List of Connections Boxes

Connection no.	Title
I.1	The Carbon Cycle and the Earth's Climate
1.1	Diamond, Graphite, and "Buckyballs"
2.1	Petroleum
3.1	Oral Contraceptives
3.2	Chemical Communication in Nature
3.3	Geometric Isomerism and Vision
4.1	Chlorofluorocarbons and the Ozone Layer
4.2	General Anesthetics
5.1	Serendipity in the Discovery of Polymers
5.2	Recycling Plastics
5.3	Terpenes
5.4	The Treatment of Atherosclerosis
6.1	Cancer and Carcinogens
6.2	Gasoline
6.3	Herbicides
7.1	Stereoisomerism in the Biological World
8.1	Drug Design
8.2	Thyroid Hormone
9.1	Methyl, Ethyl, and Isopropyl Alcohols
9.2	Neurotransmitters—The Heart of the Matter
9.3	Insecticides and Nerve Gases
9.4	Measuring Blood Alcohol
9.5	Methanol and Ethylene Glycol Poisoning
10.1	Nasal Decongestants, Diet Pills, and Stimulants
10.2	Local Anesthetics and Cocaine
10.3	Acetylcholine and Neuromuscular Blockade
10.4	Sulfa Drugs
10.5	Dyes and Dyeing
11.1	Formaldehyde and Synthetic Polymers
12.1	Food Preservatives
13.1	Aspirin and Other Analgesics
13.2	Barbiturates
14.1	Diabetes
14.2	Prevention of Disease and Detoxification
14.3	Low-Calorie Sweeteners
14.4	Nitrocellulose and Rayon
15.1	Errors in the Metabolism of Fatty Acids—Lorenzo's Oil
15.2	RU-486
16.1	Sickle Cell Anemia—A Biochemical Disease
16.2	Mad Cow Disease
16.3	Testing for Drugs, Pregnancy, and AIDS
16.4	Growth Hormone
17.1	The Human Genome Project
17.2	Acquired Immune Deficiency Syndrome (AIDS)
17.3	DNA Fingerprinting
18.1	MRI: Magnetic Resonance Imaging

SIXTH EDITION

Organic Chemistry

A Brief Survey of Concepts and Applications

Philip S. Bailey, Jr.
Christina A. Bailey

California Polytechnic State University, San Luis Obispo

PRENTICE HALL
Upper Saddle River, New Jersey 07458

Library of Congress Cataloging-in-Publication Data

Bailey, Philip S.
Organic chemistry : a brief survey of concepts and applications/
Philip S. Bailey, Jr., and Christina A. Bailey. — 6th ed.
p. cm.
Includes index.
ISBN 0-13-924119-1
1. Chemistry, Organic I. Bailey, Christina A. II. Title
QD251.2.B34 2000 99-052987
547—dc21

Executive Editor: *John Challice*
Editorial Assistants: *Gillian Buonanno, Sean Hale*
Executive Managing Editor: *Kathleen Schiaparelli*
Assistant Managing Editor: *Lisa Kinne*
Senior Marketing Manager: *Steve Sartori*
Director of Marketing: *John Tweeddale*
Manufacturing Manager: *Trudy Pisciotti*
Manufacturing Buyer: *Michael Bell*
Cover Designer: *Steven Gagliostro*
Interior Designer: *Judith A. Matz-Coniglio*
Art Manager: *Gus Vibal*
Art Editor: *Karen Branson*
Art Director: *Joseph Sengotta*
Photo Researcher: *Mary Ann Price*
Production Supervision/Composition: *Accu-color/NK Graphics*

Printed in the United States of America

10 9 8 7 6 5 4 3 2 1

ISBN 0-13-924119-1

Prentice-Hall International (UK) Limited, *London*
Prentice-Hall of Australia Pty. Limited, *Sydney*
Prentice-Hall Canada Inc., *Toronto*
Prentice-Hall Hispanoamericana, S.A., *Mexico*
Prentice-Hall of India Private Limited, *New Delhi*
Prentice-Hall of Japan, Inc., *Tokyo*
Pearson Education Asia Pte. Ltd.
Editora Prentice-Hall do Brasil, Ltda., *Rio de Janeiro*

Contents

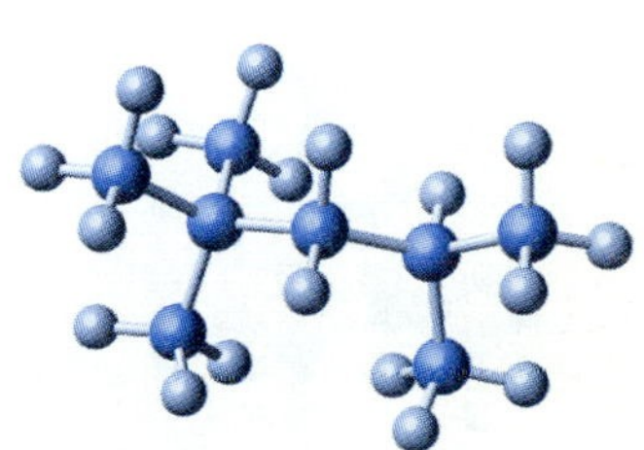

C H A P T E R 3

Alkenes and Alkynes: Structure and Nomenclature 70

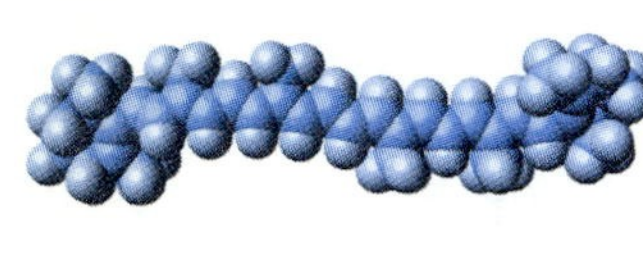

C H A P T E R 4

An Introduction to Organic Reactions 93

C H A P T E R 5

Reactions of Alkenes and Alkynes 117

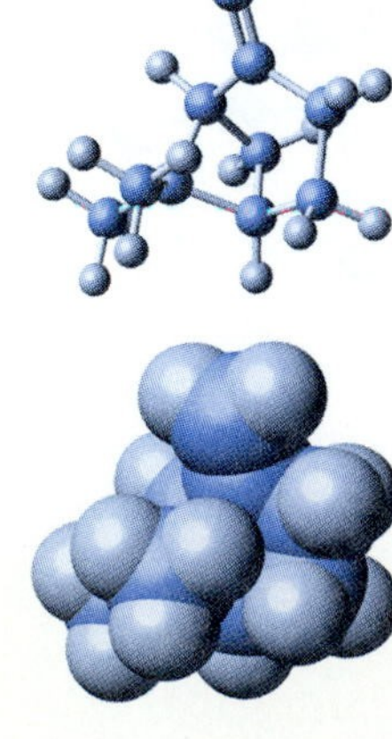

CHAPTER 6

Aromatic Compounds 153

CHAPTER 7

Stereochemistry 187

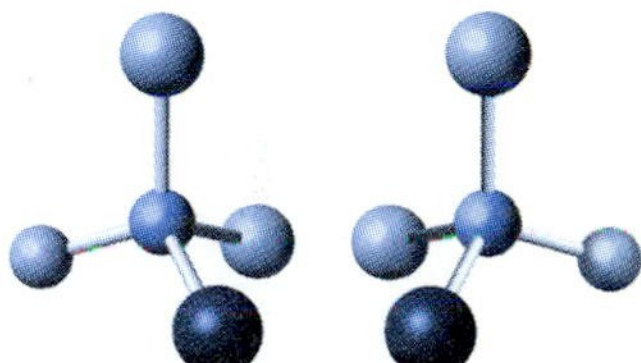

CHAPTER 8

Organic Halogen Compounds 219

CHAPTER 9

Alcohols, Phenols, and Ethers 251

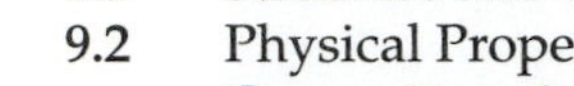

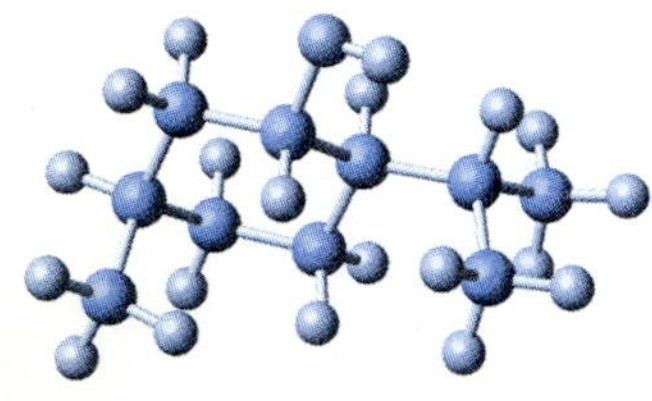

CHAPTER 10

Amines 292

CHAPTER 11

Aldehydes and Ketones 325

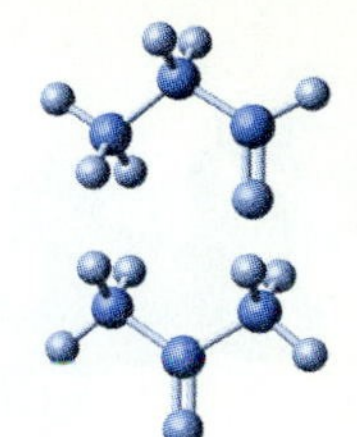

CHAPTER 12

Carboxylic Acids 358

CHAPTER 13

Derivatives of Carboxylic Acids 375

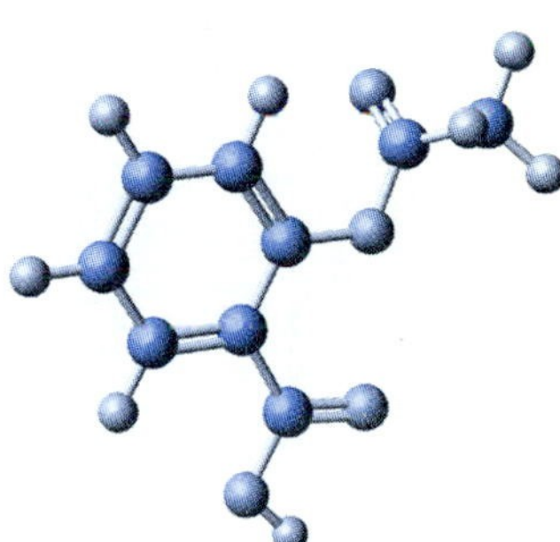

CHAPTER 14

Carbohydrates 408

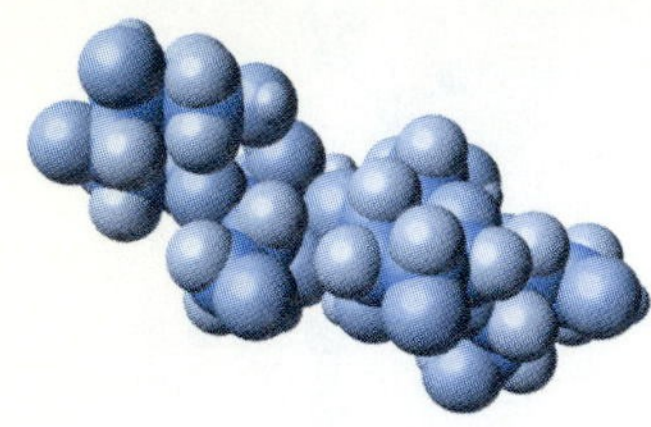

CHAPTER 15

Lipids 433

CHAPTER 16

Proteins 460

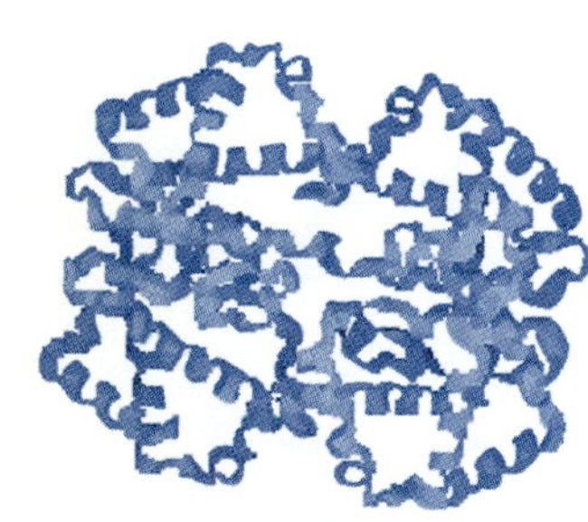

CHAPTER 17

Nucleic Acids 495

CHAPTER 18

Spectroscopy 515

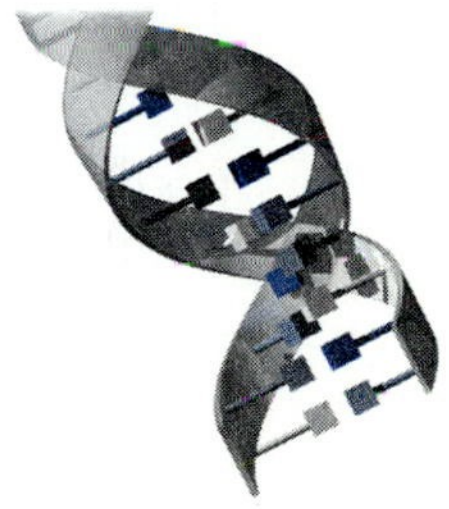

APPENDIX

Summary of IUPAC Nomenclature A1

Glossary G1

Index I1

Preface

It is a privilege to present the sixth edition of our introductory organic chemistry textbook coincidentally with the beginning of the twenty-first century. We have been honored by opportunities for revision, translations into other languages, and the use of this textbook in many parts of the world as well as the United States. Our appreciation goes to the thousands of students and faculty members who have used this book during its 22-year history. Thank you; this experience has been a special part of our professional and personal lives.

The following is a description of the focus and approach of *Organic Chemistry,* as well as how students and their instructors might best use the textbook.

This treatment of organic chemistry is designed for a short course that consists of one quarter, one semester, or two quarters. It is written especially for students who are not chemistry majors but are in fields for which chemistry is a core course. Accordingly, we have limited the scope of the presentation and have introduced applications related to agriculture, life sciences, health and medicine, and consumer, environmental, and societal issues. The selection and organization of material is consistent with this short course focus. For each topic and concept we have tried to write clear and thorough explanations that promote understanding and critical thinking. We have tried to focus on the learning process, and to write in a way to engage the student intellectually. The facts learned in a college course are important, but the promotion of active learning and the appreciation for the material will outlast memorization.

Most introductory organic chemistry courses are organized around a functional group approach in which molecules with specific structural units (functional groups) are classified in families that share similar physical and chemical properties and rules of nomenclature. Using the functional group approach we have organized the chapters as follows:

Organization

Chapters 1 to 3: The basis for structure and nomenclature. Structure and nomenclature of organic compounds are inextricably related. Competency in these areas is essential if students are to understand and appreciate the physical and chemical properties of organic compounds or to develop a practical understanding of applications. These topics are a major portion of every chapter of the book as each functional group is presented.

In Chapter 1, structure and bonding are organized around just a few ideas, and each is clearly presented with the interrelationships shown. Chapter 2 covers the structure and nomenclature of alkanes including cycloalkanes and conformational isomers. Chapter 3 presents the structure and nomenclature of alkenes and includes a discussion of geometric isomerism. A wealth of exercises accompany each chapter to allow the student a hands-on experience with structure and nomenclature.

Chapter 4: An Introduction to Organic Reactions. The basic simplicity of organic chemistry is illustrated here: three types of reactions—substitution, addition, and elimination; three types of mechanisms—carbocation, carbanion, and free radical; and a rather short list of structural features that are considered reaction sites. A free radical substitution reaction of alkanes and a carbocation elimination reaction for the preparation of alkenes conclude the chapter.

Chapters 5 and 6: Reactions of Alkenes, Alkynes, and Aromatic Compounds. Based on structural concepts covered in Chapters 1 to 3 and the principles of organic reactions from Chapter 4, addition and substitution reactions, both with carbocation mechanisms, are presented in these chapters.

Chapters 7 and 8: Stereochemistry and Organic Halides. Chirality, enantiomerism, diastereomerism, and *R,S* configurations are presented in Chapter 7; a concluding short section relates these concepts to reactions already presented. Chapter 8 then uses the principles of organic reactions and stereochemistry in the study of nucleophilic substitution and elimination reactions.

Chapters 9 and 10: Alcohols and Ethers; Amines. There are two reasons we place these chapters here. The concepts of acidity and basicity provide a conceptual thread that is clearly evident in both chapters, and the characteristic reactions of nucleophilic substitution and elimination continue nicely from Chapter 8.

Chapters 11 to 13: Aldehydes and Ketones; Carboxylic Acids and Their Derivatives. After three chapters involving nucleophilic substitution reactions, we can now concentrate on nucleophilic addition using aldehydes and ketones (Chapter 11). Chapter 12 focuses on the nomenclature and acidity of carboxylic acids. Then in Chapter 13, nucleophilic acyl substitution reactions of carboxylic acids and their derivatives are presented. The concepts of acidity and basicity—often difficult for students—as well as the nucleophilic reactions begun in Chapters 8, 9, and 10 are reiterated and reinforced.

Chapters 14 to 17: Biochemistry—Carbohydrates, Fats and Oils, Proteins, and Nucleic Acids. The functional groups and reactions introduced in Chapters 9 to 13 lead logically to chapters on biomolecules. In keeping with the interests of the students in this course, we cover each of these topics in some detail.

Chapter 18: Spectroscopy. Infrared ultraviolet-visible, proton and C-13 nuclear magnetic resonance, and mass spectrometry are presented concisely but in a way that allows significant experiences in problem solving. We realize that many instructors do not cover this chapter while others cover it at different places in the course. We have organized the material so this chapter can be introduced any time after Chapter 6 based on the student's knowledge of structure (Chapter 13 is another logical place for introduction because all the functional groups have been presented).

Connections boxes: There are more than 40 essays throughout the book written to connect organic chemistry to the real world. They include topics related to issues of energy, the environment, consumer goods, living organisms, health, disease, and medicine.

Pedagogy

The following are learning tools available in this textbook:

NEW: "Getting Involved" modules allow students to evaluate their understanding. At least one is found in each major section, allowing for better-paced learning and greater instructor flexibility.

- **Getting Involved:** These special entries follow all major sections of each chapter and many of the subsections. Within these we try to engage the student in her/his own learning process. Each **Getting Involved** module will have one or more sections in the following order:
  - **Short questions:** These questions are a first opportunity for a student to self-evaluate his or her comprehension of what was just read.
  - **Worked examples:** Often there will be a worked example in this section. Other times there will be sufficient examples in the preceding reading.

NEW: Many more worked examples this edition.

- **Practice problems:** These provide an opportunity to work immediately with the material just presented.
- **Reference to end-of-chapter problems:** The internal textbook problems are immediate opportunities to "get involved." Appropriate, related problems from the end of the chapter are referenced here.

NEW: Cross references to related end-of-chapter problems make it easier to cover only parts of a chapter.

- **Clear and complete explanations:** We have emphasized a depth of understanding rather than breadth in this text. An emphasis on systematic, logical, and critical thinking minimizes memorization and maximizes real learning.
- **Problems:** Organic chemistry is a hands-on subject. A student has to work problems to really become adept with the material. This book has approximately 800 numbered problems distributed between in-chapter and end-of-chapter exercises. Included with these problems are some molecular model-building exercises that can be done with a variety of different model kits or with electronic model programs like Chem3D (Cambridge Soft). If you are interested, Prentice Hall can make model kits of Chem3D available with this text at a dramatic discount.

NEW: 30% of problems have been revised this edition.

NEW: Model-building exercises.

- **Marginal definitions:** A brief definition of each important term is printed in the margin next to its introduction.
- **Reaction summaries:** Summaries of all reactions are presented at the end of each chapter.
- **Summary of IUPAC nomenclature:** See the Appendix for this comprehensive, integrated summary.
- **Glossary of terms:** This appendix is an alphabetical listing of the terms defined in the margins in each chapter.

Other special features of this textbook include the following:

- **Soft cover, a second color:** This textbook is significantly lighter and less expensive than hard-cover versions, and therefore is not as heavy a burden physically or economically. We are offering an alternative, especially when the cost of textbooks is such an issue on college campuses. We believe we have the explanations, illustrations, use of a second color, and exercises to provide a substantial intellectual experience and effectively promote learning.
- **Subdivisions:** We realize that instructors have their own creative organizational methods for teaching organic chemistry and that no textbook can exactly match the preferences of many, if any, instructors. We have included subheadings within each chapter to make assignment of topics and exercises, development of a course syllabus, and adaptation of this book to many teaching styles as easy as possible.
- **Solutions Manual and Study Guide (ISBN 0-13-015480-6):** An ancillary Solutions Manual and Study Guide is available that contains chapter summaries (outlined just as are the chapters) and solutions to each problem in the textbook. To ensure a consistent voice and style, we wrote this manual ourselves.
- **Website at www.prenhall.com/bailey:** We have a companion Website developed by us and specifically designed for this textbook. We suspect it will always be a work in progress, but initially it will have the following features:

NEW: Webside allows students to access additional tutorial material and updates/current topics, all prepared by the authors.

1. Links to interesting Websites providing additional information and insight for each of the Connections essays.
2. A tutorial on the use of Chem3D.
3. Demonstration exercises on structure, isomerism, and stereochemistry.

We wish to thank our students and other users of the previous five editions of this text. We have appreciated the comments and suggestions of students and instructors and invite the same for this edition.

Drs. Philip and Christina Bailey
Department of Chemistry and Biochemistry
California Polytechnic State University
San Luis Obispo, California 93407

e-mail: *pbailey@calpoly.edu*
cbailey@calpoly.edu

Acknowledgments

We wish to acknowledge the many people involved in bringing the sixth edition of this textbook to a successful conclusion. We owe a special debt of gratitude to our excellent editor, John Challice. We also wish to thank our production editor, Celeste Clingan of Accu-color, and our picture consultant, Mary Ann Price. The work of Karl Bailey and Nhu Y Tran Stressman in checking the solutions in the Solutions Manual is greatly appreciated. With much appreciation, we acknowledge the reviewers of the manuscript; they were exceedingly helpful.

Stuart R. Berryhill, California State University, Long Beach
Clair J. Cheer, San Jose State University
Loretta T. Dorn, Fort Hays State University
Jeffrey E. Elbert, South Dakota State University
Christopher M. Hadad, Ohio State University
Phillip D. Hampton, University of New Mexico
Cliff Harris, Albion College
Thomas R. Hays, Texas A & M University, Kingsville
James R. Hermanson, Wittenberg University
John C. Hogan, Louisiana State University
Tamera S. Janke, Southwest Missouri State University
Kenneth Johnson, Emporia State University
Francis M. Klein, Creighton University
James G. Macmillan, University of Northern Iowa
Rita S. Majerle, South Dakota State University
William A. Meena, Rock Valley College
Todd A. Richmond, The Claremont Colleges
Ralph Shaw, Southeastern Louisiana University
Jason M. Stenzel, Southern Connecticut State University
Richard T. Taylor, Miami University
Anthony P. Toste, Southwest Missouri State University
Maria Vogt, Bloomfield College
Mark S. Workentin, University of Western Ontario

We also wish to thank our students and colleagues at Cal Poly. Finally, we express our gratitude for the important work and contributions of Prentice Hall regional representatives.

Biographies of the Authors

The authors of this textbook, Drs. Phil and Tina Bailey, received their Ph.D.'s in chemistry from Purdue University in 1969 and 1970, respectively. Phil specialized in organic chemistry and Tina in biochemistry. Before attending Purdue, Tina earned her B.S. in chemistry from the College of Saint Elizabeth in New Jersey and Phil from the University of Texas at Austin. Both are on the faculty of California Polytechnic State University (Cal Poly) in San Luis Obispo. Cal Poly is an institution of about 16,000 students and is located on the rural and beautiful central California coast, just 10 minutes east of the Pacific Ocean. As a polytechnic institution, Cal Poly has Colleges of Agriculture, Architecture, and Engineering as well as Business, Liberal Arts, and Science and Mathematics. The University has become one of the more selective and acclaimed public institutions in the United States.

Tina is a professor of chemistry at Cal Poly. She is the recipient of the University's Distinguished Teaching Award and was Cal Poly's 1993 nominee for Outstanding Professor of the 23-campus California State University System. Her current interests are in new pedagogies for teaching chemistry, especially general chemistry. She has given presentations at national American Chemical Society meetings and has been an invited speaker in other venues on the "studio classroom" she developed at Cal Poly for teaching general chemistry for engineers. This pedagogy integrates lecture and wet chemistry laboratory in a 64-student classroom. Clusters of students and computers allow the use of technology to collect and interpret experimental data as well as use Internet resources for regular classwork and enhancement studies.

Phil is Dean of Cal Poly's College of Science and Mathematics, a position he has held for the past 16 years, except for one year as interim Vice President for Academic Affairs. During this time he has taught approximately one-third time each term in the Department of Chemistry and Biochemistry. He received an Advisor of the Year award as well as university recognition for his efforts in educational equity.

In addition to this textbook, the Baileys have collaborated in their private lives. As the parents of four children, Tina and Phil were Girl Scout and Boy Scout leaders, officers on the local swim team, and volunteers with the PTA. In addition, the Baileys have presented chemistry magic shows to the public for over 30 years; more than 125,000 have enjoyed their performances. Now, with their children as adults living away from home, they act as Cal Poly mom and dad to many students.

A Student's Guide to Using this Text

You and your classmates come to this course with a variety of backgrounds and interests. Most of you plan to be professionals in a biological, technical, or allied health field. Knowledge of chemistry is essential to a true understanding of everything from DNA replication to drug discovery to nutrition to engineering. Indeed, the chemical properties and principles you learn in this course will pervade almost every aspect of your private and professional lives. In this text, we provide you with both the principles and applications of chemistry that will help you in your professional practice and enrich your everyday life as well.

This text is rich in pedagogical aids, both within and at the ends of the chapters. We present this "user's guide" to the text to help you get the most out of this book and your course.

Applications

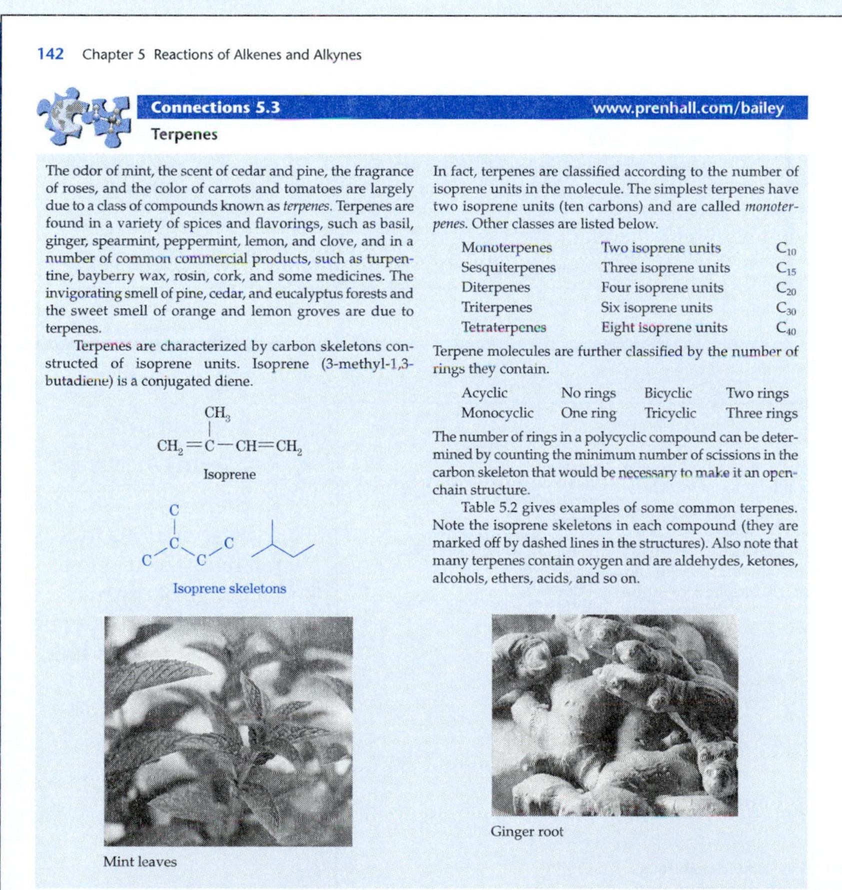

142 Chapter 5 Reactions of Alkenes and Alkynes

Connections 5.3 www.prenhall.com/bailey

Terpenes

The odor of mint, the scent of cedar and pine, the fragrance of roses, and the color of carrots and tomatoes are largely due to a class of compounds known as *terpenes.* Terpenes are found in a variety of spices and flavorings, such as basil, ginger, spearmint, peppermint, lemon, and clove, and in a number of common commercial products, such as turpentine, bayberry wax, rosin, cork, and some medicines. The invigorating smell of pine, cedar, and eucalyptus forests and the sweet smell of orange and lemon groves are due to terpenes.

Terpenes are characterized by carbon skeletons constructed of isoprene units. Isoprene (3-methyl-1,3-butadiene) is a conjugated diene.

$$CH_2{=}C(CH_3){-}CH{=}CH_2$$

Isoprene

Isoprene skeletons

In fact, terpenes are classified according to the number of isoprene units in the molecule. The simplest terpenes have two isoprene units (ten carbons) and are called *monoterpenes*. Other classes are listed below.

Monoterpenes	Two isoprene units	C_{10}
Sesquiterpenes	Three isoprene units	C_{15}
Diterpenes	Four isoprene units	C_{20}
Triterpenes	Six isoprene units	C_{30}
Tetraterpenes	Eight isoprene units	C_{40}

Terpene molecules are further classified by the number of rings they contain.

Acyclic	No rings	Bicyclic	Two rings
Monocyclic	One ring	Tricyclic	Three rings

The number of rings in a polycyclic compound can be determined by counting the minimum number of scissions in the carbon skeleton that would be necessary to make it an open-chain structure.

Table 5.2 gives examples of some common terpenes. Note the isoprene skeletons in each compound (they are marked off by dashed lines in the structures). Also note that many terpenes contain oxygen and are aldehydes, ketones, alcohols, ethers, acids, and so on.

Mint leaves

Ginger root

Connections Boxes

There are many boxes in this text that focus on how we apply our chemical knowledge to solve real-life problems. These readings will help you see how chemistry affects everyday life and how we arrived at our current understanding of chemistry.

Learning Tools

Getting Involved

Getting Involved sections give you the opportunity to practice what you have just learned. They include a number of parts:

- Concept Checks: Make sure you can answer these questions before reading any further.
- Worked Examples: Read through the problem we have solved for you and understand each step.
- Practice Problems: Test your understanding of the material by seeing if you can solve the Practice Problems.
- Cross-references to related problems: If you want more practice on this topic, solve these problems at the end of the chapter.

GETTING INVOLVED

✓ Describe the relative proportions of reactants needed to promote monohalogenation and polyhalogenation. Explain the difference.

✓ Why does a symmetrical alkane give fewer monohalogenation isomers than the alkane's unsymmetrical isomers?

Example 4.1

Write the monochlorination products of pentane and dimethylpropane.

Solution

In pentane there are three places a chlorine can replace a hydrogen and form a different product; in dimethylpropane, all sites are equivalent.

$$\underset{\text{Excess}}{CH_3CH_2CH_2CH_2CH_3} + Cl_2 \xrightarrow{\text{Light}}$$

$$\underset{|\\Cl}{CH_2}CH_2CH_2CH_2CH_3 + CH_3\underset{|\\Cl}{CH}CH_2CH_2CH_3 + CH_3CH_2\underset{|\\Cl}{CH}CH_2CH_3$$

Three possible monochlorination products

$$\underset{\text{Excess}}{CH_3-\overset{CH_3}{\underset{CH_3}{C}}-CH_3} + Cl_2 \xrightarrow{\text{Light}} CH_3-\overset{CH_3}{\underset{CH_3}{C}}-CH_2Cl + HCl$$

One monochlorination product

Problem 4.14

Consider the chlorination of 2-methylbutane, an isomer of pentane and dimethylpropane discussed in Example 4.1, for this problem. **(a)** What is the relative ratio of alkane to chlorine to have total chlorination occur? **(b)** What is the relative ratio to promote monochlorination? **(c)** Draw the possible products formed by the monochlorination reaction.

See related problems 4.27–4.29.

Key Terms in Margin

Key terms appear in the chapter margins for easy reference. Learning the language of organic chemistry is an essential part of understanding this material.

11.1 Structure of Aldehydes and Ketones

Aldehydes and **ketones** are structurally very similar; both have a carbon-oxygen double bond called a **carbonyl** group. They differ in that aldehydes have at least one hydrogen atom bonded to the carbonyl group, whereas in ketones the carbonyl is bonded to two carbons. The biological preservative, formaldehyde, is the simplest aldehyde; benzaldehyde, oil of bitter almond, is the simplest aromatic aldehyde. Acetone is the simplest ketone; it is an important industrial solvent and a principal ingredient in some fingernail polish removers. Acetophenone is the simplest aromatic ketone; it is used in perfumery.

aldehyde
functional group in which at least one H is bonded to a carbonyl

ketone
functional group in which two organic substituents are bonded to a carbonyl

carbonyl
the carbon-oxygen double bond, C=O

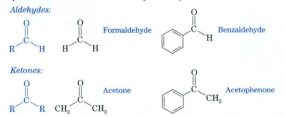

Aldehydes:

$R-\overset{O}{\overset{\|}{C}}-H$ $\quad H-\overset{O}{\overset{\|}{C}}-H$ Formaldehyde $\quad C_6H_5-\overset{O}{\overset{\|}{C}}-H$ Benzaldehyde

Ketones:

$R-\overset{O}{\overset{\|}{C}}-R$ $\quad CH_3-\overset{O}{\overset{\|}{C}}-CH_3$ Acetone $\quad C_6H_5-\overset{O}{\overset{\|}{C}}-CH_3$ Acetophenone

Aldehydes and ketones are quite prevalent in nature. They occur as natural fragrances and flavorings. In addition, carbonyl groups and their derivatives are the main structural features of carbohydrates and appear in other natural compounds, including dyes, vitamins, and hormones.

The carbonyl group is exceedingly important in organic chemistry. In addition to aldehydes and ketones, it is found in carboxylic acids and carboxylic acid

Learning Tools

Reaction Summaries

Study the illustrations and organic chemistry can seem like a vast collection of reactions; by understanding the mechanisms of only a few reactions, however, you can understand most of organic chemistry. For easy reference and review, refer to the reaction summaries at the end of most chapters.

REACTION SUMMARY

1. Formation of Carboxylic Acid Salts

Section 12.4.A; Problems 12.6–12.7, 12.30.

$$RCOH\ (C{=}O) + M^+OH^- \longrightarrow RCO^-M^+\ (C{=}O) + H_2O$$

2. Preparations of Carboxylic Acids

Section 12.5; Problems 12.12–12.17, 12.25–12.26.

A. Oxidation of Alkylbenzenes

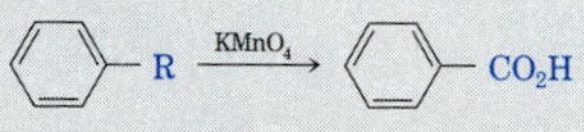

$$C_6H_5\text{–}R \xrightarrow{KMnO_4} C_6H_5\text{–}CO_2H$$

B. Oxidation of Primary Alcohols

$$RCH_2OH \xrightarrow{CrO_3/H^+} RCO_2H$$

C. Hydrolysis of Nitriles

$$RC{\equiv}N \xrightarrow{H_2O/H^+} RCO_2H$$

D. Carbonation of Grignard Reagents

$$RX \xrightarrow[\text{Ether}]{Mg} RMgX \xrightarrow[2)\ H_2O/H^+]{1)\ CO_2} RCO_2H$$

End-of-Chapter Problems

End-of-chapter problems offer you an opportunity to further test your understanding. Each question is categorized so you can quickly find questions related to specific subtopics.

Problems

14.17 **Terms:** Distinguish between the members of the following pairs of terms:

(a) hexose, pentose
(b) aldose, ketose
(c) reducing sugar, nonreducing sugar
(d) monosaccharide, polysaccharide
(e) α-D-glucose, β-D-glucose
(f) Haworth formula, Fischer projection
(g) amylose, amylopectin
(h) glycogen, cellulose
(i) Type 1 diabetes, Type 2 diabetes
(j) viscose rayon, acetate rayon
(k) Fehling's and Tollens' tests

14.18 **Structure:** How are the members of the following pairs of saccharides different from each other structurally? Which are reducing, and which are nonreducing? Explain

(a) cellobiose, maltose
(b) lactose, sucrose
(c) α-D-glucose, α-D-galactose
(d) α-D-glucose, α-D-fructose
(e) α-D-xylose, β-D-ribose
(f) maltose, lactose
(g) cellulose, starch

14.19 **Structure:** Draw Haworth formulas for the six-membered ring structures (pyranose forms) of the following:

(a) β-D-fructose (b) α-D-idose
(c) β-D-talose (d) α-D-lyxose

14.20 **Structure** D-2-deoxyribose is found in DNA, our genetic code. Draw the structure of this monosaccharide.

14.21 **Reactions:** Deoxyribose and ribose (RNA) form esters with phosphoric acid in DNA and RNA. Can both of these monosaccharides react to form the same number of ester combinations? Explain.

14.22 **Structure:** Draw the open-chain forms of the following cyclic saccharides:

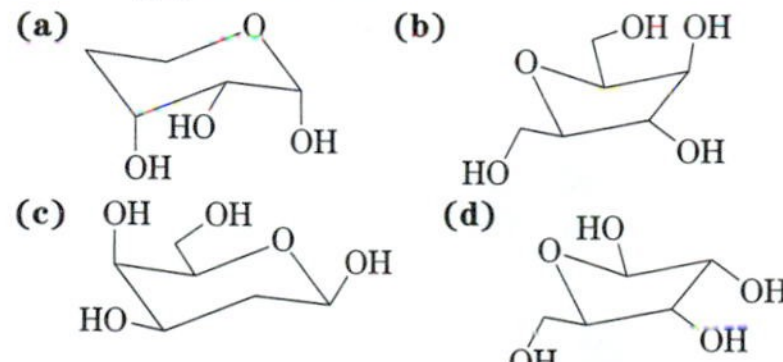

14.23 **Stereoisomers:** Draw the stereoisomers of 3-ketopentose. Which are the enantiomers, diastereomers, and meso compounds?

14.24 **Reactions:** Pure α-D-glucose or pure β-D-glucose in the presence of methanol (CH_3OH) and acid will give a mixture of α- and β-methyl glucosides. Why?

14.25 **Structure:** Specify the type of glycosidic bond that appears in each of the following disaccharides. Also identify the general type of monosaccharide units that appear in each, such as aldopentose.

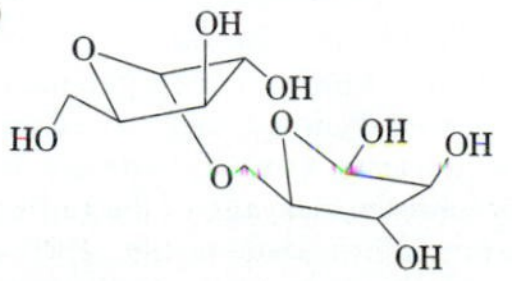

(a) OH OH O OH OH OH HO OH HO O O OH

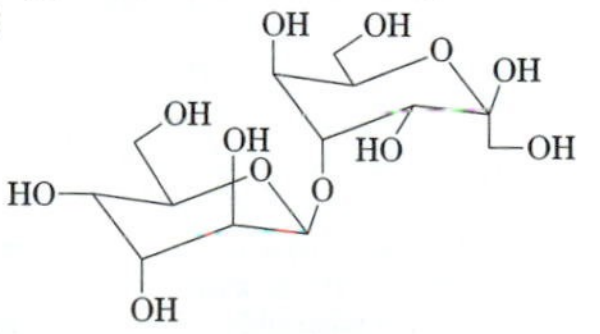

(b) OH O OH HO O OH OH O OH

Visualization

Illustrations

Study the illustrations and graphics carefully. Chemistry is a visual science, and the art will help you to visualize atoms, molecules, and chemical processes that cannot be seen with the unaided eye.

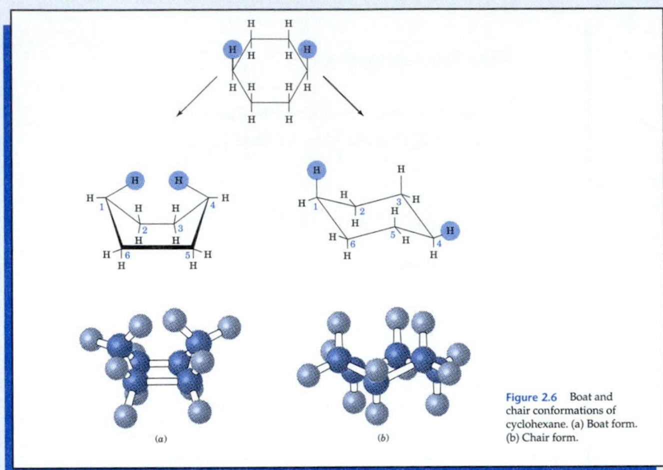

Figure 2.6 Boat and chair conformations of cyclohexane. (a) Boat form. (b) Chair form.

Activities with Molecular Models

Organic chemistry is a three-dimensional science. Perform these exercises using your model kit and you'll improve your ability to think about molecules in three dimesions.

Activities with Molecular Models

1. Make a molecular model with one carbon and four different attached groups; this is a chiral carbon atom. Manipulate the two models to convince yourself that they are not superimposable. Make a similar model but with only three different groups. Show that these are superimposable.

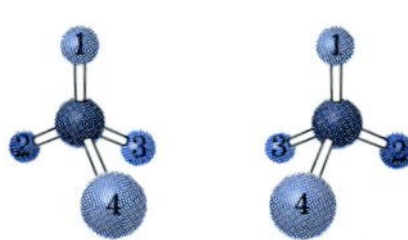

2. Make a molecular model of 1,2-dibromobutane. Put the model in the eclipsed conformation around carbons 2 and 3; this is a representation of the Fischer projection. A depiction is shown below for the meso compound. Exchange two groups on one of the carbon-bearing atoms and you will have one of the enantiomers. Make a mirror image to get the other. Notice in Fischer projections we have a totally staggered conformation that seems to curl toward the rear of the viewer.
3. You may wish to use the molecular models with the chiral carbon atom from exercise 1 to assist you in visualizing R and S configurations.

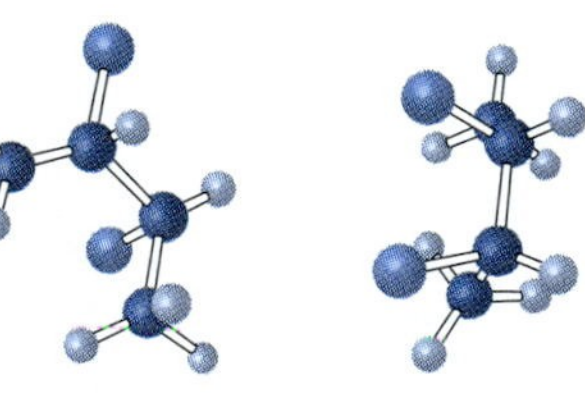

Voice Balloons

Voice balloons help you understand each step in a reaction mechanism. Make sure you understand where each product comes from; don't just memorize them.

tion of any hydrogen halide to any alkene. In all cases, the mechanism is a two-step one: first, bonding of the electrophile, the hydrogen ion; second, neutralization of the carbocation intermediate by a **nucleophilic** halide ion.

nucleophile
species with electron availability that donates electrons to electrophiles in a chemical reaction. Nucleophiles are Lewis bases

General Mechanism for the Electrophilic Addition of Hydrogen Halides to Alkenes

$$\gt C{=}C\lt \;+\; H^+ \longrightarrow -\overset{|}{\underset{H}{C}}-\overset{|}{\underset{+}{C}}- \;+\; :\ddot{X}:^- \longrightarrow -\overset{|}{\underset{H}{C}}-\overset{|}{\underset{X}{C}}-$$

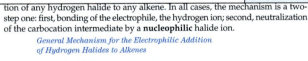

HX = HCl, HBr, HI

Step 1: π electrons bond to electrophile H^+, forming a new C–H σ bond and a carbocation.

Step 2: Negative halide ion donates electron pair to carbocation, producing the final addition product.

The electrophilic addition of hydrogen halides to alkenes can be effectively described by a potential energy diagram (section 4.1C) as shown in Figure 5.3. In the first transition state, a carbon-hydrogen bond is in the process of forming simultaneously with the cleavage of the carbon-carbon double bond. A high-energy carbocation intermediate results. The carbon-chlorine bond is forming in the second transition state with the final product being the result. Similar energy diagrams describe most electrophilic addition reactions.

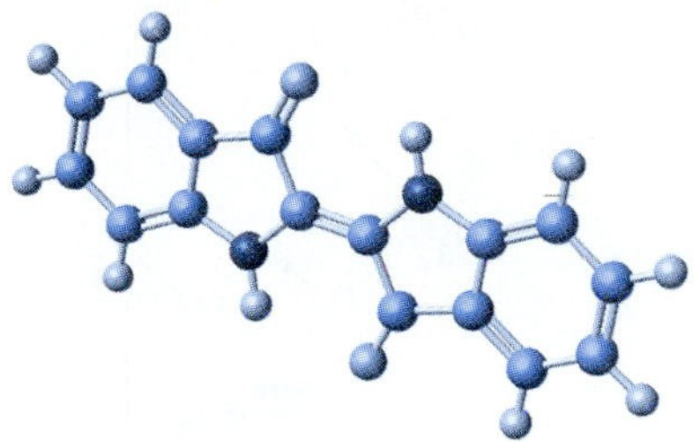

Organic Chemistry: An Introduction and a Message to Students

The Carbon Cycle: An Introduction to Organic Chemistry

Every day for billions of years the Sun has bathed our planet with generous amounts of energy. From their first appearance on Earth, green plants have lived in the Sun, using its energy to convert carbon dioxide and water into structural materials, principally starch and cellulose. By this process of photosynthesis, solar energy is stored in myriad chemical bonds, many of them carbon-carbon bonds. The by-product of photosynthesis, oxygen, is released to the atmosphere. Animals eat plants, the source of the carbon compounds vital to their existence. They use oxygen from the air to convert glycogen, their own "animal starch," to carbon dioxide, water, and the energy they need to live, function, and reproduce.

This cycle that starts with carbon dioxide, water, and energy from the Sun and "ends" with carbon dioxide, water, and energy produced by living organisms to sustain life is called the *carbon cycle* (Figure I.1). Carbon is the basis for life on Earth. Of course, plants and animals die, and, under certain conditions, large masses of plant and animal matter are transformed over time into vast deposits of coal and petroleum. Coal and petroleum are basically large deposits of hydrocarbons, which are organic compounds containing only carbon and hydrogen. Hydrocarbons are used by humans for a variety of purposes, including the manufacture of plastics, fibers, drugs, dyes, and detergents. Primarily hydrocarbons are burned to produce energy. This energy of combustion powers our cars, heats our homes, operates factories, and generates electricity. It is the same energy that was originally absorbed from sunlight and stored in chemical bonds by green plants millions of years ago. The chemical products of combustion are carbon dioxide and water, the components of the ever-continuing carbon cycle.

Carbon: The Element of Life, An Element of Civilization

Since life on Earth is based on carbon, carbon has a special place in the science of chemistry. In fact, two branches of chemistry—organic and biochemistry—are rooted in this element. *Organic chemistry* is the study of compounds containing carbon. *Biochemistry* is the study of the chemicals and processes that sustain life, the life on Earth that is based on carbon.

That it is the element of life is reason enough to give carbon its own special branch of chemistry. But there are other reasons, too. First, the number of known organic compounds is approaching 20 million. This is at least ten times the number of inorganic compounds. In fact, the number of possible organic compounds is virtually limitless. In addition, the occurrences and uses of organic

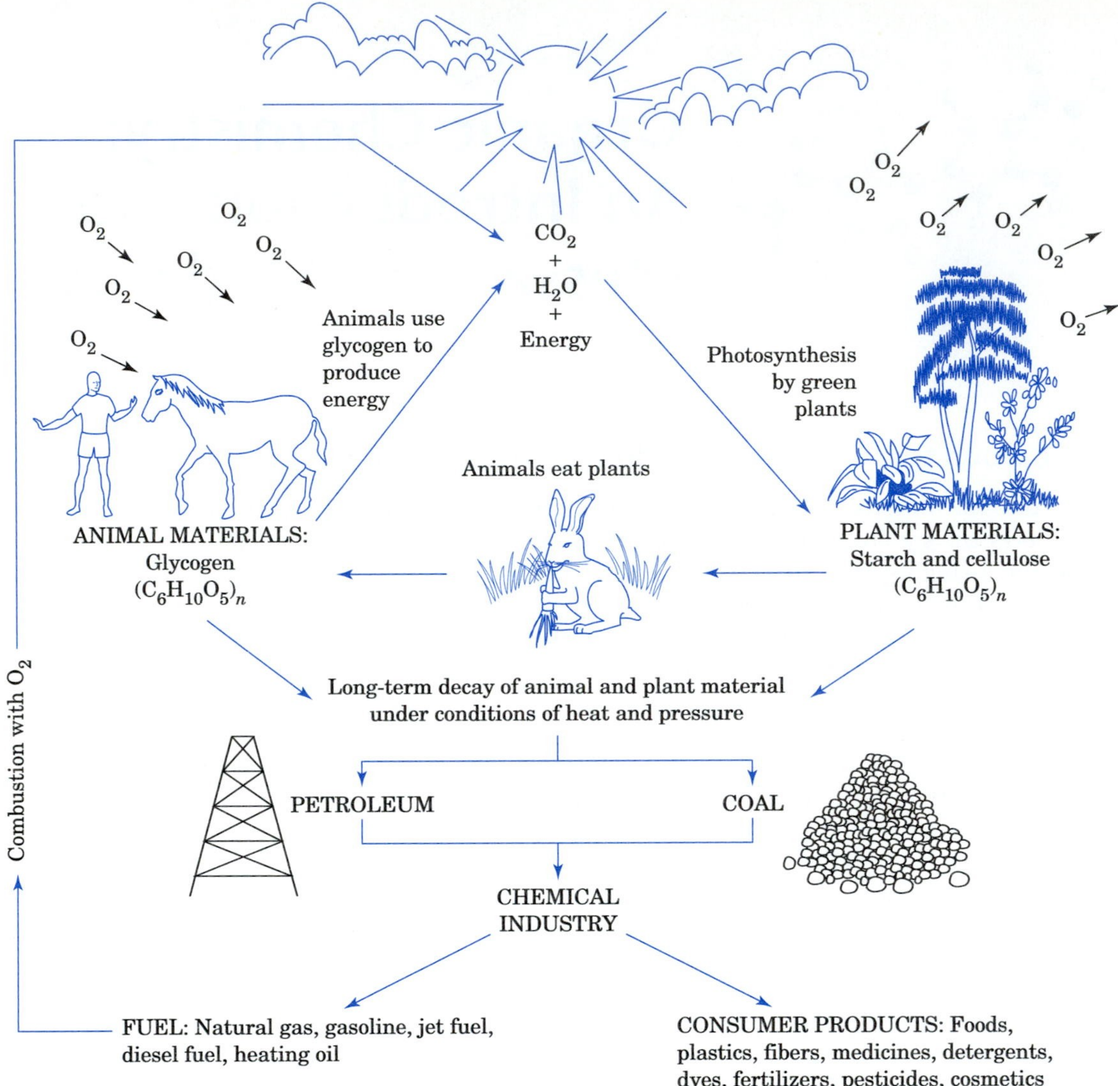

Figure I.1 The carbon cycle.

compounds are truly incredible and are not only essential to the presence of life, but they have become ubiquitous influences on life and civilization.

Take a minute to think of your personal contact with organic compounds. The foods you eat—carbohydrates, fats, proteins—as well as many food additives, are simple to highly complex organic compounds. Fertilizers, herbicides, and pesticides include organic compounds that have become almost indispensable to agriculture. Drugs and medicines, hygiene and beauty products, are all composed of organic substances that alleviate the symptoms of headaches and colds, combat serious illnesses, act as disinfectants, and are the basis of the gigantic pharmaceutical, soap, detergent, and cosmetics industries. The clothes we wear, whether made from natural fibers such as cotton, silk, wool, or from synthetic fibers such as polyester, nylon, rayon, and also the dyes that color them, are all composed of organic compounds that we will study in this book. Plastics such as PVC, Saran, Teflon, polyethylene, Styrofoam, Lucite, Melmac, and so on are enormous organic molecules called polymers. Finally, fuel, coal, and petroleum, which are composed of carbon and hydrogen, have had a profound influence on our personal lives, and the politics and economies of our planet.

The pace of development of organic chemistry, like so many other scientific and technical fields, has been phenomenal. Until 1828, when Frederick Wöhler converted the inorganic compounds ammonium hydroxide and lead cyanate to urea, it was thought that all organic compounds had to be produced from living organisms, a theory called the vital force theory. The first oil well drilled in 1859 opened up a seemingly endless treasure trove of organic compounds. The petrochemical industry eventually resulted and with it the medicines, plastics, synthetic fibers, and other incredible materials that seem to continually amaze us. How, when, and why do we use these materials, and what do we do with them when we are finished? These and other questions challenge all of us to become informed consumers of the ideas and products of our scientific and technological society in which organic chemistry plays a major role.

Tiny Molecules, Big Ideas: Understanding Organic Chemistry

It is difficult to imagine the extremely tiny size of molecules. You couldn't count the raindrops falling in a driving rainstorm or the grains of sand on a beach. But grab one of those raindrops and imagine that you could get inside it and count the molecules of water. At a rate of five molecules per second, day and night, it would take over 10 trillion years to count the water molecules in that one tiny raindrop. Now imagine that each of the molecules in the raindrop was a fine grain of sand. These grains of sand would be enough to fill the Grand Canyon, all 200 miles of it, with an average depth of 1 mile and width of 12 miles. San Luis Obispo, California, (the authors' home town), would have to have its annual average rainfall of 20 inches for more than a million years before as many raindrops hit the city as there are molecules of water in just one of those raindrops. The challenge of organic chemistry lies in becoming intimately familiar with these very tiny molecules, but not losing sight of how minute these molecules truly are while you are discovering the relationships between molecular structure and chemical and physical properties of organic compounds.

As you study organic chemistry you will find a thread of unifying principles. All organic compounds contain carbon; only a few other elements are commonly found—hydrogen, nitrogen, oxygen, sulfur, phosphorus, and the halogens. Organic compounds are constructed of covalent bonds—single, double, and triple bonds. Depending on the types of bonds to a carbon atom, the surrounding groups will be arranged in tetrahedral, triangular, or linear geometries. Organic compounds undergo three common types of reactions—addition, elimination, and substitution. These reactions proceed through three common reactive intermediates—free radicals, carbocations, and carbanions.

Three types of bonds, three geometries of bonded carbons, three types of chemical reactions, three reactive intermediates: the ideas are few, but big and important, and they lead to understanding. We will follow a logical path to develop this understanding. To begin, we will review atomic structure and how atoms combine to form molecules. Next we will study molecular structure, progress to drawing organic molecules, and then learn to name them. With the confidence of structure and nomenclature, the study of physical properties and chemical change is possible, and we will apply the concepts of structure to understand these properties.

Atoms, bonding, molecular structure, and nomenclature, followed by the study of physical and chemical change . . . this order makes sense. The chemical properties of organic compounds are also logically considered using a functional

Connection I-1 www.prenhall.com/bailey

The Carbon Cycle and the Earth's Climate

Carbon-based life on Earth has been influenced dramatically by the climatic conditions on the planet. Fluctuations have occurred over millenia from warm periods (during some of which the planet may have been entirely free of ice) to glacial periods in which large, continental areas were covered in deep layers of ice and snow. In the warm periods 55 million years past, tropical conditions extended far into today's temperate latitudes. Trees and other vegetation existed in the Arctic and Antarctic regions; alligators inhabited Ellesmere Island, located at the northernmost tip of the North American continent, 500 miles from the North Pole. Ice ages were caused by a complex variety of factors, including the distance of the Earth from the sun, solar energy output, heights and positions of the continents, and the chemical composition of the atmosphere. Carbon dioxide in the atmosphere may have been a significant contributor to the warm and cold cycles experienced by the planet.

Short-wavelength solar energy enters the atmosphere, whereas the energy that radiates back from the Earth is of longer wavelengths (heat). Carbon dioxide, among other atmospheric gases, absorbs some of this longer wavelength radiation, "trapping" it and thereby warming the planet. Carbon dioxide has been called a "greenhouse" gas for this reason.

A major source of the Earth's atmospheric carbon dioxide comes from volcanic activity. Photosynthesizing plants and bacteria absorb the carbon dioxide and metabolically exchange it for the oxygen required by more complex organisms. As these simpler organisms die, their chemically decomposed structures deposit the carbon in runoff from the land and in the sediments of the ocean floors. Eventually these become sources of the volcanic carbon dioxide, which begins the carbon cycle anew.

There is evidence to the effect that the most recent cooling trend, about 50 million years ago, was due to the meeting of the land masses of the Asian and Indian continents. The collision, which is still happening, resulted in the uplifting of the land that today we refer to as the Himalayas and Tibetan Plateau. As the mountains rose, rainfall began to erode the newly exposed mineral deposits. Carbon dioxide dissolved in rain aided in the solution process, forming mineral salts such as calcium carbonate, limestone. This diverted the recycling of carbon for a period of time and diminished the level of carbon dioxide in the atmosphere. Less heat was reabsorbed to the planet. Planetary cooling and glaciation, an Ice Age, occurred.

You can see that the carbon cycle is neither simple nor entirely understood. It is a cycle of life that extends to and unites the organic and inorganic worlds. As the intelligent caretakers of the planet, it is important to have good practices for the management and maintainence of carbon. The burning of the Amazon rainforest not only destroys oxygen-giving vegetation but also upsets the balance of gas consumption and production in the atmosphere, increases global warming through the excess production of carbon dioxide, and annihilates the potential utilization of natural products that have been and could be invaluable to humanity. Excessive consumption of carbon-based fuels, coal and petroleum, unnaturally depletes the sources of carbon cycle materials and contributes to the greenhouse effect. The carbon cycle is both natural and precarious in its own right. We must live in harmony with the Earth and use its resources in thoughtful and intelligent ways.

The Himalayas

Amazon fires as seen from space

group approach. We start with the hydrocarbons—alkanes, alkenes, alkynes, and aromatics—compounds with single, double, and triple bonds composed of only carbon and hydrogen. Basic concepts learned from this study can be applied to the widely diverse organic compounds containing oxygen and nitrogen. These are the amines, alcohols, aldehydes, ketones, and carboxylic acids, the functional groups that are the basis of biological molecules: carbohydrates, proteins, lipids, and nucleic acids.

From a single carbon atom to the complex molecules of life: this is the journey you will take in your study of organic chemistry.

Getting Involved: Learning Organic Chemistry

Tell me and I will forget,
Show me and I will remember,
Engage me and I will understand.

A Lakota Sioux saying

Your instructors and the authors of textbooks can each tell you about things, show you some, and lay an encouraging foundation for engagement. It is the student's responsibility, however, to get involved, to become engaged, to truly learn and understand.

Here is how you can get involved, become engaged, and learn organic chemistry with the help of this book and your instructor.

Keep up: This is crucial. Organic chemistry builds upon itself. How can you draw organic compounds without understanding bonding, or name compounds if you can't draw them, or understand chemical change if you don't understand molecular structure?

Read for comprehension: Organic chemistry is not light reading. Do it with pencil and paper in hand. Ask yourself questions about what you read. Avoid rote memorization, go for understanding.

Work problems: You can really become involved here. It's hands on, you are doing organic chemistry. This is probably the most important aspect of learning organic chemistry. For example, Chapter 2 has a lot of problems on drawing organic compounds. You *must* do these to become competent with organic structure! Without structure, little else is possible. Keep up with the problems as you read the book; don't wait until the night before an exam.

Study for deep and lasting understanding: Give yourself time. A rule of thumb in universities states that for every unit of coursework one is taking, 2 hours per week of outside study is needed. This would be a minimum for organic chemistry. If you are in a course with three lectures and one lab, for example, you need to be studying at least 10 hours per week. Avoid superficiality. Avoid rote memorization of things you don't understand. Exercise your mind, enjoy learning.

Preparing for exams: Test yourself. Never let the instructor be the first to test your knowledge. If you can talk about something, if you can explain it well to others, there is a good chance you know it.

This textbook can help: Here are a few of the features that will help you become engaged in learning.

- **Getting Involved:** Throughout the book you will find sections called ***Getting Involved.*** These will usually consist of the following. **(1) *Questions:*** First there will be some questions about what you just read. Take these seriously. Answer them with understanding. Re-read the preceding section if you can't answer them. Work with the material until you feel comfortable with these questions. **(2)** ***Worked examples:*** When appropriate, a worked example problem will be included to guide your approach to the exercises. **(3)** ***Problems:*** Following the questions will be some introductory problems on the material. You can become engaged, actually work with the material while it is fresh and new. **(4)** ***End-of-chapter problems:*** Finally, these "Getting Involved" sections will refer you to additional end-of-chapter problems. You can work them now or later, but you need the practice that comes with these additional problems. There is a separate solutions manual that accompanies this textbook. Work the problems first and then use the solutions manual to check your work. Don't just look at the answer and think "I understand, I could do that." This is still at the telling and showing stage of learning . . . you must become engaged by actually working the problems. The solutions manual also has a chapter summary that you might use to test yourself on the material. Read the summary sections and then explain the concepts to yourself.

- **Glossary:** Key terms are defined in the margins and again in a glossary as an appendix at the back of the book.
- **Reaction summary:** Each chapter with reactions has a summary of these reactions at the end.
- **Activities with Molecular Models:** The first 13 chapters have molecular model exercises you can do with a model kit or with computer modeling programs.
- **Nomenclature summary:** Nomenclature is covered by functional group throughout the book. A summary in the appendix integrates all of nomenclature.
- **Web site:** Prentice Hall has an Internet Web site for this book
  www.prenhall.com/bailey
  with links to other materials and learning aids.

Good luck. We hope you enjoy learning organic chemistry.

Philip S. Bailey, Jr.
Christina A. Bailey

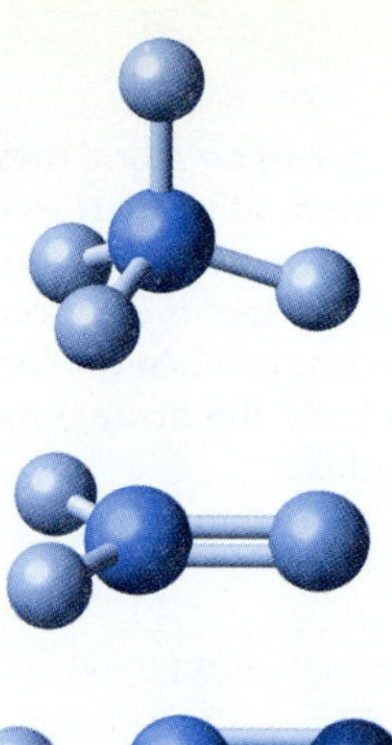

CHAPTER 1

Bonding in Organic Compounds

Organic chemistry is the study of compounds of carbon. Life on Earth is based on the element carbon and thrives because living organisms are able to synthesize marvelously large and complex as well as incredibly small and simple organic compounds. These compounds allow organisms to grow, to reproduce, and to occupy their places in nature. As mysterious as life itself is, the chemicals of life are simply combinations of atoms of only a handful of elements held together by chemical bonds. In this chapter you will see how these chemical bonds are formed and of what they are composed.

organic chemistry
the chemistry of the compounds of carbon

1.1 Elements and Compounds: Atoms and Molecules

Elements are the fundamental building units of all substances, living and nonliving, in our known universe. They are composed of remarkably tiny particles called **atoms,** the smallest particles of an element that still retain the chemical properties of that element. Elements—actually their atoms—enter into a seemingly infinite variety of chemical combinations to form compounds. The result of the bonding together of a group of atoms is a **molecule,** which is the smallest particle of a compound that still exhibits the chemical properties of that compound.

atom
smallest particle of an element

molecule
smallest particle of a compound; a bonded group of atoms

Atoms are composed of **neutrons,** which are electrically neutral, **protons,** which are positively charged, and **electrons,** which carry a negative charge. Protons and neutrons account for virtually the entire mass of an atom (Table 1.1), yet they reside in the infinitesimally small atomic **nucleus.** The nucleus is surrounded by electrons that are equal in number to the protons in the nucleus, thereby imparting neutrality of charge to the atom itself. These electrons occupy a tremendous volume, approximately a billion times the volume of the nucleus, yet they themselves are almost without mass. If all the space occupied by the electrons in a nickel could be filled with neutrons and protons, the coin would increase in weight from 5.5 grams to 100 million tons. A marble hung in a gigantic indoor sports arena gives a visual image of the relationship between an atom's nucleus and the space occupied by its surrounding electrons.

neutron
neutral subatomic particle with mass = 1 amu

proton
positively charged subatomic particle with mass = 1 amu

electron
negatively charged subatomic particle with negligible mass

nucleus
center of atom; contains protons and neutrons

Table 1.1 Comparison of Subatomic Particles

	Neutron	Proton	Electron
Charge	0	+1	−1
Mass 1	1 amu*	1 amu	$\frac{1}{1840}$

* amu = atomic mass unit.

Because electrons comprise virtually all of the volume of an atom, they play a predominant role in determining the chemical and physical properties of elements and compounds. Chemical compounds are formed by the transfer of electrons from one atom to another to form ions or by the sharing of electron pairs between atoms to form molecules. To understand the importance of electrons, you must be aware of electron configuration, particularly the arrangement of electrons at the outer extremities of an atom or molecule.

GETTING INVOLVED

- ✓ How do elements and compounds differ? What is the relationship of atoms and molecules to elements and compounds?
- ✓ How do electrons, protons, and neutrons compare in terms of mass and charge? What are their relative positions in an atom? Why is an atom neutral?

1.2 Electron Configuration

A. Atomic Number and Atomic Mass

atomic number
number of protons (or electrons) in an atom

mass number
number of protons plus neutrons in an atom

isotope
atoms of an element that differ in number of neutrons

atomic mass
weighted average of an element's naturally occurring isotopes

The **atomic number** of an atom is the number of protons in the nucleus. Since, in a neutral atom, the number of electrons and protons is equal, the atomic number also is the number of electrons surrounding the nucleus.

The **mass number** is the number of protons plus neutrons in the nucleus of an atom. Electrons have negligible mass and are generally disregarded in describing the mass of an atom. **Isotopes** are atoms with the same numbers of protons and of electrons, but different numbers of neutrons; thus they have the same atomic number but different mass numbers. The **atomic mass** of an element is the weighted average of the naturally occurring isotopes. Because of this it is usually not an integer. For example, the two most abundant isotopes of bromine have mass numbers of 79 (35 protons and 44 neutrons) and 81 (35 protons and 46 neutrons); both isotopes have 35 electrons, the atomic number. Since in nature these two isotopes occur in almost equal proportions, the atomic weight of bromine is 79.9. (Which isotope is slightly more abundant?) Look at the periodic table on the inside back cover; the atomic number is always an integer, but the atomic mass is not.

GETTING INVOLVED

- ✓ What is the difference between atomic number and mass number?
- ✓ How are isotopes of an element similar and different?
- ✓ How is the atomic mass of an element related to its isotopes?

Example 1.1

How many electrons, protons, and neutrons compose the oxygen isotope with mass number = 18.

Solution

Oxygen has an atomic number of 8 and thus has 8 protons and 8 electrons. To obtain a mass number of 18, there must be 10 neutrons.

Problem 1.1

How many electrons, protons, and neutrons do each of the following atoms have? **(a)** carbon: mass number = 12 and mass number = 13; **(b)** chlorine: mass number = 35 and mass number = 37. Give the atomic number and atomic mass of the following elements: **(c)** F; **(d)** S; **(e)** Al.

See related problem 1.16.

B. Atomic Orbitals

An atom's electrons, equal to the atomic number, do not exist randomly around the nucleus. They are described as residing in shells or energy levels that have been assigned the numbers 1 through 7, starting at the one nearest to the nucleus. Each shell has a certain capacity for electrons, as described in Table 1.2. As an electron's distance from the nucleus increases, its energy level also increases, and it is held less strongly by the positive pull of the nucleus.

Within each shell, electrons occupy specific atomic **orbitals.** An orbital describes an electron's distance from the nucleus and the shape and geometric orientation of the volume it occupies. An atomic orbital can be occupied by zero, one, or two electrons of opposite spin.

orbital
a defined region in space with the capacity to be occupied by up to two electrons

The shape and geometric orientation of the space occupied by electrons are described by s, p, d, and f orbitals. In organic chemistry, we will be involved almost exclusively with the s and p orbitals. Their shapes are easily described.

An **s orbital** is spherical, with the atom's nucleus located at its center (Figure 1.1); s orbitals in succeeding shells are all spherical but of increasing size. If an s orbital in the first shell is imagined to be a small marble, the s orbital in the second shell would be a Ping Pong ball, the one in the third shell a tennis ball, the fourth a softball, the fifth a soccer ball, the sixth a basketball, and the seventh a beach ball. Each succeeding orbital encompasses the previous ones.

s orbital
a spherical atomic orbital

A **p orbital** has two lobes with the atom's nucleus located between the two lobes (Figure 1.2a). Each shell, except the first, has three p orbitals identical in energy, size and shape and perpendicular to one another. On a three-dimensional coordinate system, one is oriented along the x-axis (p_x), another along the y-axis (p_y), and the third along the z-axis (p_z). These are pictured in Figure 1.2b.

p orbital
a double-lobed atomic orbital

Table 1.2 Electron Shells

Shell Number	Maximum Number of Electrons	Orbitals
1	2	1s
2	8	2s 2p
3	18	3s 3p 3d
4	32	4s 4p 4d 4f

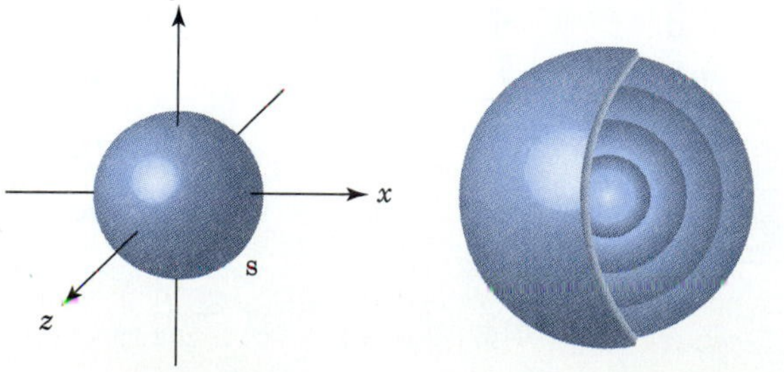

Figure 1.1 Spherical s orbitals. The cutaway model on the right indicates the concentric spherical s orbitals of succeeding energy levels.

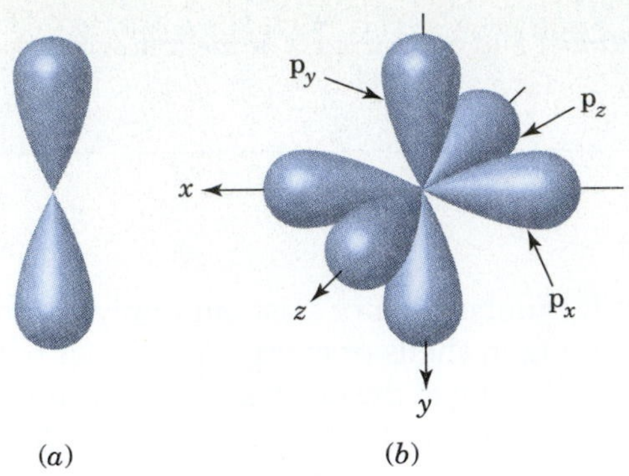

FIGURE 1.2
(a) Representation of a double-lobed p atomic orbital. (b) The three p orbitals within a particular shell are mutually perpendicular to one another (p_x, p_y, p_z).

GETTING INVOLVED

- ✓ What is an atomic orbital? What are the shapes of s and p orbitals?
- ✓ How does an s or p orbital in one energy level (shell) differ from one in the next energy level?
- ✓ Within a shell, how do p_x, p_y, and p_z orbitals differ from one another?

C. Filling Atomic Orbitals

Any orbital, in any shell, can accommodate a maximum of two electrons, which must have opposite spins. The first shell of electrons in an atom is composed of a single s orbital and has a maximum capacity of two electrons. The second shell has a single s orbital and a set of three p orbitals that can hold a total of six electrons. Thus the maximum electron capacity for the second shell is eight (two s electrons and six p electrons). The third shell can accommodate 18 electrons—two s electrons, six p electrons, and ten d electrons in five d orbitals. In addition to the s, p, and d orbitals, the fourth shell of electrons has a set of seven f orbitals that can accommodate 14 electrons, giving a maximum shell capacity of 32. See Figure 1.3 and Table 1.2.

Aufbau principle
the described order of filling atomic orbitals from lowest to highest energy with electrons

electron configuration
description of orbital occupancy by electrons of an atom or ion by energy level and number of electrons

In filling atomic orbitals, electrons occupy the orbitals nearest the atomic nucleus first—the lowest-energy orbitals—and then proceed to orbitals of higher energy. This principle is known as the **Aufbau principle;** the exact order of filling orbitals is illustrated in Figure 1.3. **Electron configurations** can be written for the elements by (1) filling atomic orbitals from lowest energy to highest energy (Aufbau principle); (2) placing one electron at a time in the orbitals of a given set

Type of Orbital	Number of Orbitals	Number of Electrons
s	1	2
p	3	6
d	5	10
f	7	14

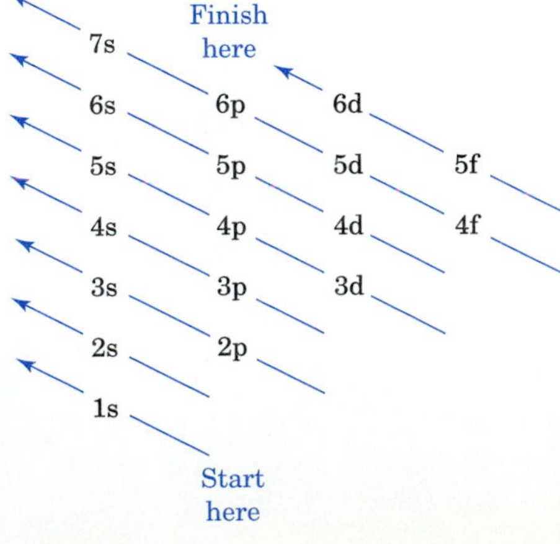

Figure 1.3 Aufbau principle. Atomic orbitals are filled beginning with the lowest-energy orbitals and proceeding to higher-energy ones. Follow the diagonal arrows to determine the order of priority for filling atomic orbitals.

Table 1.3 Electron Configurations

	Atomic Number	Number of Electrons	1s	2s	$2p_x$	$2p_y$	$2p_z$	Electron Configuration
H	1	1	↑					$1s^1$
He	2	2	↑↓					$1s^2$
Li	3	3	↑↓	↑				$1s^22s^1$
Be	4	4	↑↓	↑↓				$1s^22s^2$
B	5	5	↑↓	↑↓	↑			$1s^22s^22p^1$
C	6	6	↑↓	↑↓	↑	↑		$1s^22s^22p^2$
N	7	7	↑↓	↑↓	↑	↑	↑	$1s^22s^22p^3$
O	8	8	↑↓	↑↓	↑↓	↑	↑	$1s^22s^22p^4$
F	9	9	↑↓	↑↓	↑↓	↑↓	↑	$1s^22s^22p^5$
Ne	10	10	↑↓	↑↓	↑↓	↑↓	↑↓	$1s^22s^22p^6$

(orbitals of the same energy) until each is half full; and (3) placing remaining electrons in these half-filled orbitals, so that no more than two electrons, of opposite spin, occupy a single orbital. This process is illustrated in Table 1.3 for elements in the first two periods of the periodic table.

GETTING INVOLVED

✓ What is the Aufbau principle and how is it used in writing electron configurations?

✓ How many electrons can a set of s orbitals and a set of p orbitals in a shell accommodate?

Example 1.2

Write the complete electron configuration for Ca.

Solution

$1s^2\ 2s^2\ 2p^6\ 3s^2\ 3p^6\ 4s^2$

Problem 1.2

Write electron configurations as shown in Table 1.3 for all elements in period 3 of the periodic table.

D. Electron Configuration and the Periodic Table

Elements are organized into groups in the periodic table (see the inside back cover for a complete periodic table). Most of the elements that are of interest to us in organic chemistry are found in the first three periods and have only s and p electrons. Using the Aufbau principle, let us consider the logical development of the first three periods of the periodic table while taking special note of each element's outer-shell electron configuration.

Period 1 contains two elements: $_1H$ ($1s^1$) and $_2He$ ($1s^2$). These represent the orderly completion of the first energy level and its 1s orbital. The next element, $_3Li$, begins period 2, since according to Figure 1.3 the second energy level is next to be filled. Lithium and beryllium complete the 2s orbital. With the next six elements, B, C, N, O, F, and Ne, the three p orbitals (six total electrons) in the second shell are occupied (see Figure 1.4). In period 3, the 3s (two electrons, two elements) and 3p (six electrons, six elements) are being filled. This period

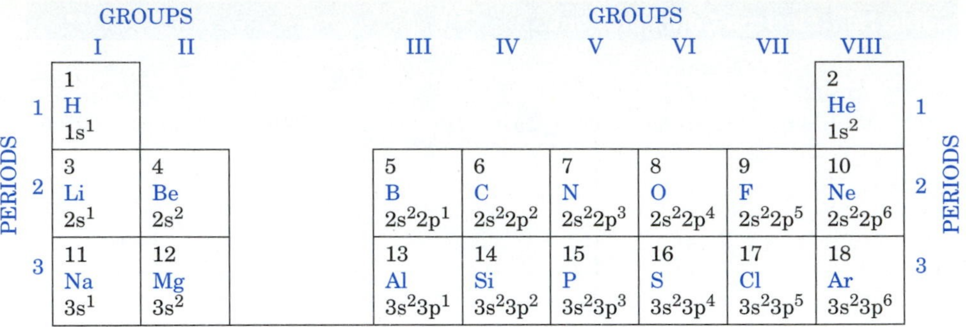

Figure 1.4 Periods 1 to 3 of the periodic table. Note that only outer-shell electron configurations are shown here. See the inside back cover for the complete table.

begins with ${}_{11}Na$ ($1s^22s^22p^63s^1$) and ends with ${}_{18}Ar$ ($1s^22s^22p^63s^23p^6$). Notice that only the outer-shell electron configuration of each element is shown in Figure 1.4.

If one continues, period by period, following the Aufbau principle and the periodic table, the following observations can be made. Groups I and II result from filling s orbitals, groups III to VIII from filling p orbitals, the three transition series from filling the 3d, 4d, and 5d orbitals, and the lanthanide and actinide series from filling the 4f and 5f orbitals (14 electrons, 14 elements). Notice that the elements in each vertical group (I to VIII) have the same general outer-shell electron configuration but that the electrons are located in different shells (principal energy levels). As a result, the elements within a particular group show similar chemical properties. Figure 1.4 shows this for part of the periodic table; also see Table 1.4.

GETTING INVOLVED

✓ On the periodic table what types of orbitals are being occupied in Groups I-II, Groups III-VIII, and the transition series? Why are there 10 elements in each transition series?

✓ Within the elements of a group, how are the outer shell configurations the same and how are they different?

Example 1.3

Write the outer-shell electron configurations for each member of the halogen group (group VII).

Solution

All have outer shells of s^2p^5. Only the shell number differs and it is the same as the period number. F, $2s^22p^5$; Cl, $3s^23p^5$; Br, $4s^24p^5$; I, $5s^25p^5$; At, $6s^26p^5$.

Problem 1.3

Take a look at Figure 1.4. Extend it by writing the probable outer shell electron configurations of the elements in those groups in periods 4 and 5.

Problem 1.4

How many outer shell electrons does each of the following atoms have?

(a) H (b) Al (c) C (d) N (e) S (f) Br

See related problems 1.17–1.18.

Table 1.4 Outer-Shell Configuration within Groups

	Groups							
	I	II	III	IV	V	VI	VII	VIII
Number of electrons in outer shell	1	2	3	4	5	6	7	8
Outer-shell electron configuration	s^1	s^2	s^2p^1	s^2p^2	s^2p^3	s^2p^4	s^2p^5	s^2p^6

E. Stable Octets

The elements in group VIII, the noble gases, are the most stable and least reactive in the periodic table, entering into chemical combination infrequently and with difficulty. All these elements except helium have the same outer-shell or valence-shell configuration, s^2p^6, which is known as a **stable octet.** Helium has a $1s^2$ configuration. We shall find that elements, in forming chemical compounds, tend to achieve a stable configuration, that is, the same electron configuration possessed by a noble gas of group VIII.

stable octet
an outer-shell electron configuration of eight electrons (s^2p^6)

GETTING INVOLVED

✓ In terms of outer shell electron configuration, what is a stable octet and what is its significance?

Problem 1.5

Write the outer-shell electron configuration (including principal energy level) of every element in group VIII.

1.3 Ionic Bonding

A. Ionic Bonding, Electronegativity, Electron Configuration, and the Periodic Table

Ionic bonding involves the complete transfer of electrons between two atoms of widely different electronegativities to form ions. The atom losing electrons becomes positive, a *cation,* and the one gaining electrons becomes negative, an *anion.* The ionic bond results from the electrostatic attraction between these two oppositely charged species.

ionic bond
bond between two atoms caused by electrostatic attraction of plus and minus charged ions

What elements form ionic bonds with one another? For a complete transfer of one or more electrons from one atom to another to occur, one atom must have a very strong attraction for electrons and the other a very weak attraction. **Electronegativity** is defined as the attraction of an atom for its outer-shell electrons. Electronegativity increases from left to right within a period, since the number of protons per nucleus increases and the electrons are entering the same main energy level (outer shell). The attraction between the increasingly positive nucleus and electrons thus becomes stronger. Electronegativity decreases from top to bottom within a group, even though the nuclear charge increases, since the outer shell is farther and farther from the nucleus and is shielded from the nucleus by inner-shell electrons. As a result of their electronegativities, elements on the far left of the periodic table (low electronegativities) tend to lose electrons, and elements on the far right (high electronegativities) tend to gain electrons in ionic bonding.

electronegativity
ability of an atom to attract its outer-shell electrons and electrons in general

Salt flat

Elements gain and lose electrons to obtain a stable outer-shell configuration, in most instances a stable octet ($1s^2$ or s^2p^6). Consider as an example the reaction between sodium and chlorine atoms to form sodium chloride (table salt). In principle, sodium can obtain a complete outer shell of eight electrons by gaining seven electrons, or by losing one and leaving the underlying complete inner shell to become the outer shell. Similarly, in principle, chlorine can achieve an octet of electrons by gaining one electron, thus completing its outer shell, or by losing seven electrons. The simpler of these two possibilities occurs if one electron is transferred from the sodium (low electronegativity) to the chlorine (high electronegativity), resulting in a noble gas outer shell for each (NaCl).

$Na\cdot$	+	$:\ddot{\underset{\cdot\cdot}{Cl}}\cdot$	⟶	Na^+	+	$:\ddot{\underset{\cdot\cdot}{Cl}}:^-$
$1s^22s^22p^63s^1$		$1s^22s^22p^63s^23p^5$		$1s^2 2s^22p^6$		$1s^22s^22p^6 3s^23p^6$
One outer-shell electron		Seven outer-shell electrons		($_{10}$Ne configuration)		($_{18}$Ar configuration)

Elements on the far left of the periodic table (such as sodium in the example) achieve a stable octet by releasing the small number of electrons in their outer shells, thereby leaving the underlying complete inner shell. These are called **electropositive** elements; they form ions with a positive charge, **cations.** The acquisition of these electrons by elements on the far right of the periodic table (such as chlorine) completes their outer shells, resulting in a noble gas configuration. These elements are **electronegative** and form ions with a negative charge, **anions.** Elements on the left of the periodic table have low electronegativities (are electropositive) and can lose electrons easily; those on the right have high electronegativities and easily gain electrons. The intrinsic logic of the periodic table is evident. You should review the **charges** of the ions commonly found in ionic compounds, as presented in Table 1.5.

electropositive
element with electron-donating capabilities

cation
positively charged ion

electronegative
element with electron-attracting capabilities

anion
negatively charged ion

ionic charge
sign and magnitude of the charge on an ion

Ionic compounds, such as sodium chloride, do not exist as molecules. In the crystalline state of sodium chloride, each sodium ion is surrounded by six chloride ions and each chloride ion by six sodium ions (Figure 1.5). The number of sodium ions equals the number of chloride ions, but no one sodium can be identified as belonging to a particular chloride and vice versa.

Table 1.5 Ionic Charges

Group Charges		Other Elements				
Group I	1+	H^+	Zn^{2+}	Cu^+, Cu^{2+}	Fe^{2+}, Fe^{3+}	Pb^{2+}, Pb^{4+}
Group II	2+	Ag^+	Cd^{2+}	Hg^+, Hg^{2+}		Sn^{2+}, Sn^{4+}
Group III	3+		Ni^{2+}			
Group VI	2−		Co^{2+}			
Group VII	1−					

Polyatomic Ions							
NH_4^+	Ammonium	HCO_3^-	Bicarbonate	SO_4^{2-}	Sulfate	PO_4^{3-}	Phosphate
OH^-	Hydroxide	ClO_3^-	Chlorate	CO_3^{2-}	Carbonate		
NO_2^-	Nitrite	MnO_4^-	Permanganate	CrO_4^{2-}	Chromate		
NO_3^-	Nitrate	CN^-	Cyanide	$Cr_2O_7^{2-}$	Dichromate		

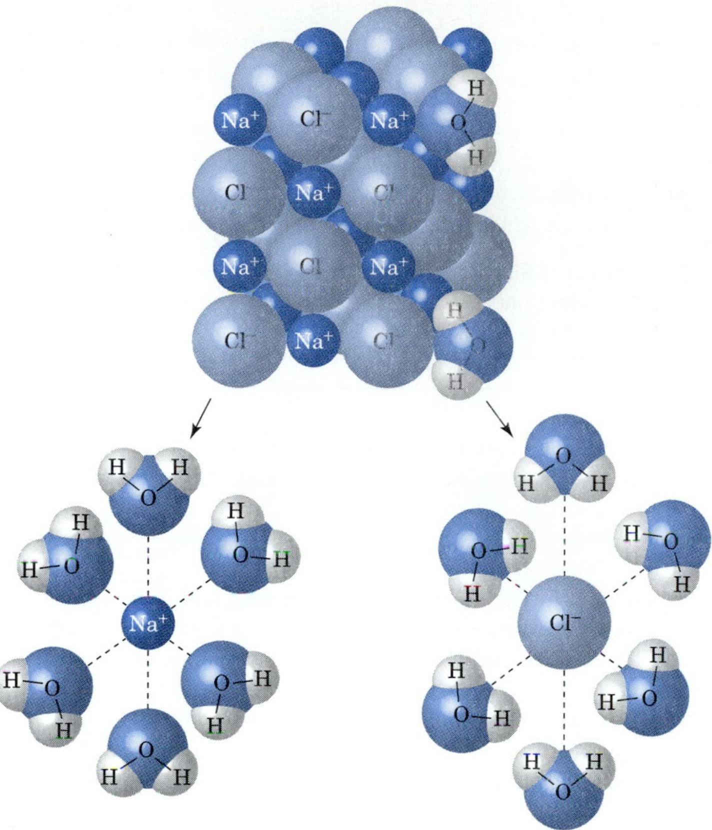

Figure 1.5 Crystalline structure of sodium chloride. Dark circles represent Na^+ and light circles Cl^-. When the ions are dissolved in water, they become independent and solvated (surrounded) by water molecules.

B. Electron Dot Representation of Ions

Ion formation can be illustrated very simply with the aid of electron dot formulas, in which the outer-shell electrons of the involved atoms are represented by dots. When a positive ion is formed, the outer-shell electrons of the atom forming the cation are lost, thereby leaving a complete inner shell, a stable octet, as the new outer shell (this shell is usually not shown in the ion). The atom forming the anion gains electrons until there are eight in the outer shell. The following examples, showing the formation of the salt substitute potassium chloride and lime (calcium oxide), illustrate the electron dot method.

$$\mathrm{K\cdot} + \cdot\ddot{\underset{\cdot\cdot}{\mathrm{Cl}}}\mathrm{:} \longrightarrow \mathrm{K}^+ + \mathrm{:}\ddot{\underset{\cdot\cdot}{\mathrm{Cl}}}\mathrm{:}^-$$

$$\mathrm{Ca:} + \cdot\ddot{\underset{\cdot}{\mathrm{O}}}\mathrm{:} \longrightarrow \mathrm{Ca}^{2+} + \mathrm{:}\ddot{\underset{\cdot\cdot}{\mathrm{O}}}\mathrm{:}^{2-}$$

GETTING INVOLVED

- ✓ How do cations and anions differ? How is each formed from a neutral atom?
- ✓ Why does electronegativity increase left to right in a period and decrease top to bottom within a group on the periodic table?
- ✓ Would anions form from atoms with high or low electronegativity? What about cations? Why are they different?

✓ How is an ionic bond formed? What causes the attraction in an ionic bond? What does the electronegativity of an atom and the concept of a stable octet have to do with ionic bond formation?

Example 1.4

Write the formula for the stomach antacid aluminum hydroxide.

Solution

The chemical formulas of ionic compounds can be written from their names by combining the ions of the components in a ratio that will produce a neutral compound. Aluminum has a valence of +3 and hydroxide of −1. They must be in a 1:3 ratio to produce a neutral compound: $Al(OH)_3$

Problem 1.6

Using both electron configurations as in section 1.3.A and electron dot formulas as in section 1.3.B, show the ionic reactions between the elements of the following pairs: **(a)** Li and F; **(b)** Mg and O.

Problem 1.7

Write formulas for the following familiar compounds: **(a)** sodium fluoride (in some fluoride toothpastes); **(b)** magnesium hydroxide (milk of magnesia); **(c)** ammonium sulfate (fertilizer); **(d)** lithium carbonate (used in the treatment of manic psychosis; **(e)** calcium oxide (lime); **(f)** calcium carbonate (limestone, stomach antacid); **(g)** sodium nitrite (food preservative); **(h)** potassium chlorate (component of match heads).

See related problem 1.19.

1.4 Covalent Bonding

A. Covalent Bonding, Electron Configuration, and the Periodic Table

covalent bond
bond formed by the sharing of electrons (in pairs) between two atoms

Unlike ionic bonds, which are formed by the complete transfer of electrons, **covalent bonds** involve the sharing of electron pairs between atoms of similar electronegativity. Because of the similarity in electronegativity, neither atom can relieve the other of its outer-shell electrons, as is the case in ionic bonding. Instead, each atom's nucleus is electrically attracted to the mutually shared electron pair, and a bond is the result.

electron dot formula
molecular representation using dots to show each atom's outer-shell electrons, both bonding and nonbonding pairs

bonding electron pair
outer-shell electron pair involved in a covalent bond

nonbonding electron pair
a lone outer-shell electron pair not involved in a bond

In the simplest kind of covalent bond formation, each atom involved provides one electron to the bond. A shared pair of electrons results, becoming part of the outer shell of each atom. Consider, for example, hydrogen, chlorine, and hydrogen chloride, which are represented below by **electron dot formulas** showing both **bonding** and **nonbonding** outer-shell electrons.

Atoms:	$H\cdot \quad \cdot H$	$:\overset{\cdot\cdot}{\underset{\cdot\cdot}{Cl}}\cdot \quad \cdot\overset{\cdot\cdot}{\underset{\cdot\cdot}{Cl}}:$	$H\cdot \quad \cdot\overset{\cdot\cdot}{\underset{\cdot\cdot}{Cl}}:$
Molecules:	$H:H$	$:\overset{\cdot\cdot}{\underset{\cdot\cdot}{Cl}}:\overset{\cdot\cdot}{\underset{\cdot\cdot}{Cl}}:$	$H:\overset{\cdot\cdot}{\underset{\cdot\cdot}{Cl}}:$

Usually, electron sharing occurs in a way that provides one or both atoms in the bond with the outer-shell configuration of a noble gas (in these cases two outer-shell electrons for hydrogen and eight for chlorine).

In Table 1.6, bond formation is related to groups in the periodic table. The number of covalent bonds an element may commonly form, the valence, is illustrated in this table by bonding each of the elements in each group to hydrogen. Hydrogen forms only one covalent bond, since it has only one electron ($H\cdot$).

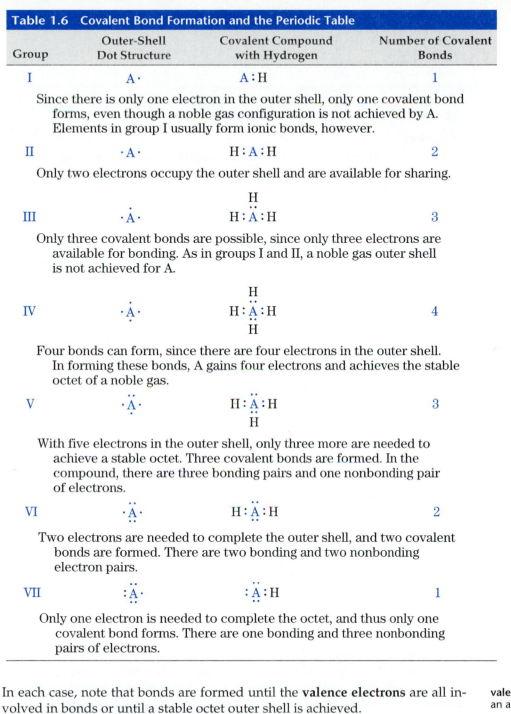

<table>
<tr><th colspan="4">Table 1.6 Covalent Bond Formation and the Periodic Table</th></tr>
<tr><th>Group</th><th>Outer-Shell Dot Structure</th><th>Covalent Compound with Hydrogen</th><th>Number of Covalent Bonds</th></tr>
<tr><td>I</td><td>A·</td><td>A:H</td><td>1</td></tr>
<tr><td colspan="4">Since there is only one electron in the outer shell, only one covalent bond forms, even though a noble gas configuration is not achieved by A. Elements in group I usually form ionic bonds, however.</td></tr>
<tr><td>II</td><td>·A·</td><td>H:A:H</td><td>2</td></tr>
<tr><td colspan="4">Only two electrons occupy the outer shell and are available for sharing.</td></tr>
<tr><td>III</td><td>·
·A·</td><td>H
··
H:A:H</td><td>3</td></tr>
<tr><td colspan="4">Only three covalent bonds are possible, since only three electrons are available for bonding. As in groups I and II, a noble gas outer shell is not achieved for A.</td></tr>
<tr><td>IV</td><td>·
·A·
·</td><td>H
··
H:A:H
··
H</td><td>4</td></tr>
<tr><td colspan="4">Four bonds can form, since there are four electrons in the outer shell. In forming these bonds, A gains four electrons and achieves the stable octet of a noble gas.</td></tr>
<tr><td>V</td><td>··
·A·
·</td><td>··
H:A:H
··
H</td><td>3</td></tr>
<tr><td colspan="4">With five electrons in the outer shell, only three more are needed to achieve a stable octet. Three covalent bonds are formed. In the compound, there are three bonding pairs and one nonbonding pair of electrons.</td></tr>
<tr><td>VI</td><td>··
·A·
··</td><td>··
H:A:H
··</td><td>2</td></tr>
<tr><td colspan="4">Two electrons are needed to complete the outer shell, and two covalent bonds are formed. There are two bonding and two nonbonding electron pairs.</td></tr>
<tr><td>VII</td><td>··
:A·
··</td><td>··
:A:H
··</td><td>1</td></tr>
<tr><td colspan="4">Only one electron is needed to complete the octet, and thus only one covalent bond forms. There are one bonding and three nonbonding pairs of electrons.</td></tr>
</table>

In each case, note that bonds are formed until the **valence electrons** are all involved in bonds or until a stable octet outer shell is achieved.

valence electrons an atom's outer-shell electrons

GETTING INVOLVED

- ✓ How do covalent bonds and ionic bonds differ? What causes the attraction between atoms in each?
- ✓ What is the difference between a bonding electron pair and a nonbonding electron pair?
- ✓ Elements in Groups I to IV on the periodic table form the same number of covalent bonds (when they do so) as their numbers of outer-shell electrons. What determines

the number of covalent bonds formed by elements in each of Groups V to VII? Which groups form stable octets upon bonding?

✓ Why do elements in Groups V to VII have nonbonding electrons in covalent compounds whereas those in Groups I to IV do not?

Example 1.5

How many covalent bonds do Al and Br normally form?

Solution

Al is in Group III, has three outer shell electrons, and forms three covalent bonds; a stable octet is not achieved. Br is in Group VII, has seven outer shell electrons, and forms one covalent bond to gain the additional electron for a stable octet.

Problem 1.8

How many covalent bonds would each of the following elements commonly form?
(a) B; **(b)** C; **(c)** Si; **(d)** N; **(e)** O; **(f)** S; **(g)** Cl; **(h)** I.

B. Covalent Bonding in Organic Compounds

In organic compounds, carbon, a member of group IV, forms four covalent bonds because it has four electrons in its outer shell and needs four more to attain a stable octet. More than any other element, carbon tends to share electrons with atoms of its own kind. An infinite variety of compounds results, ranging from single carbon molecules to gigantic chains of dozens or hundreds of carbon atoms.

```
    ·
   ·C·
 · ··  ·  ·  ·
·C:C:C:C:C·
 ·  ·  · ··  ·
         ·C·
          ·
```

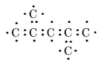

In this multicarbon structure, electrons are still available for bonding with carbon or other elements.

Elements other than carbon are also found in organic compounds. The following list of the elements most commonly found in organic compounds includes the number of covalent bonds each usually forms:

C	4
N	3
O, S	2
H	1
F, Cl, Br, I	1

valence
the number of covalent bonds an atom usually forms

single bond
bond with one shared pair of electrons

double bond
bond with two shared pairs of electrons

triple bond
bond with three shared pairs of electrons

The required **valence** of an atom does not have to be achieved using only single covalent bonds. Any combination of **single bonds** (one electron pair shared), **double bonds** (two electron pairs shared), and **triple bonds** (three electron pairs shared) can be used as long as the total adds up to the required valence.

Single Bond	Double Bond	Triple Bond
A:B	A: :B	A: : :B
A—B	A=B	A≡B

The four covalent bonds necessary for a carbon atom, for example, can be achieved in the following ways:

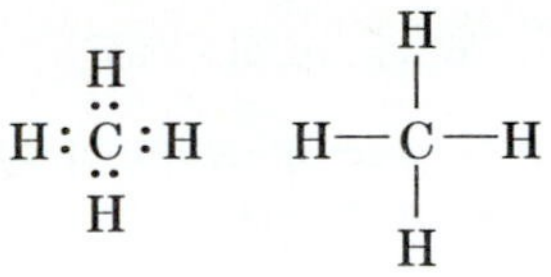

Four single bonds

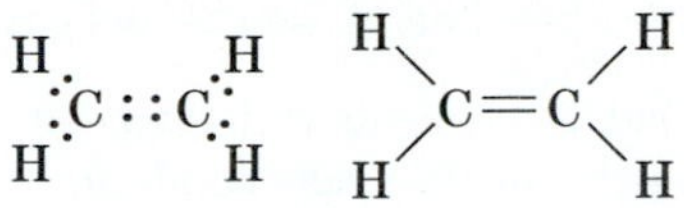

One double and two single bonds

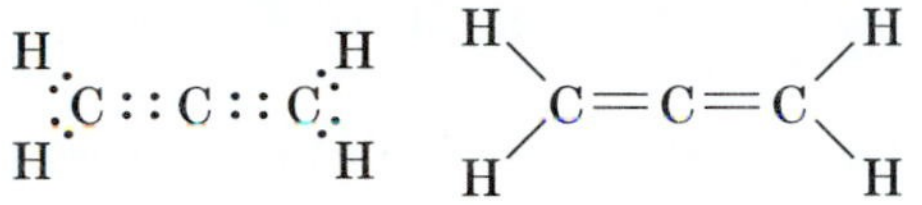

Two double bonds

One triple and one single bond

These represent two common ways to depict covalent bonds in molecules: electron dot formulas, in which all bonding and nonbonding electron pairs are shown (also see Table 1.6), and **line-bond formulas,** in which lines are used for bonds (each line represents a bonding pair of electrons). Electron dot formulas are often called **Lewis structures** after the American chemist G. N. Lewis, who, in 1915, proposed theories describing covalent bonding and what today is known as the *octet rule.*

line-bond formula
molecular representation in which bonding electron pairs are represented by lines

Lewis structure
another term for electron dot formula

GETTING INVOLVED

✓ What are the valences of the elements commonly found in organic compounds?
✓ How do single, double, and triple bonds differ?
✓ What are Lewis electron dot formulas?

C. Drawing Electron Dot Formulas

The following two rules must be followed in writing electron dot formulas of covalent compounds:

1. Use every atom in the molecular formula.
2. Satisfy the valence (the number of covalent bonds formed) of each atom: C-4; N-3; O, S-2; H, F, Cl, Br, and I-1.

The following procedure is useful:

1. Write each atom, showing its valence electrons. Assign a valence to each.
2. Bond together continuously with single bonds (one electron pair) all atoms with a valence greater than one. In bonding two atoms in a single bond, use one electron from each.
3. Attach the monovalent atoms to the polyvalent atoms until all valences are satisfied.
4. If there are insufficient monovalent atoms to accomplish step 3, insert double and triple bonds between the polyvalent atoms until all valences can be satisfied. Making cyclic structures may also be helpful.

GETTING INVOLVED

Example 1.6

Draw electron dot formulas for the following, showing all bonding and nonbonding electron pairs.

(a) C_3H_8, propane, is used as fuel gas for camping and in rural areas.

Bond the elements with the greatest valences (valences greater than 1) together by the single bonds and then fill in with hydrogens.

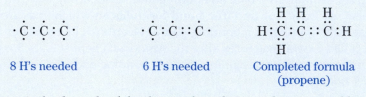

·C· (Valence = 4) H· (Valence = 1) ·C:C:C· H:C:C:C:H (with H above and below each C)

(b) C_3H_6, propene, (precursor of polypropylene plastic) and cyclopropane (an inhalation anesthetic).

See example **(a)**. Note that there are only six hydrogens here. Insertion of a double bond is helpful.

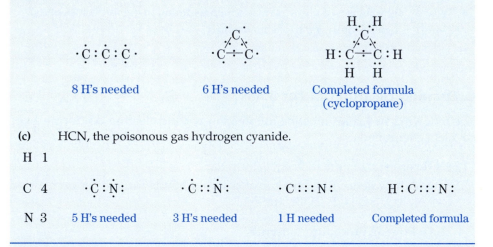

·C:C:C· (8 H's needed) ·C:C::C· (6 H's needed) H:C:C::C:H with H above each C and H below the first C (Completed formula (propene))

Alternatively, the ends of the three-carbon chain can be connected by a single bond.

·C:C:C· (8 H's needed) ring of three C's joined by single bonds (6 H's needed) ring with two H's on each C, H:C–C:H (Completed formula (cyclopropane))

(c) HCN, the poisonous gas hydrogen cyanide.

Element	Valence				
H	1				
C	4	·C:N:	·C::N:	·C:::N:	H:C:::N:
N	3	5 H's needed	3 H's needed	1 H needed	Completed formula

Problem 1.9

Draw electron dot formulas for the following compounds, showing all bonding and nonbonding electron pairs: **(a)** $CHCl_3$ (chloroform); **(b)** CH_2O (formaldehyde, a biological preservative); **(c)** CO_2 (carbon dioxide); **(d)** C_2HCl (chloroacetylene).

See related problems 1.20–1.25.

D. Structural Nature of Compounds

Ionic compounds are composed of positive and negative ions in a formula ratio providing for an electrically neutral compound. If an ionic compound is melted or dissolved in water, the positive and negative ions go into independent motion. For example, crystalline or liquid salt or salt in solution contains no molecules of sodium chloride, merely a conglomeration of positive and negative ions, as in Figure 1.5.

Covalent compounds are composed of molecules. The atoms in a molecule belong exclusively to that particular molecule and travel together as a fixed unit. In the solid state of sugar, $C_{12}H_{22}O_{11}$, each unit of the crystal lattice is a complete sugar molecule. When the sugar crystal is dissolved in water, it disperses into

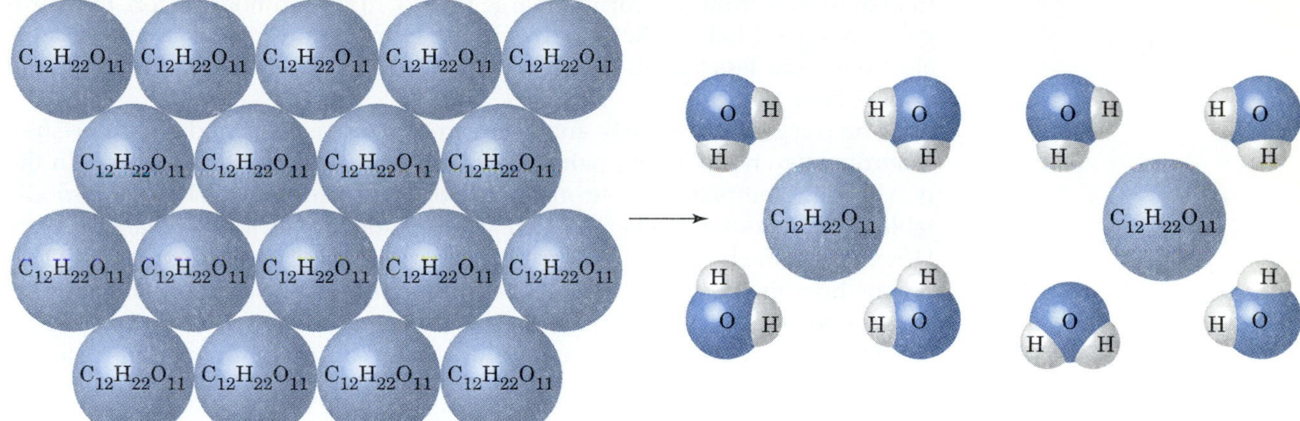

Figure 1.6 Sugar molecules dissolving in water.

sugar molecules with each molecule containing 12 carbons, 22 hydrogens, and 11 oxygens (Figure 1.6). Not only are the atoms of a covalent molecule exclusive to that particular molecule, but they are arranged in a specific pattern. The molecule has a definite shape and three-dimensional geometry, with definite bond lengths and angles.

E. Polyatomic Ions and Formal Charge

In studying chemical reactions, we will encounter polyatomic charged species in which the atoms are connected by covalent bonds. It is useful to know how to draw these **polyatomic ions** and determine the specific atom or atoms on which the charge resides, the location of the **formal charge.** The formal charge on an atom is equal to the difference between the number of valence electrons it has as a neutral free atom and the number assigned to it when it is bonded to other atoms. As we have seen, the number of outer-shell electrons in a neutral unbonded atom is equal to the atom's group number in the periodic table (review Table 1.4).

polyatomic ion
ion composed of several atoms

formal charge
difference between the number of outer-shell electrons "owned" by a neutral free atom and the same atom in a compound

A covalently bonded atom is assigned all of its outer-shell unshared, nonbonding electrons and half of the shared, bonding electrons. As an illustration, let's take a look at the ammonium and hydroxide ions compared to ammonia and water.

Ammonia	Ammonium ion	Water	Hydroxide ion
$H:\ddot{N}:H$ with H below	$H:\ddot{N}^{+}:H$ with H above and below	$H:\ddot{O}:H$ (lone pair below)	$H:\ddot{O}:^{-}$ (lone pair below)

How can we tell which of these species are positive, negative, or neutral? And in the case of charged species, how do we identify the atom or atoms bearing charges?

Let us look at the constituent atoms individually, starting with hydrogen, which has a formal charge of zero in all four species. Neutral hydrogen has one electron. In each of these cases, hydrogen shares a pair of bonding electrons and has formal possession of one of them.

Nitrogen is in group V of the periodic table and will have complete possession of five outer-shell electrons when neutral. In ammonia, nitrogen is assigned the nonbonding pair and half of the electrons in the three bonding pairs for a total of five electrons. Its formal charge is zero. In the ammonium ion, the nitrogen is assigned half of the electrons in the four bonding pairs—a total of four electrons—one less than its atomic outer shell. Thus nitrogen has a formal charge of plus one.

The oxygen is similarly analyzed. In water it is assigned six outer-shell electrons, two nonbonding pairs (four electrons), and half the electrons in the two bonding pairs (two electrons). Since oxygen is in group VI of the periodic table (with six outer-shell electrons), we conclude that its formal charge is zero. However, in the hydroxide ion, oxygen owns seven outer-shell electrons—half on one bonding pair and three nonbonding pairs—and has a formal charge of minus one.

The method for determining formal charge can be summarized by the following equation.

$$\text{Formal charge} = \begin{bmatrix}\text{group number,}\\ \text{i.e., number of}\\ \text{electrons in}\\ \text{outer shell of}\\ \text{the neutral atom}\end{bmatrix} \text{minus} \begin{bmatrix}\text{unshared}\\ \text{(nonbonding)}\\ \text{electrons}\end{bmatrix} \text{minus} \begin{bmatrix}\text{half of the}\\ \text{shared}\\ \text{(bonding)}\\ \text{electrons}\end{bmatrix}$$

Let us apply this to the following ion shown in Example 1.7 to determine the location of the positive charge in the following example.

GETTING INVOLVED

✓ In terms of valence electrons, what is the difference between atoms that have no formal charge, those with a positive formal charge, and those with a negative formal charge?

Example 1.7

Assign formal charges to each atom in the following ion.

$$\left[\begin{array}{c} \text{H}\ \ \text{H} \\ \text{H}:\overset{..}{\underset{..}{\text{C}}}:\overset{..}{\underset{..}{\text{O}}}:\text{H} \\ \text{H} \end{array}\right]^{+}$$

Hydrogens $= (1) - (0) - \frac{1}{2}(2) = 0$ formal charge

Carbon $= (4) - (0) - \frac{1}{2}(8) = 0$ formal charge

Oxygen $= (6) - (2) - \frac{1}{2}(6) = +1$ formal charge

Problem 1.10

Write electron dot formulas for the $CH_3O^{\ominus}$ and $CH_3NH_3^{\oplus}$ ions and determine which atoms have a formal charge.

See related problems 1.26–1.27.

F. Polar Covalent Bonds

An ionic bond is formed by the transfer of electrons between atoms of widely different electronegativities. In a covalent bond, electrons are shared between atoms of identical or similar electronegativities. Covalent bonds in which the electronegativities of the atoms are dissimilar, but not sufficiently different to cause complete transfer of electrons, are called **polar covalent bonds.** In polar covalent bonds, the shared electrons are pulled closer to the more electronegative atom so that the electrons are unequally shared. As a result, since electrons are negative, the more electronegative atom develops a partially negative charge and the other atom develops a partially positive charge.

polar bond
covalent bond between two atoms of different electronegativities, causing one atom to have a greater attraction for the bonding pair(s) and thus charge separation within the bond

A polar covalent bond can be represented by

$$\overset{\delta+}{A}-\overset{\delta-}{B}$$

where atom B is more electronegative than atom A. The symbol δ (delta) is used to signify that only partial charges are formed, not full ones as in ionic compounds.

An oversimplified but useful way to use electronegativity for predicting polarity is to consider that carbon and hydrogen have almost identical electronegativities. Of the atoms commonly found in organic compounds, those to the right of carbon and hydrogen in the periodic table are more electronegative, and those to the left are less electronegative.

	H			
B	C	N	O	F
		P	S	Cl
				Br
				I

Electronegativity increases →

Most carbon-carbon and carbon-hydrogen bonds are nonpolar. The following examples show how polarity is predicted using the electronegativity gradient.

$$\overset{\delta+}{H}-\overset{\delta-}{O} \qquad \overset{\delta+}{C}-\overset{\delta-}{Cl} \qquad \overset{\delta+}{N}=\overset{\delta-}{O} \qquad \overset{\delta-}{C}-\overset{\delta+}{B} \qquad \overset{\delta+}{C}\equiv\overset{\delta-}{N}$$

The concept of polar bonds will be used frequently to predict and explain reactions of organic compounds.

GETTING INVOLVED

✓ What are the similarities and differences among ionic bonds, covalent bonds, and polar covalent bonds?

Problem 1.11

Using δ+ and δ−, illustrate the polarity of the following bonds: **(a)** C—Br; **(b)** C=O; **(c)** N—H; **(d)** C=N; **(e)** C—O; **(f)** C—S.

See related problem 1.28.

1.5 An Orbital Approach to Covalent Bonding

A. Sigma and Pi Covalent Bonds

How are electron pairs shared to form covalent bonds? This is accomplished by the overlap of atomic orbitals (each with one electron) to form a **covalent bond** consisting of two spin-paired electrons. There are two important types of covalent bonds—sigma bonds and pi bonds.

covalent bond
a covalent bond results from the overlap of two atomic orbitals, each with one electron

sigma bond
a covalent bond formed by the head-to-head overlap of atomic orbitals

pi bond
a covalent bond formed by the overlap of parallel p orbitals at both lobes

A **sigma bond (σ bond)** is formed by the head-to-head overlap of atomic orbitals in one position. As shown in Figure 1.7, such a bond can be formed by the overlap of s orbitals, as in hydrogen; end-to-end overlap of p orbitals, as in chlorine; or s-p overlap, as in hydrogen chloride.

A **pi bond (π bond)** is formed when parallel p orbitals, each with one electron, overlap in two positions (Figure 1.8).

GETTING INVOLVED

- ✓ How are covalent bonds formed from atomic orbitals?
- ✓ What are the two types of covalent bonds and how do they differ?
- ✓ For practice, draw a sigma bond between two carbons and then draw a pi bond between two carbons.

B. Electron Configuration of Carbon

The chemical properties of an element depend on the electron configuration of the outer shell. Carbon has four electrons in its outer shell, two in the 2s orbital, and one each in the $2p_x$ and $2p_y$ orbitals. One would expect carbon, with this configuration, to be divalent, since the 2s orbital is filled and only the $2p_x$ and $2p_y$ orbitals have an unpaired electron to share. Carbon's tetravalence is explained by promoting one 2s electron to a 2p orbital, creating four unpaired electrons during bonding (Figure 1.9). Since bond formation is an energy-releasing process and since the formation of four bonds creates a stable octet in carbon's outer shell, the total process is energetically favorable. Figure 1.10 shows the shapes and geometric orientations of the four half-filled orbitals in carbon's outer shell.

C. Shapes of Organic Molecules

Let us consider a carbon with two atoms bonded to it by a triple bond and a single bond. How would these two atoms orient themselves around the carbon? There are two extreme possibilities: one in which the two atoms are as close to

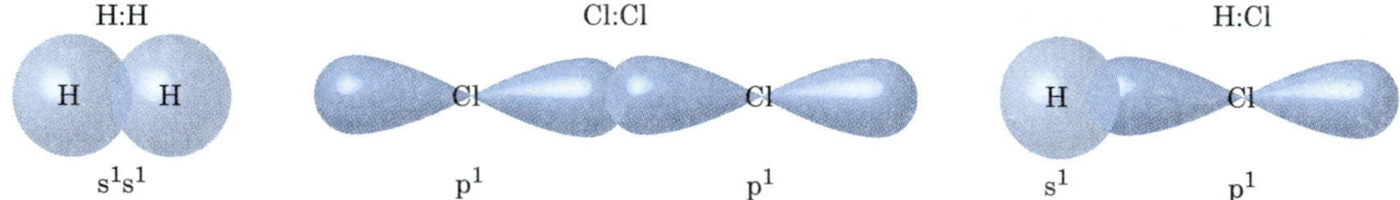

Figure 1.7 Sigma (σ) bonds. Atomic orbitals overlap in one position to form σ bonds.

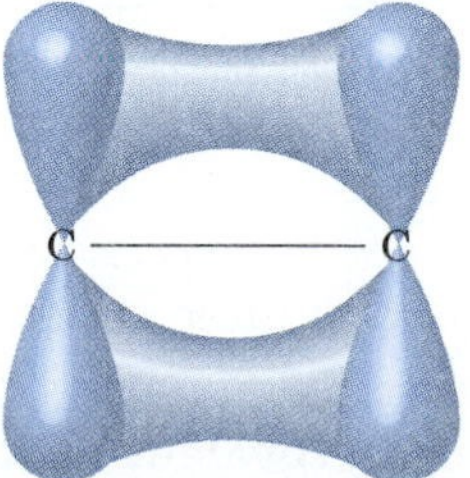

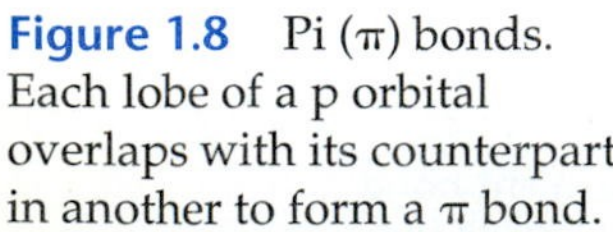

Figure 1.8 Pi (π) bonds. Each lobe of a p orbital overlaps with its counterpart in another to form a π bond.

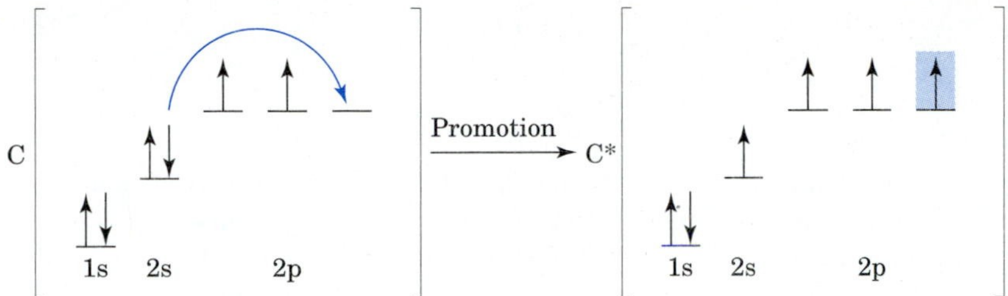

Figure 1.9 Promotion of an electron in carbon, allowing formation of four covalent bonds. It will be seen later that this does not describe the whole story; orbital hybridization must occur also.

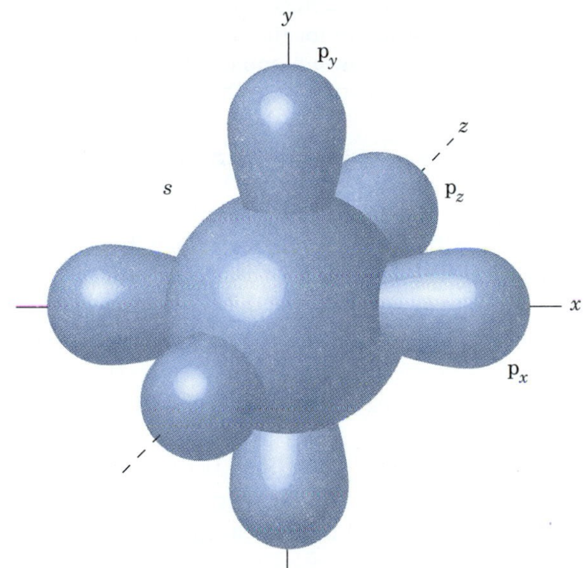

Figure 1.10 Outer-shell atomic orbitals of carbon. The carbon nucleus is at the origin. It is surrounded by one spherical s orbital and three double-lobed orbitals (p_x, p_y, and p_z), which are identical except in geometric orientation. Each orbital possesses one electron after promotion of one of the 2s electrons.

one another as possible, and one in which they are as far apart as possible. Common sense would lead us to choose the latter case, in which the two atoms are a maximum distance from each other. A linear arrangement in which the two atoms are on opposite sides of the central carbon would allow this (for example, hydrogen cyanide, H—C≡N).

In most cases, atoms are oriented in a molecule so that repulsion between electron pairs (either bonding or nonbonding) around an atom is minimized. The following simple principle is useful for predicting the shape of a molecule or the geometry of a portion of a molecule: *atoms (and nonbonding electron pairs) attached to a common central atom are arranged as far apart as possible in space.* Depending on the types of bonds involved, a carbon will have four, three, or two atoms bonded to it (see section 1.4.B). If there are four bonded groups, the geometric orientation that will position these atoms around the central carbon as far from one another as possible is a **tetrahedron** (four-cornered pyramid). The geometric orientation will be a planar triangle if there are three bonded groups and a straight line if there are two.

tetrahedron
four-cornered pyramid

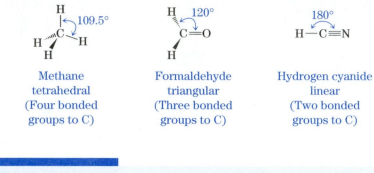

109.5°	120°	180° H—C≡N
Methane tetrahedral (Four bonded groups to C)	Formaldehyde triangular (Three bonded groups to C)	Hydrogen cyanide linear (Two bonded groups to C)

GETTING INVOLVED

- ✓ Why does carbon form four covalent bonds? Why must an outer-shell electron be promoted from a 2s to a 2p orbital to allow this?
- ✓ What is the difference in geometry of carbons with four bonded groups, three bonded groups, and two bonded groups? What combination of single, double, and triple bonds are necessary for these compounds (there are two possibilities for carbon with two bonded groups and one for each of the others)?

D. Carbon Bonded to Four Atoms

The simplest example of an organic compound with a carbon bonded to four atoms is methane, or natural gas, CH_4. To satisfy the valence of all five atoms, each of the hydrogens must be bonded to the carbon by a single bond. The most stable molecular geometry calls for the four hydrogens to be a maximum distance from one another. Placing the hydrogens at the four corners of a tetrahedron with the carbon in the center accomplishes this (Figure 1.11). The bond angle between any two hydrogens is 109.5°, and all the carbon-hydrogen bonds are equivalent.

If the hydrogens, with their 1s orbitals, were to bond to carbon's outer-shell atomic orbitals, as pictured in Figure 1.10, this stable tetrahedral molecule could not be formed. Recall that the angles between the p orbitals are 90°, not 109.5°. Furthermore, the carbon-hydrogen bonds would not be equivalent, since there are two types of atomic orbitals, s and p, in carbon's outer shell. To establish the more stable tetrahedral geometry, the outer-shell orbitals (2s, $2p_x$, $2p_y$, and $2p_z$) **hybridize,** or blend, to form four new orbitals that are equivalent and at the ideal 109.5° angle from one another. The four new orbitals, called sp^3 hybrid orbitals because they were formed from one s and three p orbitals, are directed toward the corners of a tetrahedron. Four σ bonds will form by the overlap of the four

hybridization
combination of atomic orbitals to form new orbitals of different shapes and orientations

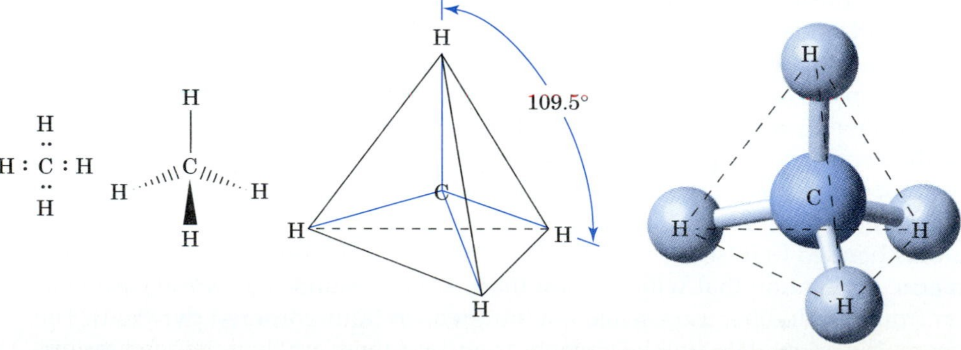

Figure 1.11 Methane, CH_4, has a tetrahedral geometry with bond angles of 109.5°.

raindrop-shaped sp^3 orbitals of carbon and the spherical s orbitals of four hydrogens. Methane, CH_4, is **sp^3-hybridized** and tetrahedral, with four equivalent σ bonds, that is, four equivalent hydrogens with **bond angles** equal to 109.5° (Figure 1.12). It is important to understand, however, that bond formation in organic compounds is not a sequential process of electron promotion, orbital hybridization, and bonding. Rather, it is an integrated event in which the electrons assume the most stable configuration.

sp^3-hybridization
combination of one s and three p orbitals to form four sp^3 hybrid orbitals that are tetrahedrally oriented

bond angle
angle between two adjacent bonds

GETTING INVOLVED

- ✓ Why does sp^3 hybridization occur in a carbon with four bonded groups? Why aren't the original s and three p orbitals used? Why are each of the four hybridized orbitals called sp^3?
- ✓ What are the geometry and bond angles of the bonded carbon?
- ✓ What kinds of bonds (σ, π) are formed when an sp^3 hybridized carbon engages in bonding?

Problem 1.12

Draw a bonding picture for propane, $CH_3CH_2CH_3$, (campstove and rural fuel gas), showing all σ bonds. Indicate the shape, bond angles, and hybridization of the carbons.

See related problems 1.29a, 1.30a, 1.31a.

E. Carbon Bonded to Three Atoms

Ethene, $CH_2{=}CH_2$ (from which the plastic polyethylene is made), has three atoms bonded to each carbon: two hydrogens and the other carbon. The geometric arrangement that allows three atoms bonded to a central carbon atom to be as far apart in space as possible is triangular, or **trigonal.** In ethene, each

trigonal
geometric arrangement in which a central atom has three bonds directed to the corners of a triangle

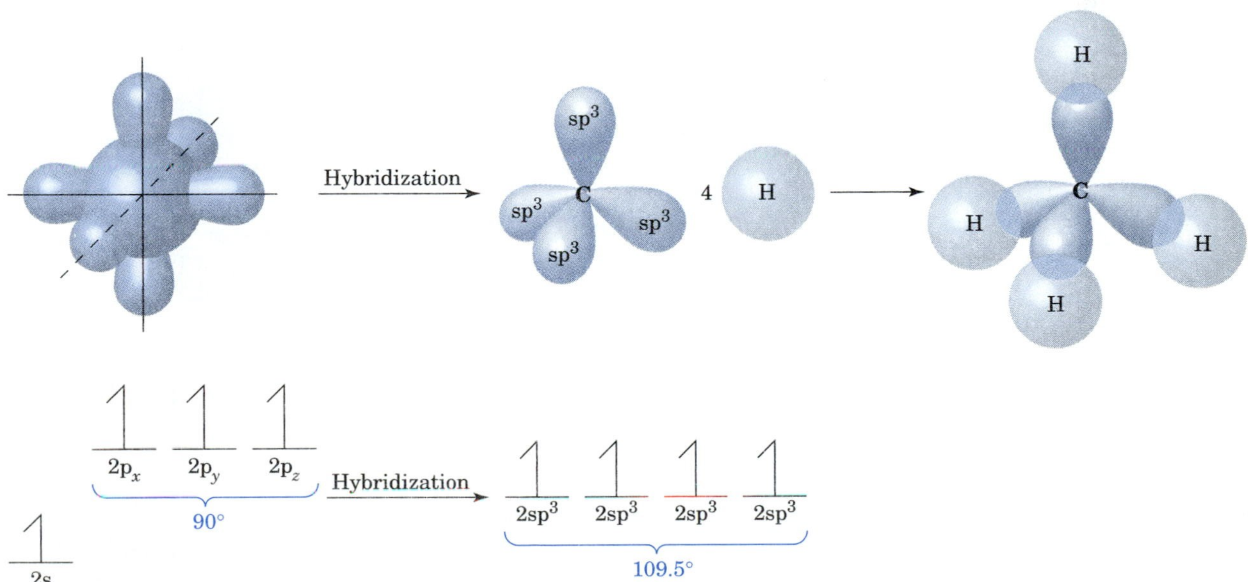

Figure 1.12
Hybridization and bonding in methane. Carbon's four outer-shell atomic orbitals (s, p_x, p_y, p_z) are converted into four new sp^3 hybrid orbitals with angles of 109.5° separating them. Each sp^3 orbital overlaps the s orbital of a hydrogen atom to form a σ bond.

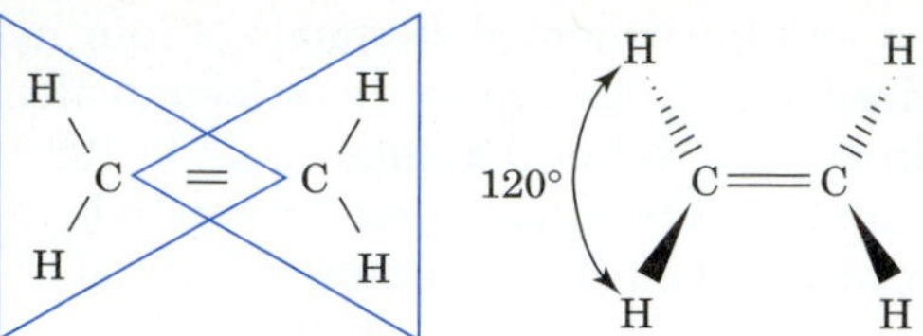

Figure 1.13 Trigonal structure of ethene, with 120° bond angles. The molecule is completely flat.

carbon is at the center of a triangle with the two hydrogens and the other carbon occupying the three corners. The bond angles are each approximately 120° (Figure 1.13).

Once again, the electron configuration of carbon as shown in Figure 1.10 would not allow a trigonal arrangement, because the three p orbitals are perpendicular to one another. The outer-shell orbitals must be hybridized to create an orbital geometry consistent with the preferred triangular shape. In this case, only three of the four orbitals have to be hybridized, since only three bonded atoms must be arranged in space. The s and two of the p orbitals are combined to form three new **sp^2 hybrid orbitals.** These sp^2 orbitals are directed toward the corners of an equilateral triangle (Figure 1.14). An unhybridized p orbital remains unchanged on each carbon, perpendicular to the hybridized orbitals, whose axes all lie in one plane.

sp^2-hybridization combination of one s and two p orbitals to form three sp^2 hybrid orbitals that are trigonally oriented

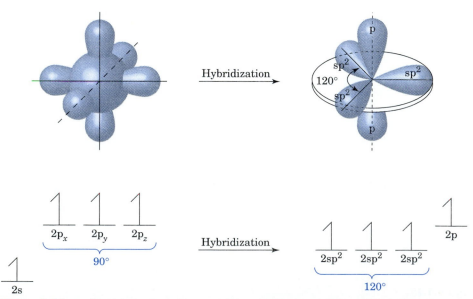

If the two hybridized carbons are now brought together, a σ bond can form between them by the overlap of two sp^2 hybrid orbitals. Both carbons also have unhybridized p orbitals, which can be oriented parallel to one another and can thereby overlap. Both lobes of the p orbitals merge above and below the σ bond, forming a pi bond (Figure 1.15). Thus a double bond is composed of a σ bond and a π bond. The molecule is completed when σ bonds are formed by overlapping each remaining sp^2 hybrid orbital of the carbons with a spherical s orbital of hydrogen. The two carbons involved in the double bond and the four attached atoms all lie in a single plane, with the π bond overlap occurring above and below the plane.

Hybridization

p

sp^2

120°

sp^2

sp^2

p

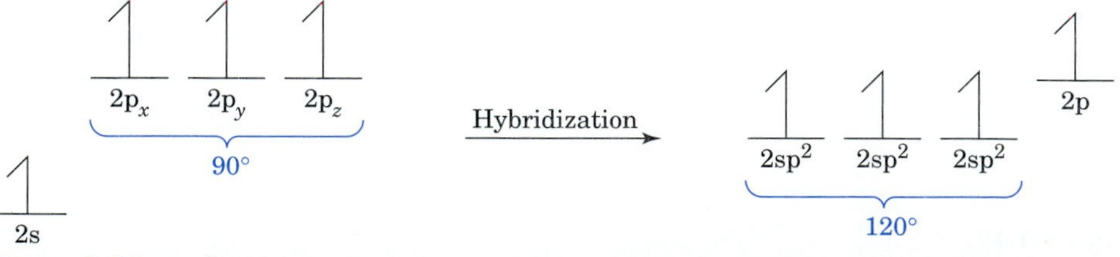

Figure 1.14 sp^2-hybridization. The s and two p orbitals hybridize to form three new sp^2 orbitals that are directed to the corners of a triangle. A p orbital remains unhybridized and is perpendicular to the plane defined by the sp^2 orbitals.

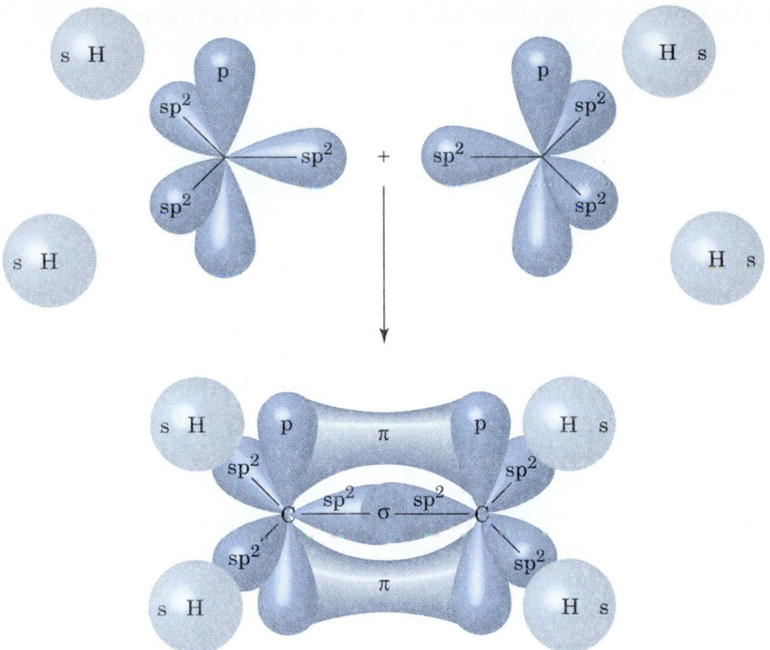

Figure 1.15 Orbital overlap of two sp^2-hybridized carbons and four hydrogens with spherical s orbitals to form ethene. The carbon-hydrogen bonds are s-sp^2 σ bonds, and the carbon-carbon double bond is composed of one sp^2-sp^2 σ bond and one p-p π bond. Ethene is a planar molecule, with the p orbitals overlapping to form a π bond above and below the plane.

GETTING INVOLVED

- ✓ Why are only three orbitals hybridized in sp^2 hybridization? Why are the three new orbitals called sp^2? What happens to the unhybridized p orbital?
- ✓ What are the geometry and bond angles associated with an sp^2 hybridized carbon?
- ✓ What kinds of bonds (σ, π) comprise the two single bonds and one double bond?

Problem 1.13

Draw a bonding picture for propene (from which polypropylene is made), showing all σ and π bonds. Indicate the shape, bond angles, and hybridization of each carbon.

```
    H   H   H
    |   |   |
H—C—C=C—H
    |
    H
```

See related problems 1.29b, 1.30b, and 1.31b.

F. Carbon Bonded to Two Atoms

Each carbon in acetylene (used in oxyacetylene welding torches) is bonded to only two other atoms, a hydrogen and the other carbon. These atoms are positioned as follows:

$$\underset{180°}{\overset{180°}{H{-}C{\equiv}C{-}H}}\quad \text{Acetylene}$$

The two bonded atoms are on opposite sides of the central carbon at a maximum distance from one another. The molecule is linear, with 180° bond angles. To

sp-hybridization combination of one s and one p orbital to form two sp hybrid orbitals that are linearly oriented

produce two equivalent orbitals directed 180° from one another, the 2s orbital and a 2p orbital on each carbon hybridize, forming two **sp hybrid orbitals.**

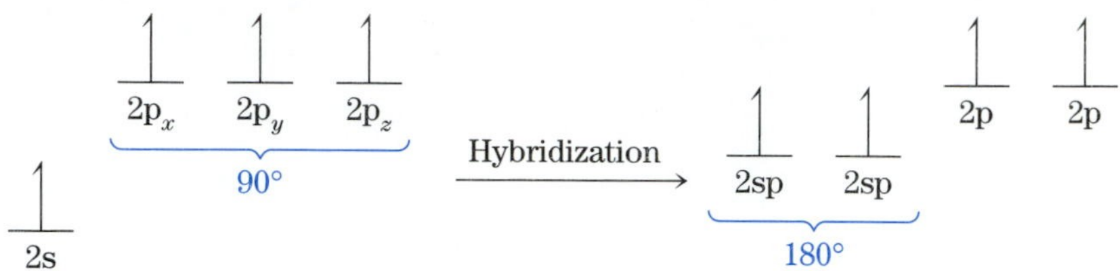

The overlapping of an sp orbital on each carbon joins the two carbons by a σ bond. The hydrogens are connected on the other sides of the carbons by σ bond formation between the remaining sp orbitals and the s orbitals of hydrogen. As in methane and ethene, the geometry of the molecule is determined by σ bond formation (Figure 1.16).

Two unhybridized, perpendicular p orbitals still remain on each carbon. The carbons are oriented so that the p orbitals on one carbon are parallel to the corresponding ones on the other carbon. These orbitals overlap to form two π bonds, one above and below and the other in front of and behind the σ bond. Thus a triple bond is composed of a σ bond and two π bonds. A virtual cylinder of electrons surrounds the two carbons sharing the triple bond (Figure 1.17).

GETTING INVOLVED

- ✓ Why are only two orbitals hybridized in sp hybridization? Why are the two new orbitals called sp? What happens to the two unhybridized p orbitals?
- ✓ What are the geometry and bond angles associated with an sp hybridized carbon?
- ✓ What kinds of covalent bonds (σ, π) comprise the single and triple bonds (problem 1.14a)?
- ✓ A carbon involved in two double bonds exhibits sp hybridization. Answer the questions in the first two checks for this case and practice with problem 1.14b.

Problem 1.14

Draw a bonding picture for the following molecules, showing all σ and π bonds. Indicate the shapes, bond angles, and hybridization of each carbon.

(a) $H_3C{-}C{\equiv}C{-}H$

(b) $H_2C{=}C{=}CH_2$

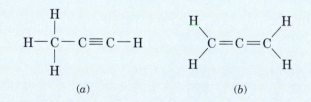

See related problems 1.29c, 1.30c, and 1.31c.

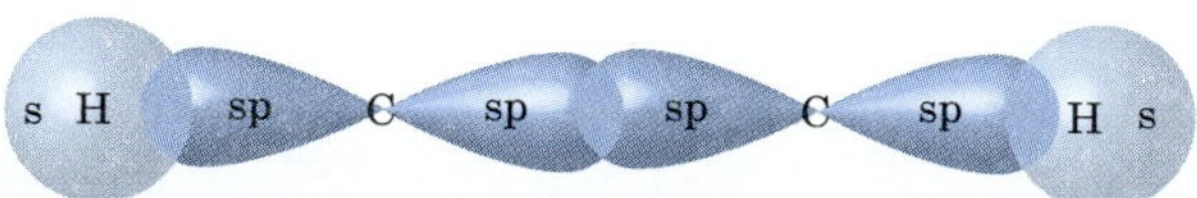

Figure 1.16 The linear geometry of acetylene is determined by the orientation of the two sp hybrid orbitals that engage in σ bond formation.

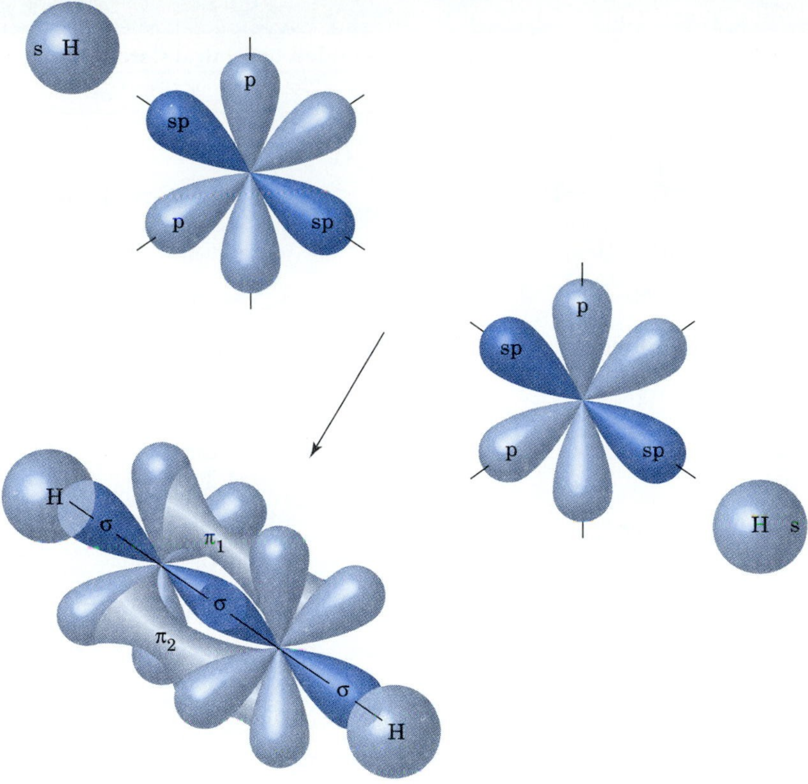

Figure 1.17 Bonding in acetylene. The carbon-hydrogen bonds are s-sp σ bonds, and the carbon-carbon triple bond is composed of one sp-sp σ bond and two p-p π bonds. The π bonds are perpendicular to one another.

G. Bonding in Organic Compounds: A Summary

1. **Geometry and hybridization:** a carbon with
   a. four bonded groups is tetrahedral, sp^3-hybridized, and has 109.5° bond angles.
   b. three bonded groups is trigonal, sp^2-hybridized, and has 120° bond angles.
   c. two bonded groups is linear, sp-hybridized, and has a 180° bond angle.
2. **Types of bonds:**
   a. All single bonds are σ bonds.
   b. A double bond is made up of a σ and a π bond.
   c. A triple bond is a σ and two π bonds (see Table 1.7).
3. **Bond strength:**

$$C{\equiv}C > C{=}C > C{-}C$$

bond strength energy required to break a covalent bond (usually in kJ/mole)

4. **Bond length:**

$$C{-}C > C{=}C > C{\equiv}C$$

bond length distance between atoms in a covalent bond (usually in angstroms, 10^{-10} meters)

Table 1.7 presents bond lengths and bond energies (the energy necessary to break a bond). In Example 1.8 we present the three types of hybridization in a single molecule. Although at first this molecule may seem complex, looking at it

Table 1.7 Bonding in Organic Compounds

	Number of Atoms Bonded to Central Carbon			
	4	3	2 or	2
Example	$H_3C{-}CH_3$ (H—C—C—H with three H on each C)	$H_2C{=}CH_2$	H—C≡C—H	O=C=O
Hybridization	sp^3	sp^2	sp	sp
Geometry	Tetrahedral	Trigonal	Linear	Linear
Bond angles	109.5°	120°	180°	180°
Types of bonds around carbon	4 Single	2 Single 1 Double	1 Single 1 Triple	2 Double
Covalent bonds	4σ	2σ $1\sigma, 1\pi$	1σ $1\sigma, 2\pi$	$1\sigma, 1\pi$ $1\sigma, 1\pi$
Carbon-carbon bond length	1.54 Å	1.34 Å	1.20 Å	
Carbon-carbon bond energy	83 kcal/mol (347 kJ)	146 kcal/mol (611 kJ)	200 kcal/mol (837 kJ)	

on a carbon-by-carbon, bond-by-bond basis will reveal the simple application of principles discussed in this section.

GETTING INVOLVED

- ✓ Let's bring our understanding of hybridization of carbon together. On a piece of paper draw the s and three p unhybridized orbitals of carbon, showing correct geometric placement. Below this write the electron configuration, i.e., each orbital has one electron. Now draw three arrows pointing to where you will describe sp^3, sp^2, and sp hybridization. In each area write the electron configuration using the hybridized orbitals and those left unhybridized. Now draw these orbitals around each carbon showing correct geometry.
- ✓ What are the differences in these three types of hybridization in terms of geometry of bonded groups and bond angles?
- ✓ Also, describe what types of bonds (single, double, and triple) are needed to cause these three hybridizations and how these bonds are constructed with sigma and pi bonds.
- ✓ Finally, do Example 1.8 on your own and check your work against the solution presented.
- ✓ Why is a carbon-carbon triple bond shorter and stronger than a carbon-carbon double or single bond?

See related problems 1.29d, 1.30d, and 1.31d.

Example 1.8

Draw a bonding picture for $CH_3CH{=}CH{-}C{\equiv}CH$ showing all σ and π bonds.

Solution

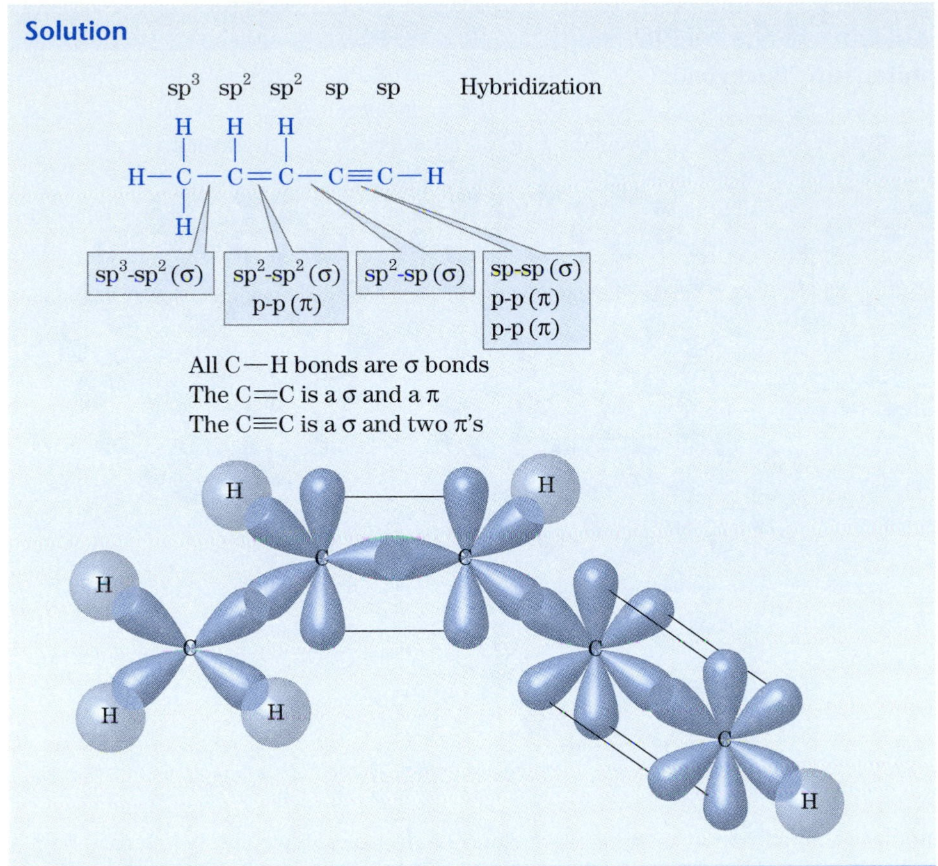

1.6 Bonding to Oxygen and Nitrogen

Like carbon, oxygen and nitrogen can participate in single and multiple bonds composed of σ and π bonds. The atoms hybridize and show molecular geometries consistent with the number of surrounding space-occupying groups. Oxygen and nitrogen differ from carbon, however, in that in chemical compounds each has nonbonding electron pairs (nitrogen has one pair; oxygen, two). These nonbonding electron pairs occupy space just as bonded atoms do and consequently influence the geometry of the molecule.

Consider methanol (wood alcohol) and formaldehyde (a biological preservative), shown in Figure 1.18. The oxygen in methanol has two single bonds and two nonbonding electron pairs. This constitutes four space-occupying groups surrounding the atom, and thus the oxygen is sp^3-hybridized. The four sp^3 hybrid orbitals are oriented toward the corners of a tetrahedron. Two of the orbitals possess one electron each and overlap with orbitals from carbon and hydrogen to form σ bonds. The other two sp^3 orbitals are each occupied by a nonbonding electron pair.

In formaldehyde, there are a carbon (double bond) and two nonbonding electron pairs to occupy the space around the oxygen. The oxygen is sp^2-hybridized, with two of the three sp^2 orbitals occupied by nonbonding electron pairs. The remaining sp^2 orbital and the unhybridized p orbital overlap with their counterparts on carbon to form the σ and π molecular orbitals of the double bond.

Connections 1.1

www.prenhall.com/bailey

Diamond, Graphite, and "Buckyballs"

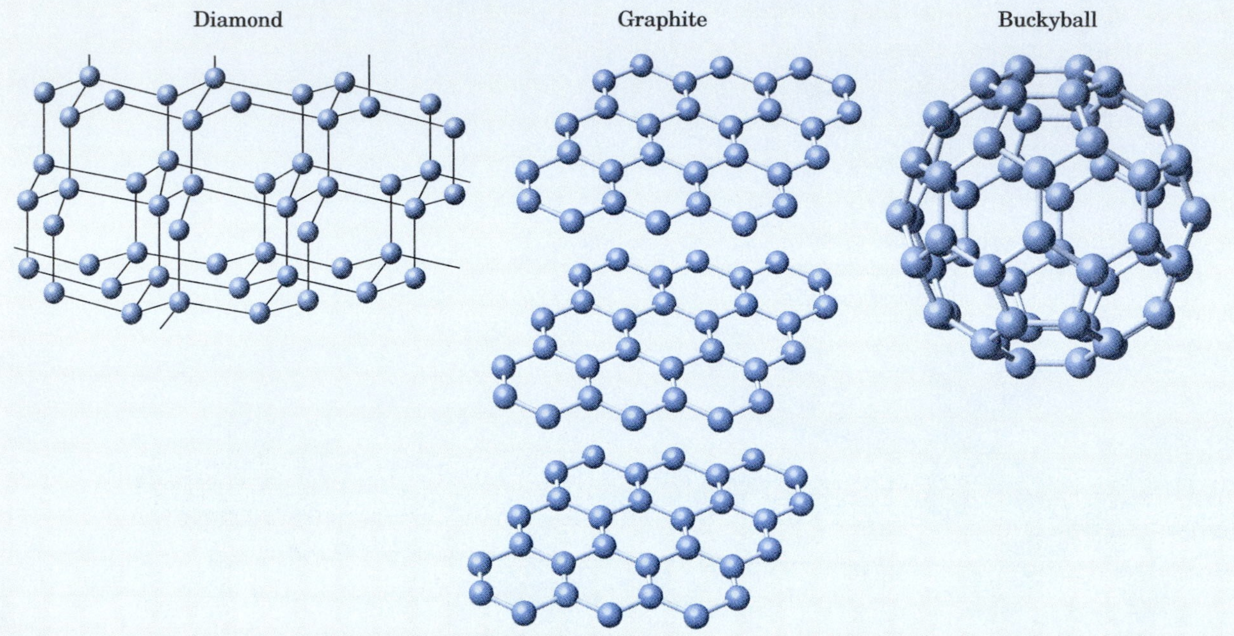

Diamond and graphite are two familiar crystalline forms of elemental carbon. Diamond is the hardest, most abrasive mineral known; graphite is soft, slippery, and often used as a lubricant, for example, to relieve stiff locks. Diamonds can be colorless and transparent, whereas graphite is a black, opaque material (as found in pencils). Graphite can conduct an electric current; diamond cannot. And the difference in price is tremendous, even though each is merely a collection of carbon atoms.

To explain these diverse properties, let us delve into the crystalline structure on an atomic scale. Both substances are covalently bonded. In diamond, each carbon is singly bonded to four other carbons in a tetrahedral arrangement (sp^3-hybridized) with 109.5° bond angles. This pattern extends continuously throughout a vast network. The strength and hardness of diamond are a result of this crystalline structure. Breaking a diamond involves not merely the cleavage of the substance between molecules but the actual cracking of the molecule along innumerable strong covalent bonds.

In graphite, each carbon is bonded to three other carbon atoms. The geometry around each carbon is that of a planar equilateral triangle with 120° bond angles (sp^2-hybridized). Like all sp^2-hybridized carbons, those in graphite have an unhybridized p orbital, in this case possessing one electron. Because of this geometry, all carbons in a molecule of graphite are necessarily in the same plane. The p orbitals can thereby overlap continuously, creating a mobile π cloud of electrons above and below each large graphite molecule. These large hexagonal sheets are layered on one another, cushioned by the π electron cloud. The loosely held electron mass is responsible for the electrical conductivity of graphite. Furthermore, gas molecules can be absorbed between the atomic layers, where they act like ball bearings. This allows the carbon sheets to slide by one another—hence the lubricating properties of graphite.

Recently, a new form of elemental carbon has been discovered consisting of molecules of 60 carbon atoms, each sp^2-hybridized. The spherical molecules are called *buckminsterfullerenes* (or just fullerenes), because they resemble geodesic domes designed by the late inventor Buckminster Fuller. The caged molecules consist of carefully arranged hexagonal and pentagonal rings resulting in a molecule that resembles a soccer ball, hence the fashionable name *buckyball.* The C_{60} molecule has 20 hexagons and 12 pentagons; each vertex is a carbon atom that can be thought of as having one double and two single bonds. It has no edges, no charges, and is the most spherical molecule known to date. The future holds the potential of exciting research as imaginative chemists attach molecular decorations to the sphere's exterior, make larger (C_{70} looks like a rugby ball) and smaller fullerenes, replace some of the carbons with other atoms, and trap smaller atoms and molecules inside the cage. There are high hopes that the discovery of useful applications will result.

Unhybridized oxygen

2s 2p$_x$ 2p$_y$ 2p$_z$ Outer shell

Methanol

$$H-\underset{\displaystyle H}{\overset{\displaystyle H}{\underset{|}{\overset{|}{C}}}}-\ddot{\underset{\cdot\cdot}{O}}-H$$

Hybridized oxygens

Formaldehyde

$$\begin{matrix}H \\ & \diagdown \\ & & C=\ddot{O}: \\ & \diagup \\ H\end{matrix}$$

sp^3 sp^3 sp^3 sp^3

sp^2 sp^2 sp^2 p

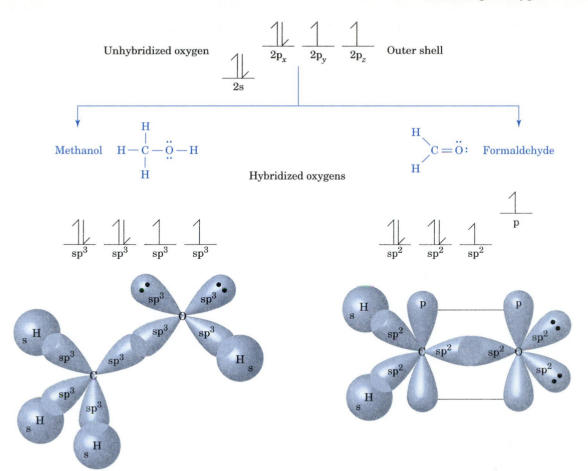

Figure 1.18 Bonding pictures of methanol and formaldehyde.

Analogous reasoning can be used for nitrogen, as shown in Figure 1.19. In addition to its nonbonding electron pair, nitrogen can have three, two, or one bonded atom(s). In effect, nitrogen can have four, three, or two space-occupying groups and shows sp^3-, sp^2-, or sp-hybridization, respectively.

GETTING INVOLVED

✓ How are sp^3 carbons, oxygens, and nitrogens different? How many bonding and nonbonding electron pairs does each have? Answer the same questions for the sp^2 hybridized atoms.

✓ Compare sp hybridized carbons and nitrogens.

Problem 1.15

Draw a bonding picture for the following molecule, showing all sigma and pi bonds and orbitals with nonbonding electrons. Indicate the hybridization of each atom.

$$N\equiv C-\underset{\displaystyle H}{\overset{\displaystyle H}{\underset{|}{\overset{|}{C}}}}-\overset{\displaystyle O}{\overset{\|}{C}}-O-H$$

See related problems 1.32–1.34.

Diamond

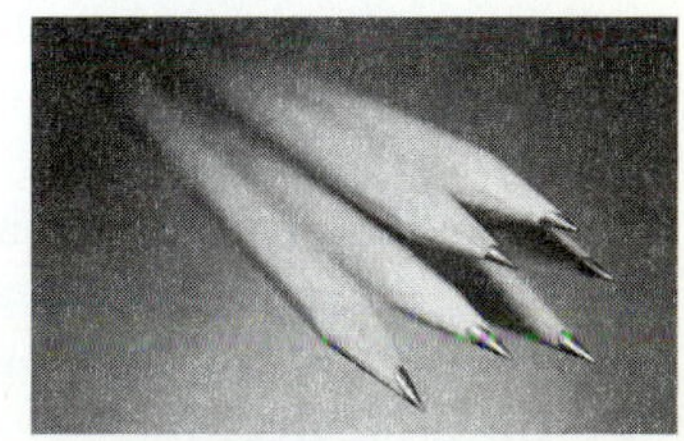

Graphite pencils

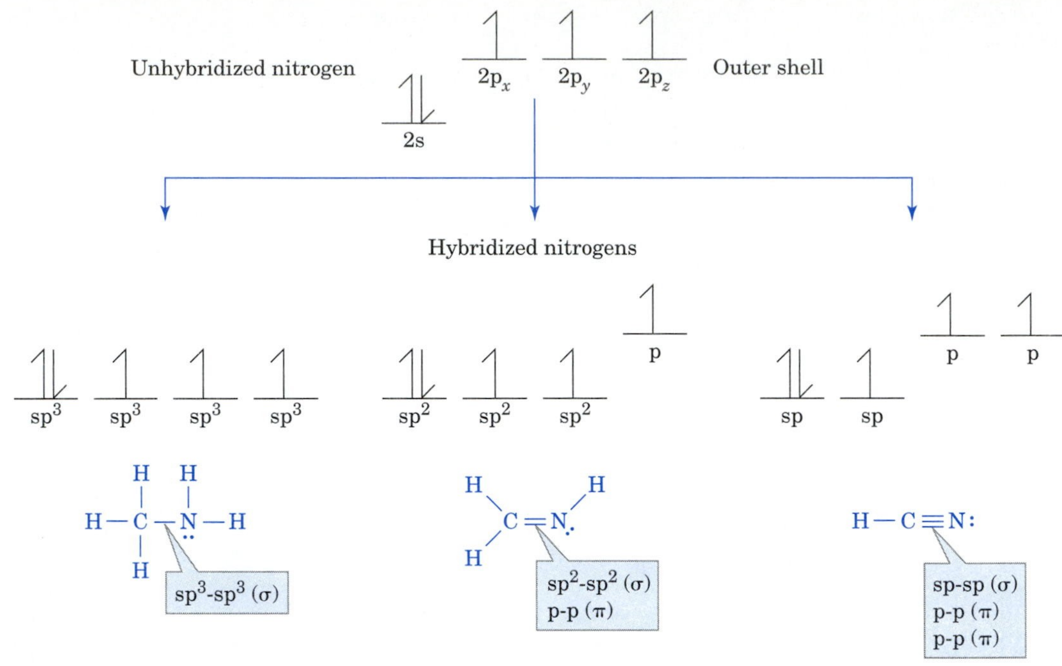

Figure 1.19 Bonding patterns in nitrogen compounds.

Problems

1.16 **Atomic and Mass Numbers:** Write the number of protons, electrons, and neutrons in each of the following atoms.

(a) ^{127}I **(b)** ^{27}Al **(c)** ^{58}Ni **(d)** ^{208}Pb

1.17 **Electron Configurations:** Using Table 1.4, Figure 1.4, and the periodic table, write the specific outer-shell electron configuration for each of the following elements from their positions on the periodic table. For example, Li is in period 2, group I and is thus $2s^1$.

(a) Na **(b)** Mg **(c)** B **(d)** Ge
(e) P **(f)** O **(g)** I **(h)** Kr

1.18 **Electron Configurations:** Identify the elements that have the following outer-shell electron configurations. Use a periodic table without electron configurations.

(a) $7s^1$ **(b)** $5s^25p^2$ **(c)** $3s^23p^5$
(d) $3s^2$ **(e)** $2s^22p^1$ **(f)** $4s^24p^4$
(g) $1s^2$ **(h)** $5s^25p^6$ **(i)** $4s^24p^3$

1.19 **Ionic Reactions:** Using electron dot formulas as illustrated in section 1.3.B, write balanced ionic reaction equations for the reactions between

(a) calcium and fluorine **(b)** sodium and oxygen

1.20 **Electron Dot Formulas:** Write electron dot formulas for the following covalent molecules, showing all bonding and nonbonding electron pairs:

(a) CF_2Cl_2 (a freon)
(b) CH_4O (wood alcohol)
(c) CH_5N (odor of fish)
(d) H_2S (hydrogen sulfide, odor of rotten eggs)
(e) C_2H_6 (ethane, component of natural gas)
(f) C_2Cl_4 (a dry-cleaning agent)
(g) CS_2 (carbon disulfide, used in the manufacture of rayon)
(h) $COCl_2$ (the nerve gas phosgene)
(i) HCl (hydrogen chloride gas)
(j) BH_3O_3 (boric acid, eyewash, veterinary antiseptic; no O—O bonds in formula)
(k) CH_2Cl_2 (methylene chloride, a degreasing solvent)
(l) CH_4S (added to natural gas to provide a warning odor)
(m) N_2 (nitrogen gas)
(n) N_2H_4 (hydrazine, rocket fuel)
(o) Cl_2 (chlorine gas)
(p) $AlCl_3$ (aluminum trichloride)
(q) HONO (nitrous acid)

1.21 **Electron Dot Formulas:** There are two possible covalent compounds for each of the following molecular formulas. Write electron dot formulas for each, showing all bonding and nonbonding electron pairs.

(a) C_4H_{10} **(b)** C_2H_6O **(c)** C_3H_7Br
(d) $C_2H_4Cl_2$ **(e)** C_2H_7N **(f)** C_3H_6

1.22 **Electron Dot Formulas:** Following are condensed formulas of some common compounds. Write electron dot formulas for each, showing all bonding and nonbonding electron pairs. To assist you in interpretation, in problem (a) the first carbon has three hydrogens bonded to it and it is connected to the next carbon by a single bond. That carbon has two bonded hydrogens and an OH. Problems (c), (e), (g), and (i) each have a carbon-oxygen double bond.

(a) CH_3CH_2OH (beverage alcohol)
(b) $CH_2{=}CHCl$ (precursor of PVC)
(c) CH_3CO_2H (acetic acid, the sour-tasting substance in vinegar)
(d) $CH_2{=}CHC{\equiv}N$ (acrylonitrile, from which Orlon is made)
(e) H_2NCONH_2 (urea, by means of which nitrogen is excreted in urine)
(f) $CH_3N{=}C{=}O$ (MIC, methyl isocyanate, the gas that caused the tragedy in Bhopal, India)
(g) $HCO_2CH_2CH_3$ (artificial rum flavor)
(h) $CH_3CH{=}CHCH_2SH$ (a constituent of skunk scent)
(i) $CH_3CHOHCO_2H$ (lactic acid, found in sour milk and sore muscles)
(j) $CH_3CH_2CH_2SH$ (in fresh onions)

1.23 **Electron Dot Formulas:** Place nonbonding electron pairs where ever they exist in the following molecules.

(a) 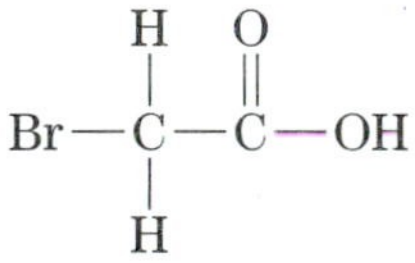
```
      H  O
      |  ||
  Br—C—C—OH
      |
      H
```
(b) 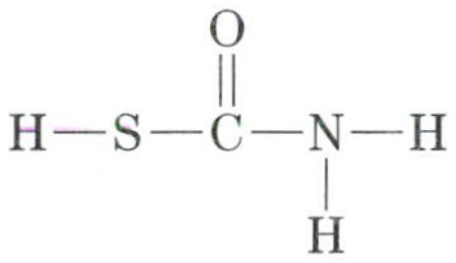
```
          O
          ||
  H—S—C—N—H
              |
              H
```

1.24 **Electron Dot Formulas:** What is the maximum number of double bonds possible in an electron dot formula of C_5H_6? Draw an example. What is the maximum number of triple bonds? Draw an example.

1.25 **Electron Dot Formulas and Formal Charge:** Write an electron dot formula for ozone, O_3, in which there are no cyclic structures and only oxygen-oxygen single bonds. Identify any formal charges.

1.26 **Formal Charge:** Determine if any atoms in the following species are charged. If so, indicate the formal charge.

(a) CH_3 **(b)** $\cdot CH_3$
(c) $:CH_3$ **(d)** $CH_3\overset{H}{\underset{\cdot\cdot}{O}}CH_3$
(e) $(CH_3)_4N$ **(f)** $CH_3{-}N{\equiv}N:$
(g) $CH_3\overset{\cdot\cdot}{\underset{\cdot\cdot}{O}}:$ **(h)** $:\overset{\cdot\cdot}{\underset{\cdot\cdot}{Br}}$

1.27 **Electron Dot Formulas and Formal Charge:** Draw electron dot formulas for the carbonate ion (CO_3^{2-}) and bicarbonate ion (HCO_3^-). Show all bonding and nonbonding electron pairs and formal charges. There are no O—O bonds.

1.28 **Polar Covalent Bonds:** Identify and show the polarity of the polar covalent bonds in the amino acid cysteine, which is found in hair protein and undergoes transformations during permanents. Use the δ+ and δ− symbols shown in section 1.4.F. (Amino acids actually exist in a dipolar salt form.)

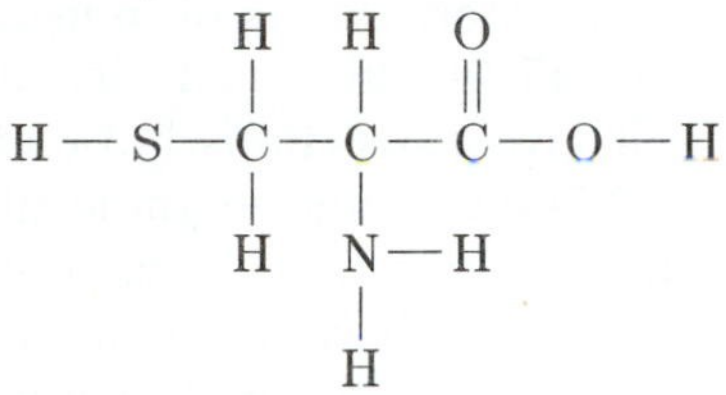

```
         H   H   O
         |   |   ||
  H—S—C—C—C—O—H
         |   |
         H   N—H
             |
             H
```

1.29 **Bond Angles, Geometry, Hybridization of Carbon:** Indicate for each carbon in the following molecules the hybridization, geometric shape, and bond angles:

(a) CH_3CH_3
(b) $CH_3CH{=}CHCH_3$
(c) $CH_3C{\equiv}CCH_3$
(d) $CH_2{=}CHCH_2C{\equiv}CH$

1.30 **σ and π Bonds:** For the compounds in problem 1.29, identify σ and π bond locations.

1.31 **Bonding Pictures:** Construct a bonding picture for the compounds in problem 1.29, showing in your drawing all σ and π bonds.

1.32 **Bond Angles, Geometry, Hybridization:** For each carbon, oxygen, or nitrogen, indicate the geometric shape, hybridization, and bond angles.

(a) $CH_3CH_2NH_2$ **(b)** $CH_3CH{=}NH$
(c) $CH_3C{\equiv}N$ **(d)** CH_3CH_2OH
(e) 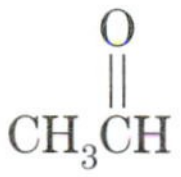 $CH_3\overset{O}{\overset{\|}{C}}H$

1.33 **σ and π Bonds:** For the compounds in problem 1.32, identify σ and π bond locations.

1.34 **Bonding Pictures:** Construct a bonding picture for each of the compounds in problem 1.32, showing all σ and π bonds in your drawing.

1.35 **Silicon:** Life on Earth is based on the element carbon. One of the episodes of the television series *Star Trek* centered around an alien life form based on the element silicon. Do you think the author picked this element randomly, or was there logic involved in the choice? Explain.

1.36 **Molecular Shape: (a)** Why does NH_3 have bond angles of 107° and BF_3 angles of 120°? **(b)** In terms of s and p, describe the hybridization in the two compounds in (a). Explain.

1.37 **Bond Angles:** CH_4 has bond angles of 109°, NH_3 of 107°, and H_2O of 105°. Explain the decreasing angles.

1.38 **Reactivity:** CH_4 is a relatively unreactive molecule. However, NH_3 and BH_3 are very reactive for different reasons. Offer an explanation for the different reactivities of these three compounds.

Activities with Molecular Models

In these sections of the book, we offer opportunities for you to see molecules more intimately using molecular models. There are many kinds of molecular models that you can build and manipulate with your hands including simple ball and stick, just stick, or space-filling models; all are shown in exercise 1 of this section. Any kit you use will have instructions, but you will soon find that making models is simple and self-evident. You can even generate 3-D simulations of molecules and view them from different perspectives using computers. To see examples of this you can visit this textbook on the Prentice-Hall website **www.prenhall.com/bailey** where you will be able to work with models made using CS Chem3D and CS ChemDraw; these programs also were used for the structures and models in the solutions manual for this text. If you have access to these programs you can quickly learn to make computer-generated molecular models yourself.

1. Make models of ethane (C_2H_6), ethene (C_2H_4), and ethyne (C_2H_2). These molecules illustrate sp^3 (tetrahedral), sp^2 (trigonal), and sp (linear) hybridizations, respectively. Note the geometries and bond angles as you look at your models.

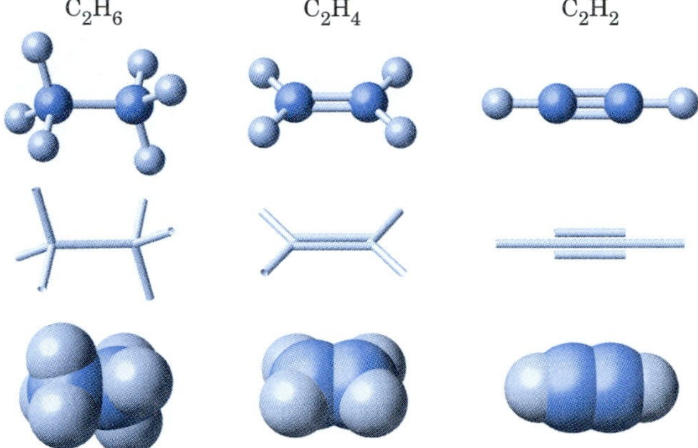

2. Make models of methane (CH_4), formaldehyde (CH_2O), and hydrogen cyanide (HCN). Observe the geometries and bond angles at each carbon.

3. Make models of methanol (CH_4O) and formaldehyde (CH_2O). Note the geometries and bond angles of both the carbons and oxygens in these molecules.

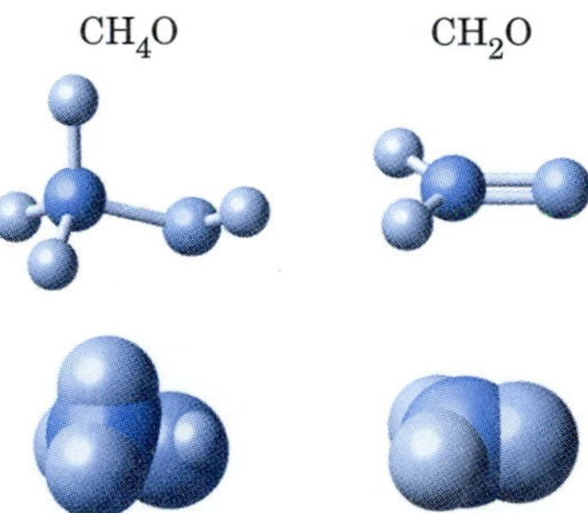

4. Make models of CH_5N, CH_3N, and HCN. Note the geometries and bond angles at both the carbons and the nitrogens.

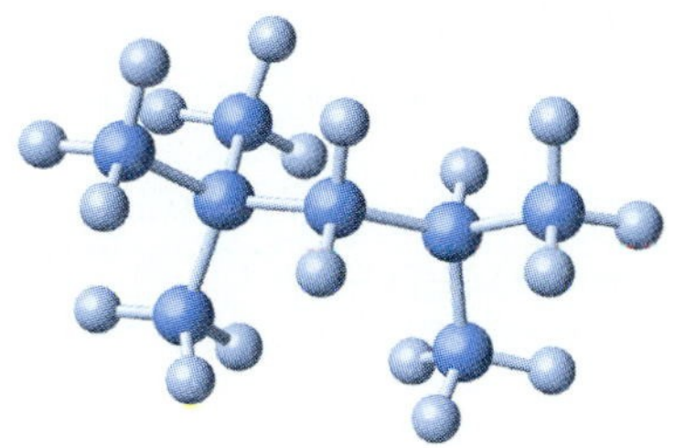

CHAPTER 2 The Alkanes: Structure and Nomenclature of Simple Hydrocarbons

Organic compounds are classified according to common structural features that impart similar chemical and physical properties to the compounds within each group or family. Studying organic chemistry in terms of these families simplifies the task of understanding the reactions that compounds undergo; this textbook is organized accordingly.

2.1 Hydrocarbons: An Introduction

In this chapter, we will begin a thorough coverage of organic molecular structure and nomenclature using **hydrocarbons,** which are in many respects the simplest of organic compounds because they are composed of only carbon and hydrogen. First we shall look at **alkanes,** which are hydrocarbons whose carbon-carbon bonds are all single bonds. Learning the relationship of structure to the nomenclature of alkanes should help you master both.

hydrocarbon
compound composed of only carbon and hydrogen

alkane
compound composed of only carbon and hydrogen and single bonds

Petroleum and coal are the major sources of hydrocarbons. They are complex mixtures of literally thousands of compounds, most of them hydrocarbons, formed by the decay and degradation of marine plants and animals. Some of these compounds are probably already familiar to you: methane (natural gas), propane and butane (bottled fuel gas used in rural communities and camping equipment).

```
   H             H   H   H              H   H   H   H
   |             |   |   |              |   |   |   |
H--C--H       H--C---C---C--H        H--C---C---C---C--H
   |             |   |   |              |   |   |   |
   H             H   H   H              H   H   H   H
```

Methane | Propane | Butane

Mostly, the hydrocarbons in petroleum are burned as fuel, but a small portion is converted into petrochemicals such as plastics, fibers, dyes, detergents, medicines, pesticides, and other products.

saturated compound
compound with only single bonds

unsaturated compound
compound with at least one double or triple bond

alkene
compound composed of carbon and hydrogen and at least one double bond

alkyne
compound composed of carbon and hydrogen and at least one triple bond

Hydrocarbons fall into two major classes: **saturated** hydrocarbons (alkanes) in which all carbon-carbon bonds are single bonds and **unsaturated** hydrocarbons in which the molecules have at least one carbon-carbon double bond **(alkenes)** or triple bond **(alkynes).** Aromatic compounds, originally named for their aromas, also fall into the unsaturated designation; their structures will be discussed in a later chapter. Following are simple examples of each class of hydrocarbons: ethane, a minor component of natural gas; ethylene, precursor of the plastic polyethylene; acetylene, the fuel of oxyacetylene torches; and benzene, a gasoline component.

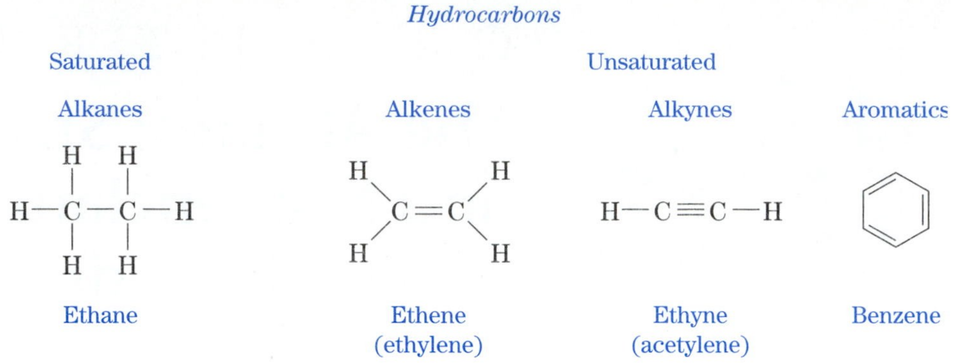

Hydrocarbons

Saturated	Unsaturated		
Alkanes	Alkenes	Alkynes	Aromatics
H–C–C–H (each C bearing two further H)	$H_2C{=}CH_2$	H–C≡C–H	(benzene ring)
Ethane	Ethene (ethylene)	Ethyne (acetylene)	Benzene

GETTING INVOLVED

✓ What are the similarities and differences between saturated and unsaturated hydrocarbons; among alkanes, alkenes, and alkynes?

Problem 2.1

Using the previous examples of alkanes, alkenes, and alkynes, with two carbons see if you can draw an alkane (all single bonds) of C_3H_8, an alkene (has a C=C) of C_3H_6, and an alkyne (has a C≡C) of C_3H_4.

2.2 Molecular and Structural Formulas: Isomerism

molecular formula
formula that gives the number of each kind of atom in a compound

structural formula
formula that provides the bonding arrangement of atoms in a molecule

Theoretically, the number of possible organic compounds is infinite. Several million have already been synthesized or isolated from their natural sources. These compounds are commonly represented by either molecular or structural formulas. A **molecular formula** describes the exact number of each kind of atom in a compound. The three simplest alkanes have molecular formulas of CH_4, C_2H_6, and C_3H_8. Although this is important information, **structural formulas** are more useful. They not only provide the exact number of each kind of atom in a molecule but also the bonding arrangement of these atoms—that is, which atoms are bonded to each other and by what kind of bond. The structural formulas of the three simplest alkanes are shown in Figure 2.1 along with molecular model representations.

isomers
compounds with the same molecular formula but different structural formulas

structural isomers
isomers that vary in the bonding attachments of atoms

constitutional isomers
isomers that vary in the bonding attachments of atoms

stereoisomers
isomers with same bonding attachments of atoms but different spatial orientations

The greater the number of atoms in a molecular formula, the greater the number of possible compounds with that molecular formula. For example, although only one compound is possible for alkanes with the formulas CH_4, C_2H_6, and C_3H_8, two are possible for C_4H_{10}; three for C_5H_{12}; 75 for $C_{10}H_{22}$; 366,319 for $C_{20}H_{42}$; and 62,491,178,805,831 for $C_{40}H_{82}$. These different compounds with the same molecular formula but different structural formulas are called **isomers.** Very few of the isomers of $C_{20}H_{42}$ or $C_{40}H_{82}$ have been synthesized, isolated from natural sources, or characterized. Yet the possibility of their existence aptly illustrates the enormous scope of organic chemistry.

There are five types of isomerism, that we will cover in this and the next chapter. Skeletal, positional, and functional isomers fall under the general heading of **structural isomerism** (sometimes called **constitutional isomers**). In structural isomers, different atoms are attached to one another. In **stereoisomerism**—geometric and conformational—the same atoms are bonded to one another, but their orientation in space differs. We will investigate skeletal isomerism in alkanes in the next section.

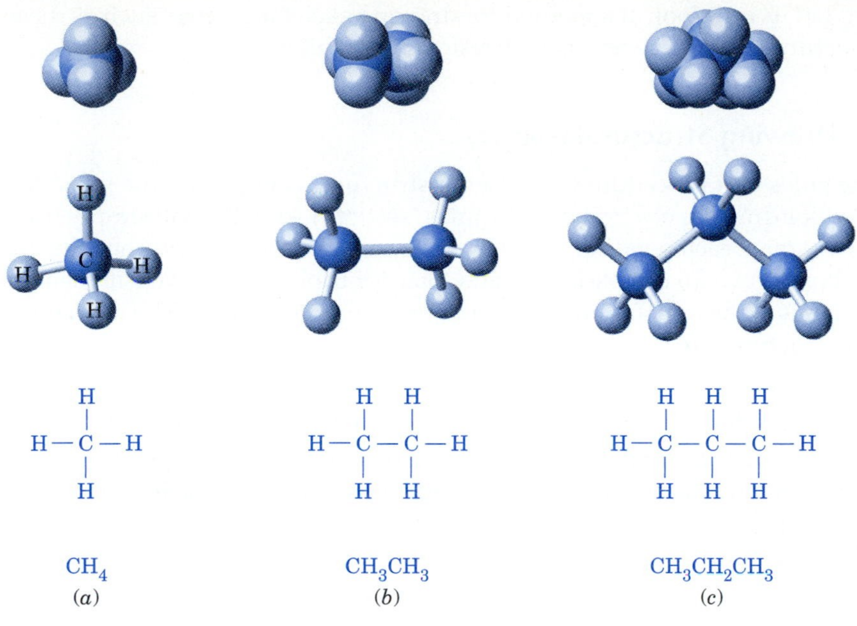

Figure 2.1 Representations of (a) methane, (b) ethane, and (c) propane.

GETTING INVOLVED

✓ What is the difference between a molecular and structural formula?

✓ What are isomers? How are isomers similar and different? What is the difference between structural isomers and stereoisomers?

Problem 2.2

The first illustrative example of isomers in the next section is the two isomers of C_4H_{10}. Before continuing, see if you can draw these using the third type of representation in Figure 2.1. You can check your work in Figure 2.2.

2.3 Skeletal Isomerism in Alkanes

A. Isomers

Isomers are compounds with the same molecular formula but different structural formulas. Alkanes demonstrate a type of isomerism known as **skeletal isomerism.** Since all noncyclic alkanes have the general formula C_nH_{2n+2} (CH_4, C_2H_6, C_3H_8, C_4H_{10}, C_5H_{12}, and so on), they can have only carbon-carbon single bonds. The only structural variations possible are in the arrangements of the carbons, the carbon skeleton. For example, the simplest member of the alkane family that demonstrates isomerism has the molecular formula C_4H_{10}. There are two possible structural formulas, one in which the four carbons are arranged in a continuous chain and another in which the chain is branched.

skeletal isomers isomers that differ in the arrangement of the carbon chain

```
                                   H
                                   |
                                 H-C-H
   H   H   H   H               H   |   H
   |   |   |   |               |   |   |
 H-C---C---C---C-H           H-C---C---C-H
   |   |   |   |               |   |   |
   H   H   H   H               H   H   H
```

Let us now look at a method for drawing skeletal isomers such as these and structural isomers generally in a systematic fashion.

B. Drawing Structural Isomers

The rules and procedure for drawing structural isomers are the same that we used for drawing electron dot formulas (section 1.4.C). We will simply use a line to designate each bonding pair of electrons instead of two dots (one line for single bonds, two for double bonds, and three for triple bonds). We will then quickly progress to the use of even more condensed representations. The rules for drawing structures are

1. Every atom in the molecular formula must be used—no more, no less.
2. The valence (number of bonds) of every atom must be satisfied. The valences of atoms commonly found in organic compounds are as follows:

C	4
N	3
O, S	2
H, F, Cl, Br, I	1

When you are just beginning to learn organic structure, it is good to have a procedure for drawing isomers. The following procedure, in which polyvalent atoms are considered first, is useful.

1. Bond together continuously with single bonds all atoms with valences greater than one.
2. Attach monovalent atoms to the polyvalent atoms until all the valences have been satisfied.
3. If there are insufficient monovalent atoms in the molecular formula to accomplish step 2, insert double or triple bonds between the polyvalent atoms until it is possible to satisfy all valences. Drawing cyclic structures may also be helpful.
4. To construct the isomers of a molecular formula, vary the arrangement of atoms and bonds in the molecules.

Let us apply these rules and this procedure to the simple example discussed in Part A of this section, C_4H_{10} (see Figure 2.2). First bond the four carbons to one another by single bonds. Then attach hydrogens to the carbons one at a time until each carbon has four bonds. You will see that all ten hydrogens are used and no **multiple bonds are required.**

multiple bond
a double bond or triple bond

```
                         H   H   H   H
                         |   |   |   |
C—C—C—C               H—C — C — C — C—H
                         |   |   |   |
                         H   H   H   H
```

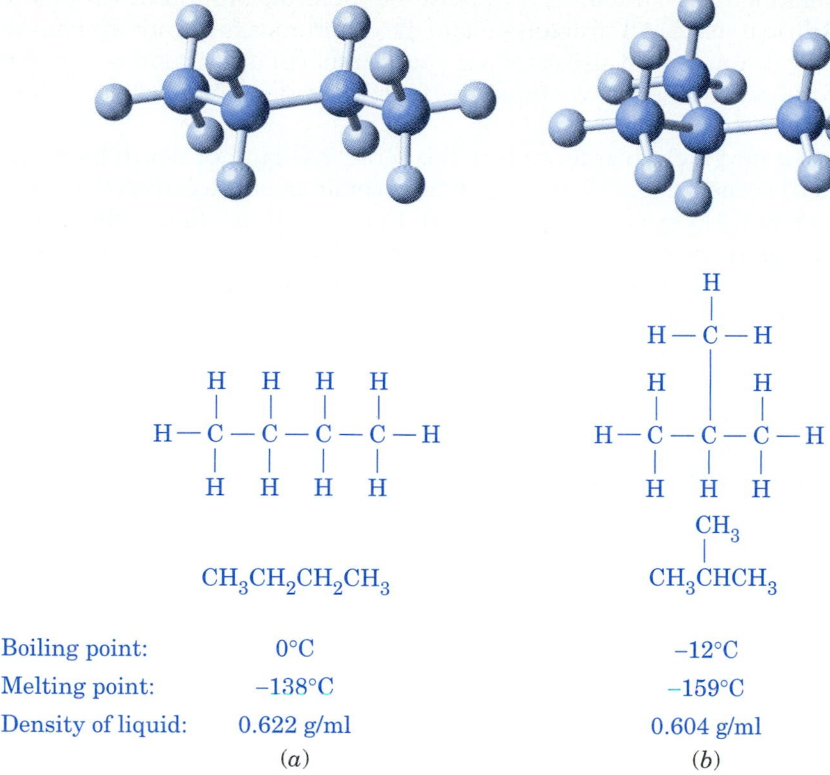

Figure 2.2 Isomers of C_4H_{10}. (a) Butane. (b) Methylpropane. Each has the same molecular formula but a unique structure.

Now vary the arrangement of the polyvalent atoms—the four carbons—and attach hydrogens as before.

```
    C                    H
    |                    |
C — C — C         H  H — C — H  H
                  |      |      |
              H — C ———— C ———— C — H
                  |      |      |
                  H      H      H
```

To a person just learning organic chemistry, it may seem that there should be quite a few isomers of C_4H_{10}. For example, why aren't the following different from our first isomer, butane?

```
    H   H   H   H             H                  H   H   H              H
    |   |   |   |             |                  |   |   |              |
H — C — C — C — C — H     H — C — H          H — C — C — C — H      H — C — H   H
    |   |   |   |             |                  |   |   |              |       |
    H   H   H   H             |   H   H          H   H   |          H — C ——————C — H
        Butane                |   |   |                  |              |       |
                          H — C — C — C — H          H — C — H          H   H — C — H
                              |   |   |                  |                      |
                              H   H   H                  H                      H
```

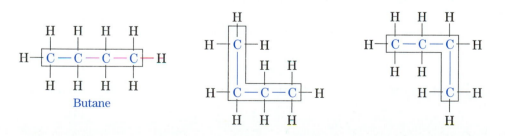

Close examination reveals that each of these structures has a continuous chain of four carbons and is therefore identical to butane. All that has been done in generating these structures is to twist the molecule around in various contortions like a snake. All structures retain the continuous four-carbon chain and are identical. You should also convince yourself that in our second isomer it makes no difference whether we bond the fourth carbon above or below the three-carbon chain.

The next molecular formula in this series is C_5H_{12}, for which there are three skeletal isomers (Figure 2.3). The first has a continuous five-carbon chain (only one isomer of C_5H_{12} can have such a chain). The second has a four-carbon chain with a one-carbon branch on the second carbon. Note that it makes no difference on which of the two interior carbons the one-carbon branch is placed. In either case, it will be on the second carbon from the end of a four-carbon chain. The third isomer has two one-carbon branches on the middle carbon of a three-carbon chain. Note that if the two one-carbon branches had been put on the first carbon of the chain, or if one two-carbon chain had been put on the second carbon of the three-carbon chain, we would have repeated the second isomer, since the longest chain would be extended to four carbons.

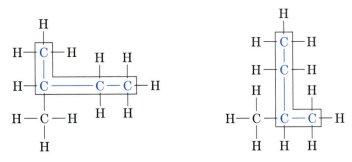

Figure 2.3 shows condensed structural formulas for the compounds under discussion. Since this type of formula is the most frequently used, you should

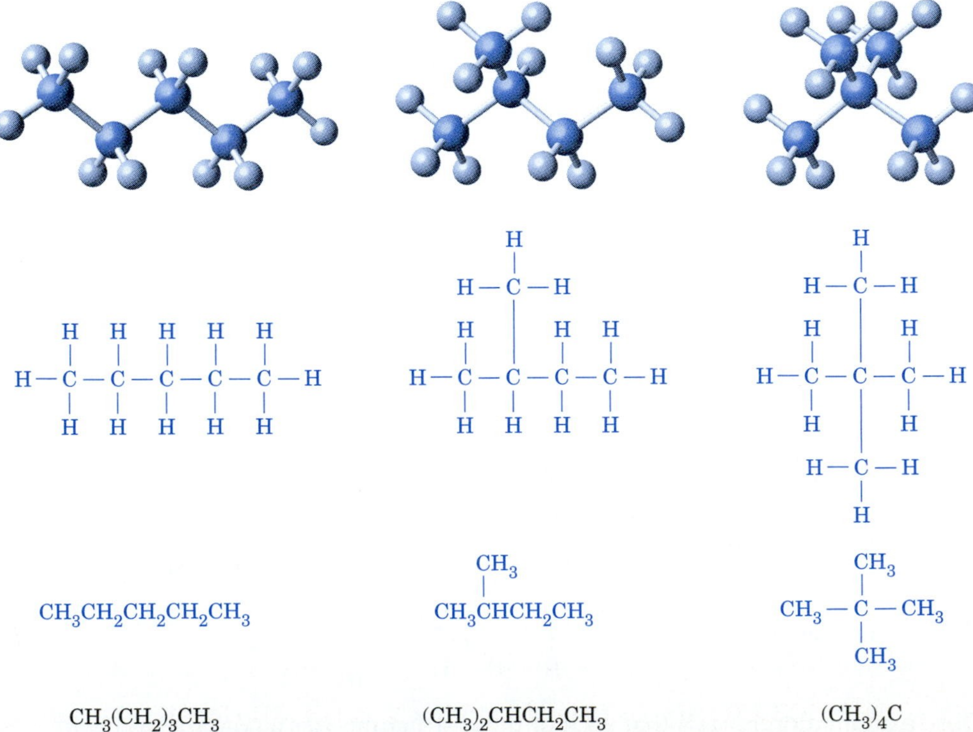

Figure 2.3 Structural formula representations for the isomers of C_5H_{12}.

compare the expanded and condensed formulas to make certain of their meaning.

Example 2.1 illustrates the use of **condensed formulas** in drawing the skeletal isomers of C_6H_{14}.

condensed formula structural formula in which not all the bonds or atoms are individually shown

GETTING INVOLVED

Example 2.1

Draw the five skeletal isomers having the molecular formula C_6H_{14}.

Solution

A systematic approach should be followed in drawing isomers of molecular formulas to avoid repetition or omission of an isomer. Begin by arranging the carbons in one continuous chain:

$$CH_3CH_2CH_2CH_2CH_2CH_3$$

Now reduce the length of the longest chain by one carbon, and place the remaining one-carbon chain in as many different locations as possible (carbons 2 and 3). Placing the CH_3 on the fourth carbon is the same as placing it on the second; drawing it up or down is the same.

$$\begin{array}{c} CH_3 \\ | \\ CH_3CHCH_2CH_2CH_3 \end{array} \qquad \begin{array}{c} CH_3 \\ | \\ CH_3CH_2CHCH_2CH_3 \end{array}$$

After forming as many isomers as we can with a five-carbon chain, we now reduce the longest chain length to four carbons and consider the remaining carbons as one two-carbon branch ($-CH_2CH_3$) and as two one-carbon branches ($-CH_3$, $-CH_3$). There is no place a two-carbon branch can be placed without extending the length of the longest chain. If it were attached to the end, the longest chain would become six carbons. If it were placed on an interior carbon, the chain would be extended to five. There are, however, two arrangements of two one-carbon branches on a four-carbon chain.

$$\begin{array}{c} CH_3 \\ | \\ CH_3CCH_2CH_3 \\ | \\ CH_3 \end{array} \qquad \begin{array}{c} CH_3 \quad CH_3 \\ | \qquad | \\ CH_3CH - CHCH_3 \end{array}$$

Problem 2.3

Determine which, if any, compounds in the following sets are the same.

$$\begin{array}{c} CH_3 \\ | \\ CH_3CHCH_2CH_2CH_2CH_3 \end{array}, \quad \begin{array}{c} CH_3CH_2CH_2CH_2CHCH_3 \\ | \\ CH_3 \end{array}, \quad \begin{array}{c} CH_3 \\ | \\ CHCH_2CH_2 \\ | \qquad | \\ CH_3 \quad CH_2CH_3 \end{array}$$

(a)

$$\begin{array}{c} CH_2CH_3 \\ | \\ CH_3CHCH_2CH_2CH_3 \end{array}, \quad \begin{array}{c} CH_2CH_3 \\ | \\ CH_3CH_2CHCH_2CH_3 \end{array}, \quad \begin{array}{c} CH_3 \\ | \\ CH_3CH_2CHCH_2CH_2CH_3 \end{array}$$

(b)

Problem 2.4

Draw the nine skeletal isomers with the formula C_7H_{16}.

See related problem 2.19 a–h.

C. Cycloalkanes

cycloalkane
cyclic compound containing only carbon and hydrogen

Cycloalkanes have the general molecular formula C_nH_{2n}. They have two fewer hydrogens than the corresponding alkanes. The simplest cycloalkane has three carbons, C_3H_6. There is one possible structural formula, a three-membered ring. There are two skeletal isomers that are cycloalkanes with the formula C_4H_8.

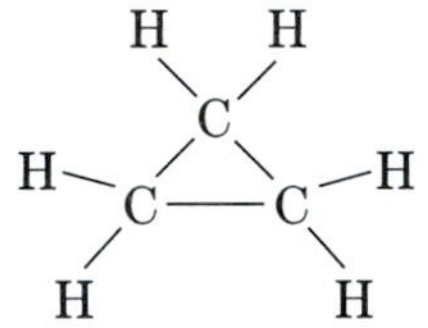

C_3H_6 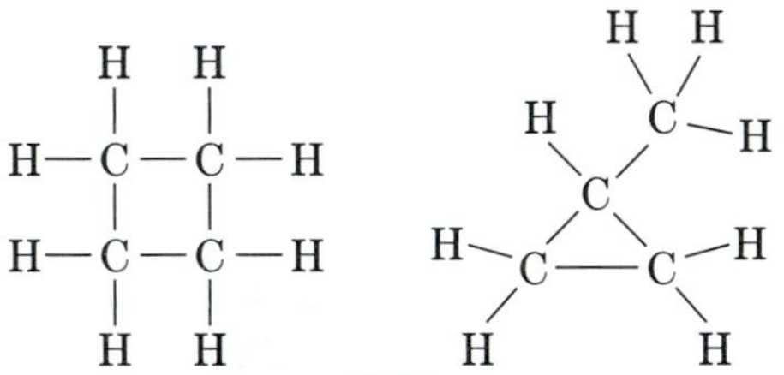 $\qquad$ C_4H_8

GETTING INVOLVED

✓ How does the general formula C_nH_{2n} for cycloalkanes compare to general formulas for alkanes and for alkenes.

Problem 2.5

Draw the five skeletal isomers of cycloalkanes with the molecular formula C_5H_{10}.

See related problem 2.19 i–k.

2.4 Representations of Structural Formulas

You began writing structural formulas in Chapter 1 when you learned to write electron dot formulas. Most of the structural formulas we have used so far in this chapter have been like electron dot formulas except that the bonding pairs of electrons are illustrated with lines rather than dots. Until you are absolutely sure of what you are doing, these are the best formulas to use. But as your proficiency increases, you will want to find shorter methods for representing compounds. This is possible with condensed formulas.

Methods for condensing formulas vary from grouping hydrogens on a given carbon together to using stick diagrams. For example,

$CH_3CH(CH_3)CH_2CH_2CH_3$ $\qquad$ $(CH_3)_2CH(CH_2)_2CH_3$

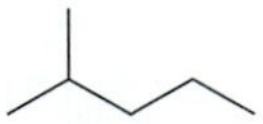

are all representations of the compound

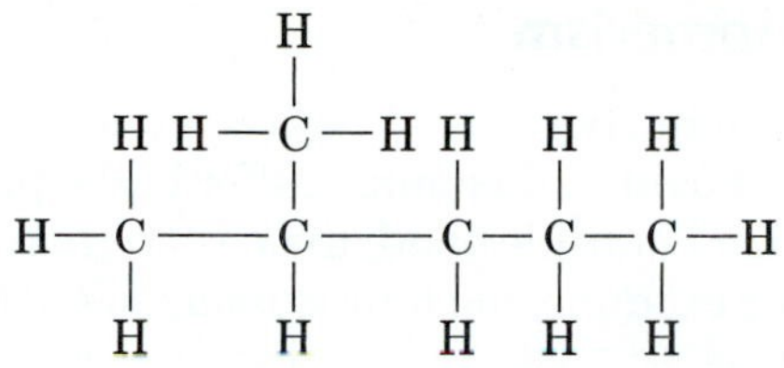

In the first condensation, the hydrogens on each carbon are grouped. In the second, CH_3's and CH_2's are grouped in parentheses as appropriate. Again, examine Figures 2.1 to 2.3 for examples of these methods. The framework formula uses lines, with the intersections of lines and end of lines understood to be carbons unless otherwise specified. The carbons have sufficient hydrogens to satisfy their valences.

Framework formulas are especially useful in representing cyclic (or ring) compounds. Each corner of the polygon represents a carbon unless otherwise specified. Double lines mean double bonds, and three lines designate triple bonds.

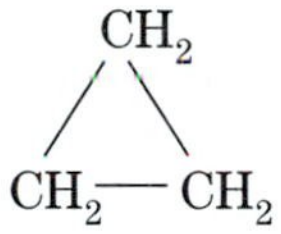

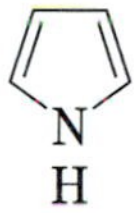

GETTING INVOLVED

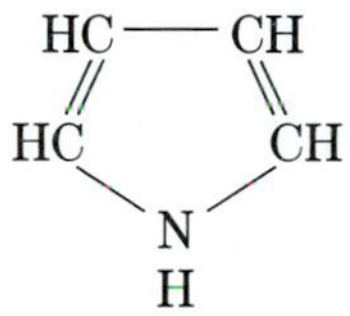

Example 2.2

Draw an expanded structural formula for the following saturated hydrocarbon:

$$(CH_3)_2CH(CH_2)_3CH(CH_3)CH_2C(CH_3)_3.$$

Solution

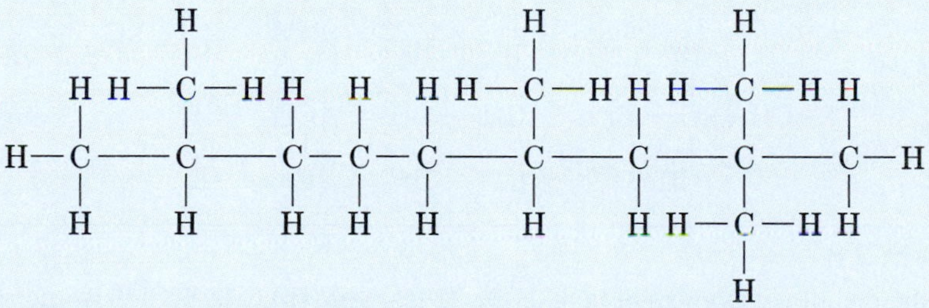

Problem 2.6

Draw the stick representation and a condensed representation using C's and H's equivalent to the stick drawing of the compound in Example 2.2.

Problem 2.7

Determine the molecular formulas (i.e., how many carbons, hydrogens, nitrogens, oxygens, bromines) for each of the following compounds.

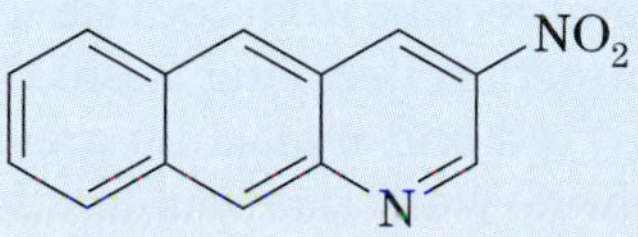

NO_2 N (a)

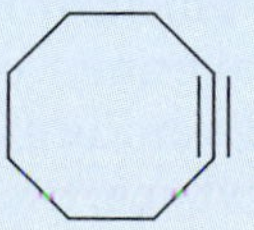

(b)

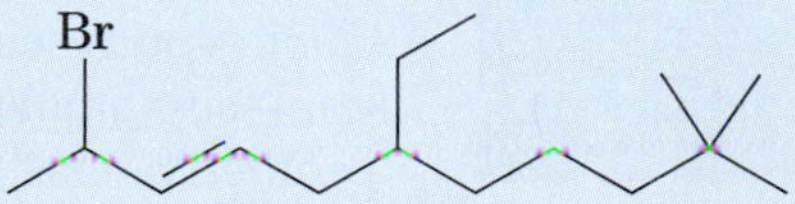

Br (c)

2.5 Positional Isomerism

positional isomers
isomers that differ in the location of a noncarbon group or a double or triple bond

Skeletal isomers differ in the position of carbon atoms, that is, the arrangement of the carbon skeleton. **Positional isomers** differ in the position of a noncarbon group or of a double bond or triple bond; there is no change in the carbon skeleton. Let us consider, for example, the four isomers of C_4H_9Br.

Skeletal isomers:

$CH_3CH_2CH_2CH_2Br$ and $CH_3CH_2CH(Br)CH_3$ — Positional isomers

$CH_3CH(CH_3)CH_2Br$ and $CH_3C(CH_3)(Br)CH_3$ — Positional isomers

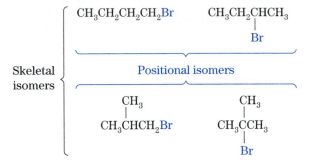

The members of the upper pair of compounds are positional isomers because each has a four-carbon continuous chain and they differ only in the position of the bromine. Likewise, the members of the lower pair have identical carbon skeletons and differ only in the position of the bromine. The two pairs of compounds differ, however, in the carbon skeleton and are therefore related as skeletal isomers.

GETTING INVOLVED

Example 2.3
Draw the positional isomers of $C_2H_3Br_2Cl$.

Solution

$CH_3{-}C(Br)_2{-}Cl$ (i.e., $CH_3{-}CBr_2Cl$) $CH_2(Cl){-}CH(Br)_2$ $CH_2(Br){-}CH(Br)(Cl)$

Problem 2.8
Draw the positional isomers of **(a)** $C_2H_4Br_2$ (2 isomers) **(b)** C_3H_6BrCl (5 isomers).
See related problems 2.21–2.22.

2.6 IUPAC Nomenclature of Alkanes

A. An Introduction to IUPAC Nomenclature

Pine forest

Trivial or common names have been assigned to organic compounds for a variety of reasons, including pioneering chemists' simple ignorance of the structures of the compounds they were investigating. The rationales for many of these names are obvious: source—pinene from pine trees and cocaine from coca leaves; smell—putrescine and cadaverine, compounds formed in decaying flesh; flavor—cinnamaldehyde and vanillin; color—Congo red and malachite green; geometry—basketane and cubane; trade names—Nutrasweet®; and public adulation—recently buckyballs or fullerenes (see Connections 1.1).

The potential for discovery or synthesis of great numbers of chemical compounds showed the need for a method of systematically naming millions of compounds. Early in the 20th century, the International Union of Pure and Applied Chemists (IUPAC) developed the IUPAC system of nomenclature, which relates names of compounds to molecular structure. For example, the name of a simple hydrocarbon is based on the number of carbons in the longest continuous chain of carbons in the molecule. Double bonds and triple bonds are identified, and their locations within the longest chain are described numerically.

We have just considered structure and isomerism of alkanes and are now prepared to relate their structures to nomenclature. Nomenclature of other classes of organic compounds will be considered in individual chapters as the chemistry of each functional group is presented. A summary of IUPAC nomenclature appears in the appendix. Some common names are still used; the more persistent of these will be introduced throughout the book.

B. Nomenclature of Continuous-Chain, Unbranched Alkanes: Basis of Organic Nomenclature

Compounds containing only carbon and hydrogen, with continuous, unbranched carbon chains and with only single bonds, are named with the Greek name for the number of carbons followed by the suffix *-ane*. For example, a compound with a five carbon chain is named *pent*, the Greek denoting five, followed by the suffix *-ane*, which indicates that all carbon-carbon bonds are single bonds.

Ring, or cyclic, hydrocarbons follow the same naming scheme except that the prefix *cyclo-* is used to indicate that the chain is a ring.

Cyclopropane Cyclobutane Cyclopentane Cyclohexane Cyclooctane

Table 2.1 presents structural formulas and names for the first ten saturated hydrocarbons. Since the nomenclature of alkanes is the basis of organic nomenclature in general, it is important that you learn the names of at least the first ten hydrocarbons. Notice that the first four hydrocarbons in Table 2.1 have trivial names that were incorporated in the IUPAC system because of their extensive prior use. Analyze the names in Table 2.1, realizing that the prefixes (*pent-*, *hex-*, *hept-*, *oct-*, *non-*, and *dec-*) describe the number of carbons (five through ten, respectively), and that the *-ane* suffix signifies the presence of nothing but carbon-carbon single bonds.

Table 2.1 Continuous-Chain Hydrocarbons

First Ten Hydrocarbons			
CH_4	Methane	$CH_3(CH_2)_4CH_3$	Hexane
CH_3CH_3	Ethane	$CH_3(CH_2)_5CH_3$	Heptane
$CH_3CH_2CH_3$	Propane	$CH_3(CH_2)_6CH_3$	Octane
$CH_3(CH_2)_2CH_3$	Butane	$CH_3(CH_2)_7CH_3$	Nonane
$CH_3(CH_2)_3CH_3$	Pentane	$CH_3(CH_2)_8CH_3$	Decane

C. Nomenclature of Branched-Chain Alkanes

alkyl group
hydrocarbon chain with one open point of attachment
R—: sometimes your instructor may use the symbol "R" for a general alkyl group.

Branched-chain alkanes are compounds in which shorter carbon chains (alkyl groups) are attached to longer carbon skeletons. Their names are based on the name of the longest continuous carbon chain; the names of the attached hydrocarbon substituents **(alkyl groups)** are derived by changing the ending of the appropriate hydrocarbon name from *-ane* to *-yl* (Table 2.2). These compounds are named according to the following procedure:

1. Find the longest carbon chain in the molecule and name it according to Table 2.1 with the Greek for the number of carbons followed by the suffix *-ane*. For a cyclic compound, the ring is usually the base of the name regardless of the longest continuous chain.
2. Name the attached shorter chains (alkyl groups). See Table 2.2.
3. To locate the positions of the alkyl groups, number the longest carbon chain consecutively from one end to the other, starting at the end that will give the lowest number to the first substituent. This step is used concurrently with step 2.

If more than one of a particular alkyl group appears in a molecule, a Greek prefix indicating the number of identical groups is used. For example, if a compound has two, three, four, five, or eight methyl groups, we indicate this by using dimethyl, trimethyl, tetramethyl, pentamethyl, and octamethyl. The location of each methyl group is described with its own number.

The alkyl groups with three- and four-carbon chains deserve special mention. A three-carbon alkyl group could be attached to a longer chain at either of the outside carbons or at the middle carbon; the groups are called propyl and isopropyl, respectively.

$CH_3CH_2CH_2—$ Propyl $CH_3\underset{|}{C}HCH_3$ Isopropyl

There are two structural isomers of a four-carbon alkyl group, each of which has two different points of connection.

Table 2.2 Alkyl Groups

$CH_3—$	Methyl	$CH_3(CH_2)_4CH_2—$	Hexyl
$CH_3CH_2—$	Ethyl	$CH_3(CH_2)_5CH_2—$	Heptyl
$CH_3CH_2CH_2—$	Propyl	$CH_3(CH_2)_6CH_2—$	Octyl
$CH_3(CH_2)_2CH_2—$	Butyl	$CH_3(CH_2)_7CH_2—$	Nonyl
$CH_3(CH_2)_3CH_2—$	Pentyl	$CH_3(CH_2)_8CH_2—$	Decyl

Branched Alkyl Groups

$CH_3\underset{|}{C}HCH_3$	$CH_3\underset{|}{C}HCH_2CH_3$	$CH_3\overset{CH_3}{\overset{|}{C}}HCH_2—$	$CH_3\underset{|}{\overset{CH_3}{\overset{|}{C}}}CH_3$
Isopropyl	Secondary butyl (*sec-butyl*)	Isobutyl	Tertiary butyl (*tert-*, or *t-butyl*)

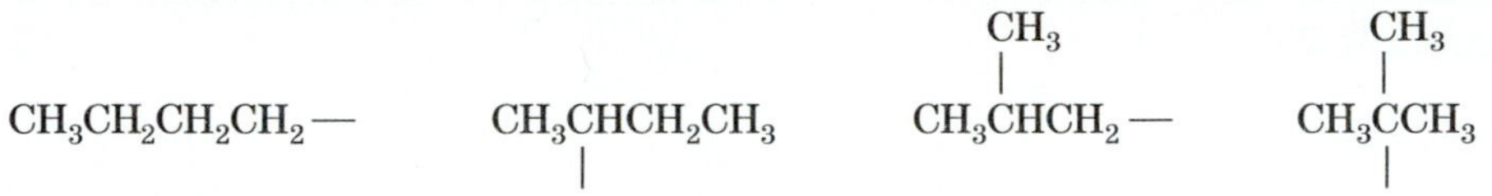

```
                                                CH3            CH3
                                                 |              |
CH3CH2CH2CH2—        CH3CHCH2CH3           CH3CHCH2—         CH3CCH3
                        |                                       |
    Butyl          Secondary butyl          Isobutyl        Tertiary butyl
```

GETTING INVOLVED

✓ Take a minute to think through what you need to know to name alkanes: knowing the names of the first ten hydrocarbons, finding and naming the longest chain, identifying and naming branches on the longest chain, numbering the longest chain, and knowing the names of alkyl groups with three and four carbons.

✓ Now take a close look at Examples 2.4 through 2.7.

Example 2.4

Name

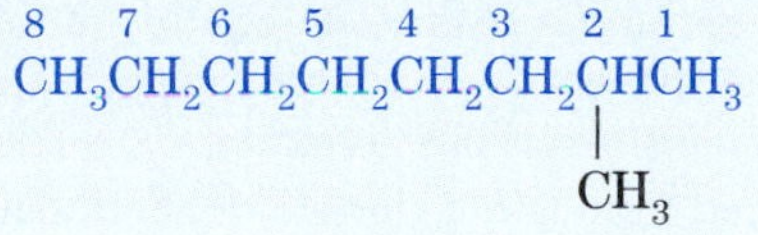

```
 8   7   6   5   4   3   2  1
CH3CH2CH2CH2CH2CH2CHCH3
                    |
                   CH3
```

Solution

1. The longest continuous chain has eight carbons, and thus the base of the name is octane.
2. A one-carbon substituent, a methyl group, is attached to the longest chain. The compound is a methyloctane.
3. Numbering left to right locates the methyl group on carbon-7. Conversely, numbering right to left puts it on carbon-2. The second alternative gives the lowest number to the substituent. The complete name is 2-methyloctane.

Example 2.5

Name

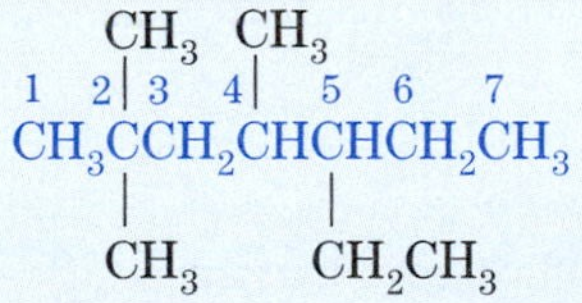

```
     CH3  CH3
 1  2| 3  4|  5   6   7
CH3CCH2CHCHCH2CH3
     |       |
    CH3    CH2CH3
```

Solution

1. The longest continuous chain has seven carbons: a heptane.
2. There are four shorter branches on the heptane chain: three one-carbon chains and one two-carbon chain. One-carbon substituents are called methyl groups. Since there are three methyl groups, part of the name must be trimethyl. The two-carbon chain is called an ethyl group. The alkyl groups are named alphabetically. The parent compound is an ethyl trimethylheptane.
3. The substituents must now be placed on the longest chain. Numbering from left to right allows the lowest designations. The three methyl groups are on carbon-2 and carbon-4 (2,2,4-trimethyl), and the ethyl group is on carbon-5 (5-ethyl). Each substituent gets its own number. The complete name is 5-ethyl-2,2,4-trimethylheptane.

Example 2.6

Name

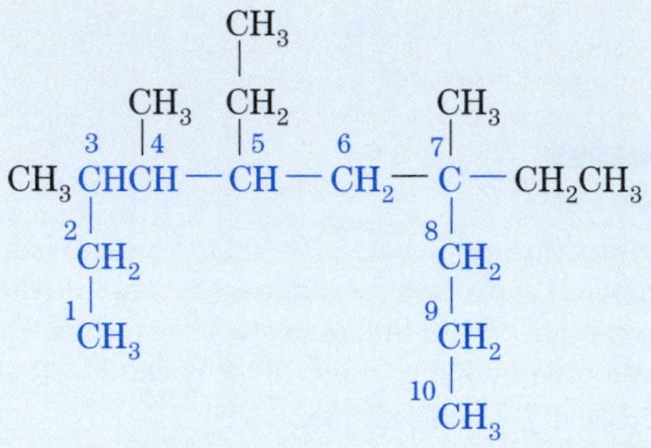

```
                 CH3
                  |
          CH3    CH2          CH3
        3  |4     |5     6    7|
CH3 CHCH ——— CH ——— CH2 ——— C ——— CH2CH3
     2|                        8|
     CH2                       CH2
     1|                        9|
     CH3                       CH2
                              10|
                               CH3
```

Solution

1. Finding the longest continuous chain may require careful inspection. We are looking not for the longest straight chain but rather for the longest continuous carbon chain. In this case, it has ten carbons: a decane.
2. The chain is numbered from left to right to obtain the lowest numbering for the location of the substituents.
3. Identify the substituents on the decane chain. On the third, fourth, and seventh carbons, there are methyl groups (3,4,7-trimethyl). Attached to the fifth and seventh carbons are ethyl groups (5,7-diethyl). Arranging the substituents in alphabetical order, we get the name 5,7-diethyl-3,4,7-trimethyldecane.

Example 2.7

Name

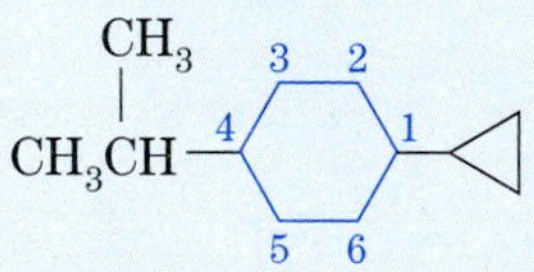

```
  CH3      3  2
   |      /‾‾\
CH3CH —4 <    > 1— ◁
          \__/
          5  6
```

Solution

1. The cyclohexane ring is the base of the name.
2. Numbering in this case follows the alphabetical order of the substituents. The substituent on carbon-1 should be designated cyclopropyl- and the one on carbon-4, isopropyl.
3. The name is 1-cyclopropyl-4-isopropylcyclohexane.

Problem 2.9

(a) Name the five isomers drawn in Example 2.1 by the IUPAC system of nomenclature.

(b) Name the nine isomers in Problem 2.4.

(c) Write structures for the following compounds: (a) 1-isobutyl-3-isopropylcyclopentane; (b) 5,6-diethyl-2,2,4,8-tetramethylnonane.

See related problems 2.20, 2.24, and 2.25.

D. Nomenclature of Halogenated Hydrocarbons (Alkyl Halides)

We shall learn in Chapters 4 and 8 that halogenated hydrocarbons play an important role in organic reactions. Their nomenclature follows standard rules. Halogens attached to a hydrocarbon chain are named by the prefixes *fluoro-* (F), *chloro-* (Cl), *bromo-* (Br), and *iodo-* (I).

alkyl halide
alkane possessing at least one F, Cl, Br, or I. Your instructor may occasionally use the symbol X to represent halogen in general examples.

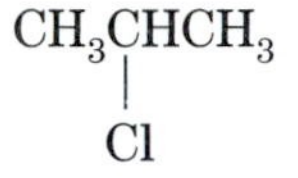

```
CH3CHCH3
   |
   Cl
```

2-chloropropane

```
   CH3  CH3
   |    |
CH3CCH2CCH2CH3
   |    |
   Br   I
```

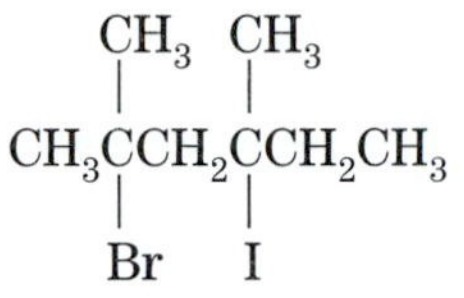

2-bromo-4-iodo-2,4-dimethylhexane

GETTING INVOLVED

Problem 2.10

In section 2.5, we drew the four isomers of C_4H_9Br. Name these by the IUPAC system of nomenclature.

Problem 2.11

Name the (a) two isomers of $C_2H_4Br_2$ and (b) five isomers of C_3H_6BrCl that you drew in problem 2.8.

Problem 2.12

There are eight isomers with the formula $C_5H_{11}Cl$. Draw and name these. If you need a little help, consult Figure 2.3 for the possible carbon skeletons. Then place the Cl in place of a hydrogen in as many *different* locations as you can.

See related problems 2.23, 2.26, and 2.27.

2.7 Conformational Isomerism

Stereoisomerism refers to isomer variations in spatial or three-dimensional orientation of atoms. One type of stereoisomerism is conformational isomerism. This is a more subtle form of isomerism than skeletal and positional isomerism, which we covered earlier in this chapter. In these structural isomers the actual bonding arrangement of atoms differs, with variations of the carbon skeleton or the positions of noncarbon atoms. In **conformational isomerism,** the bonding arrangement of atoms remains constant, but the relationship of the atoms in space differs as a result of rotation around carbon-carbon single bonds. As we shall see, this rotation occurs readily, with easy interconversion of the conformational isomers (conformers). Thus they are not isolatable and not isomers in the same sense as those we have covered so far.

conformational isomers
isomers that differ as a result of the degree of rotation around a carbon-carbon single bond

Let us take ethane (CH_3CH_3) as a simple example. The two carbons are connected by a single bond, a σ bond. Sigma bonds overlap in only one position, and consequently rotation of the carbons around the single bond does not affect the degree of overlap. As a result such rotation is more or less unrestricted. As the carbons in ethane rotate, the relationship of the hydrogens on the adjacent carbons changes continually; theoretically there is an infinite number of conformational isomers of ethane. However, since bond rotation is continual, conformational isomers are interconverting rapidly. None is an independent entity or a compound that can be isolated.

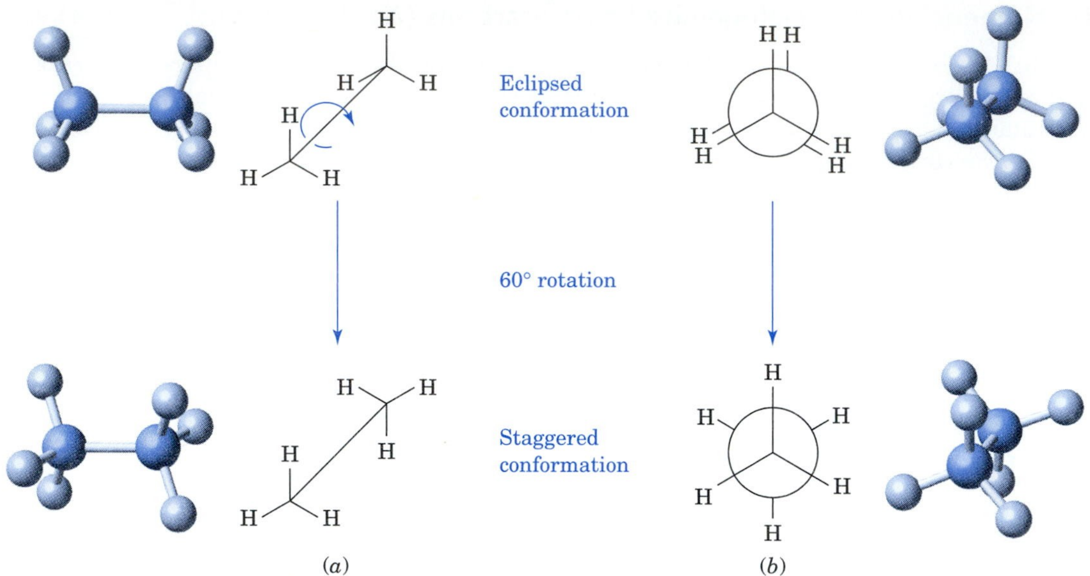

Figure 2.4 The conformations of ethane, $CH_3—CH_3$. (a) Sawhorse diagram. (b) Newman projection.

eclipsed
conformation around a carbon-carbon single bond in which attached atoms are as close together as possible

staggered
conformation around a carbon-carbon single bond in which attached atoms are as far apart as possible

sawhorse diagram
a way of representing conformational isomers with stick drawings

Newman projection
a way of representing conformational isomers using an end-on projection of a carbon-carbon bond

There are two extreme forms of ethane we can easily visualize (Figure 2.4). In one, the hydrogens on the adjacent carbons are lined up with one another and are therefore as close together as possible. As a result, this conformation, called the **eclipsed** conformation, is the least stable (highest energy) of all possibilities and is not very abundant in a sample of ethane. In the other extreme, the hydrogens on adjacent carbons are **staggered** with one another and thus are as far apart as possible. This conformation, called the staggered conformation, is the most stable (lowest energy). These conformations can be represented by either **sawhorse diagrams** or **Newman projections,** as illustrated in Figure 2.4. The sawhorse diagrams are self-explanatory, but let us take a closer look at the Newman projections.

In the Newman projection, one is viewing the carbon-carbon bond end-on along the axis of connection. The point of intersection of the three lines represents the front carbon, and the perimeter of the circle represents the rear carbon. The projection shows that the hydrogens are as near to each other as possible in the eclipsed conformation. This leads to maximum repulsion between the bonding pairs of electrons and accounts for the instability of this conformation. A 60° rotation around the carbon-carbon bond axis places the hydrogens and the bonding pairs a maximum distance from each other, thus minimizing repulsion. This is the staggered conformation. Although there are differences in stability among the conformers of ethane, the energy differences are not great. For this reason, rotation around the carbon-carbon bond occurs with almost no restriction, making it impossible to isolate the different conformers—they are constantly interconverting.

In 1,2-dibromoethane ($BrCH_2—CH_2Br$), two staggered and two eclipsed conformations are possible, with all four having different stabilities (Figure 2.5). Because of the large size and high electron density of the bromines, their proximity to one another is the prime determinant of conformer stability. The two staggered conformations are more stable than the two eclipsed; the staggered conformation with the bromines maximally separated is the most stable. By analogous reasoning, the eclipsed conformation in which the two bromines are eclipsed with each other (rather than with hydrogens) is the least stable. All sam-

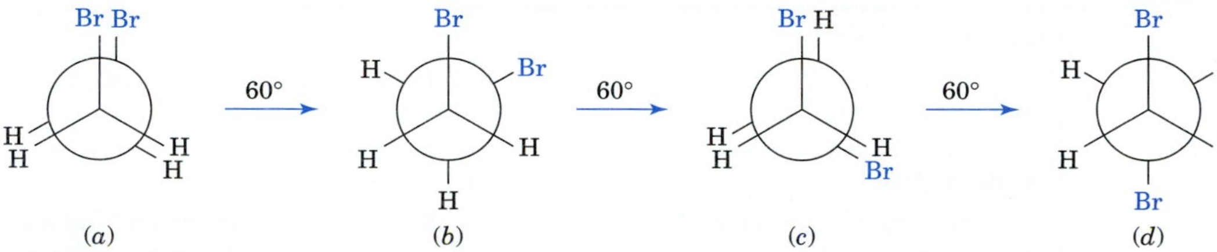

Br Br / H H / H H (a) — 60° → Br / H Br / H H / H (b) — 60° → Br H / H H / H Br (c) — 60° → Br / H H / H H / Br (d)

Figure 2.5 Newman projections of the conformations of 1,2-dibromoethane ($BrCH_2—CH_2Br$). (a) Eclipsed (least stable). (b) Staggered. (c) Eclipsed. (d) Staggered (most stable).

ples of 1,2-dibromoethane are identical, however, and are composed of these four and other intermediate conformers in concentrations approximately related to their stabilities. The conformations cannot be separated or isolated; the sample is merely a dynamic mixture of these conformers.

GETTING INVOLVED

- ✓ How do structural isomers and conformational isomers differ? What is the difference between eclipsed and staggered conformations? Which is more stable? How does an eclipsed conformation change to a staggered conformation?
- ✓ Draw the general representations of staggered and eclipsed conformations using Newman projections and sawhorse diagrams several times until you become facile with them.

Example 2.8

Draw the staggered and eclipsed conformational isomers of butane, $CH_3CH_2CH_2CH_3$, looking down the $C_1—C_2$ bond. Use Newman projections.

Solution

First determine what atoms are on C_1 and C_2.

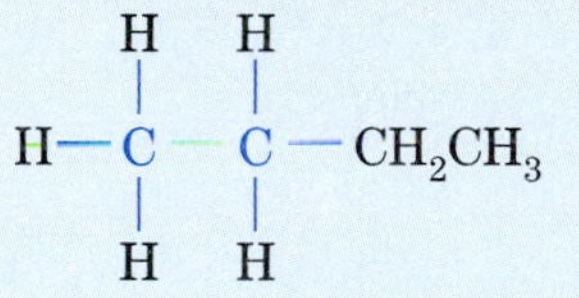

$$H{-}\underset{H}{\overset{H}{C}}{-}\underset{H}{\overset{H}{C}}{-}CH_2CH_3$$

C_1 H, H, H
C_2 H, H, CH_2CH_3

Draw a circle with three lines at 120° angles emanating from the center and three more lines, staggered with the others, coming from the perimeter. The lines coming from the center are from C_1; put the three hydrogens on them. Those from the perimeter are from C_2 (behind); put the two H's and the CH_2CH_3 on them. Now rotate one of the carbons 60° to get the eclipsed conformation.

H / H H / H H / CH_2CH_3 — 60° → H_H / CH_3CH_2 H / H H H

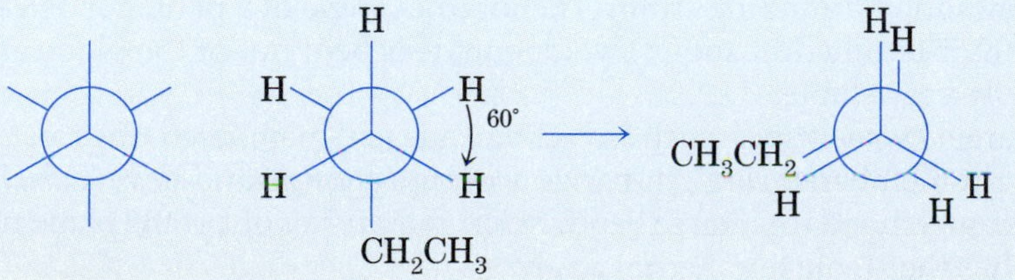

Problem 2.13

Using Newman projections draw the staggered and eclipsed conformations of propane. Which is more stable? For guidance, take a look at Example 2.8.

Problem 2.14

Draw the two staggered and two eclipsed conformational isomers of butane, $CH_3CH_2CH_2CH_3$, formed by 60° rotations about the single bond between the second and third carbons. Use Newman projections. Identify the most and least stable conformers. If you need assistance, check out Example 2.8 and Figure 2.5.

See related problem 2.28.

2.8 Cycloalkanes—Conformational and Geometric Isomerism

A. Structure and Stability

Saturated hydrocarbons possessing one or more rings are called *cycloalkanes*. They are often described by regular polygons, each corner of which represents a carbon with enough hydrogens to satisfy the valence. The smallest member of this class, cyclopropane, has three carbons. Cyclopropane and cyclobutane (Four-carbon ring) have been shown to be less stable than larger-ring cycloalkanes.

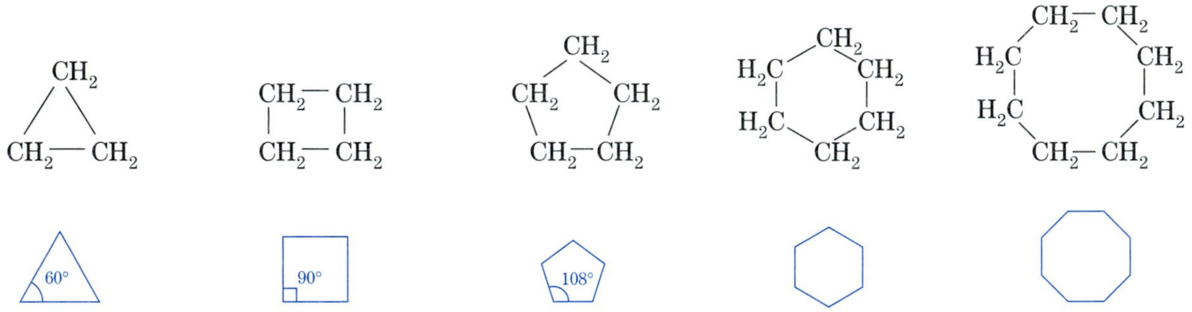

60° Cyclopropane

90° Cyclobutane

108° Cyclopentane

Cyclohexane

Cyclooctane

The source of this relative instability is the internal angles of each ring. Each carbon in a ring has four bonded atoms, is sp^3-hybridized, and should be tetrahedral, with 109.5° bond angles. Since cyclopropane has a three-membered ring, and three points define a plane, the molecule as a whole must have the geometry of an equilateral triangle with internal angles of 60°. This angle differs significantly from the preferred tetrahedral angle, causing decreased orbital overlap in the σ bonds and internal angle strain. Although cyclobutane is not planar, it still geometrically approximates a square, with internal bond angles close to 90°; it thereby suffers from ring strain. The internal angle of a pentagon is very close to the 109.5° tetrahedral angle. Cyclopentane is bent out of the plane and is energetically very stable.

In larger cycloalkanes, such as cyclohexane (six-membered ring) and cyclooctane (eight-membered ring), the rings are large enough and have sufficient flexibility through bond rotation to bend, twist, and pucker out of the plane until each carbon has the stable tetrahedral angle.

GETTING INVOLVED

✓ Explain why cyclopropane and cyclobutane are less stable than their larger ring counterparts. How do larger rings avoid internal angle strain?

B. Conformational Isomerism in Cyclohexane

To visualize the puckering in cyclohexane that provides for 109.5° bond angles, let us begin with cyclohexane as a regular hexagon. The simplest way to make this molecule nonplanar, minus internal strain, is to bend the two "end" carbons (carbons 1 and 4) out of the plane of the ring. Both carbons can be "puckered" in the same direction to form the **boat conformation** (Figure 2.7a). Or one carbon can be pulled above the plane of the ring and the other below, producing the **chair conformation** (Figure 2.7b). In both conformations, each carbon is tetrahedral and all bond angles are 109.5°. The two conformations are not of equal stability, however. The chair form is more stable and by far the predominant conformer of cyclohexane. The difference in stability is evident if one compares the structures of the boat and chair forms.

boat conformation
an unstable conformation of cyclohexane with 109.5° bond angles but in which most bonds are eclipsed

chair conformation
the most stable conformation of cyclohexane in which all bonds are staggered and bond angles are 109.5°

In the boat form, the carbons on opposite ends (carbon-1 and carbon-4) are pulled toward each other, causing steric interactions between the "flagpole" hydrogens because of their proximity (see Figures 2.6 and 2.7). In the chair form, these same two carbons are bent away from each other—one up and one down—and

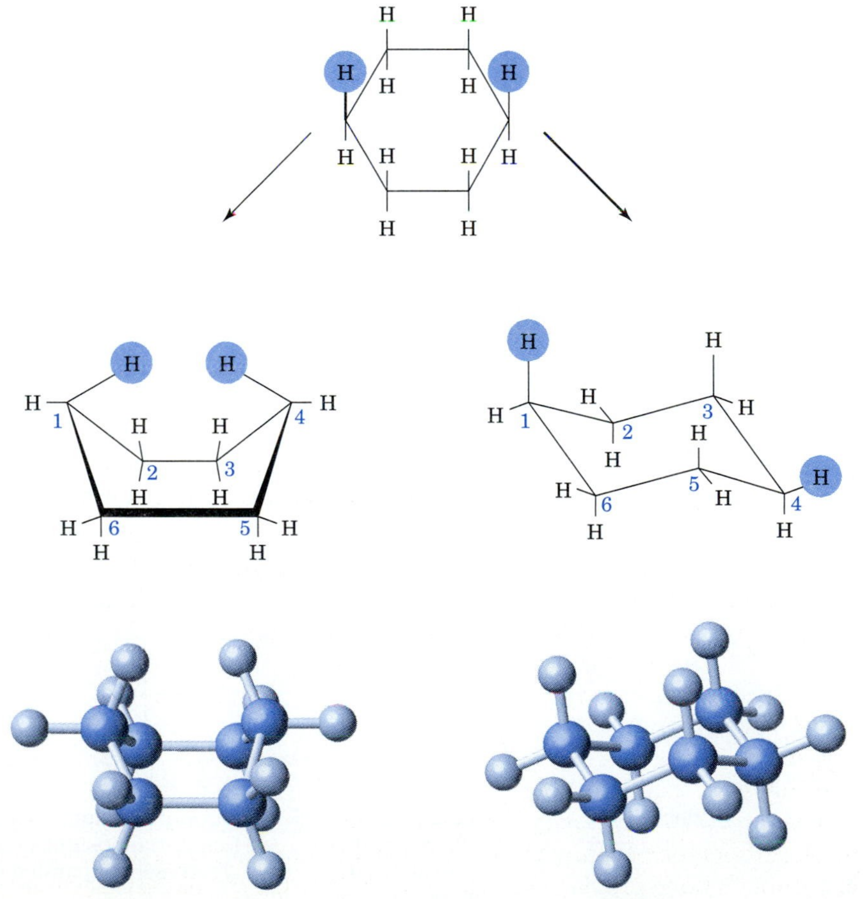

Figure 2.6 Boat and chair conformations of cyclohexane. (a) Boat form. (b) Chair form.

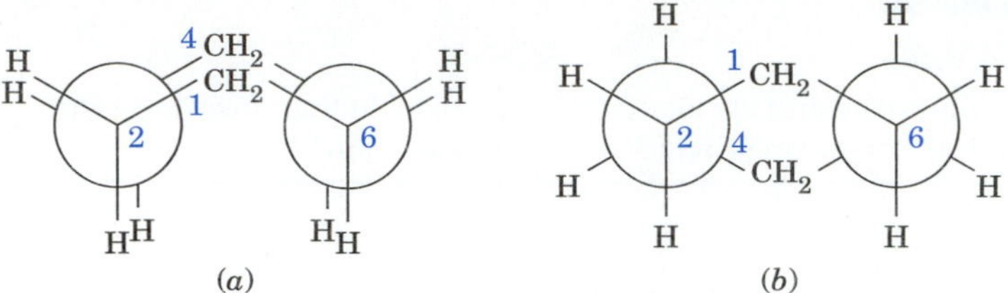

Figure 2.7 Newman projections of the (a) boat and (b) chair conformations of cyclohexane. Compare the numbered carbons to those in the boat and chair forms in Figure 2.6.

thus are not subject to mutual repulsion. A second destabilizing factor can be found by viewing the $C_2—C_3$ and $C_5—C_6$ bonds end-on, using Newman projections (Figure 2.8). In the boat form, the bonded atoms are in the less stable eclipsed conformation, whereas in the chair form, they are staggered (see section 2.6). For these two reasons, the chair form is the more stable, predominant conformation.

Close examination of the chair form of cyclohexane reveals that there are two basic orientations of the hydrogens (Figure 2.8). Six of the hydrogens are approximately perpendicular to the ring and are called **axial hydrogens.** There are three above and three below the ring on alternate carbons (Figures 2.6 and 2.8). The other six hydrogens, one on each carbon, lie in the average "plane" of the ring and protrude outward from it; these are called **equatorial hydrogens.** The axial hydrogens are nearer to one another than the equatorial hydrogens are. Substituted cyclohexanes thus exist predominantly in conformations in which the group or groups that have replaced hydrogens are in the roomier equatorial positions (section 2.8.D).

axial bonds
bonds on cyclohexane chair perpendicular to the ring with three up and three down on alternating carbons

equatorial bonds
bonds on cyclohexane chair parallel to the ring

C. Drawing the Cyclohexane Chair

Six-membered rings are very common in organic chemistry. Thus you should take some time to learn how to draw the chair form of cyclohexane. First, draw the chair framework on scratch paper over and over until you feel comfortable doing so and are getting good reproductions. In drawing the chair, note that there are four carbons in a plane. Attached to these is an "end" carbon that is above the plane and, on the other "end," a carbon that is down below the plane.

Now, on the "end" carbon that is up, draw a vertical line straight up. Do the same on alternate carbons to produce a total of three lines straight up. On each of the other three carbons, draw a vertical line down. These six lines represent the axial positions.

Finally, place on each of the six carbons a line that is neither up nor down but that is coming off the perimeter of the chair. These are the equatorial positions. This process is summarized below.

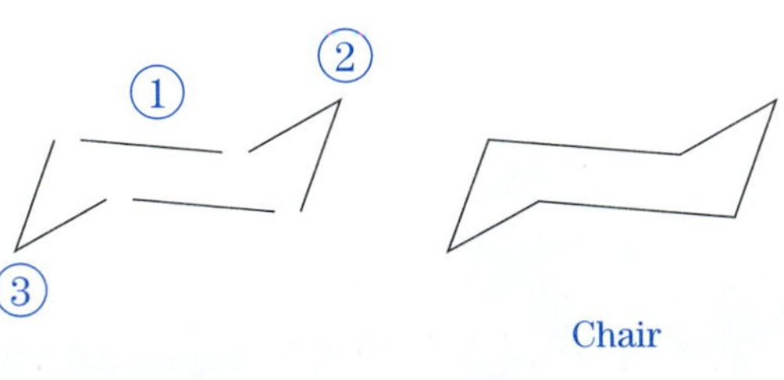

Chair

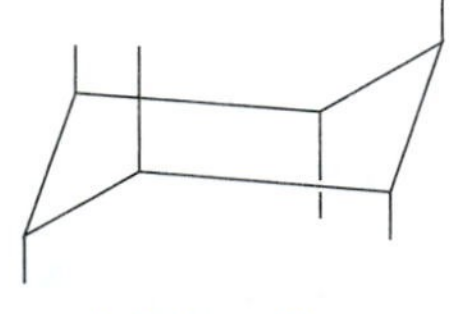

Axial positions all are parallel and alternate up or down on alternate carbons

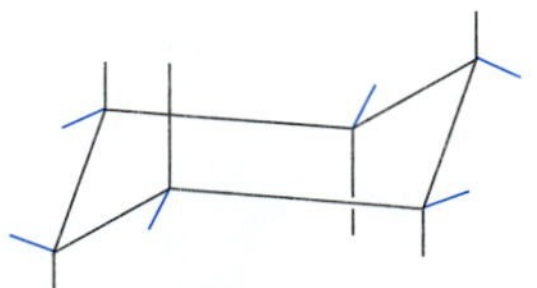

Axial positions (black)
Equatorial positions (color)
Three sets of parallel lines

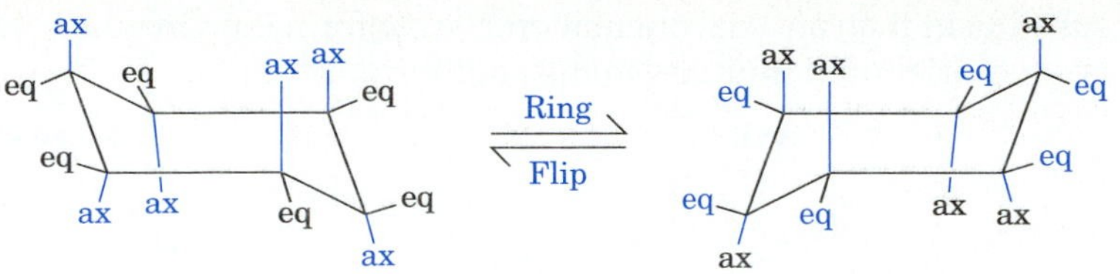

ax eq eq ax ax ax eq eq ax ax eq eq ax eq
Ring Flip
ax ax eq eq eq eq eq eq ax ax ax ax eq eq

Figure 2.8 Axial and equatorial positions on the chair form of cyclohexane. The chair form is constantly flipping between two identical conformers. As this happens, all axial hydrogens become equatorial and all equatorial hydrogens axial.

GETTING INVOLVED

✓ Become proficient at drawing the cyclohexane chair. First draw the chair over and over until you feel comfortable. Then take some of those better drawings you did and draw in the three axial positions above and the three below the ring. Now that you have the axials shown, put in the six equatorial positions.

Example 2.9

Draw 1,3-dimethylcyclohexane with both methyl's axial, both equatorial, and one axial and one equatorial.

Solution

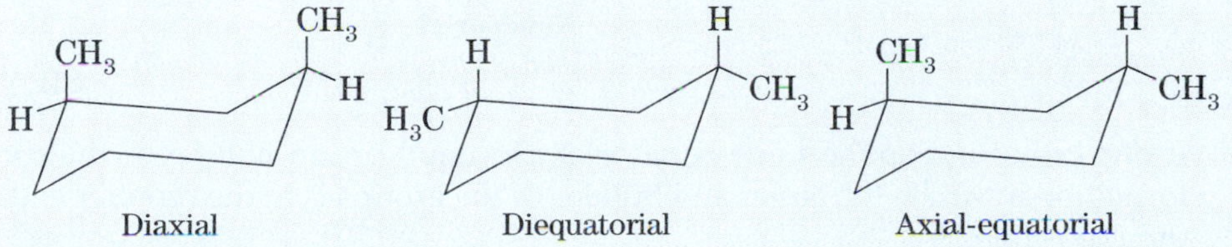

CH_3 H CH_3 H — Diaxial

H H_3C H CH_3 — Diequatorial

CH_3 H H CH_3 — Axial-equatorial

Problem 2.15

To develop a feel for the chair conformation of cyclohexane and axial and equatorial positions, draw chair forms of cyclohexane with the following substituents: **(a)** axial CH_3; **(b)** equatorial CH_3; **(c)** Br's on carbon-1 and carbon-2, both axial; **(d)** Br's on carbon-1 and carbon-4, one axial and one equatorial; **(e)** Br's on carbon-1 and carbon-3, both equatorial.

D. Conformational Isomerism in Substituted Cyclohexanes

Axial positions in the chair conformation of cyclohexane are more crowded than the equatorial. Axial hydrogens protrude directly above or below the ring and are closer to one another than equatorial hydrogens, which are situated around the perimeter and directed away from the ring (see the molecular model representations in Figure 2.6). Because the equatorial positions are more spacious, a substituent bonded to the ring in place of hydrogen will form a more stable compound if it is equatorial rather than axial.

Consider methylcyclohexane, for example. An equilibrium exists between two conformations, one in which the methyl group is axial and another where it is equatorial. The equilibrium exists because the cyclohexane ring is constantly flipping between the two conformations; every time a flip occurs, all axial positions become equatorial and all equatorial positions become axial (see Figure 2.8). Since the equatorial position offers a roomier environment for the methyl substituent than an axial orientation, the conformation in which the methyl group is equatorial predominates in the equilibrium.

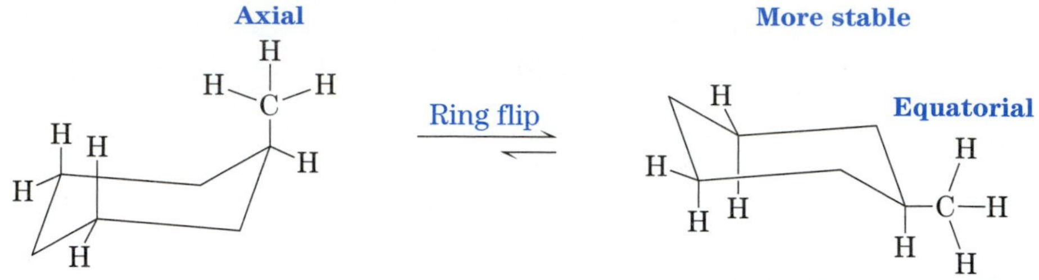

If two groups are attached to a cyclohexane ring, two equilibrium possibilities exist: axial-axial in equilibrium with equatorial-equatorial, and axial-equatorial in equilibrium with equatorial-axial. In the first of these possibilities it is clear that the equilibrium will favor the equatorial-equatorial conformation, where the extreme crowding that would result from a diaxial arrangement is avoided. This is dramatically illustrated with 1-isopropyl-3-methylcyclohexane.

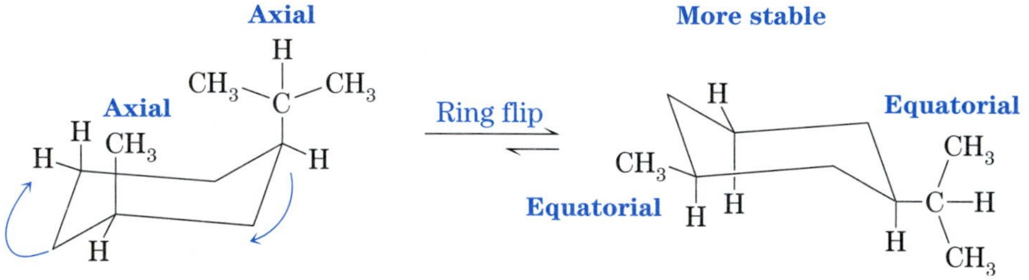

If one group is axial and one equatorial, the equilibrium will favor the conformation in which the larger and bulkier of the groups is in the roomier equatorial position.

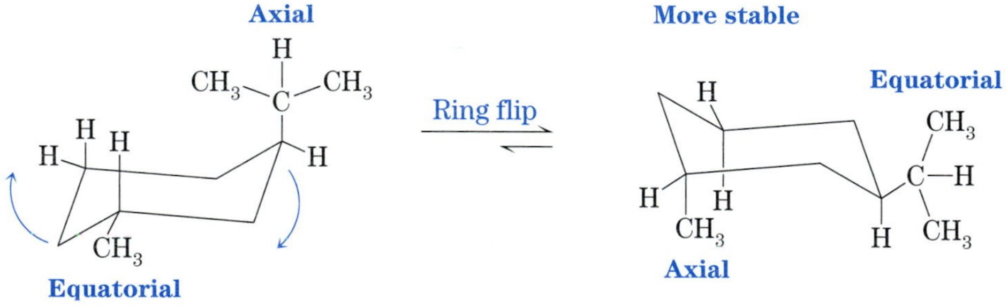

GETTING INVOLVED

Problem 2.16

Draw the equilibria described and in each case determine which conformation is more stable. **(a)** Draw the chair forms of bromocyclohexane to illustrate the equilibrium between the conformations in which bromine is axial and in which bromine is equatorial. **(b)** Illustrate the axial-axial/equatorial-equatorial equilibrium, using 1-isopropyl-2-methylcyclohexane. Note that in this example the diaxial groups are on opposite sides of the ring as opposed to the 1-isopropyl-3-methyl example shown in this section. The same would be true for diaxial in the 1,4 positions. **(c)** Illustrate the axial-equatorial/equatorial-axial equilibrium, using 1-isopropyl-4-methylcyclohexane.

See related problems 2.29–2.32.

E. Geometric Isomerism in Cyclic Compounds

Although relatively free rotation around carbon-carbon single bonds exists in open chain alkanes, it does not exist in cycloalkanes. For example, in small ring compounds like cyclopropane, if two carbons began to rotate in opposite directions, the third carbon would be forced to break its attachments because it is bonded to both and obviously cannot follow the opposing rotations. Because of this and the nature of the cyclic structure, cycloalkanes can be thought of as having sides. In appropriately substituted cycloalkanes, this results in a type of stereoisomerism called **geometric isomerism.**

geometric isomers
cis and *trans* isomers; a type of stereoisomerism in which atoms or groups display orientation differences around a double bond or ring

***cis* isomer**
geometric isomer in which groups are on same side of ring or double bond

***trans* isomer**
geometric isomer in which groups are on opposite sides of ring or double bond

Consider, for example, 1,2-dibromocyclopropane. The two bromines can be on the same side of the ring; this is called the ***cis* isomer.** Or the two bromines can be on opposite sides of the ring, the ***trans* isomer.**

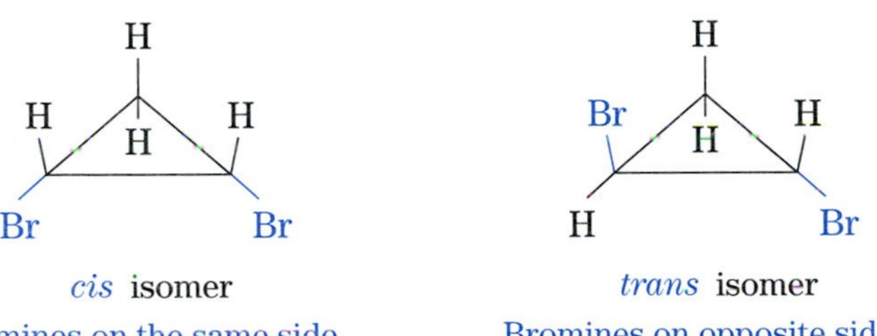

cis isomer — Bromines on the same side

trans isomer — Bromines on opposite sides

Structural isomers, such as skeletal and positional isomers, have different bonding arrangements of atoms. Notice in the above examples, however, that the same atoms are connected to one another, a characteristic of stereoisomerism. The compounds differ in the spatial orientation of the atoms; the bromines are on the same side *(cis)* or opposite sides *(trans)* but still connected to the same carbons.

GETTING INVOLVED

✓ Describe geometric isomerism and why it occurs in cyclic compounds. How are *cis* and *trans* isomers different?

Example 2.10

Draw the geometric isomers of 1,2- and 1,3-dimethylcyclobutane.

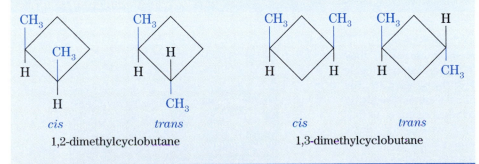

cis *trans*
1,2-dimethylcyclobutane

cis *trans*
1,3-dimethylcyclobutane

Problem 2.17

Draw the geometric isomers of 1-bromo-2-methylcyclopentane and 1-bromo-3-methylcyclopentane.

See related problems 2.33–2.35.

solid
state of matter with constant volume and shape; strong attractive forces between immobile molecules in crystal lattice

liquid
state of matter with constant volume but variable shape; molecules in random motion but with intermolecular attractions

gas
state of matter with variable volume and shape; molecules are independent, in random motion, and without intermolecular attractions

2.9 Hydrocarbons: Relation of Structure and Physical Properties

How does the molecular structure we have discussed relate to physical properties of compounds? The **solid, liquid,** and **gaseous** states of a compound do not represent differences in the structure of the individual molecules. They represent, rather, variations in the arrangement of the molecules. In a solid, the molecules are arranged very compactly and are relatively immobilized in an orderly crystal lattice. Attractive forces between molecules are at a maximum (Figure 2.9). In the liquid state, the intermolecular attractions still exist, but the molecules are mobile; they have greater kinetic energy. Molecular mobility in the vapor phase is so great that intermolecular attractions are practically nonexistent and each molecule is theoretically independent of the others. Energy, in the form of heat, is required to provide molecules with the impetus and mobility to break

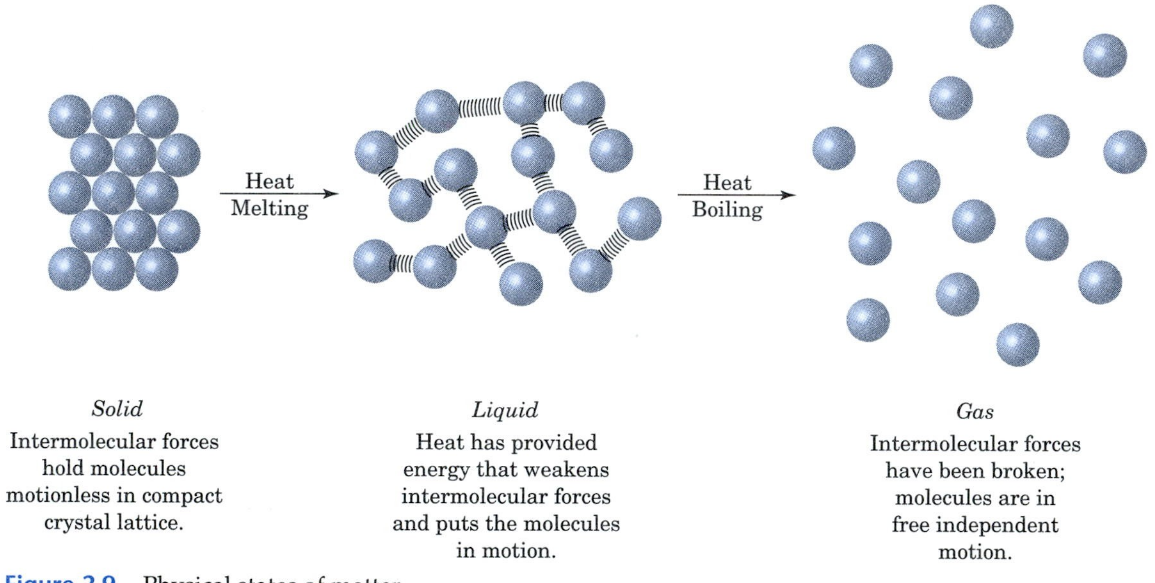
Heat
Melting
Heat
Boiling
Solid
Intermolecular forces hold molecules motionless in compact crystal lattice.
Liquid
Heat has provided energy that weakens intermolecular forces and puts the molecules in motion.
Gas
Intermolecular forces have been broken; molecules are in free independent motion.

Figure 2.9 Physical states of matter.

Connections 2.1 www.prenhall.com/bailey

Petroleum

From the time green plants appeared on this planet, they have been using the sun's energy to convert carbon dioxide and water into oxygen and the organic chemicals of life. The process is called *photosynthesis.* Animals, either directly by eating plants or indirectly, acquire these organic materials from plants and modify them for their own existence. Some are used for body structure and others for energy. When "burning" organic compounds for energy, animals use oxygen from the air and return carbon dioxide and water. The carbon cycle (see Figure I.1) is thus completed.

As normal life cycles of growth, death, and decay continued century after century, the remains of prehistoric plant and animal life settled to the bottoms of lakes, marshes, and oceans. Sediments of these organic wastes accumulated over the ages; under some conditions they were converted into a complex mixture of organic compounds called petro-

Connections 2.1 *(cont.)*

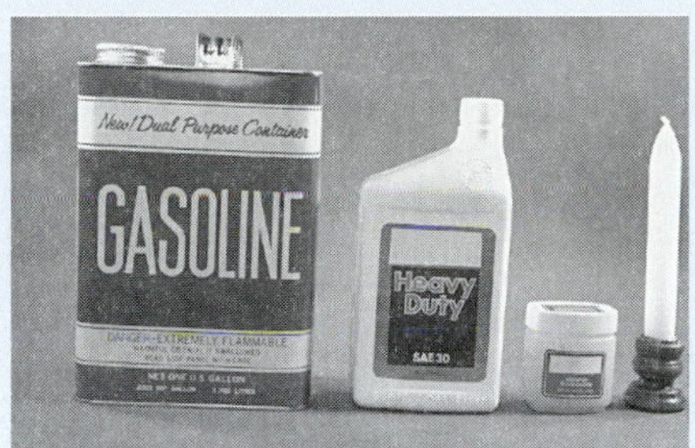

leum, and under other conditions into massive deposits of coal. The evolution of coal occurred in several stages, with the latter stages having a high elementary carbon content and a concentration of volatile organic compounds (largely aromatic).

Coal had long been valued as an energy resource, but it wasn't until 1859, when Edwin Drake drilled a hole 69.5 feet deep next to a surface oil seep, that the ancient legacy of unaccountable generations of plants and animals—petroleum—was released.

Crude oil is usually a black, viscous, foul-smelling liquid composed primarily of hundreds of different hydrocarbon molecules. To convert crude oil into usable components, it must be separated into various fractions. Since the boiling points of hydrocarbons increase with molecular weight (section 2.9), crude oil can be separated into its components by fractional distillation using gigantic distillation towers such as that schematically depicted in Figure 2.10. Low-boiling, low-molecular-weight compounds are collected high on the column, and those of increasing molecu-

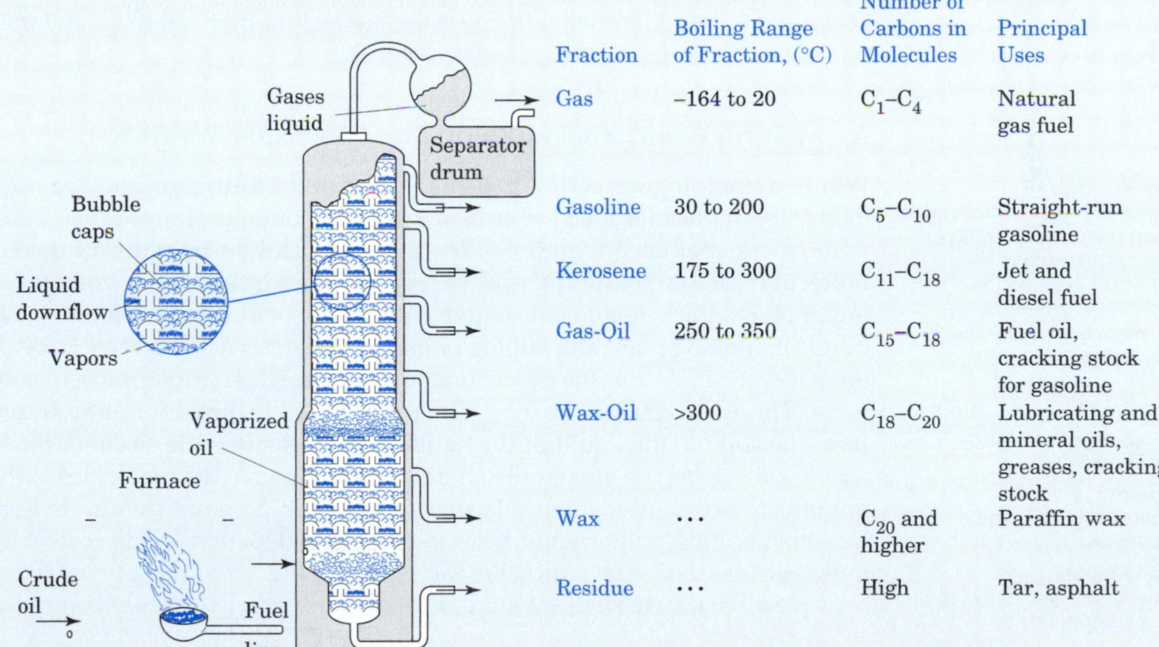

Fraction	Boiling Range of Fraction, (°C)	Number of Carbons in Molecules	Principal Uses
Gas	−164 to 20	C_1–C_4	Natural gas fuel
Gasoline	30 to 200	C_5–C_{10}	Straight-run gasoline
Kerosene	175 to 300	C_{11}–C_{18}	Jet and diesel fuel
Gas-Oil	250 to 350	C_{15}–C_{18}	Fuel oil, cracking stock for gasoline
Wax-Oil	>300	C_{18}–C_{20}	Lubricating and mineral oils, greases, cracking stock
Wax	...	C_{20} and higher	Paraffin wax
Residue	...	High	Tar, asphalt

Figure 2.10 Fractional distillation of crude petroleum. A bubble-cap distillation tower and the principal fractions of petroleum, their properties, and uses are illustrated. High-boiling, high-molecular-weight materials are collected at the bottom of the column and low-boiling, low-molecular-weight materials at upper regions of the column. Natural gas and liquefied natural gas (LNG) are primarily methane, CH_4. Liquefied petroleum gas (LPG)—used for heating in rural areas and in camping stoves, lighters, and torches—is mainly propane with lesser amounts of ethane and butane. Straight-run gasoline represents neither the quality nor the quantity for today's needs. Consequently, gasoline is also produced from the refining of other petroleum fractions (see Connections 6.2).

Connections 2.1 *(cont.)*

lar weight and boiling point are separated at various stages lower on the column. These different fractions are then commercially developed into several thousand consumable products, from life-giving medicines to deadly pesticides, from tar for roads to delicate synthetic fibers and plastics.

Most of a barrel (1 barrel = 42 gallons) of crude oil is converted into fuels: over 40% into gasoline; 30% to 35% to fuel oil for heating and other purposes; and 7% to 10% to jet fuel; much of what remains is sold as aviation gasoline, liquefied petroleum gas, lubricating oils, greases, and asphalt. Only a small portion, around 5%, of this rich supply of organic materials is used in the petrochemical industry to manufacture drugs, dyes, plastics, and other products commonly used in today's culture.

Most of this treasure of hydrocarbon energy originally obtained from ancient sunlight and stored in chemical bonds by prehistoric organisms is reconverted by combustion to carbon dioxide and water, from which it came.

Combustion:

$$\text{Hydrocarbon} + O_2 \longrightarrow CO_2 + H_2O$$

Petroleum is a dwindling resource and the prospect of its depletion sometime in the twenty-first century is causing political problems throughout the world. Methods to enhance the amount of oil recoverable from a well are being improved. Extraction of oil from shale and tar sands and the conversion of coal into synthetic liquid fuels are possibilities for the future. Nonpetroleum energy sources such as nuclear fission and fusion, solar and geothermal energy, biomass energy, hydroelectric energy, tidal and wind power, and energy from thermal gradients in the ocean are in use or under consideration.

out of a crystal lattice and form a liquid or to sever all attractive forces and become a vapor. Here we will consider the factors that influence the melting points and boiling points of alkanes (Table 2.3).

A. Melting Point, Boiling Point, and Molecular Weight

homologous series
a series in which each compound differs from the one preceding by a constant factor; each of the members of the homologous series—methane, ethane, propane, butane, pentane and so on—differs from the preceding by a $-CH_2-$ group

melting point
temperature at which a solid becomes a liquid

boiling point
temperature at which a liquid becomes a gas

molecular weight
sum of the atomic weights of the atoms in a compound

Within a **homologous series, melting points** and **boiling points** increase with increasing molecular weight for most classes of organic compounds. Consider your own experience with the following hydrocarbon fractions of increasing **molecular weight:** natural gas is, of course, a gas; gasoline is a volatile liquid; motor oil is a thick, nonvolatile liquid; and paraffin wax (candles) is a solid. These trends in melting point and boiling point are understandable on two bases. First, the larger the molecule, the more numerous are the sites for intermolecular attractions. These attractions must be either weakened or broken in any transition from the solid to the liquid or the liquid to the gaseous state. Second, the heavier the substance, the greater the energy needed to give the molecules sufficient impetus to break these intermolecular forces. Melting point trends are less regular than boiling point trends, since melting also depends on the correct fit of a molecule into its crystal lattice. We can see in Table 2.3 that these generalizations hold for alkanes. We shall see similar trends with other classes of compounds.

B. Melting Point, Boiling Point, and Molecular Structure

Boiling points decrease with chain branching in hydrocarbons. Branching makes a molecule more compact and decreases the surface area. The smaller the surface area, the fewer the opportunities for intermolecular attraction; consequently, branched molecules have lower boiling points than unbranched ones of comparable molecular weight. On the other hand, compactness or molecular symmetry usually increases the melting point of a compound. Such molecules fit

Table 2.3 Melting Points and Boiling Points of Alkanes

Name	Formula	Molecular Weight	Melting Point, °C	Boiling Point, °C
Methane	CH_4	16	−182	−164
Ethane	CH_3CH_3	30	−183	−89
Propane	$CH_3CH_2CH_3$	44	−190	−42
Butane	$CH_3(CH_2)_2CH_3$	58	−138	−1
Pentane	$CH_3(CH_2)_3CH_3$	72	−130	36
Hexane	$CH_3(CH_2)_4CH_3$	86	−95	69
Heptane	$CH_3(CH_2)_5CH_3$	100	−91	98
Octane	$CH_3(CH_2)_6CH_3$	114	−57	126
Nonane	$CH_3(CH_2)_7CH_3$	128	−51	151
Decane	$CH_3(CH_2)_8CH_3$	142	−30	174
Pentadecane	$CH_3(CH_2)_{13}CH_3$	212	10	271
Eicosane	$CH_3(CH_2)_{18}CH_3$	282	37	343

more easily into a crystal lattice. A more stable crystal lattice requires a larger energy to disrupt. Consequently, the melting points are higher for highly branched compounds than for compounds with a longer, straighter chain.

Consider, for example, the isomers of C_5H_{12} (all with molecular weight 72). Note the progressive decrease in boiling point with branching. Also observe the large difference in melting point between pentane and the highly compact, symmetrical dimethylpropane.

	$CH_3CH_2CH_2CH_2CH_3$	$CH_3CH(CH_3)CH_2CH_3$	$CH_3C(CH_3)_2CH_3$
Melting point:	−160°C	−130°C	−17°C
Boiling point:	36°C	28°C	9.5°C

A dramatic difference in melting point occurs between the isomers octane and 2,2,3,3-tetramethylbutane.

	$CH_3CH_2CH_2CH_2CH_2CH_2CH_2CH_3$	$CH_3C(CH_3)_2—C(CH_3)_2CH_3$
Melting point:	−57°C	101°C
Boiling point:	126°C	107°C

Fitting octane into an orderly crystal lattice would be like trying to stack wet spaghetti, whereas arranging 2,2,3,3-tetramethylbutane would be analogous to stacking wooden blocks. However, octane, with more surface area, has a higher boiling point.

C. Solubility and Density

From experience, we know that hydrocarbons such as alkanes are water insoluble. This is because water is a polar solvent (polar O—H bonds) and alkanes are nonpolar (composed of nonpolar carbon-carbon and carbon-hydrogen bonds). Hydrocarbons are also less dense than water and will float on its surface. (Oil spills remain for the most part on the ocean's surface, for example.)

solubility
the amount of material that will dissolve in a solvent and produce a stable solution

density
weight per unit volume of a substance

GETTING INVOLVED

✓ What is a homologous series? Why do melting points and boiling points generally increase within a homologous series of hydrocarbons?

✓ Within a series of hydrocarbon isomers, what influence does chain branching have on melting points and boiling points?

✓ Why does gasoline float on water and not dissolve in it?
See problem 2.38.

Problems

2.18 **Molecular Weights:** Using atomic weights from the periodic table, calculate the molecular weights of the following compounds:

(a) CH_4—methane, natural gas
(b) CH_3CH_2OH—beverage alcohol
(c) $C_{12}H_{22}O_{11}$—table sugar
(d) $C_{11}H_{17}N_2O_2SNa$—sodium pentothal

2.19 **Skeletal Isomerism:** Draw the skeletal isomers with the following formulas. Disregard isomers that display isomerism other than skeletal. You may wish to do problem 2.20 concurrently.

(a) 18 isomers with the formula C_8H_{18}
(b) 3 isomers of C_9H_{20} with eight carbons in the longest chain.
(c) 11 isomers of C_9H_{20} with seven carbons in the longest chain
(d) 8 isomers of C_9H_{20} with five carbons in the longest chain
(e) 4 isomers of $C_{10}H_{22}$ with nine carbons in the longest chain
(f) 2 isomers of $C_{10}H_{22}$ in which there are only two alkyl groups on a six-carbon chain
(g) 6 isomers of $C_{10}H_{22}$ with five carbons in the longest chain
(h) the isomer of $C_{13}H_{28}$ with the shortest possible longest chain of carbons
(i) 5 cyclic compounds of C_5H_{10}
(j) 12 cyclic compounds of C_6H_{12}
(k) 5 compounds of C_8H_{16} that have a six-membered ring
(l) 12 isomers of C_9H_{18} with a six-membered ring.

2.20 **Nomenclature of Alkanes:** Name the compounds drawn in problem 2.19.

2.21 **Positional Isomers:** Suppose you have a method by which you can remove a single hydrogen from a molecule and replace it with a chlorine. For each of the following molecules, determine how many different isomers, each possessing one chlorine, could be made by replacing a single hydrogen with a chlorine.

(a) $CH_3CH_2CH_2CH_2CH_2CH_3$

(b)
$$\begin{array}{rcccl} & CH_3 & & CH_3 & \\ & | & & | & \\ CH_3 & C & CH_2CH_2 & CH & CH_3 \\ & | & & & \\ & CH_3 & & & \end{array}$$

(c)
$$\begin{array}{rcccccl} & CH_3 & & CH_3 & & CH_3 & \\ & | & & | & & | & \\ CH_3 & CH & CH_2 & CH & CH_2 & CH & CH_3 \end{array}$$

(d) $\triangle$

2.22 **Skeletal and Positional Isomerism:** Draw the isomers described. It may be helpful to construct the possible carbon skeletons first and then vary the positions of the halogens on each skeleton. You may wish to do problem 2.23 concurrently.

(a) 3 isomers of $C_2H_3Br_2F$
(b) 4 isomers of $C_3H_6Br_2$
(c) 12 isomers of C_4H_8BrF
(d) 6 isomers of $C_4H_8Br_2$ with four carbons in the longest chain.
(e) 9 isomers of $C_5H_{10}Br_2$ with four carbons in the longest chain
(f) 5 isomers of $C_6H_{13}Cl$ with four carbons in the longest chain

2.23 **Nomenclature of Halogenated Alkanes:** Name the compounds in problem 2.22 by the IUPAC system of nomenclature.

2.24 **Nomenclature of Alkanes:** Name the following by the IUPAC system of nomenclature:

(a) $CH_3CH_2CH_3$
(b) $CH_3(CH_2)_8CH_3$
(c) $CH_3(CH_2)_6CH_3$

2.25 **Nomenclature of Alkanes:** Name the following by the IUPAC system of nomenclature:

(a)
$$\begin{array}{rcl} & CH_3 & \\ & | & \\ CH_3CH_2CH_2 & CH & CH_2CH_2CH_2CH_2CH_3 \end{array}$$

(b)
$$\begin{array}{rcl} & CH_3CH_2CH_2 & \\ & | & \\ CH_3CH_2CH_2CH_2 & CH & CH_2CH_2CH_2CH_2CH_3 \end{array}$$

(c)
$$\begin{array}{rcccl} & CH_3 & & CH_3 & \\ & | & & | & \\ CH_3 & CH & CH_2CH_2 & CH & CH_3 \end{array}$$

(d) 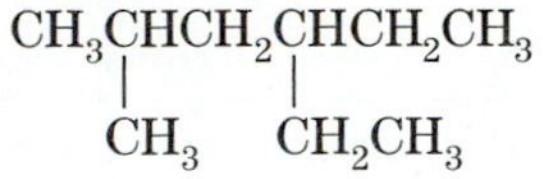

(e) 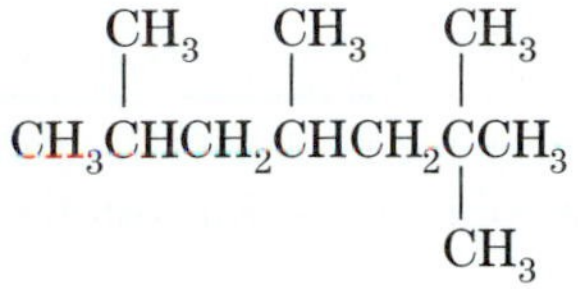

(f) 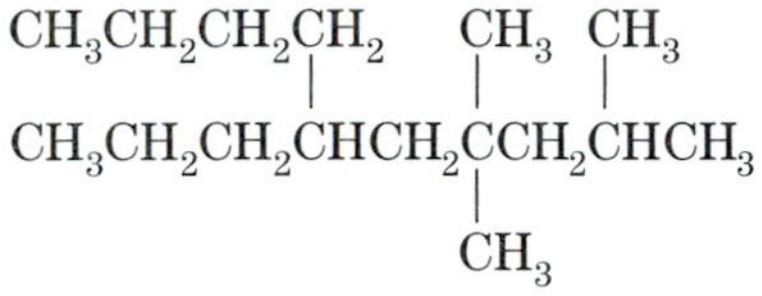

(g) 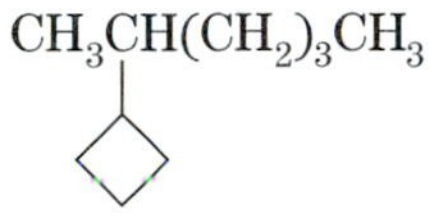

(h) 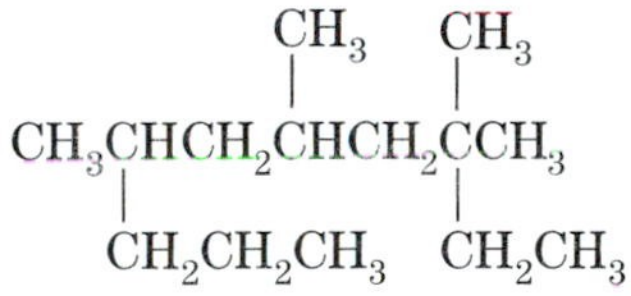

(i) 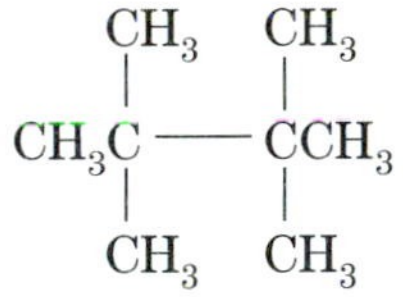

(j) 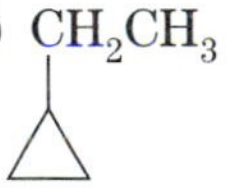

(k) 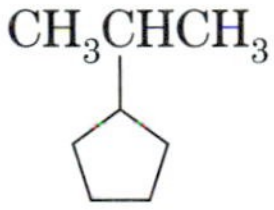

(l) 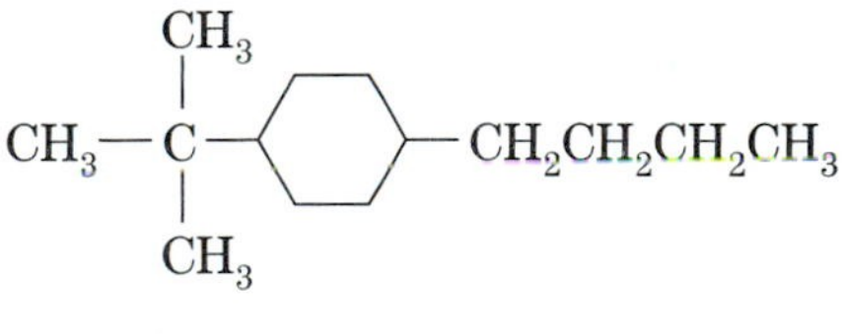

(m) $(CH_3)_3CCH_2CH_3$

(n) $(CH_3)_2CHCH_2CH(CH_2CH_3)CH_3$

(o) $(CH_3CH_2)_2CHCH_2C(CH_3)_3$

2.26 **Nomenclature of Halogenated Alkanes:** Name the following compounds by the IUPAC system of nomenclature:

(a) CHI_3

(b) 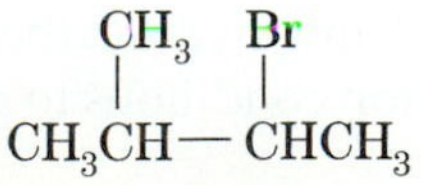

(c) 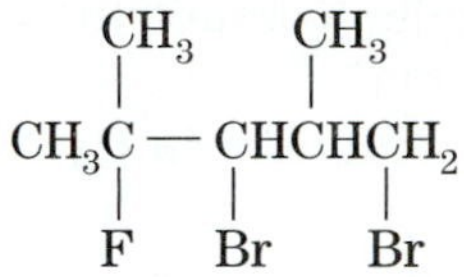

2.27 **IUPAC Nomenclature:** Draw the following compounds:

(a) dichlorodifluoromethane (a freon)

(b) 2,2,4-trimethylpentane (a component of high-octane gasoline)

(c) 1,2,3,4,5,6-hexachlorocyclohexane (an insecticide)

(d) 4-ethyl-2,2,6-trimethyl-5-propyloctane

2.28 **Conformational Isomers:** Using Newman projections or sawhorse diagrams, draw the staggered and eclipsed conformational isomers of

(a) ethyl alcohol, CH_3CH_2OH (beverage alcohol).

(b) ethylene glycol, $HOCH_2CH_2OH$ (antifreeze; production of Dacron).

(c) isopropyl alcohol, $CH_3CHOHCH_3$ (rubbing alcohol).

2.29 **Conformational Isomerism in Substituted Cyclohexanes:**

(a) Draw the most stable chair form of ethylcyclohexane.

(b) Draw the most stable chair forms of 1,2-dimethylcyclohexane; 1,3-dimethylcyclohexane; and 1,4-dimethylcyclohexane.

(c) Draw the least stable chair forms of the compounds in part (b).

(d) Draw 1,2-dimethylcyclohexane with one group axial and one equatorial.

(e) Draw the most stable chair form of 1-butyl-3-methylcyclohexane in which one group is axial and the other is equatorial.

2.30 **Conformational Isomerism in Substituted Cyclohexanes:**

(a) Draw the two possible chair forms of cyclohexane substituted with a single bromine. Which is more stable?

(b) Draw the chair form of cyclohexane with two bromines on adjacent carbons, both axial. Do a ring flip to obtain the more stable form with both equatorial. Now draw the chair with one axial and one equatorial.

(c) Draw the chair form of cyclohexane with a $-CH_3$ and a $-CH_2CH_3$ on carbons 1 and 3, respectively, both axial. Do a ring flip to get the more stable equilibrium conformation. Now draw the pair of conformers so that one group is axial and the other equatorial. Which is more stable?

(d) Repeat part **(c)** with the two groups on carbons 1 and 4.

(e) Draw the chair form of cyclohexane with bromines on carbons 1, 3, and 5 and with one of the bromines axial. Do a ring flip to obtain the other conformer. Which is more stable?

2.31 **Conformational Isomerism in Cyclohexane:** Camphor, a compound long respected for its medicinal qualities (actually it has been found to have little or no value), exists in the boat conformation. Draw the molecule in this conformation, and explain why it cannot exist in the chair form.

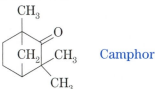

CH_3, O, CH_2, CH_3, CH_3 Camphor

2.32 **Conformational Isomers:** You can buy a bottle of the compound on the left but there is no way to bottle the conformer on the right. Explain.

CH_3, H, H_3C, H CH_3, H, H, CH_3

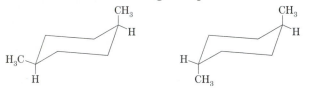

2.33 **Geometric Isomerism:** Draw the *cis* and *trans* isomers of each of the following compounds:

(a) 1,2-dimethylcyclopropane

(b) 1-bromo-3-chlorocyclobutane

2.34 **Geometric Isomerism:** Draw the *cis* and *trans* isomers of each of the following compounds. In each case, draw the cyclohexane ring as a planar hexagon lying on its side so that the plane appears to be going in front of and behind the paper.

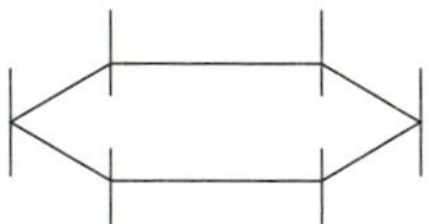

(a) 1,2-dimethylcyclohexane

(b) 1,3-dimethylcyclohexane

(c) 1,4-dimethylcyclohexane

2.35 **Geometric Isomerism:** Cycloalkanes with more than two substituents sometimes provide more than two possibilities of geometric isomerism (the terms *cis* and *trans* are not applicable in these compounds). Writing the ring as a planar polygon, draw the isomers described.

(a) the three geometric isomers of 1,2-dibromo-3-chlorocyclopropane

(b) the two geometric isomers of 1,2,3-tribromocyclopropane

(c) the three geometric isomers of 1,2,4-tribromocyclopentane

(d) the four geometric isomers of 1-chloro-2,4-dibromocyclopentane

2.36 **Geometric Isomerism in the Cyclohexane Chair:** The following table summarizes representations of geometric isomerism in a disubstituted cyclohexane chair.

	cis	*trans*
1,2-disubstituted	ax-eq or eq-ax	ax-ax or eq-eq
1,3-disubstituted	ax-ax or eq-eq	ax-eq or eq-ax
1,4-disubstituted	ax-eq or eq-ax	ax-ax or eq-eq

Draw the most stable chair form of each of the following disubstituted cyclohexanes. Rationalize the table presented.

(a) *cis* 1,2-dimethylcyclohexane

(b) *cis* 1-bromo-3-chlorocyclohexane

(c) *trans* 1,4-diethylcyclohexane

(d) *cis* 1-ethyl-4-methylcyclohexane

(e) *trans* 1-ethyl-3-methylcyclohexane

2.37 **Molecular formulas:** A compound has seven carbons, two nitrogens, one oxygen, one sulfur, two bromines, one triple bond, two double bonds and one ring. The rest of the atoms are hydrogen. How many hydrogens are there?

2.38 **Physical Properties:** Explain the difference in melting point or boiling point, as indicated, for each of the following sets of compounds:

(a) boiling point

CH_4 −164°C CH_3CH_3 −89°C

(b) boiling point

$CH_3CH_2CH_2CH_2CH_2CH_3$ 69°C $CH_3CH(CH_3)—CH(CH_3)CH_3$ 58°C

(c) boiling point

CBr_4 190°C CCl_4 77°C

(d) melting point

$CH_3(CH_2)_4CH_3$ −95°C   (cyclohexane) 6.5°C

(e) melting point

CH_3CH_3 −183°C $CH_3(CH_2)_{18}CH_3$ 37°C

2.39 **Combustion:** Write balanced chemical equations for the total combustion of

(a) methane—natural gas

(b) propane—rural gas

(c) 2,2,4-trimethylpentane—a high-octane gasoline component.

See Connections 2.1 on petroleum; hydrocarbons react with oxygen under combustion conditions to produce carbon dioxide and water.

2.40 **Petroleum Fractions:** Write the structures of one or two molecules that could be representative of the following petroleum fractions:

(a) gas
(b) gasoline
(c) kerosene
(d) gas-oil
(e) wax-oil
(f) wax

Activities with Molecular Models

1. Make molecular models of the three skeletal isomers of C_5H_{12}. Note the tetrahedral geometry of each carbon. Also observe the increasing compactness of the molecules with branching.

2. For each of the three isomers you made in exercise 1, see how many different places you can remove a hydrogen and attach a bromine. In each different case, you have made a positional isomer. How many can you make for each? Does symmetry within each molecule influence the number of possible positional isomers?

3. Make a model of ethane, C_2H_6. Rotate the two carbons relative to each other around the carbon-carbon bond. Make the staggered and eclipsed conformations.

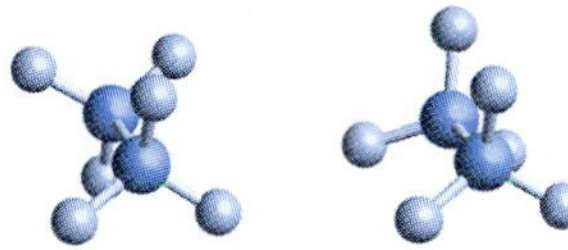

4. Using your models from exercise 3, remove one hydrogen and replace it with a bromine. Find the one staggered and one eclipsed conformation. Now replace a second hydrogen, on the other carbon, with a bromine. Find the two staggered and two eclipsed conformations.

5. Make a model of the chair form of cyclohexane. Observe that all the carbon-carbon bonds are in the staggered conformation. Identify the axial and equatorial hydrogens.

6. Using the model from exercise 5, replace an axial hydrogen with a methyl group and then replace an equatorial hydrogen with a methyl group. Which conformer do you think is more stable?

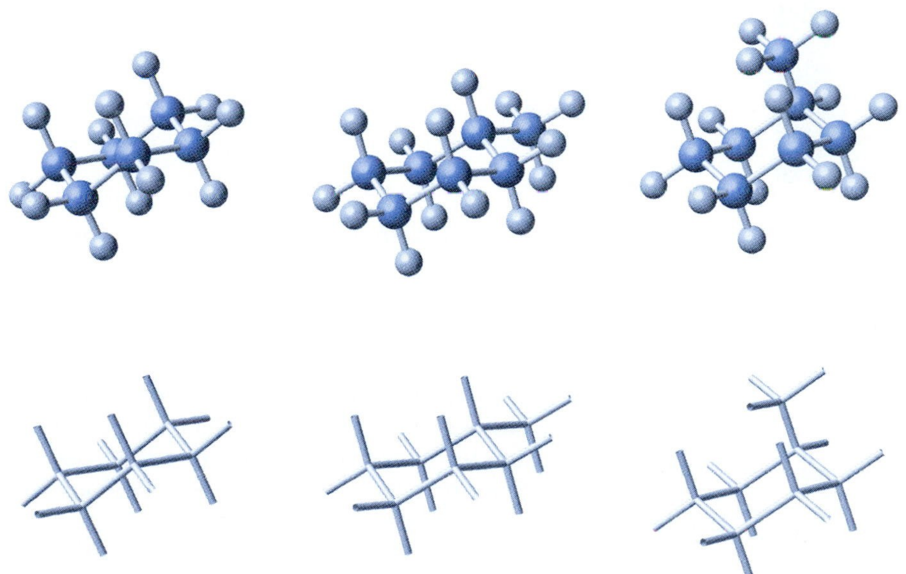

7. Using the model from exercise 6, make 1,2-dimethylcyclohexane with both methyl groups axial, one axial and one equatorial, and both equatorial. Which of these do you think is the most stable? Are any of them interconvertible with a ring flip? Now do the same with 1,3-dimethylcyclohexane. Notice the difference in diaxial between the 1,2 and 1,3 isomers. Finally, make the same models of 1,4-dimethylcyclohexane and answer the same questions.

CHAPTER 3

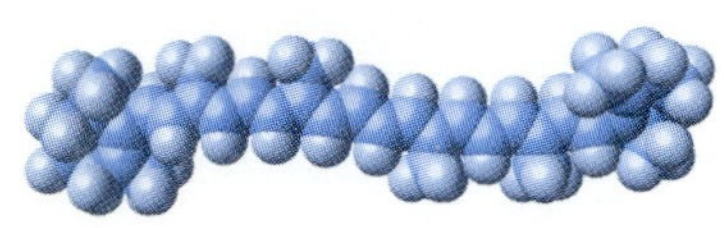

Alkenes and Alkynes: Structure and Nomenclature

3.1 Introduction to Alkenes and Alkynes

alkane
compound composed of only carbon and hydrogen and single bonds

alkene
hydrocarbon with at least one carbon-carbon double bond

alkyne
hydrocarbon with at least one carbon-carbon triple bond

In Chapter 2 we studied the structure and nomenclature of **alkanes.** Like all hydrocarbons, alkanes consist of carbon and hydrogen only. However, unlike other hydrocarbons, alkanes have only carbon-carbon single bonds. In contrast, **alkenes** have at least one carbon-carbon double bond and **alkynes** at least one carbon-carbon triple bond. The simplest examples of alkenes and alkynes are ethene (commonly known as ethylene, the precursor to polyethylene plastic) and ethyne (acetylene, used in welding). Compare these with ethane:

Welder with an oxyacetylene torch

$H_3C{-}CH_3$ (H—C—C—H, each C bearing two H)	$H_2C{=}CH_2$	H—C≡C—H
Ethane	Ethene	Ethyne
C_2H_6	C_2H_4	C_2H_2
C_nH_{2n+2}	C_nH_{2n}	C_nH_{2n-2}
saturated	unsaturated	unsaturated
Alkane	Alkene	Alkyne

saturated
a saturated molecule has all single bonds; each atom has the maximum number of attached atoms possible

unsaturated
an unsaturated molecule has at least one double bond or triple bond

Note that ethane has the maximum number of hydrogens that can be accommodated by two carbons in a molecule, six, whereas ethene and ethyne have fewer. Ethane and the other alkanes are called **saturated** hydrocarbons because all carbons have four attached atoms. The term **unsaturated** is applied to alkenes and alkynes, because not all of their carbons have the maximum number of bonded atoms. A carbon involved in a carbon-carbon double bond has only three attached atoms, and one involved in a carbon-carbon triple bond has only two. Though each carbon has four bonds (a double and two singles in the first case and a triple and one single in the second), these carbons have fewer than four bonded atoms, the maximum possible.

As we saw in Chapter 2, alkanes have the general formula C_nH_{2n+2}; that is, they have twice as many hydrogens plus two as carbons. This is evident in the molecular formulas for the first few alkanes, CH_4, C_2H_6, C_3H_8, and C_4H_{10} (components of the gas fraction of petroleum). To insert a double bond between two carbons of an alkane (as shown in ethane compared with ethene), two hydrogens must be removed from adjacent carbons. The general formula for noncyclic alkenes with one double bond then is C_nH_{2n}. Since four hydrogens must be removed from adjacent carbons to convert alkane to an alkyne (or two more hydrogens removed from the double bond of an alkene), the general formula for noncyclic alkynes with one triple bond is C_nH_{2n-2}. Convince yourself of this with the ethane, ethene, and ethyne structures previously shown.

GETTING INVOLVED

✓ What are the differences between alkanes, alkenes, and alkynes in structure and general molecular formula?

✓ What is the difference between a saturated and unsaturated compound?

Example 3.1

Draw an example of a simple alkatriene, a hydrocarbon with three double bonds, and a cycloalkadiene, a cyclic hydrocarbon with two double bonds. Determine the general molecular formulas for each.

C_6H_8 C_5H_6

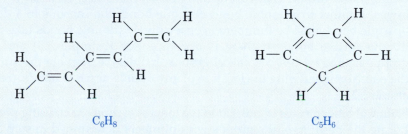

In both formulas there are four less hydrogens than twice the number of carbons. The general molecular formula is C_nH_{2n-4}.

Problem 3.1

What are the general molecular formulas (comparable to C_nH_{2n+2} for alkanes) for cycloalkanes, cycloalkenes, dienes (two double bonds), and cycloalkynes?

3.2 Nomenclature of Alkenes and Alkynes

A. IUPAC Nomenclature

The names of unsaturated hydrocarbons, and of organic compounds in general, follow the same conventions we saw with the alkanes. The number of carbons in the longest continuous chain of carbons is indicated by Greek-based prefixes. If the carbon chain has only carbon-carbon single bonds the suffix *-ane* is affixed to the Greek for the number of carbons. If there is a double bond, the suffix *-ene* (alkenes) is used, and for a triple bond, the suffix *-yne* (alkynes) is used. To indicate more than one double or triple bond, *di, tri, tetra* and so on are placed in front of the *-ene* or *-yne.* Double and triple bonds must be specifically identified and their positions determined in order to name a compound. All other carbon-carbon bonds are assumed to be single bonds. The following examples illustrate these points.

$CH_3CH_2CH_2CH_3$	$CH_2{=}CHCH_2CH_3$	$CH_3C{\equiv}CCH_3$	$CH_2{=}CHCH{=}CH_2$
Butane	1-butene	2-butyne	1,3-butadiene

B. Procedure for Naming Alkenes and Alkynes

Before considering additional examples, let us develop a procedure that we can follow when naming hydrocarbons.

1. Identify and name the longest continuous chain of carbon atoms (see Table 2.1 in the previous chapter). If the longest chain excludes multiple bonds, select the longest continuous chain with the maximum number of double and triple bonds.
2. Indicate the presence of a double bond with the suffix *-ene* and a triple bond with *-yne.* If there is neither, retain the suffix *-ane.*

C—C	C=C	C≡C
-ane	-ene	-yne
Alkanes	Alkenes	Alkynes

3. Number the carbon chain from one end to the other, so as to give the lowest number to a double bond or triple bond; the carbons in a double or triple bond must be numbered consecutively (double bonds take priority over triple bonds if there is a choice; see Example 3.5) and then groups named by prefixes (alkyl groups, halides). Complete the suffix of the name by identifying the location of any double or triple bonds. (See Example 3.2.)
4. Name all other groups connected to the longest carbon chain, using prefixes. Locate the position of the groups with numbers determined in step 3 and list them at the beginning of the name (alphabetical order is used in the chemical literature). (See Example 3.3.)

GETTING INVOLVED

✓ Look over the four-step procedure for nomenclature and be sure you understand the logic of it.

✓ What is the first step in naming any organic compound? What functional groups are always named with suffixes and what are those suffixes? How are carbon chains numbered?

Example 3.2

Name the following positional isomers.

$CH_3CH_2CH_2CH_2CH{=}CH_2$	$CH_3CH_2CH_2CH{=}CHCH_3$	$CH_3CH_2CH{=}CHCH_2CH_3$
6 5 4 3 2 1	6 5 4 3 2 1	1 2 3 4 5 6
A	**B**	**C**

1. All three compounds have six carbon continuous chains, and thus the Greek *hex* is used.
2. In each case the double bond is indicated by the suffix *-ene:* hexene.
3. Number the carbon chains from the ends that will give the lowest numbers to the double bonds. The position of the double bond is described by the lowest-numbered carbon involved. A is 1-hexene, B is 2-hexene, and C is 3-hexene.

Example 3.3

Name

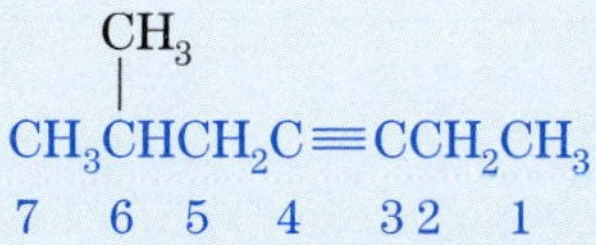

$$\begin{array}{c} CH_3 \\ | \\ CH_3CHCH_2C{\equiv}CCH_2CH_3 \\ 7\ \ 6\ \ 5\ \ 4\ \ \ 3\ 2\ \ 1 \end{array}$$

Solution

1. There are seven carbons in the longest continuous chain: hept-.
2. The triple bond is designated by the suffix *-yne:* heptyne.
3. Number so as to give the multiple bond the lowest number, right to left: 3-heptyne.
4. Name the branched methyl group with a prefix. The complete name is 6-methyl-3-heptyne.

Example 3.4

Name

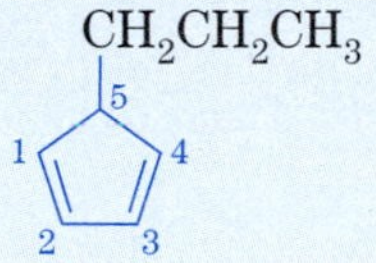

$CH_2CH_2CH_3$ (on carbon 5 of a cyclopentadiene ring numbered 1, 2, 3, 4, 5)

Solution

1. There are five carbons in the ring. This is designated by cyclopent-.
2. The two double bonds are indicated by *-diene* (*di* means "two"), cyclopentadiene (the syllable *a* is added for smoother pronunciation).
3. The ring is numbered, giving the lowest possible numbers to the carbons involved in the double bonds.
4. The propyl group is named with a prefix. The complete name is 5-propyl-1,3-cyclopentadiene.

Problem 3.2

Name the following by the IUPAC system of nomenclature:

(a) $CH_3CH_2CH_2CH_2CH_2CH{=}CH_2$ **(b)** $CH_3C{\equiv}CCH_2CH_2CH_2CH_3$

(c) 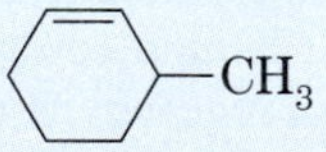 (cyclohexene ring)—CH_3 **(d)** $CH_3C{\equiv}C{-}C{\equiv}CCH_2CH_3$

See related problems 3.21–3.22.

C. Naming Compounds with Both Double and Triple Bonds

Consider the following examples of the numbering of carbon chains containing both double and triple bonds. The lowest numbers are given to multiple bonds whether double or triple. When there is a choice, however, the double bond takes precedence.

$\overset{1}{C}H_2{=}\overset{2}{C}H\overset{3}{C}{\equiv}\overset{4}{C}\overset{5}{C}H_3$	$\overset{5}{C}H_3\overset{4}{C}H{=}\overset{3}{C}H\overset{2}{C}{\equiv}\overset{1}{C}H$	$\overset{1}{C}H_2{=}\overset{2}{C}H\overset{3}{C}H_2\overset{4}{C}{\equiv}\overset{5}{C}H$
1-penten-3-yne	3-penten-1-yne	1-penten-4-yne

GETTING INVOLVED

Example 3.5

Name

$$\overset{4}{\mathrm{H}\mathrm{C}}\equiv\overset{3}{\mathrm{C}}-\overset{2}{\mathrm{C}\mathrm{H}}=\overset{1}{\mathrm{C}\mathrm{H}_2}$$

Solution

1. The longest chain is four carbons: but-.
2. The double bond is designated by the suffix *-ene,* and the triple bond by the suffix *-yne:* buten-yne.
3. Number from the end that gives the lowest number to the double bond (right to left). The complete name is 1-buten-3-yne.

Problem 3.3

Name by the IUPAC system of nomenclature:

(a) $CH_3C\equiv C-CH=CHCH_3$ **(b)** $HC\equiv CCH_2CH=CH\underset{\displaystyle CH_2CH_3}{\underset{|}{C}}HCH_2CH_3$

See related problems 3.24–3.25.

D. Common Nomenclature

Some alkenes are commonly referred to with an "alkylene"-type nomenclature, such as ethylene and propylene, from which the plastics polyethylene and polypropylene are made. The simplest alkyne is commonly called *acetylene* (as

Connections 3.1 www.prenhall.com/bailey

Oral Contraceptives

Complex cyclic hydrocarbon derivatives known as *steroids* are responsible for many of the hormonal responses in our bodies. Three classes of steroids, classified as the androgens, estrogens, and progestins, govern reproduction and the sexual characteristics of men and women. It is a complex interplay of estradiol and progesterone, for example, that is involved in the maturation, release, implantation, and maintenance of a fertilized egg in the uterus.

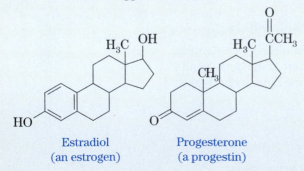

Estradiol (an estrogen)

Progesterone (a progestin)

Progesterone is indispensable to the integrity of a pregnancy in part by telling the brain, in a process called "feedback," that no more eggs are needed for fertilization.

Therefore, the most effective female contraceptive should be some type of progestin. Progesterone itself can be injected or dispensed by a vaginal insert for continuous delivery and birth control. These two methods of administration are not always convenient for use by large populations of women. In order to produce a drug that could be taken orally, the five-membered ring of the steroid had to be modified with an ethyne group. Norethindrone is one of the most commonly used progestins in "the minipill." Combining an estrogen with a progestin increases the efficiency of contraception to more than 99%. Two commonly used estrogens are mestranol and ethinyl estradiol.

At the end of 1991 the Food and Drug Administration (FDA) approved the use of implantable progestin pellets (Norplant®), which provide contraception for up to five

Connections 3.1 *(cont.)*

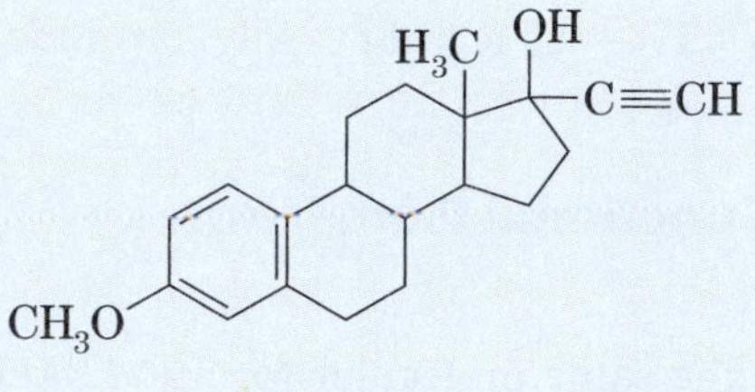

Mestranol

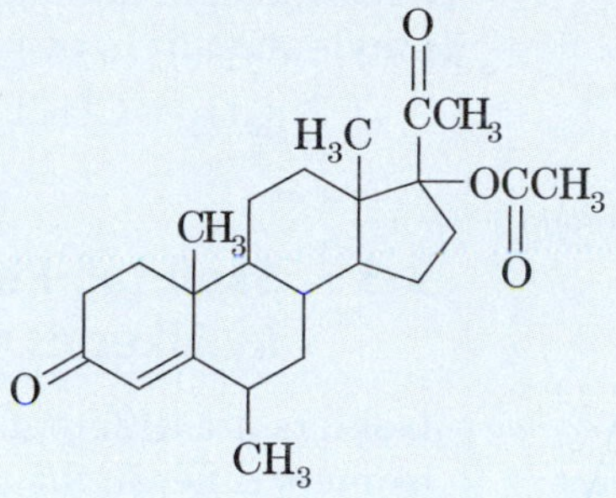

Medroxyprogesterone acetate (Depo-Provera®)

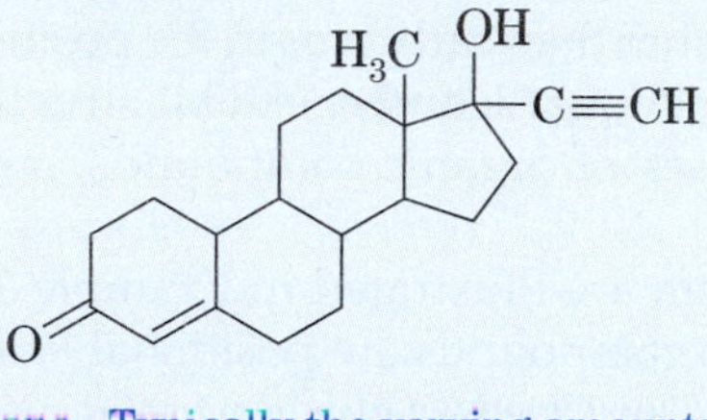

Norethindrone Typically the varying amounts of progestin are given with estrogen for 21 days. Then the drug is withdrawn for 7 days. Menstruation occurs, and the schedule is started again.

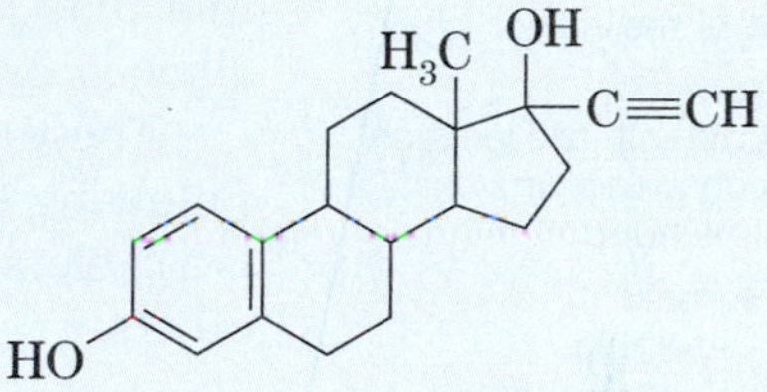

Ethinyl estradiol A typical preparation of "the pill" would contain 0.035 mg of ethinyl estradiol and 0.5–0.75 mg of norethindrone.

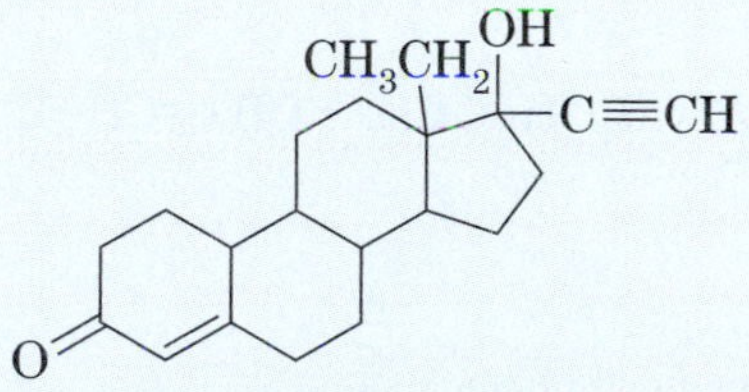

Levonorgestrel (Norplant®)

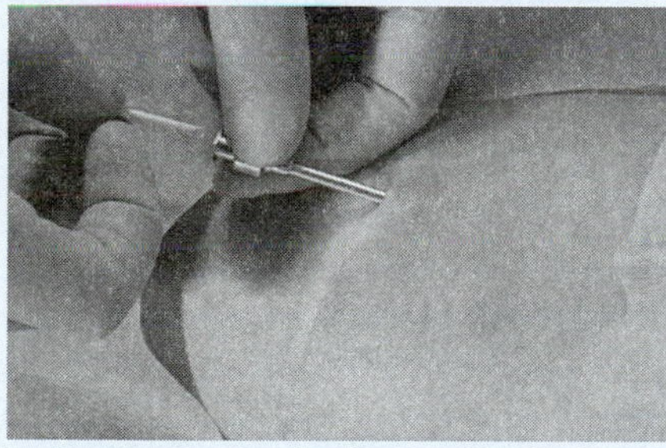

years. In 1992 the FDA approved the use of an injected dose of progestin (Depo-Provera®), offering a three-month-long method of reversible birth control.

In addition to contraception, these compounds can be used to help relieve amenorrhea (the absence of menstruation) and dysmenorrhea (irregular and/or painful menstrual cycles). As little as 0.020 mg of ethinyl estradiol can alleviate the symptoms and discomfort of menopause. This may be taken orally or through a removable patch applied to the skin.

The story of the chemical synthesis and development of "the pill" is described by its discoverer and self-proclaimed "mother," Dr. Carl Djerassi, in his book *The Pill, Pigmy Chimps, and Degas' Horse.*

As controversial as birth control is throughout the world, even more argument has been generated by the introduction of RU-486, a "contragestive" or antiprogesterone medication. It blocks the action of progesterone, thereby interrupting and preventing the implantation of the fertilized egg and/or preventing its development. RU-486 also has potential use in the treatment of progesterone-sensitive cancers, as well as other hormonal disorders.

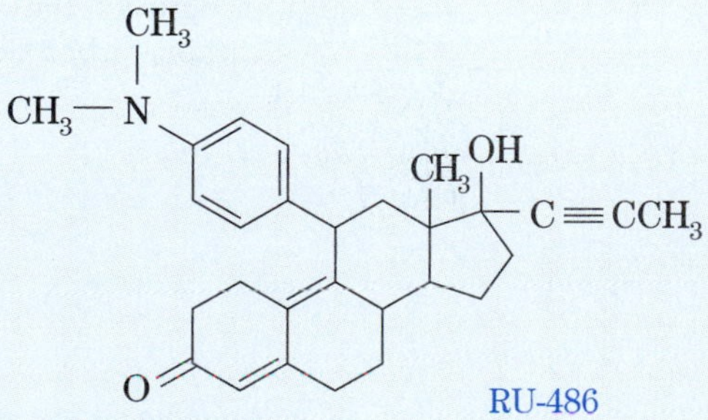

RU-486

in oxyacetylene welding torches), and more complex alkynes are sometimes named as derivatives of acetylene.

$H_2C{=}CH_2$	$CH_3CH{=}CH_2$	$HC{\equiv}CH$	$CH_3C{\equiv}CH$
Ethylene	Propylene	Acetylene	Methylacetylene

vinyl
$CH_2{=}CH{-}$ is the vinyl group

allyl
$CH_2{=}CHCH_2{-}$ is the allyl group

Occasionally ethylene and propylene groups are referred to with the prefixes **vinyl-** and **allyl-,** respectively (the plastic PVC is made from vinyl chloride).

$CH_2{=}CHCl$ Vinyl chloride $CH_2{=}CHCH_2Br$ Allyl bromide

3.3 Skeletal, Positional, and Functional Isomerism in Alkenes and Alkynes

isomers
compounds with the same molecular formula but different structural formulas

skeletal isomers
isomers that differ in the arrangement of the carbon chain

positional isomers
isomers that differ in the location of a noncarbon group or a double or triple bond

functional isomers
isomers with structural differences that place them in different classes of organic compounds

Isomers are different compounds with the same molecular formula. **Skeletal isomers** differ in the arrangement of carbon atoms. **Positional isomers** vary in the position of an atom or group other than carbon or in the position of a carbon-carbon double or triple bond. **Functional isomers** exhibit structural variations that place them in different classes of organic compounds, such as alkanes, alkenes, or alkynes.

These three types of isomerism are illustrated in Example 3.6, and by the following examples. The first two compounds are positional isomers, the third is a skeletal isomer of the first two, and the last one is a functional isomer of the others.

$CH_3CH_2CH_2CH_2C{\equiv}CH$ $CH_3CH_2CH_2C{\equiv}CCH_3$

$CH_3CH_2CH(CH_3)C{\equiv}CH$ $CH_3CH{=}CH{-}CH{=}CHCH_3$

GETTING INVOLVED

✓ What are the differences among skeletal, positional, and functional isomers?

Example 3.6

Draw the five compounds with the formula C_4H_8. Identify skeletal, positional, and functional isomers.

Solution

First let us draw compounds with four carbons in the longest chain. To satisfy the valences of the carbons, a double bond must be included between the first two carbons or between the middle two carbons. These compounds differ in the position of the carbon-carbon double bond and are *positional isomers.*

$CH_3CH_2CH{=}CH_2$ 1-butene $CH_3CH{=}CHCH_3$ 2-butene

Now let us change the carbon skeleton to the branched one. There is only one possible isomer with this skeleton; it is a *skeletal isomer* of both 1-butene and 2-butene.

$CH_3C(CH_3){=}CH_2$ Methylpropene

There are two cyclic compounds with the formula C_4H_8, cyclobutane and methylcyclopropane.

$$\begin{array}{c} CH_2-CH_2 \\ | \quad\quad | \\ CH_2-CH_2 \end{array} \quad \text{Cyclobutane} \qquad \begin{array}{c} CH_3 \\ | \\ CH \\ / \quad \backslash \\ CH_2-CH_2 \end{array} \quad \text{Methylcyclopropane}$$

These two molecules belong to the class of compounds called *cycloalkanes* (section 2.8) and they are functional isomers of the three alkenes.

Problem 3.4

Draw a positional, skeletal, and functional isomer of 1-heptene. To what class of compounds does the functional isomer belong?

Problem 3.5

Draw a positional, skeletal, and three functional isomers of 1-heptyne. To what classes of compounds do the three functional isomers belong?

Problem 3.6

There are three isomers of C_3H_4, an alkyne, a diene, and a cycloalkene. Each is technically a functional isomer of the others. Draw and name these compounds.

Problem 3.7

There are three alkynes with the formula C_5H_8. Draw and name these compounds and identify which are positional and which are skeletal isomers of each other.

Problem 3.8

There are five alkenes with the formula C_5H_{10} that are positional or skeletal isomers. Draw and name these compounds and identify which are positional and which are skeletal isomers of each other. Draw one compound that is a functional isomer of the alkenes you drew.

See related problems 3.20, 3.26, 3.29–3.30.

3.4 Functional Isomerism in Organic Chemistry

Let us take a broader look at functional isomerism in order to gain an appreciation for the variety of compounds possible in organic chemistry.

Functional isomers belong to different classes of organic compounds because they possess different functional groups. A **functional group** is usually the site of the characteristic reactions of a particular class of compounds. Consider, for example, the two isomers with the molecular formula C_2H_6O.

functional group
a structural unit (grouping of atoms) in a molecule that characterizes a class of organic compounds and causes the molecule to display the characteristic chemical and physical properties of the class of compounds

CH_3CH_2OH — An alcohol (beverage alcohol)

CH_3OCH_3 — An ether

Each member of the class of compounds called alcohols possesses a saturated carbon bonded to a hydroxy group (C—O—H), whereas ethers possess a unit of two saturated carbons separated by an oxygen (C—O—C). These characteristic structural units are the functional groups of their respective compounds. They illustrate an especially important type of isomerism. Different compounds possessing the same functional group have similar chemical properties; those with different functional groups often undergo distinctively different chemical reactions. The functional group is often the basis for the naming of an organic compound. Table 3.1 summarizes some of the major classes of organic compounds. A more complete list appears inside the front cover.

Following are some examples of functional isomers with the formula C_4H_8O (not all isomers are shown):

$CH_3CH_2CH_2\overset{\overset{\displaystyle O}{\|}}{C}H$ — Aldehyde

$CH_3CH_2\overset{\overset{\displaystyle O}{\|}}{C}CH_3$ — Ketone

$CH_2{=}CHCH_2CH_2OH$ — Alkene–alcohol

$CH_2{=}CHCH_2OCH_3$ — Alkene–ether

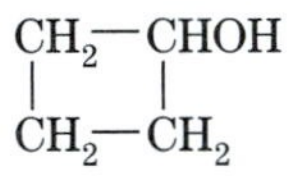

$\begin{array}{l} CH_2{-}CHOH \\ |\quad\quad | \\ CH_2{-}CH_2 \end{array}$ — Alcohol

$\begin{array}{c} O \\ CH_2 \quad\; CH_2 \\ CH_2{-}CH_2 \end{array}$ — Ether

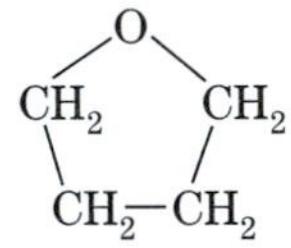

GETTING INVOLVED

✓ You will become very familiar with the functional groups as we first cover their nomenclature and then reactions in the chapters devoted to each. Still, at this point it would be well to learn the structural characteristics of each of those in Table 3.1 at this time as you will probably see them prior to the individual chapters.

Problem 3.9

Draw: **(a)** four carboxylic acids with the formula $C_5H_{10}O_2$; **(b)** two alcohols and one ether with the formula C_3H_8O; **(c)** four amines with the formula C_3H_9N; **(d)** three straight chain aldehydes and ketones with the formula $C_6H_{12}O$.

Problem 3.10

Draw a specific example containing three carbons of the following types of compounds: **(a)** alkene; **(b)** alkyne; **(c)** carboxylic acid; **(d)** aldehyde; **(e)** ketone; **(f)** alcohol; **(g)** amine; **(h)** ether.

See related problems 3.27–3.28, 3.31–3.33.

Problem 3.11

There are 19 compounds shown in Connections 3.2, "Chemical Communication in Nature." As you read the Connections material, try to locate the functional groups present in each structure as described below each compound. Don't be concerned with the complexity of the molecules: they are not presented to learn or memorize.

3.5 Geometric Isomerism in Alkenes

A. Cis-trans Isomerism

A carbon-carbon single bond is composed of a σ bond in which there is only one position of overlap. As a result, there is more or less free rotation about single bonds, and this rotation is capable of producing an infinite number of confor-

Table 3.1 Major Classes of Organic Compounds

Functional Group Name	Functional Group Structure	Example	Name and Application
Alkane	$-\overset{\vert}{\underset{\vert}{C}}-\overset{\vert}{\underset{\vert}{C}}-$	$CH_3CH_2CH_3$	Propane (rural or camping gas)
Alkene	$>C=C<$	$CH_2=CH_2$	Ethene (precursor of polyethylene)
Alkyne	$-C\equiv C-$	$HC\equiv CH$	Acetylene (used in oxyacetylene torches)
Aromatic hydrocarbon	 (benzene ring)	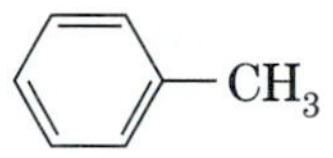 (benzene ring)$-CH_3$	Toluene (high-octane gasoline component)
Carboxylic acid	$-\overset{O}{\overset{\Vert}{C}}OH$	$CH_3\overset{O}{\overset{\Vert}{C}}OH$	Acetic acid (vinegar acid)
Aldehyde	$-\overset{O}{\overset{\Vert}{C}}H$	$H\overset{O}{\overset{\Vert}{C}}H$	Formaldehyde (biological preservative)
Ketone	$-\overset{\vert}{\underset{\vert}{C}}-\overset{O}{\overset{\Vert}{C}}-\overset{\vert}{\underset{\vert}{C}}-$	$CH_3\overset{O}{\overset{\Vert}{C}}CH_3$	Acetone (fingernail polish remover)
Alcohol	$-\overset{\vert}{\underset{\vert}{C}}-OH$	CH_3CH_2OH	Ethanol (beverage alcohol)
Ether	$-\overset{\vert}{\underset{\vert}{C}}-O-\overset{\vert}{\underset{\vert}{C}}-$	$CH_3CH_2OCH_2CH_3$	Diethyl ether (general anaesthetic)
Amine	$-\overset{\vert}{N}-$	CH_3NH_2	Methylamine (fishy odor of herring brine)

mational isomers (section 2.7). For example, in 1,2-dibromoethane we can picture the two bromines on the same side of the carbon chain (eclipsed) or on opposite sides (staggered) as possible conformations. However, rotation around the carbon-carbon single bond is unrestricted, and thus these and other conformers of 1,2-dibromoethane cannot be separated or isolated because they are constantly interconverting.

Eclipsed

Less stable Br's sterically close

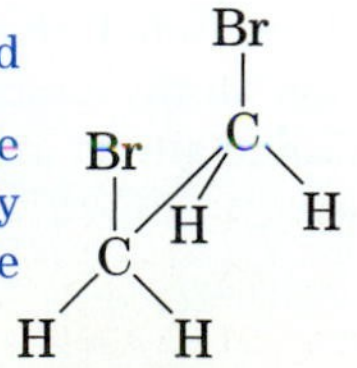

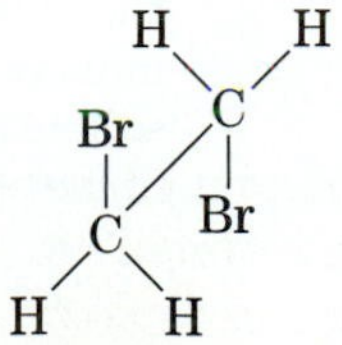

Br, C, H, H; Br, C, H, H

Staggered

More stable Br's sterically separated

H, H, C, Br; Br, C, H, H

Connections 3.2

www.prenhall.com/bailey

Chemical Communication in Nature

Writing, talking, music, art, telephones, radio and television, computers—these are means of human communication. But how is communication accomplished by less intelligent organisms? How do armies of household ants follow the same trail to spilled food? How do bees know to follow their queen, build a hive, defend their community, and reproduce? These types of behavior are elicited by substances called *pheromones*. A pheromone is a chemical substance that, when secreted by an individual of a species, can elicit a certain type of behavior or psychological change in other individuals. Pheromones have been classified according to the type of behavior they produce; alarm, recruiting, primer, and sex pheromones are examples.

Alarm pheromones warn of danger. The alarm pheromone of the honeybee found in the sting gland is isoamyl acetate (this compound has a distinct banana odor). Several species of ants secrete simple aldehydes and ketones such as 2-hexenal. Aphids produce a tetraene.

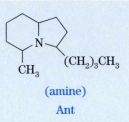

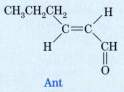

$CH_3COCH_2CH_2CHCH_3$ (with C=O and CH_3 branch)

Honeybee

$CH_3CH_2CH_2CH{=}CHCHO$ (H, H on the double bond)

Ant
(alkene, aldehyde)

CH_3, CH_3, CH_2, CH_3, CH_2

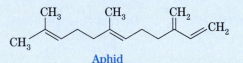

Aphid
(alkene–tetraene)

Recruiting, aggregating, or trail pheromones direct others of a species to a food source. Citral and geraniol attract honeybees, and a heterocyclic compound containing nitrogen is used by the common household ant to form trails to food.

CH_3, CH, O, H_3C, CH_3

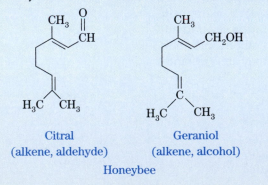

Citral
(alkene, aldehyde)

CH_3, CH_2OH, C, H_3C, CH_3

Geraniol
(alkene, alcohol)

Honeybee

N, CH_3, $(CH_2)_3CH_3$

(amine)
Ant

Sex pheromones (sometimes called sex attractants) attract the opposite sex and promote sexual behavior. Extremely small amounts of these pheromones can cause such responses. For example, 0.01 gram of periplanone B, the sex pheromone of the American cockroach, is enough to excite 100 billion cockroaches with a combined weight of approximately 10,000 tons. The female East African tick attracts male ticks with phenol and *p*-methylphenol. The common housefly *(Musca domestica)* uses a long-chain alkene commonly called muscalure as a sex attractant.

O, O, H_2C, H, H, O, CH_3, CH, H, CH_3

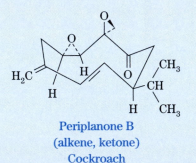

Periplanone B
(alkene, ketone)
Cockroach

A moth alights near a small amount of cotton treated with pheromones, a chemical secreted by animals that influences behavior. Moths typically secrete the chemical to attract members of the opposite sex

Connections 3.2 *(cont.)*

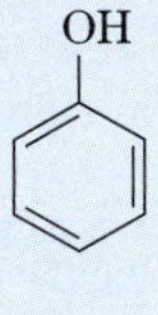

Phenol

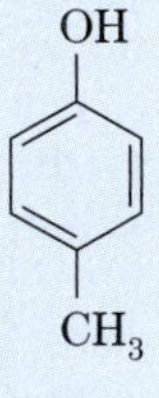

p-methylphenol
(aromatic)
Tick

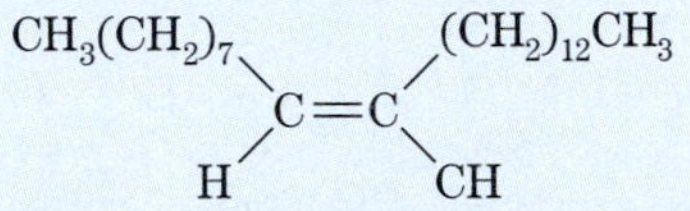

$CH_3(CH_2)_7CH{=}CH(CH_2)_{12}CH_3$

Muscalure
(alkene)
Housefly

$CH_3C(O)(CH_2)_5CH{=}CHCO_2H$

trans-9-oxo-2-decenoic acid
(ketone, alkene, carboxylic acid)
Queen honeybee

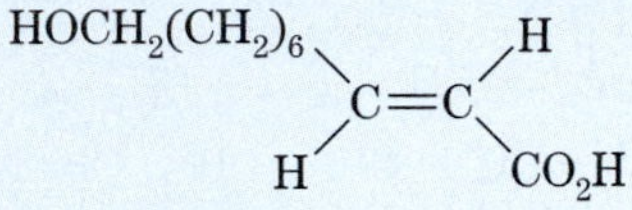

$HOCH_2(CH_2)_6CH{=}CHCO_2H$

trans-10-hydroxy-2-decenoic acid
(alcohol, alkene, carboxylic acid)
Worker honeybee

Sex pheromones can be quite useful for insect control. They can be used as bait in traps to attract large numbers of insects, which can then be efficiently destroyed with chemical insecticides. Alternatively, sex pheromones sprayed in the air can be so confusing to male insects that they find it impossible to locate females to mate with.

It is believed that mammals (possibly even humans) may also communicate chemically, for example, methyl p-hydroxybenzoate, secreted in the vaginas of female dogs in heat, sexually arouses male dogs. Dimethyl disulfide appears to be a sex attractant for male hamsters.

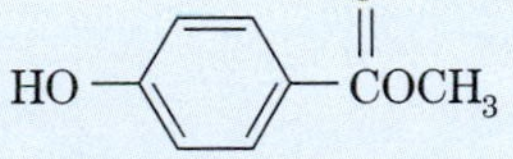

$HO{-}C_6H_4{-}C(O)OCH_3$

Methyl p-hydroxybenzoate
(aromatic)
Dog

CH_3SSCH_3

Dimethyl disulfide
Hamster

Evidence suggests that sex pheromones even play a role in communication between the male and female genders in plants during reproduction.

Primer pheromones are used by social insects to regulate their caste system. Much of the regulatory power enjoyed by the queen honeybee is attributed to *trans*-9-oxo-2-decenoic acid. If a queen dies or leaves the hive, the absence of this royal pheromone causes worker bees to produce a new queen. The worker honeybee pheromone has a similar structure.

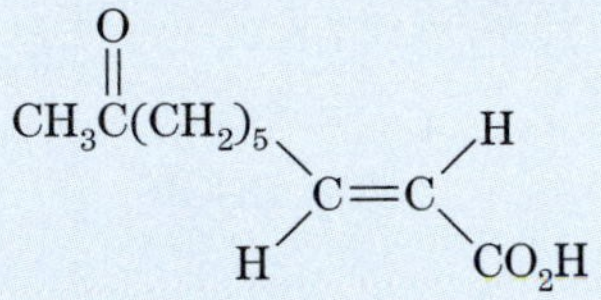

Allomones are chemicals produced by organisms for defensive purposes or for an advantage in adapting to their environment. Juglone, a product of the decay of fallen walnut tree leaves, poisons underlying plants, thereby protecting the walnut tree's nutrient and water supply. Plants also produce allomones to protect themselves from insects. Two familiar examples are nicotine and the pyrethrins, both of which are used commercially as insecticides. Some plants and insects protect themselves by producing chemicals that are unpalatable to predators. Other insects secrete allomones that are fatal to predators. Quinones, such as p-benzoquinone secreted by African termites, are common chemicals of insect warfare.

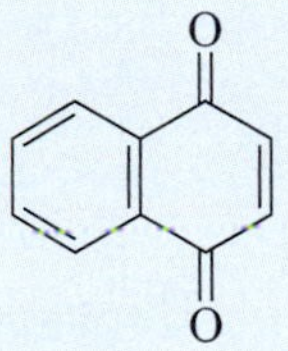

Juglone
(aromatic, ketone,
alkene)

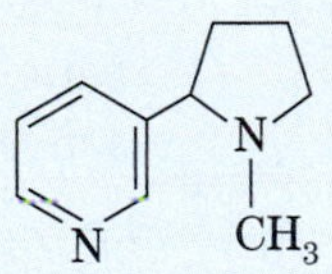

Nicotine
(amine)

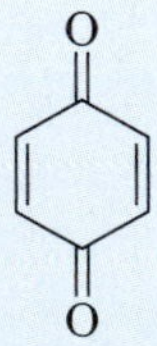

p-benzoquinone
(ketone, alkene)

Even microorganisms have developed very sophisticated antibiotic molecules for defensive purposes.

Kairomones are chemicals that are produced by one

Connections 3.2 *(cont.)*

$$\begin{array}{c} O \\ \| \\ CH_3CHCOH \\ | \\ OH \end{array}$$

Lactic acid
(alcohol, carboxylic acid)

$$\begin{array}{cccccccccccc} & CH_3 & & & CH_3 & & & CH_3 & & \\ & | & & & | & & & | & & \\ & C & CH_2 & & C & CH_2 & & C & & CH_2 \\ / & \backslash\backslash & / & \backslash & \backslash\backslash & / & \backslash & // & \backslash & // \\ CH_3 & & CH & CH_2 & & CH & CH & & CH & \end{array}$$

α-farnesene
(alkene–tetraene)

organism which give an advantage to the different organism they excite. For example, lactic acid produced by humans attracts the mosquito that carries yellow fever, and α-farnesene in apple skins attracts the codling moth.

A substance may be both a kairomone and an allomone. Chemicals that give flowers their scents attract pollinating insects and are allomones from the flowering plant's point of view. For the insects, they point to a source of nectar and are kairomones.

In contrast, free rotation around carbon-carbon double bonds is not possible, because a double bond is constructed of both a σ and a π bond. Although rotation can occur around a σ bond without diminishing orbital overlap, this is not possible with a π bond because it is formed by the overlap of parallel p orbitals in two positions. For rotation to occur, the π bond would have to be broken, a process that is not energetically favored (Figure 3.1). As a result, a compound like 1,2,-dibromoethene has two distinct isomers that can be separated and isolated and that do not interconvert under normal conditions. One isomer, in which the two bromines are on the same side, is termed ***cis*** (Latin, "same side"), and the other, in which the bromines are on opposite sides, is ***trans*** (Latin, "across").

***cis* isomer**
geometric isomer in which groups are on the same side of a ring or double bond

***trans* isomer**
geometric isomer in which groups are on opposite sides of a ring or double bond

1,2-dibromoethene

$$\begin{array}{ccc} H & & H \\ & C{=}C & \\ Br & & Br \end{array}$$

cis
Less stable
Br's sterically close

$$\begin{array}{ccc} H & & Br \\ & C{=}C & \\ Br & & H \end{array}$$

trans
More stable
Br's sterically separated

1,1-dibromoethene

$$\begin{array}{ccc} H & & Br \\ & C{=}C & \\ H & & Br \end{array}$$

Does not exhibit geometric isomerism

geometric isomers
cis and *trans* isomers; a type of stereoisomerism in which atoms or groups display orientation differences around a double bond or ring

These *cis/trans* isomers are called **geometric isomers** because they differ in the geometric orientation of atoms, not in the structural (atom-to-atom) arrange-

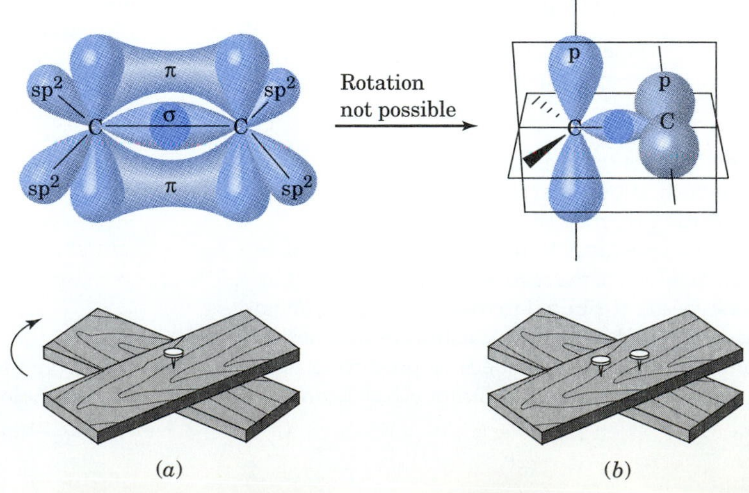

Figure 3.1
Rotation around a double bond is not usually possible, since the π bond, which has two positions of overlap, must be broken. The difference between rotation about a carbon-carbon single bond and rotation about a carbon-carbon double bond is analogous to the difference in the ability to rotate (a) two pieces of wood connected by one nail versus (b) two pieces connected by two nails.

ment. For geometric *(cis/trans)* isomerism to be possible, each carbon involved in a carbon-carbon double bond must have two different groups attached. For example, 1,2-dibromoethene exhibits *cis-trans* isomerism, as illustrated, but 1,1-dibromoethene ($CH_2{=}CBr_2$) does not. (See section 2.8E for geometric isomerism in cyclic compounds.)

Just as the staggered conformation is more stable than the eclipsed in conformational isomerism in 1,2-dibromoethane, the *trans* configuration is more stable than the *cis* in 1,2-dibromoethene. In both the staggered and eclipsed structures, the relatively large and bulky bromines are maximally separated from each other thus minimizing steric hindrance that occurs in eclipsed and *cis* structures.

In compounds such as the geometric isomers illustrated in Example 3.7 the *cis-trans* designation refers to the configuration of the double bonds as it relates to the longest continuous carbon chain, that is, whether the chain proceeds across the double bond in a *cis* or *trans* fashion.

GETTING INVOLVED

- ✓ What is the difference between conformational isomers and geometric isomers?
- ✓ What is the difference between *cis* and *trans* isomers?
- ✓ Why is geometric isomerism possible around some double bonds and not others?
- ✓ Why is rotation restricted in a double bond?
- ✓ Explain the relative stabilities of simple *cis* and *trans* isomers; compare with staggered and eclipsed conformations.
- ✓ Practice drawing geometric isomers with the following problems. To get started, follow the method in Example 3.7.

Example 3.7

Draw the two geometric isomers of $CH_3CH{=}CBrCH_2CH_3$.

Solution

Visually isolate the doubly bonded carbons and identify the two groups connected to each carbon: CH_3 and H; Br and CH_2CH_3. To get the first isomer, place two groups on each carbon randomly but oriented to the corners of a triangle. Interchange the two groups on one of the carbons to get the other isomer.

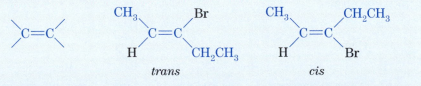

C=C

CH_3 Br
C=C
H CH_2CH_3
trans

CH_3 CH_2CH_3
C=C
H Br
cis

Problem 3.12

(a) Draw the geometric isomers of 2-butene, $CH_3CH{=}CHCH_3$. Label them *cis* and *trans*. **(b)** Does 1-butene, $CH_3CH_2CH{=}CH_2$, have geometric isomers? Explain.

Problem 3.13

In each of the following sets of compounds, identify the one that can exhibit geometric isomerism. **(a)** 1-bromopropene, 2-bromopropene, 3-bromopropene; **(b)** 1-pentene, 2-pentene; **(c)** 2-methyl-2-pentene, 3-methyl-2-pentene.

Problem 3.14

Draw the two geometric isomers possible for each of the following compounds: **(a)** 3-methyl-3-hexene; **(b)** 1-bromo-1-pentene.

Problem 3.15

Draw the following compounds: **(a)** trans 2,3-dibromo-2-butene; **(b)** *cis* 2-heptene.

Problem 3.16

Draw the geometric isomer of 1,4-dibromo-1,3-butadiene in which one double bond is *cis* and the other *trans*. Be sure to show both simultaneously in the molecule and draw each double bond like the one in Example 3.7.

See related problems 3.34, 3.37–3.38.

B. The E-Z System for Designating Configuration of Geometric Isomers

The *cis-trans* designation of configuration in geometric isomers does not always work. For example, the following compounds are geometric isomers but cannot be designated as *cis* or *trans;* there are no like groups to observe or a carbon chain to follow to determine *cis* or *trans.*

F(H)C=C(Cl)(Br)   H(F)C=C(Cl)(Br)

If we could assign a priority to each group on each carbon of the double bond, we would have a method for designating configuration; this is the basis of the E-Z method. If our two higher priority groups are on opposite sides of the double bond, the configuration is **E** (entgegen, German for opposite) and if on the same side, the configuration is **Z** (zusammen, German for together).

HIGHER PRIORITY (upper left), lower priority (upper right), lower priority (lower left), HIGHER PRIORITY (lower right) — C=C

E (opposite)

lower priority (upper left), lower priority (upper right), HIGHER PRIORITY (lower left), HIGHER PRIORITY (lower right) — C=C

Z (together)

In our example, a readily apparent way to distinguish the atoms into high and low priorities is by atomic number. F has a higher atomic number than H and is of higher priority; on the other carbon, Br has a higher atomic number than Cl and likewise is of higher priority.

F (HIGHER PRIORITY), H on one carbon; Cl, Br (HIGHER PRIORITY) on the other — C=C

E (opposite)

H, F (HIGHER PRIORITY) on one carbon; Cl, Br (HIGHER PRIORITY) on the other — C=C

Z (together)

The same procedure can be used on molecules that can be designated as *cis* or *trans.* In the E-Z method we compare the atoms directly bonded to the carbon-carbon double bond such as the carbon in methyl. (In Chapter 7, Section 7.6, there are a set of priority rules for distinguishing more complex bonded groups.) Which system one uses is optional except where a double bond cannot be clearly designated as *cis* or *trans.* The E-Z method always works.

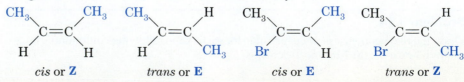

CH_3(H)C=C(CH_3)(H) — *cis* or **Z**

CH_3(H)C=C(H)(CH_3) — *trans* or **E**

CH_3(Br)C=C(CH_3)(H) — *cis* or **E**

CH_3(Br)C=C(H)(CH_3) — *trans* or **Z**

Connections 3.3

www.prenhall.com/bailey

Geometric Isomerism and Vision

Geometric isomerism is extremely important in some biological processes, especially vision. The rod cell in the retina of the eye contains a protein called *opsin*, which combines with the organic molecule retinal to form a complex known as *rhodopsin*. Retinal has five double bonds, one inside a cyclohexene ring and four outside. Of the four double bonds in the carbon chain attached to the ring, three are permanently *trans*, while one can undergo a reversible isomerism from *cis* to *trans*. The transition from *cis* to *trans* is triggered by light energy. This, in turn, stimulates nerve cells to the brain, which records the data from the light. The process is reversed through a type of protein known as an *enzyme* (a biological catalyst), and the *cis*-retinal is ready to be changed again.

cis

Light

Enzyme

11-*cis*-retinal

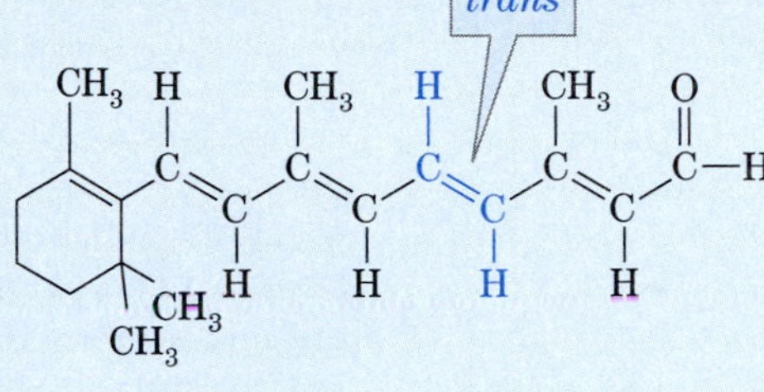

All *trans*-retinal

Retinal can be made in the body from beta-carotene, a yellow pigment found in many vegetables, especially carrots and sweet potatoes. Beta-carotene is cleaved in half by enzymes to form retinol, or vitamin A. The retinol is then oxidized to retinal and combines with opsin.

Deficiencies in vitamin A, which is called a fat soluble vitamin because of its very nonpolar, hydrocarbon nature, can cause conditions ranging from dry skin and mucous membranes to night blindness in adults, and physical and mental retardation in children.

Excesses of vitamin A, which are stored in the liver and fat tissue of the body, can also be toxic, leading to yellowing and peeling of the skin, headaches, and vomiting.

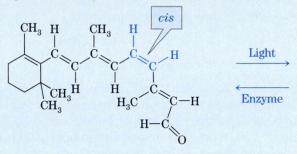

Point of cleavage

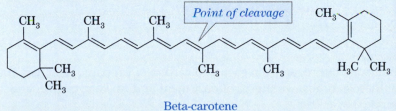

Beta-carotene

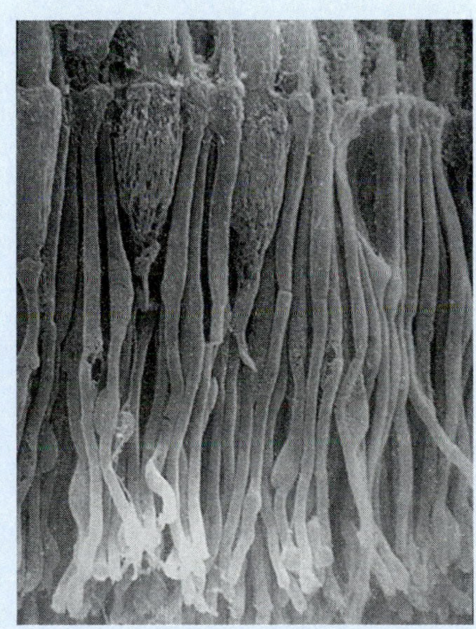

Rods of the eye

GETTING INVOLVED

✓ In the E-Z designation of configuration, what exactly do E and Z represent?

✓ How are the relative priorities of two groups determined?

Problem 3.17

Designate the configurations of the following alkenes as Z or E.

(a) $CH_3(H)C{=}C(H)CH_2CH_2OH$ (CH₃ and CH₂CH₂OH on the same side) **(b)** $Cl(H)C{=}C(Br)CH_3$ (Cl and Br on the same side) **(c)** $CH_3(Br)C{=}C(H)CH_2CH_3$ (CH₃ and CH₂CH₃ on the same side)

See additional problem 3.36.

3.6 Units of Unsaturation

Throughout this chapter we have seen compounds with double bonds, triple bonds, and rings as well as the more common single bonds. We have developed general molecular formulas for saturated and unsaturated hydrocarbons. But how can we tell from a specific molecular formula whether the possible isomers have double bonds, triple bonds, or rings and how many of each? The answer becomes evident once we determine the number of **units of unsaturation** present in a molecular formula. We shall see in this section that units of unsaturation can be expressed in the following ways:

units of unsaturation a unit of unsaturation is expressed as a ring or double bond. A triple bond is two units of unsaturation

double bond: one unit of unsaturation

triple bond: two units of unsaturation

ring: one unit of unsaturation

First, consider those formulas for which there can be no multiple bonds. This simplest class of organic compounds is the alkanes. We have seen that they are entirely composed of carbon and hydrogen and contain only single bonds. The general formula, C_nH_{2n+2}, allows them the maximum possible number of hydrogens in a molecule with n carbons, and therefore these compounds are said to be saturated. Consider propane, for example, the single compound with the formula C_3H_8.

```
   H   H   H
   |   |   |
H—C—C—C—H      Propane
   |   |   |
   H   H   H
```

There is no way to associate more than eight hydrogens with three carbons; this compound is *saturated* with hydrogens.

There are ways to associate fewer than eight hydrogens with three carbons. If we remove hydrogens in pairs from propane, we come up with other possible structures. For example, in drawing C_3H_6, we can remove one hydrogen from each of two adjacent carbons. To satisfy the covalence of carbon, we must insert a double bond, which gives propene.

```
   H   H   H              H   H   H
   |   |   |              |   |   |
H—C—C—C—H          H—C—C=C—H      Propene
   |   •   •              |
   H                      H
```

Or we can remove a hydrogen from each end carbon of propane. To satisfy the covalence of these carbons, we must connect them and make cyclopropane, a ring compound.

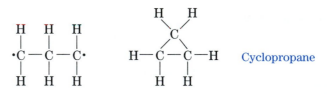

```
                         H   H
                          \ /
  H   H   H                C
  |   |   |               / \
•C—C—C•              H—C—C—H      Cyclopropane
  |   |   |               |   |
  H   H   H               H   H
```

Both propene and cyclopropane have two fewer hydrogens than are maximally possible with three carbons. They are considered to have one unit of unsaturation. A single unit of unsaturation can be expressed as either one double bond or one ring. Note that cyclopropane is "saturated" in the sense that it has no double or triple bonds but has one unit of unsaturation because of the one ring.

Now let us consider the molecular formula C_4H_6. Completely saturated hydrocarbons with four carbons have the formula C_4H_{10} (C_nH_{2n+2}). C_4H_6 has four fewer hydrogens or two units of unsaturation (for every two hydrogens fewer than the maximum, there is one unit of unsaturation).
The two units of unsaturation can be expressed as

```
   H  H  H
   |  |  |  |
H—C—C—C—C—
   |  |  |  |
   H  H
```

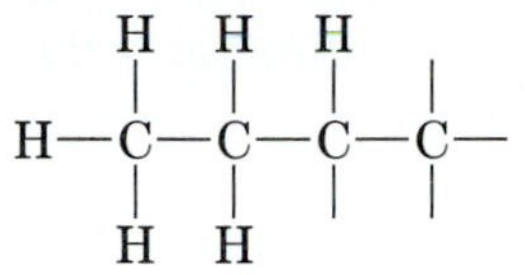

1. *One triple bond*

$$CH_3CH_2C{\equiv}CH \quad CH_3C{\equiv}CCH_3$$

2. *Two double bonds*

$$CH_3CH{=}C{=}CH_2 \quad CH_2{=}CH{-}CH{=}CH_2$$

3. *One double bond and one ring*

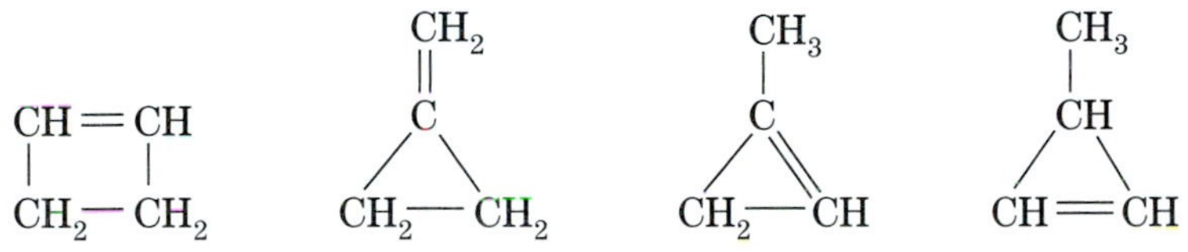

```
                      CH2            CH3           CH3
                      ||             |             |
CH=CH                 C              C             CH
|   |                / \            / \\          /  \
CH2—CH2          CH2—CH2         CH2—CH         CH=CH
```

4. *Two rings*

```
CH—CH2
| \  |
CH2—CH
```

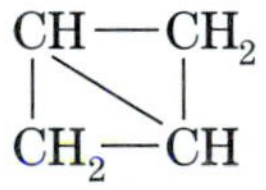

How do we handle compounds that have other elements in addition to carbon and hydrogen? First, monovalent elements (F, Cl, Br, I) should be considered to be equivalent to hydrogen. For example, $C_3H_6Br_2$ has eight monovalent elements, two times plus two as many ($2n + 2$) monovalent atoms as the number of carbons. There are no units of unsaturation.

Oxygen has no effect on the calculation of units of unsaturation since it is divalent and can be inserted without disrupting the carbon : hydrogen ratio. Compare the isomers of C_2H_6O with ethane (CH_3CH_3).

$$CH_3CH_3 \quad CH_3CH_2OH \quad CH_3OCH_3$$

Disregarding the oxygen, C_2H_6O has twice as many plus two monovalent elements (hydrogens) as carbons and has no units of unsaturation.

Nitrogen, however, is trivalent and has the effect of adding a hydrogen. Insert a nitrogen atom between the C—H or C—C bond of ethane.

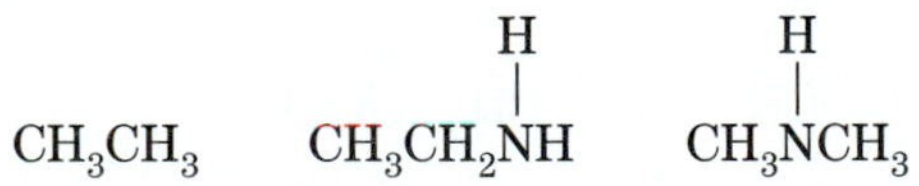

```
                    H          H
                    |          |
CH3CH3        CH3CH2NH     CH3NCH3
```

Compare the molecular formulas C_2H_6 and C_2H_7N. To satisfy nitrogen's covalence, we have to add a hydrogen. Thus in calculations we should ignore the nitrogen and one hydrogen.

We can summarize the calculation of units of unsaturation very concisely. A unit of unsaturation can be expressed as a multiple bond (a double bond is one unit; a triple bond, two) or as a ring (one). To calculate the number of units of unsaturation, compare the number of monovalent atoms (H, F, Cl, Br, I) to the number of carbons. Ignore oxygen. Ignore nitrogen, but subtract one hydrogen or monovalent atom from the formula for each nitrogen. If there are two times plus two as many monovalent atoms as carbons (C_nX_{2n+2}), there are no units of unsaturation. For every two monovalent atoms fewer than $2n + 2$, there is one unit of unsaturation. Example 3.8 illustrates this and two other methods for determining units of unsaturation.

GETTING INVOLVED

✓ What is the difference between a hydrocarbon with no units of unsaturation and one with units of unsaturation in terms of the numbers of carbons and hydrogens? What are the three ways in which units of unsaturation can be expressed?

✓ Describe how to determine the number of units of unsaturation from a molecular formula?

Example 3.8

Calculate the number of units of unsaturation in $C_4H_4BrClN_2O_2$.

Solution

(a) There are four carbons, and, for the compound to be saturated, there should be $2n + 2$ or ten monovalent elements. Ignore the two oxygens. Ignore the nitrogens and one monovalent element for each (say two of the hydrogens). The formula then is C_4H_2BrCl. There are four monovalent elements, six fewer than the ten needed for saturation. Thus there are three units of unsaturation (6/2 = 3).

(b) Another method is to bond all the polyvalent atoms by single bonds and then count the number of monovalent atoms required to satisfy valences.

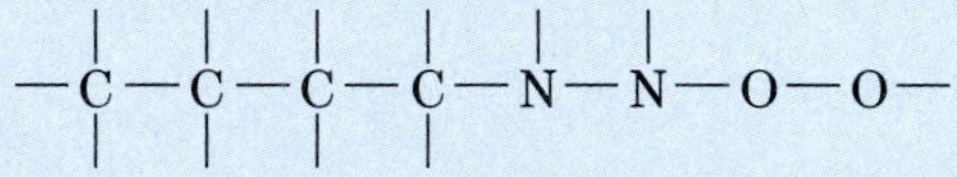

$$-\overset{|}{\underset{|}{C}}-\overset{|}{\underset{|}{C}}-\overset{|}{\underset{|}{C}}-\overset{|}{\underset{|}{C}}-\overset{|}{N}-\overset{|}{N}-O-O-$$

In this case, 12 monovalent atoms would produce saturation, but only six such atoms are available. Each unit of unsaturation is achieved by removal of two monovalent atoms. This system is six monovalent atoms short of being saturated. Thus there are three units of unsaturation [(12 − 6)/2 = 3].

(c) A third alternative is to bond the polyvalent atoms together by single bonds and then change some of the single bonds to double and/or triple bonds until the valences can be satisfied with the available monovalent atoms. For this case, three double bonds (or one triple bond and one double bond) are required to accomplish this, and thus there are three units of unsaturation.

$$\text{Br}-\overset{\text{Cl}}{\overset{|}{C}}=\overset{\text{H}}{\overset{|}{C}}-\overset{\text{H}}{\overset{|}{C}}=\overset{\text{H}}{\overset{|}{C}}-N=N-O-O-H$$

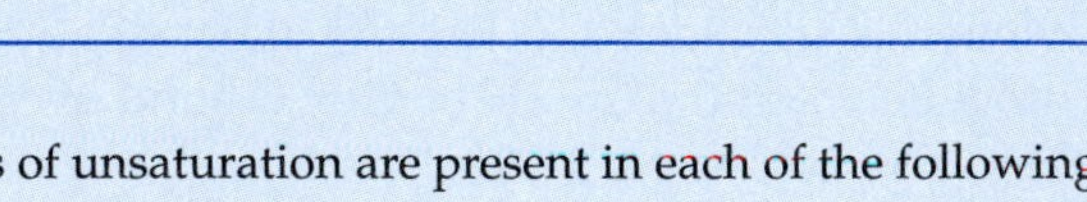

Problem 3.18

How many units of unsaturation are present in each of the following molecules?

(a) $HC{\equiv}C-CH{=}CH-CH{=}CH_2$ **(b)** 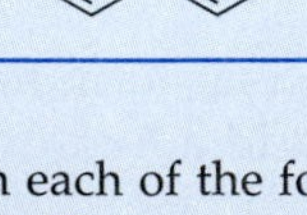 [naphthalene structure] **(c)** 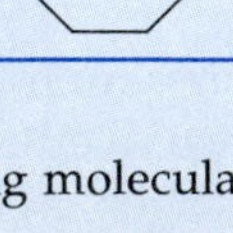 [cyclooctyne structure]

Problem 3.19

Determine the number of units of unsaturation in each of the following molecular formulas: **(a)** C_5H_{10}; **(b)** $C_8H_{12}Br_2$; **(c)** C_6H_6; **(d)** $C_7H_7NO_3$.

See related problems 3.40–3.43 and additional problems 3.45–3.46.

Problems

3.20 **Skeletal and Positional Isomerism:** Draw the skeletal and positional isomers described. You may wish to do this exercise in conjunction with problem 3.21.

(a) the 13 alkenes with the formula C_6H_{12}

(b) the 12 cycloalkanes with the formula C_6H_{12} (disregard geometric isomerism)

(c) the seven alkynes with the formula C_6H_{10}

3.21 **Nomenclature of Alkenes, Alkynes, and Cycloalkanes:** Name the compounds you drew in problem 3.20 by the IUPAC system of nomenclature.

3.22 **Nomenclature of Alkenes:** Name the following compounds by the IUPAC system of nomenclature.

(a) $CH_3(CH_2)_4CH{=}CH_2$

(b)
$$\begin{array}{c} \quad\quad\quad\quad CH_3\ CH_3 \\ \quad\quad\quad\quad |\quad\ \ | \\ CH_3CH_2CH_2C{=}CCH_2CH_3 \end{array}$$

(c)
$$\begin{array}{c} \quad CH_3 \\ \quad | \\ CH_3CCH{=}CHCH_3 \\ \quad | \\ \quad CH_3 \end{array}$$

(d) CH_2CH_3 on a cyclopentene ring

(e)
$$\begin{array}{c} \triangledown \quad\quad\quad\quad CH_2CH_2CH_3 \\ | \quad\quad\quad\quad\quad | \\ CH_3CHCH{=}CHCHCH_2CH_2CH_3 \end{array}$$

(f)
$$\begin{array}{c} CH_3CH_2 \quad CH_2CH_3 \quad CH_3 \\ | \quad\quad\quad\quad | \quad\quad\quad | \\ CH_3CH_2C{=}CHCHCH_2CH_2CHCH_3 \end{array}$$

(g) $CH_3CH_2CH_2CH{=}CH{-}CH{=}CHCH_3$

(h) (cyclooctatetraene ring)

(i)
$$\begin{array}{c} CH_3 \quad\quad\quad Br \quad Br \\ | \quad\quad\quad\quad | \quad\ \ | \\ CH_3C{=}CH{-}C{=}C{-}CH{=}CHCH_2CH_3 \end{array}$$

3.23 **Nomenclature of Alkynes:** Name the following compounds by the IUPAC system

(a) $CH_3CH_2C{\equiv}CH$

(b) $CH_3CBr_2C{\equiv}CCH_2CH_2CH(CH_3)_2$

(c)
$$\begin{array}{c} \quad CH_3 \\ \quad | \\ CH_3CHC{\equiv}CCH_3 \end{array}$$

(d)
$$\begin{array}{c} \quad\quad CH_3 \\ \quad\quad | \\ HC{\equiv}CCCH_2CH_3 \\ \quad\quad | \\ \quad\quad CH_2CH_3 \end{array}$$

(e)

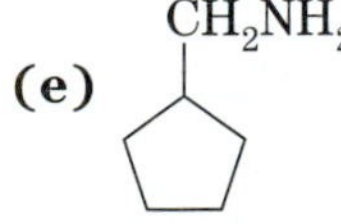

(f)
$$\begin{array}{c} \quad CH_3 \\ \quad | \\ CH_3CHCH_2C{\equiv}C{-}C{\equiv}C{-}C{\equiv}CCH_3 \end{array}$$

3.24 **Nomenclature of Alkenes and Alkynes:** Name the following compounds by the IUPAC system of nomenclature.

(a) $(CH_3)_2C{=}CHCH_2CH{=}CH(CH_2)_2C(CH_3)_3$

(b) $CH_3(C{\equiv}C)_4CH_3$

(c) $CH_3CH{=}CHCH_2C{\equiv}CH$

(d) $CH_3CH{=}CHC{\equiv}CCH_3$

(e)
$$\begin{array}{c} \quad\quad CH_3 \\ \quad\quad | \\ H_2C{=}C{-}CH{=}CHC{\equiv}CC{\equiv}CC{\equiv}CH \end{array}$$

3.25 **IUPAC Nomenclature:** Draw the following compounds.

(a) 3,7-dimethyl-1,3,6-octatriene (ocimene, a component of basil)

(b) tetrafluoroethene (precursor of Teflon)

(c) 2-chloro-1,3-butadiene (precursor of Neoprene rubber)

(d) 1,3,11-tridecatrien-5,7,9-triyne (a rare occurrence of alkynes in nature; this compound is found in some plants and fungi)

(e) 1,1-dichloroethene (precursor of the plastic Saran)

(f) 2,2,4-trimethylpentane (a high-octane gasoline component)

3.26 **Positional Isomerism:** Draw the positional isomers of the following compounds. In each maintain the carbon skeleton and functional group shown.

(a) $CH_3CH_2CH_2CH_2CH_2CH_2CH{=}CH_2$

(b)
$$\begin{array}{c} \quad CH_3 \\ \quad | \\ CH_3CHCH_2CH_2OH \end{array}$$

(c)
$$\begin{array}{c} \quad O \\ \quad \| \\ CH_3CCH_2CH_2CH_2CH_2CH_3 \end{array}$$

(d)
$$\begin{array}{c} \quad CH_3 \\ \quad | \\ CH_3CCH_2CH_2CH_2C{\equiv}CH \\ \quad | \\ \quad CH_3 \end{array}$$

(e) CH_2NH_2 on a cyclopentane ring

(f)
$$\begin{array}{c} \quad CH_3 \quad\ \ CH_3 \\ \quad | \quad\quad\quad | \\ CH_3CHCH_2CHCH_2Br \end{array}$$

3.27 **Functional Groups:** Identify the one functional group in each of the following molecules.

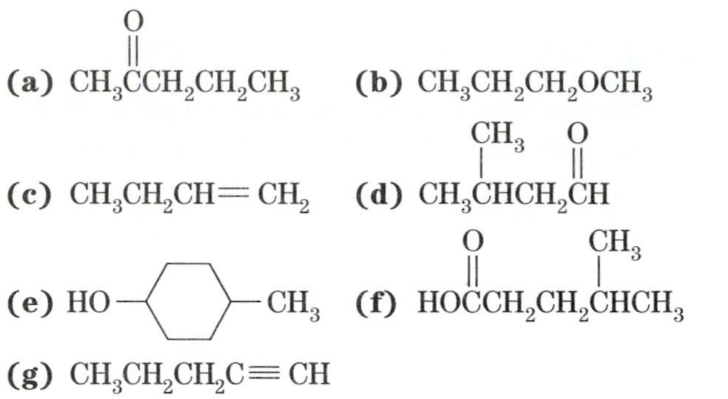

(a) $CH_3\overset{\overset{O}{\|}}{C}CH_2CH_2CH_3$ **(b)** $CH_3CH_2CH_2OCH_3$

(c) $CH_3CH_2CH{=}CH_2$ **(d)** $CH_3\underset{}{\overset{\overset{CH_3}{|}}{C}}HCH_2\overset{\overset{O}{\|}}{C}H$

(e) HO—(cyclohexane ring)—CH_3 **(f)** $HO\overset{\overset{O}{\|}}{C}CH_2CH_2\overset{\overset{CH_3}{|}}{C}HCH_3$

(g) $CH_3CH_2CH_2C{\equiv}CH$

3.28 **Functional Isomerism:** Draw a functional isomer of each compound in problem 3.27.

3.29 **Positional Isomerism:** Draw a positional isomer of each compound in problem 3.27.

3.30 **Skeletal Isomerism:** Draw a skeletal isomer of each compound in problem 3.27.

3.31 **Functional Isomers:** Using the formula $C_4H_8O_2$, draw a(n)

(a) carboxylic acid	**(b)** alcohol–aldehyde
(c) alcohol–ketone	**(d)** ether–aldehyde
(e) ether–ketone	**(f)** alkene–dialcohol
(g) alkene–diether	**(h)** alcohol–ether
(i) dialcohol	**(j)** diether

3.32 **Skeletal, Positional, and Functional Isomers** Draw the isomers described.

(a) aldehydes with the formula C_4H_8O

(b) ketones with the formula $C_6H_{12}O$

(c) aldehydes or ketones with the formula $C_5H_{10}O$ (8 total)

(d) carboxylic acids with the formula $C_6H_{12}O_2$ (7 total)

(e) alcohols or ethers with the formula $C_4H_{10}O$ (7 total)

(f) alcohols or ethers with the formula $C_5H_{12}O$ (14 total)

(g) amines with the formula $C_4H_{11}N$ (8 total).

3.33 **Functional Isomerism:** Draw six functional isomers with the formula $C_5H_{10}O$. Identify the functional group in each.

3.34 **Geometric Isomerism in Alkenes:** Draw the geometric isomers of the following compounds.

(a) $BrCH{=}CHCl$

(b) $BrFC{=}CHCl$

(c) $CH_3CH_2CH{=}CHCH_3$

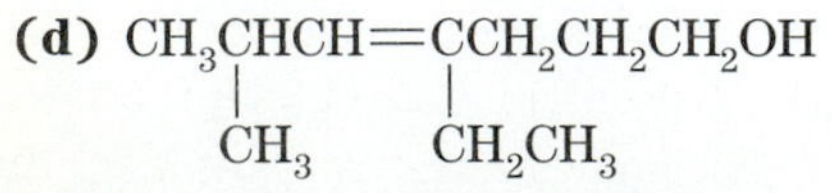

(d) $CH_3\underset{\underset{CH_3}{|}}{C}HCH{=}\underset{\underset{CH_2CH_3}{|}}{C}CH_2CH_2CH_2OH$

3.35 **Geometric Isomerism:** Draw the *cis* and *trans* isomers of 3-methyl-2-pentene

3.36 **The E-Z Method:** Draw the following compounds.

(a) E 2-octene

(b) E 3,6-dibromo-2-hexene

(c) Z 3-chloro-3-hexene

(d) Z 1-bromo-2-chloro-2-butene

3.37 **Geometric Isomerism in Alkenes:** Draw the geometric isomers of the following dienes. Be sure to show clearly the geometric isomerism at each double bond; remember, each can be *cis* or *trans.* As a result, there are four combinations and four isomers of the first example. Why are there only three possibilities for the second example?

(a) $BrCH{=}CH{-}CH{=}CHCl$

(b) $CH_3CH{=}CH{-}CH{=}CHCH_3$

(c) How many geometric isomers are possible for the following?

$CH_3CH{=}CH{-}CH{=}CH{-}CH{=}CHCH_2CH_3$

Identify in words the combinations (such as *cis, cis, cis*). Draw the isomer in which all three double bonds are *cis* and the one in which all the double bonds are *trans.*

3.38 **Geometric Isomerism:** Draw the following compounds.

(a) *cis-trans-cis* 2,4,6-octatriene

(b) *cis-cis* 3,5-octadiene

3.39 Which compound of each pair is more stable? Explain.

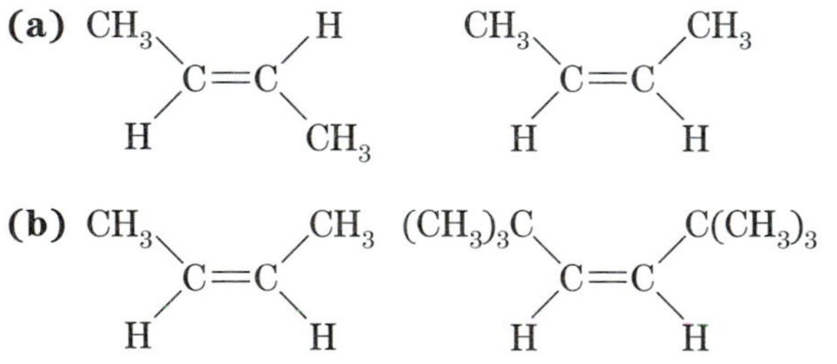

(a) $(CH_3)(H)C{=}C(H)(CH_3)$ [CH_3 and H on one carbon; H and CH_3 on the other, CH_3 groups on opposite sides] and $(CH_3)(H)C{=}C(CH_3)(H)$ [CH_3 groups on the same side]

(b) $(CH_3)(H)C{=}C(CH_3)(H)$ [CH_3 groups on the same side] and $((CH_3)_3C)(H)C{=}C(C(CH_3)_3)(H)$ [$(CH_3)_3C$ groups on the same side]

3.40 **Expressing Units of Unsaturation:** Determine the units of unsaturation in the following molecular formulas, and then draw the isomers described.

(a) Using the formula C_8H_{10}, draw three compounds, one with as many triple bonds as possible, one with as many double bonds as possible, and one with as many rings as possible.

(b) Draw six isomers of C_6H_8 so that each differs from the others in the number of triple bonds or double bonds or rings.

(c) Describe in words how four units of unsaturation can be expressed in terms of triple bonds, double bonds, and rings.

3.41 **Units of Unsaturation:** Consider the molecular formula $C_{11}H_{14}$.

(a) What is the greatest number of triple bonds possible in a compound with this formula?
(b) What is the maximum number of double bonds possible?
(c) If a compound has a triple bond, what is the maximum number of rings it can have?

3.42 **Units of Unsaturation:**
(a) A compound has 13 carbons, one triple bond, one double bond, no rings, and three bromines, and the remainder of the atoms are hydrogen. How many hydrogens are there?
(b) A compound has seven carbons, five hydrogens, one oxygen, two double bonds, one triple bond, and one ring. The remaining atoms are chlorine. How many chlorines are there?

3.43 **Units of Unsaturation:** How many units of unsaturation are present in the following compounds?

(a) 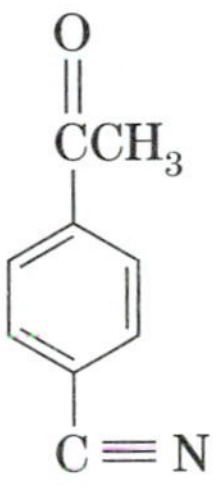 $C(=O)CH_3$ on a benzene ring, para to $C\equiv N$

(b) [bicyclic anhydride structure]

(c) [anthracene structure]

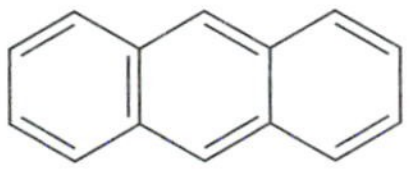

3.44 **Isomers:** Draw the isomers of the following molecular formulas.
(a) C_4H_8 (6 isomers)
(b) C_3H_5Br (5 isomers)

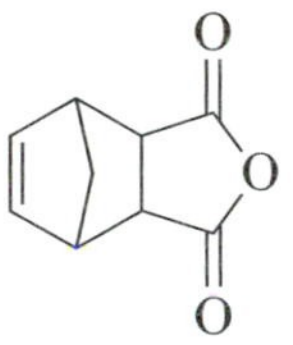

3.45 **Types of Isomerism:** Indicate the type of isomerism (skeletal, positional, functional, conformational, geometric) displayed by each of the following pairs.

(a) $CH_3CH_2CH_2CH_2CH_3$ $\quad$ $CH_3-C(CH_3)_2-CH_3$

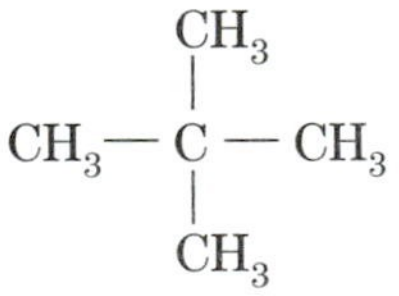

(b) $CH_3C(=O)CH_3$ $\quad$ $CH_3CH_2C(=O)H$

(c) $(CH_3CH_2)(H)C=C(CH_3)(Br)$ $\quad$ $(CH_3CH_2)(H)C=C(Br)(CH_3)$

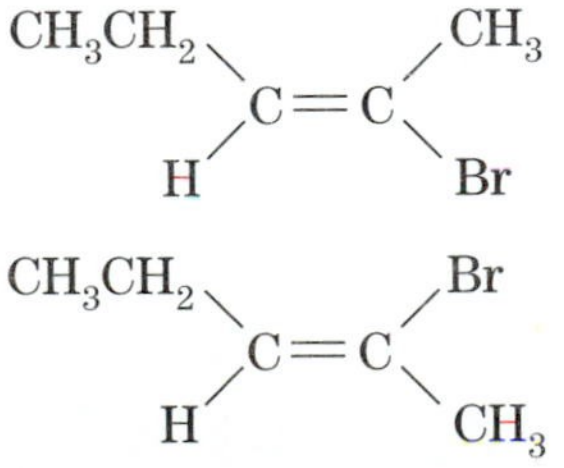

(d) 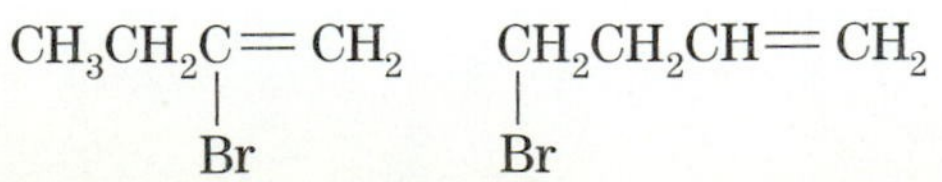 $CH_3CH_2C(Br)=CH_2$ $\quad$ $CH_2(Br)CH_2CH=CH_2$

(e) 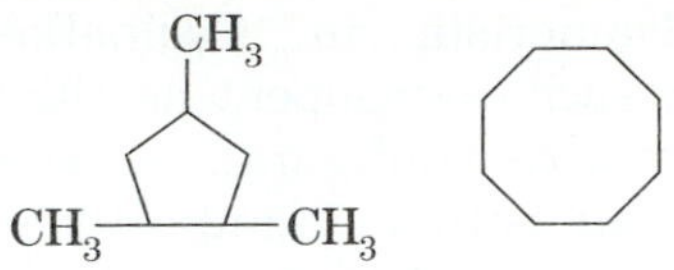 [1,2,4-trimethylcyclopentane] $\quad$ [cyclooctane]

(f) 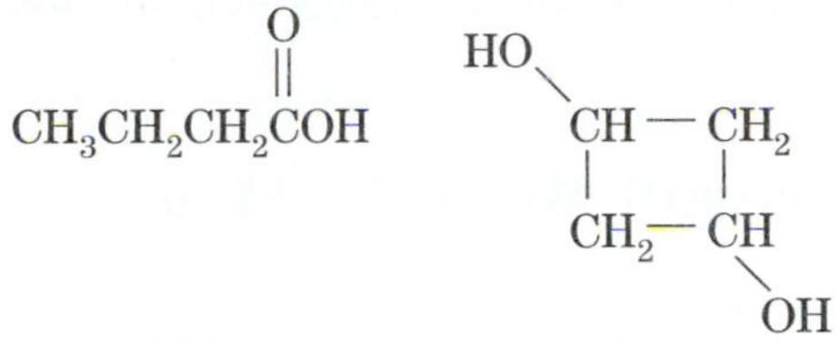 $CH_3CH_2CH_2C(=O)OH$ $\quad$ [cyclobutane-1,3-diol: HO–CH–CH₂ / CH₂–CH–OH]

(g) 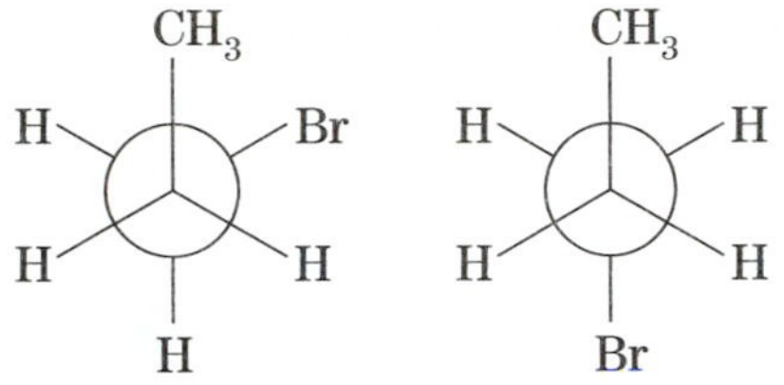 [Newman projections: front carbon CH_3, H, H; back carbon Br, H, H — CH_3 and Br gauche] $\quad$ [CH_3 and Br anti]

(h) $CH_3CH_2C\equiv CCH_2CH_3$ $\quad$ $CH_3C\equiv CCH_2CH_2CH_3$

(i) CH_3CH_2OH $\quad$ CH_3OCH_3

(j) 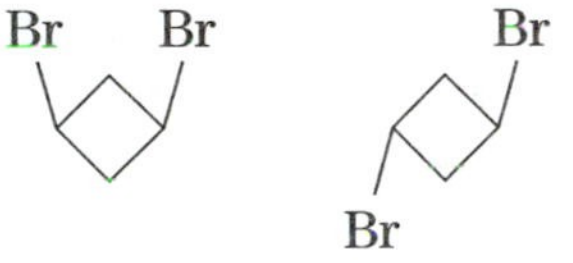 [cis-1,3-dibromocyclobutane] $\quad$ [trans-1,3-dibromocyclobutane]

(k) 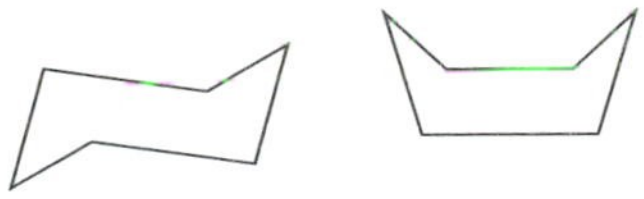 [two cyclohexane chair conformations]

3.46 **Isomers:** Draw the isomers described.
(a) The isomer of C_8H_{18} with the shortest carbon chain
(b) The isomer of $C_{12}H_{26}$ with the longest carbon chain
(c) An isomer of $C_{10}H_{20}O_2$ with the longest carbon chain
(d) A ketone with the formula C_5H_4O
(e) An example of geometric isomerism around a carbon-carbon double bond in a compound with the formula C_5H_4O
(f) An example of geometric isomerism in a cyclic compound with the formula C_5H_{10}
(g) A cyclic alcohol with the formula C_5H_8O
(h) An isomer of C_9H_6 with as many triple bonds as possible
(i) An isomer of $C_3H_8O_2$ with as few carbon-carbon bonds as possible
(j) An isomer of C_3H_9N with as many carbon-nitrogen bonds as possible
(k) A chair form of a compound with the formula C_9H_{18} and one side chain axial and one equatorial

3.47 **Geometric Isomerism:** Is it possible that $CH_3CH=NOH$ could exhibit geometric isomerism? If so, explain and draw the isomers.

3.48 **Geometric Isomerism in Cycloalkenes:** In small ring alkenes such as cyclopentene, the double bond must have a *cis* configuration, whereas in cyclooctene there are both a *cis* and a *trans* isomer. Draw the three isomers described and explain the difference in ability to exhibit *cis-trans* isomerism.

3.49 **Geometric Isomerism in cycloalkenes:** Which do you think is more stable, *cis* or *trans* cyclooctene? *Trans* cyclooctene or *trans* cyclodecene? Explain.

Activities with Molecular Models

1. Make molecular models of ethane and ethene. Notice the tetrahedral shape and 109° bond angles around each carbon of ethane and the trigonal shape and 120° bond angles around each carbon of ethene. Also notice that you can rotate the single bond of ethane but not the double bond of ethene.

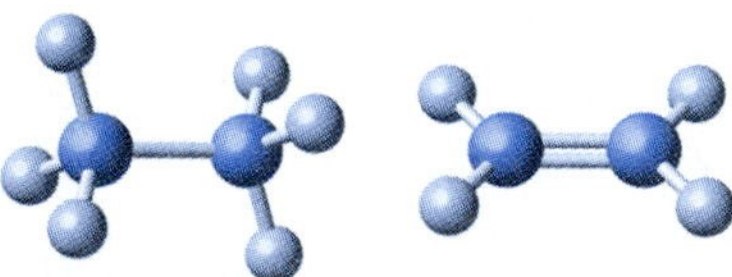

2. Using the models you made in exercise 1, replace a hydrogen on each carbon with a bromine. Notice in 1,2-dibromoethane that you can rotate around the carbon-carbon single bond to get all of the conformations and that they are interconvertible. However, notice that there is no rotation around the carbon-carbon double bond. If you put the two bromines on the same side you have the *cis* geometric isomer, and if you put them on opposite sides you have *trans*. The *cis* and *trans* isomers are not interconvertible.

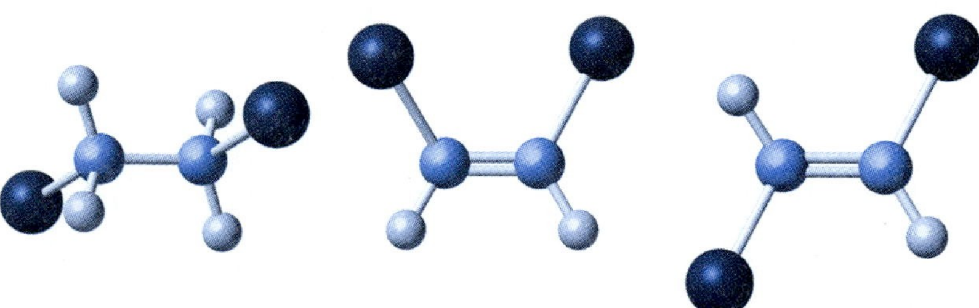

3. Make a model of ethyne. Now replace the hydrogens with methyl groups to get 2-butyne. Compare to 2-butene which exhibits *cis-trans* isomerism. Why does 2-butene show geometric isomerism but not 2-butyne?

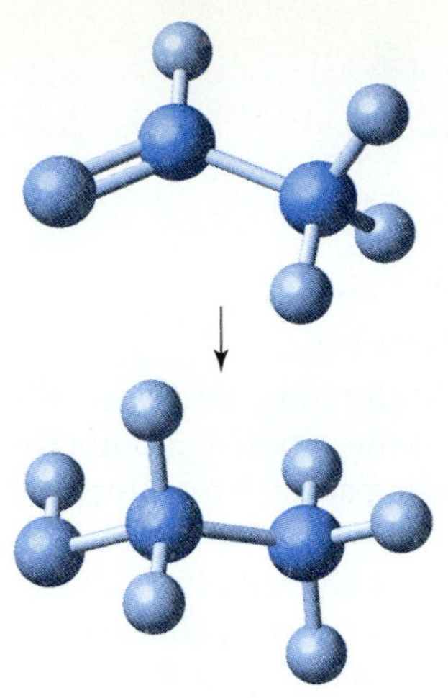

CHAPTER 4

An Introduction to Organic Reactions

We live on a planet abundant in life forms, and as a result organic reactions are going on around us—and within our own bodies—at all times. Basic and applied research also has introduced a wealth of reactions used in synthetic organic chemistry which have led to the many commercial products—including plastics, fibers, and medicines—that so dramatically influence our lives. Although there are an incredible number and variety of reactions, all follow a few fundamental principles that are based on the molecular structure of organic compounds. This allows us to study reactions within a logical framework based on reaction type and mechanism of occurrence.

Before proceeding, let us take a moment to put in perspective what we have learned to this point. We started with a review of atomic structure and electron configuration and used this knowledge to look at the bond-forming characteristics of elements commonly found in organic compounds. We looked at atoms and learned how their hybridization and bonding determine molecular geometry. The enormous breadth of organic chemistry is the result of many types of isomers—structural or constitutional isomers and stereoisomers. Our ability to draw complex structures, using electron dot formulas and condensed representations of molecules, is important both for understanding organic chemistry and for communicating our knowledge. Our ability to communicate is enhanced by our learning a systematic method of nomenclature.

We have gone from atoms and electrons to bonds and molecules to sophisticated structural representations and nomenclature of organic molecules. Each step has been rooted in and dependent on the previous one. We have already begun to apply these principles to hydrocarbons. Now that we are becoming competent with organic structure and nomenclature, we are prepared to embrace the world of organic chemical reactions.

4.1 General Principles of Organic Reactions

A. Types of Reactions; The Reaction Equation

In an organic reaction, one organic compound is converted into another. Bonds are broken in the reactants and new bonds are formed in the products. A **reaction equation** describes what happens in such a transformation by showing the reactants and products. For example, consider the following reaction equation describing the addition of HCl to propene.

reaction equation
an equation that shows what happens in a chemical reaction by showing reactants and products

On the left, the reactant side of the equation, the double bond of propene breaks, leaving a single bond, and the single bond of H — Cl breaks. On the right side of the equation, the product side, a new carbon-hydrogen bond and a new

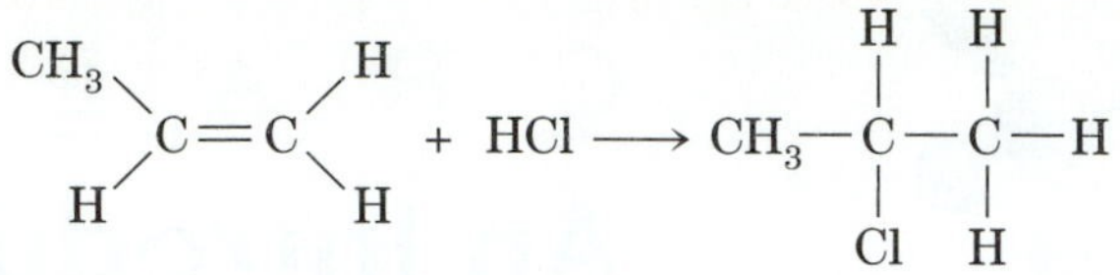

$$CH_3(H)C{=}CH_2 + HCl \longrightarrow CH_3{-}CH(Cl){-}CH_2{-}H$$

The reactants — The product

carbon-chlorine bond are formed. The equation specifically indicates where the hydrogen and chlorine bond, that is, 2-chloropropane is the product and not the positional isomer, 1-chloropropane. The changes in bonding and the reactants and products are clearly shown in the reaction equation.

Although the variety of chemical reactions in organic chemistry is immense, fortunately most fall into one of three reaction types—substitution, elimination, and addition. As you study the reactions presented in this text, you should try to organize them according to reaction type.

substitution reaction
a reaction in which an atom or group on a molecule is replaced by another atom or group

1. *Substitution.* In a **substitution reaction,** an atom or group of atoms is replaced by another species.

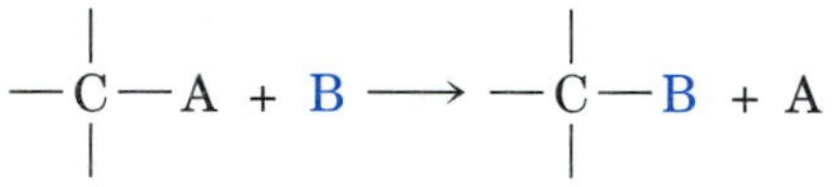

$$-\overset{|}{\underset{|}{C}}-A + B \longrightarrow -\overset{|}{\underset{|}{C}}-B + A$$

elimination reaction
a reaction in which atoms or groups are removed from adjacent atoms to form a double or triple bond

2. *Elimination.* An **elimination reaction** involves the removal of a pair of atoms or groups from adjacent carbon atoms. This necessarily results in the formation of a multiple bond.

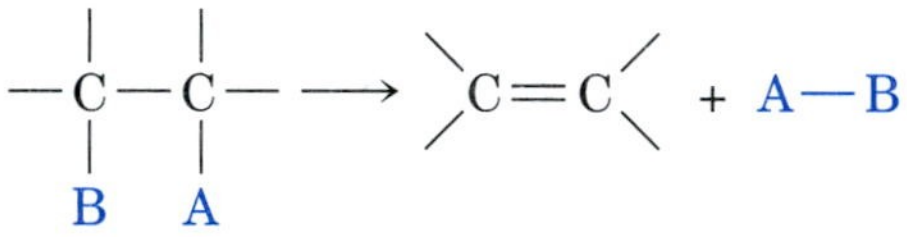

$$-\overset{|}{\underset{B}{C}}-\overset{|}{\underset{A}{C}}- \longrightarrow \;>C{=}C< \; + A{-}B$$

addition reaction
a reaction in which atoms or groups add to adjacent atoms of a multiple bond

3. *Addition.* In an **addition reaction,** atoms or groups add to the adjacent carbons of a multiple bond. To maintain the proper valence, the multiplicity of the bond decreases.

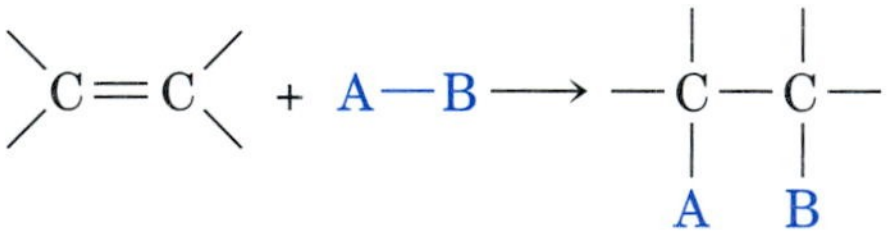

$$>C{=}C< \; + A{-}B \longrightarrow -\overset{|}{\underset{A}{C}}-\overset{|}{\underset{B}{C}}-$$

GETTING INVOLVED

✓ What information does a reaction equation convey?

✓ Define substitution, elimination, and addition reactions. Write general equations for each.

✓ Describe in words what is happening in terms of bond breaking and bond formation in substitution, elimination, and addition reactions.

Problem 4.1

In the general equation for an elimination reaction, A-B is eliminated to produce a double bond. Write a general equation in which A-B is eliminated twice to generate a triple bond. Now write the reverse of this to illustrate the addition of A-B twice to a triple bond.

Problem 4.2

Write specific reaction equations for the following reactions: **(a)** addition of HBr to propene to form 2-bromopropane; **(b)** elimination of HBr from 1-bromopropane;

(c) substitution reaction in which propane reacts with Br_2 to form 2-bromopropane and HBr; **(d)** addition of Br_2 to propyne.

Problem 4.3

Classify the following reactions, which involve either a preparation or a reaction of ethanol (beverage alcohol), as substitution, elimination, or addition:

(a) $CH_3CH_2Br + NaOH \longrightarrow CH_3CH_2OH + NaBr$

(b) $CH_3CH_2OH \xrightarrow[\text{catalyst}]{H_2SO_4} CH_2{=}CH_2 + H_2O$

(c) $CH_3\overset{\overset{\displaystyle O}{\|}}{C}H + H_2 \xrightarrow[\text{catalyst}]{Ni} CH_3CH_2OH$

(d) $CH_2{=}CH_2 + H_2O \xrightarrow[\text{catalyst}]{H_2SO_4} CH_3CH_2OH$

(e) $CH_3CH_2OH \xrightarrow[\text{catalyst}]{\text{Heat, Cu}} CH_3\overset{\overset{\displaystyle O}{\|}}{C}H + H_2$

(f) $CH_3CH_2OH + HCl \xrightarrow{ZnCl_2} CH_3CH_2Cl + H_2O$

B. Reaction Mechanisms

reaction mechanism a step-by-step description of how a chemical reaction occurs

The reaction equation describes what happens in a chemical reaction. The reaction mechanism tells how it happens. The **reaction mechanism** is a step-by-step description of how the chemical reaction occurs. Consider, for example, the possible mechanisms by which hydrogen chloride could add to propene.

$$(CH_3)(H)C{=}CH_2 + HCl \longrightarrow CH_3{-}\underset{Cl}{\overset{H}{C}}{-}\underset{H}{\overset{H}{C}}{-}H$$

Do the hydrogen and chlorine add simultaneously in a one-step mechanism? Perhaps the hydrogen bonds first, followed by the chlorine in a two-step mechanism? Or is it possible that the chlorine bonds first, followed by the hydrogen? Do the hydrogen and chlorine add as charged or neutral species? Are any short-lived intermediate species formed during the steps of the reaction? A reaction mechanism answers these questions.

For example, this reaction proceeds by a two-step mechanism. The hydrogen adds first as a positive ion to form a short-lived intermediate called a *carbocation*. The carbocation is neutralized in the second step by a negative chloride ion.

$$(CH_3)(H)C{=}CH_2 \xrightarrow[\text{Step 1}]{H^+} CH_3{-}\underset{+}{\overset{H}{C}}{-}\underset{H}{\overset{H}{C}}{-}H \xrightarrow[\text{Step 2}]{:\ddot{\underset{..}{Cl}}:^-} CH_3{-}\underset{Cl}{\overset{H}{C}}{-}\underset{H}{\overset{H}{C}}{-}H$$

The organic reactant — A carbocation intermediate — The product

The carbocation is known as a reaction intermediate. There are three very common reaction intermediates, which are discussed in section 4.1D.

GETTING INVOLVED

✓ How do a reaction equation and reaction mechanism differ? What kind of information does each give? Describe each.

Problem 4.4

For the reaction equation illustrating the addition of HCl to propene in the previous section, describe what bonds break on the left of the equation and what bonds form on the right. Now look at the mechanism and give the order in which these bonds break and form.

C. Reaction Mechanisms and Potential Energy Diagrams

potential energy diagram a graphical depiction of energy changes during a chemical reaction

transition state a dynamic process of change in which bonds are being broken and formed in a reaction

energy of activation the energy difference between reactants and a transition state

heat of reaction the difference in energy between the reactants and the products

exothermic reaction a reaction in which energy is released

endothermic reaction a reaction in which energy is absorbed

Potential energy diagrams are used to depict energy changes during chemical reactions. Energy is required to break bonds and potential energy increases as bonds break during the initial stages of a reaction. As new bonds form and a reaction comes to a conclusion, energy is released.

Let's apply this idea to the reaction we just presented, the addition of HCl to propene. If the reaction occurs in one step (it does not, it is a two-step reaction), the hydrogen-chlorine and the carbon-carbon double bond break simultaneously with the formation of the new carbon-hydrogen and carbon-chlorine bond. The **transition state,** where all of this is occuring, is of high energy, high enough to cause the bonds to break (Figure 4.1). The rate of reaction depends on the difference in energy between that of the starting materials and the transition state; this is the **energy of activation.** As the new bonds form and the product is produced, energy is released. The difference in energy between the starting materials and the products is called the **heat of reaction.** If the product is of lower energy, as in our example, heat is released and the reaction is **exothermic** (more energy is released in new bond formation than was consumed in breaking bonds in the reactants). The opposite is called an **endothermic** reaction.

This addition reaction actually occurs in two steps with a carbocation **intermediate** (Figure 4.1). In the first transition state, the HCl bond and the carbon-carbon double bond are breaking as a new carbon-hydrogen bond is forming. The result is a carbocation intermediate that is positively charged and of high energy. The carbon-chlorine bond is forming in the second transition state and consid-

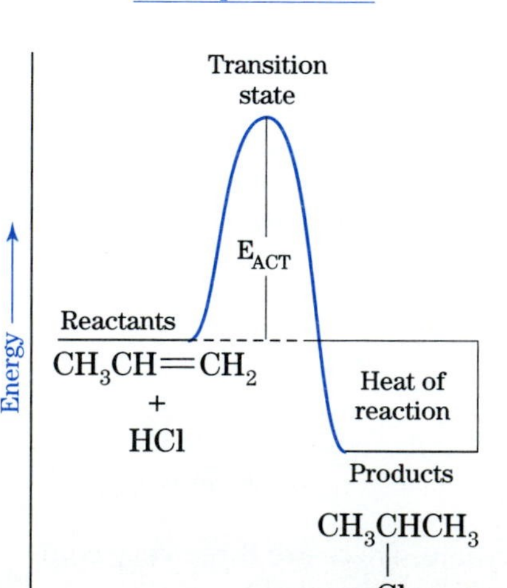

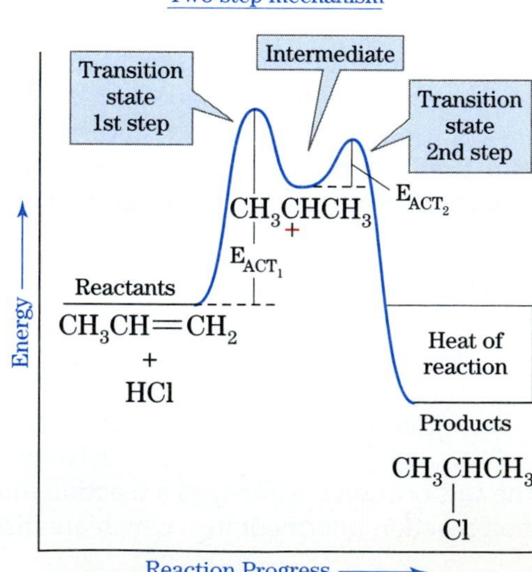

Figure 4.1 Potential energy diagrams.

erable energy is released as the final product develops; the reaction is exothermic. The reaction rate depends on the step with the higher energy of activation, the slower step; in this case it is the first step.

How do a transition state and intermediate differ? A transition state is a dynamic process of change. Bonds are in the process of being broken and formed. Because change is occuring in a transition state, it cannot be thought of as an isolable entity. An intermediate is the result of a transition. Nothing is happening in the intermediate stage and it is theoretically isolable.

GETTING INVOLVED

- ✓ What is a potential energy diagram? Why is energy required to break bonds and released when bonds are formed?
- ✓ What is the difference between a transition state and a reaction intermediate?
- ✓ What is the difference between energy of activation and heat of reaction?
- ✓ How do an exothermic and endothermic reaction differ?

D. Reaction Intermediates

When an organic reaction occurs, some covalent bonds must break and new bonds must form. In multistep reaction mechanisms, as bonds break, unstable, short-lived species called **reaction intermediates** form. There are three major types: **carbocations, free radicals,** and **carbanions.**

$$-\overset{|}{\underset{+}{C}}- \qquad -\overset{|}{\underset{\cdot}{C}}- \qquad -\overset{|}{\underset{\ddot{-}}{C}}-$$

Carbocation   Free radical   Carbanion

Each of these species is unstable for one or both of the following reasons: (1) the particle is charged (carbocation, carbanion); (2) the particle does not have an octet of electrons in the outer shell (carbocation, free radical). In the carbocation, the charged carbon "owns" half of the bonding pairs of electrons, a total of three electrons. Since it is in group IV of the periodic table, to be neutral it must formally own four outer-shell electrons. Consequently, carbocations are positive. Free-radical carbons have one additional electron and are neutral, and carbanion carbons have formal ownership of five outer-shell electrons and are negative.

These intermediates can be formed in a variety of ways. Here we will consider their formation from the **homolytic** or **heterolytic cleavage** of a single bond.

$$-\overset{|}{\underset{|}{C}}:L \xrightarrow[\text{cleavage}]{\text{Heterolytic}} -\overset{|}{\underset{|}{C}}{}^{+} + :L^{-} \qquad \text{Carbocation}$$

$$-\overset{|}{\underset{|}{C}}\cdot\!\cdot L \xrightarrow[\text{cleavage}]{\text{Homolytic}} -\overset{|}{\underset{|}{C}}\cdot + \cdot L \qquad \text{Free radical}$$

$$-\overset{|}{\underset{|}{C}}:L \xrightarrow[\text{cleavage}]{\text{Heterolytic}} -\overset{|}{\underset{|}{C}}:^{-} + {}^{+}L \qquad \text{Carbanion}$$

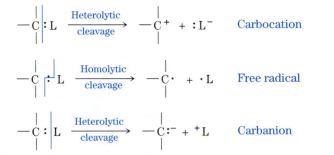

In heterolytic cleavage, as illustrated by the formation of carbocations or carbanions, both electrons involved in the single bond remain with one of the atoms. Two charged species result. In homolytic cleavage, the electrons are parted; each atom retains one. Neutral free radicals result.

Once these intermediates are formed, they are neutralized very quickly. Carbocations are neutralized by negative species, carbanions by positive species, and free radicals by other free radicals.

reaction intermediate
an unstable, short-lived species formed during a chemical reaction; examples are carbocations, free radicals, and carbanions

carbocation
a species with a carbon that has only three bonds, six outer-shell electrons, and a positive charge

free radical
a neutral species with a carbon that has only three bonds and seven outer-shell electrons, one of which is unpaired

carbanion
a species with a carbon that has only three bonds, eight outer-shell electrons including one nonbonding pair, and a negative charge

homolytic cleavage
bond cleavage in which the bonding electrons are evenly divided between the two parting atoms

heterolytic cleavage
bond cleavage in which the bonding electrons are unevenly divided between the two parting atoms

GETTING INVOLVED

✓ What is a reaction intermediate? In terms of outer shell electrons and charge, what are the differences between carbocations, free radicals, and carbanions?

✓ What is the difference between homolytic and heterolytic cleavage of a bond? Which type is most likely to form charged reaction intermediates?

Problem 4.5

In chemical reactions, bromine, Br_2, can cleave homolytically and heterolytically depending on the conditions. Give the electron dot formulas of each bromine atom after cleavage and any charges.

Problem 4.6

Write the structures of two carbocations, two free radicals, and two carbanions that have three carbons and seven hydrogens.

See related problems 4.25–4.26.

4.2 Sites of Organic Reactions

The reactivity of an organic compound is determined by its structure. Specific reaction sites tend to be atoms or groups in which there is a special availability or deficiency of electrons. The electron-rich site of one reactant may then react with the electron-deficient area of another.

electrophile
an electron-deficient species that accepts electrons from nucleophiles in a chemical reaction. Electrophiles are Lewis acids

nucleophile
a species with electron availability that donates electrons to electrophiles in a chemical reaction. Nucleophiles are Lewis bases

Regions of a compound or ion that are deficient in electrons, or positive, tend to attract negative or electron-rich species and to accept electrons in a chemical reaction. Compounds or ions with these properties are called **electrophiles** (from the Greek, "electron-loving"). Their counterparts with high electron density or availability attract positive species and are called **nucleophiles** (from the Greek, "nucleus-loving"). They provide electrons in a chemical reaction. The following sections present four types of reaction sites: multiple bonds, polar bonds, and Lewis acids and bases. As you investigate these sites, keep in mind the concepts of electron availability (nucleophilicity) or deficiency (electrophilicity).

GETTING INVOLVED

✓ What is the difference between an electrophile and nucleophile? Why are they important in organic reactions?

Problem 4.7

Classify the following as either electrophiles or nucleophiles.

(a) $:\ddot{\underset{..}{Cl}}^{+}$ **(b)** $H\ddot{\underset{..}{O}}:^{-}$ **(c)** $H_3C:^{-}$ **(d)** $CH_3CH_2\ddot{\underset{..}{S}}:^{-}$ **(e)** $CH_3\overset{+}{C}HCH_3$ **(f)** $CH_3\ddot{N}H_2$

See related problem 4.24.

A. Multiple Bonds

multiple bond
a double bond or triple bond

Multiple bonds, double and triple bonds, are usually more active reaction sites than single bonds because their electrons are more readily available to an attacking species. In a single bond, σ bond, the electrons are concentrated between the two atoms and are not easily accessed. Double bonds and triple bonds, however, also have π bonds in which electron density is above and below the involved atoms, very accessible to species seeking electrons.

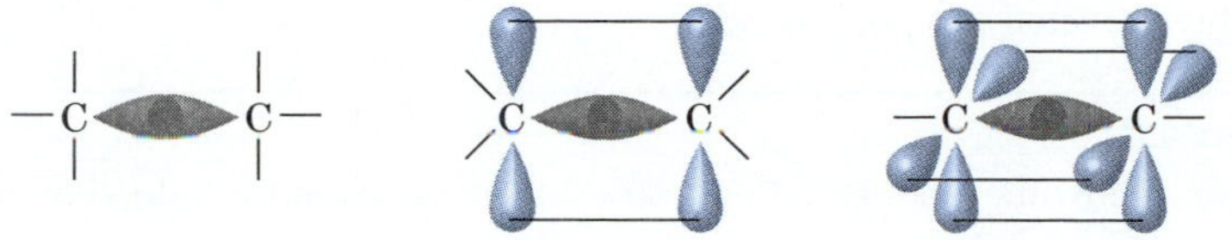

Single bond Double bond Triple bond

As an example of the reactivity of double bonds, consider again the reaction between propene and HCl described in section 4.1.B, in particular the mechanism in which a positive hydrogen ion adds first. As a positive species (electrophile), the hydrogen is attracted to the electron-rich π cloud of the double bond. It bonds to one of the carbons, using the two π electrons of the double bond (remember, electrophiles accept electrons), and forms a carbocation intermediate. The carbocation is short-lived and quickly neutralized by the negative chloride ion (a nucleophile in this case), which provides electrons for the new carbon-chlorine bond.

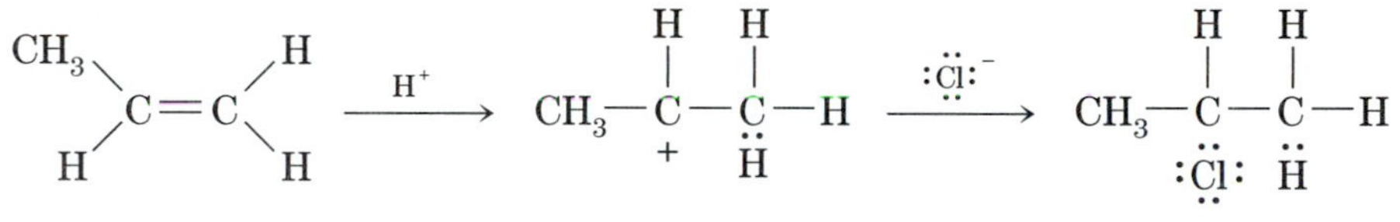

$$CH_3CH{=}CH_2 \xrightarrow{H^+} CH_3-\overset{+}{C}H-CH_3 \xrightarrow{:\ddot{Cl}:^-} CH_3-CH(Cl)-CH_3$$

Carbocation

GETTING INVOLVED

✓ Why are carbon-carbon double bonds and triple bonds generally more reactive than single bonds?

B. Polar Bonds

Polar bonds (section 1.4.F) are covalent bonds in which there is an uneven sharing of electron pairs between atoms of different electronegativities. One atom of the bond is partially positive and the other partially negative, since electrons are distorted toward the more electronegative atom. As a result, the molecule has a region of high electron density that might attract electrophiles and a region of low electron density that might attract nucleophiles.

polar bond
a covalent bond between two atoms of different electronegativities causing one atom to have a greater attraction for the bonding pair(s) and thus charge separation

Basically, opposite charges attract each other. Thus if two molecules, each with polar bonds, were brought together, an attraction between the opposite charges could result in reaction. For example, formaldehyde, with a polar carbon-oxygen double bond, and hydrogen cyanide (in sodium cyanide solution) react by an addition reaction. In simple terms, first the negative cyanide (a nucleophile) is attracted to the positive carbon, and then the positive hydrogen (an electrophile) bonds to the negative oxygen.

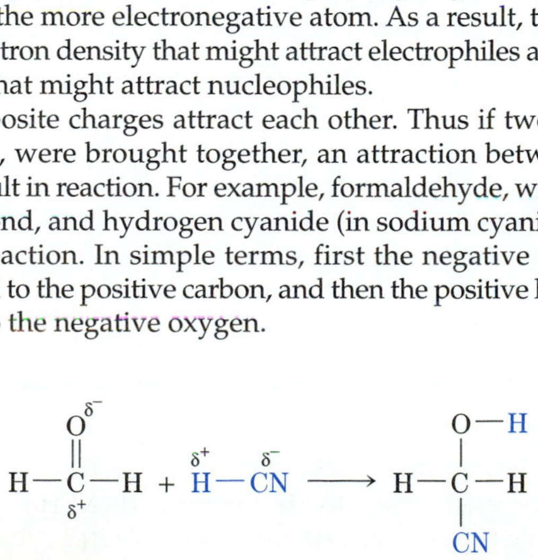

$$H-\overset{\overset{O^{\delta-}}{\|}}{\underset{\delta^+}{C}}-H + \overset{\delta^+}{H}-\overset{\delta^-}{CN} \longrightarrow H-\overset{O-H}{\underset{CN}{C}}-H$$

GETTING INVOLVED

✓ Why are polar bonds reaction sites? Why are both electrophiles and nucleophiles possibly attracted to a polar bond?

Problem 4.8

Identify and show the polarity of the polar covalent bonds in the following molecules:

(a) $ClCH_2C(=O)OCH_3$ (b) $H_2NCH_2CH_2OH$ (c) $CH_3C{\equiv}N$

See related problem 4.20.

C. Lewis Acids and Bases

Lewis base
a substance with an outer-shell nonbonding electron pair that it can share in a chemical reaction with a Lewis acid. Nucleophiles are Lewis bases

Lewis acid
a substance that can accept a pair of electrons for sharing from a Lewis base in a chemical reaction. Electrophiles are Lewis acids

A **Lewis base** is a species that has a nonbonding pair of outer-shell electrons that it can share in a chemical reaction. Because a Lewis base donates electrons in a chemical reaction, it falls under the definition of nucleophile. A **Lewis acid** is a substance that can accept a pair of electrons for sharing in a chemical reaction, falling under the general category of electrophile.

Let us apply these concepts to a simple acid-base neutralization reaction involving a hydrogen ion and hydroxide ion. In the reaction of hydrochloric acid with sodium hydroxide, the hydroxide ion provides a nonbonding pair of electrons for sharing and is a Lewis base. The hydrogen ion, which has no electrons at all, accepts this pair as a Lewis acid. Water is the product.

$$HCl + NaOH \longrightarrow H_2O + NaCl$$

$$\underset{\text{Lewis acid}}{H^+} + \underset{\text{Lewis base}}{{}^{-}\!:\ddot{\underset{..}{O}}\,H} \longrightarrow H:\ddot{\underset{..}{O}}:H$$

We are also familiar from general chemistry with the formation of the hydronium ion in aqueous solutions of acids and the reaction of ammonia with acids to form ammonium salts. Both are Lewis acid-base reactions. The hydronium ion forms as the hydrogen ion from an acid bonds to a lone pair of electrons on the oxygen of water, a Lewis base. Oxygen in the hydronium ion is positively charged, since it can be assigned only five outer-shell electrons (half of the three bonding pairs and all of the nonbonding pair). As a member of group VI of the periodic table, it requires six to be neutral (see section 1.4.E on formal charge).

$$HCl + H_2O \longrightarrow H_3O^+ + Cl^-$$

$$\underset{\text{Lewis acid}}{H^+} + \underset{\text{Lewis base}}{H:\ddot{\underset{..}{O}}:H} \longrightarrow \underset{\text{Hydronium ion}}{H:\overset{\overset{H}{..}}{\underset{..}{O}}:H^{+}}$$

The ammonium ion results from the reaction of the lone pair of electrons on the nitrogen of ammonia (a Lewis base) with a hydrogen ion (Lewis acid).

$$\underset{\text{Nitric acid}}{HNO_3} + \underset{\text{Ammonia}}{NH_3} \longrightarrow \underset{\text{Ammonium nitrate}}{NH_4NO_3}$$

$$\underset{\text{Lewis acid}}{H^+} + \underset{\text{Lewis base}}{:\overset{\overset{H}{..}}{\underset{\underset{H}{..}}{N}}:H} \longrightarrow \underset{\text{Ammonium ion}}{H:\overset{\overset{H}{..}}{\underset{\underset{H}{..}}{N}}:H^{+}}$$

Why does the nitrogen have a nonbonding pair of electrons? Remember, an atom tends, when possible, to complete its outer shell and achieve a stable octet (section 1.4.B–C). In its uncombined state, nitrogen has five electrons in its outer shell (group V) and needs only three more for the octet. In ammonia, each of the three hydrogens provides one electron to the three bonding pairs, and a nonbonding pair remains. Similar reasoning is used to explain the two nonbonding electron pairs in water.

Most organic compounds that are Lewis bases can be thought of as derivatives of water or ammonia, such as alcohols and amines. In these compounds, one can visualize that one or more hydrogens of water or ammonia are replaced by alkyl groups. The oxygen still has two nonbonding electron pairs and the nitrogen, one. Both remain Lewis bases in structure and react with hydrogen ions in a manner analogous to that of water and ammonia.

$$CH_3CH_2\ddot{\underset{\cdot\cdot}{O}}H + H^+Cl^- \longrightarrow CH_3CH_2\underset{\cdot\cdot}{\overset{H}{\overset{|}{\dot{\dot{O}}}}}{}^+H + Cl^-$$

Ethyl alcohol

$$CH_3\ddot{N}CH_3\ (\text{with } CH_3 \text{ on N}) + H^+Cl^- \longrightarrow CH_3\overset{H}{\overset{+}{N}}CH_3\ (\text{with } CH_3 \text{ on N}) + Cl^-$$

Trimethylamine

Some group III compounds, such as BF_3 and $AlCl_3$, are common Lewis acids. In these compounds, neither the boron nor aluminum atom has a stable octet in its outer shell. However, the acquisition of two more electrons from a Lewis base will complete the outer shell, and as a result these compounds are Lewis acids. Consider the reaction of aluminum trichloride (Lewis acid) with the following ether (Lewis base, since the oxygen can share a nonbonding electron pair):

$$AlCl_3 + (CH_3)_2\ddot{\underset{\cdot\cdot}{O}} \longrightarrow Cl_3\overset{-}{Al}:\overset{+}{\underset{\cdot\cdot}{O}}(CH_3)_2$$

Lewis acid Lewis base

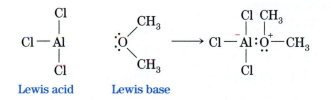

The aluminum has gained electrons and is negative (minus one, since it now owns four outer-shell electrons). The oxygen has decreased in assigned outer-shell electrons from six to five and is thus plus one. The overall product is neutral, however.

GETTING INVOLVED

- ✓ Describe Lewis acids and bases. How do they differ in terms of outer shell electrons at the acid or base site?
- ✓ Why are oxygen and nitrogen derivatives the most common Lewis bases in organic compounds?
- ✓ Why is hydrogen ion a Lewis acid? Why are water and alcohols Lewis bases? Why are ammonia and amines Lewis bases?
- ✓ Why is aluminum trichloride a Lewis acid?

Problem 4.9

(a) Identify Lewis base sites in the following molecules and explain why the site is a Lewis base. **(b)** Write the Lewis acid, Lewis base reaction of these compounds with HCl, in which hydrogen ion is the Lewis acid as illustrated by examples in this section. Be sure to show formal charges in the products.

(a) $CH_3CH_2\ddot{N}H_2$ **(b)** 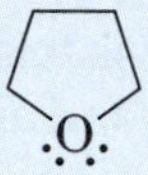 (tetrahydrofuran ring with $:\ddot{O}:$) **(c)** 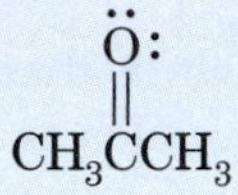 $CH_3\overset{\overset{\Large ..}{O}:}{\overset{\|}{C}}CH_3$ **(d)** $CH_3C{\equiv}N:$

Problem 4.10

Explain why boron trifluoride is a Lewis acid and ammonia is a Lewis base. Write the Lewis acid, Lewis base reaction involving these two compounds. Be sure to show formal charges in the product.

See related problems 4.21–4.22.

D. Combination of Site Types

Some reaction sites contain two or more of the structural concepts discussed in this introduction. As a simple example, consider the carbon-oxygen double bond in formaldehyde. This reaction site is (a) a polar bond because the oxygen is more electronegative than carbon, (b) a Lewis base since there are unshared pairs of electrons on the oxygen that can be shared with a Lewis acid, and (c) a multiple bond with a π bond in which electrons are very available to incoming species.

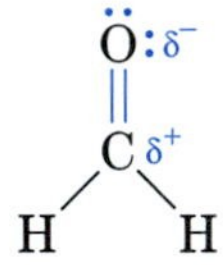

$$H_2C^{\delta+}{=}\ddot{O}:^{\delta-}$$

Formaldehyde

GETTING INVOLVED

Problem 4.11

Identify the possible reaction sites in the following molecules:

(a) $HOCH_2CH{=}CH_2$ **(b)** 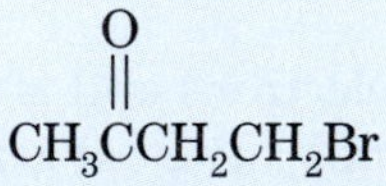 $CH_3\overset{\overset{O}{\|}}{C}CH_2CH_2Br$ **(c)** $CH_3CH_2CH_2NH_2$

See related problem 4.23.

4.3 Getting Started

Chemical reactions will be presented in this text in an organized manner to facilitate learning. Usually a new reaction will be presented in a generalized equation, followed by a few specific examples. In some reactions, there will clearly be a choice of products that could arise from a given starting material. A method for determining the predominant product will be presented when possible, followed again by specific examples. In many reactions, the mechanism will also be covered. We suggest you organize your study of each reaction as follows:

1. *General reaction equation.* Learn the general equation and identify the reaction as substitution, elimination, or addition.
2. *Predominant product.* Learn to determine which product predominates when more than one is possible.
3. *Reaction mechanism.* Learn the step-by-step mechanism in a general form so that you can apply it to specific examples; classify an intermediate as a carbocation, carbanion, or free radical.
4. *Specific examples and practice problems.* Be sure to include specific examples in your summaries, and make sure that you can write reactions by doing practice problems.

So far in this chapter we have introduced reaction types and intermediates and have illustrated mechanisms by an addition reaction. We now introduce two reactions in detail. First is a substitution reaction with a free-radical mechanism, halogenation of alkanes. Then we look at an elimination reaction with a carbocation mechanism under the general heading of preparation of alkenes and alkynes. Chapter 5 will build on this by introducing addition reactions of alkenes and alkynes, most of which occur with a carbocation mechanism; this is called *electrophilic addition*. Electrophilic aromatic substitution will be covered in Chapter 6 (again a reaction with a carbocation mechanism). Nucleophilic substitution will be introduced in Chapter 8. The principles you learn in these reactions will be seen over and over throughout this text and in organic chemistry in general.

4.4 Halogenation of Alkanes: Chlorination and Bromination

A. General Reaction

In the general equation for a reaction, we show only the bonds involved in the transformation. **Halogenation** is a substitution reaction in which a hydrogen on an alkane is replaced by a halogen. The reaction occurs when an alkane is combined with either chlorine or bromine in the presence of heat or light. On the reactant side of the equation for halogenation, a carbon-hydrogen and a halogen-halogen bond break, and on the product side, a carbon-halogen and a hydrogen-halogen bond form. This is illustrated by the following general equation:

halogenation introduction of a halogen (chlorination, bromination, and the like) into a molecule

General Reaction Equation for Halogenation of Alkanes

$$-\overset{|}{\underset{|}{C}}-H \quad + \quad X_2 \quad \xrightarrow[\text{heat}]{\text{Light or}} \quad -\overset{|}{\underset{|}{C}}-X \quad + \quad HX$$

An alkane X_2 is Cl_2 or Br_2 An alkyl halide   HX is HCl or HBr

GETTING INVOLVED

✓ In halogenation of alkanes, what bonds break in the reactants and what bonds form in the products? Which of the three reaction types is illustrated?

Problem 4.12

Write the equations for the following halogenation reactions: **(a)** methane and bromine to form bromomethane and HBr. **(b)** ethane and chlorine to form chloroethane and HCl.

B. Chlorination of Methane: An Example of Halogenation

Methane, the simplest alkane, has four carbon-hydrogen bonds that could potentially react in a halogenation reaction. In the simplest expression of the reaction, one of the hydrogens is replaced by a halogen, producing chloromethane. But as the chloromethane forms, it too can compete with the unreacted methane for the chlorine and, as a result, some dichloromethane is also formed. Notice that dichloromethane still has two carbon-hydrogen bonds and can react with the remaining chlorine along with the methane and chloromethane. Trichloromethane results, which can further react to form tetrachloromethane. The entire process is illustrated below.

	IUPAC Name	Common Name
$CH_4 + Cl_2 \longrightarrow CH_3Cl + HCl$	Chloromethane	Methyl chloride
$CH_3Cl + Cl_2 \longrightarrow CH_2Cl_2 + HCl$	Dichloromethane	Methylene chloride
$CH_2Cl_2 + Cl_2 \longrightarrow CHCl_3 + HCl$	Trichloromethane	Chloroform
$CHCl_3 + Cl_2 \longrightarrow CCl_4 + HCl$	Tetrachloromethane	Carbon tetrachloride

Even when methane is mixed with chlorine in a 1:1 molar ratio, all four organic products form in which from one to four hydrogens have been replaced. At the conclusion of the reaction, when all the chlorine has been consumed, the reaction vessel contains the four chlorination products, one mole of HCl, and some unreacted methane. This polyhalogenation is a serious detriment to the synthetic use of alkane halogenation.

GETTING INVOLVED

✓ Describe to yourself what is happening in the chlorination of methane and why mono, di, tri, and tetrachlorinated products are formed.

Problem 4.13

Show all the possible products that could result from the chlorination of ethane in light. Use the chlorination of methane as an example. If positional isomerism among the polyhalogenated products is considered, there are nine possible isomers.

C. Control of the Halogenation Reaction

As with many potentially useful reactions, the challenge of halogenation lies in controlling it, so that an acceptable yield of desired product is obtained. How could the halogenation reaction conditions be adjusted so as to obtain, say, predominantly tetrachloromethane (CCl_4)? The formation of tetrachloromethane from methane involves replacement of all four hydrogens. This requires 4 moles of chlorine for each mole of methane. Thus, if the reactants methane and chlorine are combined in at least a 1:4 ratio, tetrachloromethane can be obtained almost exclusively.

$$CH_4 + 4Cl_2 \xrightarrow{\text{Light}} CCl_4 + 4HCl$$

For maximal polyhalogenation, an alkane should be exposed to at least as many moles of halogen as there are hydrogen atoms in the molecule.

On the other hand, how could the predominance of chloromethane (CH_3Cl) be assured? For such an outcome, conditions have to be such that chlorine is more likely to encounter a methane molecule than it is any of the chloromethane that forms. This can be accomplished by running the reaction in a large excess of methane (for example, $10CH_4$:$1Cl_2$).

$$\underset{\text{Large excess}}{CH_4} + Cl_2 \xrightarrow{\text{Light}} CH_3Cl + HCl$$

To favor monohalogenation, the alkane is introduced in excess so that the halogen molecules will always have a higher probability of reacting with the alkane than with the monohalogenated product. Even when monohalogenation is predominant, however, the less symmetrical the alkane, the greater the number of products possible. Compare the possible monochlorination products of pentane and its symmetrical skeletal isomer, dimethylpropane, in Example 4.1.

GETTING INVOLVED

- ✓ Describe the relative proportions of reactants needed to promote monohalogenation and polyhalogenation. Explain the difference.
- ✓ Why does a symmetrical alkane give fewer monohalogenation isomers than the alkane's unsymmetrical isomers?

Example 4.1

Write the monochlorination products of pentane and dimethylpropane.

Solution

In pentane there are three places a chlorine can replace a hydrogen and form a different product; in dimethylpropane, all sites are equivalent.

$$\underset{\text{Excess}}{CH_3CH_2CH_2CH_2CH_3} + Cl_2 \xrightarrow{\text{Light}}$$

$$\underset{\substack{|\\Cl}}{CH_2}CH_2CH_2CH_2CH_3 + CH_3\underset{\substack{|\\Cl}}{CH}CH_2CH_2CH_3 + CH_3CH_2\underset{\substack{|\\Cl}}{CH}CH_2CH_3$$

Three possible monochlorination products

$$\underset{\text{Excess}}{CH_3-\overset{\substack{CH_3\\|}}{\underset{\substack{|\\CH_3}}{C}}-CH_3} + Cl_2 \xrightarrow{\text{Light}} \underset{\text{One monochlorination product}}{CH_3-\overset{\substack{CH_3\\|}}{\underset{\substack{|\\CH_3}}{C}}-CH_2Cl} + HCl$$

Problem 4.14

Consider the chlorination of 2-methylbutane, an isomer of pentane and dimethylpropane discussed in Example 4.1, for this problem. **(a)** What is the relative ratio of alkane to chlorine to have total chlorination occur? **(b)** What is the relative ratio to promote monochlorination? **(c)** Draw the possible products formed by the monochlorination reaction.

See related problems 4.27–4.29.

D. Mechanism of Halogenation

To illustrate the mechanism of halogenation, let us consider the monochlorination of methane:

$$CH_4 + Cl_2 \xrightarrow{\text{Light}} CH_3Cl + HCl$$

For the reaction to occur, a C—H and Cl—Cl bond must be broken and a C—Cl and H—Cl bond must be formed. The Cl—Cl bond (bond energy = 58 kcal/mol) is weaker than the C—H bond (bond energy = 102 kcal/mol) and is cleaved by heat or light to form two chlorine atoms (free radicals).

$$:\ddot{\underset{..}{Cl}}:\ddot{\underset{..}{Cl}}: \xrightarrow[\text{heat}]{\text{Light or}} 2\ :\ddot{\underset{..}{Cl}}\cdot \quad \text{Chlorine atom (free radical)}$$

The chlorine atoms will immediately seek a method for completing their octets. This can be accomplished by abstracting a hydrogen atom from methane, thus cleaving a C—H bond and forming one of the reaction products, HCl.

$$\mathrm{H}:\overset{\mathrm{H}}{\underset{\mathrm{H}}{\ddot{\underset{..}{C}}}}:\mathrm{H} + \cdot\ddot{\underset{..}{Cl}}: \longrightarrow \mathrm{H}:\overset{\mathrm{H}}{\underset{\mathrm{H}}{\ddot{\underset{..}{C}}}}\cdot + \mathrm{H}:\ddot{\underset{..}{Cl}}:$$

Free radical

Now the carbon lacks on octet of electrons—a methyl free radical. It can change this by abstracting a chlorine from an undissociated chlorine molecule to form the other reaction product, CH_3Cl.

$$\mathrm{H}:\overset{\mathrm{H}}{\underset{\mathrm{H}}{\ddot{\underset{..}{C}}}}\cdot + :\ddot{\underset{..}{Cl}}:\ddot{\underset{..}{Cl}}: \longrightarrow \mathrm{H}:\overset{\mathrm{H}}{\underset{\mathrm{H}}{\ddot{\underset{..}{C}}}}:\ddot{\underset{..}{Cl}}: + :\ddot{\underset{..}{Cl}}\cdot$$

Another chlorine free radical that can attack yet another methane molecule is formed in this step, thus continuing the process.

chain reaction
a reaction that sustains itself through repeating chain-propagation steps

The chlorination of methane (and halogenation of alkanes in general) occurs by a free-radical **chain reaction.** It is initiated by the light- or heat-induced cleavage of a chlorine molecule. Once initiated, the reaction will proceed in the absence of light or heat. This is due to the alternate formation of methyl and chlorine free radicals in the two propagation steps just described. Each step produces a product and a reactive intermediate which participates in the other step.

The reaction will not proceed indefinitely, however, since chain termination steps, although not as statistically probable as propagation ones, do occur. These result in the consumption of free radicals without producing new ones to continue the chain. The entire mechanism is summarized as follows:

$$Cl_2 \xrightarrow[\text{light}]{\text{Heat or}} 2Cl\cdot$$

$$CH_4 + Cl\cdot \longrightarrow CH_3\cdot + HCl$$

$$CH_3\cdot + Cl_2 \longrightarrow CH_3Cl + Cl\cdot$$

$$Cl\cdot + Cl\cdot \longrightarrow Cl_2$$

$$CH_3\cdot + Cl\cdot \longrightarrow CH_3Cl$$

$$CH_3\cdot + CH_3\cdot \longrightarrow CH_3CH_3$$

Connections 4.1

www.prenhall.com/bailey

Chlorofluorocarbons and the Ozone Layer

A knowledge of free-radical chemistry as just described is important to the understanding of the effects of some chemicals on the environment. An example is the damage to the earth's ozone layer caused by chlorofluorocarbons.

Chlorofluorocarbons (CFC's, freons) are small gaseous molecules containing carbon, fluorine, and chlorine. They were developed in the early 1930s by chemists searching for a new refrigerant to replace the toxic ammonia and sulfur dioxide then in use. Chlorofluorocarbons became widely used as dry-cleaning solvents, as refrigerants for freezers, refrigerators, and air-conditioning units, and as propellants in aerosol cans for dispensing many consumer products, including deodorants, hair sprays, whipped cream, metered-dose inhalants, and window cleaners. Two examples of CFC's are shown below.

CCl_3F	CCl_2F_2
Freon 11 (bp 24°C)	Freon 12 (bp −30°C)

Because of their stability, chlorofluorocarbons are not readily biodegraded or chemically destroyed after use on the earth's surface. Instead, they slowly diffuse toward the upper atmosphere. In the middle of the stratosphere, about 8 to 30 miles above the earth's surface, is a layer of ozone (O_3), a form of elemental oxygen, that is approximately 20 miles thick. This ozone absorbs certain levels of ultraviolet radiation and in doing so shields the earth from the harmful effects of ultraviolet rays. As it absorbs the UV light, ozone is converted to molecular oxygen (O_2) and oxygen atoms. Recombination of these and other naturally occurring oxygen atoms regenerates ozone.

$$O_3 \xrightarrow{\text{Ultraviolet light}} O_2 + \cdot\ddot{\underset{\cdot\cdot}{O}}\cdot$$

Ultraviolet light also causes chlorofluorocarbons to dissociate, a process that produces chlorine atoms, chlorine free radicals. In the stratosphere, these chlorine atoms react with ozone to form chlorine monoxide and molecular oxygen; a molecule of ozone is destroyed. This would not cause a serious ozone depletion if it weren't for a subsequent reaction in which the chlorine monoxide reacts with naturally occurring oxygen atoms to regenerate the original chlorine radical and molecular oxygen. This new chlorine atom can attack and destroy yet another ozone molecule. A free-radical chain reaction is initiated by each CFC molecule that is dissociated, and each chain results in the destruction of thousands of molecules of ozone.

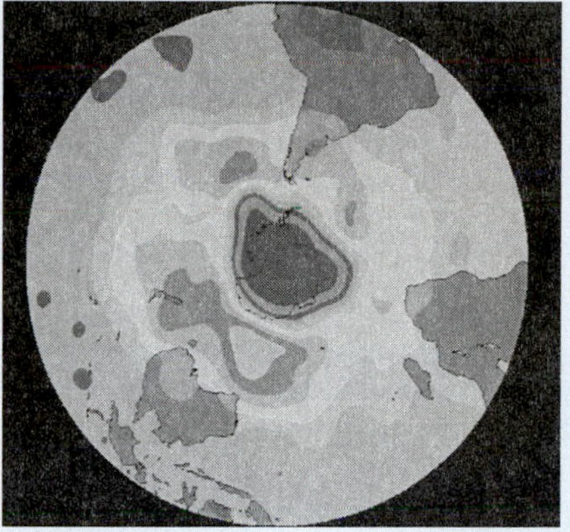

Hole in ozone layer as seen from space

Chain Initiation:

$$CF_2Cl_2 \longrightarrow \cdot CF_2Cl + :\ddot{\underset{\cdot\cdot}{Cl}}\cdot$$

Chain Propagation:

$$:\ddot{\underset{\cdot\cdot}{Cl}}\cdot + O_3 \longrightarrow ClO + O_2$$

$$ClO + \cdot\ddot{\underset{\cdot\cdot}{O}}\cdot \longrightarrow :\ddot{\underset{\cdot\cdot}{Cl}}\cdot + O_2$$

Even small depletions in the ozone layer result in increasing levels of ultraviolet radiation reaching the earth's surface; this can cause significant increases in sunburn, skin cancer, and eye disease. There are also serious implications for plant and aquatic life as well as climatological changes. Measurable effects of damage to the ozone layer include a decrease in some frog populations because increased UV radiation destroys their eggs, which float near the surface of water, and regional holes in the ozone shield that appear over the Arctic and Antarctic.

International agreements and accelerated efforts by some nations and industries are in effect to phase out the use of CFC's and other substances that cause depletion of the ozone layer. Other substances include the agricultural fumigant, methyl bromide; halons, bromine-containing fluorocarbons used in fire protections systems; and hydrochlorofluorocarbons (HCFC's), a less-threatening replacement of CFC's (the hydrogens on HCFC's make them more susceptible to oxidation and destruction in the lower atmosphere). Common aerosol replacements for CFC's include gaseous hydrocarbons such as propane and butane. Even with these efforts underway, it is uncertain how long it will take the environment to recover from damage already done.

GETTING INVOLVED

✓ Describe the initiation step for chlorination. What type of species is formed? Why is it reactive? What does it do next?

✓ Describe the two propagation steps for the chlorination of methane. What type of reaction intermediate is formed? Why do the two steps alternate and create a chain reaction? Each step results in one of the products and one of the reactive intermediates; describe these in each case.

✓ Describe the three possible termination steps. Why do they stop the chain reaction? How many chains does each step break?

Connections 4.2 www.prenhall.com/bailey

General Anesthetics

A person undergoing major surgery must be kept unconscious, without perception of sensations, for a controlled period of time without undue danger of death or toxic side effects. The halogenated general anesthetics are a group of relatively nontoxic, nonflammable, easily vaporized organic liquids used for this purpose.

Since the membranes of our bodies, including those of the nerve cells in our brains, are largely hydrocarbon in structure, these anesthetics can pass into our cells rapidly and exit just as quickly. In their passage they render the person unconscious by mechanisms yet to be discovered.

The different anesthetics shown below have slightly different properties; the type of surgery as well as the patient's physical state will partly determine which one will be used.

Use of a general anesthetic is commonly preceded by the injection of a barbiturate in order to put the patient to sleep quickly. In addition, an opiate pain reliever such as morphine may be administered, as well as a muscle relaxant, so that incisions can be made more easily by the surgeon. Oxygen must be given along with the anesthetic, and the gas pressures of both must be closely monitored and regulated in order to prevent adverse effects.

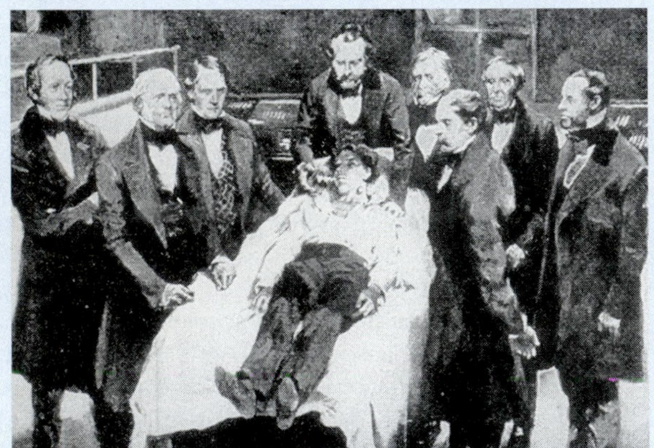

The first ether-assisted surgical operation

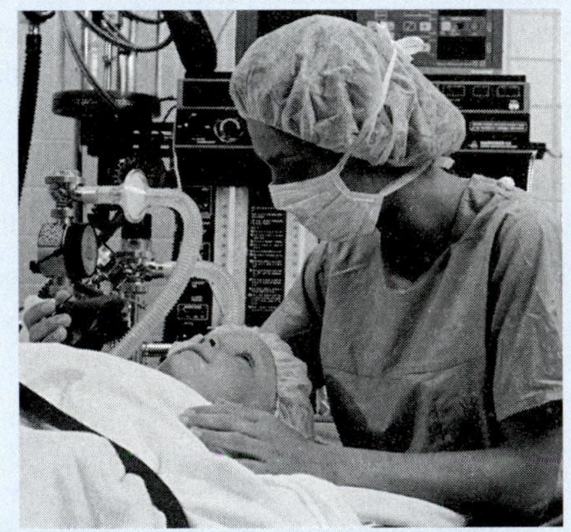

Modern anesthesiologist

$$\begin{array}{ccccc} & \mathrm{F} & \mathrm{Br} & \\ & | & | & \\ \mathrm{F}- & \mathrm{C} - & \mathrm{C} & -\mathrm{H} \\ & | & | & \\ & \mathrm{F} & \mathrm{Cl} & \end{array}$$

Halothane

$$\begin{array}{cccccc} & \mathrm{F} & \mathrm{Br} & & \mathrm{F} & \\ & | & | & & | & \\ \mathrm{F}- & \mathrm{C} - & \mathrm{C} - & \mathrm{O} - & \mathrm{C} & -\mathrm{H} \\ & | & | & & | & \\ & \mathrm{F} & \mathrm{Cl} & & \mathrm{F} & \end{array}$$

Enflurane

$$\begin{array}{cccccc} & \mathrm{Cl} & \mathrm{F} & & \mathrm{H} & \\ & | & | & & | & \\ \mathrm{F}- & \mathrm{C} - & \mathrm{C} - & \mathrm{O} - & \mathrm{C} & -\mathrm{H} \\ & | & | & & | & \\ & \mathrm{Cl} & \mathrm{F} & & \mathrm{H} & \end{array}$$

Methoxyflurane

$$\begin{array}{cccccc} & \mathrm{F} & \mathrm{H} & & \mathrm{F} & \\ & | & | & & | & \\ \mathrm{F}- & \mathrm{C} - & \mathrm{C} - & \mathrm{O} - & \mathrm{C} & -\mathrm{H} \\ & | & | & & | & \\ & \mathrm{F} & \mathrm{Cl} & & \mathrm{F} & \end{array}$$

Isoflurane

Problem 4.15

Write a step-by-step free radical chain reaction for the light-induced monobromination of ethane. Use the monochlorination of methane as a guide.

See related problems 4.30–4.31.

4.5 Preparation of Alkenes and Alkynes: Elimination Reactions

A. General Reaction Equations

Elimination reactions are used to introduce carbon-carbon double or triple bonds into a molecule. To generate a carbon-carbon double bond, two atoms or groups, one from each of two adjacent carbons, must be eliminated. It follows that to generate a triple bond, four atoms or groups, two from each of two adjacent single-bonded carbons, must be eliminated. The following general reaction equations illustrate these concepts.

General Equations of Elimination Reactions for Preparing Alkenes and Alkynes

Alkenes
$$-\underset{B}{\overset{|}{\underset{|}{C}}}-\underset{A}{\overset{|}{\underset{|}{C}}}- \longrightarrow \;\rangle C{=}C\langle \;+\; A{-}B$$

Alkynes
$$-\underset{B}{\overset{B}{C}}-\underset{A}{\overset{A}{C}}- \longrightarrow -C{\equiv}C- \;+\; 2\,A{-}B$$

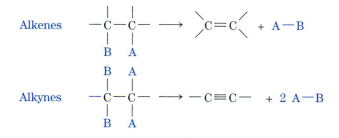

The general concept is as follows: eliminate once to produce a double bond; eliminate twice to give a triple bond. But we need to be more specific. What do the symbols A and B stand for, and what reagents will cause their elimination to occur? We will consider two types of elimination reactions: dehydrohalogenation, in which A and B are hydrogen (H) and halogen (Cl, Br, I), and dehydration, where A and B are the elements of water (H and OH).

In **dehydrohalogenation** reactions, the elements of a hydrogen halide (HCl, HBr, HI) must be removed. Because hydrogen halides are acidic in nature (hydrochloric, hydrobromic, and hydroiodic acids), bases are effective dehydrohalogenating reagents. Many bases are useful, but we will simplify our study by considering only potassium hydroxide for the preparation of alkenes and sodium amide, a stronger base, for the preparation of alkynes. Aqueous alcohol (CH_3CH_2OH in water) is often the solvent in dehydrohalogenation reactions using potassium hydroxide because it will dissolve both the KOH (soluble in water) and alkyl halide (soluble in alcohol).

dehydrohalogenation
a reaction in which hydrogen and halogen are eliminated from a molecule

General Equations for Dehydrohalogenation Reactions

Preparation of alkenes
$$-\underset{X}{\overset{|}{\underset{|}{C}}}-\underset{H}{\overset{|}{\underset{|}{C}}}- \;+\; KOH \xrightarrow[\text{in aqueous alcohol}]{} \;\rangle C{=}C\langle \;+\; KX \;+\; H_2O$$

Preparation of alkynes
$$-\underset{X}{\overset{X}{C}}-\underset{H}{\overset{H}{C}}- \;\text{or}\; -\underset{H}{\overset{X}{C}}-\underset{X}{\overset{H}{C}}- \;+\; 2NaNH_2 \longrightarrow -C{\equiv}C- \;+\; 2NaX \;+\; 2NH_3$$

(X is Cl, Br, or I in these reactions)

Following are examples illustrating the preparations of a specific alkene, propene, and an alkyne, propyne, by dehydrohalogenation. Direct your attention initially to the halogens, as they are eliminated. Then remove an equal number of hydrogens from the adjacent carbon to form the double or triple bond.

$$CH_3CH_2CH_2Cl + KOH \longrightarrow CH_3CH{=}CH_2 + KCl + H_2O$$

$$CH_3CH_2CHBr_2 + 2NaNH_2 \longrightarrow CH_3C{\equiv}CH + 2NaBr + 2NH_3$$

dehydration
a reaction in which the elements of water (H and OH) are eliminated from a molecule

In **dehydration** reactions, the elements of water are eliminated from adjacent carbons of an alcohol. Acids are generally effective dehydrating agents. We will use sulfuric acid to illustrate the reaction. It acts as a catalyst and is not consumed in the reaction. Alkenes can be prepared by dehydration reactions involving alcohols, but dehydration is not an effective method for making alkynes.

General Reaction Equation for Preparation of Alkenes by Dehydration

$$-\underset{H}{\overset{|}{\underset{|}{C}}}-\underset{OH}{\overset{|}{\underset{|}{C}}}- \xrightarrow{H_2SO_4} \;\rangle C{=}C\langle\; + H_2O$$

Consider again the preparation of propene, this time by dehydration. Note that the middle carbon loses the OH and the hydrogen can come from either of the adjacent carbons. In either case, propene is the product.

$$CH_3\underset{OH}{\underset{|}{C}}HCH_3 \xrightarrow{H_2SO_4} CH_3CH{=}CH_2 + H_2O$$

GETTING INVOLVED

- ✓ Describe an elimination reaction. What is the difference in an elimination reaction that forms a carbon-carbon double bond to one that forms a triple bond? What bonds break and form in an elimination reaction?
- ✓ What specifically is dehydrohalogenation? What atoms are eliminated in the reaction? Why are bases effective reagents? What base is commonly used in the preparation of alkenes and what is used to prepare alkynes? Can you write a general reactions equations for both?
- ✓ What specifically is dehydration? What is eliminated in the reaction? What reagent is commonly used? Can you write a general reaction equation?

Problem 4.16

Complete the following reactions illustrating the preparations of alkenes. **(a)** $CH_3CH_2CH_2OH$ and H_2SO_4; **(b)** 1-bromobutane and KOH; **(c)** 2-bromobutane and KOH (note that two alkenes are possible here, draw both; in the next section we will learn how to predict the predominant product).

Problem 4.17

Complete the following reactions illustrating the preparations of alkynes. **(a)** 1,2 dichloroethane and 2 $NaNH_2$; **(b)** 1,1-dibromobutane and 2 $NaNH_2$; **(c)** 2,2-dibromopropane and 2 $NaNH_2$ (notice here that elimination can occur to form an alkyne or a diene, two double bonds; draw both).

B. Orientation of Elimination

In all of the examples in the previous section, only one elimination product was possible. How do we determine whether more than one is possible and then predict the predominant one? Consider the dehydration of the following alcohols,

1-butanol and 2-butanol, which are positional isomers of each other. Only one product is possible from 1-butanol.

$$CH_3CH_2CH_2CH_2OH \xrightarrow{H_2SO_4} CH_3CH_2CH{=}CH_2 + H_2O$$

Two products are formed from 2-butanol because the hydrogen that leaves after the hydroxy can come from either carbon-1 or carbon-3. The alkenes are not formed in equal amounts, however; 2-butene occurs in the greater amount.

$$\underset{\displaystyle OH}{CH_3CH_2\underset{|}{C}HCH_3} \xrightarrow{H_2SO_4} CH_3CH_2CH{=}CH_2 + \underset{\text{Predominant alkene}}{CH_3CH{=}CHCH_3} + H_2O$$

The more stable alkene is the predominant product in these elimination reactions; stability can be predicted by the following method.

Prediction of Orientation of Elimination

In dehydration and dehydrohalogenation, if elimination can result in the formation of more than one alkene, the most stable alkene is formed predominantly. The most stable alkene is the one most highly substituted with alkyl groups. This is known as **Zaitsev's** (sometimes spelled **Saytzeff) rule,** after the Russian chemist Alexander Zaitsev.

Zaitsev's rule in applicable elimination reactions, the most substituted alkene (with alkyl groups) will predominate

Stability and Ease of Formation of Alkenes

$$CH_2{=}CH_2 < RCH{=}CH_2 < \begin{matrix} R_2C{=}CH_2 \\ RCH{=}CHR \end{matrix} < R_2C{=}CHR < R_2C{=}CR_2$$

Least substituted, least stable — Most substituted, most stable

R = general symbol for alkyl group

To determine the degree of substitution of an alkene, count the number of carbons directly attached to the two carbons involved in the double bond. In 1-butene, the first carbon has two attached hydrogens, and the second has a hydrogen and a carbon; 1-butene has a monosubstituted double bond. Each carbon of 2-butene has a hydrogen and a carbon attached; 2-butene has a disubstituted double bond. Since 2-butene is more highly substituted, it is formed predominantly.

GETTING INVOLVED

✓ What is the Zaitsev's rule and what does it say about stability of alkenes and the predominant product of elimination when more than one alkene is possible? How is the degree of substitution and the stability of an alkene determined?

Example 4.2

Write structures for the three alkenes that may result from the dehydrobromination of 3-bromo-2,3-dimethylpentane. Determine the degree of substitution of each and the predominant product.

Solution

Focus your attention on the halogen (or OH in dehydration), since you know this atom will be eliminated. After eliminating the bromine in your mind, remove a hydrogen from an adjacent carbon to form the double bond. There are three directly adjacent carbons. Elimination of a hydrogen from the one with the fewest hydrogens forms the most substituted alkene.

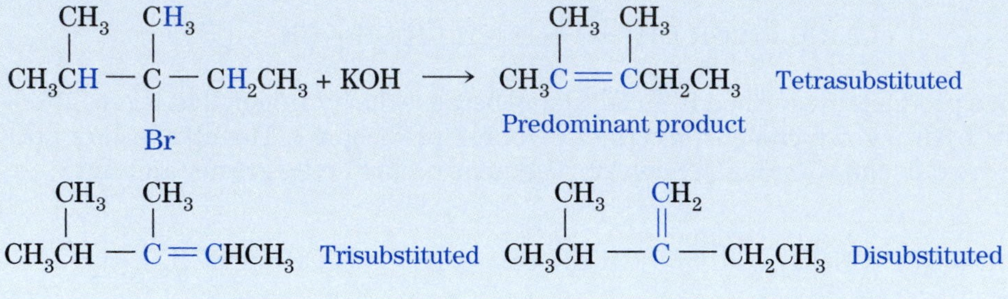

$$CH_3CH(CH_3)-C(CH_3)(Br)-CH_2CH_3 + KOH \longrightarrow CH_3C(CH_3)=C(CH_3)CH_2CH_3 \quad \text{Tetrasubstituted (Predominant product)}$$

$$CH_3CH(CH_3)-C(CH_3)=CHCH_3 \quad \text{Trisubstituted} \qquad CH_3CH(CH_3)-C(=CH_2)-CH_2CH_3 \quad \text{Disubstituted}$$

Problem 4.18

Complete the following reactions, showing the major organic product:

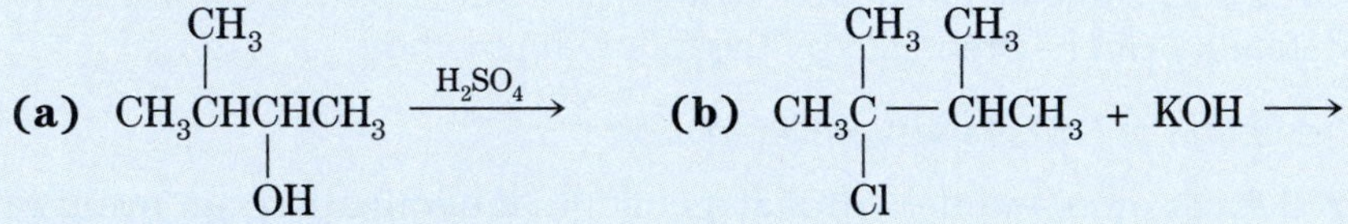

$$\textbf{(a)}\ CH_3CH(CH_3)CH(OH)CH_3 \xrightarrow{H_2SO_4} \qquad \textbf{(b)}\ CH_3C(CH_3)(Cl)-CH(CH_3)CH_3 + KOH \longrightarrow$$

See related problems 4.33–4.36.

C. Mechanism of the Dehydration Reaction

We will expand on the mechanism of the dehydrohalogenation reaction in the chapter on organic halogen compounds (Chapter 8). Here let us look at the mechanism of dehydration, using the reaction of isopropyl alcohol (rubbing alcohol) with sulfuric acid to produce propene (from which the plastic polypropylene is made).

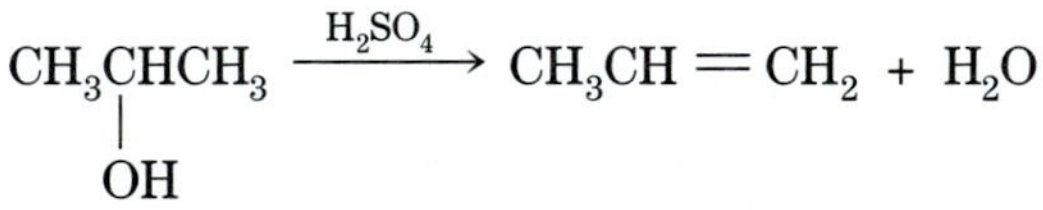

$$CH_3CH(OH)CH_3 \xrightarrow{H_2SO_4} CH_3CH=CH_2 + H_2O$$

For the reaction to occur there must be interaction between isopropyl alcohol and sulfuric acid. What attractions might there be between these substances? First, consider that sulfuric acid is a strong acid and is able to provide a positive hydrogen ion (often called a *proton*) in a chemical reaction. We have seen in this chapter that alcohols are Lewis bases (section 4.2.C). The oxygen of isopropyl alcohol has two nonbonding electron pairs to which the hydrogen ion is attracted. The reaction is analogous to the ionization of a strong acid in water to form the hydronium ion ($H_2O + H_2SO_4 \longrightarrow H_3O^+ + HSO_4^-$).

Step 1: $$CH_3CH(\ddot{O}H)CH_3 + H^+ \longrightarrow CH_3CH(\ddot{O}H_2^+)CH_3$$

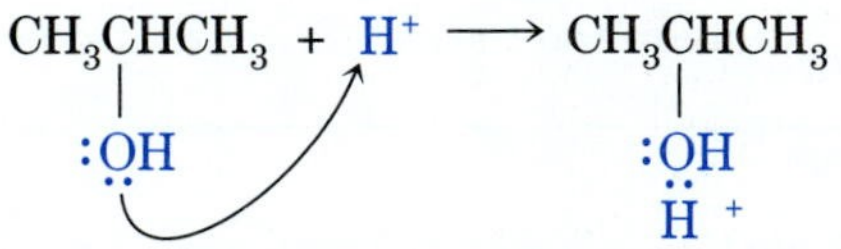

In the second step of the reaction, the protonated alcohol loses a molecule of water, producing a carbocation; the OH is thereby eliminated.

Step 2: $$CH_3CH(\ddot{O}H_2^+)CH_3 \longrightarrow CH_3\overset{+}{C}HCH_3 + H_2\ddot{O}:$$

A carbocation

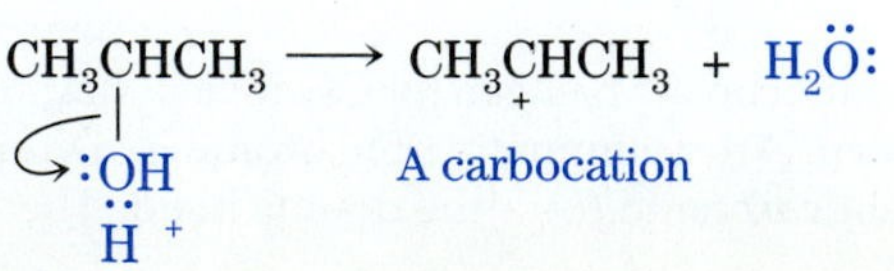

Note that the oxygen retained the pair of electrons in the carbon-oxygen bond, creating a neutral water molecule and leaving a positive charge on the carbon. If you look closely, you will see that the carbocation has only one more hydrogen than the product. In the final step of the mechanism, the carbocation eliminates a positive hydrogen ion to form the neutral product, propene. The hydrogen ion provided by sulfuric acid in the first step is returned to it in the last; sulfuric acid is truly a **catalyst.**

catalyst
a reagent that influences the course and rate of a reaction without being consumed

Step 3: $CH_3\overset{+}{C}H-CH_2(H) \longrightarrow CH_3CH{=}CH_2 + H^+$

We can write mechanisms in a general form just as reactions can be written in a general form. Both are illustrated below for the dehydration of alcohols, showing only the atoms actually involved in the reaction.

General Reaction for the Dehydration of Alcohols

$$-\underset{OH}{\overset{|}{C}}-\underset{H}{\overset{|}{C}}- \xrightarrow{H_2SO_4} \rangle C{=}C\langle + H_2O$$

General Mechanism for the Dehydration of Alcohols

$$-\underset{:\ddot{O}H}{\overset{|}{C}}-\underset{H}{\overset{|}{C}}- \xrightarrow{H^+} -\underset{:\overset{+}{\ddot{O}}H_2}{\overset{|}{C}}-\underset{H}{\overset{|}{C}}- \xrightarrow{-H_2\ddot{O}:} -\overset{|}{\underset{+}{C}}-\underset{H}{\overset{|}{C}}- \xrightarrow{-H^+} \rangle C{=}C\langle$$

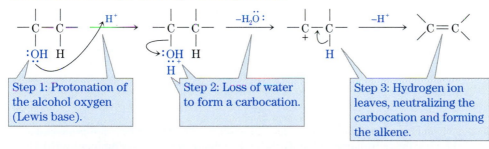

Step 1: Protonation of the alcohol oxygen (Lewis base).

Step 2: Loss of water to form a carbocation.

Step 3: Hydrogen ion leaves, neutralizing the carbocation and forming the alkene.

Use of Curved Arrows

Curved arrows are used by organic chemists to show the movement of electrons in a reaction mechanism. The arrow begins at the electrons that move and the head points to the atom or bond where they move. Examine this in the mechanism for dehydration of isopropyl alcohol and the general mechanism for dehydration of alcohols. A full arrow is used to depict movement of electron pairs and "fish hooks" are used to show movement of single electrons.

GETTING INVOLVED

Let's ask ourselves questions about the dehydration mechanism one step at a time:

- ✓ Step 1: What happens? Describe the attraction of the hydrogen ion to the alcohol in terms of Lewis acids and bases. Explain the formal charge in the oxygen.
- ✓ Step 2: What happens? Where do the electrons in the C—O bond go when water leaves? What is the reactive intermediate called that remains and why is it positively charged? How far along are we in the elimination?
- ✓ Step 3: What happens? Where do the electrons in the C—H bond go when hydrogen ion leaves? Why is sulfuric acid truly a catalyst?

Example 4.3

Write a reaction mechanism illustrating the formation of the predominant dehydration product from 2-butanol (structure below).

Solution

First determine the predominant product, the more highly substituted, of the two possible.

$$CH_3CH_2CHCH_3 \text{ (with OH on C-2)} \xrightarrow{H_2SO_4} CH_3CH_2CH{=}CH_2 + CH_3CH{=}CHCH_3 + H_2O$$

Predominant product ($CH_3CH{=}CHCH_3$)

Now write the three-step mechanism for the predominant product: (1) protonation of the oxygen; (2) loss of water to produce a carbocation; and (3) loss of a proton to form the double bond.

$$CH_3CH_2CHCH_3 \text{ (with } \ddot{O}H\text{)} \xrightarrow{H^+} CH_3CH_2CHCH_3 \text{ (with } :\overset{+}{O}H_2\text{)} \xrightarrow{-H_2\ddot{O}:}$$

$$CH_3CH(H)\overset{+}{C}HCH_3 \xrightarrow{-H^+} CH_3CH{=}CHCH_3$$

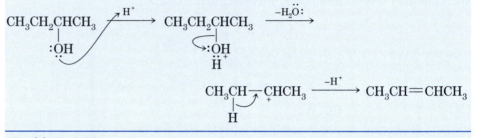

Problem 4.19

The following alcohol can produce two possible alkenes:

$$CH_3C(CH_3)(OH)CH_2CH_3 \xrightarrow{H_2SO_4}$$

(a) First write a reaction equation illustrating the dehydration of this alcohol and showing the two alkenes. Identify the alkene that is formed predominantly. **(b)** Then write a step-by-step mechanism for the dehydration. Steps 1 and 2 are the same for the two products. The last step will produce different products, depending on the position of the hydrogen that is lost.

See related problem 4.32.

REACTION SUMMARY

A. Halogenation of Alkanes

Section 4.4; Example 4.1; Problems 4.12–4.14, 4.27–4.31.

$$-\overset{|}{\underset{|}{C}}-H + X_2 \xrightarrow[\text{heat}]{\text{Light or}} -\overset{|}{\underset{|}{C}}-X + HX \quad (X = Cl, Br)$$

B. General Equations for Elimination Reactions for Preparing Alkenes and Alkynes

Section 4.5; Examples 4.2–4.3; Problems 4.15–4.17, 4.32–4.36.

Alkenes

$$-\overset{|}{\underset{B}{C}}-\overset{|}{\underset{A}{C}}- \longrightarrow -\overset{|}{C}{=}\overset{|}{C}- + A-B$$

Alkynes

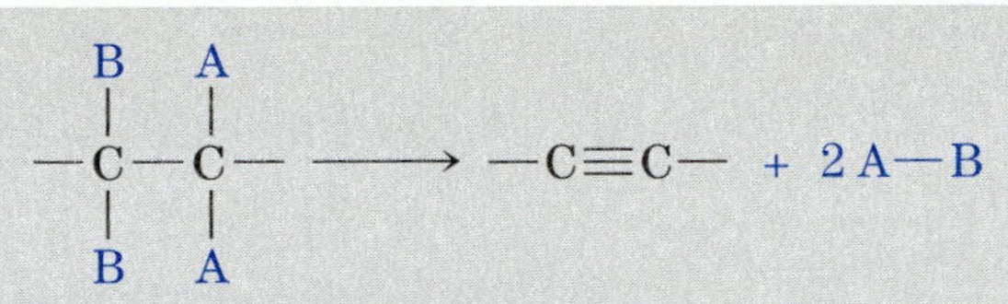

$$-\overset{\overset{\large B}{|}}{\underset{\underset{\large B}{|}}{C}}-\overset{\overset{\large A}{|}}{\underset{\underset{\large A}{|}}{C}}- \longrightarrow -C{\equiv}C- + 2\,A{-}B$$

Dehydrohalogenation: AB = HCl, HBr, HI
Elimination Reagent is KOH for Alkenes
$2NaNH_2$ for Alkynes

Dehydration: AB = H—OH (water)
Elimination Catalyst is H_2SO_4
Used to prepare alkenes only

Problems

4.20 **Polar Bonds:** Identify and show the polarity using δ^+ and δ^- of the polar bonds in the following molecule (sodium pentothal, a barbiturate):

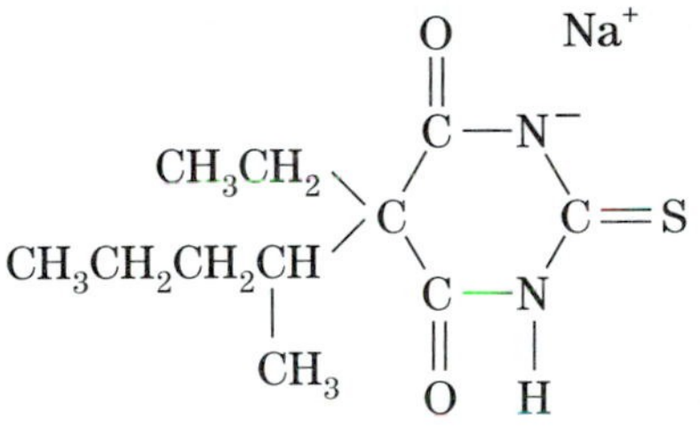

CH_3CH_2 and $CH_3CH_2CH_2CH(CH_3)$ on C of a ring: C(=O)—N^- (Na^+)—C=S—N(H)—C(=O)

4.21 **Lewis Acids and Bases:** Determine whether the following are Lewis acids or Lewis bases:

(a) $AlBr_3$ (b) CH_3NHCH_3
(c) CH_3OH (d) BH_3

4.22 **Lewis Acids and Bases:** Write, using electron dot formulas, Lewis acid–Lewis base reactions between the following species:

(a) H^+, CH_3CH_2OH (b) $AlCl_3, H_2O$
(c) CH_3NH_2, HCl (d) $BF_3, (CH_3)_3N$

4.23 **Reaction Sites:** Identify the reaction sites—polar bonds, multiple bonds, Lewis acids/bases—in the following molecule, which is produced by queen honeybees and is largely responsible for their regulatory powers:

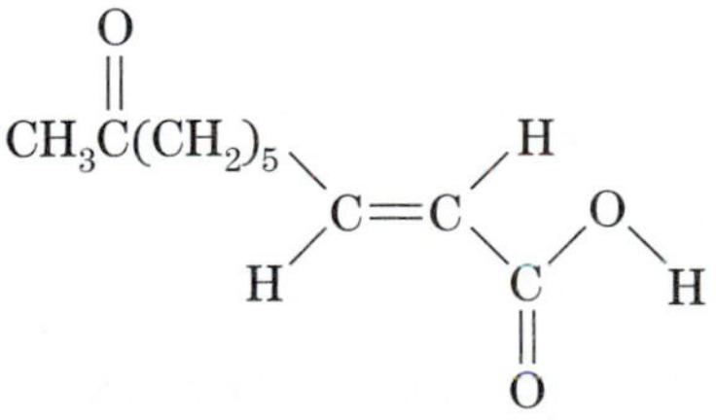

$CH_3C(=O)(CH_2)_5$(H)C=C(H)—C(=O)—O—H

4.24 **Electrophiles and Nucleophiles:** Classify the following as either electrophiles or nucleophiles:

(a) $^-{:}C{\equiv}N{:}$ (b) $:\ddot{Br}^+$ (c) $CH_3\ddot{O}H$
(d) $(CH_3)_3N{:}$ (e) $CH_3CH_2\ddot{O}{:}^-$
(f) $(CH_3)_2\overset{+}{C}H$ (g) $CH_3CH_2{:}^-$ (h) $^-{:}\ddot{S}H$

4.25 **Reactive Intermediates:** Draw a tertiary butyl carbocation, free radical, and carbanion. Show how the electrons in the C—A bond in the following general compound must be distributed in the bond cleavage to form each:

$$(CH_3)_3C{-}A \longrightarrow$$

4.26 **Reactive Intermediates:** Show the formation of a carbocation, free radical, and carbanion by addition of A^+, $A\cdot$, and $A{:}^-$ to a carbon-carbon double bond.

4.27 **Halogenation of Alkanes:** How many different monochlorination isomers could be formed by the light-induced monochlorination of the following compounds? In other words, in how many different places could one hydrogen be replaced by one chlorine? Write the compounds formed.

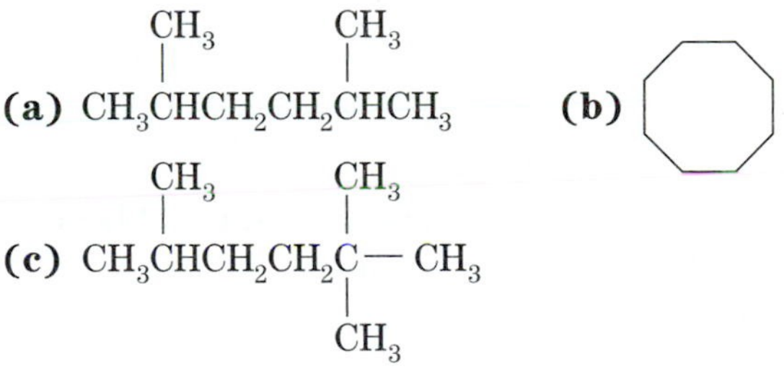

(a) $CH_3CH(CH_3)CH_2CH_2CH(CH_3)CH_3$ (b) cyclooctane

(c) $CH_3CH(CH_3)CH_2CH_2C(CH_3)_2—CH_3$

4.28 **Halogenation of Alkanes:** Write the structural formula for the alkane with each of the following molecular formulas that gives only one monobromination isomer:

(a) C_5H_{12} (b) C_8H_{18}

4.29 **Halogenation of Alkanes:** Describe the reaction conditions that would favor the formation of bromoethane and hexabromoethane when ethane is treated with bromine in the presence of light.

4.30 **Halogenation of Alkanes—Reaction Mechanism:** Write a step-by-step reaction mechanism for the light-induced monobromination of methane.

4.31 **Halogenation of Alkanes—Reaction Mechanism:** Tetraethyllead (an antiknock agent in leaded gasoline) decomposes to elemental lead and ethyl free radicals at 140°C.

$$Pb(CH_2CH_3)_4 \xrightarrow{140°C} Pb + 4CH_3CH_2\cdot$$

Although methane and chlorine, in the absence of light, react at 250°C, in the presence of minute quantities of tetraethyllead they can be made to react at 140°C. Write a mechanism showing how tetraethyllead catalyzes the chlorination of methane.

4.32 **Dehydration of Alcohols—Reaction Mechanism:** Write a step-by-step reaction mechanism for the acid-catalyzed dehydration of $CH_3CH_2CHOHCH_2CH_3$.

4.33 **Elimination Reactions to Produce Alkenes:** Complete the following reactions, showing the predominant organic products:

(a) $CH_3CH_2CH_2Br + KOH \longrightarrow$

(b) 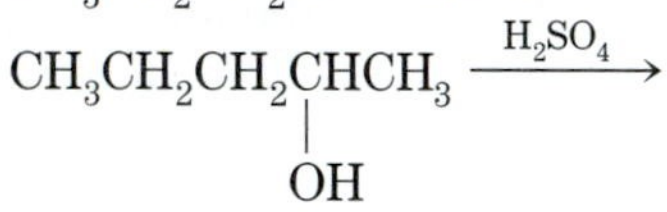 $CH_3CH_2CH_2CH(OH)CH_3 \xrightarrow{H_2SO_4}$

(c) 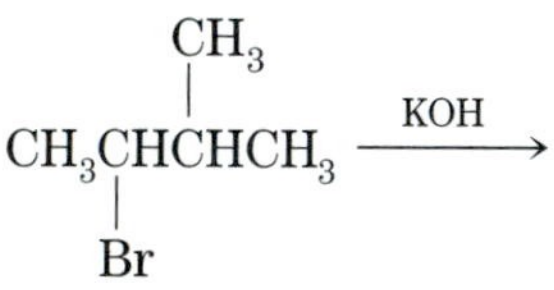 $CH_3CH(Br)CH(CH_3)CH_3 \xrightarrow{KOH}$

(d) 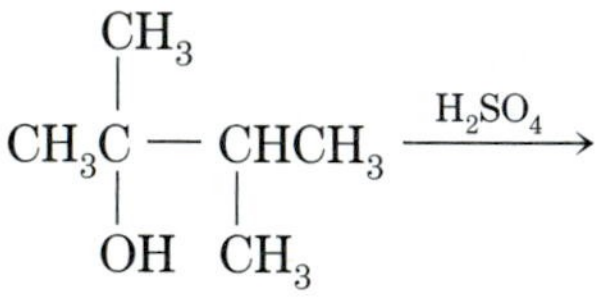 $CH_3C(CH_3)(OH)—CH(CH_3)CH_3 \xrightarrow{H_2SO_4}$

(e) 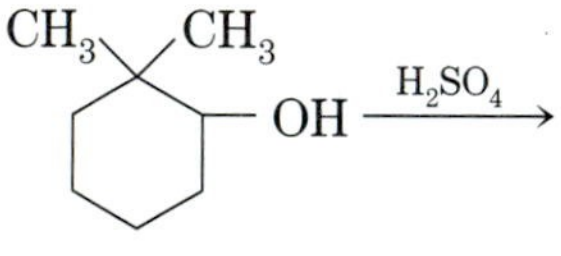 1-(CH₃)-1-...

(f) $HOCH_2CH_2CH_2CH_2OH \xrightarrow{H_2SO_4}$

4.34 **Elimination Reactions to Produce Alkynes:** Complete the following reactions:

(a) $CH_3CH_2CHBr_2 + 2NaNH_2 \longrightarrow$

(b) 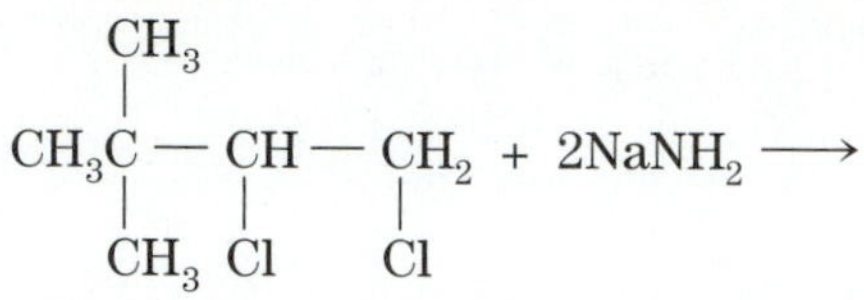 $CH_3C(CH_3)_2—CH(Cl)—CH_2Cl + 2NaNH_2 \longrightarrow$

4.35 **Preparation of Alkenes and Alkynes:** Write the structure of a starting compound and necessary reagents for preparing the following in one step:

(a) 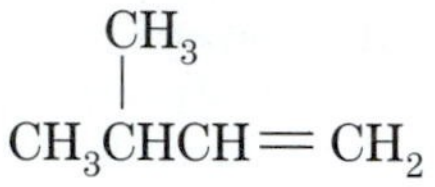 $CH_3CH(CH_3)CH{=}CH_2$ **(b)** 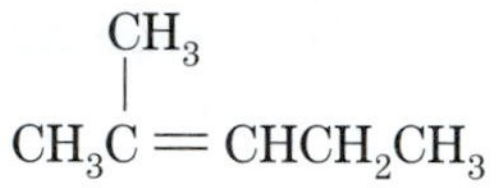 $CH_3C(CH_3){=}CHCH_2CH_3$

(c) $CH_3CH_2C{\equiv}CH$

4.36 **Preparation of Alkenes and Alkynes:** In each case select the better method for preparing the compound desired. Explain your choice.

(a) 1-bromopentane or 2-chloropentane for the preparation of 2-pentene, using KOH in aqueous alcohol

(b) 1,1-dichloropropane or 2,2-dichloropropane for preparing propyne, using sodium amide.

(c) 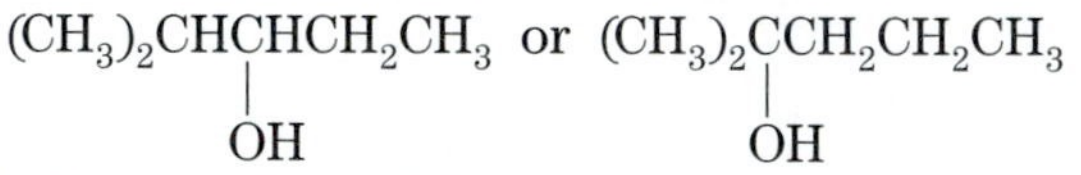 $(CH_3)_2CHCH(OH)CH_2CH_3$ or $(CH_3)_2C(OH)CH_2CH_2CH_3$

4.37 **Carbocations:** What are the shape, bond angles, and hybridization of a carbocation carbon? There is an atomic orbital containing no electrons. What kind of orbital is it?

4.38 **Carbanions:** What are the shape, bond angles, and hybridization of a carbanion carbon? What is the orbital possessing the nonbonding electron pair?

4.39 **Lewis Acid, Lewis Base Reactions:** In section 4.2C an example reaction between $AlCl_3$ and an ether is shown. Determine the hybridization, geometry, and bond angles of the aluminum and oxygen before and after reaction.

4.40 **Reaction Mechanisms:** Isopropyl alcohol reacts with HBr to produce 2-bromopropane by a mechanism similar to that of dehydration presented in section 4.5C. Write this mechanism.

Activities with Molecular Models

1. Make a model of butane. How many different monobromination products are possible? Make a model of each.
2. Using the models you made in exercise 1 of 1-bromo and 2-bromobutane, demonstrate the result of dehydrobromination. To do so, remove the bromine and a hydrogen from an adjacent carbon; insert a double bond between these two carbons. How many isomers are possible from each compound? Which is the more stable in the case where two are possible.
3. Make a model of ethanol and its dehydration product ethene.

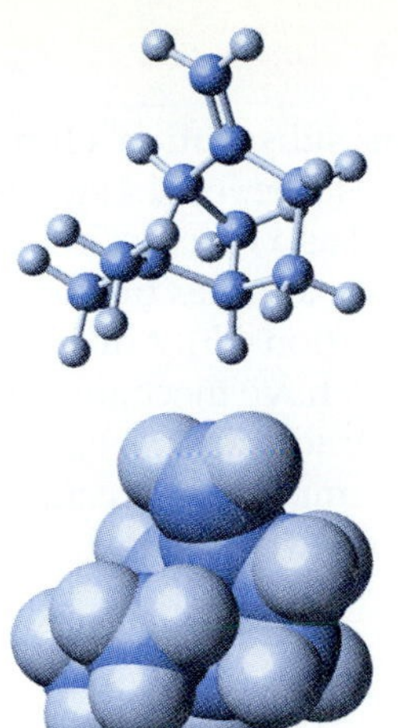

C H A P T E R 5

Reactions of Alkenes and Alkynes

Alkenes and alkynes are very reactive hydrocarbons especially compared to their saturated counterparts, the alkanes. They readily engage in addition reactions. Why are alkenes and alkynes reactive and why is addition the characteristic reaction?

First, carbon-carbon double and triple bonds are composed of π bonds in addition to a σ bond; the double bond has one π bond, the triple bond has two (Figure 5.1). In σ bonds, the only type of bond in alkanes, the shared electron pair is concentrated between the bonded atoms and tightly held. It is not especially accessible to passing reagents. Pi bonds, in contrast, are formed by p-orbitals that overlap above and below the σ bond (section 1.5D–F). Their electron density is like a loosely held, negative cloud projected away from the bonded carbons and is quite susceptible to attack, especially by electrophilic (electron-loving) reagents.

Second, alkenes and alkynes are unsaturated hydrocarbons. Not only are the double and triple bonds attractive to reactive species, but also the two carbons do not have the maximum possible number of bonded atoms or groups. Each carbon in a double bond can accommodate one more atom, and each in a triple bond can accommodate two. Addition reactions can occur. Double bonds can undergo addition once, and triple bonds can under go addition twice, given enough reagent.

Addition to alkenes

$$>C{=}C< + E{-}A \longrightarrow -\overset{|}{\underset{E}{C}}-\overset{|}{\underset{A}{C}}-$$

Addition to alkynes

$$-C{\equiv}C- + 2E{-}A \longrightarrow -\overset{E}{\underset{E}{C}}-\overset{A}{\underset{A}{C}}-$$

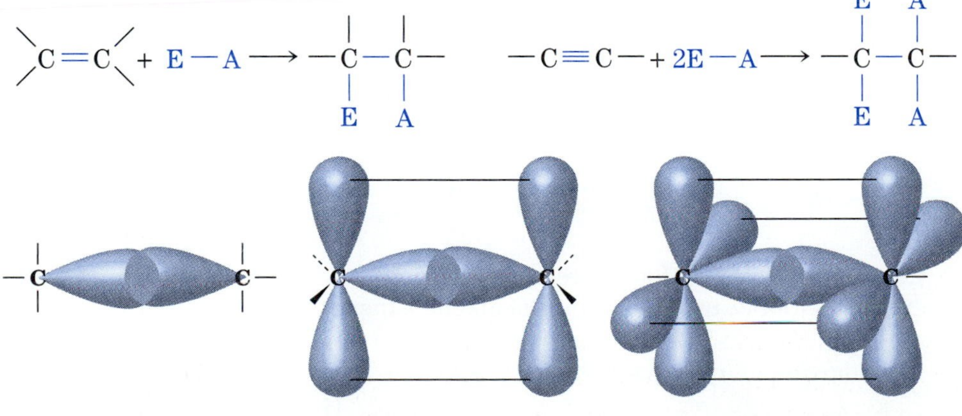

Figure 5.1 Bonding in hydrocarbons.

In Chapter 4, we introduced the three types of reactions: substitution, elimination, and addition (section 4.1A). Halogenation of alkanes (section 4.4) gave us an example of substitution and introduced the free radical as a reactive intermediate. In dehydration and dehydrohalogenation, we saw examples of elimination reactions in the preparation of alkenes and alkynes (section 4.5). Addition reactions are the opposite of elimination and characteristically have mechanisms involving carbocations (we saw carbocations in dehydration); addition reactions are covered in the next section. Later we will investigate an important practical application of this reaction in the formation of addition polymers, which are familiar to us as plastics, synthetic fibers, and resins.

GETTING INVOLVED

✓ Take a couple of minutes here to review the definitions and differences among substitution, elimination, and addition reactions. Also do you remember what carbocations, free radicals, and carbanions are?

Problem 5.1

Addition and elimination reactions are essentially the opposite of each other. To illustrate this: **(a)** write a reaction equation illustrating the dehydrohalogenation of bromoethane with KOH to form ethene and then the addition of HBr to ethene to form bromoethane; **(b)** the dehydration of ethanol, CH_3CH_2OH, to form ethene (sulfuric acid catalyst) and the addition of water to ethene to form ethanol (again, sulfuric acid is the catalyst).

5.1 Addition Reactions of Alkenes

Addition reactions of alkenes will be introduced with a general reaction equation followed by examples. We will then examine the step-by-step reaction mechanism. Finally we will use the mechanism to predict orientation of addition when more than one product is possible.

A. General Reaction Equation for Addition to Alkenes

Carbon forms four bonds; if they are all single bonds, the maximum number of attached atoms is four. Each carbon involved in a double bond has only three attached atoms and potentially can add one atom. This is what occurs in an addition reaction. Reagents add to the carbon-carbon double bond of an alkene to form a saturated compound. In the following equation involving a general reagent EA, E bonds to one carbon, A bonds to the other carbon, and the double bond becomes a single bond; addition occurs.

General Reaction for Addition to Alkenes

$$\mathrm{>C{=}C<} + \mathrm{E{-}A} \longrightarrow -\underset{\mathrm{E}}{\overset{|}{\underset{|}{\mathrm{C}}}}-\underset{\mathrm{A}}{\overset{|}{\underset{|}{\mathrm{C}}}}-$$

Common possibilities for EA are hydrogen, halogen, hydrogen halide, and water; examples follow, using the simplest alkene, ethene.

1. ***Addition of hydrogen halides.*** (E = H, A= X; HX = HCl, HBr, HI)

$$\mathrm{H_2C{=}CH_2} + \mathrm{HCl} \longrightarrow \mathrm{H}-\underset{\mathrm{H}}{\overset{\mathrm{H}}{\mathrm{C}}}-\underset{\mathrm{Cl}}{\overset{\mathrm{H}}{\mathrm{C}}}-\mathrm{H}$$

2. ***Halogenation.*** ($E = X$, $A = X$; $X_2 = Cl_2$ or Br_2, F_2 is too reactive; I_2 is too unreactive.)

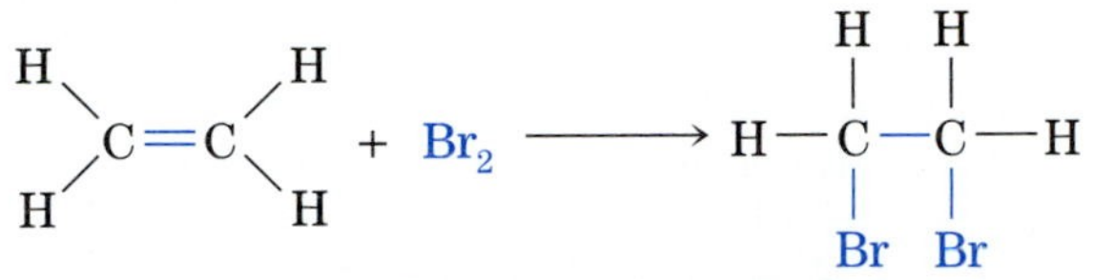

$$\begin{matrix} H & & H \\ & C{=}C & \\ H & & H \end{matrix} + Br_2 \longrightarrow H-\overset{\displaystyle H}{\underset{\displaystyle Br}{C}}-\overset{\displaystyle H}{\underset{\displaystyle Br}{C}}-H$$

halogenation
reaction in which halogen is introduced into a molecule

3. ***Hydration.*** ($E = H$, $A = OH$; H_2SO_4 is the catalyst.)

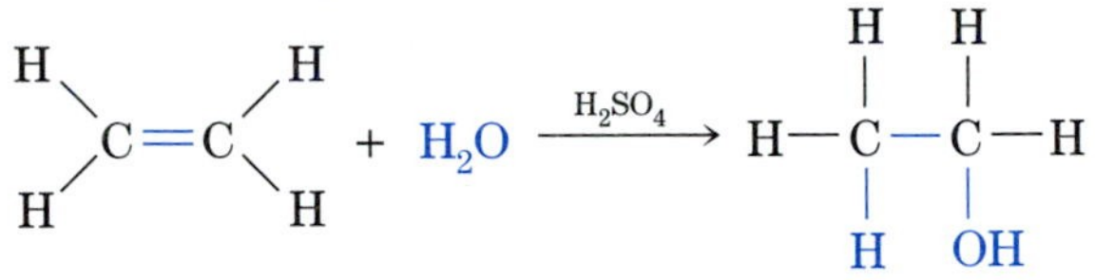

$$\begin{matrix} H & & H \\ & C{=}C & \\ H & & H \end{matrix} + H_2O \xrightarrow{H_2SO_4} H-\overset{\displaystyle H}{\underset{\displaystyle H}{C}}-\overset{\displaystyle H}{\underset{\displaystyle OH}{C}}-H$$

hydration
reaction in which the elements of water (H and OH) are introduced into a molecule

4. ***Hydrogenation.*** ($E = H$, $A = H$; metal catalyst such as nickel, platinum, or palladium; reaction conducted under pressure.)

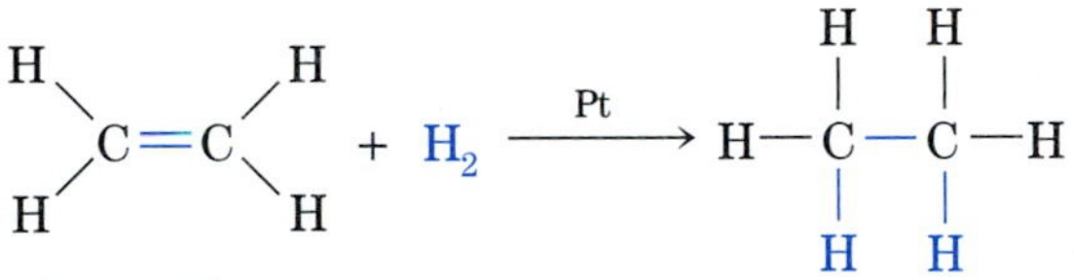

$$\begin{matrix} H & & H \\ & C{=}C & \\ H & & H \end{matrix} + H_2 \xrightarrow{Pt} H-\overset{\displaystyle H}{\underset{\displaystyle H}{C}}-\overset{\displaystyle H}{\underset{\displaystyle H}{C}}-H$$

hydrogenation
reaction in which the elements of hydrogen (H_2) are introduced into a molecule

GETTING INVOLVED

✓ In terms of addition reactions of alkenes, define hydrogenation, halogenation, hydrohalogenation, and hydration. What groups/atoms are adding to the alkene in each case and what happens to the carbon-carbon double bond?

Example 5.1

Write reaction equations for the reactions of propene with hydrogen (Pt catalyst) and bromine, and the reaction of cyclohexene with HCl, and with water (H_2SO_4 catalyst).

Solution

Write the structure of the alkene, focusing your attention on the carbon-carbon double bond; this is where the reaction occurs. Now determine the two parts of the adding reagent (they are emphasized in the examples shown; H and H in the first, H and OH in the last). Place one on each carbon of the double bond and change the double bond to a single bond.

Hydrogenation:

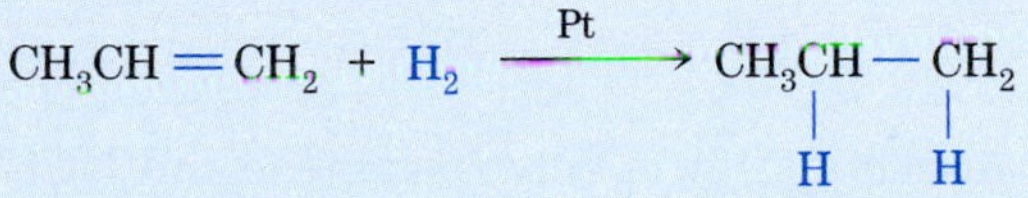

$$CH_3CH{=}CH_2 + H_2 \xrightarrow{Pt} CH_3\underset{\displaystyle H}{CH}-\underset{\displaystyle H}{CH_2}$$

Halogenation:

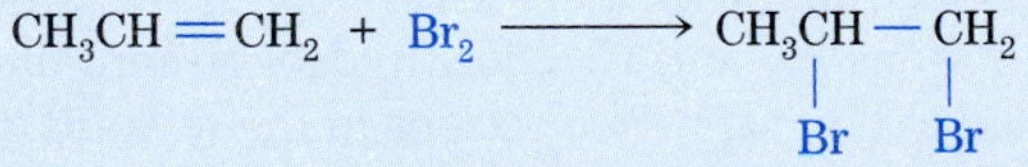

$$CH_3CH{=}CH_2 + Br_2 \longrightarrow CH_3\underset{\displaystyle Br}{CH}-\underset{\displaystyle Br}{CH_2}$$

Hydrohalogenation:

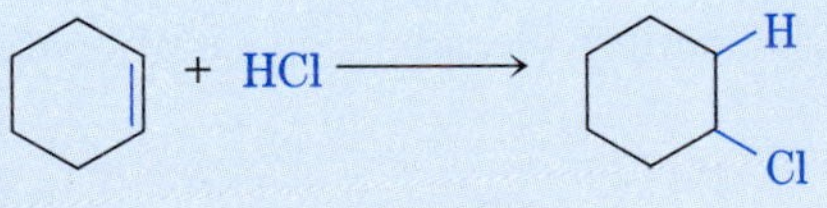

cyclohexene + HCl ⟶ cyclohexane with H and Cl on adjacent carbons

Hydration:

cyclohexene + H_2O $\xrightarrow{H_2SO_4}$ cyclohexane with H and OH on adjacent carbons

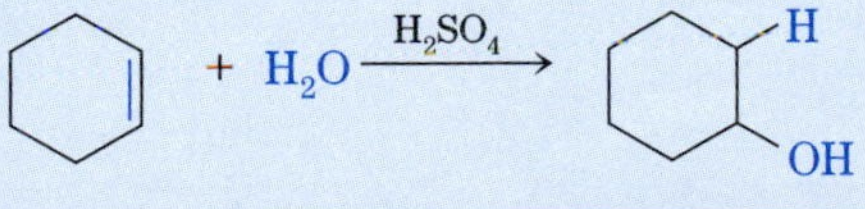

Problem 5.2

Write equations showing the reaction of 2-butene with each of the following reagents: **(a)** H_2/Pt; **(b)** Cl_2; **(c)** HBr; **(d)** H_2O/H_2SO_4.

B. Mechanism of Electrophilic Addition

electrophilic addition
addition reaction initiated by an electron-deficient species (electrophile)

carbocation
species with a carbon that has only three bonds, six outer-shell electrons, and a positive charge

The reaction equations in section 5.1.A show only reactants and products, not mechanism. With the exception of hydrogenation, the addition reactions described here for alkenes occur by a mechanism called **electrophilic addition.** In this mechanism, an electron-deficient species called an *electrophile* (remember that electrophile means "electron-loving") seeks electrons and is attracted to the electron-rich double bond of an alkene. The double bond is an accessible source of electrons because the electrons of the component π bond are in p orbitals that overlap above and below the σ bond (Figure 5.2). The π bond can be imagined as a loosely held cloud of negative charge that attracts the positive electrophile and initiates the reaction (hence the term electrophilic addition). The electrophile uses the electron pair in the pi-bond of the double bond for bonding to one of the carbons. A **carbocation** results on the other carbon, which is quickly neutralized in the second step of the mechanism.

Step 1: $>C{=}C< \; + \; \overset{+}{E}{-}\overset{-}{A} \longrightarrow -\underset{\underset{E}{|}}{\overset{|}{C}}-\underset{+}{\overset{|}{C}}- \; + \; :A^-$

Carbocation

Step 2: $-\underset{\underset{E}{|}}{\overset{|}{C}}-\underset{+}{\overset{|}{C}}- \; + \; :A^- \longrightarrow -\underset{\underset{E}{|}}{\overset{|}{C}}-\underset{\underset{A}{|}}{\overset{|}{C}}-$

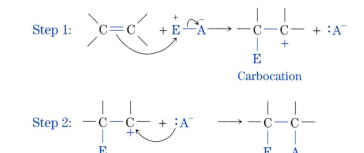

We will examine specific examples of electrophilic addition mechanisms in the following sections.

1. ***Mechanism for Addition of Hydrogen Halides** (HX).* (HX = HCl, HBr, and HI.)

As an example, let us examine the addition of HCl to ethene. First, we should write the reaction equation, showing all bonds in the reaction vicinity.

$$H_2C{=}CH_2 + HCl \longrightarrow H-\underset{\underset{H}{|}}{\overset{\overset{H}{|}}{C}}-\underset{\underset{Cl}{|}}{\overset{\overset{H}{|}}{C}}-H$$

What attraction is there between ethene and HCl? The double bond of ethene is rich in electrons, which are accessible in the π bond. HCl is polar and can be thought of as H^+ and Cl^-. The hydrogen ion, the electrophile, is positive; it is attracted to the negative electrons in the π cloud of the double bond. It bonds to one of the carbons, using the two electrons of the π bond. This leaves the other carbon deficient in electrons, so a carbocation is formed (section 4.1.B–C).

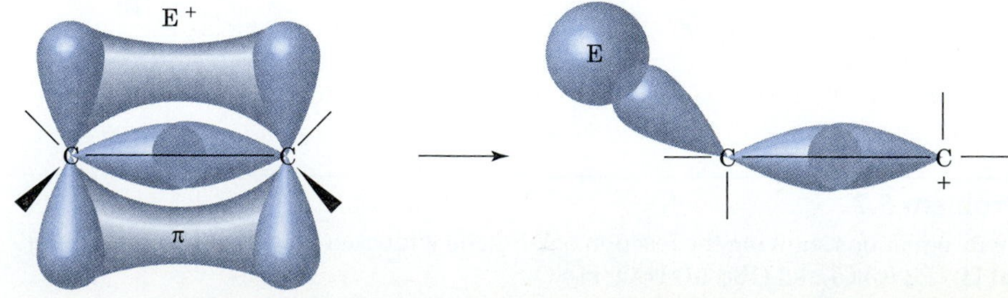

Figure 5.2 Bonding picture of an alkene showing σ and π bonds of the carbon-carbon double bond. The electrophile E^+ becomes embedded in the π cloud and thereby initiates the addition reaction.

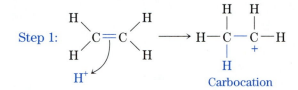

Step 1: $H_2C{=}CH_2 + H^+ \longrightarrow H_3C{-}\overset{+}{C}H_2$

Carbocation

With only six outer-shell electrons, a carbocation is a short-lived, unstable, reactive intermediate whose positive charge is susceptible to immediate neutralization. The negative chloride ion (a nucleophile) is attracted to the positive charge and uses one of its lone pairs of electrons to bond to the positive carbon. This is the second and concluding step of the mechanism.

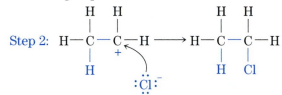

Step 2: $H_3C{-}\overset{+}{C}H_2 + :\ddot{Cl}:^- \longrightarrow H_3C{-}CH_2Cl$

This mechanism for the addition of HCl to ethene can be generalized for the addition of any hydrogen halide to any alkene. In all cases, the mechanism is a two-step one: first, bonding of the electrophile, the hydrogen ion; second, neutralization of the carbocation intermediate by a **nucleophilic** halide ion.

nucleophile
species with electron availability that donates electrons to electrophiles in a chemical reaction. Nucleophiles are Lewis bases

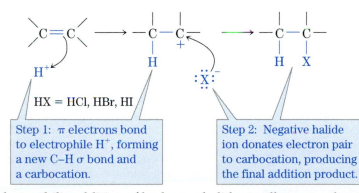

General Mechanism for the Electrophilic Addition of Hydrogen Halides to Alkenes

$>C{=}C< + H^+ \longrightarrow -\underset{H}{C}-\overset{+}{C}- \xrightarrow{:\ddot{X}:^-} -\underset{H}{C}-\underset{X}{C}-$

HX = HCl, HBr, HI

Step 1: π electrons bond to electrophile H^+, forming a new C–H σ bond and a carbocation.

Step 2: Negative halide ion donates electron pair to carbocation, producing the final addition product.

The electrophilic addition of hydrogen halides to alkenes can be effectively described by a potential energy diagram (section 4.1C) as shown in Figure 5.3. In the first transition state, a carbon-hydrogen bond is in the process of forming simultaneously with the cleavage of the carbon-carbon double bond. A high-energy carbocation intermediate results. The carbon-chlorine bond is forming in the second transition state with the final product being the result. Similar energy diagrams describe most electrophilic addition reactions.

GETTING INVOLVED

Let's describe the two steps of electrophile addition of hydrogen halides:

✓ Step 1: What is the electrophile that is attracted to the π cloud? When it bonds, where does it get the needed electrons? What reactive intermediate results?

✓ Step 2: How is the carbocation neutralized?

Example 5.2

Write the electrophilic addition reaction mechanism illustrating the addition of HCl to propene to form 2-chloropropane.

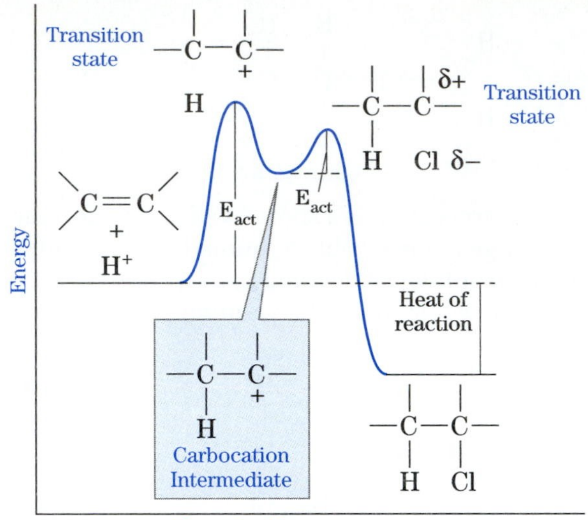

Figure 5.3
Potential energy diagram for electrophilic addition of HCl to alkenes.

Solution

First, be sure that you know what the product of the reaction is. Focus your attention on the double bond; add the hydrogen of HCl to one carbon and the chloride to the other. The double bond becomes a single bond.

$$CH_3CH{=}CH_2 + HCl \longrightarrow \underset{\displaystyle Cl}{CH_3CH}{-}\underset{\displaystyle H}{CH_2}$$

Next show how the reaction occurs—the mechanism. As you write the mechanism, remember that all you have to do is show how the hydrogen became bonded to one carbon and the chlorine to the other. The H^+ is attracted to the negative electrons in the double bond. It acquires two of them to form a bond to the outside carbon. Since the inside carbon was involved in the sharing of these electrons, it becomes positive; a carbocation results. The carbocation is quickly neutralized by the negative chloride ion to give the final product.

$$CH_3CH{=}CH_2 \xrightarrow{H^+} CH_3\overset{+}{C}H{-}\underset{\displaystyle H}{CH_2} \xrightarrow{:\ddot{Cl}:^-} \underset{\displaystyle :\ddot{Cl}:}{CH_3CH}{-}\underset{\displaystyle H}{CH_2}$$

Carbocation

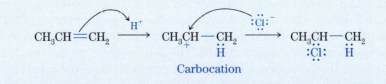

Problem 5.3

Write equations for the following reactions: **(a)** cyclopentene and HCl; **(b)** 1-butene and HBr (Do you notice that two addition products are possible? Write both.).

Problem 5.4

(a) Using Example 5.2 as a guide, write the electrophilic addition mechanism for the addition of HBr to 1-butene to form 2-bromobutane. **(b)** In **(a),** you bonded the hydrogen ion to the external carbon of the double bond. If it bonds to the internal carbon, 1-bromobutane eventually results. Write this mechanism. In section 4.2C, we will learn how to predict which product predominates.

2. ***Mechanism for the Addition of Halogen (X_2).*** ($X_2{=}Cl_2$, Br_2). The reaction mechanism for the halogenation of alkenes is similar to the addition of hydrogen halides. Although Cl_2 and Br_2 are nonpolar, as they approach the π cloud of the double bond, the repulsion between their outer-shell electrons and the π cloud momentarily polarizes the halogen molecule ($X^+ X^-$). As this happens, the positive halogen is instantaneously attracted to the π cloud and bonds to one of the carbons, forming a carbocation on the other, which is neutralized by the negative halide ion.

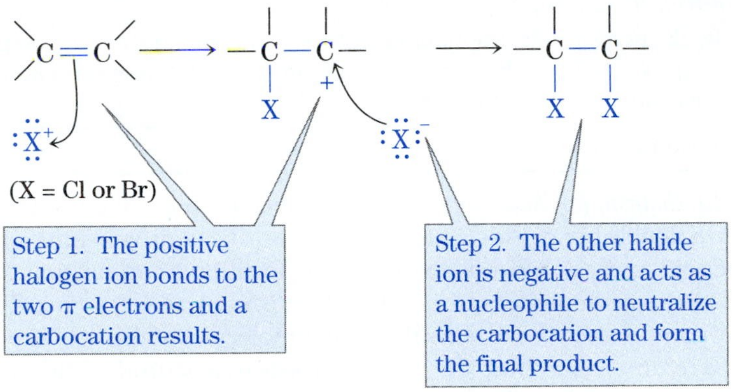

(X = Cl or Br)

Step 1. The positive halogen ion bonds to the two π electrons and a carbocation results.

Step 2. The other halide ion is negative and acts as a nucleophile to neutralize the carbocation and form the final product.

Actually, this mechanism for halogenation is not entirely correct and does not account for the products formed in some reactions. For example, bromination of cyclopentene gives the *trans* product entirely. If a simple carbocation were the intermediate, one would expect some *cis* product, since it would appear that the bromide ion should be able to neutralize the planar carbocation from either above or below.

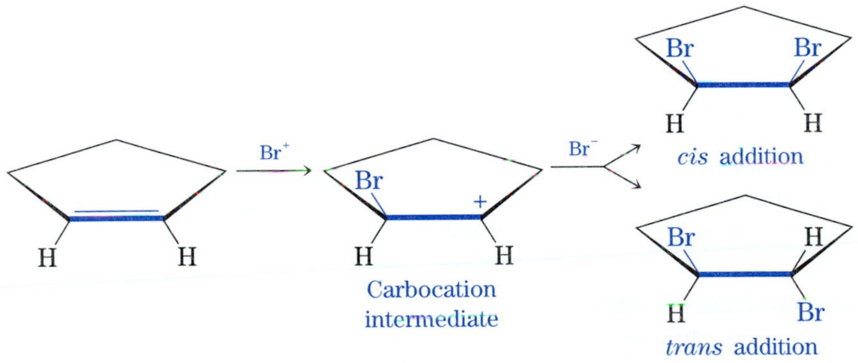

Carbocation intermediate

cis addition

trans addition

The exclusive formation of the *trans* product is explained by the formation of a special type of cation called a *bromonium ion.* In this ion, a lone pair of electrons on the bromine overlaps with the vacant p orbital of the carbocation. The ring is shielded from attack from above, and the bromonium ion is neutralized exclusively from below the ring, giving *trans* 1,2-dibromocyclopentane.

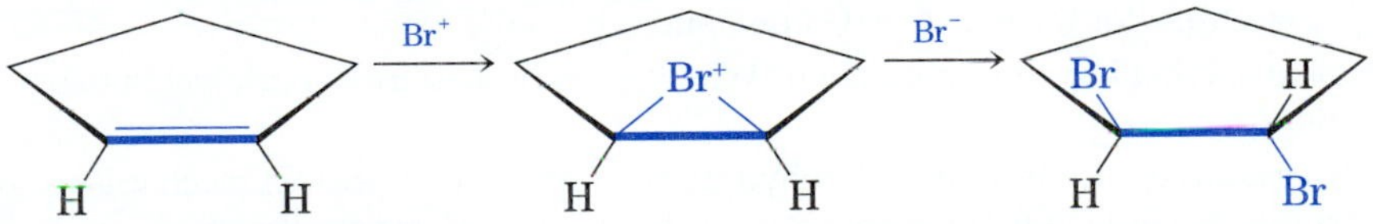

GETTING INVOLVED

✓ Describe the two-step mechanism for bromination of alkenes. Why does a Br^+ form? What is the reactive intermediate generated and how is it neutralized?

✓ What is a bromonium ion and why does it direct the addition of bromine to be *trans?*

Problem 5.5

Write equations for the following halogenation reactions: **(a)** propene and Cl_2; **(b)** 1,2-dimethylcyclohexene and Br_2 (be sure to show the *trans* addition stereochemistry in this example).

Problem 5.6

(a) Write the two-step electrophilic addition mechanism for the reaction propene and chlorine (problem 5.5a). **(b)** Draw the structure of the bromonium ion that forms in the mechanism of the reaction in problem 5.5b.

See related problem 5.29.

3. *Mechanism for Addition of Water.* In hydration of alkenes, the addition of water to a double bond, the electrophile is a positive hydrogen ion provided by the sulfuric acid catalyst. It bonds to the alkene, forming a carbocation. Since there is no OH^- in an acid medium, the carbocation is neutralized by bonding to one of the lone pairs of electrons on a neutral water molecule, which acts as a nucleophile. Loss of a hydrogen ion from the oxygen produces a neutral addition product and regenerates the catalyst. The mechanism is exactly the reverse of the mechanism for dehydration of alcohols to form alkenes (section 4.5.A and C).

General Reaction Mechanism for the Hydration of Alkenes

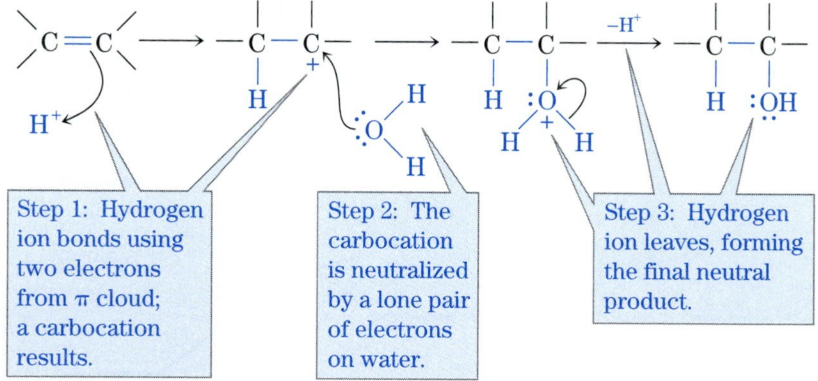

$-H^+$
Step 1: Hydrogen ion bonds using two electrons from π cloud; a carbocation results.
Step 2: The carbocation is neutralized by a lone pair of electrons on water.
Step 3: Hydrogen ion leaves, forming the final neutral product.

GETTING INVOLVED

Let's take a close look at the hydration mechanism.

- ✓ Step 1: Hydrogen ion is the electrophile, what is its source? What reactive intermediate is formed when it bonds to the alkene? What is the source of electrons for the new C—H bond?
- ✓ Step 2: Why is water and not hydroxide ion the species that neutralizes the carbocation? What role does the Lewis base character of water play? What is the source of electrons for the new C—O bond and why is the oxygen positive?
- ✓ Step 3: What happens in this step? Why is sulfuric acid truly a catalyst in this reaction?
- ✓ Compare the hydration and dehydration (section 4.5A and C) mechanisms and convince yourself that they are the exact opposites of one another.

Problem 5.7

(a) Write the equation for the hydration of 2-butene with water and sulfuric acid. **(b)** Now write the three-step mechanism for this reaction.

4. *Summary of Electrophilic Addition.* Electrophilic addition basically involves two steps. In step one, an electrophile (H^+ or X^+ in the cases we examined) attacks the π bond of a double bond, extracts the π electrons, and uses them to form a bond to one carbon; a carbocation results, with the positive charge on the other carbon. In the second step, the carbocation is neutralized by the nucleophilic species of the adding reagent.

General Reaction Mechanism for the Electrophilic Addition

$$>C{=}C< \;\xrightarrow{E^+}\; -\overset{|}{\underset{E}{\underset{|}{C}}}-\overset{|}{\overset{+}{C}}- \;\xrightarrow{:A^-}\; -\overset{|}{\underset{E}{\underset{|}{C}}}-\overset{|}{\underset{A}{\underset{|}{C}}}-$$

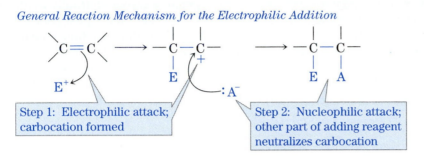

Step 1: Electrophilic attack; carbocation formed

Step 2: Nucleophilic attack; other part of adding reagent neutralizes carbocation

A different mechanism occurs in the hydrogenation of alkenes. It is discussed in section 5.2.B.

C. Orientation of Addition

When a symmetrical reagent adds to a symmetrical alkene, only one positional isomer is possible.

$$\underset{\text{Symmetrical}}{CH_3CH{=}CHCH_3} + \underset{\text{Symmetrical}}{Br_2} \longrightarrow CH_3\underset{Br}{\underset{|}{C}}H-\underset{Br}{\underset{|}{C}}HCH_3$$

A symmetrical reagent E—A is one in which E and A are identical, such as the bromine molecule. Hydrogen bromide has atoms that are different and is therefore unsymmetrical. An alkene is symmetrical if bisection of the double bond gives two identical halves (as in 2-butene) and is unsymmetrical if different halves result (1-butene). If either the alkene or the added reagent is symmetrical, only one positional isomer is possible from the addition.

$$\underset{\text{Symmetrical}}{CH_3CH{=}CHCH_3} + \underset{\text{Unsymmetrical}}{HBr} \longrightarrow CH_3CH_2\underset{Br}{\underset{|}{C}}HCH_3$$

$$\underset{\text{Unsymmetrical}}{CH_3CH_2CH{=}CH_2} + \underset{\text{Symmetrical}}{Br_2} \longrightarrow CH_3CH_2\underset{Br}{\underset{|}{C}}H-\underset{Br}{\underset{|}{C}}H_2$$

If both the alkene and the added reagent are unsymmetrical, however, two positional isomers are possible. They are usually formed in unequal amounts.

$$\underset{\text{Unsymmetrical}}{CH_3CH_2CH{=}CH_2} + \underset{\text{Unsymmetrical}}{HBr} \longrightarrow \underset{\text{Major}}{CH_3CH_2\underset{Br}{\underset{|}{C}}HCH_3} + \underset{\text{Minor}}{CH_3CH_2CH_2CH_2Br}$$

The proportion of each isomer depends on the relative stabilities of the intermediate carbocations formed in the reaction mechanism. To be able to predict which will dominate, one must understand carbocation stability.

carbocation stability
order of stability is 3° > 2° > 1°

R-group
R is a generic symbol for an alkyl group

tertiary atom
atom with three directly attached carbons (alkyl groups)

secondary atom
atom with two directly attached carbons (alkyl groups)

primary atom
atom with one directly attached carbon (alkyl group)

1. ***Carbocation Stability.*** Structurally, a carbocation is a carbon with three bonds, six outer-shell electrons (in the three bonds), and a positive charge. The simplest is the methyl carbocation, CH_3^+. Groups directly attached to the positive carbon that can partially neutralize or disperse the positive charge stabilize the carbocation. Alkyl groups (**R groups:** methyl, ethyl, propyl, and the like) are electron-releasing groups and stabilize carbocations. The electron density of the adjacent σ bonds "spreads over" to the positive carbon, partially neutralizing the charge. The more alkyl groups directly attached to the positive carbon, the more stable the carbocation. A **tertiary** (3°) carbocation has three attached alkyl groups and is more stable than a **secondary** (2°) carbocation with two or a **primary** (1°) carbocation with only one. To determine the number of directly attached alkyl groups, count the number of carbons directly bonded to the positive carbon. The order of carbocation stability follows:

**Carbocation Stability:*

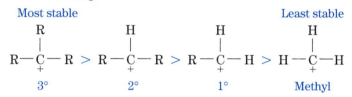

Most stable → Least stable

$$\underset{3°}{R_3C^+} > \underset{2°}{R_2CH^+} > \underset{1°}{RCH_2^+} > \underset{\text{Methyl}}{CH_3^+}$$

GETTING INVOLVED

- ✓ What is a symmetrical and unsymmetrical alkene; a symmetrical and unsymmetrical adding reagent?
- ✓ Describe the carbocation in terms of number of bonds, number of outer shell electrons, and charge. Why is a carbocation charged?
- ✓ What is the difference among primary, secondary, and tertiary carbocations?
- ✓ Describe and explain the order of stability of carbocations.

Problem 5.8

Using $C_4H_9^+$, draw a primary, secondary, and tertiary carbocation and arrange them in order of stability.

2. *Predicting Addition Products.* The reaction of 1-butene (an unsymmetrical alkene) with hydrogen bromide (an unsymmetrical reagent) gives two positional isomers as addition products, 1-bromobutane and 2-bromobutane. To predict which predominates, we write a reaction mechanism (electrophilic addition, section 5.1.B) leading to each isomer and determine which carbocation is more stable.

$$CH_3CH_2CH{=}CH_2 \xrightarrow{HBr}$$

$$\xrightarrow{H^+} CH_3CH_2\overset{+}{C}HCH_3 \text{ (More stable, 2° carbocation)} \xrightarrow{Br^-} CH_3CH_2CH(Br)CH_3 \text{ (Major product)}$$

$$\xrightarrow{H^+} CH_3CH_2CH_2\overset{+}{C}H_2 \text{ (Least stable, 1° carbocation)} \xrightarrow{Br^-} CH_3CH_2CH_2CH_2Br \text{ (Minor product)}$$

If the hydrogen ion from hydrogen bromide bonds to the first carbon, the second carbon becomes positive (the two π electrons are pulled away from it). Since

*R stands for an alkyl group (methyl, ethyl, propyl, and so on).

there are two attached alkyl groups, this is a secondary carbocation. Neutralization by bromide ion forms 2-bromobutane. Because this secondary carbocation is more stable than the primary carbocation formed if the hydrogen ion bonds to the second carbon (leading to 1-bromobutane upon neutralization by Br^-), it is formed more readily, and 2-bromobutane is the predominant product.

Reactions such as these in which one product predominates are termed **regioselective.** If one product is formed exclusively, the reaction is **regiospecific.**

Pioneering work in this field of organic chemistry was performed by the Russian chemist Vladimir Markovnikov. Appropriately, the rule for predicting orientation of addition is known as Markovnikov's rule, which can be stated in modern terms as follows: *When an unsymmetrical reagent adds to an unsymmetrical alkene, the positive portion of the reagent adds to the carbon that results in the formation of the more stable carbocation.* **Markovnikov's rule** actually referred to the addition of a reagent such as HX to a carbon-carbon double bond and stated that the H bonds to the carbon with the most hydrogens (or fewest alkyl groups). We can identify the most stable carbocation in this same way.

regioselective
in addition reactions, a reaction in which one of two possible positional isomers predominates

regiospecific
in addition reactions, a reaction in which one of two possible positional isomers is formed exclusively

Markovnikov's rule
rule for predicting orientation of addition of unsymmetrical reagents to unsymmetrical alkenes

GETTING INVOLVED

✓ How many products are possible if an unsymmetrical reagent adds to an unsymmetrical alkene? How is the predominant product predicted considering stability of the intermediate carbocations?

Example 5.3

Using the mechanism of electrophilic addition, predict the predominant product of the reaction between methylpropene and water (H_2SO_4 as catalyst).

Solution

Hydrogen ion from H_2SO_4 adds first, bonding to one of the carbons involved in the double bond. Addition to carbon-1 produces the more stable 3° carbocation, and the resulting product predominates.

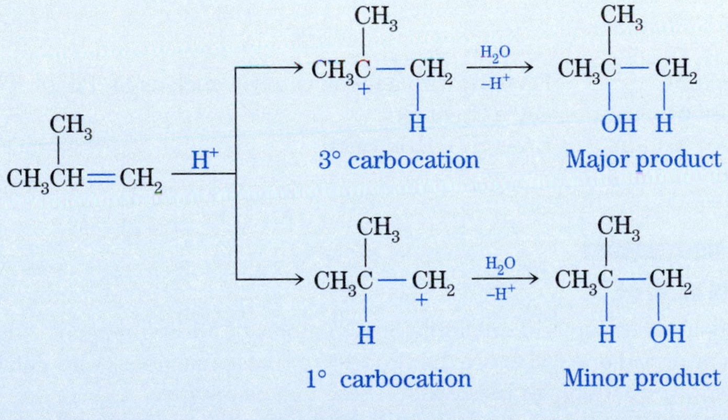

$$CH_3\underset{|}{\overset{CH_3}{C}}H{=}CH_2 \xrightarrow{H^+} CH_3\overset{CH_3}{\underset{+}{C}}-\underset{H}{CH_2} \xrightarrow[-H^+]{H_2O} CH_3\overset{CH_3}{\underset{OH}{C}}-\underset{H}{CH_2}$$

3° carbocation — Major product

$$CH_3\underset{|}{\overset{CH_3}{C}}H{=}CH_2 \xrightarrow{H^+} CH_3\overset{CH_3}{\underset{H}{C}}-\overset{+}{CH_2} \xrightarrow[-H^+]{H_2O} CH_3\overset{CH_3}{\underset{H}{C}}-\underset{OH}{CH_2}$$

1° carbocation — Minor product

Problem 5.9

Predict the major product of addition in the following reactions by writing the two possible intermediate carbocations formed in the mechanism of electrophilic addition and determining which is more stable.

(a) $CH_3CH_2\overset{CH_3}{\overset{|}{C}}{=}CH_2 + HCl \longrightarrow$ (b) 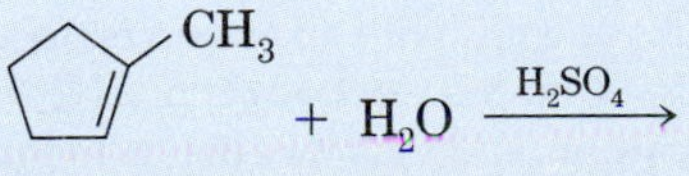 1-methylcyclopentene (CH_3 on ring double-bond carbon) $+ H_2O \xrightarrow{H_2SO_4}$

See related problems 5.25, 5.27, 5.29.

5.2 Addition Reactions of Alkynes

A. General Reaction Equation for Addition to Alkynes

Alkynes have structural similarities to alkenes, and we would predict that they might have similar chemical properties. Like alkenes, they are unsaturated and are capable of addition reactions. Since a triple bond is composed of a σ and two π bonds, we would expect it to be very attractive to electrophiles. The characteristic reaction of alkynes is the same as that of alkenes, electrophilic addition.

Unlike alkenes, however, alkynes can add either one or two equivalents of reagent, converting the carbon-carbon triple bond to a double bond or single bond. Hydrogen, halogen, and hydrogen halide readily add to alkynes. For example, one mole of hydrogen chloride adds to one mole of acetylene to produce vinyl chloride (from which PVC is made), or two moles add to form 1,1-dichloroethane.

$$\mathrm{HC{\equiv}CH} \xrightarrow{\mathrm{HCl}} \underset{\mathrm{Cl}\quad\ \mathrm{H}}{\mathrm{HC{=}CH}} \xrightarrow{\mathrm{HCl}} \overset{\mathrm{Cl}\quad\ \mathrm{H}}{\underset{\mathrm{Cl}\quad\ \mathrm{H}}{\mathrm{HC{-}CH}}}$$

Addition reactions of alkynes can be generalized as follows:

General Equations for Addition Reactions of Alkynes

$$-\mathrm{C{\equiv}C}- + 1\mathrm{E{-}A} \longrightarrow \underset{\mathrm{E}\qquad\quad \mathrm{A}}{\mathrm{C{=}C}}$$

$$-\mathrm{C{\equiv}C}- + 2\mathrm{E{-}A} \longrightarrow \overset{\mathrm{E}\quad \mathrm{A}}{\underset{\mathrm{E}\quad \mathrm{A}}{-\mathrm{C{-}C}-}}$$

Hydrogenation: EA is H_2 with a metal catalyst such as Pt, Pd, or Ni
Halogenation: EA is Cl_2 or Br_2
Hydrohalogenation: EA is HCl, HBr, or HI

GETTING INVOLVED

✓ Explain why alkynes can add one mole or two moles of adding reagent, whereas alkenes can only add one. What are the products of these reactions? Can you write general reaction equations for the addition reactions of alkynes?

Example 5.4

Write equations showing the reactions of propyne with two moles H_2 (Pd catalyst), one mole Br_2, and two moles HCl.

Solution

As you did with alkenes, first focus on the carbon-carbon triple bond of the alkyne; this is where the reaction occurs. Now determine what the reagent is; one atom bonds to one carbon, and the other bonds to the other carbon. If one mole of reagent is used, the triple bond is converted to a double bond, and if two moles are used, it becomes a single bond.

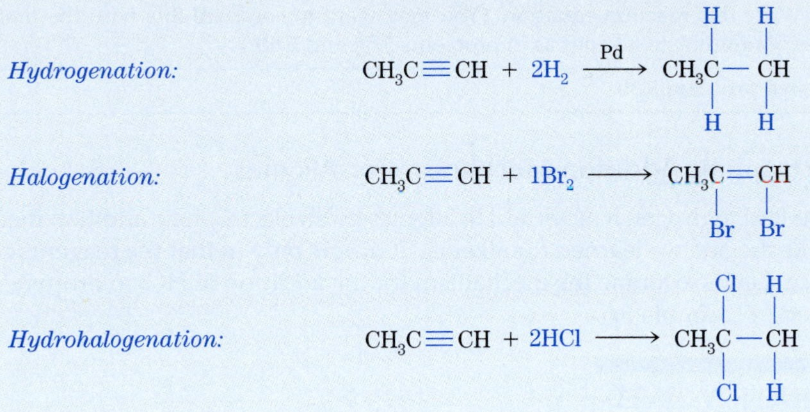

Hydrogenation: $CH_3C{\equiv}CH + 2H_2 \xrightarrow{Pd} CH_3CH_2-CH_3$ (product drawn as $CH_3C(H)(H)-CH(H)(H)$)

Halogenation: $CH_3C{\equiv}CH + 1Br_2 \longrightarrow CH_3C(Br){=}CH(Br)$

Hydrohalogenation: $CH_3C{\equiv}CH + 2HCl \longrightarrow CH_3CCl_2-CH_3$ (product drawn as $CH_3C(Cl)(Cl)-CH(H)(H)$)

Problem 5.10

Write equations showing the reaction of 2-butyne with each of the following: **(a)** $1Br_2$; **(b)** $2Br_2$; **(c)** $1Cl_2$; **(d)** $2H_2/Ni$.

See related problem 5.26.

B. Mechanism of Catalytic Hydrogenation of Alkenes and Alkynes

In hydrogenation, an alkene or alkyne is combined wit hydrogen gas in the presence of a metal catalyst such as nickel, palladium or platinum. In this heterogeneous (involving more than one phase) system, the alkene or alkyne and the hydrogen are adsorbed on the metallic surface. Both the multiple bond and the hydrogen-hydrogen bond are weakened as a result of this adsorption and the hydrogens add to the multiple bond from the catalyst surface. Because both reactants are on the catalyst surface, the hydrogens add to the same side of the multiple bond: *cis*-addition. This is illustrated below for hydrogenation of an alkyne; in this case a *cis* alkene necessarily results.

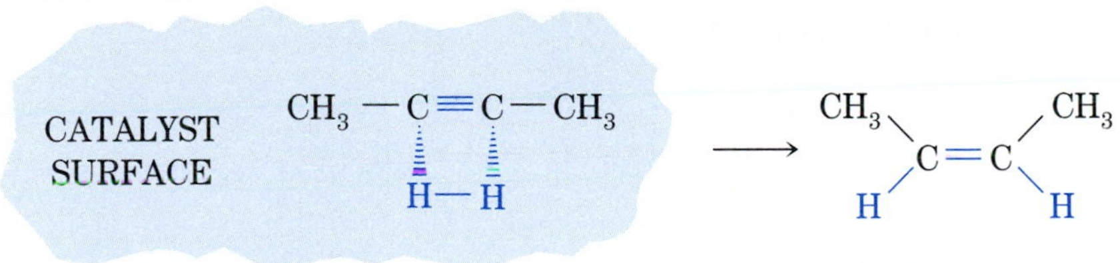

CATALYST SURFACE: $CH_3-C{\equiv}C-CH_3$ with $H-H$ associated beneath $\longrightarrow$ *cis*-$CH_3(H)C{=}C(H)CH_3$ (both CH_3 on the same side, both H on the same side)

GETTING INVOLVED

Problem 5.11

Write equations showing the reaction of 2-pentyne with hydrogen in the presence of a metal catalyst. Write one in which one mole of hydrogen adds and another showing the addition of two moles. In the first, be sure to show which geometric isomer is formed from the addition.

Problem 5.12

Hydrogen adds *cis* to alkenes also though in most alkenes this would not be obvious in the product. For example, write the product of hydrogen addition to cyclohexene. However, if we hydrogenate 1,2-dimethylcyclohexene, we can see that *cis* addition occurs. Write this reaction equation. (You may want to compare this with the *trans* addition of bromine to alkenes as in problems 5.5b and 5.6b.)

See related problem 5.30.

C. Electrophilic Addition Mechanism for Alkynes

Halogens and hydrogen halides add to alkynes by an electrophilic addition mechanism like the one we learned for alkenes. It differs only in that the reagents can add twice. Let us examine the mechanism for the addition of HCl to propyne as illustrated in Example 5.5.

GETTING INVOLVED

Example 5.5

Write a step-by-step reaction mechanism illustrating the addition of two moles of HCl to propyne by an electrophilic addition mechanism.

Solution

First, let us write the reaction equation. The first mole of HCl will add to the triple bond of propyne to form a double bond. Then the second mole will add to give the final saturated product.

$$CH_3C{\equiv}CH \xrightarrow{HCl} CH_3\underset{\substack{|\\Cl}}{C}{=}\underset{\substack{|\\H}}{CH} \xrightarrow{HCl} CH_3\overset{\substack{Cl\\|}}{\underset{\substack{|\\Cl}}{C}}{-}\overset{\substack{H\\|}}{\underset{\substack{|\\H}}{CH}}$$

Basically, we need to write the same electrophilic addition mechanisms we have been writing, but we must do it twice. The reaction is initiated by a hydrogen ion, which bonds to the outer carbon of propyne to form the more stable secondary carbocation (bonding to the interior carbon would have formed a primary carbocation on carbon-1), which is neutralized by a chloride ion. A hydrogen ion now adds to the double bond, forming the more stable carbocation on the interior carbon. As before, this carbocation is neutralized by a chloride ion.

$$CH_3C{\equiv}CH \xrightarrow{H^+} CH_3\overset{+}{C}{=}\overset{\substack{H\\|}}{CH} \xrightarrow{:\ddot{Cl}:^-} CH_3\overset{\substack{:\ddot{Cl}:\\|}}{C}{=}\overset{\substack{H\\|}}{CH}$$

$$CH_3\overset{\substack{Cl\\|}}{C}{=}\overset{\substack{H\\|}}{CH} \xrightarrow{H^+} CH_3\overset{\substack{Cl\\|}}{\underset{+}{C}}{-}\overset{\substack{H\\|}}{\underset{\substack{|\\H}}{CH}} \xrightarrow{:\ddot{Cl}:^-} CH_3\overset{\substack{Cl\\|}}{\underset{\substack{|\\:\ddot{Cl}:}}{C}}{-}\overset{\substack{H\\|}}{\underset{\substack{|\\H}}{CH}}$$

Problem 5.13

Write a reaction equation showing the reaction of 1-pentyne with two moles of HBr. Using Example 5.5 as a guide, write a reaction mechanism illustrating this reaction.

See related problems 5.26 and 5.28.

D. Addition of Water to Alkynes

Water adds to alkynes but not in the way we have described for other addition reactions of alkynes. The difference lies in the formation of an intermediate enol (a compound with both a double bond and an alcohol function) from the addition of one mole of reagent. Enols are unstable and rearrange to aldehydes and ketones. This can be illustrated by adding water to acetylene in the presence of sulfuric acid and mercury (II) sulfate as catalysts.

$$\mathrm{HC{\equiv}CH} + \mathrm{H_2O} \xrightarrow[\mathrm{HgSO_4}]{\mathrm{H_2SO_4,}} \underset{\text{Enol}}{\mathrm{HC(H){=}CH(OH)}} \longrightarrow \underset{\text{Aldehyde}}{\mathrm{H_2HC{-}CH{=}O}}$$

GETTING INVOLVED

Problem 5.14

When water adds to an unsymmetrical alkyne, the hydrogen ion adds first to form the more stable carbocation which is then neutralized by water. Rearrangement of the resulting enol then occurs. Write the equation for the addition of water to propyne showing the predominant product.

See related problem 5.32.

5.3 Addition Polymers

A **polymer** is a giant molecule composed of a repeating structural unit called a **monomer** (*poly* means many, *mono* means one, and *mer* means unit). Monomeric units are repeated hundreds, even thousands, of times in a polymer molecule.

polymer
a giant molecule composed of a repeating structural unit

monomer
compound(s) from which a polymer is made

Polymers are part of our daily lives. Examples include Teflon®, PVC, Dacron®, nylon, polyethylene, styrofoam, and rubber. Polymers are used for toys, dishes, clothes, credit cards, computer disks, artificial organs, furniture coverings, machine parts, combs and brushes, shower curtains, and garden hose, and many other products.

Polymers resulting from the addition of alkene molecules to each other are called **addition polymers** and are the major products of the gigantic plastics industry. Some typical addition polymers and their uses are shown in Table 5.1. A general reaction equation showing the formation of an addition polymer from an alkene monomer follows. As the alkene molecules add over and over to each other, a long polymer chain develops and the double bonds are converted to single bonds by the addition reaction, using mechanisms we have already considered.

addition polymer
polymer that results from polymerization of alkenes

$$n\ \underset{\text{Monomer}}{\mathrm{>C{=}C<}} \xrightarrow{\text{Catalyst}} \underset{\text{Polymer}}{\left(\mathrm{-\overset{|}{\underset{|}{C}}-\overset{|}{\underset{|}{C}}-}\right)_n}$$

GETTING INVOLVED

✓ What is the difference between a monomer and a polymer? What specifically is an addition polymer?

Table 5.1 Commercial Applications of Addition Polymers

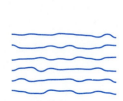

Polymer	Reaction	Uses
Polyethylene	$CH_2{=}CH_2 \longrightarrow -(CH_2-CH_2)_n-$ High Density Low Density	*High density:* composed of long, unbranched, closely packed chains; strong plastic used for large drums, pipes, conduits, tanks, crates, and baby bottles. *Low density:* composed of highly branched, loosely packed polymer chains; lower-melting, softer plastic used for packaging of foods, garments, and dry cleaning, and for other uses such as garbage bags and disposable diaper liners.
Polypropylene	$CH_2{=}CH(CH_3) \longrightarrow -(CH_2-CH(CH_3))_n-$	Filaments and fibers for indoor-outdoor carpeting; used in appliances, luggage, packing crates, and car and truck parts.
Polystyrene	$CH_2{=}CH(C_6H_5) \longrightarrow -(CH_2-CH(C_6H_5))_n-$	*Foam:* Styrofoam coolers, disposable drinking cups, and protective packaging; good insulating properties. *Hard solid:* children's toys, plastic picnic eating utensils, lighting fixtures, wallcoverings, and plastic furniture.
Polymethyl methacrylate	$CH_2{=}C(CH_3)(CO_2CH_3) \longrightarrow -(CH_2-C(CH_3)(CO_2CH_3))_n-$	Also called Lucite or Plexiglas; safety glass (Lucite plus glass) for automobile windshields, plastic coating in stucco and some paints, advertising signs and displays. This polymer and similar derivatives are used in hard and soft contact lenses.
Orlon	$CH_2{=}CH(CN) \longrightarrow -(CH_2-CH(CN))_n-$	Wearing apparel, blankets, and carpeting.
Polyvinyl chloride	$CH_2{=}CH(Cl) \longrightarrow -(CH_2-CH(Cl))_n-$	PVC; used as a rubber substitute in raincoats, shower curtains, garden hose, baby pants, swimming pool liners, weatherstripping, outdoor siding, and gutters. Electrical conduit and pipe for plumbing. Also plastic bottles, phonograph records, credit cards, toys, auto mats, and upholstery.
Polyvinylidene chloride	$CH_2{=}CCl_2 \longrightarrow -(CH_2-CCl_2)_n-$	Saran; packaging film for foods.
Teflon	$CF_2{=}CF_2 \longrightarrow -(CF_2-CF_2)_n-$	Nonstick surface coating in cooking utensils, valves, and gaskets.

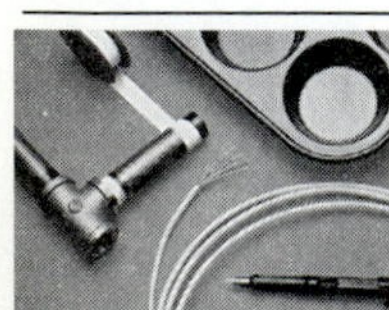

Teflon

Contact lens

Plastic bottles

Connections 5.1 www.prenhall.com/bailey

Serendipity in the Discovery of Polymers

Serendipity has played a large role in discoveries that enabled the advancements in science and technology affecting life today. In each case an inquiring and imaginative mind rescued an accidental discovery from oblivion and developed it into something of value. The following anecdotes and countless other stories of serendipity, luck, determination, and persistence show the roots of today's gigantic polymer industry.

One day in 1846, while performing some experiments in the kitchen of his home, Christian Schoenbein, a professor of chemistry at the University of Basel in Switzerland, cleaned up a spill of nitric and sulfuric acids with his wife's cotton apron. He rinsed the apron and hung it in front of the hot stove to dry. To his amazement, the apron flash-burned and disappeared. Nitrocellulose had been discovered, and in 1869 it was converted by John Wesley Hyatt, a New York printer, into the first plastic, celluloid (movie film used to be made from celluloid—it was quite flammable and dangerous).

Just a few years earlier, Charles Goodyear had spilled a mixture of latex rubber (a natural polymer) and sulfur on his hot stove. During the aggravation of cleaning the material off the stove, he found that the properties of the rubber had greatly improved; he had accidentally come upon the process of vulcanization.

In 1907 a chemist named Leo Baekeland was experimenting with a chemical reaction that tended to foul his glassware so badly it had to be discarded. Unlike his predecessors, he investigated the material rather than throwing it away in disgust and as a result discovered the first truly synthetic polymer (plastic), Bakelite (among its uses, light switches). Earlier in his career, Baekeland invented Velox, the first photographic paper that could be exposed with artificial light. George Eastman (inventor of the Kodak camera) offered Baekeland one million dollars for his Velox patent just moments before Baekeland was preparing to take an initial bargaining stance of $25,000.

In the late 1920s Arnold Collins, a member of Wallace Carothers' group that developed nylon for DuPont, turned off a distillation apparatus late one Friday and left for the weekend. When he returned on Monday morning, he found that the distilled material had solidified. Almost as a reflex to its appearance, he threw the material against a bench; it bounced in a lively fashion. Sales of neoprene rubber (gas pump hose) began in 1932.

One morning in 1938, Roy Plunkett, also a DuPont chemist, found the pressure gauge on a cylinder recently filled with tetrafluoroethylene gas reading zero. Since the seals were tight, he weighed the cylinder and found that it weighed the same as it had just after he had filled it. Rather than simply filling another cylinder and going on with his experiments, Plunkett opted to investigate the zero pressure reading by cutting the cylinder open. A waxy white solid fell out; Teflon was born.

A. Cationic Polymerization by Electrophilic Addition

Cationic polymerization is a form of electrophilic addition. Let us consider the acid-initiated polymerization of isobutylene to form the adhesive polyisobutylene.

cationic polymerization addition polymerization of alkenes initiated by an electrophile

$$nCH_2{=}C(CH_3)_2 \xrightarrow[\text{catalyst}]{H^+} \sim\!\left(CH_2{-}C(CH_3)_2\right)_n\!\sim$$

Isobutylene Polyisobutylene

For simplicity, we will use a generic acid, HA, present only in trace amounts. In the initiation step, hydrogen ion attacks the electron-rich π cloud of the double bond and bonds to one of the carbons, forming the more stable tertiary carbocation.

Initiation:

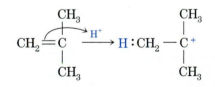

$$CH_2{=}C(CH_3)_2 \xrightarrow{H^+} H{:}CH_2{-}\overset{+}{C}(CH_3)_2$$

Since there is only a trace amount of acid relative to the abundant isobutylene, the carbocation is unlikely to encounter an A^- and be neutralized. Rather, it is attracted to the electron-rich double bond of another isobutylene molecule. The carbocation extracts the π electrons for bonding and forms a new carbocation; two isobutylene molecules are now connected. This new carbocation can continue the process with the formation of new carbocations many times; the reaction is propagated and the polymer grows.

Propagation:

$$H{:}CH_2-\overset{CH_3}{\underset{CH_3}{C^+}}\ \ CH_2{=}\overset{CH_3}{\underset{CH_3}{C}} \longrightarrow H{:}CH_2-\overset{CH_3}{\underset{CH_3}{C}}{:}CH_2-\overset{CH_3}{\underset{CH_3}{C^+}}$$

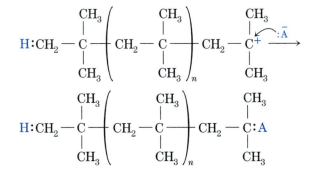

Eventually the polymerization is terminated by a reaction that does not produce a carbocation, for example, the capturing of the anion and the resulting neutralization of the carbocation.

Termination:

$$H{:}CH_2-\overset{CH_3}{\underset{CH_3}{C}}\left(CH_2-\overset{CH_3}{\underset{CH_3}{C}}\right)_n CH_2-\overset{CH_3}{\underset{CH_3}{C^+}}\ \ {:}\bar{A} \longrightarrow$$

$$H{:}CH_2-\overset{CH_3}{\underset{CH_3}{C}}\left(CH_2-\overset{CH_3}{\underset{CH_3}{C}}\right)_n CH_2-\overset{CH_3}{\underset{CH_3}{C}}{:}A$$

The net result is an addition polymer in which the double bonds of the alkene essentially are added to each other over and over again, producing long polymer chains.

GETTING INVOLVED

Problem 5.15

(a) Write the formula of the polymer formed by addition polymerization of propene.
(b) Now write the cationic polymerization mechanism for this process.
See related problem 5.36.

free-radical polymerization addition polymerization of alkenes initiated by a free radical

B. Polymerization by a Free-Radical Chain Reaction

Consider the polymerization of the monomer ethylene to form polyethylene. The reaction can be induced by combining a small amount of peroxide (ROOR) with a large volume of ethylene. When peroxides are heated, they decompose readily to form highly reactive free radicals.

Peroxide Decomposition:

$$\underset{\text{A peroxide}}{R\ddot{\underset{..}{O}}{:}\ddot{\underset{..}{O}}R} \xrightarrow{\text{Heat}} \underset{\text{A free radical}}{2R\ddot{\underset{..}{O}}\cdot}$$

Connections 5.2 www.prenhall.com/bailey

Recycling Plastics

Plastics comprise a significant volume of household and office trash, but society has been slow in recycling these materials compared with aluminum, glass, and paper. One problem in plastics recycling is that several dozen different polymer formulations are in common use for making products as different as plastic wrap, soft drink bottles, and protective hard hats. Unless the different plastic polymers can be separated by type, recycling has limited practical value. And, unlike metal and glass, but like paper, plastics have a limited reprocessing life, since the polymer chains are degraded somewhat during each cycle and undesirable coloration can become a problem.

Methods are being devised for sorting plastics into various types by making use of differences in density and melting points or spectral characteristics (such as interaction with infrared light). High-density plastics can be separated from low-density ones in gigantic centrifuge-like machines. Careful heating of plastic mixtures can separate low-melting polymers from higher-melting materials. Yet another method is sorting prior to recycling, using the identification codes commonly found on recyclable resins. These familiar codes and their meanings are presented in this essay.

Many items can be made from recycled plastics. Among them are flowerpots, detergent bottles, drainage pipes, fiberfill for pillows, shower stalls, carpeting, videocassette boxes, desk supplies, and park benches. One focus in plastics recycling is to convert products with short first-use lives, like styrofoam, into something with a relatively long second use, like a drainage pipe, cassette holder, or picnic table.

Plastics Recycling Codes

Polyethylene terephthalate, a polyester condensation polymer used in large soft-drink bottles.

High-density polyethylene used in containers for detergents, liquid bleaches, motor oil, shampoos, body powders, and milk; also, plastic grocery bags.

Polyvinyl chloride, PVC, used in a variety of containers, clear or opaque, such as those for liquid body soaps, mouthwashes, shampoos, and conditioners.

Low-density polyethylene used for food wrappers, garment bags, department store shopping bags, and shrink wrap packaging.

Polypropylene used in plastic tubs for sour cream, party dips, yogurt, sauces, and margarine.

Polystyrene: styrofoam cartons and cups; clear plastic salad containers; plastic eating utensils.

Other: mixed materials.

These free radicals will immediately seek a source of electrons to complete their octets. The π bond in an ethylene molecule supplies an electron to pair with the radical, but in the process a new free radical is formed.

Initiation:

$$R\ddot{\underset{\cdot\cdot}{O}}\cdot \; CH_2{=}CH_2 \longrightarrow R\ddot{\underset{\cdot\cdot}{O}}{:}CH_2{-}CH_2\cdot$$

As before, this new radical will seek a pairing electron by attacking another ethylene molecule, in turn producing yet another free radical. The process repeats itself many times as the polymer is built.

Propagation:

$$RO{:}CH_2{-}CH_2\cdot \; CH_2{=}CH_2 \; CH_2{=}CH_2 \; CH_2{=}CH_2 \cdots \longrightarrow$$
$$RO{:}CH_2{-}CH_2{:}CH_2{-}CH_2{:}CH_2{-}CH_2{:}CH_2{-}CH_2\cdot \sim$$

The polymerization proceeds by a free-radical chain reaction analogous to the chlorination of alkanes (section 4.4.D). Attack by the peroxide is the initiation step, and the repeated additions are propagation steps. Occasionally, two propagating chains meet end to end to conclude the reaction and form a complete polymer molecule—a termination step.

Termination:

$$RO\sim\!\!\left(CH_2CH_2\right)_x\cdot + \cdot\sim\!\!\left(CH_2CH_2\right)_y OR \longrightarrow RO\sim\!\!\left(CH_2CH_2\right)_x : \sim\!\!\left(CH_2CH_2\right)_y OR$$

GETTING INVOLVED

Problem 5.16

(a) Write the formula of the polymer formed by addition polymerization of 1,1-dichloroethene. **(b)** Now write the free radical polymerization mechanism showing initiation, propagation, and termination steps.

5.4 Electrophilic Addition to Conjugated Dienes

conjugation alternating double and single bonds in a molecule

Dienes are compounds with two carbon-carbon double bonds. When the double bonds are separated by one single bond, the diene is termed **conjugated,** as in the case of 1,3-butadiene. Conjugated systems have alternating double and single bonds, and this can lead to some interesting chemical properties.

$$CH_2{=}CH{-}CH{=}CH_2$$ 1,3-butadiene, a conjugated diene

If 1,3-butadiene were treated with one mole of bromine, you might expect that bromine would add to one of the double bonds and produce 3,4-dibromo-1-butene. However, two products are formed when one mole of bromine is added to 1,3-butadiene. One of them is the 3,4-dibromo product formed by what is called 1,2 addition, but the other is a perplexing 1,4-dibromo-2-butene product, formed by 1,4 addition of the bromine.

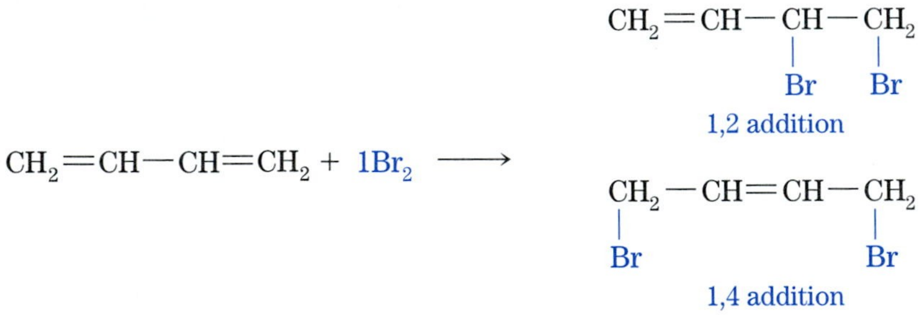

$$CH_2{=}CH{-}CH{=}CH_2 + 1Br_2 \longrightarrow CH_2{=}CH{-}CH(Br){-}CH_2Br \quad \text{(1,2 addition)}$$

$$\phantom{CH_2{=}CH{-}CH{=}CH_2 + 1Br_2} \longrightarrow BrCH_2{-}CH{=}CH{-}CH_2Br \quad \text{(1,4 addition)}$$

1,2 addition

1,4 addition

allylic carbocation carbocation in which positive carbon is adjacent to a carbon-carbon double bond

To understand this seemingly unusual behavior, we need to look at the very special type of carbocation formed as this reaction proceeds, an **allylic carbocation.** An allylic carbocation is one in which the carbon bearing the positive charge is directly adjacent to a carbon-carbon double bond. In this reaction, a positive bromine attacks the diene, preferentially forming the more stable allylic carbocation.

$$CH_2{=}CH{-}CH{=}CH_2 \xrightarrow{Br^+} CH_2{=}CH{-}CH(Br){-}\overset{+}{C}H_2$$

Less stable primary carbocation

$$CH_2{=}CH{-}CH{=}CH_2 \xrightarrow{Br^+} CH_2{=}CH{-}\overset{+}{C}H{-}CH_2Br$$

More stable allylic carbocation

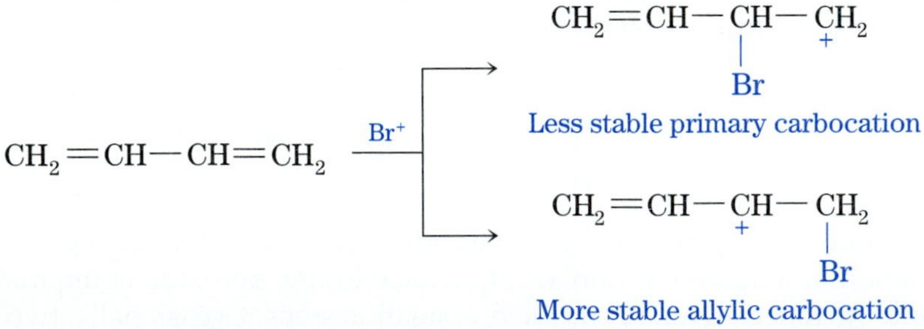

Allylic carbocations are among the most stable carbocations due to a phenomenon called resonance. One can write structures where the electrons in the double bond seem to migrate to the positively charged carbon and neutralize it, with the charge transferred to another carbon. These two structures are called **resonance forms.** Neither actually exists—they merely represent a symbolic way for showing how delocalization of the π electrons in the double bond can stabilize the carbocation. A **resonance hybrid,** which one can think of as an average of the resonance forms, is the actual structure of the carbocation.

resonance forms
symbolic, nonexistent structures, differing only in position of electrons, which are used to describe an actual molecule or ion

resonance hybrid
"average" of the resonance forms used to describe a molecule or ion that cannot be described by a single structure

Resonance forms:

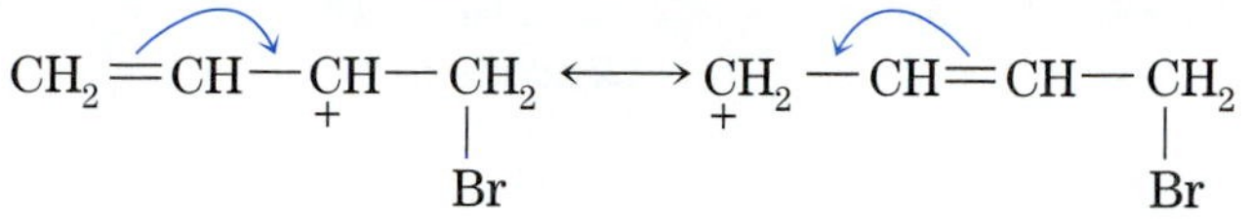

$$CH_2{=}CH{-}\overset{+}{C}H{-}CH_2(Br) \longleftrightarrow \overset{+}{C}H_2{-}CH{=}CH{-}CH_2(Br)$$

Resonance hybrid:

$$\underbrace{CH_2 \overset{...}{=} CH \overset{...}{=} CH}_{+}{-}CH_2(Br)$$

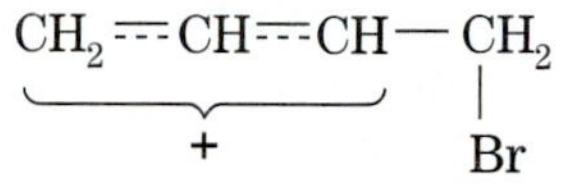

The presence of resonance in a species is always a stabilizing influence. To illustrate this and the drawing of resonance forms, let us take a look at the allylic carbocation under consideration. First, note that the resonance forms differ in the position of electrons, π electrons in particular, not in the position of atoms. We can convert one form into the other merely by moving the pair of π electrons (symbolized by one line of the double bond) as shown by the arrows. The double-headed arrow separating the two structures indicates that these are resonance forms. This arrow distinguishes resonance forms from species that are in equilibrium with one another. The two resonance forms do not even exist, so they cannot be in equilibrium. They are merely useful representations of the actual allylic carbocation, which is more accurately represented by the resonance hybrid.

The resonance hybrid is an average of the resonance forms. Look at the three atoms involved in the resonance. Wherever there is a double bond in one, there is a single bond in the other. As a result, we draw something more like one and one-half bonds in the resonance hybrid, as shown by the dashed lines. In the resonance forms, the positive charge appears to be concentrated on individual atoms. In the hybrid, we show it delocalized across the three-atom system. It is this delocalization that stabilizes the allylic carbocation. The positive charge, normally a destabilizing influence, is accommodated in a more stable manner if it is spread out over several atoms rather than concentrated on one. This charge delocalization and stability are analogous to a gentle wind blowing all over town as opposed to a vicious tornado touching down in a small area; the gentle wind is delocalized energy, which is less destabilizing to the community.

We often use resonance forms in writing reaction mechanisms, with the understanding that they themselves do not exist, because they clearly show where the actual ion can be neutralized. The mechanism for electrophilic addition of bromine to 1,3-butadiene follows. In drawing the resonance forms of the allylic carbocation, we illustrate two sites for neutralization, resulting in both 1,2 and 1,4 addition products.

Step 1: Electrophile Br^+ is attracted to π cloud and uses two π electrons to bond; a more stable allylic carbocation results

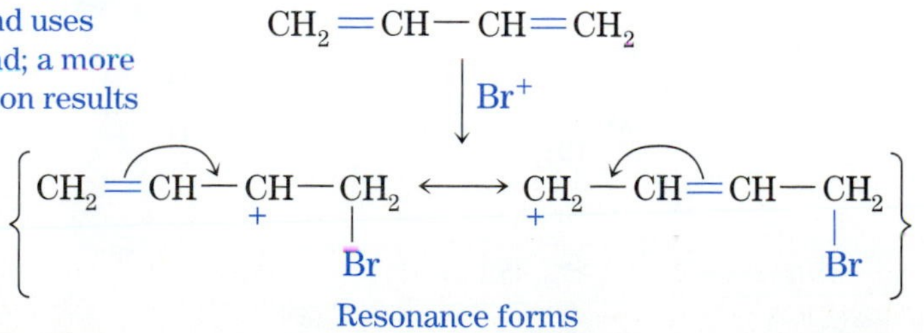

$$CH_2{=}CH{-}CH{=}CH_2 \xrightarrow{Br^+}$$

$$\left\{ CH_2{=}CH{-}\overset{+}{C}H{-}CH_2(Br) \longleftrightarrow \overset{+}{C}H_2{-}CH{=}CH{-}CH_2(Br) \right\}$$

Resonance forms

Step 2: The allylic carbocation is resonance stabilized; resonance forms show the two places it can be neutralized by bromide ion

Resonance forms

$\downarrow$ Br^-

$CH_2{=}CH{-}CH(Br){-}CH_2Br$ + $CH_2Br{-}CH{=}CH{-}CH_2Br$

1,2 addition — 1,4 addition

If we draw the π-bonding picture of the resonance forms, we can understand how delocalization of electrons and the positive charge occurs (Figure 5.3). In each of the two resonance forms, there is overlap of two p orbitals to form the π bond (σ bonds are represented by lines). The positive carbocation carbon has an empty p orbital. Now, if you look at these more closely you can see that the only difference between the two resonance forms as we have drawn them is which p orbitals overlap. Further thought suggests that the middle p orbital must unavoidably overlap with those on either side because of their size and proximity. Thus the actual structure of the allylic carbocation is the resonance hybrid in which there is continuous overlap of the three p orbitals and in which the positive charge is delocalized throughout the three-atom system.

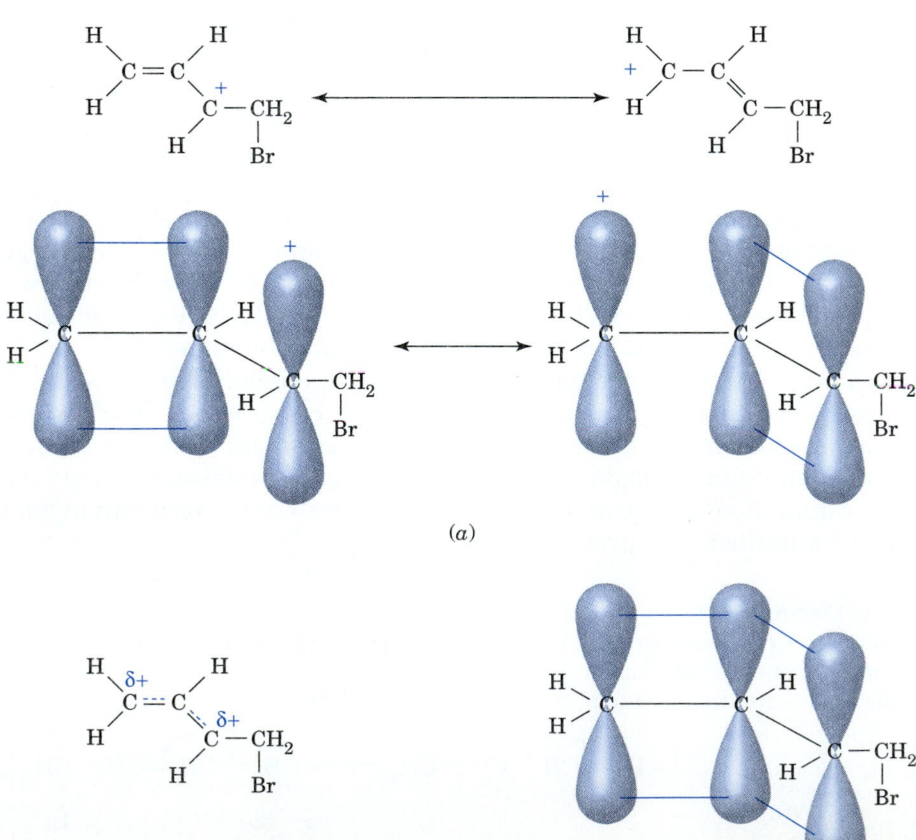

Figure 5.3
π-bonding picture of the allylic carbocation formed during bromination of 1,3-butadiene. (a) Resonance forms. (b) Resonance hybrid.

GETTING INVOLVED

✓ What is a conjugated diene? What is meant by 1,2 and 1,4 addition to such a diene?

✓ What is an allylic carbocation?

✓ Distinguish between resonance forms and resonance hybrid. How does the π-bonding picture of an allylic carbocation help explain the distinction?

✓ Can you write the mechanism for the 1,2 and 1,4 addition of bromine to 1,3-butadiene?

Problem 5.17

Write the 1,2 and 1,4 addition products resulting from the following reactions: **(a)** 1,3-butadiene and HBr; **(b)** 1,3-cyclopentadiene and Br_2.

Problem 5.18

Write electrophilic addition mechanisms for the two reactions in problem 5.17. Be sure to properly show the two resonance forms of the intermediate allylic carbocation with a double-headed arrow.

See related problem 5.33.

5.5 Resonance Stabilization of Reactive Intermediates

We have just examined the allylic carbocation and the resonance stabilization associated with it. Allylic free radicals and carbanions also are resonance-stabilized. An allylic reactive intermediate is one in which the plus charge of the carbocation, the unpaired electron of the free radical, or the nonbonding electron pair and negative charge of the carbanion are on a carbon directly adjacent to a carbon-carbon double bond. Since the double bond has two p orbitals involved in a π bond and the carbon of the reactive intermediate has a p orbital with zero (carbocation), one (free radical), or two electrons (carbanion), the three-atom allylic system has a p orbital on each carbon. Because of the continuous overlap of these p orbitals, the charge or unpaired electron of the reactive intermediate is delocalized over the three-atom system, and the intermediate is stabilized.

To use resonance effectively, one should be able to draw resonance structures and resonance hybrids. Resonance structures differ only in the positions of electrons, not atoms, as shown for the allylic carbocation, free radical, and carbanion in Figure 5.4. Drawing resonance forms essentially involves depicting electrons within these three-atom systems to show delocalization of the charge or unpaired electrons. As you examine Figure 5.4, note that the resonance hybrid is drawn as an average of the resonance structures. Also note in the bonding picture that each carbon in the system possesses a p orbital and that there is continuous overlap among them. Finally, note that the intermediate can be neutralized at either the first or third carbon of the system.

GETTING INVOLVED

✓ What is an allylic carbocation, an allylic free radical, and an allylic carbanion? Can you draw resonance forms, resonance hybrids, and π-bonding pictures for each?

Problem 5.19

Using Figure 5.4 as a guide draw the resonance forms, resonance hybrid, and π-bonding picture for each of the following allylic systems.

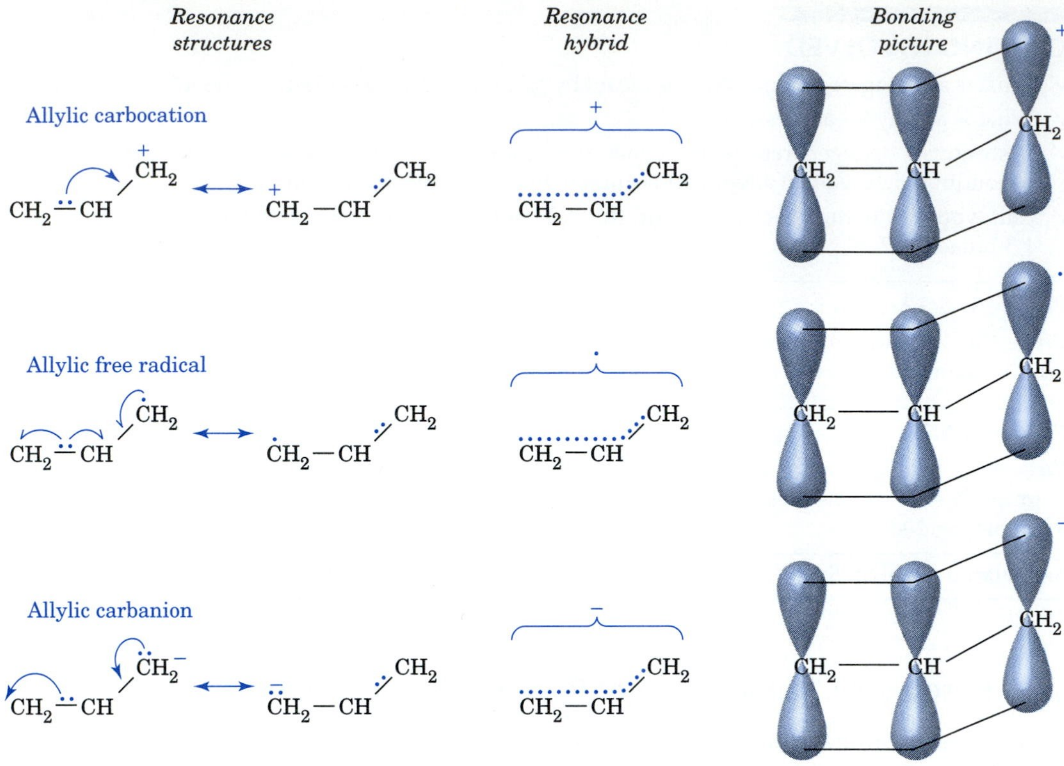

Figure 5.4 Resonance structures for the allylic system ($CH_2{=}CH{-}CH_2{-}$).

(a) $CH_3CH{=}CH\overset{+}{C}H_2$ **(b)** $CH_3\dot{C}H{-}CH{=}CH_2$ **(c)** $CH_3CH{=}CH\overset{\ddot{-}}{C}H_2$ **(d)** (3-methylcyclopentenyl cation, CH_3)

See related problems 5.34–5.35.

Tapping a rubber tree

5.6 Natural and Synthetic Rubber

A. Natural Rubber

Natural rubber is produced from a milky-white colloidal latex found in the stems of some plants (even in the common dandelion and goldenrod). The commercial source is the rubber tree, which can yield as much as a ton of rubber per acre. The term *rubber* was coined by Joseph Priestley, who used it to "rub out" pencil marks. Structurally, natural rubber is a polyterpene (See Connections 5.3 on terpenes) composed of many recurring isoprene (2-methyl-1,3-butadiene) skeletons.

$$CH_2{=}\underset{\text{Isoprene}}{\overset{CH_3}{\overset{|}{C}}}{-}CH{=}CH_2 \qquad \sim\!\!\left(CH_2\overset{CH_3}{\overset{|}{C}}{=}CHCH_2\right)_n\!\!\sim$$

Isoprene

B. Synthetic Polyisoprene Rubber

Polyisoprene rubber can also be produced synthetically by the addition polymerization of isoprene (2-methyl-1,3-butadiene), a conjugated diene, by 1,4 addition (section 5.4).

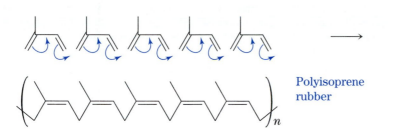

Polyisoprene rubber

In addition to natural and synthetic polyisoprene rubber, several synthetic rubbers play a large part in the present-day rubber industry. By far the most important of these is SBR, a styrene-butadiene copolymer with elastomer properties. (A copolymer is a polymer prepared from two or more different monomer units.) This rubber, once called GRS (government rubber styrene), was the first and most widely produced synthetic rubber of World War II. Composed of approximately 75% 1,3-butadiene and 25% styrene, it is similar in chemistry of formation to other addition polymers. Styrene undergoes normal 1,2 addition, and the butadiene mainly 1,4 addition:

$$n\mathrm{CH_2{=}CH{-}CH{=}CH_2} + n\mathrm{CH_2{=}CH(C_6H_5)} \longrightarrow \sim\!\left(\mathrm{CH_2CH{=}CHCH_2{-}CH_2CH(C_6H_5)}\right)_n\!\sim$$

The synthetic rubbers polybutadiene, polyisoprene, and neoprene are all made by the 1,4 addition polymerization of dienes:

$$n\mathrm{CH_2{=}C(G){-}CH{=}CH_2} \longrightarrow \sim\!\left(\mathrm{CH_2C(G){=}CHCH_2}\right)_n\!\sim$$

where G = CH_3 for polyisoprene,
= H for polybutadiene, and
= Cl for neoprene.

Like most natural and synthetic rubbers, polybutadiene and polyisoprene have their predominant use in tire manufacture. Neoprene, however, is more resistant to oils, chemicals, heat, and air than most rubbers and thereby finds specialty uses, as in gasoline pump hoses and rubber tubing in automobile engines.

Although rubber was introduced in Europe shortly after Columbus sailed to the New World, it had limited uses until 1839, when Charles Goodyear accidentally discovered **vulcanization.** Natural rubber tends to be sticky when warm and brittle when cold and, though elastic, does not regain its shape quickly or completely when stretched. One day in 1839, Charles Goodyear accidentally

Charles Goodyear (far right) and colleagues

vulcanization
process in which rubber is treated with sulfur to improve its properties

Connections 5.3 www.prenhall.com/bailey

Terpenes

The odor of mint, the scent of cedar and pine, the fragrance of roses, and the color of carrots and tomatoes are largely due to a class of compounds known as *terpenes.* Terpenes are found in a variety of spices and flavorings, such as basil, ginger, spearmint, peppermint, lemon, and clove, and in a number of common commercial products, such as turpentine, bayberry wax, rosin, cork, and some medicines. The invigorating smell of pine, cedar, and eucalyptus forests and the sweet smell of orange and lemon groves are due to terpenes.

Terpenes are characterized by carbon skeletons constructed of isoprene units. Isoprene (2-methyl-1,3-butadiene) is a conjugated diene.

$$CH_2{=}C(CH_3){-}CH{=}CH_2$$

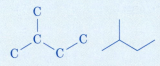

Isoprene

Isoprene skeletons

In fact, terpenes are classified according to the number of isoprene units in the molecule. The simplest terpenes have two isoprene units (ten carbons) and are called *monoterpenes.* Other classes are listed below.

Monoterpenes	Two isoprene units	C_{10}
Sesquiterpenes	Three isoprene units	C_{15}
Diterpenes	Four isoprene units	C_{20}
Triterpenes	Six isoprene units	C_{30}
Tetraterpenes	Eight isoprene units	C_{40}

Terpene molecules are further classified by the number of rings they contain.

Acyclic	No rings	Bicyclic	Two rings
Monocyclic	One ring	Tricyclic	Three rings

The number of rings in a polycyclic compound can be determined by counting the minimum number of scissions in the carbon skeleton that would be necessary to make it an open-chain structure.

Table 5.2 gives examples of some common terpenes. Note the isoprene skeletons in each compound (they are marked off by dashed lines in the structures). Also note that many terpenes contain oxygen and are aldehydes, ketones, alcohols, ethers, acids, and so on.

Mint leaves

Ginger root

spilled one of his experiments, a mixture of rubber and sulfur, on a hot stove. The substance he scraped off was not sticky and exhibited a greatly increased elasticity. Today, the process is called vulcanization.

In vulcanization, sulfur adds to the double bonds in rubber, constructing cross-links between polymer chains (Figure 5.5). The sulfur bridges are strained when the rubber is stretched; the strain is relieved only when the rubber is allowed to assume the original conformation. Soft rubber has 1%–3% sulfur by weight, whereas hard rubber (as in rubber mallets) has as much as 20%–30% sulfur.

Connections 5.3 *(cont.)*

Table 5.2 Terpenes

Monoterpenes

Acyclic

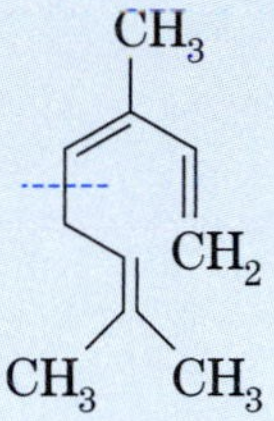

CH_3 CH_2 CH_3 CH_3

Ocimene (basil)

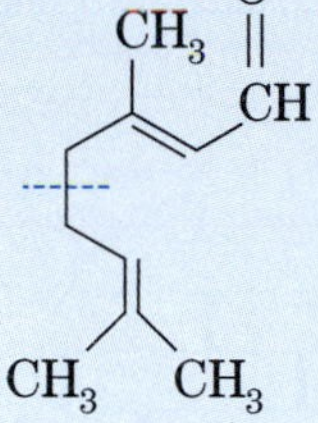

O CH_3 CH CH_3 CH_3

Citral (lemon oil)

Monocyclic

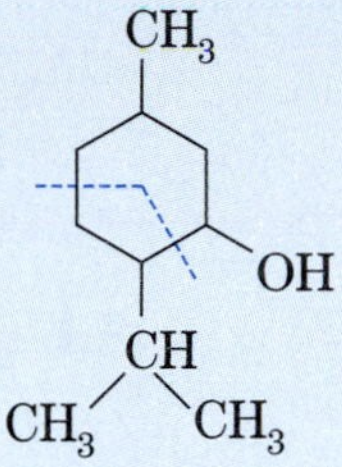

CH_3 OH CH CH_3 CH_3

Menthol (mint flavor)

Bicyclic

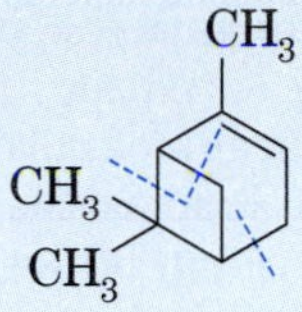

CH_3 CH_3 CH_3

α-pinene (oil of turpentine)

Sesquiterpenes

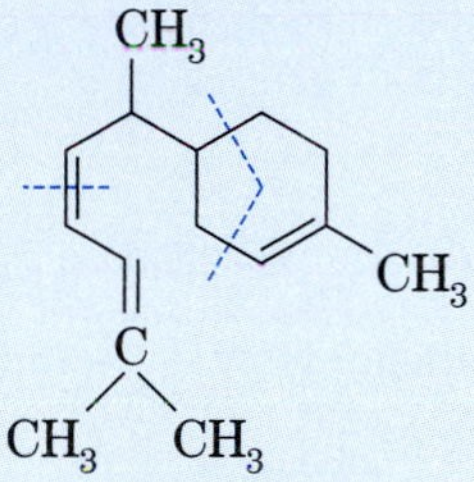

CH_3 CH_3 C CH_3 CH_3

Zingiberene (oil of ginger)

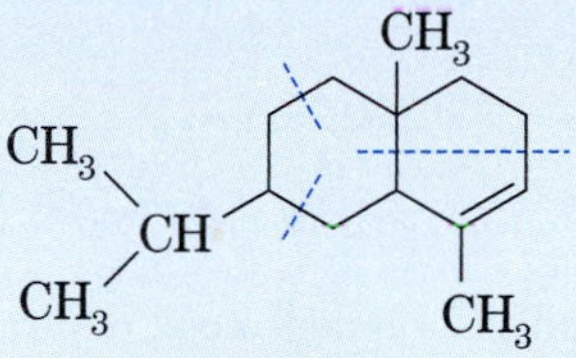

CH_3 CH_3 CH CH_3 CH_3

Selinene (oil of celery)

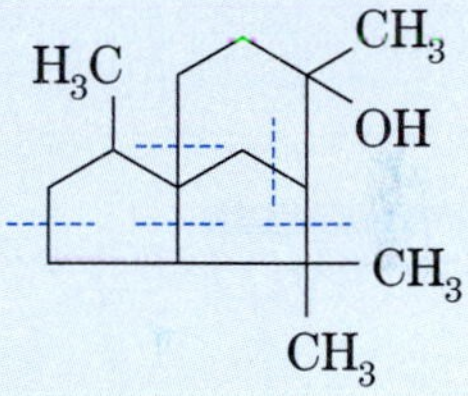

H_3C CH_3 OH CH_3 CH_3

Cedrol (oil of cedar)

Diterpene

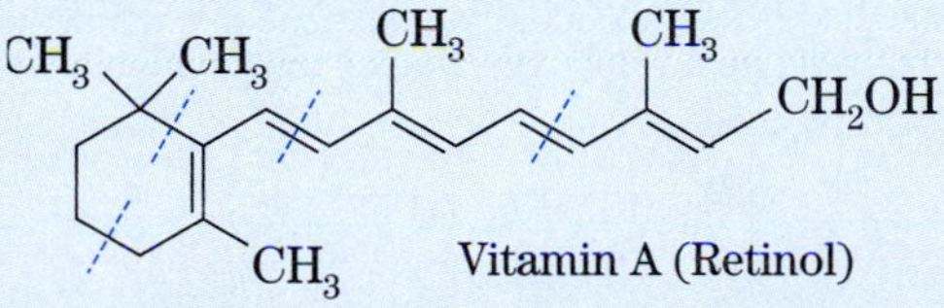

CH_3 CH_3 CH_3 CH_3 CH_2OH CH_3

Vitamin A (Retinol)

Triterpene

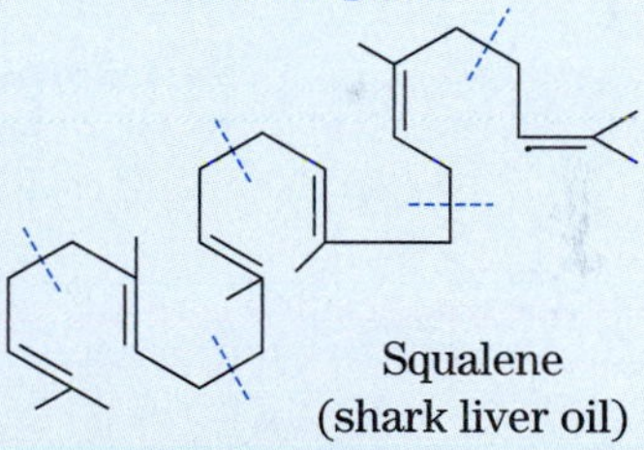

Squalene (shark liver oil)

Tetraterpene

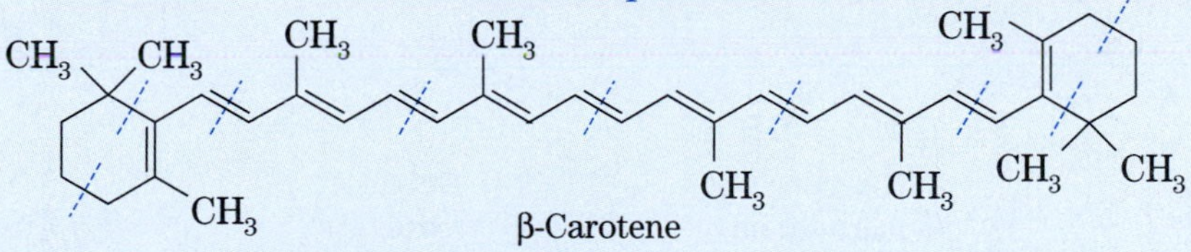

CH_3 CH_3 CH_3 CH_3 CH_3 CH_3 CH_3 CH_3 CH_3 CH_3

β-Carotene (carrots)

Polyterpene

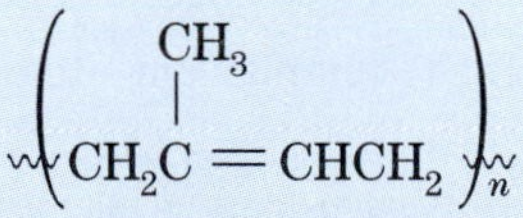

$$\left(CH_2\overset{CH_3}{\overset{|}{C}}=CHCH_2 \right)_n$$

Natural rubber

Although his discovery made possible the enormous growth of the rubber industry, Goodyear did not personally profit from it. He died in 1860 after years of court battles over patents, leaving his wife and six children with debts of more than $200,000.

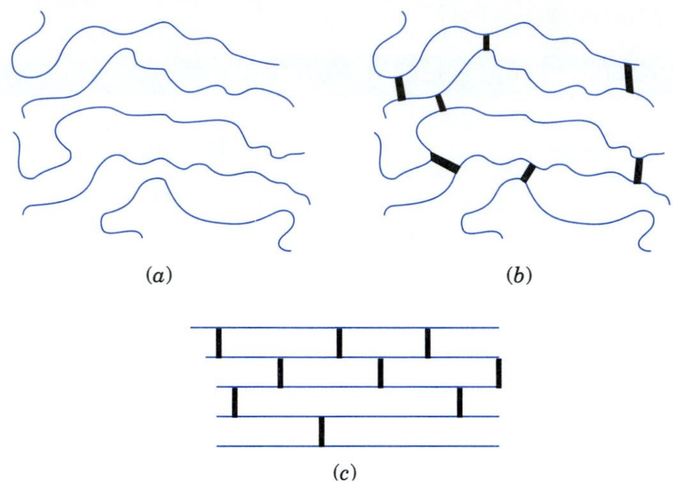

Figure 5.5 (a) Untreated rubber molecules are bent and convoluted. (b) Vulcanized rubber has many sulfur bridges linking the polymer chains, thus providing a "chemical memory" of the original shape before stretching. (c) When stretched, the rubber molecules align. This puts a strain on the system, since the sulfur bridges tend to hold the molecules in the original conformation. The strain is relieved when the rubber is allowed to snap back into the original conformation.

GETTING INVOLVED

- ✓ What is polyisoprene rubber, natural and synthetic? How is 1,4 addition involved in producing the synthetic rubber?
- ✓ What is SBR and neoprene rubber?
- ✓ What is vulcanization? What effect does it have on the properties of rubber and why?

Problem 5.20

The following are structures of common terpenes. Classify each compound as a mono-, di-, tri, or tetraterpene and as acyclic, mono-, bi-, tri-, or tetracyclic. Using dashed lines or circles, identify each isoprene unit.

(a) Carvone (spearmint oil)

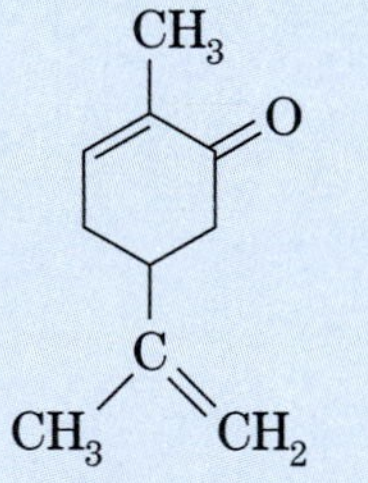

(b) Geraniol (rose oil)

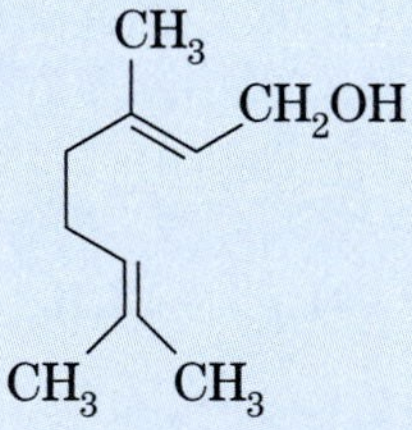

(c) Eudesmol (from eucalyptus oils)

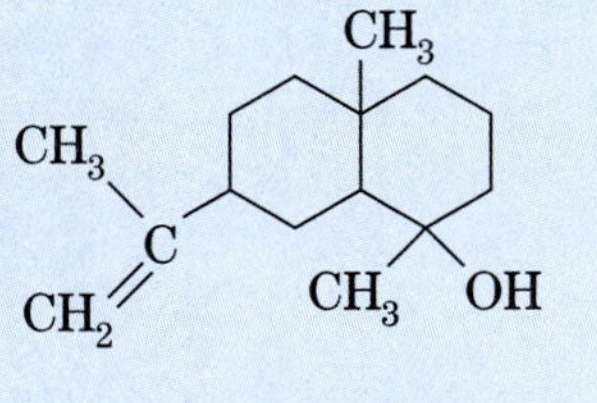

(d) $CH_3\overset{CH_3}{\overset{|}{C}}{=}CHCH_2CH_2\sim\left(\overset{CH_3}{\overset{|}{C}}{=}CH{-}CH{=}CH\right)_3\sim\left(CH{=}\overset{CH_3}{\overset{|}{C}}{-}CH{=}CH\right)_2\sim CH{=}\overset{CH_3}{\overset{|}{C}}CH_2CH_2CH{=}\overset{CH_3}{\overset{|}{C}}CH_3$

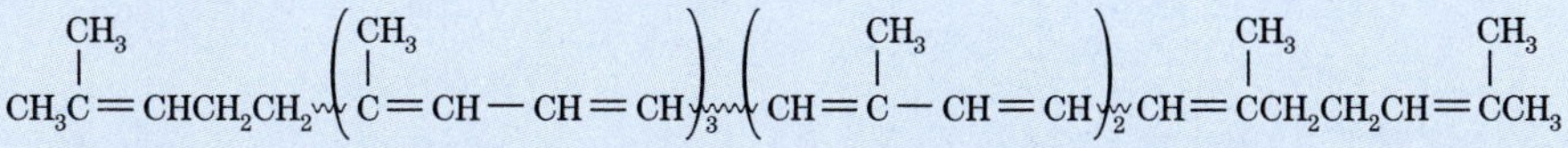

Lycopene (plant pigment—tomatoes, pyracantha; all trans double bonds)

(e) Abietic acid (chief constituent of rosin)

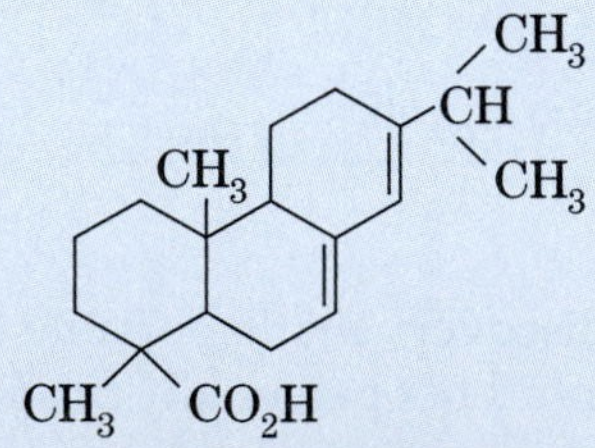

(f) Limonene (lemon and orange oil)

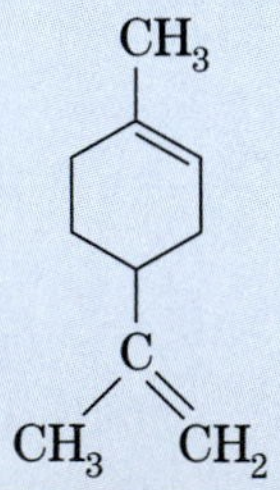

(g) Myrcene (bayberry)

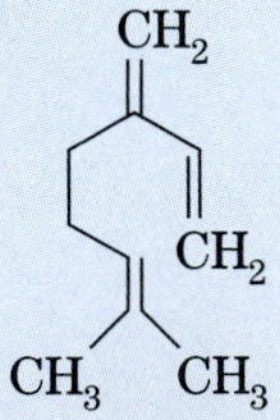

5.7 Oxidation of Alkenes

A. Hydroxylation with Potassium Permanganate

Alkenes react with potassium permanganate to form 1,2-diols.

$$3\ \gt C{=}C\lt \ + \underset{\text{(Purple)}}{2KMnO_4} + 4H_2O \longrightarrow 3{-}\underset{OH}{\overset{|}{C}}{-}\underset{OH}{\overset{|}{C}}{-} + \underset{\text{(Brown)}}{2MnO_2} + 2KOH$$

This reaction, known as the Baeyer test, is useful in distinguishing alkenes from alkanes. Alkanes do not undergo reaction with potassium permanganate. A positive test is easy to detect visually because potassium permanganate solutions are deep purple. When such a solution is added to an alkene, the purple quickly disappears, leaving a murky brown precipitate of manganese oxide.

In this reaction, the permanganate ion provides both oxygens simultaneously and as a result *cis* (or *syn*) addition occurs.

$$\text{cyclohexene} \xrightarrow[H_2O]{KMnO_4} \text{cis-1,2-cyclohexanediol (H, H; OH OH)} \quad \text{via} \quad \text{cyclic manganate ester (H, H; O, O; Mn; O, O}^-)$$

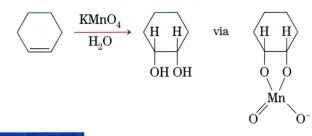

GETTING INVOLVED

✓ Write a general reaction equation for potassium permanganate reaction with alkenes. Why does *cis* addition occur?

Problem 5.21

Write equations showing the reaction of the following with potassium permanganate: **(a)** propene; **(b)** 1-methylcyclohexene.

See additional problems 5.31 and 5.37.

B. Ozonolysis

Double bonds are easily cleaved oxidatively by reaction with ozone followed by zinc and water.

$$\gt C{=}C\lt \ + O_3 \xrightarrow[Zn]{H_2O,} \gt C{=} \ + \ O{=}C\lt$$

The reaction products are aldehydes and ketones. Ozone, prepared by passing oxygen gas through an electric discharge, is bubbled into a solution of the alkene in an inert solvent such as carbon tetrachloride. Mechanistically, the ozone adds to the double bond, and a molozonide is formed. This rearranges to an ozonide in which the carbon-carbon bond is completely cleaved. Reduction and hydrolysis of the ozonide produces aldehydes and ketones.

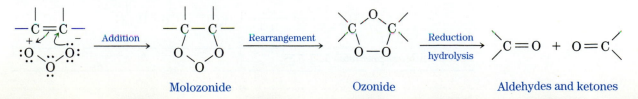

$$-C{=}C- \ + \ O_3 \xrightarrow{\text{Addition}} \underset{\text{Molozonide}}{-C{-}C-\ (O{-}O{-}O\ \text{ring})} \xrightarrow{\text{Rearrangement}} \underset{\text{Ozonide}}{C{-}O{-}C\ (O{-}O)} \xrightarrow[\text{hydrolysis}]{\text{Reduction}} \underset{\text{Aldehydes and ketones}}{\gt C{=}O + O{=}C\lt}$$

Ozonolysis is particularly useful for elucidating the location of double bonds in alkenes. An unknown alkene is cleaved to smaller, more easily identifiable aldehydes and ketones. The aldehydes and ketones are then pieced back together like a puzzle. Wherever a carbon-oxygen double bond occurs, originally that carbon was involved in a carbon-carbon double bond (see Example 5.6).

GETTING INVOLVED

✓ What happens to a carbon-carbon double bond in ozonolysis? What types of products are formed? How can ozonolysis be useful in elucidating the structure of an alkene? Can you write a general equation for the reaction?

Example 5.6

Suppose we have an unknown alkene with the molecular formula C_4H_8. After ozonolysis, it has been converted to the following compounds:

$$\underset{\text{Unknown alkene}}{C_4H_8} \xrightarrow{O_3} \xrightarrow[Zn]{H_2O,} CH_3\underset{\underset{O}{\|}}{C}CH_3 + H\underset{\underset{O}{\|}}{C}H$$

What is the alkene?

Solution

The carbon-oxygen double bonds identify the carbons involved in the alkene linkage. Connecting these two carbons gives us the structure of the unknown alkene, 2-methylpropene.

$$CH_3\underset{\underset{CH_2}{\|}}{C}CH_3$$

The other two isomeric alkenes with the formula C_4H_8 give quite different ozonolysis products.

$$CH_3CH_2CH = CH_2 \xrightarrow{O_3} \xrightarrow[Zn]{H_2O,} CH_3CH_2\overset{\overset{O}{\|}}{C}H + H\overset{\overset{O}{\|}}{C}H$$

$$CH_3CH = CHCH_3 \xrightarrow{O_3} \xrightarrow[Zn]{H_2O,} 2CH_3\overset{\overset{O}{\|}}{C}H$$

Problem 5.22

Write the products formed by ozonolysis of each of the following alkenes: **(a)** 2-methyl-2-pentene; **(b)** 3,7-dimethyl-1,3,6-octatriene (ocimene, a component of the herb basil); **(c)** cyclohexene.

Problem 5.23

Determine the structure of the alkene with the formula C_5H_{10} from the ozonolysis products.

$$C_5H_{10} \xrightarrow{O_3} \xrightarrow[H_2O]{Zn,} CH_3\overset{\overset{O}{\|}}{C}H + CH_3CH_2\overset{\overset{O}{\|}}{C}H$$

See additional problems 5.38–5.39.

Connections 5.4

www.prenhall.com/bailey

The Treatment of Atherosclerosis

The Cholesterol Connection

The formation of terpenes (Connections 5.3) is a polymerization process that can be continued in biological organisms to the point of some very complex molecules. One such process is the formation of cholesterol, a multicyclic precursor to metabolic and sex hormones made from isoprene-like units.

Some persons have a genetic disposition to producing and circulating large quantities of cholesterol. They suffer from a condition known as *hypercholesterolemia*. Others lead a life style that includes fat- and cholesterol-laden diets, which can be just as deadly as a genetic fault. A change in diet can help in some cases, but others require either drug or surgical intervention.

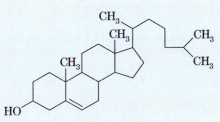

Cholesterol
part of membrane structures

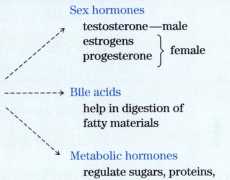

Sex hormones
testosterone—male
estrogens, progesterone — female

Bile acids
help in digestion of fatty materials

Metabolic hormones
regulate sugars, proteins, salt, and water

Cholesterol has received some very bad press in the past, in that too much cholesterol circulating in the bloodstream can be deposited on the walls of arteries. These deposits, known as *atherosclerotic plaque*, grow until they can completely block the flow of blood. When this occurs in the heart, oxygen cannot reach the cells in that organ and a heart attack, or myocardial infarction, occurs. If the blockage occurs in a blood vessel leading to the brain, a stroke results. The consequences can range from the relatively mild disability of hypertension or high blood pressure to mental and physical incapacity and/or to death.

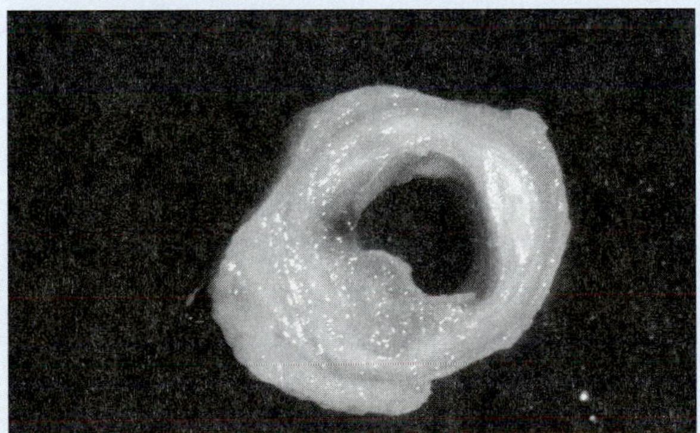

A partially clogged artery

The most recent drug treatment developed to slow down the biosynthesis of cholesterol involves stopping (inhibiting) the enzyme that starts the polymerization process. The enzyme has the tongue-twisting title of 3-hydroxy-3-methyl glutaryl coenzyme A reductase; HMG-CoA reductase for short. The drugs are called *HMG-reductase inhibitors;* lovastatin was the first marketed.

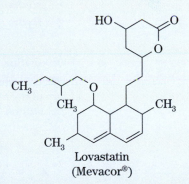

Lovastatin
(Mevacor®)

Ingestible resins are also used to "soak up" cholesterol and fats in the intestine, and other drugs can alter the ratio of high-density lipoproteins (fat-cholesterol-protein complexes), or HDLs, to low-density lipoproteins, LDLs. The HDLs are what could be referred to as a "good" form of circulating cholesterol, which gets metabolized out of the system, while LDLs are "bad" cholesterol, which can eventually end up deposited on the walls of arteries, causing atherosclerosis.

5.8 Acidity of Terminal Alkynes

Terminal alkynes, in which the triple bond is at the end of a carbon chain, are very weakly acidic. The hydrogen can be abstracted by strong bases such as sodium amide ($:NH_2^-$ is a stronger base than $:\ddot{O}H^-$).

$$RC{\equiv}CH + NaNH_2 \longrightarrow RC{\equiv}C{:}^-Na^+ + NH_3$$

Alkanes and alkenes are not acidic and do not undergo this reaction. These sodium salts of alkynes are useful in producing higher alkynes by nucleophilic substitution reactions (section 8.4.A).

GETTING INVOLVED

Problem 5.24

Write an equation showing the reaction between 1-butyne and sodium amide.

REACTION SUMMARY

A. Reactions of Alkenes

Addition Reactions

Section 5.1; Examples 5.1–5.3; Problems 5.2–5.9, 5.25, 5.27, 5.29.

$$>C{=}C< + E{-}A \longrightarrow -\underset{E}{\overset{|}{\underset{|}{C}}}-\underset{A}{\overset{|}{\underset{|}{C}}}-$$

- **Hydrohalogenation:** EA = HX; E = H and X = Cl, Br, or I
- **Halogenation:** EA = X_2; X_2 = Cl_2 or Br_2; E = X and A = X
- **Hydration:** EA = H_2O with H_2SO_4 catalyst; E = H and A = OH
- **Hydrogenation:** EA = H_2 with Ni, Pt, or Pd catalyst; E = H and A = H

Orientation of Addition

When an unsymmetrical reagent adds to an unsymmetrical alkene, the positive portion of the reagent adds to the carbon that results in the formation of the more stable carbocation.

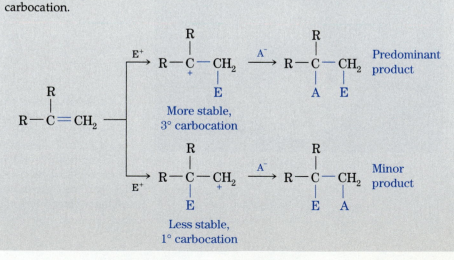

$$R_2C{=}CH_2 \xrightarrow{E^+} R_2\overset{+}{C}{-}CH_2E \xrightarrow{A^-} R_2C(A){-}CH_2E$$

More stable, 3° carbocation — Predominant product

$$R_2C{=}CH_2 \xrightarrow{E^+} R_2C(E){-}\overset{+}{C}H_2 \xrightarrow{A^-} R_2C(E){-}CH_2A$$

Less stable, 1° carbocation — Minor product

Carbocation Stability

Most stable 3° > 2° > 1° > CH_3^+ least stable

Oxidation of Alkenes

Section 5.7; Example 5.6; Problems 5.21–5.23, 5.37–5.39.

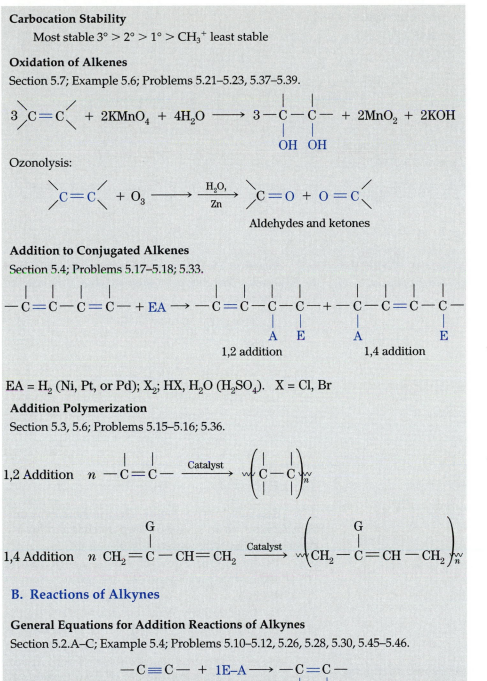

$$3\ \rangle C{=}C\langle\ + 2KMnO_4 + 4H_2O \longrightarrow 3\ {-}\underset{OH}{\overset{|}{\underset{|}{C}}}{-}\underset{OH}{\overset{|}{\underset{|}{C}}}{-} + 2MnO_2 + 2KOH$$

Ozonolysis:

$$\rangle C{=}C\langle\ + O_3 \longrightarrow \xrightarrow[Zn]{H_2O,} \rangle C{=}O + O{=}C\langle$$

Aldehydes and ketones

Addition to Conjugated Alkenes

Section 5.4; Problems 5.17–5.18; 5.33.

$$-\overset{|}{C}{=}\overset{|}{C}-\overset{|}{C}{=}\overset{|}{C}- + EA \longrightarrow \underset{\text{1,2 addition}}{-\overset{|}{C}{=}\overset{|}{C}-\underset{A}{\overset{|}{\underset{|}{C}}}-\underset{E}{\overset{|}{\underset{|}{C}}}-} + \underset{\text{1,4 addition}}{-\underset{A}{\overset{|}{\underset{|}{C}}}-\overset{|}{C}{=}\overset{|}{C}-\underset{E}{\overset{|}{\underset{|}{C}}}-}$$

EA = H_2 (Ni, Pt, or Pd); X_2; HX, H_2O (H_2SO_4). X = Cl, Br

Addition Polymerization

Section 5.3, 5.6; Problems 5.15–5.16; 5.36.

1,2 Addition $$n\ -\overset{|}{C}{=}\overset{|}{C}- \xrightarrow{\text{Catalyst}} \sim\!\left(\underset{|}{\overset{|}{C}}-\underset{|}{\overset{|}{C}}\right)_n\!\sim$$

1,4 Addition $$n\ CH_2{=}\overset{G}{\overset{|}{C}}-CH{=}CH_2 \xrightarrow{\text{Catalyst}} \sim\!\left(CH_2-\overset{G}{\overset{|}{C}}{=}CH-CH_2\right)_n\!\sim$$

B. Reactions of Alkynes

General Equations for Addition Reactions of Alkynes

Section 5.2.A–C; Example 5.4; Problems 5.10–5.12, 5.26, 5.28, 5.30, 5.45–5.46.

$$-C{\equiv}C- + 1E\text{–}A \longrightarrow -\underset{E}{\underset{|}{C}}{=}\underset{A}{\underset{|}{C}}-$$

$$-C{\equiv}C- + 2E\text{–}A \longrightarrow -\underset{E}{\underset{|}{\overset{E}{\overset{|}{C}}}}-\underset{A}{\underset{|}{\overset{A}{\overset{|}{C}}}}-$$

Hydrogenation: EA is H_2 with a metal catalyst such as Pt, Pd, or Ni
Halogenation: EA is Cl_2 or Br_2
Hydrohalogenation: EA is HCl, HBr, or HI

Addition of Water to Alkynes

Section 5.2.D; Problem 5.14, 5.32.

$$-C\equiv C- + H_2O \xrightarrow[HgSO_4]{H_2SO_4} -\underset{H}{\overset{H}{C}}-\underset{O}{\overset{\|}{C}}- \quad \left(\text{via } -\underset{H}{C}=\underset{OH}{C}-\right)$$

Acidity of Terminal Alkynes

Section 5.8; Problem 5.24.

$$R-C\equiv C-H + NaNH_2 \longrightarrow R-C\equiv CNa + NH_3$$

Problems

5.25 **Addition Reactions of Alkenes:** Write the products of the following addition reactions:

(a) $CH_3(CH_2)_3CH{=}CH_2 + Br_2 \longrightarrow$

(b) cyclopentene $+ H_2 \xrightarrow{Ni}$

(c) 1-methylcyclohexene (CH_3 on ring) $+ HBr \longrightarrow$

(d) $CH_3CH_2\overset{CH_3}{\overset{|}{C}}{=}CH_2 + HCl \longrightarrow$

(e) $CH_3\overset{CH_3}{\overset{|}{C}}{=}CHCH_3 + HI \longrightarrow$

(f) $CH_3\overset{CH_3}{\overset{|}{C}}HCH{=}\overset{CH_3}{\overset{|}{C}}CH_3 + H_2O \xrightarrow{H_2SO_4}$

(g) $CH_2{=}CH-CH{=}CH_2 + 2Cl_2 \longrightarrow$

(h) cyclohexenyl$-CH_2CH_3 + H_2O \xrightarrow{H_2SO_4}$

5.26 **Addition Reactions of Alkynes:** Write the products of the following addition reactions:

(a) $CH_3CH_2C\equiv CH + 1Cl_2 \longrightarrow$

(b) $CH_3CH_2C\equiv CCH_3 + 2Br_2 \longrightarrow$

(c) $CH_3CH_2CH_2CH_2C\equiv CH + 1H_2 \xrightarrow{Ni}$

(d) $CH_3CH_2C\equiv CCH_2CH_3 + 2H_2 \xrightarrow{Ni}$

(e) $CH_3CH_2CH_2C\equiv CH + 1HBr \longrightarrow$

(f) $CH_3CH_2CH_2C\equiv CH + 2HBr \longrightarrow$

5.27 **Reaction Mechanisms—Electrophilic Addition to Alkenes:** Write step-by-step mechanisms for the following reactions:

(a) $CH_3CH{=}CH_2 + Br_2 \longrightarrow$

(b) $CH_3\overset{CH_3}{\overset{|}{C}}{=}CHCH_3 + HCl \longrightarrow$

(c) 1-methylcyclohexene (CH_3 on ring) $+ HBr \longrightarrow$

(d) $CH_3\overset{CH_3}{\overset{|}{C}}{=}CH_2 + H_2O \xrightarrow{H_2SO_4}$

5.28 **Reaction Mechanisms—Electrophilic Addition to Alkynes:** Write a step-by-step mechanism for the reaction of 2-butyne with two moles of HCl.

5.29 **Bromination:** Show the product (including stereochemistry; *cis* or *trans*) of the bromination of cyclohexene.

5.30 **Hydrogenation:** Write equations showing the reaction of one mole of hydrogen with a metal catalyst such as Pd with each of the following compounds. In each case show the product formed from both hydrogens adding to the same side of the multiple bond. **(a)** 4-methyl-2-pentyne; **(b)** 1,2-dimethylcyclopentene.

5.31 **Reaction with $KMnO_4$:** Write the reaction of methylcyclopentene with potassium permanganate showing *cis* (syn) addition.

5.32 **Hydration of Alkynes:** Write equations showing the reaction of the following alkynes with water in the presence of sulfuric acid and mercury(II) sulfate: **(a)** 1-butyne; **(b)** 2-butyne.

5.33 **Electrophilic Addition to Conjugated Dienes** Write reaction equations showing 1,2 and 1,4 addition of the reagents shown to the conjugated dienes. For each, draw the two resonance forms of the intermediate allylic carbocation.

(a) $CH_2{=}C(CH_3){-}CH{=}CH_2 + H_2O \xrightarrow{H_2SO_4}$

(b)  cyclopentadiene $+ Cl_2 \longrightarrow$

5.34 **Resonance Forms and Resonance Hybrids:** Draw the resonance forms and resonance hybrids of the following reaction intermediates. Examples (a) and (b) each have two resonance forms and example (c) has three. In each case, the first one is shown.

(a) $CH_3CH{=}CH\dot{C}H_2$ (b) $\overset{..}{\underset{-}{C}}H_2{-}\overset{O}{\overset{\|}{C}}H$

(c) 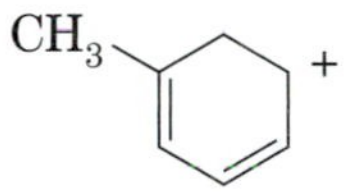 CH_3-substituted cyclohexenyl cation (+) (d) 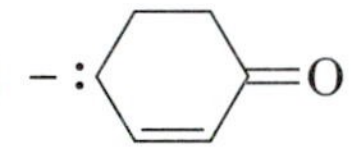 $-{:}$ cyclohexenone anion ($=O$)

5.35 **Resonance Forms and Resonance Hybrids:** The carbonate ion has three resonance forms (one is shown below). Draw these resonance forms and the resonance hybrid. All carbon-oxygen bonds are identical in the hybrid. Finally, draw a π-bonding picture.

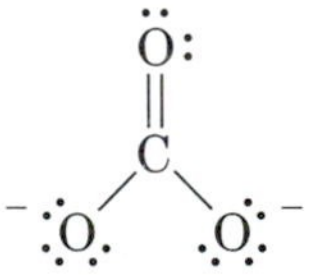

$O{=}C(O^-)(O^-)$ Carbonate ion

5.36 **Addition Polymers:** Write structures for the polymers produced from the following monomers:
(a) $CH_2{=}CF_2$ Its polymer is polyvinylidene fluoride, used to make rubberlike articles.
(b) $CH_2{=}CHBr$ Vinyl bromide is used to produce flame-retardant polymers.

5.37 **Oxidation of Alkenes:** Write equations showing the reactions of the following alkenes with a solution of potassium permanganate: **(a)** propene; **(b)** cyclopentene.

5.38 **Ozonolysis:** Write the products of ozonolysis of the following compounds with O_3 followed by Zn/H_2O:

(a) $CH_3\overset{CH_3}{\overset{|}{C}}{=}CHCH_2CH_2\overset{CH_2}{\overset{\|}{C}}CH{=}CH_2$

Myrcene, bayberry

(b) 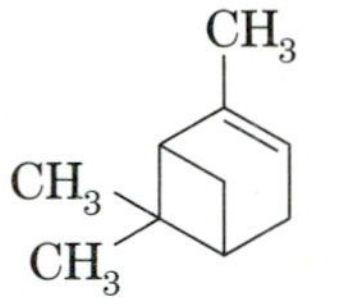 Pinene (bicyclic, CH_3 on ring double bond; gem-CH_3, CH_3)

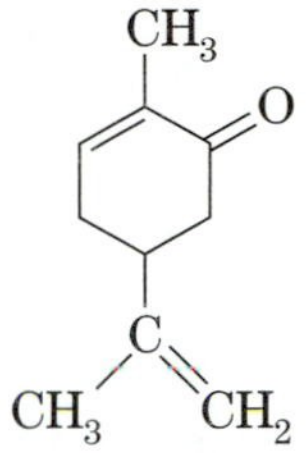

(c) Carvone (cyclohexenone with CH_3; substituent $C(CH_3){=}CH_2$)

5.39 **Ozonolysis** Write molecular structures for the following compounds based on the molecular formulas and ozonolysis products shown:

(a) $C_8H_{16} \xrightarrow{O_3} \xrightarrow[H_2O]{Zn} 2CH_3\overset{O}{\overset{\|}{C}}CH_2CH_3$

(b) $C_7H_{14} \xrightarrow{O_3} \xrightarrow[H_2O]{Zn} CH_3CH_2CH_2\overset{O}{\overset{\|}{C}}H + CH_3\overset{O}{\overset{\|}{C}}CH_3$

(c) $C_{10}H_{18} \xrightarrow{O_3} \xrightarrow[H_2O]{Zn} H\overset{O}{\overset{\|}{C}}CH_2CH_2\overset{O}{\overset{\|}{C}}H + 2CH_3\overset{O}{\overset{\|}{C}}CH_3$

(d) $C_9H_{14} \xrightarrow{O_3} \xrightarrow[H_2O]{Zn} CH_3\overset{O}{\overset{\|}{C}}{-}C_5H_8{-}\overset{O}{\overset{\|}{C}}CH_3$ (1,3-cyclopentane)

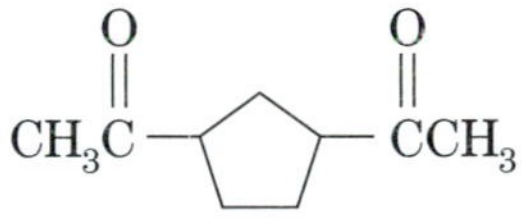

5.40 **Acidity of Terminal Alkynes:** Write equations showing the reactions of the following alkynes with sodium amide: **(a)** propyne; **(b)** 1-pentyne; **(c)** 2-pentyne.

5.41 **Synthesis:** In formulating a synthesis for an organic compound, one often works backwards from the product desired to a starting material that is available. Try some of the following by working backward and proposing structures for the unknown compounds A, B, C, ... The reactions you will need are the elimination reactions to prepare alkenes and alkynes (section 4.5) and the addition reactions of alkenes and alkynes (sections 5.1–5.2).

(a) $A + H_2 \xrightarrow{Ni} CH_3CH_2CH_2CH_2CH_3$

(b) $B \xrightarrow{2NaNH_2} CH_3C{\equiv}CH$

(c) $C \xrightarrow{H_2SO_4} CH_3\overset{CH_3}{\overset{|}{C}}HCH{=}CHCH_3$

(d) $D \xrightarrow{HCl} CH_3\overset{CH_3}{\overset{|}{C}}H\underset{Cl}{\underset{|}{C}}HCH_3$

(e) $E \xrightarrow{H_2SO_4} F \xrightarrow[Ni]{H_2}$ (cyclohexane)

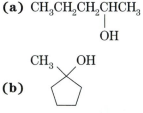

(f) $G \xrightarrow{KOH} H \xrightarrow{Br_2} CH_3CHBrCH_2Br$

(g) $I \xrightarrow{HBr} J \xrightarrow{KOH} K \xrightarrow{Cl_2} CH_3CHClCHClCH_3$

5.42 **Hydration:** What alkenes upon hydration would produce the following alcohols?

(a) $CH_3CH_2CH_2CHCH_3$ (OH on C-2)

(b) 1-methylcyclopentanol (CH_3 and OH on the same ring carbon)

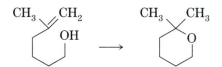

5.43 **Reaction Mechanism:** The following reaction occurs under acid conditions. Propose a mechanism. Start with hydrogen ion adding to the double bond to produce a carbocation.

(cyclic structure with CH_3, $=CH_2$, OH) → (cyclic ether with CH_3, CH_3, O)

5.44 **Hydrogenation:** Write reaction equations showing the preparation of the following compounds from hydrogenation of alkynes: **(a)** *cis* 2-pentene; **(b)** *cis* 4-methyl-2-hexene.

5.45 **Reactions of Alkynes:** Write reaction equations showing the preparation of the following compounds from alkynes: **(a)** 2,2-dibromopentane; **(b)** 3,3-dichloropentane; **(c)** 2,2,3,3-tetrachloropentane.

5.46 **Units of Unsaturation:** How many units of unsaturation are there in 1-buten-3-yne? How many moles of bromine could add to one mole of the compound?

5.47 **Units of Unsaturation:** A noncyclic hydrocarbon with eight carbons when completely hydrogenated consumes four mole equivalents of hydrogen. Give the molecular formula of both the starting material and the hydrogenation product.

5.48 **1,4 Addition:** Draw the diene from which 1,4-dibromo-2-cyclohexene would be formed upon addition of one mole of bromine.

5.49 **Allylic Carbocations:** Identify the different carbons to which a bromide ion could bond in neutralizing the following carbocation. Write the structures of the possible products. (Hint: draw the resonance forms.)

$$CH_3CH{=}CH{-}CH{=}CH{-}\overset{+}{C}HCH_2CH_3$$

Activities with Molecular Models

1. Make molecular models of ethene and ethyne. Now convert these to the products formed when bromine (Br_2) adds to the double bonds and triple bonds to form single bonds. How many bromines are needed to convert a double bond to a single bond and a triple bond to a single bond? How many bromines are in your products and to which carbons did they add?

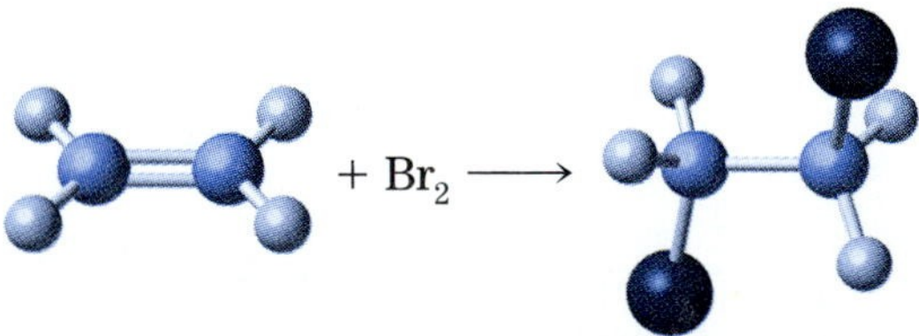

2. Make molecular models of 1-butene and 2-butene *(cis* or *trans)*. Make models of the one product formed from the addition of HBr to 2-butene and the two products formed from 1-butene. Why is there a difference in the number of addition products. Which product predominates in the addition to 1-butene?

3. Make a model of 2-butyne and the product *cis* addition of hydrogen.

4. Make a model of cyclopentene and the product of *trans* addition of bromine.

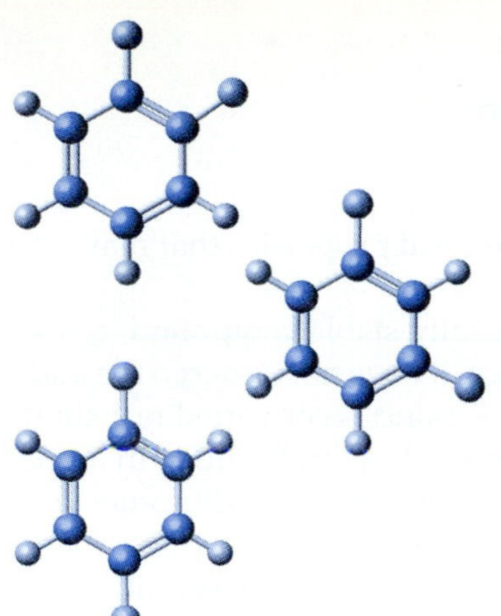

CHAPTER 6

Aromatic Compounds

6.1 Introduction to Aromatic Compounds

The term **aromatic** is used to refer to benzene and compounds similar to benzene in structure and chemical behavior. Benzene, C_6H_6, is a cyclic compound commonly written as a hexagon with alternating double and single bonds, or with a circle drawn in the center of a hexagon. Each corner of the hexagon represents a carbon with one bonded hydrogen.

aromatic compounds compounds that resemble benzene in structure and chemical behavior

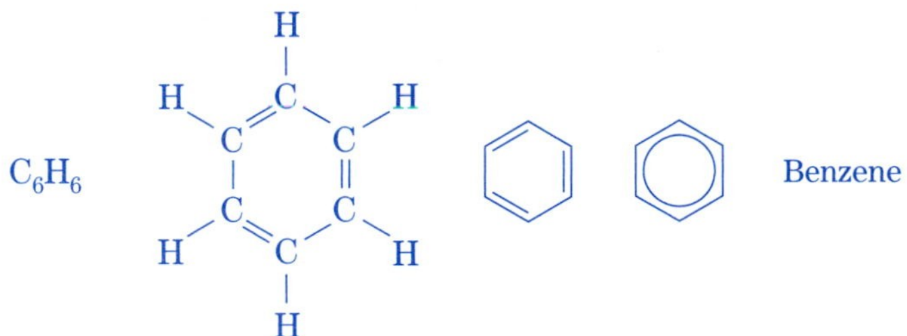

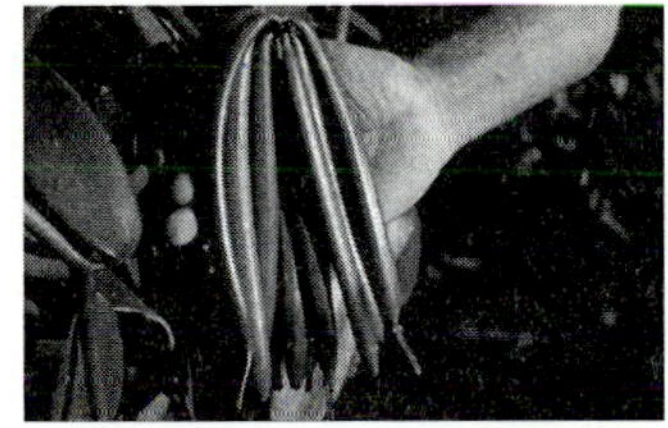
Vanilla beans

Although benzene and many other aromatic compounds are extracted from foul-smelling coal tar, they tend to have a fragrant odor, hence the term *aromatic.* Some aromatic compounds with additional functional groups impart the characteristic fragrances of wintergreen, cinnamon, cloves, vanilla, and roses.

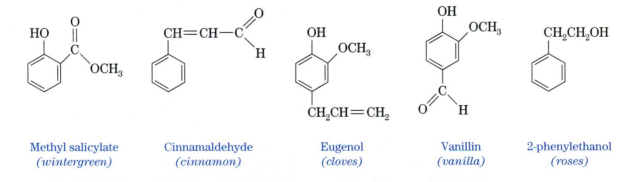

Methyl salicylate *(wintergreen)*	Cinnamaldehyde *(cinnamon)*	Eugenol *(cloves)*	Vanillin *(vanilla)*	2-phenylethanol *(roses)*

Both naturally occurring and synthetic aromatic compounds have a variety of applications, including uses as antiseptics, local anesthetics, food preservatives, gasoline components, dyes, detergents, plastics and synthetic fibers, pain relievers, and other medications.

6.2 Benzene: Structure and Bonding

A. Unusual Characteristics of Benzene

Benzene shows some unusual structural and chemical properties that make it a special type of organic compound.

1. ***Unexpected Stability.*** Benzene is an unusually stable compound, as evidenced by its relative resistance to chemical change. It does not undergo the addition reactions typical of alkenes, in which a double bond is converted to a single bond. For example, cyclohexene reacts instantaneously with bromine in a relatively dilute carbon tetrachloride solution, decolorizing it as 1,2-dibromocyclohexane forms. Benzene, with what appears to be the equivalent of three double bonds, surprisingly does not react with bromine at all under these conditions.

Cyclohexene: + Br_2/CCl_4 ⟶ (1,2-dibromocyclohexane: Br, Br)

Benzene: + Br_2/CCl_4 ⟶ No reaction

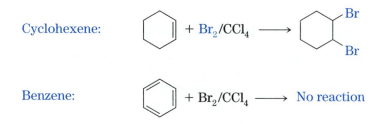

In order to react with bromine, benzene requires a catalyst ($FeBr_3$), and even then it undergoes substitution rather than addition. In substitution reactions, the integrity of the benzene ring is preserved—testimony to its unusual stability.

C_6H_6 + $Br_2/FeBr_3$ ⟶ C_6H_5Br + HBr

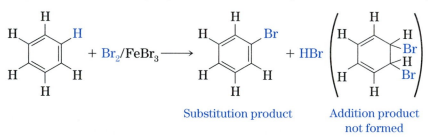

Substitution product

Addition product not formed

Benzene does undergo some addition reactions but with greater difficulty than might be expected and, again, with perplexing results. For example, the hydrogenation of cyclohexene is an exothermic reaction with 28.6 kcal of energy evolved for each mole of cyclohexene hydrogenated. 1,3-Cyclohexadiene, with twice the number of double bonds, produces almost twice as much energy upon hydrogenation, just as we might expect.

Cyclohexene + H_2 $\xrightarrow{Ni}$ cyclohexane + heat (28.6 kcal/mol)

1,3-Cyclohexadiene + $2H_2$ $\xrightarrow{Ni}$ cyclohexane + heat (55.4 kcal/mol)

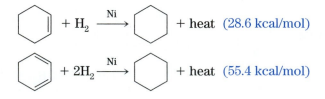

Following this same logic, we would expect benzene, with what appears to be three carbon-carbon double bonds, to produce three times the energy of cyclohexene upon hydrogenation or 85.8 kcal/mole. But hydrogenation of benzene produces 49.8 kcal/mole, even less than that produced by 1,3-cyclohexadiene.

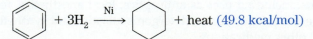

Benzene + $3H_2$ $\xrightarrow{Ni}$ cyclohexane + heat (49.8 kcal/mol)

Benzene contains 36 kcal/mole less energy than would be predicted. This energy difference is called the **resonance energy.** Benzene is 36 kcal/mole more stable than would be expected.

resonance energy
a measure of the degree to which a compound is stabilized by resonance

2. ***Carbon-Carbon Bond Lengths.*** Only one form of 1,2-dibromobenzene is known. The following structures represent the same compound even though, as written, the bromines flank a single bond in one case and a double bond in the other.

Br Br

Br Br

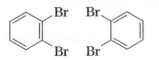

Physical measurements show that all the carbon-carbon bond lengths in benzene are identical and intermediate in length between normal carbon-carbon single and double bonds.

C—C	C⋯C	C=C
Single bonds	Bonds in benzene	Double bonds
1.54 Å	1.40 Å	1.34 Å

As a result, benzene is sometimes depicted as a hexagon with a circle drawn inside rather than with alternating double and single bonds. The circle within a hexagon is descriptively more accurate, and the alternating single- and double-bond model is better for electron bookkeeping.

GETTING INVOLVED

- ✓ What is the molecular formula for benzene? How is it represented in a structural formula? Draw the hexagon with alternating double and single bonds a few times to become comfortable with it.
- ✓ Does benzene really have alternating double and single bonds? What are the carbon-carbon bonds like? How does the study of 1,2-dibromobenzene and bond length measurements shed light on bonding in benzene?
- ✓ How does the difference in reactivity and in reaction type in the bromination of benzene relative to the bromination of cyclohexene provide evidence of benzene's unusual stability?
- ✓ How does a comparison of heats of hydrogenation of cyclohexene, 1,3-cyclohexadiene, and benzene illustrate the stability of benzene? What is resonance energy?

B. Bonding in Benzene

1. ***Resonance Hybrid Picture of Benzene.*** Benzene can be described by two resonance forms (see section 5.4 for resonance introduction).

Resonance forms (indicated by double-headed arrows) are used to describe electron delocalization. They themselves do not actually exist. Instead, the resonance forms are classical electronic structures used to describe a more complex structure. They differ only in the position of electrons, not in atoms. In the two resonance forms of benzene, the positions of the carbon-carbon double bonds change, but the carbon atoms remain stationary.

resonance forms
symbolic, nonexistent structures, differing only in positions of electrons, that are used to describe an actual molecule or ion

resonance hybrid
"average" of the resonance forms used to describe a molecule or ion that cannot be described by a single structure

However, benzene does not alternate between the two resonance structures, nor are some benzene molecules of one form and the rest of the other. The true structure of benzene is an average of the resonance forms called the **resonance hybrid.** Wherever there is a double bond in one resonance representation, there is a single bond in the other. Averaging these, we get a resonance hybrid with six identical carbon-carbon bonds all intermediate in length between carbon-carbon double and single bonds, commonly represented by the circle in a hexagon.

Resonance hybrid of benzene

Describing benzene by using two resonance forms to depict the resonance hybrid is analogous to describing a mule as a hybrid of a horse and a donkey.* The mule is not a horse part of the time and a donkey the rest, but an individual creature with characteristics of both. The analogy fails in that horses and donkeys actually exist, whereas contributing resonance structures do not. Another analogy describing a rhinoceros as a hybrid of the fictional dragon and unicorn is better.† The rhinoceros is real, but the dragon and unicorn are not.

2. ***The Orbital Picture of Benzene.*** An orbital description of benzene more satisfactorily explains the structure of the resonance hybrid. Since each carbon in the resonance forms is involved in a double bond, and we know that a double bond is composed of a σ bond and a π bond, each carbon must possess a p orbital. The only difference in the two resonance forms is in which p orbitals are shown overlapping (Figure 6.1a). However, if you could put yourself in the position of a p orbital, you would find the two adjacent p orbitals on either side of you to be identical and equidistant. Consequently, the p orbital would necessarily and unavoidably overlap with *both* of the adjacent p orbitals. This is the situation with the p orbitals in benzene. There is continuous overlap of the six p orbitals around the ring in the resonance hybrid (Figure 6.1b). This explains the fact that all carbon-carbon bonds in benzene are equivalent and intermediate in length between single and double bonds.

C. Structure of Benzene—A Summary

The following summary statements describe benzene, the parent of aromatic compounds (see Figure 6.1).

1. The molecular formula is C_6H_6.
2. The carbons exist in a flat six-membered ring with a cloud of six π electrons overlapping above and below the ring.
3. All six carbons are equivalent.
4. All carbon-carbon bond lengths are equivalent and intermediate between single and double bonds.
5. All six hydrogens are equivalent.
6. Each carbon is trigonal, sp^2 hybridized, and has 120° bond angles.

A wide variety of structures qualify as aromatic compounds, all of which are more stable than would have been predicted. All bear structural features resembling benzene: cyclic, planar, p orbital on each atom of the ring, and 2, 6, 10, 14,

*Analogy by G. W. Wheland, University of Chicago.
†Analogy by J. D. Roberts, California Institute of Technology.

(a)

(b)

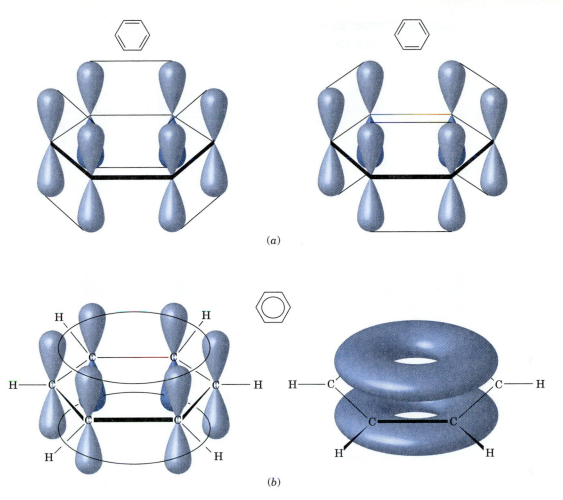

Figure 6.1 π bonding representation of benzene. (a) Resonance forms of benzene. (b) Resonance hybrid of benzene showing continuous overlap of p orbitals. The total π cloud represents six electrons with delocalization above and below the ring.

18, and so forth π electrons (benzene has six; naphthalene in Example 6.1 has 10, and anthracene in problem 6.1 has 14), which are delocalized by resonance.

Benzene is commonly represented with alternating double and single bonds (resonance forms) or with a circle in a hexagon (resonance hybrid).

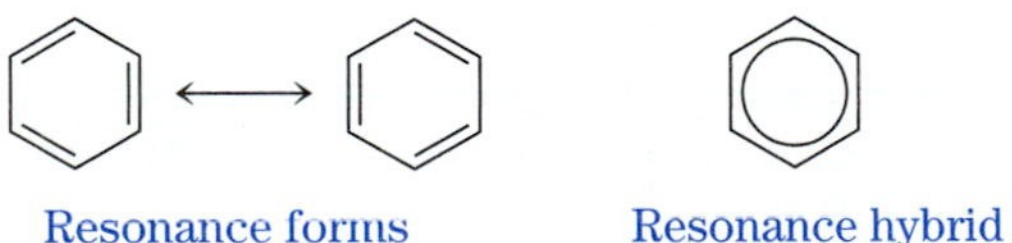

The circle in a hexagon more clearly represents that all carbon-carbon bonds are equivalent and intermediate between double and single bonds. The alternating double- and single-bond representation is good for electron-bookkeeping and for writing reaction mechanisms. For this reason, we will use it throughout the text. However, you must remember that there are no double or single bonds and that the resonance forms do not even exist; they are merely used to describe a more complex structure.

GETTING INVOLVED

- ✓ How are resonance forms and the resonance hybrid related? Which really exists? Write the resonance forms and the resonance hybrid of benzene?
- ✓ Draw the π-bonding picture of benzene and describe how it explains the equivalencies of all carbon-carbon bonds, the six carbons, and the six hydrogens.
- ✓ Describe the hybridization, geometry, and bond angles common to each carbon in benzene.

Example 6.1

A classical structure for naphthalene is shown on the next page. Draw the resonance hybrid and the π bonding picture.

Connections 6.1 www.prenhall.com/bailey

Cancer and Carcinogens

Many aromatic compounds have been indicted as cancer-causing agents. These include both naturally occurring and synthetic species.

Cancer is a term that strikes fear into the hearts of everyone. As we continue to conquer bacterial and viral infections and control heart disease, cancer remains obstinately resistant to cure. It is characterized by cell proliferation beyond an organism's normal growth and replacement needs. There are many natural and environmental agents, generically called *carcinogens,* which can cause or stimulate cancerous processes. Some of these substances are unavoidable, and others are a matter of lifestyle choice.

O O O O O OCH_3

Aflatoxin B_1

$HOCH_2-C_6H_4-NHNH\overset{O}{\overset{\|}{C}}(CH_2)_2\underset{NH_2}{\underset{|}{C}}H\overset{O}{\overset{\|}{C}}OH$

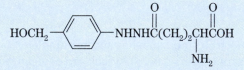

Agaritine

Although both fungi and plants can be a source of beneficial medicines, others produce aromatic compounds that are potent carcinogens. Such chemicals include aflatoxin B_1 and agaritine from fungi and a variety of toxins from tobacco.

Aflatoxin B_1 is produced by a mold, *Aspergillus flavus,* which grows on crops such as corn, dried chili peppers, and peanuts. It is both toxic and carcinogenic and can sometimes be found in homemade peanut butter, as well as in the food staples of underdeveloped nations.

Agaritine is found in mushrooms, including those that are edible. It is not as potent as aflatoxin B_1, but one would not want to consume excessive quantities of mushrooms to prove a point.

Carcinogens can also be found in plants, most notably in tobacco, which contains nitrosonornicotine and related toxins. Cigarette smoke has been shown to contain more than 4000 chemical compounds, among them a number of polycyclic aromatics such as benzo[a]pyrene. These are some of the most potent carcinogenic materials known. In a bizarre twist of nature, the polycyclic aromatics are converted to carcinogens during the natural process of biotransformation within the body. They are oxidized to epoxides, electrophiles that are attracted to the electron-rich, nitrogen-containing rings of DNA and RNA, our genetic memory. The resulting reactions alter the integrity of the genetic code.

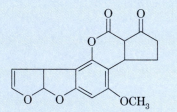

Lung cancer has been indisputably linked to smoking and is almost invariably fatal. The spouses and children of smokers also have a higher-than-normal incidence of this deadly form of cancer when passively subjected to the so-called side-stream effluent of the smoker.

NO N N

Nitrosonornicotine
(found in tobacco)

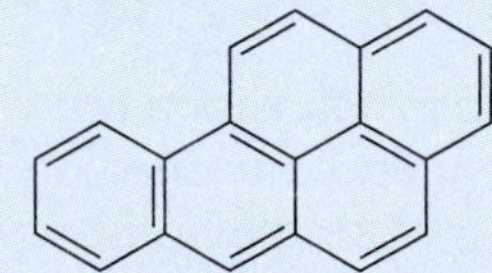

Benzo(a)pyrene
(product of tobacco smoke)

Connections 6.1 *(cont.)*

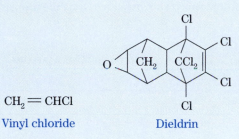

$CH_2{=}CHCl$

Vinyl chloride

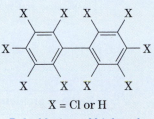

Dieldrin

X = Cl or H

Polychlorinated biphenyls
PCBs

Synthetic halogenated hydrocarbons are also potential carcinogens. This has led to the strict control or outright banning of industrial and agricultural chemicals. The monomer unit of polyvinyl chloride, vinyl chloride, is a carcinogen, as are the banned insecticide dieldrin and aromatics still found in power transformers, polychlorobiphenyls (PCBs).

A great deal of controversy surrounds safe exposure limits, such as how much of a known toxin or carcinogen is safe to consume. Dr. Bruce Ames of the University of California at Berkeley, discoverer of a widely used test for carcinogenicity, is a well-known proponent of moderation in evaluating chemicals. He has documented many naturally occurring carcinogens to which we are exposed daily in amounts comparable to industrial carcinogens with few observable effects. Many substances classified by the U.S. Environmental Protection Agency as carcinogens are actually not directly cancer-causing. They may be precursors, stimulators, or cooperative species depending upon circumstances and/or other chemicals to launch the cancerous growth process. This is an ongoing debate that requires that citizens be scientifically informed.

Mushrooms

Solution

In drawing the resonance hybrid and π bonding picture, notice that each carbon in naphthalene is involved in a double bond and therefore has a p orbital. Draw the framework of naphthalene, place a p orbital in each carbon, and connect the p orbitals continuously around the ring to obtain the bonding picture. To simplify the drawing of the bonding picture, it is best to represent naphthalene in a vertical fashion. The resonance hybrid should be shown with a circle in each ring, as was done with benzene.

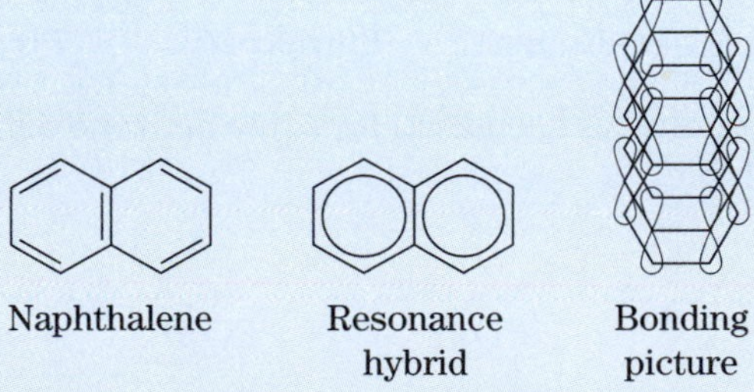

Naphthalene — Resonance hybrid — Bonding picture

Problem 6.1

Give the molecular formulas for each of the following compounds. Count the carbons and hydrogens very carefully: **(a)** bromobenzene; **(b)** naphthalene (see Example 6.1); **(c)** anthracene (see problem 6.2).

Problem 6.2

A classical structure for anthracene is shown below. Draw the resonance hybrid and a π bonding picture.

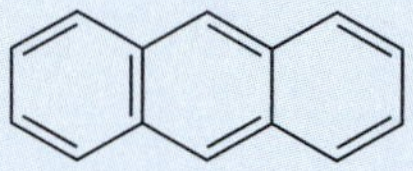

Anthracene (from coal tar)

See related problem 6.19.

6.3 Nomenclature of Aromatic Compounds

A. Aromatic Hydrocarbon Ring Systems

Benzene, C_6H_6, is the most common aromatic ring. There are a variety of fused-ring aromatic hydrocarbons of which naphthalene, anthracene, and phenanthrene are the most common. The numbering system shown is used to name derivatives of these three compounds.

Benzene

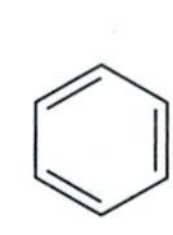

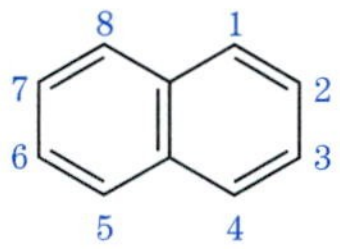

Naphthalene

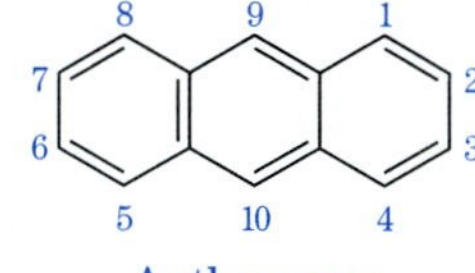

Anthracene

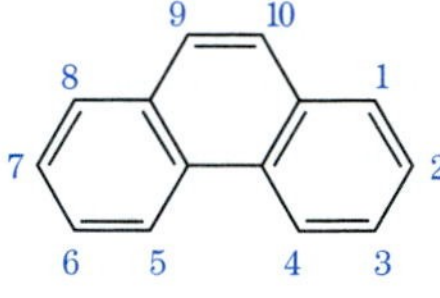

Phenanthrene

GETTING INVOLVED

Problem 6.3

Draw all the positional isomers in which a bromine can replace a hydrogen on **(a)** naphthalene, **(b)** anthracene, and **(c)** phenanthrene. Do not repeat any structures.

See related problem 6.27.

B. Monosubstituted Benzenes

Monosubstituted benzenes are named as derivatives of benzene.

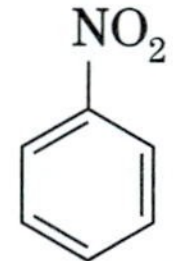

Nitrobenzene

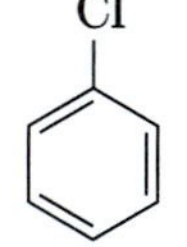

Chlorobenzene

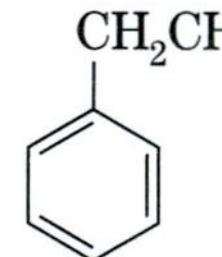

Ethylbenzene

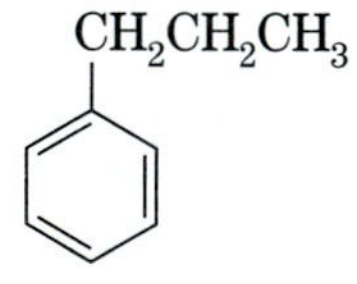

Propylbenzene

Common names have always been used for some benzene derivatives, and these are currently acceptable.

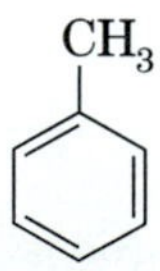

CH_3	$\overset{O}{\|\|}$ CH	CO_2H	SO_3H	OH	NH_2
Toluene	Benzaldehyde	Benzoic acid	Benzenesulfonic acid	Phenol	Aniline

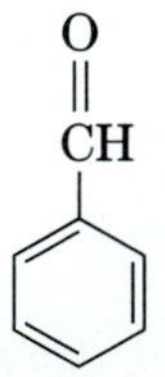

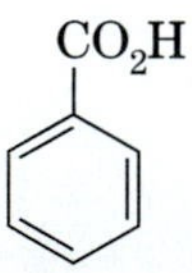

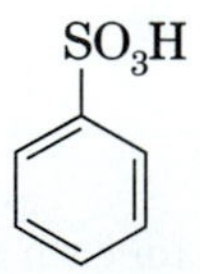

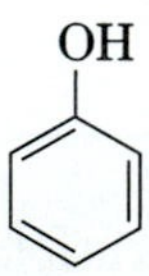

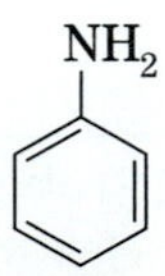

GETTING INVOLVED

✓ Be sure to learn the names and structures of the six compounds given in which common names are acceptable.

Problem 6.4

Name the following compounds:

(a) 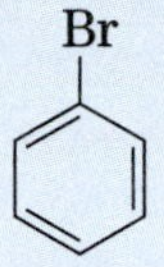 Br (b) 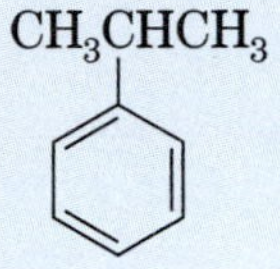 CH_3CHCH_3 (c) 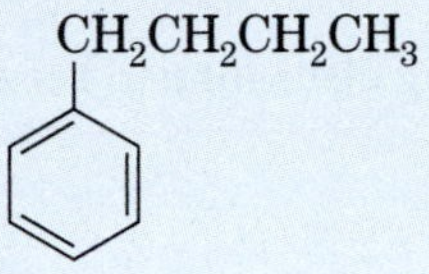 $CH_2CH_2CH_2CH_3$ (d) 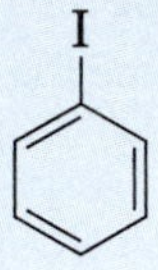 I

See related problem 6.21 a–c.

C. Disubstituted Benzenes

To name a disubstituted benzene, both groups and their relative positions must be identified. Every disubstituted benzene will have three positional isomers, as illustrated by the xylenes (dimethylbenzenes used in high-octane gasoline). If the groups are adjacent, in a 1,2 relation, they are termed *ortho* (o); if 1,3, *meta* (m); and if 1,4 *para* (p).

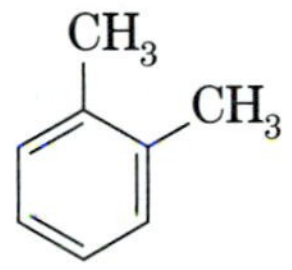

CH_3 CH_3 | CH_3 CH_3 | CH_3 CH_3

ortho-xylene, or 1,2-dimethylbenzene 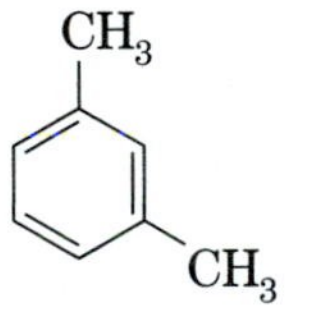 | *meta*-xylene, or 1,3-dimethylbenzene 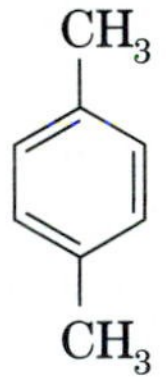 | *para*-xylene, or 1,4-dimethylbenzene

When two substituents are different, they are usually put in alphabetical order. If the compound is a derivative of a monosubstituted benzene designated by an accepted common name, it can be named as such.

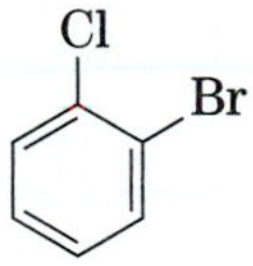

Cl Br | CO_2H NO_2 | CH_3 Cl

o-bromochlorobenzene 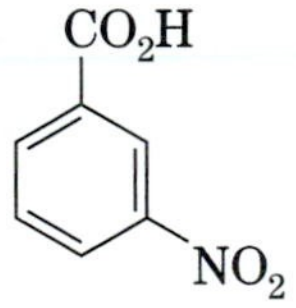 | *m*-nitrobenzoic acid 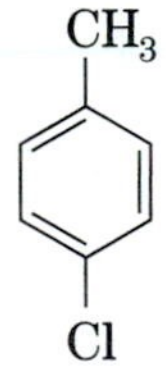 | *p*-chlorotoluene

GETTING INVOLVED

✓ What do the terms ortho, meta, and para designate?

Example 6.2

Name the following compounds:

(a) 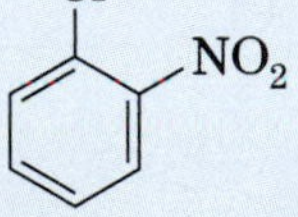 Cl NO_2 (b) 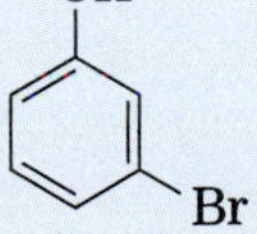 OH Br

Solution

(a) There are two substituents in a 1,2, or ortho, relationship; both are named by prefixes.
o-chloronitrobenzene, or 1-chloro-2-nitrobenzene

(b) Cover the bromine with your finger and you will see hydroxybenzene, which is usually called phenol. The two groups are *meta* (1,3).
m-bromophenol, or 3-bromophenol

Problem 6.5

Name the following disubstituted benzenes:

(a) 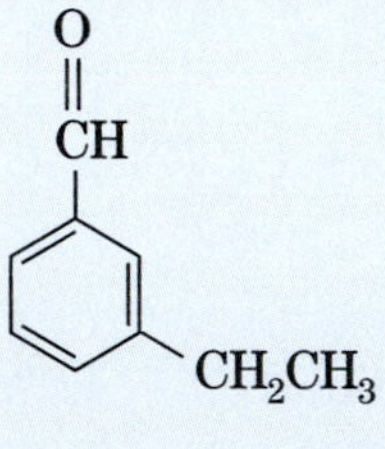 O=CH (benzene ring) with CH_2CH_3 (meta)   (b) 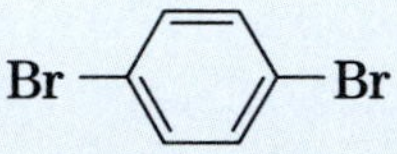 Br–(benzene ring)–Br (para)   (c) 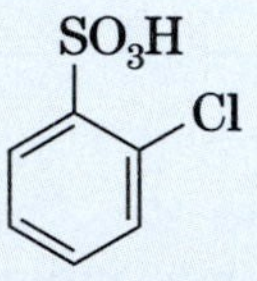 SO_3H, Cl (ortho)   (d) 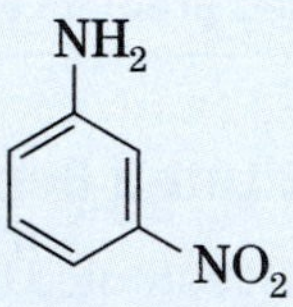 NH_2, NO_2 (meta)

See related problems 6.21 and 6.25.

D. Polysubstituted Benzenes

When more than two groups are on a benzene ring, their positions must be numbered. Ortho, meta, and para designations are not acceptable. If one of the groups is associated with a common name, the molecule can be named as a derivative of the monosubstituted compound, numbering from the group designated in the common name.

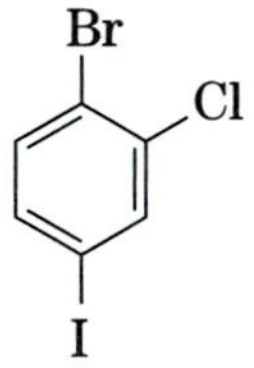

Br, Cl, I — 1-bromo-2-chloro-4-iodobenzene

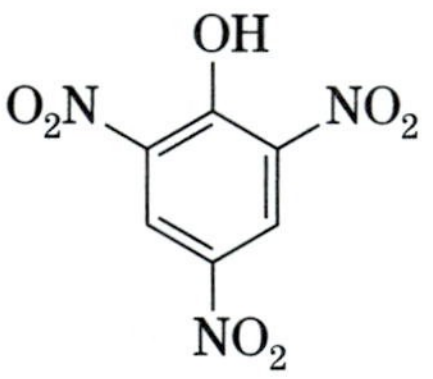

OH, O_2N, NO_2, NO_2 — 2,4,6-trinitrophenol

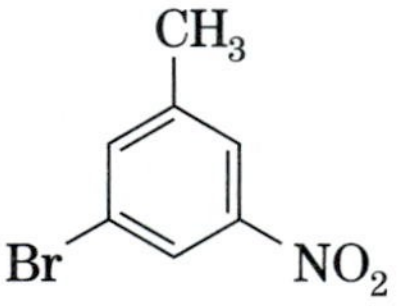

CH_3, Br, NO_2 — 3-bromo-5-nitrotoluene

GETTING INVOLVED

Problem 6.6

Name the following polysubstituted benzenes:

(a) 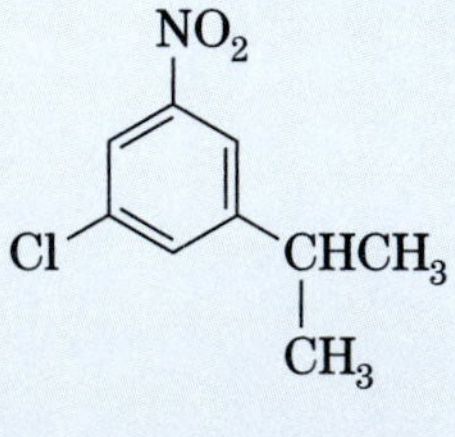 NO_2, Cl, $CHCH_3$ with CH_3 (b) 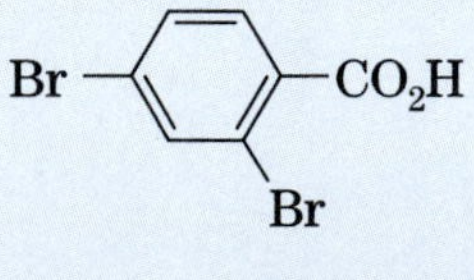 Br–(ring)–CO_2H, Br   (c) 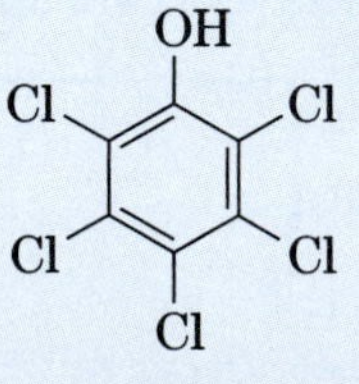 OH, Cl, Cl, Cl, Cl, Cl   (d) 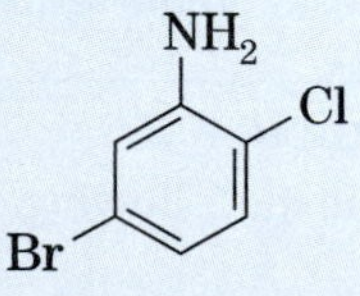 NH_2, Cl, Br

See related problem 6.22.

E. Substituted Anilines

Substituents on aniline are designated by a number (or o, m, p) if on the ring, or by N if on the nitrogen.

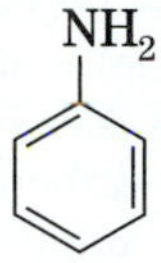

NH_2 NH_2 CH_3 $NHCH_3$ $NHCH_3$ CH_2CH_3

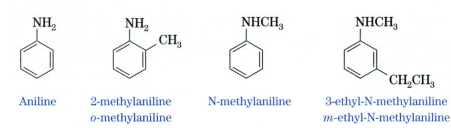

Aniline 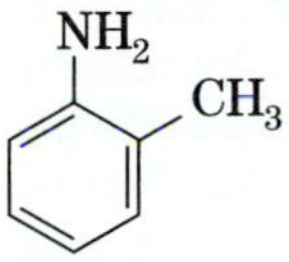 | 2-methylaniline *o*-methylaniline | N-methylaniline 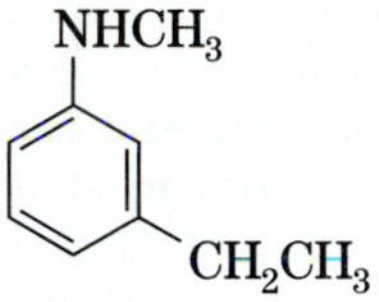 | 3-ethyl-N-methylaniline *m*-ethyl-N-methylaniline

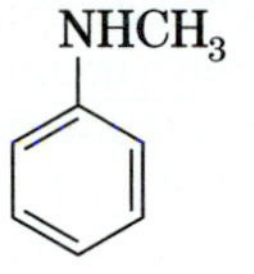

GETTING INVOLVED

Problem 6.7

Name the following substituted anilines. Notice that if you replace each substituent with H, you have aniline. To name each compound merely name and locate the substituent groups.

(a) $NHCH_2CH_2CH_2CH_3$ **(b)** CH_3 N CH_2CH_3 $CH_2CH_2CH_3$ **(c)** $N(CH_3)_2$ Cl Br

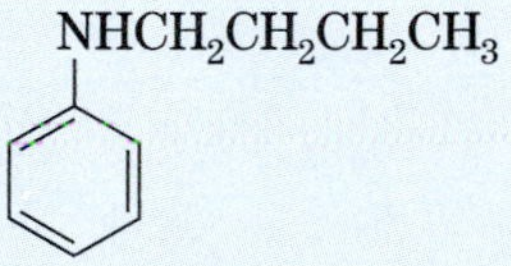

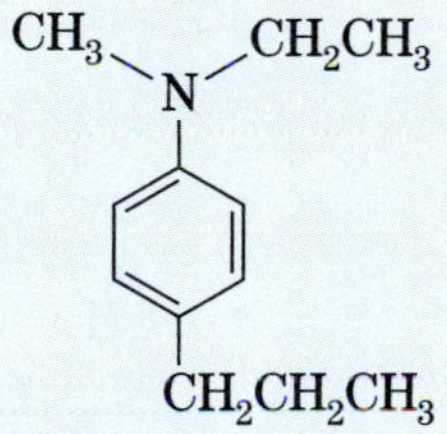

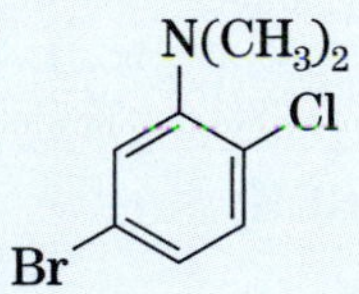

See related problem 6.23.

F. Aromatic Compounds Designated by Prefixes

Occasionally, the substituents on an aromatic ring are too complex to name conveniently with a prefix. In these cases, the aromatic ring is named with a prefix; *phenyl-* and *benzyl-* are commonly used.

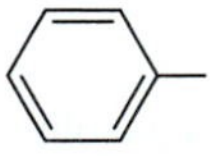

$-CH_2-$

Phenyl- 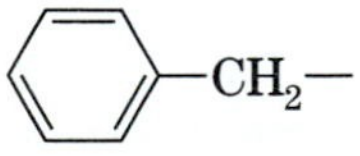 Benzyl-

In the following examples, the longest carbon chain is used as the base of the name, and the aromatic portion is identified as a substituent on the chain.

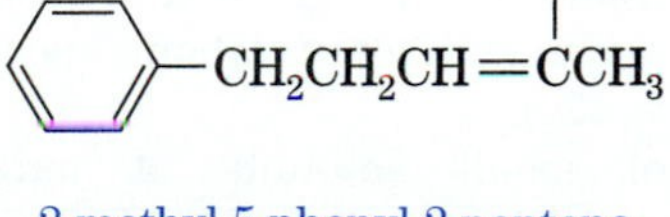

CH_3 $-CH_2CH_2CH=CCH_3$ HO CH_2

2-methyl-5-phenyl-2-pentene 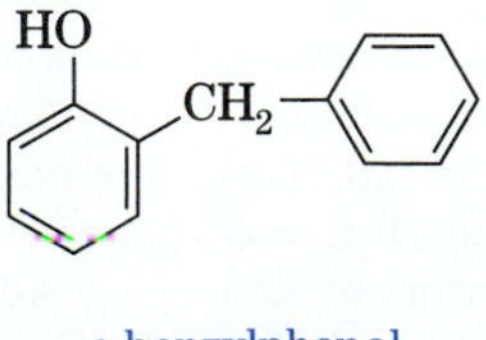 *o*-benzylphenol

Connections 6.2 — Gasoline

www.prenhall.com/bailey

Gasoline, produced from crude oil, is mainly a mixture of hydrocarbon molecules, aromatic and nonaromatic, composed of five to ten carbons. It is important that these compounds vaporize and mix with oxygen in the carburetor and then undergo a smooth and controlled combustion so that the resultant energy can steadily and evenly push against the piston and turn the wheels.

The nature of the hydrocarbons in gasoline is important. They must be easily vaporized, but not so volatile as to boil out of the gas tank. Higher molecular weight components (higher boiling points) in the summer and lower molecular weight components (lower boiling points; more easily vaporized) in the winter enhance gasoline performance.

Hydrocarbon structure also influences combustion qualities. Why is it so important how a fuel burns? A piston at the end of its compression cycle in an engine cylinder does not respond well to a sudden explosion of gasoline. Violent explosions, especially those contrary to engine timing, drive the piston through the cylinder uncontrollably, producing knocks and pinging, engine damage, and inefficient energy transfer. A good gasoline burns in a controlled fashion without knocking. Branched chain, cyclic, unsaturated, and especially aromatic hydrocarbons burn smoothly and have antiknock properties. Octane numbers are used to describe gasoline performance. When the scale was first devised, isooctane, $(CH_3)_3CCH_2CH(CH_3)_2$, a branched alkane noted for its antiknock qualities, was assigned an octane number of 100. Heptane, a straight-chained alkane and poor automotive fuel, was assigned zero. A gasoline that burns like a mixture of 90% isooctane and 10% heptane is assigned an octane number of 90 whether determined by the research octane number (RON, laboratory conditions) or motor octane number (MON, road conditions); the number posted on a gas pump is usually an average of the two. Octane numbers of some hydrocarbons are given in Table 6.1. In 1935, the research octane number for regular gasoline was 72 and 78 for premium. Octane numbers peaked around 1968 with regular at 94 and premium at 100. They have decreased since then as fuel conservation in automobiles has gained importance.

Two general refinery methods are used to enhance the quantity of gasoline obtainable from crude oil. Hydrocarbon molecules that are too large to meet gasoline requirements (more than 10 carbons) are thermally or catalytically "cracked" into molecules that are within the C_5 to C_{10} carbon range. Molecules with fewer than five carbons are combined to produce hydrocarbons with five to ten carbons in a process called *alkylation*.

Table 6.1 Research Octane Numbers for Hydrocarbon Molecules

Hydrocarbon	Unleaded RON	Hydrocarbon	Unleaded RON
$CH_3CH_2CH_3$	97		
$CH_3(CH_2)_4CH_3$	25	$C_6H_5CH_3$	120
$CH_3(CH_2)_6CH_3$	−19		
$CH_3(CH_2)_2CH{=}CHCH_3$	93		
$CH_3CH(CH_3)CH_2CH_2CH(CH_3)CH_3$	65	CH_3, CH_3 (ortho)	107
	101		
	106	CH_3, CH_3 (meta)	118

There are also two general refining processes for enhancing the quality of gasoline available from petroleum. Both fall under the heading of reforming, since in each, hydrocarbon molecules with five to ten carbons are transformed into higher-octane gasoline components. In isomerization, unbranched alkanes such as octane (octane number = −19) are converted into branched chain hydrocarbons such as isooctane (octane number = 100). In aromatization, saturated alkanes such as heptane (octane number = 0) are cyclized and dehydrogenated to aromatic compounds with the same number of carbons, such as toluene (octane number = 120).

Addition of small amounts of tetraethyllead, $Pb(CH_2CH_3)_4$, to gasoline can produce a substantial increase

Connections 6.2 *(cont.)*

in octane number. Tetraethyllead actually moderates fuel combustion so that it burns in a slow and controlled manner instead of detonating violently and causing knocking and pinging. Because lead compounds are often toxic, the amount of lead in gasoline has been regulated downward. Before the 1975 model year, leaded gasolines comprised 98% of the market. This statistic has been virtually reversed, with unleaded gasolines being almost completely dominant today. In order to maintain high octane ratings, unleaded gasolines are often blended with increased amounts of aromatic hydrocarbons, especially benzene, xylene, and toluene (BXT).

Gasohol is a mixture of 90% unleaded gasoline and 10% ethanol (beverage alcohol, CH_3CH_2OH). Alcohol is both a good fuel extender and octane booster and since it can be distilled from almost any type of crop or crop waste, it can be a profit-making by-product for farmers and a renewable energy source.

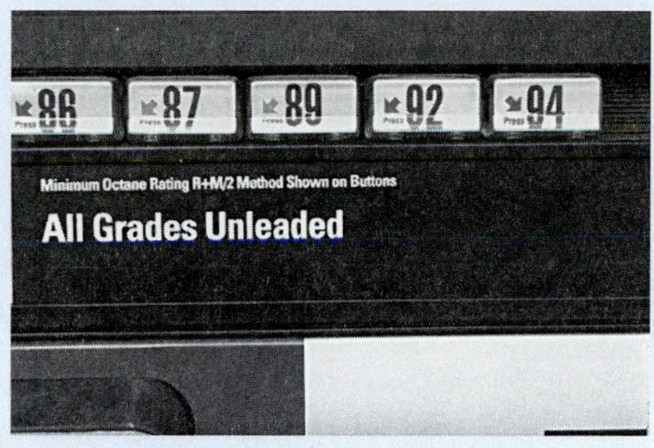

Gas pump showing octane ratings

GETTING INVOLVED

✓ What are phenyl and benzyl groups?

Example 6.3

Name the following compound:

$$(C_6H_5)_2C{=}CH{-}CH{=}CH_2$$

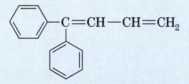

Solution

1. The longest chain is four carbons: but.
2. There are two double bonds: butadiene.
3. The chain is numbered to give the double bonds the lowest numbers: 1,3-butadiene.
4. All other substituents are named by prefixes. There are two phenyl groups on carbon-1. The complete name is

1,1,diphenyl-1,3-butadiene

Problem 6.8

Name the following compounds:

(a) $CH_3CH(C_6H_5)CH_2CH(CH_3)CH_2CH_3$ (phenyl on C-2, CH_3 on C-4)

(b) $C_6H_5{-}CH_2C{\equiv}CCH_2{-}C_6H_5$

(c) $C_6H_5{-}CH_2{-}C_6H_4{-}CO_2H$

See related problem 6.24.

6.4 Electrophilic Aromatic Substitution

We have seen that aromatic compounds are exceptionally stable and, as a consequence, are relatively resistant to chemical change. This stability is related to the unique electronic and bonding characteristics of benzene and aromatic compounds described in section 6.2. When these compounds engage in chemical reactions, they tend to retain their aromatic character, that is, the electronic structure of the benzene ring. Addition reactions would disrupt this electronic structure; as atoms or groups bonded to the ring, π bonds would be destroyed. Substitution reactions, however, preserve the integrity of the benzene ring, and the unusually stable π-bonding pattern remains intact.

Electrophilic aromatic substitution is the characteristic reaction of benzene and its derivatives. Because of its electron-rich clouds of π electrons, the benzene ring attracts electron-deficient species, electrophiles. In the course of the reaction, the electrophile replaces a hydrogen on the ring; the π-bonding pattern is retained.

These reactions will be presented according to the method we have established: (1) a summary of reactions, (2) the reaction mechanisms, and (3) a method for determining orientation of substitution.

A. Electrophilic Aromatic Substitution: The Reaction

Electrophilic aromatic substitution reactions are basically very simple—a hydrogen bonded to a benzene ring is replaced by another group or atom, as illustrated in Figure 6.2. Notice how simply the reaction can be visualized. The ring remains intact; it does not change. Something becomes bonded to the ring in place of a hydrogen. The new groups can be chlorine, bromine, alkyl groups, acyl groups, nitro, and the sulfonic acid group.

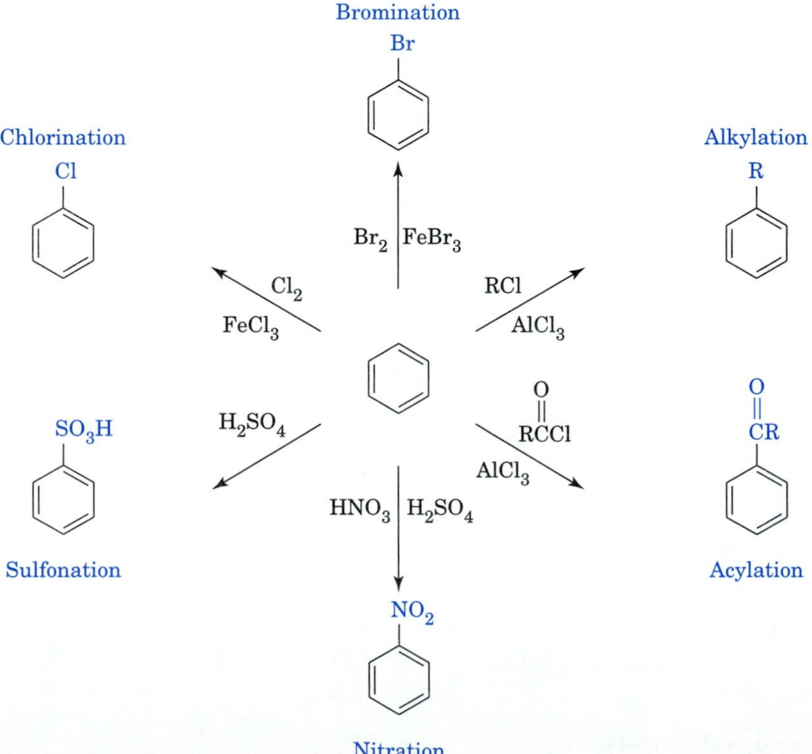

Figure 6.2 Electrophilic aromatic substitution reactions.

GETTING INVOLVED

- ✓ Why do aromatic compounds commonly undergo substitution reactions instead of addition reactions as do alkenes?
- ✓ Why are electrophiles attracted to a benzene ring? How does substitution preserve the electronic and thus aromatic character of the benzene ring; how would addition destroy it?
- ✓ Do you know the reagents and substituting groups for chlorination, bromination, alkylation, acylation, nitration, and sulfonation? Can you write reaction equations describing these reactions?

Example 6.4

Write equations illustrating the following reactions:

(a) benzene and 2-chloropropane with aluminum trichloride catalyst

(b) *p*-dibromobenzene with concentrated nitric and sulfuric acids

Solution

(a) In alkylation and acylation reactions, the carbon to which the chlorine is attached replaces the hydrogen on the benzene ring. Everything attached to that carbon goes along for the ride except the chlorine, which pairs up with the replaced hydrogen to form HCl as the by-product.

$$C_6H_6 + CH_3CH(Cl)CH_3 \xrightarrow{AlCl_3} C_6H_5CH(CH_3)CH_3 + HCl$$

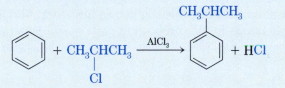

(b) These concentrated acids produce the nitro group (NO_2) as the electrophile. The same product is obtained no matter what hydrogen is replaced.

$$Br{-}C_6H_4{-}Br + HNO_3 \xrightarrow{H_2SO_4} Br{-}C_6H_3(NO_2){-}Br + H_2O$$

Problem 6.9

Write equations illustrating the reaction of *p*-xylene (1,4-dimethylbenzene) with the following reagents. All reactions are electrophilic aromatic substitution: **(a)** Cl_2, $FeCl_3$; **(b)** Br_2, $FeBr_3$; **(c)** CH_3CH_2Cl, $AlCl_3$; **(d)** $CH_3\overset{O}{\overset{\|}{C}}Cl$, $AlCl_3$; **(e)** HNO_3, H_2SO_4; **(f)** H_2SO_4.

B. Electrophilic Aromatic Substitution: The Mechanism

This is the general equation that illustrates electrophilic aromatic substitution. A hydrogen on the benzene ring is replaced by another group represented by E.

$$C_6H_5{-}H + E{-}A \xrightarrow{\text{Catalyst}} C_6H_5{-}E + HA$$

electrophile
electron-deficient species that accepts electrons from nucleophiles in a chemical reaction. Electrophiles are Lewis acids

When considering the mechanism of electrophilic aromatic substitution, we must remember that benzene is a relatively stable entity and is resistant to chemical change. It does possess an electron-rich π cloud above and below the ring. For benzene to react, it must be exposed to a species reactive enough to attract electrons from this π cloud. This species is the **electrophile;** it is positive and formed from the interaction between the reagent and catalyst, as illustrated in Table 6.2.

$$E—A + \text{catalyst} \longrightarrow \underset{\text{Electrophile}}{E^+}$$

Once formed, the positively charged electrophile is quickly drawn into benzene's π cloud of electrons. It accepts two electrons from the π system and bonds to one of the carbons; a carbocation intermediate results.

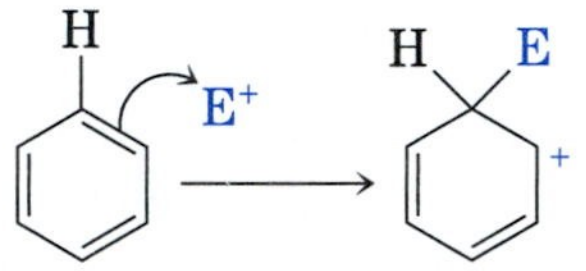

Carbocation

Benzene's aromaticity has been disrupted momentarily. However, the carbocation formed is allylic and stabilized by resonance (section 5.4). Resonance structures can be used to illustrate this stabilizing influence. Moving electron pairs from the double bonds to the carbocation carbon will generate the three resonance forms.

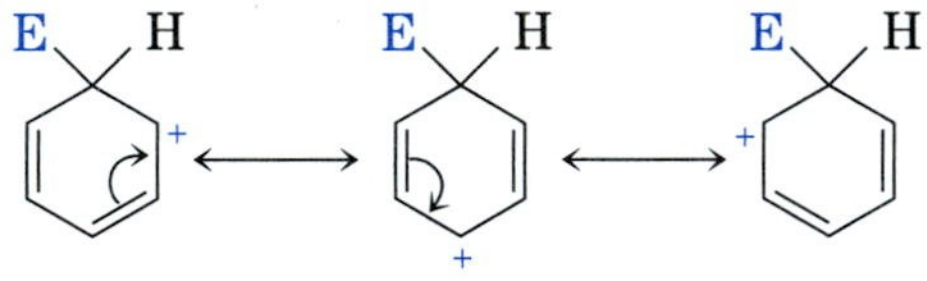

Resonance structures

Although this is a relatively stable carbocation, it is much less stable than the original benzene ring. As the hydrogen ion leaves and the two remaining electrons reconstitute the continuous ring of delocalized electrons, the aromatic benzene structure is regenerated and the substitution reaction is complete.

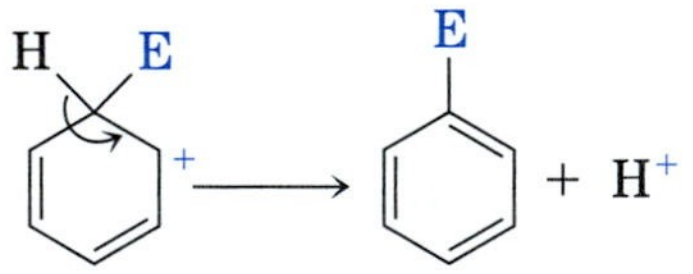

$+ H^+$

The drive to restore the stable aromatic ring is the main reason that substitution occurs instead of addition; addition would destroy the ring.

The mechanism of electrophilic aromatic substitution can be summarized as follows:

Generation of the Electrophile: $E—A + \text{catalyst} \longrightarrow E^+ + A:^-$

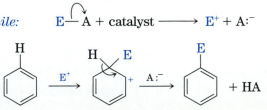

Two-Step Substitution: $\xrightarrow{E^+}$ Carbocation $\xrightarrow{A:^-}$ $+ HA$

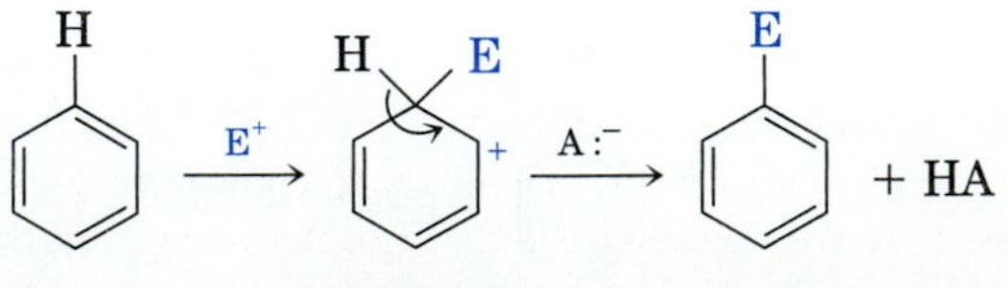

Table 6.2 Electrophilic Aromatic Substitution Reactions

Aromatic Hydrocarbon	Reagent	Catalyst	Electrophile	Product	By-product
Halogenation					
benzene (C–H)	Cl_2	Fe or $FeCl_3$	Cl^+	chlorobenzene (C–Cl)	HCl
benzene (C–H)	Br_2	Fe or $FeBr_3$	Br^+	bromobenzene (C–Br)	HBr
Friedel-Crafts Alkylation and Acylation					
benzene (C–H)	R—Cl	$AlCl_3$	R^+	alkylbenzene (C–R)	HCl
benzene (C–H)	RC(=O)—Cl	$AlCl_3$	RC^+(=O)	acylbenzene (C–C(=O)R)	HCl
Nitration					
benzene (C–H)	HNO_3	H_2SO_4	NO_2^+	nitrobenzene (C–NO_2)	H_2O
Sulfonation					
benzene (C–H)	H_2SO_4	. . .	$\overset{+}{S}O_3H$	benzenesulfonic acid (C–SO_3H)	H_2O

In the following sections, this general mechanism is applied to the specific reactions we are considering.

GETTING INVOLVED

- ✓ Describe the general mechanism for electrophilic aromatic substitution.
- ✓ Why is a positively charged electrophile attracted to the benzene ring? Where does it get electrons for bonding? What results from the bonding of the electrophile?
- ✓ Why is the carbocation formed so stable? Why is it not neutralized by a negative species directly? How is it neutralized and how does this restore the integrity of the aromatic ring?

Problem 6.10

Before we take a more detailed look at individual mechanisms of electrophilic aromatic substitution reactions, see if you can apply the general concepts to the chlorination of benzene. First, write the equation for generation of the electrophile and then the two-step substitution of the electrophile on the benzene ring.

1. ***Halogenation.*** The following general reaction equation illustrates the chlorination or bromination of benzene and its derivatives. Fluorine is too reactive and iodine too unreactive to be involved in electrophilic aromatic substitution.

$$C_6H_5H + X_2 \xrightarrow{FeX_3} C_6H_5X + HX \quad X = Cl, Br$$

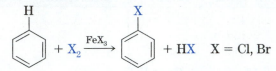

The electrophile, a positive halogen ion, is formed from the reaction of iron (III) halide, a Lewis acid, and the halogen. The electrophile is highly attracted to the electron-rich benzene ring. It bonds using a pair of electrons from the π cloud to form the carbocation intermediate previously described. Loss of a hydrogen ion reforms the benzene ring and the product, a halobenzene.

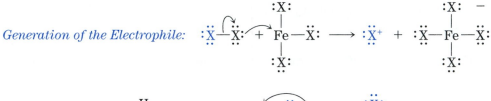

Generation of the Electrophile: $:\ddot{X}-\ddot{X}: + FeX_3 \longrightarrow :\ddot{X}^+ + FeX_4^-$

Two-Step Substitution: $C_6H_6 + :\ddot{X}^+ \longrightarrow [C_6H_6X]^+ \xrightarrow{X-FeX_3^-} C_6H_5X + HX + FeX_3$

FeX_3 is truly a catalyst. It enters the reaction to generate the electrophile and FeX_4^- but is regenerated when the FeX_4^- reacts with the departing hydrogen ion.

GETTING INVOLVED

- ✓ Describe the mechanism just presented to yourself.
- ✓ How is the electrophile generated? Why is the electrophile positive? Why is it attracted to the benzene ring?
- ✓ Where does the electrophile get the electron pair needed for bonding to the benzene ring? What reactive intermediate results and why is it relatively stable?
- ✓ How is the carbocation intermediate neutralized in the final step? How is the aromatic ring reconstituted by this neutralization?
- ✓ Is the catalyst regenerated? When? How?

Problem 6.11

Write the mechanism for the bromination of benzene with $Br_2/FeBr_3$ showing electrophile generation and a two-step substitution.

See related problem 6.32a.

2. ***Alkylation and Acylation: The Friedel-Crafts Reaction.*** Alkylation and acylation of aromatic compounds are frequently referred to as the Friedel-Crafts reaction after Charles Friedel (French) and James Craft (American), who discovered the reaction in 1877. Let us consider the reaction of benzene with 2-chloropropane as an example.

$$C_6H_5H + CH_3CH(Cl)CH_3 \xrightarrow{AlCl_3} C_6H_5CH(CH_3)CH_3 + HCl$$

Aluminum trichloride is a Lewis acid. To complete its octet, aluminum abstracts a chloride ion from 2-chloropropane. The resulting carbocation bonds to the benzene ring. Loss of the hydrogen ion reforms the benzene ring and generates the final product. The catalyst is regenerated as the hydrogen ion reacts with one of the chlorides of $AlCl_4^-$ to form HCl and $AlCl_3$.

Alkylation:

Generation of the electrophile:

$$(CH_3)_2CH{:}\ddot{Cl}{:} + AlCl_3 \longrightarrow (CH_3)_2CH^+ \; AlCl_4^-$$

Two-step substitution:

$$C_6H_6 + CH_3\overset{+}{C}HCH_3 \longrightarrow [C_6H_6{-}CH(CH_3)_2]^+ \xrightarrow{Cl-AlCl_3^-} C_6H_5CH(CH_3)_2 + HCl + AlCl_3$$

GETTING INVOLVED

✓ Describe the mechanism just presented and answer the types of questions just asked in the previous section about halogenation.

Problem 6.12

Write the mechanism for the reaction of benzene with **(a)** $CH_3\overset{O}{\overset{\|}{C}}{-}Cl/AlCl_3$ and **(b)** $CH_3CH_2Cl/AlCl_3$ showing electrophile generation and two-step substitution.

See related problems 6.32 b–c.

3. *Nitration.*

$$C_6H_6 + HNO_3 \xrightarrow{H_2SO_4} C_6H_5NO_2 + H_2O$$

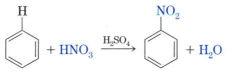

Nitration follows the same pattern we have seen. First, the electrophile is generated, a nitronium ion, from the reaction of nitric acid with sulfuric acid. The positive nitronium ion attacks the benzene ring, and the two-step substitution occurs as before.

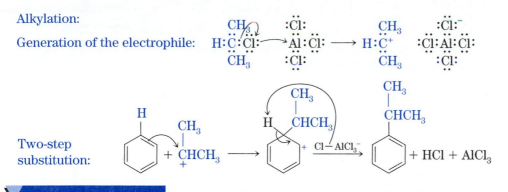

Generation of the electrophile: $HO{-}NO_2 + H^+HSO_4^- \longrightarrow H_2O + NO_2^+HSO_4^-$

Two-step substitution:

$$C_6H_6 + \overset{+}{N}O_2 \longrightarrow [C_6H_6{-}NO_2]^+ \xrightarrow{HSO_4^-} C_6H_5NO_2 + H_2SO_4$$

GETTING INVOLVED

✓ Describe the mechanism of nitration and answer the types of questions asked about halogenation.

Problem 6.13

Write the mechanism for the reaction of 1,4-dimethylbenzene with HNO_3/H_2SO_4 showing generation of the electrophile and two-step substitution.

See related problem 6.32 d.

4. ***Sulfonation.*** The reaction of benzene with concentrated sulfuric acid produces benzenesulfonic acid.

$$C_6H_5{-}H + H_2SO_4 \longrightarrow C_6H_5{-}SO_3H + H_2O$$

The mechanism involves generation of the electrophile followed by two-step substitution, as we have seen before.

Generation of the electrophile: $2H_2SO_4 \longrightarrow \overset{+}{S}O_3H\ \overset{-}{H}SO_4 + H_2O$

Two-step substitution: $C_6H_5{-}H + \overset{+}{S}O_3H \longrightarrow [C_6H_5(H)(SO_3H)]^+ \xrightarrow{HSO_4^-} C_6H_5{-}SO_3H + H_2SO_4$

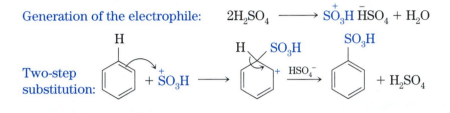

GETTING INVOLVED

✓ Describe to yourself the reaction mechanism just presented and answer the types of questions asked about halogenation.

Problem 6.14

Write the mechanism for the electrophilic aromatic substitution of 1,4-dimethylbenzene with H_2SO_4, showing generation of the electrophile and two-step substitution.

See related problem 6.32 e.

C. Orientation of Substitution

1. ***Directive Effects.*** Since benzene is a symmetrical molecule, electrophilic substitution gives only one substitution product, no matter which of the six hydrogens is replaced. Most benzene derivatives, however, are not symmetrical, and more than one substitution isomer is usually possible. For example, the nitration of chlorobenzene can give three positional isomers of chloronitrobenzene.

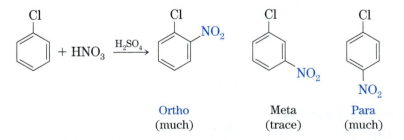

$$C_6H_5Cl + HNO_3 \xrightarrow{H_2SO_4} \text{o-}ClC_6H_4NO_2 + \text{m-}ClC_6H_4NO_2 + \text{p-}ClC_6H_4NO_2$$

Ortho (much)   Meta (trace)   Para (much)

The ortho and para isomers are formed almost to the exclusion of the meta product.

What determines the orientation of substitution, and how does one predict the predominant products? *The atom or group already present on the benzene ring directs the orientation of substitution of the incoming electrophile.* For example, in the nitration of chlorobenzene, the chlorine directs the nitro group primarily to the ortho and para positions. However, in the chlorination of nitrobenzene, the nitro group directs the incoming chlorine almost exclusively to the meta position.

$$C_6H_5NO_2 + Cl_2 \xrightarrow{FeCl_3} m\text{-}ClC_6H_4NO_2$$

Meta

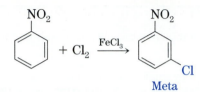

Groups already present on the benzene ring direct incoming electrophiles either to the ortho and para positions or to the meta position. Table 6.3 lists ortho-para directors and meta directors.

Table 6.3 Orientation of Substitution

Ortho-Para Directors		Meta Directors	
—OH	Hydroxy-	—C(=O)OH	Carboxylic acid
—OR	Alkoxy-		
$—NH_2$	Amino-		
—NHR	Alkylamino-	—C(=O)H	Aldehyde
$—NR_2$			
—X	Halogens		
—R	Alkyl-	—C(=O)R	Ketone
		—C≡N	Cyano-
		$—NO_2$	Nitro-
		$—SO_3H$	Sulfonic acid

To predict the orientation of substitution, analyze the effect of the groups already bonded to the benzene ring. In the sulfonation of toluene, the methyl group (an alkyl group) is an ortho-para director (Table 6.3).

$$C_6H_5CH_3 + H_2SO_4 \longrightarrow o\text{-}CH_3C_6H_4SO_3H + p\text{-}CH_3C_6H_4SO_3H$$

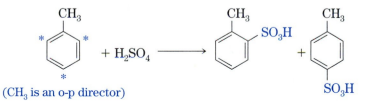

(CH_3 is an o-p director)

A carboxylic acid group is a meta director, as illustrated by the bromination of benzoic acid.

$$C_6H_5CO_2H + Br_2 \xrightarrow{FeBr_3} m\text{-}BrC_6H_4CO_2H$$

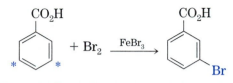

(CO_2H is an m director)

If two or more groups are already present on the benzene ring, the directive effects of each group should be analyzed individually, and prediction of the product should be based on the complete analysis.

GETTING INVOLVED

✓ What influence does a group present on a benzene ring have on electrophilic aromatic substitution? What are ortho-para directors and meta directors? Make sure you can recognize each.

Example 6.5

Predict the product of nitration of *p*-bromobenzenesulfonic acid.

Solution

Nitration is accomplished by treating an aromatic compound with a mixture of concentrated nitric and sulfuric acids. A nitro group replaces a hydrogen on the ring. The sulfonic acid group is a meta director. The bromine is an ortho-para director. The para position from bromine is already occupied by the sulfonic acid group. Direction is to the ortho positions on either side of the bromine. Both attached groups direct to the same two positions, and one product is predicted.

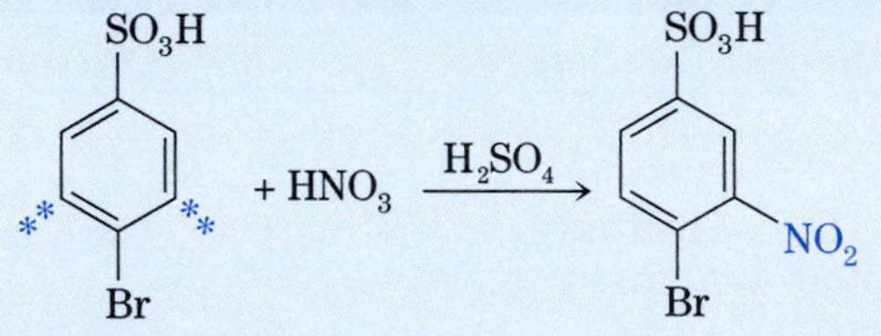

SO_3H / Br (benzene ring, ** marking positions ortho to Br) + HNO_3 $\xrightarrow{H_2SO_4}$ SO_3H / Br / NO_2

(SO_3H, m director;
Br, o-p director)

Problem 6.15

Complete the following reactions, and predict the principal substitution products:

(a) 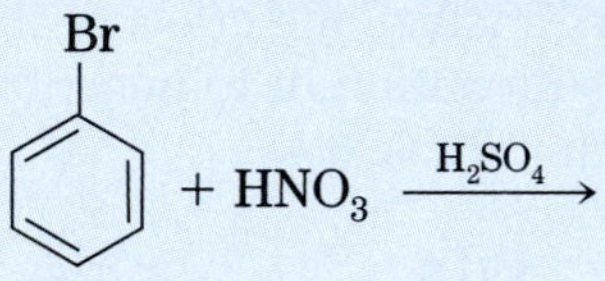 Br (bromobenzene) + HNO_3 $\xrightarrow{H_2SO_4}$

(b) 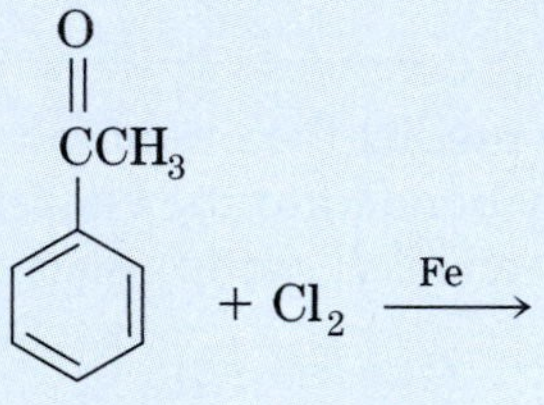 $\overset{O}{\overset{\|}{C}}CH_3$ (acetophenone) + Cl_2 $\xrightarrow{Fe}$

(c) CH_2CH_3 / CN (para) + H_2SO_4 $\longrightarrow$

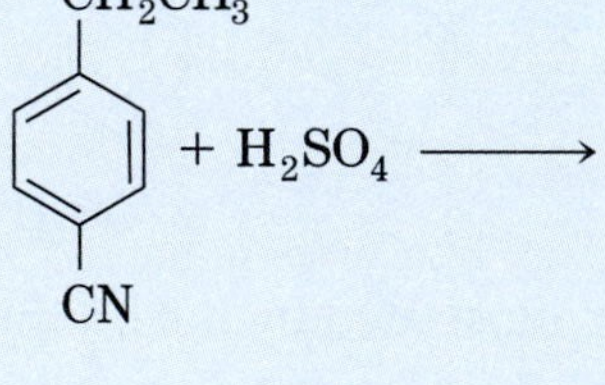

(d) 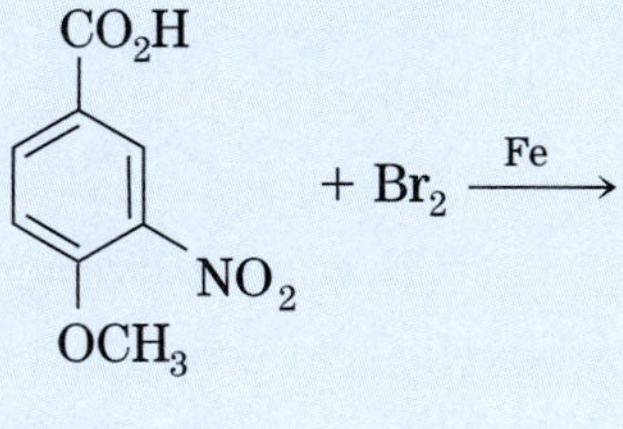 CO_2H / NO_2 / OCH_3 + Br_2 $\xrightarrow{Fe}$

See related problems 6.29–6.31 and 6.33.

2. ***Synthesis.*** In synthesizing an aromatic compound, not only do we need to know the reagents necessary for putting a particular group on a benzene ring, but we also need to determine the order in which to add the reagents. For example, consider the synthesis of *m*-bromonitrobenzene from benzene.

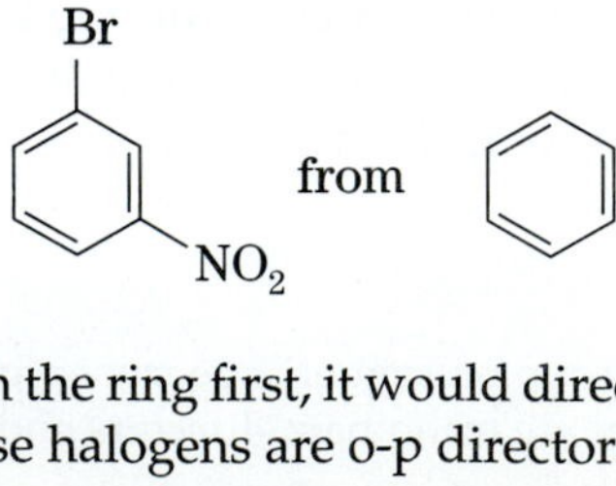

Br / NO_2 (meta) from (benzene)

If the bromine were put on the ring first, it would direct the nitration to the ortho and para positions because halogens are o-p directors. Very little of the desired product would result.

$$C_6H_6 + Br_2 \xrightarrow{Fe} C_6H_5Br \xrightarrow[H_2SO_4]{HNO_3,} o\text{-}BrC_6H_4NO_2 + p\text{-}BrC_6H_4NO_2$$

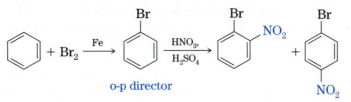

o-p director

This would be an acceptable synthesis of either *o*- or *p*-bromonitrobenzene. If the nitro group is placed on the ring first, however, since it is a meta director, bromination then gives almost exclusively *m*-bromonitrobenzene, the desired product.

$$C_6H_6 \xrightarrow[H_2SO_4]{HNO_3,} C_6H_5NO_2 \xrightarrow[Fe]{Br_2,} m\text{-}BrC_6H_4NO_2$$

m director

GETTING INVOLVED

✓ To do a synthesis you need to know the reagents necessary to introduce a group and the order in which to introduce groups (o, p and m-directors).

Example 6.6

Devise a synthesis for *p*-ethylbenzenesulfonic acid.

Solution

Draw the compound first. Cover the ethyl group with your finger to gain an impression of the situation if the sulfonic acid group were put on the ring first. The sulfonic acid group is a meta director, so it should not be introduced first. If it were, the two groups would be meta to each other in the final product. The ethyl group is an ortho-para director and should be introduced first. To do this, start with the corresponding alkyl chloride, chloroethane. This plus benzene and an aluminum chloride catalyst gives ethylbenzene. Sulfonation with concentrated sulfuric acid gives a mixture of ortho and para products from which the para product can be separated.

$$C_6H_6 \xrightarrow[AlCl_3]{CH_3CH_2Cl} C_6H_5CH_2CH_3 \xrightarrow{H_2SO_4} p\text{-}CH_3CH_2C_6H_4SO_3H + \text{ortho isomer}$$

Problem 6.16

Using reaction equations, show how the following compounds could be synthesized from benzene: **(a)** *m*-chlorobenzenesulfonic acid and **(b)** *p*-nitrotoluene.

See related problems 6.36–6.38.

3. ***Theory of Directive Effects.*** Carbocation stability was the determining factor in orientation of electrophilic addition to alkenes (section 5.1C.2) and it is also responsible for the ortho–para- and meta-directing effects in electrophilic aromatic substitution. The concept is slightly more involved because the intermediate carbocation in electrophilic aromatic substitution is resonance stabilized.

The ortho-para directors listed in Table 6.3 are electron-releasing groups and as such they stabilize carbocations. For example, let us examine ortho, meta, and para attack in the nitration of phenol. The OH group is electron-releasing and will most effectively stabilize a carbocation if, by resonance, the positive charge can be placed on the carbon to which it is attached. Notice that in ortho and para attack, the resulting carbocation has a resonance form in which the positive charge is right next to the hydroxy group. (See section 5.5 for a review of drawing resonance forms.) No such resonance form is possible for meta attack. As a result, the carbocations formed from ortho or para attack are more stable than that formed from meta attack, and ortho and para substitution are favored heavily.

Otho-para substitution favored

$+NO_2$

Ortho — Especially stable — Positive charge possible next to OH; carbocation stabilized

Meta — Positive charge not possible next to OH

Para — Especially stable — Positive charge possible next to OH; carbocation stabilized

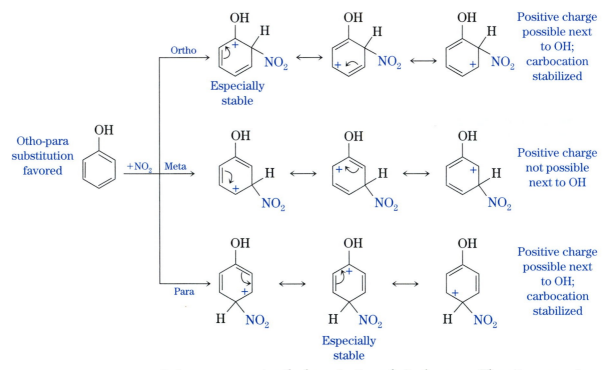

Let us now examine the bromination of nitrobenzene. The nitro group is an especially strong electron-withdrawing group and greatly destabilizes carbocations. In fact, the electron-dot formula for a nitro group shows a full formal positive charge on the nitrogen. The carbocations formed from ortho or para attack have resonance forms in which the positive charge of the carbocation resides directly on the carbon bearing the nitro group. This is not the case in meta attack. Although all three carbocations are unstable in comparison to those formed on a benzene ring with an electron-releasing group, the ones formed from ortho-para attack are especially unstable, since the nitro group can exert its maximum destabilizing effect in these cases. As a result, meta substitution is favored because the carbocation formed is least destabilized.

GETTING INVOLVED

- ✓ Do you notice that the resonance forms drawn for ortho, meta, and para attack for the nitration of phenol are the same as those drawn for bromination of nitro benzene? Can you draw all of these resonance forms?
- ✓ Why is ortho-para substitution favored in the nitration of phenol and meta substitution favored in the bromination of nitrobenzene?
- ✓ Why do electron-releasing groups stabilize and electron-withdrawing groups destabilize carbocations?

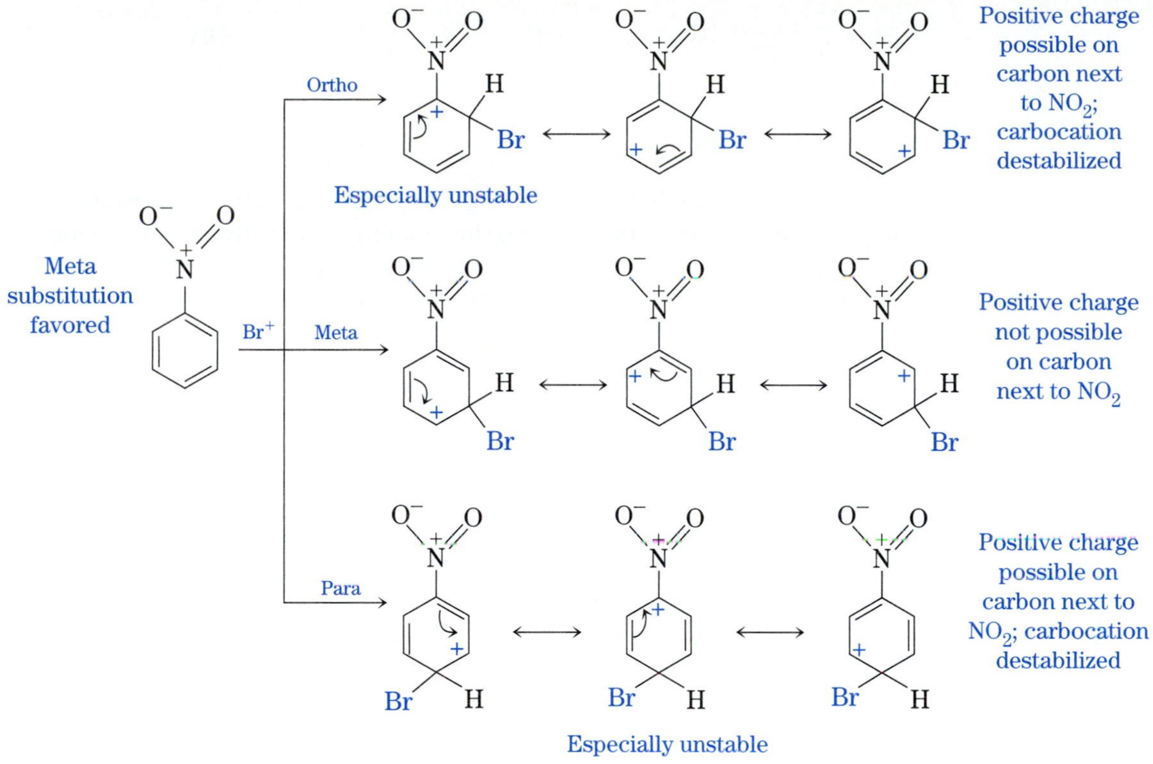

Ortho

Especially unstable

Positive charge possible on carbon next to NO_2; carbocation destabilized

Meta substitution favored

Br^+

Meta

Positive charge not possible on carbon next to NO_2

Para

Especially unstable

Positive charge possible on carbon next to NO_2; carbocation destabilized

D. Activating and Deactivating Groups

A substituent already present on a benzene ring not only directs the orientation of substitution of an incoming group (electrophile) but also influences the rate of reaction. Groups that increase the rate of electrophilic aromatic substitution are called **activating groups,** whereas those that decrease the rate are termed **deactivating groups.**

activating group group that increases the reactivity of an aromatic compound to electrophilic substitution

deactivating group group that decreases the reactivity of an aromatic compound to electrophilic substitution

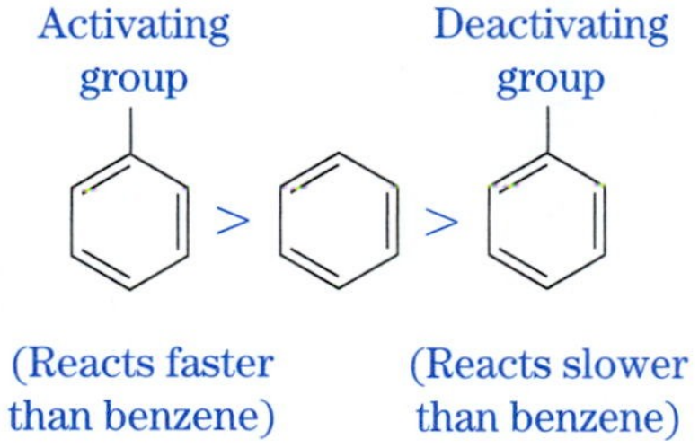

Activating group > > Deactivating group

(Reacts faster than benzene) (Reacts slower than benzene)

The rate of electrophilic substitution depends on the availability to the attacking electrophile of the π cloud of electrons above and below the benzene ring. The more electron-rich (the more negative) the cloud, the faster the electrophilic attack. Electron-releasing groups increase the electron density of the ring and are activating groups. Electron-withdrawing groups, on the other hand, decrease the electron density of the π cloud, decreasing its availability to the attacking electrophile; they are deactivating.

All ortho-para directors (Table 6.3) with the exception of the halogens are activating groups. Except for alkyl groups, all have a lone pair of electrons that is donated to the ring through resonance. This makes the ring more negative and consequently more attractive to the positive electrophiles. The hydroxy group on phenol is a good example of an activating group.

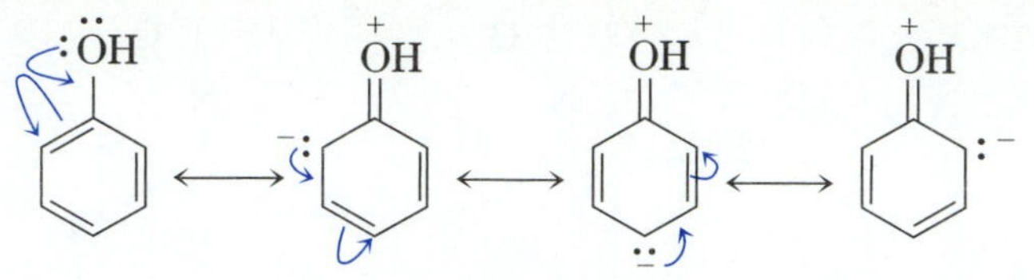

All meta directors (Table 6.3) and the halogens are deactivating groups. They withdraw electrons from the benzene ring, making it less attractive to an incoming electrophile. Because of bond polarity, many of these groups have either a full or a partial positive charge on the atom bonded to the benzene ring.

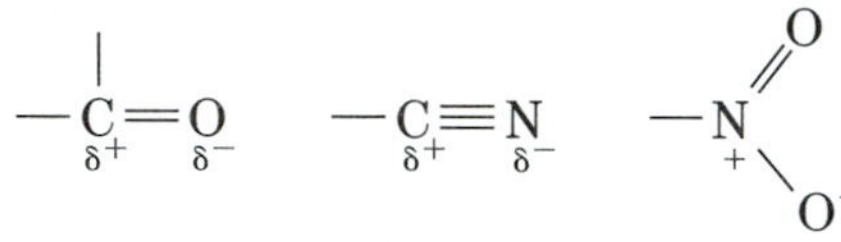

By their greater electronegativity or by resonance (as with nitro group), they withdraw electron density from the benzene ring.

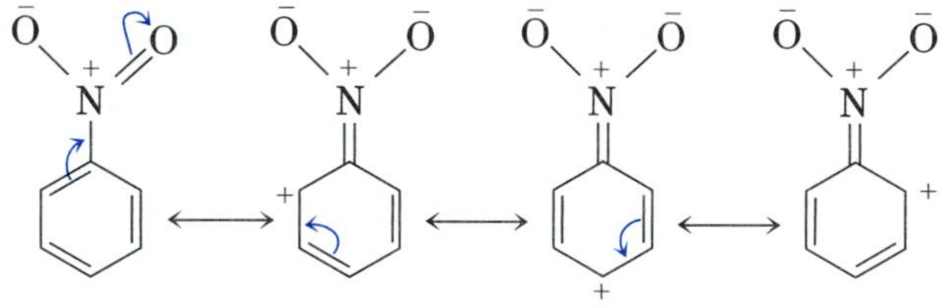

The halogens show a dual effect. Owing to their strong electronegativity, they withdraw electrons from the benzene ring, thus deactivating it. However, once an electrophile has bonded and formed a carbocation, the halogen releases a lone pair of electrons by resonance, stabilizing the positive charge. Although the halogens are deactivating, they are ortho-para directors.

GETTING INVOLVED

- ✓ What two influences do groups already bonded to a benzene ring have on electrophilic aromatic substitution and particularly the incoming electrophile?
- ✓ In what different ways do activating and deactivating groups affect electrophilic aromatic substitution? Why are most ortho-para directors activating groups and why are meta directors deactivating groups? Why do halogens have a dual effect?

Problem 6.17

One can visually experience differences in reactivity by treating an aromatic compound with bromine and noting the rate of decolorization. Arrange the following compounds in order of reactivity toward bromine: **(a)** benzene, chlorobenzene, methoxybenzene; **(b)** phenol, nitrobenzene, *p*-nitrophenol; **(c)** toluene, *p*-methylaniline, *m*-chlorotoluene.

See related problems 6.42–6.43.

6.5 Oxidation of Alkylbenzenes

When alkylbenzenes are treated with an oxidizing agent, such as potassium permanganate, alkyl groups on a benzene ring are oxidized to carboxylic acids. All primary and secondary alkyl groups are oxidized, regardless of their size or number (given enough reagent).

$$CH_3CH_2-C_6H_4-CH_3 \xrightarrow{KMnO_4} HO_2C-C_6H_4-CO_2H$$

Note in this reaction that alkyl groups, which are ortho-para directors, are changed to acid groups, which are meta directors. This must be considered in synthesis problems. For example, the following sequence is effective for producing *m*-nitrobenzoic acid from toluene.

Connections 6.3 www.prenhall.com/bailey

Herbicides

Many herbicides are aromatic compounds and can be prepared by using electrophilic substitution reactions like those described in this chapter. These include the phenoxyacetic acid herbicides introduced in 1944. The first of this class, 2,4-dichlorophenoxyacetic acid (2,4-D), continues to be one of the most useful herbicides ever developed. It and 2,4,5-T (2,4,5-trichlorophenoxyacetic acid) have seen wide use as defoliants and weed killers in agricultural and other applications. These compounds mimic plant growth hormones and affect cellular division and phosphate and nucleic acid metabolism.

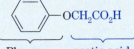

$C_6H_5-OCH_2CO_2H$ (Phenoxy | acetic acid)

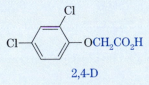

2,4-D: Cl, Cl-substituted $C_6H_3-OCH_2CO_2H$

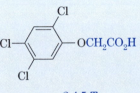

2,4,5-T: Cl, Cl, Cl-substituted $C_6H_2-OCH_2CO_2H$

Controversy concerning the effects of 2,4,5-T on humans and other mammals led the Environmental Protection Agency to curtail most of its uses in 1979. Most notable in the controversy was the use of 10 million gallons of Agent Orange, a 50/50 mixture of butyl esters of 2,4-D and 2,4,5-T, as a defoliant in war zones in Vietnam between 1965 and 1970. The formation of trace amounts of a substance known as dioxin during the production of 2,4,5-T and its alleged detrimental effects on humans were the basis of a claim by Vietnam war veterans for compensation for a variety of physical ailments following exposure to the defoliant spray.

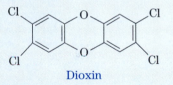

Dioxin (tetrachloro-substituted, Cl, Cl, Cl, Cl; O, O)

2,4-D and related compounds are available to home gardeners in a variety of ester and salt forms as a broadleaf-weed killer. Because of their large surface areas, broadleaf weeds quickly absorb a lethal dose, whereas thin-bladed grasses remain unaffected. Compounds similar to 2,4-D, such as *p*-chlorophenoxyacetic acid and β-naphthoxyacetic acid, are the ingredients of plant preparations used to aid tomato blossoms in setting fruit and to prevent premature dropping of fruit. In fact, even 2,4-D, in lower concentrations than in weed killers, is used in citrus culture to prevent preharvest fruit drop, to prevent fruit and leaf damage following pesticide oil sprayings, to delay fruit maturity, and to increase fruit size.

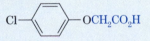

$Cl-C_6H_4-OCH_2CO_2H$

p-chlorophenoxyacetic acid

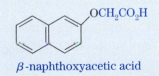

$C_{10}H_7-OCH_2CO_2H$

β-naphthoxyacetic acid

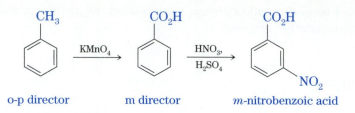

CH_3 (o-p director) $\xrightarrow{KMnO_4}$ CO_2H (m director) $\xrightarrow{HNO_3, H_2SO_4}$ *m*-nitrobenzoic acid (CO_2H, NO_2)

To produce *o*- and *p*-nitrobenzoic acid, one would reverse the sequence of reagents added—HNO_3/H_2SO_4 first, followed by $KMnO_4$.

GETTING INVOLVED

✓ What happens to alkyl groups on a benzene ring when the compound is treated with potassium permanganate?

✓ What type of directing group in electrophilic aromatic substitution is an alkyl group? A carboxylic acid group? Why is this important in synthesis schemes involving oxidation and aromatic substitution?

Example 6.7

Devise a synthesis of a disubstituted benzene having a carboxylic and a sulfonic acid group para to one another. Start with benzene.

Solution

We treat benzene with chloromethane and aluminum trichloride to introduce a methyl group. This methyl group will eventually become the carboxylic acid. The methyl group is an o-p director, and a carboxylic acid is an m director. We should therefore introduce the sulfonic acid group before oxidizing the methyl. Both *o*- and *p*-methylbenzenesulfonic acid will be formed, but we can separate the para isomer. Oxidation with potassium permanganate produces the final product.

Benzene $\xrightarrow[AlCl_3]{CH_3Cl}$ CH_3-benzene $\xrightarrow{H_2SO_4}$ CH_3, SO_3H (para) $\xrightarrow{KMnO_4}$ CO_2H, SO_3H (para)

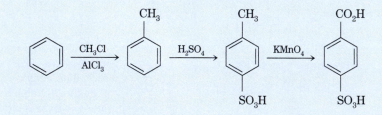

Problem 6.18

Write a reaction equation illustrating the oxidation of propylbenzene with potassium permanganate.

See related problem 6.34–6.35 and 6.37(g-h).

REACTION SUMMARY

A. Electrophilic Aromatic Substitution

Section 6.4.A–C; Figure 6.2; Tables 6.2–6.3; Examples 6.4–6.5; Problems 6.9–16, 6.29–6.33, 6.35–6.38.

Halogenation: Benzene + X_2 $\xrightarrow{FeX_3}$ X-benzene + HX (X = Cl, Br)

Friedel-Crafts Alkylation and Acylation:

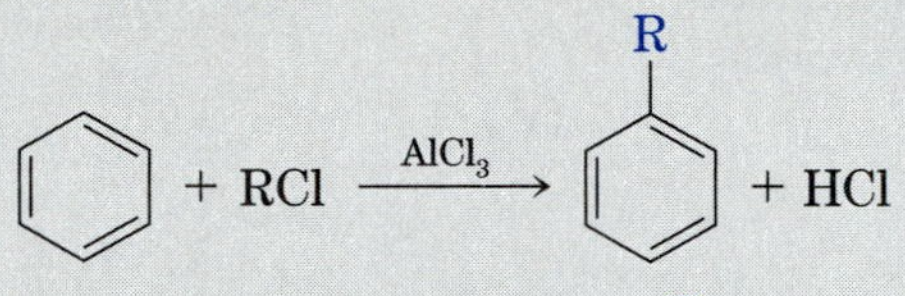

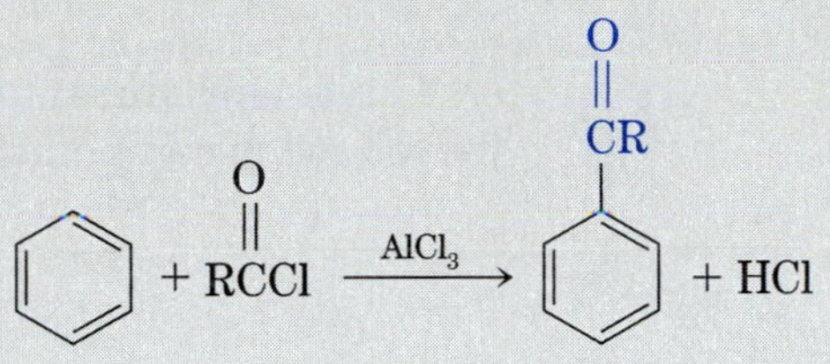

$$C_6H_6 + RCl \xrightarrow{AlCl_3} C_6H_5R + HCl$$

$$C_6H_6 + RCCl(=O) \xrightarrow{AlCl_3} C_6H_5CR(=O) + HCl$$

Nitration:

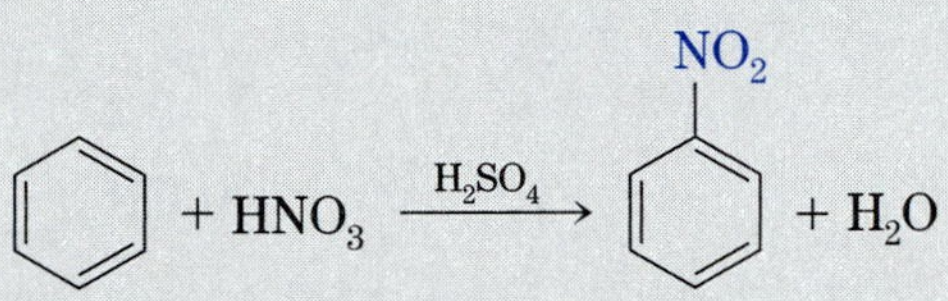

$$C_6H_6 + HNO_3 \xrightarrow{H_2SO_4} C_6H_5NO_2 + H_2O$$

Sulfonation:

$$C_6H_6 + H_2SO_4 \longrightarrow C_6H_5SO_3H + H_2O$$

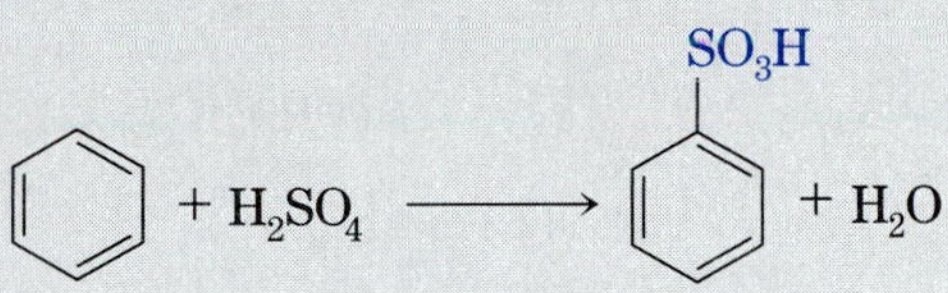

Orientation of substitution:

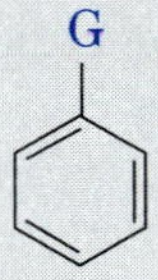

G (on benzene ring)

ortho-para director
G = OH, OR, NH_2, NHR, NR_2, X, R

meta director
G = CO_2H, CHO, CR(=O), CN, NO_2, SO_3H

B. Oxidation of Alkylbenzenes

Section 6.5; Example 6.7; Problems 6.18, 6.34–6.35.

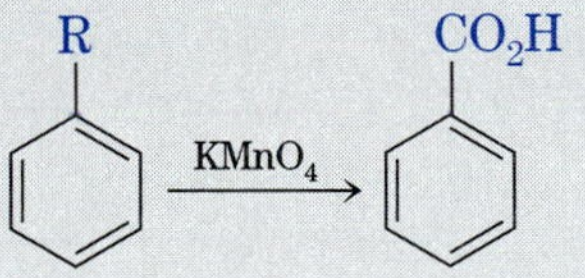

$$C_6H_5R \xrightarrow{KMnO_4} C_6H_5CO_2H$$

Problems

6.19 **Bonding Pictures:** Draw a bonding picture showing all π bonds for phenanthrene (structure in section 6.3.A).

6.20 **Molecular Formulas:** Write molecular formulas (such as C_6H_6 for benzene) for the following polynuclear aromatic compounds:

(a)

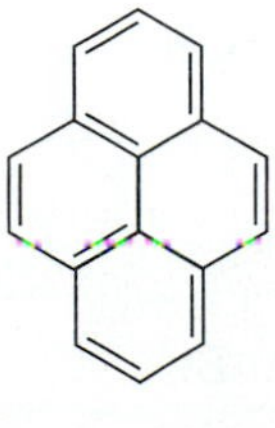

(b) 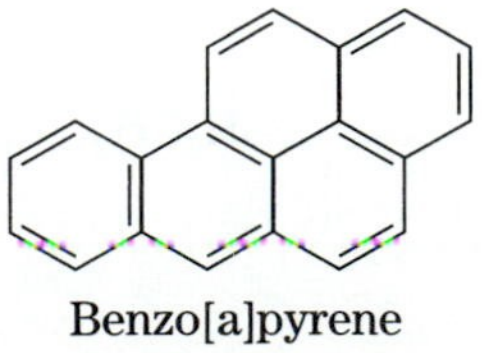 Benzo[a]pyrene

6.21 **Nomenclature of Monosubstituted and Disubstituted Benzenes:** Name the following compounds:

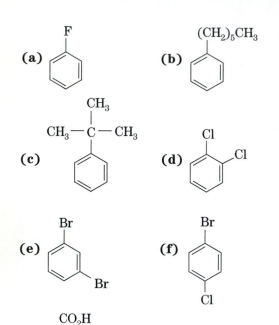

(a) 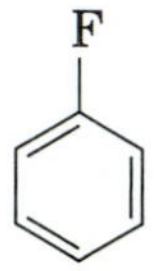 F (b) $(CH_2)_5CH_3$

(c) CH_3 / CH_3-C-CH_3 (d) Cl, Cl

(e) Br, Br (f) Br, Cl

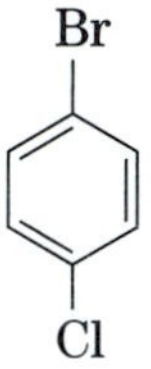

(g) 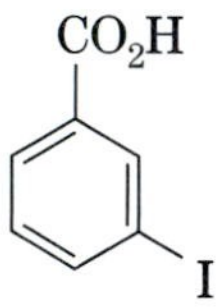 CO_2H, I (h) F, F

(i) O / ‖ / CH, CH_2CH_3 (j) 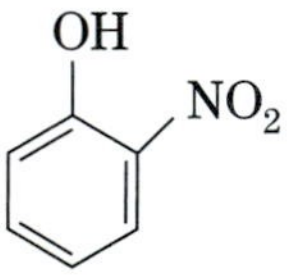 OH, NO_2

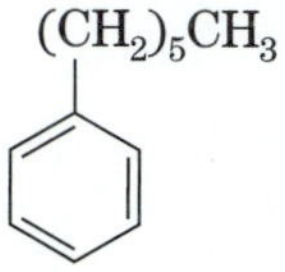

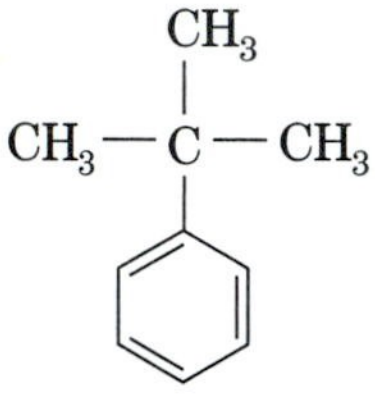

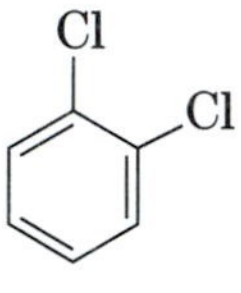

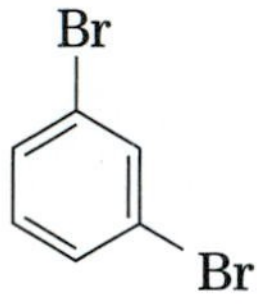

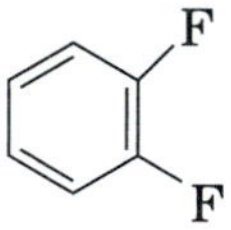

6.22 **Nomenclature of Polysubstituted Benzenes:** Name the following compounds:

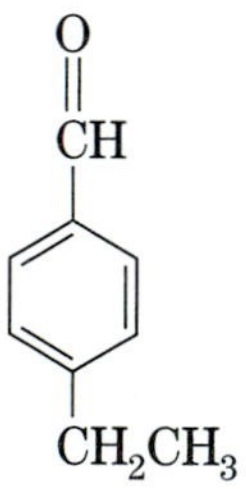

(a) 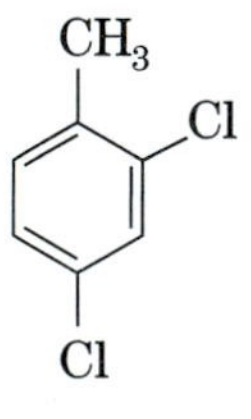 CH_3, Cl, Cl (b) 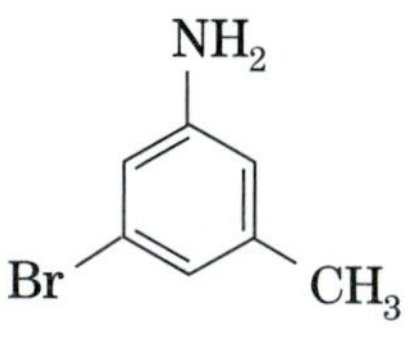 NH_2, Br, CH_3

(c) 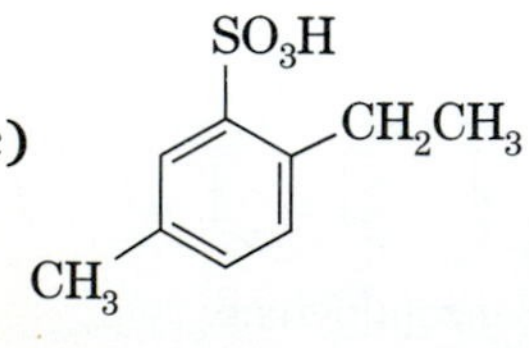 SO_3H, CH_2CH_3, CH_3 (d) 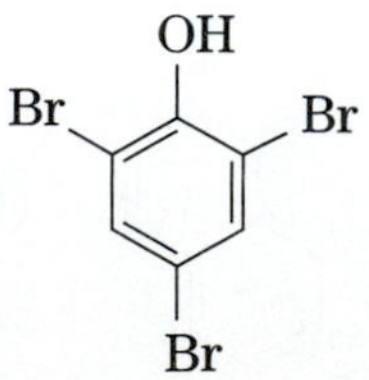 OH, Br, Br, Br

(e) 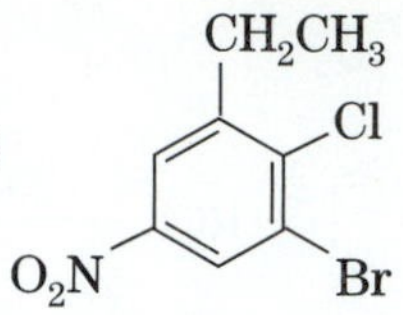 CH_2CH_3, Cl, O_2N, Br (f) 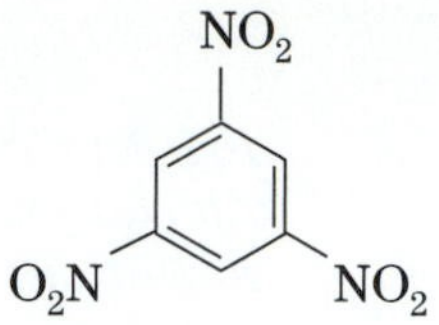 NO_2, O_2N, NO_2

6.23 **Nomenclature of Substituted Anilines:** Name the following compounds:

(a) 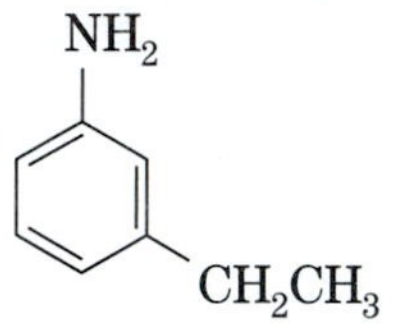 NH_2, CH_2CH_3 (b) 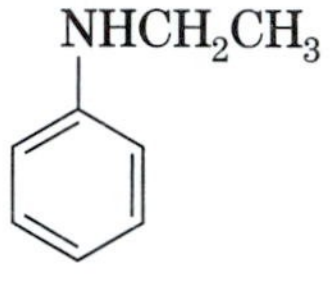 $NHCH_2CH_3$

(c) 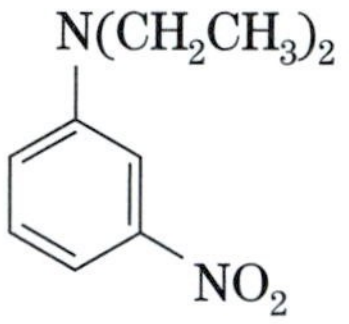 $N(CH_2CH_3)_2$, NO_2 (d) 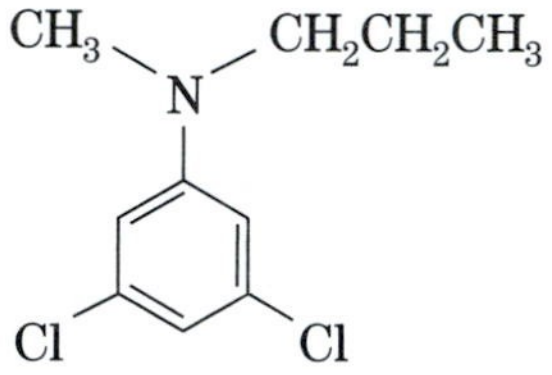 CH_3, N, $CH_2CH_2CH_3$, Cl, Cl

6.24 **Nomenclature Using Benzene as a Prefix:** Name the following compounds:

(a) 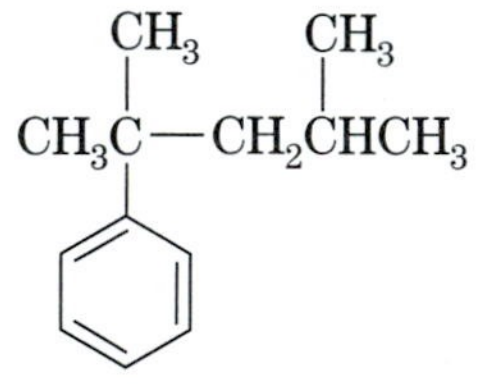 CH_3 CH_3 / $CH_3C-CH_2CHCH_3$

(b) 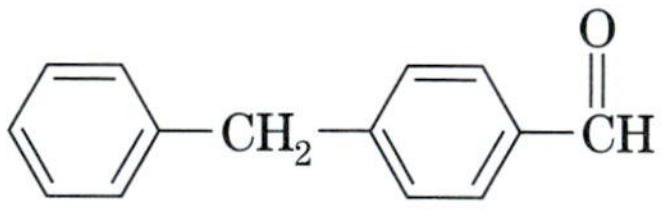 $-CH_2-$ O / ‖ / CH

(c) 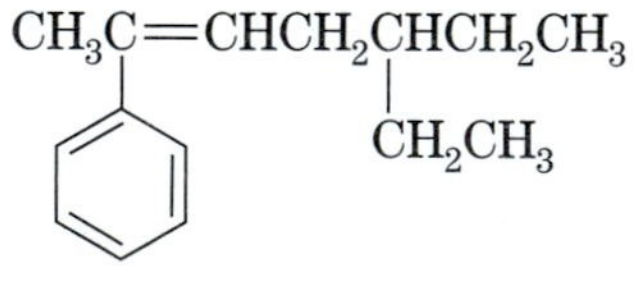 $CH_3C=CHCH_2CHCH_2CH_3$ / CH_2CH_3

6.25 **Nomenclature of Polynuclear Aromatic Compounds:** Name the following compounds:

(a) 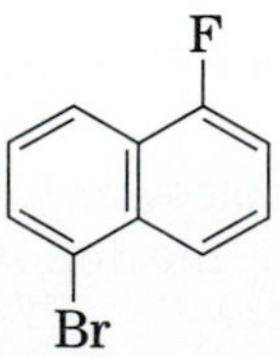 F, Br (b) 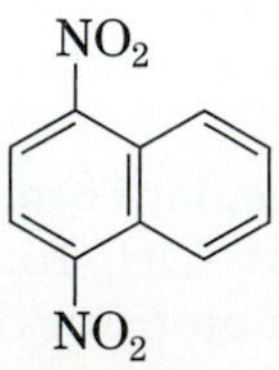 NO_2, NO_2

(c) 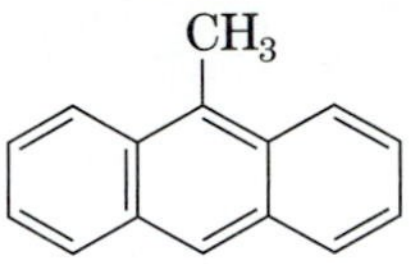
(d) 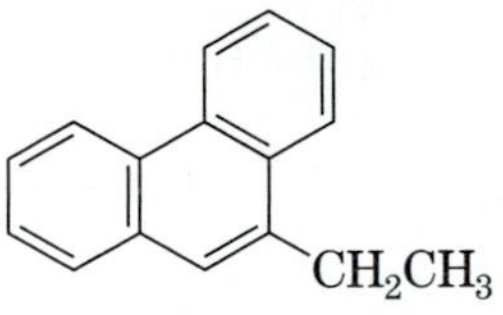

6.26 **Nomenclature:** Draw the following compounds:
(a) *p*-dichlorobenzene (mothballs)
(b) *m*-xylene (component of high-octane gasoline)
(c) 1,3,5-trinitrobenzene (an explosive, TNB)
(d) *o*-phenylphenol (a disinfectant in household deodorizers)
(e) 2,6-di-*t*-butyl-4-methylphenol (antioxidant used in gasoline)
(f) benzaldehyde (oil of bitter almonds)
(g) 2-methylnaphthalene (found in coal tar)
(h) pentachlorophenol (ant and termite killer)
(i) 2,4,6-trinitrotoluene (TNT)

6.27 **Positional Isomers:** Draw the positional isomers of the following compounds:
(a) tribromobenzenes
(b) chlorodibromobenzenes
(c) bromochlorofluorobenzenes
(d) dibromonaphthalenes
(e) dinitroanthracenes
(f) dinitrophenanthrenes

6.28 **Positional Isomers:** There are three dibromobenzenes. Their melting points are 87°C, 6°C, and −7°C. Nitration of the isomer with mp = 87°C results in only one mononitrated dibromobenzene. The isomer with mp = +6°C gives two mononitrated isomers, and the one with mp = −7°C gives three. Write the structure of each isomer.

6.29 **Positional Isomers:** Write the structure of polysubstituted benzene compounds with the following properties:
(a) formula C_9H_{12} and gives only one monochlorination isomer
(b) formula $C_{10}H_{14}$ and gives two monochlorination isomers

6.30 **Electrophilic Aromatic Substitution Reactions of Aromatic Compounds:** Predict the major product(s) of the following reactions:

(a) 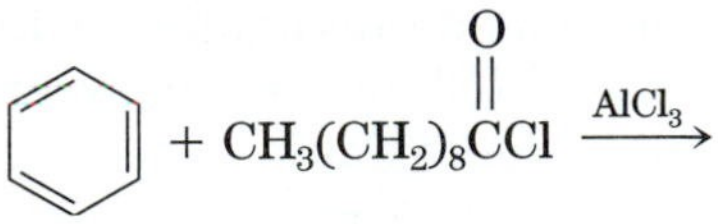

(b) 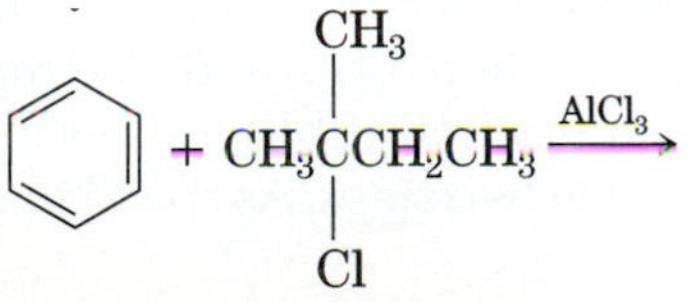

(c) 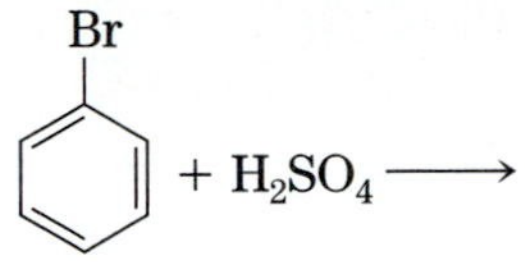

(d) 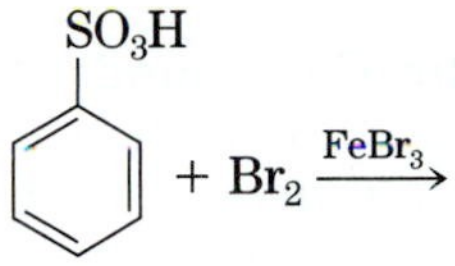

(e) 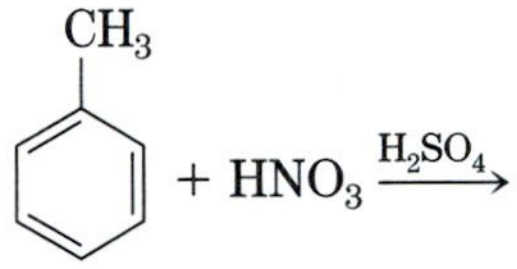

(f) 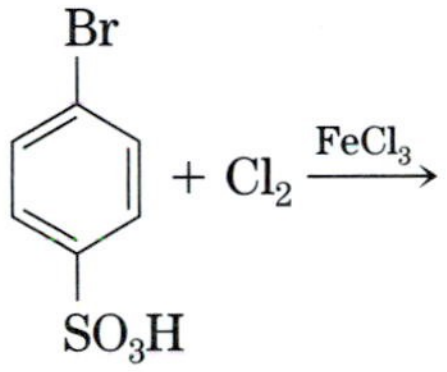

(g) 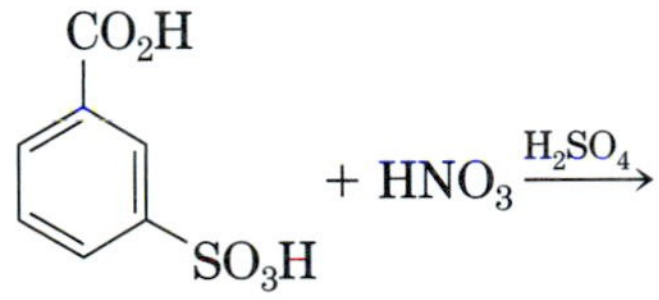

(h) 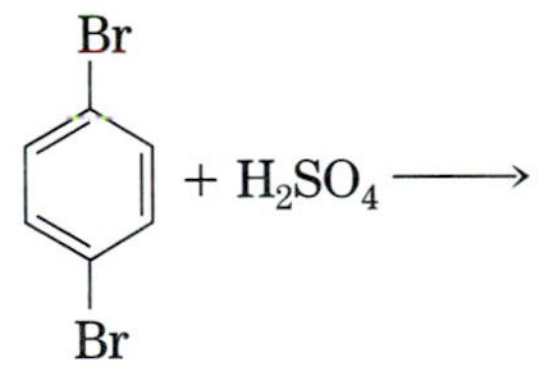

(i) 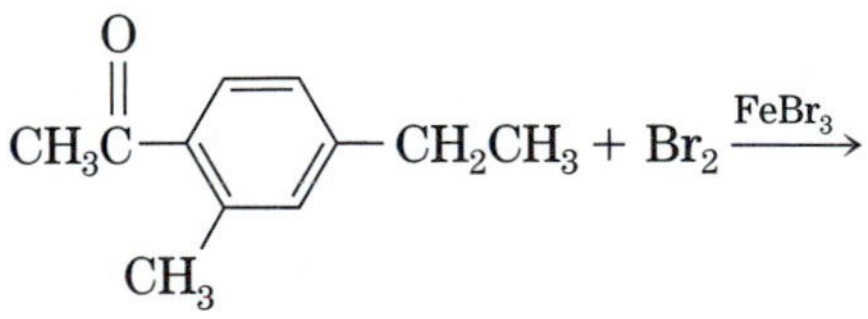

(j) 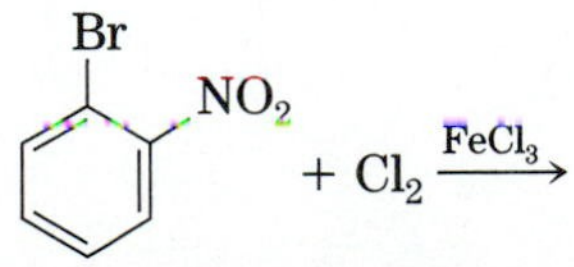

(k) 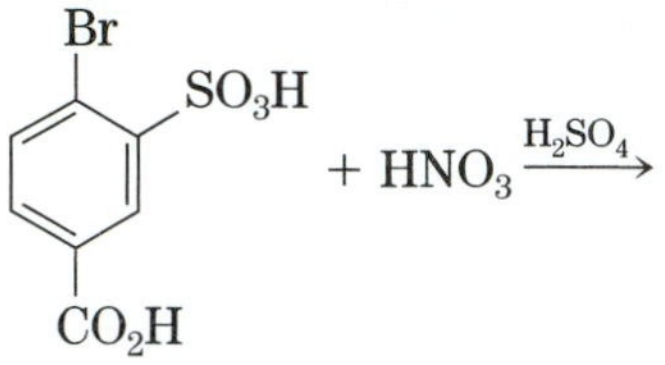 2-bromo-5-carboxybenzenesulfonic acid (Br, SO_3H, CO_2H on ring) $+ HNO_3 \xrightarrow{H_2SO_4}$

6.31 **Reactions of Aromatic Compounds:** Draw the major product(s) of monobromination of the following compounds: **(a)** nitrobenzene; **(b)** *m*-dinitrobenzene; **(c)** chlorobenzene; **(d)** *p*-methylbenzenesulfonic acid; **(e)** methoxybenzene.

6.32 **Reaction Mechanisms:** Write a step-by-step reaction mechanism for the reaction of toluene with each of the following reagents:

(a) $Cl_2/FeCl_3$ **(b)** $CH_3CH_2CCl/AlCl_3$ (with C=O)

(c) $CH_3CHCH_3/AlCl_3$ (Cl on central C) **(d)** HNO_3/H_2SO_4

(e) H_2SO_4

6.33 **Reactions of Aromatic Compounds:** Write structures for each product indicated by a letter.

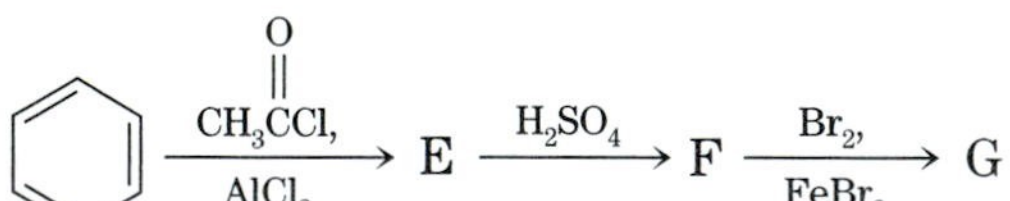

(a) 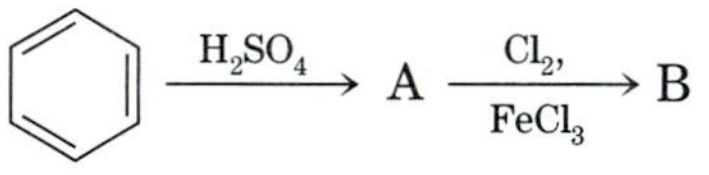 benzene $\xrightarrow{H_2SO_4}$ A $\xrightarrow[FeCl_3]{Cl_2,}$ B

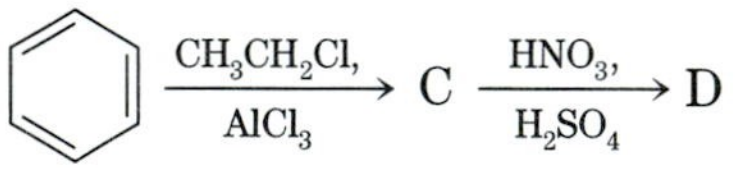

(b) benzene $\xrightarrow[AlCl_3]{CH_3CH_2Cl,}$ C $\xrightarrow[H_2SO_4]{HNO_3,}$ D

(c) benzene $\xrightarrow[AlCl_3]{CH_3CCl,}$ E $\xrightarrow{H_2SO_4}$ F $\xrightarrow[FeBr_3]{Br_2,}$ G (acyl chloride, C=O)

(d) benzene $\xrightarrow{H_2SO_4}$ H $\xrightarrow[H_2SO_4]{HNO_3,}$ I $\xrightarrow[FeCl_3]{Cl_2,}$ J

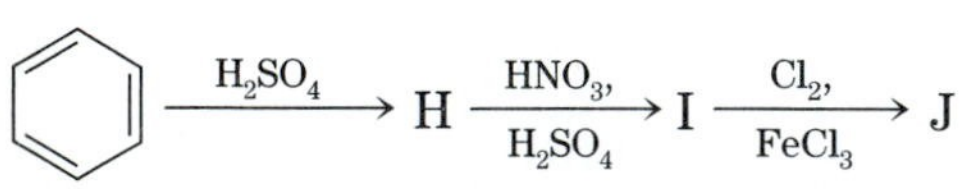

6.34 **Oxidation of Alkylbenzenes**

(a) 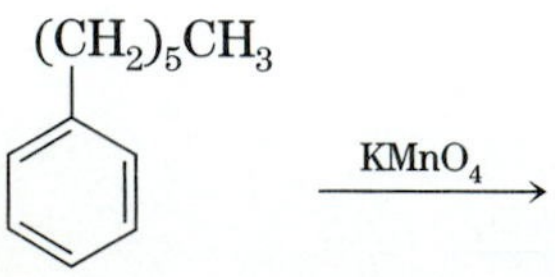 $C_6H_5(CH_2)_5CH_3 \xrightarrow{KMnO_4}$

(b) 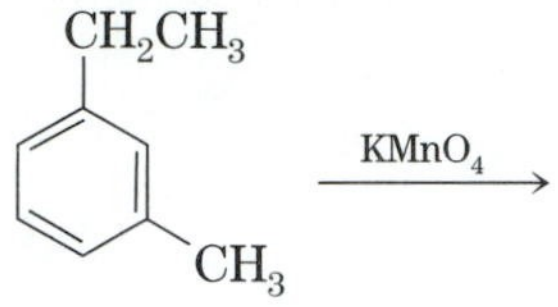 1-ethyl-3-methylbenzene (CH_2CH_3, CH_3) $\xrightarrow{KMnO_4}$

(c) benzene $\xrightarrow[AlCl_3]{CH_3Cl,}$ A $\xrightarrow{KMnO_4}$ B $\xrightarrow[FeBr_3]{Br_2,}$ C

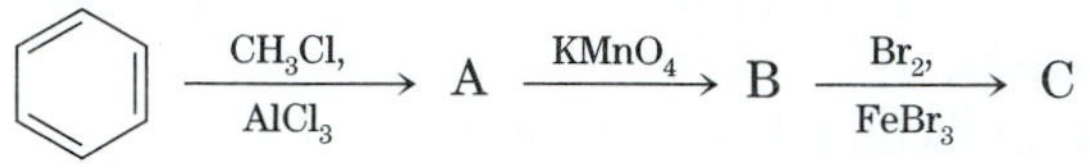

6.35 **Synthesis:** If you wished to make *m*-bromobenzoic acid from toluene, would you oxidize the methyl before or after introduction of the bromine? Why?

6.36 **Synthesis:** Which group would you introduce first in the synthesis of *p*-chlorobenzenesulfonic acid from benzene? Why?

6.37 **Synthesis:** Outline the steps in the synthesis of the following compounds from benzene (assume that ortho and para isomers can be separated):

(a) *p*-bromochlorobenzene
(b) *p*-isopropylbenzenesulfonic acid
(c) *m*-bromobenzenesulfonic acid
(d) *m*-chloronitrobenzene
(e) *p*-chloronitrobenzene
(f) 2-bromo-4-nitroethylbenzene
(g) *m*-nitrobenzoic acid
(h) *p*-nitrobenzoic acid

6.38 **Synthesis:** From the word descriptions, write reactions illustrating the preparation of the following familiar substances:

(a) Mothballs: treatment of benzene with 2 moles chlorine and $FeCl_3$ as the catalyst.

(b) TNT: trinitration of toluene with 3 moles nitric acid and concentrated sulfuric acid as a catalyst.

(c) Pentachlorophenol: a wood preservative (prevents attack by fungi and termites) used on fence posts and telephone poles; produced by pentachlorination of phenol.

(d) Synthetic detergents: benzene and 2-chlorododecane ($C_{12}H_{25}Cl$) in the presence of $AlCl_3$ as a catalyst react by a Friedel-Crafts reaction. The product is sulfonated with fuming sulfuric acid. The sulfonic acid group is neutralized with sodium hydroxide (simple acid-base reaction) to give the detergent.

(e) Sodium benzoate: a food preservative; oxidation of toluene to benzoic acid followed by neutralization of the acid with sodium hydroxide.

6.39 **Reaction Mechanisms:** The mechanisms of electrophilic addition (section 5.1.B) and electrophilic aromatic substitution (section 6.4.B) are essentially identical in the first of the two steps. As an aid to study, write general reaction mechanisms one below the other for comparison.

6.40 **Reaction Mechanisms:** Toluene can react with bromine in two different ways depending on the reaction conditions. When toluene is treated with Br_2 and $FeBr_3$, electrophilic aromatic substitution occurs on the benzene ring. If it is treated with bromine alone in the light, a free-radical chain reaction (section 4.4) occurs, involving bromination of the methyl group. These happen because toluene is both an aromatic hydrocarbon (benzene ring) and an alkane (methyl group). Write step-by-step reaction mechanisms for both reactions.

6.41 **Reaction Mechanisms:** Isopropylbenzene can be made by the Friedel-Crafts reaction in three ways. In section 6.4.B.2, we see that treating benzene with 2-chloropropane and aluminum chloride will give isopropylbenzene. Treating benzene with either 2-propanol and sulfuric acid or propene and sulfuric acid will likewise produce isopropyl benzene; the isopropyl carbocation forms as the electrophile in all three cases by reactions we have discussed. Write a step-by-step reaction mechanism for each process. See sections 6.4.B.2, 5.1.B, and 4.5.C for assistance. Note that all three processes involve the same reaction intermediates.

6.42 **Activating and Deactivating Groups:** Arrange the compounds of the following sets in order of reactivity (least to most reactive) toward electrophilic aromatic substitution:

(a) benzene, phenol, nitrobenzene
(b) benzene, chlorobenzene, aniline
(c) *p*-xylene, *p*-methylbenzoic acid, benzoic acid
(d) toluene, nitrobenzene, *p*-nitrotoluene, *p*-xylene

6.43 **Activating and Deactivating Groups:** Which benzene ring would you expect to be nitrated (HNO_3/H_2SO_4) in the following compounds? Explain your answer.

(a) 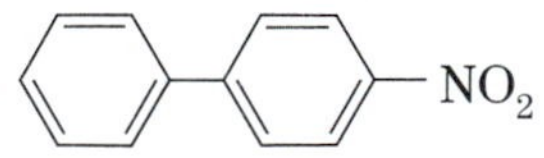 NO_2

(b) 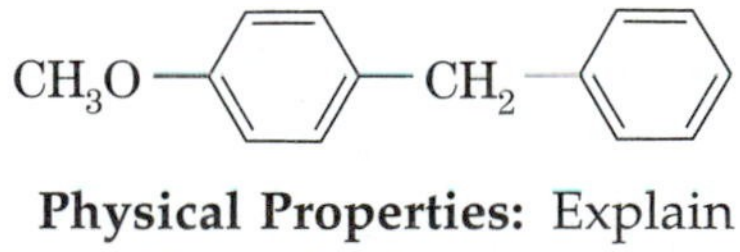 CH_3O — CH_2

6.44 **Physical Properties:** Explain the difference in melting point or boiling point, as indicated, for the compounds of the following sets:

(a) boiling point of methylbenzene (111°C) and ethylbenzene (136°C)
(b) melting point of xylenes, components of high-octane gasoline

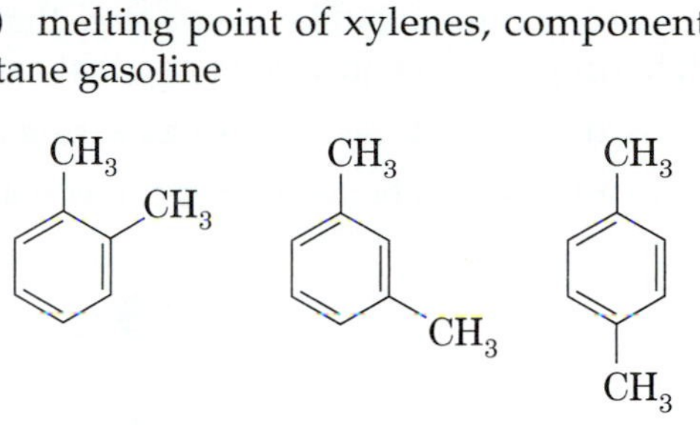

CH_3 CH_3 −25°C; CH_3 CH_3 −48°C; CH_3 CH_3 13°C

(c) melting point

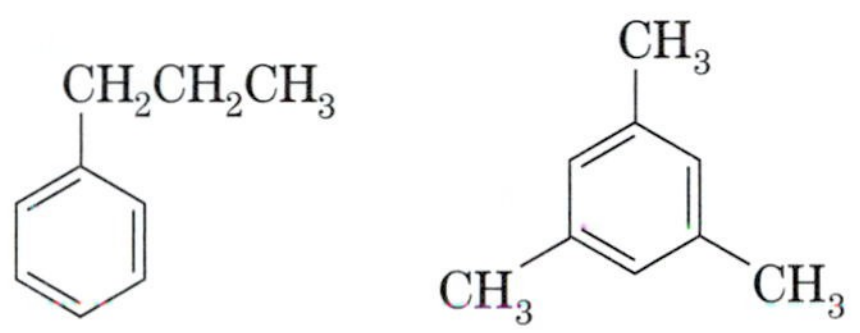

$CH_2CH_2CH_3$ −99°C; CH_3 CH_3 CH_3 −45°C

(d) melting point

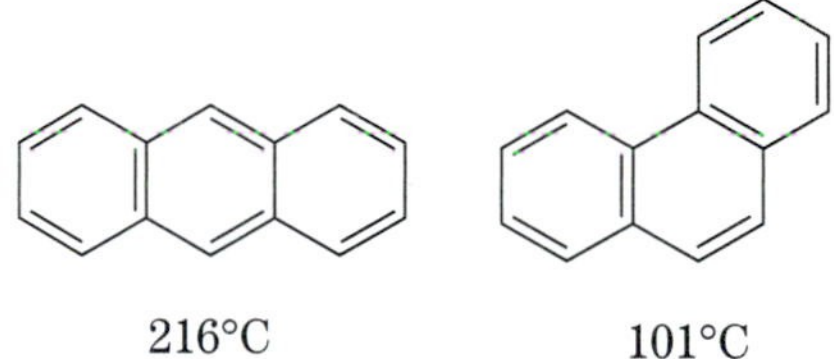

216°C 101°C

6.45 **Gasoline:** Write three structural characteristics of high-octane gasoline molecules. Draw representative molecules that possess each of the characteristics.

6.46 **Production of Gasoline:** What petroleum refining method could improve the suitability of the following compounds as gasoline components?

(a) $CH_3(CH_2)_3CH_3$ **(b)** $CH_3(CH_2)_{18}CH_3$
(c) $CH_3(CH_2)_5CH_3$ **(d)** $CH_3CH{=}CH_2$
(e) 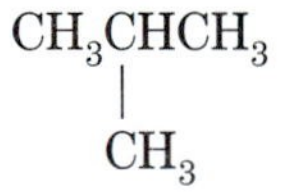 CH_3CHCH_3 (with CH_3 on the central carbon) **(f)**

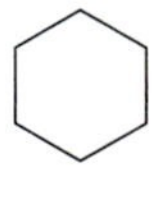

6.47 **Basicity of Aniline:** Aniline is basic because of the availability of a lone pair of electrons on the nitrogen. What effect would an electron-with-drawing group like NO_2 or an electron-releasing group like OCH_3 have on the basicity? In which position(s)—ortho, meta, or para—would these groups have their greatest effect? Explain.

Activities with Molecular Models

1. Make a molecular model of benzene. Note that the molecule is entirely planar, that all carbons are equivalent, that all hydrogens are equivalent, and that each carbon is trigonal with 120° bond angles.
2. How many different places on a benzene ring can you replace one hydrogen with a bromine?
3. How many places on a benzene ring can you substitute two bromines for two hydrogens?

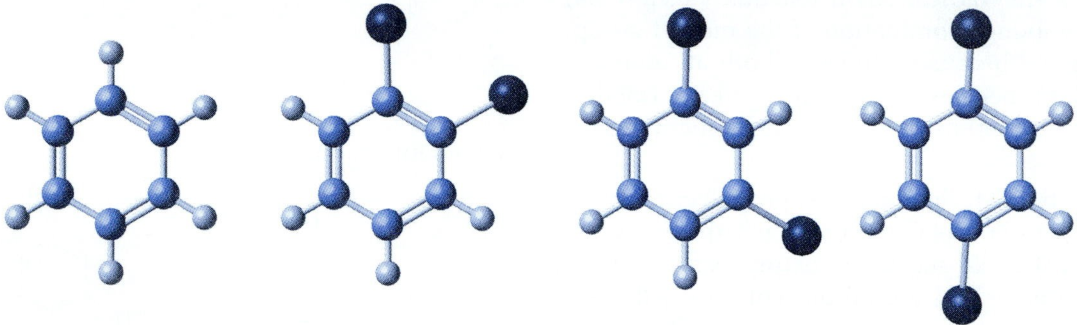

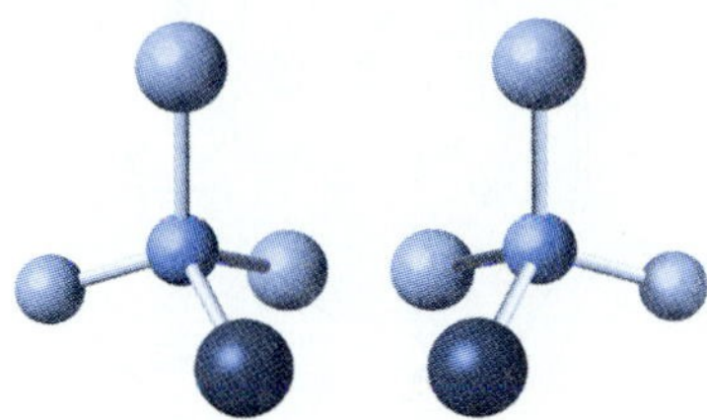

CHAPTER 7
Stereochemistry

7.1 Introduction

Following are two structural representations of lactic acid. They are written to show the three-dimensional, tetrahedral geometry of the middle carbon.

CO_2H | CO_2H
H····C—OH, CH_3 | HO—C····H, CH_3

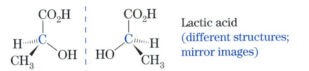

Lactic acid
(different structures; mirror images)

At first glance these two structures look very similar; you may think that they are identical. But look more closely. Try to superimpose the two. They are non-superimposable mirror images. Each form exists independently. The structure on the left is produced in muscles during exercise and is responsible for soreness. The other is found in sour milk.

Let us compare lactic acid to propanoic acid, again shown in the tetrahedral configuration.

CO_2H | CO_2H
H····C—CH_3, H | H_3C—C····H, H

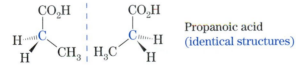

Propanoic acid
(identical structures)

These two representations are clearly identical. What is it about lactic acid that causes it to exist in two distinct and different mirror image forms, whereas a similar compound, propanoic acid, does not? Close examination of the two compounds reveals that lactic acid possesses a carbon (the one emphasized) with four different attached groups (CH_3, OH, H, CO_2H); there is no such carbon in propanoic acid. This special carbon with four different bonded groups is called a stereocenter or **chiral carbon atom** (chiral from the Greek for hand). When four different groups are placed at the corners of a tetrahedron, two different arrangements or configurations are possible that are nonsuperimposable mirror images of one another; hence the two forms of lactic acid. You can prove this to yourself using a set of molecular models. If two or more groups attached to a tetrahedral carbon are identical, as in propanoic acid, then only one structure is possible.

chiral carbon atom carbon with four different bonded groups

The similarity of the two forms of lactic acid is analogous to that of a pair of gloves. A cursory look at a pile of gloves of the same style, color, and size would not reveal whether they were left-handed, right-handed, or a mixture. With closer observation, however, you could separate the gloves into right-handed and left-handed. The difference is subtle. Like the isomers of lactic acid, right- and left-handed gloves are mirror images of each other but are not superimposable

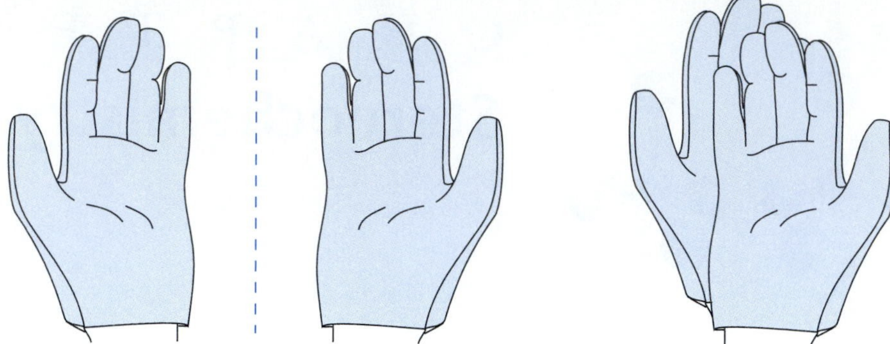

Figure 7.1
Nonsuperimposable left- and right-handed gloves.

(Figure 7.1). In contrast, plain dinner spoons, like propanoic acid, not only look alike but also have identical mirror images.

GETTING INVOLVED

Problem 7.1

Which of the following objects have a nonsuperimposable mirror image?

(a) sock	**(b)** nail	**(c)** foot	**(d)** screw
(e) fork	**(f)** spiral staircase	**(g)** pullover sweater	**(h)** scissors
(i) rubber ball	**(j)** pine cone	**(k)** key	**(l)** checkerboard
(m) coiled spring	**(n)** clock	**(o)** ocean wave	**(p)** block
(q) hammer	**(r)** ear	**(s)** golf club	**(t)** umbrella

isomers
different compounds with the same molecular formula

structural or constitutional isomers
isomers that vary in the bonding attachments of atoms

skeletal isomers
isomers that differ in the arrangement of the carbon chain

positional isomers
isomers that differ in the location of a noncarbon group or double bond or triple bond

functional isomers
isomers with structural differences that place them in different classes of organic compounds

stereoisomers
isomers with the same bonding attachments of atoms but different spatial orientations

geometric isomers
cis and *trans* isomers; atoms or groups display orientation differences around a double bond or ring

conformational isomers
isomers that differ as a result of the degree of rotation around a carbon-carbon single bond

The two mirror-image structures of lactic acid are stereoisomers of each other (specifically, they are enantiomers, a term defined in the next section). Stereoisomers differ only in the spatial orientation of atoms, not in their bonding arrangement. Let us briefly review what we have learned about isomerism.

Structural or constitutional isomers differ in the bonding arrangement of atoms; different atoms are attached to one another in the isomers. **Skeletal, positional,** and **functional** isomers are in this class.

Structural, or constitutional, isomerism:

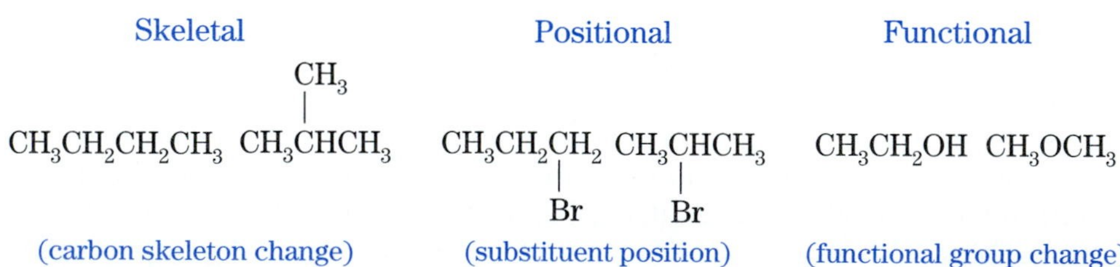

In **stereoisomerism** the same atoms are bonded to one another, but their orientation in space differs. We have already studied **geometric** and **conformational** isomerism.

GETTING INVOLVED

- ✓ What is a stereocenter (chiral carbon atom)? How many different ways can the groups on a chiral carbon atom be arranged and how are they related to one another?
- ✓ What is the difference between structural isomers and stereoisomers?
- ✓ What are the three types of structural isomers and how are they different?
- ✓ What are the three types of stereoisomers and how are they different?

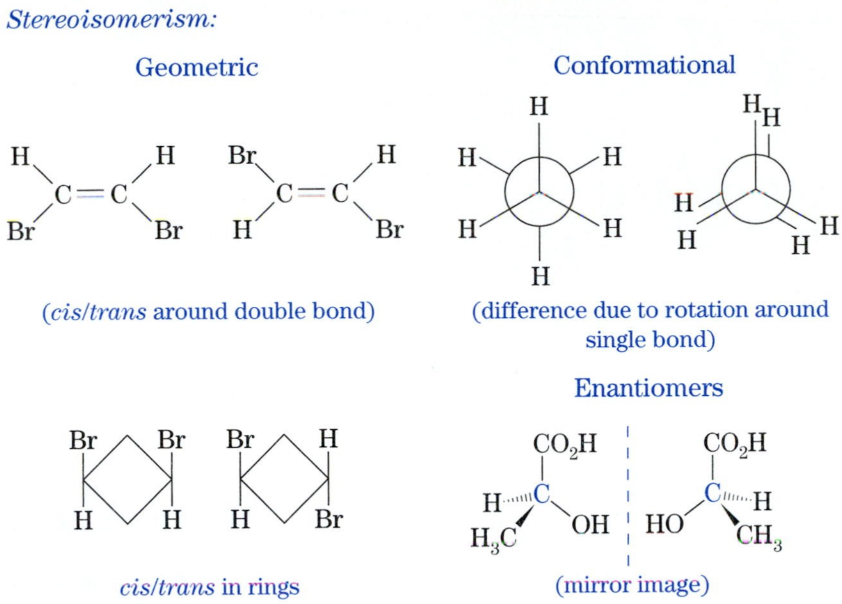

Stereoisomerism:

Geometric

(*cis/trans* around double bond)

Conformational

(difference due to rotation around single bond)

Enantiomers

cis/trans in rings

(mirror image)

GETTING INVOLVED

Problem 7.2

There are seven structural isomers of $C_4H_{10}O$. Draw a pair of skeletal, positional, and functional isomers. One of the seven isomers has a chiral carbon atom (stereocenter) and can exist in two mirror-image forms. Draw this compound and identify the chiral carbon atom.

Problem 7.3

For the formula $C_6H_{12}O$, draw a representative pair of isomers illustrating each of the following types of isomerism: **(a)** skeletal; **(b)** positional; **(c)** functional; **(d)** geometric; **(e)** conformational.

7.2 Stereoisomers with One Chiral Carbon Atom

A. Chiral Carbon Atoms, Enantiomers, and Racemic Mixtures

A carbon with four different bonded groups is called a **chiral carbon** atom (a stereocenter). The groups can be quite different, as in a carbon with four different attached halogens, or only slightly different, as in a carbon bonded to four alkyl groups of different lengths.

chiral carbon atom a carbon with four different bonded groups

chiral carbons

four different attached groups

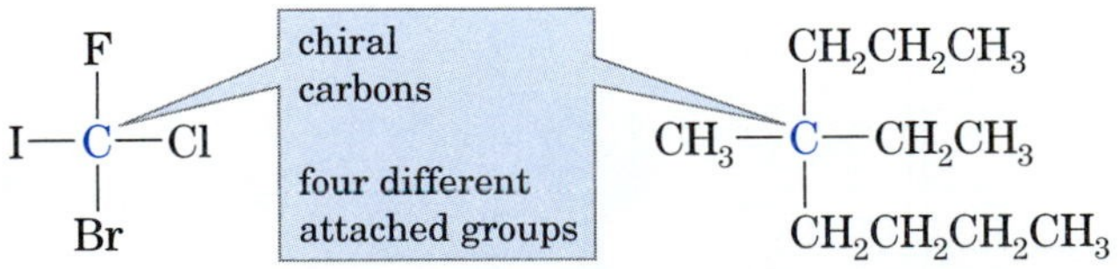

Because of its tetrahedral geometry, a chiral carbon atom can exist in either of two three-dimensional arrangements. They look very much alike just as our left and right hands look alike. But, like our hands, they are mirror images and in no way superimposable (Figure 7.2).

In the following representations, the two chiral carbon atoms just described are in their tetrahedral, nonsuperimposable, mirror-image **configurations** with three of the groups comprising the base of the pyramid and one the peak.

configuration the orientation of groups around a chiral carbon (stereocenter)

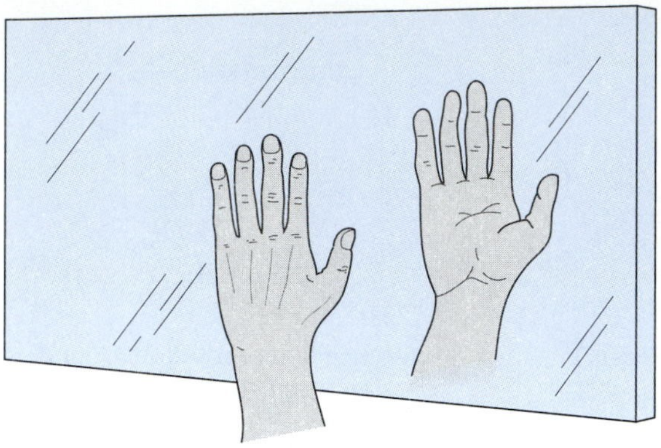

Figure 7.2 The mirror-image relations of a right and a left hand are shown.

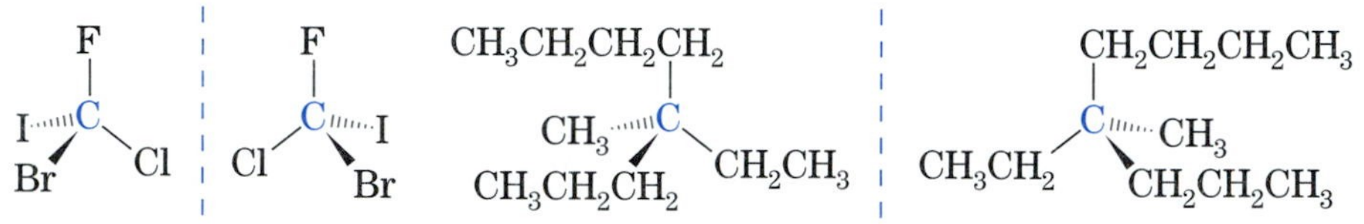

F, I, C, Br, Cl | F, C, I, Cl, Br

$CH_3CH_2CH_2CH_2$, C, CH_3, CH_2CH_3, $CH_3CH_2CH_2$ | $CH_2CH_2CH_2CH_3$, C, CH_3, CH_3CH_2, $CH_2CH_2CH_3$

Convince yourself that these mirror images are not superimposable (use molecular models). Figure 7.3a illustrates that any two groups of a chiral carbon atom can be superimposed on the mirror image, but the other two cannot. However, if any two groups on a carbon are identical, the mirror images can always be completely superimposed (Figure 7.3b).*

*Carbons with four different groups are called chiral carbon atoms, stereogenic centers, stereocenters, chirality centers, or asymmetric carbon atoms. Stereogenic centers or stereocenters are used to indicate that it is not the carbon itself that is chiral, but the resulting molecule with four different groups attached to the carbon is chiral.

The carbon is the center of chirality (in fact, chirality center is the IUPAC term) but it is the molecule that is chiral, i.e. the molecule does not have a plane of symmetry and is not superimposable on its mirror image. Since the most common stereocenters are carbons, we will use the term chiral carbon atom for a carbon with four different bonded groups.

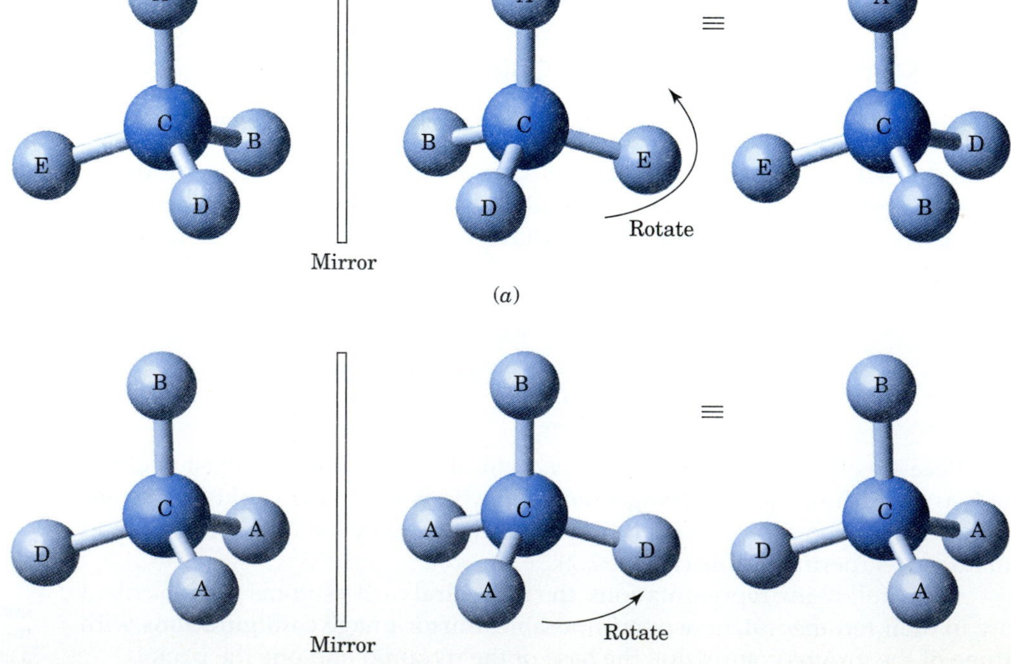

Figure 7.3 (a) A chiral carbon with four different attached groups exists in two nonsuperimposable mirror-image forms. Any two groups (in this picture A and E) can be superimposed by rotation, but the other two (B and D) will always be in conflict. (b) If a carbon has two or more identical groups, it is superimposable on its mirror image.

Stereoisomers that are nonsuperimposable mirror images, such as the two examples in this section and lactic acid in the previous section, are called **enantiomers.** Compounds with a single chiral carbon atom always have one pair of enantiomers; each enantiomer is a chiral molecule, i.e. it is not superimposable on its mirror image. We will see later that compounds with more than one chiral carbon atom have the possibility of more than one pair of enantiomers. The differences between a pair of enantiomers are subtle. Their physical properties, such as melting point, boiling point, and refractive index are identical. Enantiomers differ in only one physical property, the direction in which they rotate plane-polarized light (plane-polarized light will be discussed in section 7.3). One enantiomer rotates plane-polarized light to the right (dextrorotatory) and the other rotates it an equal amount to the left (levorotatory). A molecule that is not superimposable on its mirror image will produce a measurable rotation of plane-polarized light and is termed an optically active or **chiral molecule.**

enantiomers
stereoisomers that are mirror images

chiral molecule
a molecule that is not superimposable on its mirror image; these compounds rotate plane-polarized light

racemic mixture
a 50/50 mixture of enantiomers

A 50/50 mixture of a pair of enantiomers is called a **racemic mixture.** Because the two components rotate plane-polarized light equally but in opposite directions, a racemic mixture is optically inactive, it does not give a net rotation of plane-polarized light.

At this point you may be wondering if all of this could really be important. After all, a pair of enantiomers don't look very different, they differ only as mirror images, and their physical properties are identical with the mere exception of the direction of rotation of plane-polarized light. The answer is that stereoisomerism is crucially important since it is prevalent in organic chemistry, especially in biological molecules. Just as your right hand will fit in a right-handed glove, but not in a left-handed glove, optically active biological molecules exhibit different relationships with other optically active molecules. For example, starch and cellulose are both polymers of glucose; they differ only in the configuration of the linkage between glucose molecules and are stereoisomers. Enzymes in our mouth and intestines are also optically active and are able to digest starch such as in potatoes and bread, but not cellulose, such as in the morning newspaper. We have seen that one enantiomer of lactic acid is found in sour milk, the other enantiomer in sore muscles. The levorotatory enantiomer of epinephrine (commonly known as adrenalin) has 20 times the potency of its mirror image in raising blood pressure. One enantiomer of carvone is responsible for the odor of spearmint, and the other for the scent of caraway and dill seed; the difference in smell is determined by the interaction of our olfactory tissues, which are also composed of optically active molecules, with these two mirror-image isomers.

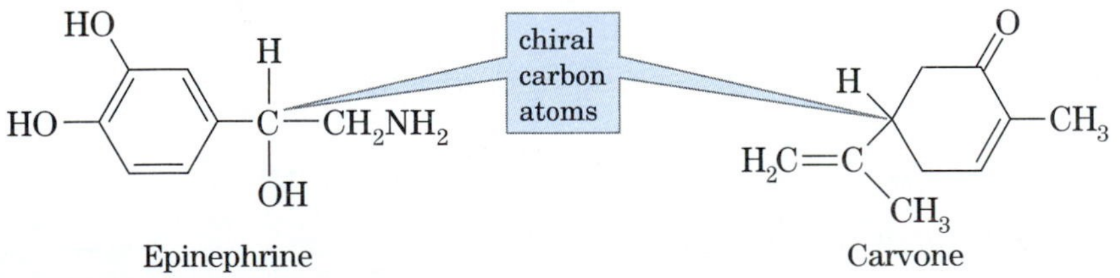

chiral carbon atoms

Epinephrine

Carvone

GETTING INVOLVED

- ✓ What is a chiral carbon atom? How are the two configurations of a chiral carbon atom related?
- ✓ What are enantiomers? How are the two compounds structurally related? How are their physical properties alike and how are they different?
- ✓ What is an optically active or chiral compound?
- ✓ What is a racemic mixture? Why is it optically inactive?

Example 7.1

Identify any chiral carbon atoms in threonine, an essential amino acid and component of protein structure.

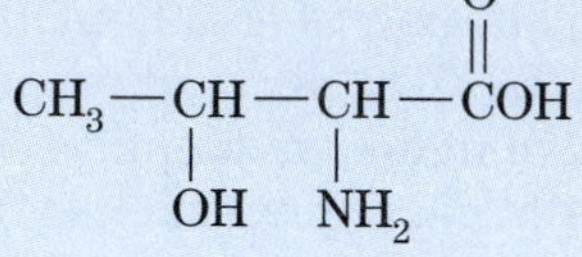

$$\mathrm{CH_3{-}\underset{\displaystyle OH}{\underset{|}{CH}}{-}\underset{\displaystyle NH_2}{\underset{|}{CH}}{-}\overset{\displaystyle O}{\overset{\|}{C}}OH}$$

A chiral carbon atom must have four different bonded groups

The first carbon from the left is a methyl group and has three hydrogens; it is not chiral. The next carbon has a methyl, hydrogen, OH, and the rest of the molecule, four different bonded groups and is a chiral atom. Likewise the next carbon is chiral because it has four different groups (H, NH_2, CO_2H, and the left side of the molecule). The last carbon, the carboxylic acid, has three different bonded groups and it is not chiral. A chiral carbon must be sp^3 hybridized; four different groups are needed for chirality.

Example 7.2

Which compound of each of the following pairs of positional isomers has a pair of enantiomers?

(a) 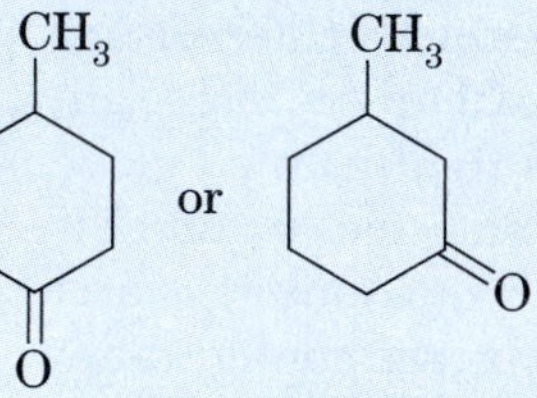 4-methylcyclohexanone (CH_3 at top, O at bottom) or 3-methylcyclohexanone (CH_3 at top, O at lower right)

(b) 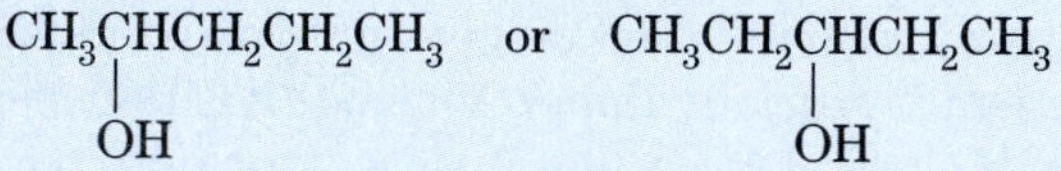 $\mathrm{CH_3\underset{\displaystyle OH}{\underset{|}{C}}HCH_2CH_2CH_3}$ or $\mathrm{CH_3CH_2\underset{\displaystyle OH}{\underset{|}{C}}HCH_2CH_3}$

If the structure has a chiral carbon atom, it will have a pair of enantiomers.

(a) The second structure has a chiral carbon atom; the carbon with the methyl group. This carbon also has a hydrogen and going around the right side of the ring is different from going around the left. There are four different attached groups. The first structure has symmetry, going around the ring from the left or right is the same so there are not four different groups.

(b) Carbon 2 of the first structure is chiral with a hydrogen, OH, methyl and propyl group; a pair of enantiomers is possible. The second structure does not have a chiral carbon atom; the carbon with the OH has two ethyl groups.

Problem 7.4

Identify chiral carbon atoms in the following compounds.

(a) 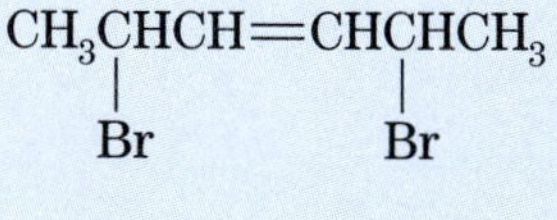 $\mathrm{CH_3\underset{\displaystyle Br}{\underset{|}{C}}HCH{=}CH\underset{\displaystyle Br}{\underset{|}{C}}HCH_3}$

(b) $\mathrm{CH_3CH_2\overset{\displaystyle CH_3}{\overset{|}{C}}HCH_2\overset{\displaystyle CH_3}{\overset{|}{C}}HCH_2CH_3}$

(c) cyclohexane-1,3-dione ring with CH_3 at C2 (top, between the two C=O) and CH_3 at C5 (bottom)

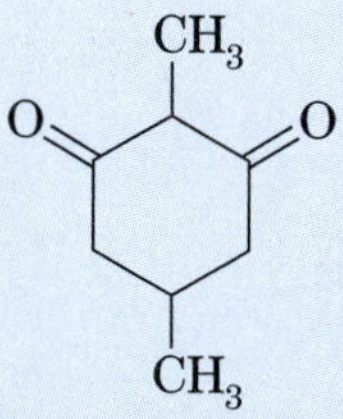

(d) 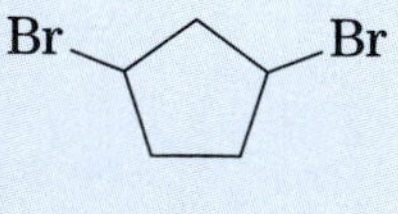 Br—cyclopentane—Br (1,3-dibromocyclopentane)

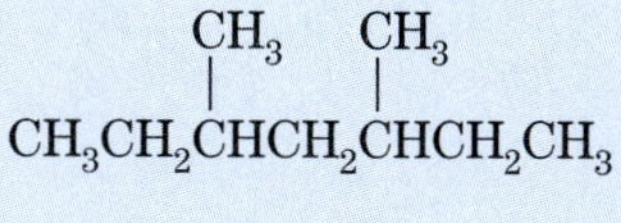

Problem 7.5

Which of the following compounds can exist as a pair of enantiomers? **(a)** methylcyclopentane; **(b)** 3-methylcyclopentene; **(c)** *trans* 2-pentene; **(d)** 4-chloro-*cis*-2-pentene; **(e)** 2-bromo-2-methylpentane.

Problem 7.6

There are nine skeletal isomers with the formula C_7H_{16}. Draw the two that have chiral carbon atoms.

See related problems 7.17–7.18.

B. Expressing the Configuration of Enantiomers in Three Dimensions

There are several excellent ways to express the three-dimensional structure of an enantiomeric pair on a two-dimensional surface. An especially effective method and one that is relatively easy to draw and manipulate is illustrated in Figure 7.4. Imagine that you have a tetrahedral ball and stick model of a chiral carbon atom and that it is sitting on a table supported by the three atoms making the "base" of the pyramid. Rotate the model upward so two atoms are sticking out toward you horizontally, and two atoms are directed vertically away from you (Figure 7.4a). You can translate this model into a wedges and dashes representation. The two horizontal atoms are represented by wedges indicating they are projecting out from the paper and the two vertical atoms by dashes showing that they projected behind the plane (Figure 7.4b). This is the type of representation we will use in this chapter.

The wedges and dashes representation can be simplified as shown in Figure 7.4c–d, but it is important to understand that horizontal bonds are projected in

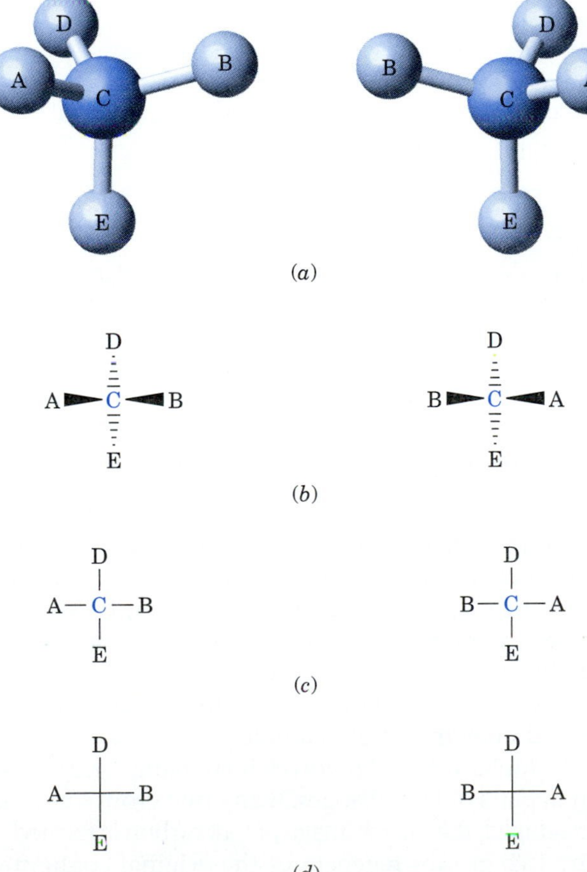

Figure 7.4 Four different representations of a chiral carbon. (a) With a tetrahedral carbon placed in this way, the horizontal bonds are in front of the plane and the vertical bonds behind. (b) The three-dimensional nature is shown by solid wedges (in front of the plane) and dashes (behind the plane). (c) No attempt is made in these formulas to show three dimensions. Vertical bonds are understood to be behind the plane, however, and horizontal bonds in front. (d) Fischer projections in the most common and simplest form, where the actual carbon atoms are assumed.

front of the plane and vertical bonds are behind. The representation shown in Figure 7.4d is called a Fischer projection and is commonly used to express configurations especially in carbohydrate chemistry.

GETTING INVOLVED

Example 7.3

Draw the enantiomers of cysteine, the amino acid found in hair protein (only one of the two is actually cysteine), and lactic acid in projections, as illustrated in Figure 7.4b.

Solution

In each case, first find the chiral carbon atom (*), the one with four different attached groups. Then draw a carbon with horizontal wedges and vertical dashes. Put the four different groups around it randomly. Finally, draw the mirror image; the two vertical groups maintain their positions, but the horizontal ones switch.

$HSCH_2\overset{*}{C}HCO_2H$ (with NH_2 on the starred carbon) → H_2N—C—H (top CO_2H, bottom CH_2SH) and H—C—NH_2 (top CO_2H, bottom CH_2SH) — Cysteine

$CH_3\overset{*}{C}HCO_2H$ (with OH on the starred carbon) → H—C—OH (top CO_2H, bottom CH_3) and HO—C—H (top CO_2H, bottom CH_3) — Lactic acid

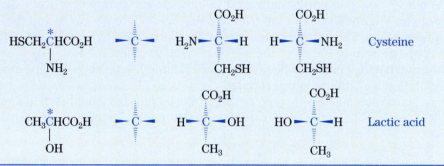

Problem 7.7

Using the wedges and dashes representations in Figure 7.4b and the method in Example 7.3 draw the mirror-image enantiomers of **(a)** the amino acid phenylalanine; **(b)** adrenalin; **(c)** amphetamine.

(a) C_6H_5—CH_2CHCO_2H (with NH_2 on CH)

(b) 3,4-(HO)$_2C_6H_3$—C(H)(OH)—CH_2NH_2

(c) CH_3CHCH_2—C_6H_5 (with NH_2 on CH)

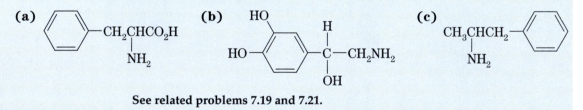

See related problems 7.19 and 7.21.

C. Comparing Representations of Enantiomers

How can we determine whether two representations of a compound are identical or mirror images? We can rotate, flip, or turn structures such as those in Figure 7.4b so long as we use wedges and dashes correctly to maintain the three-dimensional configuration. For example, a 90° rotation would result in the new vertical groups protruding out of the paper as wedges and the new horizontal groups being represented by dashes since they are directed behind the plane. A similar situation occurs if the structure is flipped. This procedure is illustrated in method 1 of Example 7.4.

A simpler method involves interchanging groups around the chiral carbon atom in pairs of interchanges. If any two groups around a chiral carbon atom are interchanged, the mirror image of that carbon is formed. Then a second interchange of any two groups regenerates the original configuration but from a different perspective. This procedure is illustrated in method 2 of the following example.

GETTING INVOLVED

Example 7.4

Determine whether the following structures are identical or mirror images (enantiomers).

D
A—C—B
E

Structure 1

B
E—C—D
A

Structure 2

Solution

Method 1: Physical maneuvering of one structure.

D
A—C—B
E
Structure 1

Rotate 90° counterclockwise →

B
D····C····E
A

Turn over, out of plane →

B
E—C—D
A
Matches structure 2

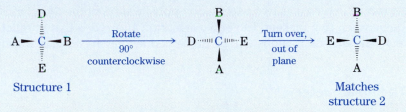

Method 2: Interchange of attached groups. To maintain the original configuration, you must perform an even number of interchanges (do them in pairs).

D
A—C—B
E
Structure 1

First, interchange B and D →

B
A—C—D
E

Second, interchange A and E →

B
E—C—D
A
Matches structure 2

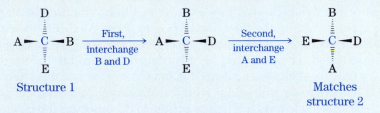

By both methods, the two structures are shown to be identical.

Problem 7.8

Determine which of the structures in (a)–(h) are identical to and which are mirror images of the following compound:

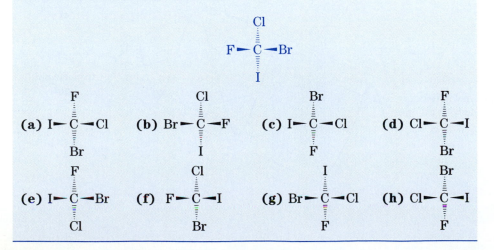

Cl
F—C—Br
I

(a) F above; I—C—Cl; Br below

(b) Cl above; Br—C—F; I below

(c) Br above; I—C—Cl; F below

(d) F above; Cl—C—I; Br below

(e) F above; I—C—Br; Cl below

(f) Cl above; F—C—I; Br below

(g) I above; Br—C—Cl; F below

(h) Br above; Cl—C—I; F below

7.3 Measurement of Optical Activity—The Polarimeter

A compound that is not superimposable on its mirror image is a chiral compound and it rotates plane-polarized light.

A. Plane-Polarized Light

plane-polarized light light oscillating in only one plane

Light can be described as a wave vibrating perpendicular to its direction of propagation. Vibration can occur in an infinite number of planes at right angles to the direction of light travel. Light vibrating in all possible planes is said to be *unpolarized.* Light oscillating in only one of the possible planes is **plane-polarized.** Plane-polarized light can be produced by passing unpolarized light through a Nicol prism (Iceland spar, a form of calcite, $CaCO_3$, used by the Scottish physicist William Nicol) or through a Polaroid sheet (specially oriented crystals embedded in plastic, invented by E. H. Land). In either case, light vibrating in only one plane is allowed to pass. Light vibrating in all other planes is rejected (Figure 7.5). A polarizer can be compared to a picket fence, and the vibrating light waves can be depicted as two people, on opposite sides of the fence, oscillating a rope between two pickets. The only oscillation allowed is that parallel to the pickets. All other oscillations will be destroyed as they try to pass through the fence.

B. The Polarimeter

polarimeter instrument used to measure the rotation of plane-polarized light

The rotation of plane-polarized light by an optically active compound is detected and measured with an instrument called a **polarimeter,** shown diagrammatically in Figure 7.6. A polarimeter has a monochromatic (single-wavelength) light source at one end, which produces unpolarized light vibrating in all possible planes perpendicular to the direction of propagation. As this light encounters the stationary polarizer, all planes but one are rejected. The light passing through is plane-polarized. The polarized light continues on through the sample tube (which for now we shall assume is empty) and reaches the variable analyzing polarizer. If this polarizer is lined up with the stationary polarizer, the polarized light will be allowed to pass and will be visible to the observer. If the variable po-

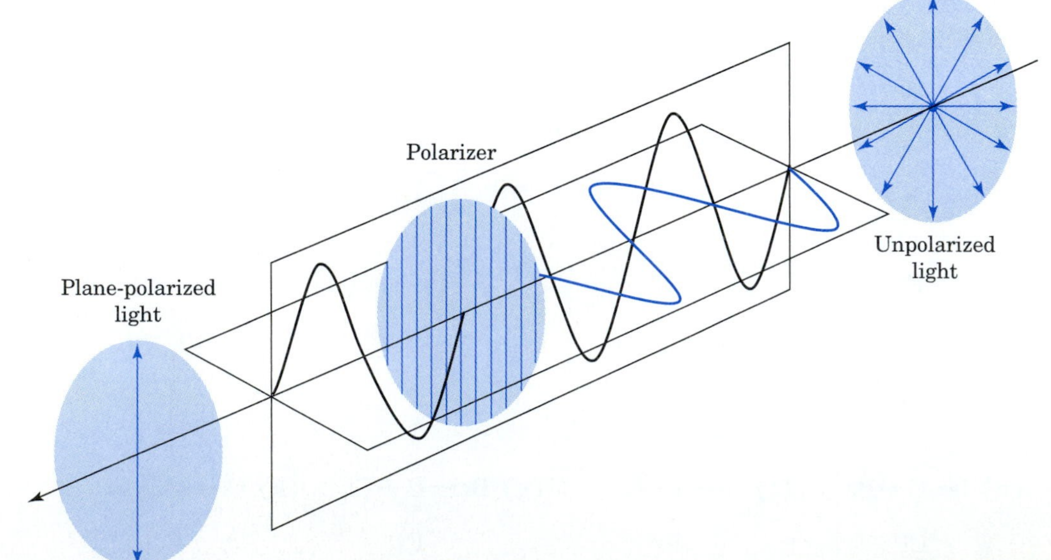

Figure 7.5 As unpolarized light encounters a polarizer, all but one plane is blocked. The resulting light that is transmitted is plane-polarized.

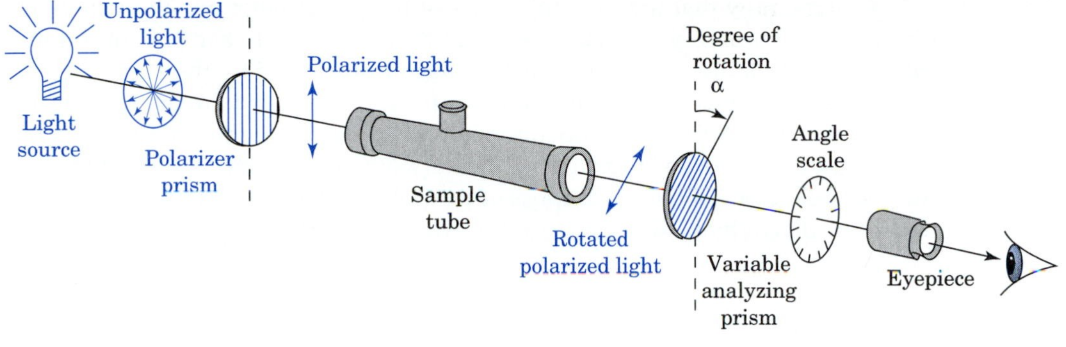

Figure 7.6 Schematic representation of a polarimeter containing an optically active sample.

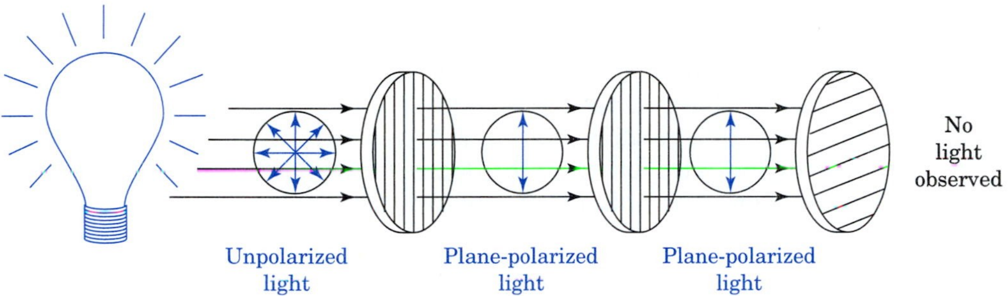

Figure 7.7 Unpolarized light from the light source is polarized by the first polarizing sheet. Since the second polarizing sheet is in alignment with the first, light is transmitted. Since the third is not, light transmission is blocked at this point. These concepts can be demonstrated by using two pairs of Polaroid sunglasses as illustrated in Figure 7.8.

larizer is rotated, however, so that its allowable planes of light transmission are 90° to the plane allowed by the stationary polarizer, the polarized light will be blocked, and the observer will perceive darkness (Figure 7.7). This can be demonstrated with two pairs of Polaroid sunglasses (Figure 7.8). If one pair is placed in front of the other so that the lenses are lined up, light will pass through both, and the pair will be transparent. If now one pair is rotated 90°, the two lenses will have their planes of allowable light transmission out of phase, and the pair will appear opaque. (Try this at a store.)*

*Three-D movies and slides make use of polarized light. To see in three dimensions, each of our eyes visualizes a scene from a slightly different perspective. These are combined by the brain to give a 3-D visual image. Two cameras are used in making 3-D movies to get two different views. The two images are projected on the screen with two different beams of polarized light (polarized in different planes). Without glasses, both of our eyes see both images, and the picture appears blurred. Using Polaroid glasses with the lenses unsynchronized, however, we get a 3-D view. Since the lenses are oriented differently, each allows a different plane of polarized light to be transmitted; the other of the two planes is rejected by each lens. Thus each eye sees only one of the originally filmed images and the two are then combined by the brain to give the 3-D effect.

Figure 7.8 Using polarized sunglasses to demonstrate the generation and detection of plane-polarized light. If the lenses are in alignment, light is transmitted. If they are not in alignment, light that passes through one will be blocked by the other as was illustrated in Figure 7.7.

Assume now that the two polarizers of the polarimeter are aligned so that there is maximum light transmission and that an optically active compound is placed in the sample tube. The plane of light transmitted by the stationary polarizer will be rotated in the sample tube and will not be maximally transmitted by the variable polarizer. The operator, however, can rotate the variable polarizer until light transmission is again maximized (when the analyzer's allowable transmission planes are the same as those of the polarized light). In this way, not only can optical activity be detected, but the angle of rotation can also be measured.

C. Specific Rotation

optically active compound a compound that rotates plane-polarized light; such compounds are not superimposable on their mirror images

dextrorotatory rotation of plane-polarized light to the right (*d* or +)

levorotatory rotation of plane-polarized light to the left (*l* or −)

specific rotation calculated degree of rotation of an optically active compound

A compound that does not rotate plane-polarized light is optically inactive, whereas a compound that does is **optically active.** A compound that rotates polarized light to the right, or clockwise, is termed **dextrorotatory,** represented by *d* or + (the variable polarizer is turned to the right to obtain maximum light transmission). If the rotation is to the left (counterclockwise), the substance is **levorotatory,** indicated as *l* or −. The degree of rotation is used to calculate the **specific rotation,** α, according to the following equation:

$$\alpha = \frac{\text{observed rotation, degrees}}{\text{length of sample tube, dm} \times \text{concentration of sample, g/cm}^3}$$

Like density, melting point, and boiling point, specific rotation is a physical property of a compound. The following are some specific rotations measured at about 20°C, using light of wavelength 5893 Å (the wavelength of the sodium D line):

Menthol	$\alpha = -50°$	Sucrose	$\alpha = +66.5°$
Cholesterol	$\alpha = -31.5°$	Vitamin C	$\alpha = +21.5°$
α-D-glucose	$\alpha = +112.2°$	Nicotine	$\alpha = -169°$

GETTING INVOLVED

- ✓ What is plane-polarized light and how is it generated?
- ✓ Describe the components of a polarimeter and how it works. When do the two polarizers allow light transmission and when is it blocked?
- ✓ What is an optically active compound? How is a dextrotatory enantiomer different from its levorotatory counterpart both structurally and in rotation of plane-polarized light? What is specific rotation?

7.4 Stereoisomers with Two Chiral Carbon Atoms

A compound possessing one chiral carbon atom can exist in two nonsuperimposable mirror-image forms called enantiomers. A compound with two chiral carbon atoms can have a maximum of four optical isomers, because each chiral carbon atom can exist in two configurations that are mirror images. The maximum number of stereoisomers possible for a compound is 2^n, where n is the number of chiral carbon atoms. This is sometimes called the *van't Hoff rule.*

A. Molecules with Two Dissimilar Chiral Carbon Atoms: Enantiomers and Diastereomers

Consider the carbohydrate molecule 2-deoxyribose, which is a structural component of the genetic material deoxyribonucleic acid (DNA).

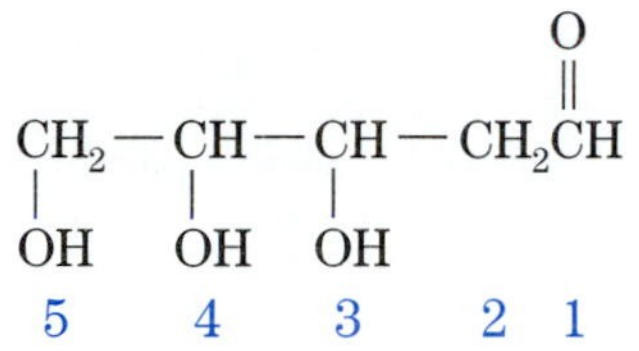

$$\underset{\underset{5}{OH}}{CH_2}-\underset{\underset{4}{OH}}{CH}-\underset{\underset{3}{OH}}{CH}-\underset{2}{CH_2}\overset{\overset{O}{\|}}{\underset{1}{CH}}$$

2-deoxyribose

This molecule has two chiral carbon atoms, carbon-3 and carbon-4. The four different groups bonded to carbon-3 are $-CH_2\overset{\overset{O}{\|}}{C}H$, $-H$, $-OH$, and $-CHOHCH_2OH$, and the four bonded to carbon-4 are $-CHOHCH_2\overset{\overset{O}{\|}}{C}H$, $-H$, $-OH$, and $-CH_2OH$. Carbon-1 has three different bonded groups but needs four to be chiral. Carbon-2 and carbon-5 each have two identical bonded groups (hydrogens) and thus cannot be chiral.

The maximum number of stereoisomers possible for this structure is four ($2^n = 2^2 = 4$). These should be drawn in a systematic fashion, with the chiral carbon atoms emphasized. It is also convenient to draw the isomers as pairs of mirror images for comparison purposes. Following are the four isomers using a form of the wedges and dashes representation. Remember that horizontal bonds are in front of the plane of the paper and vertical bonds behind. Only the two chiral carbon atoms and the bond between them are in the plane.

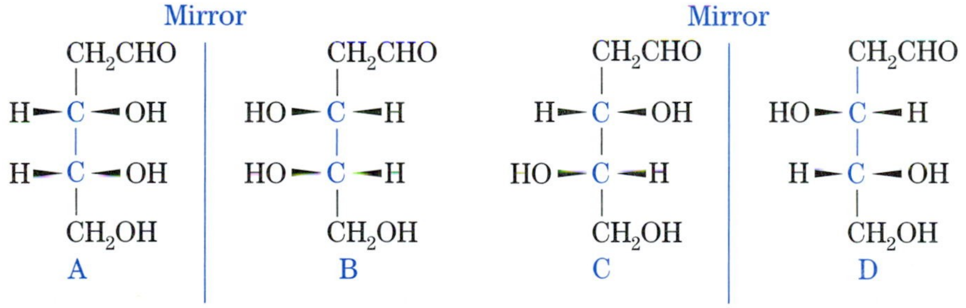

A	Mirror	B	C	Mirror	D
CH_2CHO		CH_2CHO	CH_2CHO		CH_2CHO
H—C—OH		HO—C—H	H—C—OH		HO—C—H
H—C—OH		HO—C—H	HO—C—H		H—C—OH
CH_2OH		CH_2OH	CH_2OH		CH_2OH

Structures A and B are mirror images of one another, yet in no way are they superimposable. We cannot superimpose A and B by turning either out of the plane, nor can we rotate either 180° in the plane and have them match since the top and bottom of the molecule are different. Thus A and B are an enantiomeric pair. In fact, structure A specifically is the 2-deoxyribose found in DNA. All of their physical properties are identical except the rotation of plane-polarized light. A and B rotate light in equal magnitude but in opposite directions. Likewise, structures C and D are nonsuperimposable mirror images and are a pair of enantiomers.

There are four combinations of structures that, unlike enantiomers, are not related as mirror images. These (AC, AD, BC, and BD) are called **diastereomers.** Diastereomers are stereoisomers that are not mirror images. Whereas enan-

diastereomers stereoisomers that are not mirror images

tiomers differ only in the rotation of plane-polarized light, diastereomers can differ in all physical properties. Their melting points, boiling points, densities, refractive indices, and, if they are chiral, specific rotations can differ and usually do.

GETTING INVOLVED

✓ Structurally, what is the difference between a pair of enantiomers and a pair of diastereomers? How does each differ in physical properties?

Problem 7.9

Draw the four optical isomers of the amino acid below (only one of the four is the essential amino acid threonine).

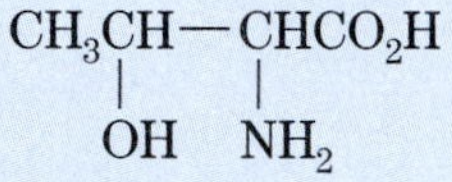

$$\begin{array}{c} CH_3CH-CHCO_2H \\ \quad | \qquad | \\ \quad OH \quad NH_2 \end{array}$$

Identify pairs of enantiomers and diastereomers.
See related problems 7.20, 7.22 a, c; 7.23 b.

B. Molecules with Two Similar Chiral Carbon Atoms: Enantiomers, Diastereomers, and Meso Compounds

Let us consider the compound tartaric acid.

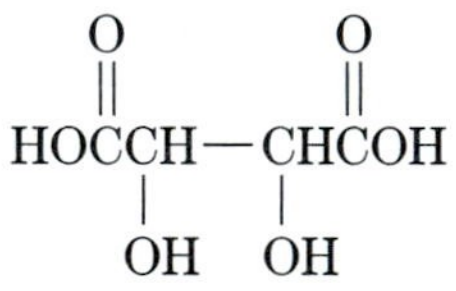

$$\begin{array}{c} O \qquad\quad O \\ \| \qquad\quad \| \\ HOCCH-CHCOH \\ | \qquad | \\ OH \quad OH \end{array}$$

Tartaric acid

Tartaric acid is found in grape juice and cream of tartar and is also the acid component of some baking powders. It has two similar chiral carbons, carbon-2 and carbon-3. Each has four different bonded groups ($-CO_2H$, $-H$, $-OH$, $-CHOHCO_2H$).

Remember that one should draw the structures systematically, in pairs of mirror images, with emphasis on the chiral carbon atoms. In the following, structures E and F are mirror images. However, by rotating either molecule 180° in the plane of the page, we can superimpose one on the other. Thus structures E and F are identical, and F should be eliminated from the list of stereoisomers. Although E has chiral carbon atoms, the overall molecule is not chiral since it is superimposable on its mirror image. Such compounds are called **meso compounds** and are optically inactive; they do not rotate plane-polarized light. Actually, meso compounds probably interact with plane-polarized light, but the rotation is undetectable due to internal compensation. Close inspection of *meso*-tartaric acid (E) reveals that the top half of the molecule is a mirror image of the bottom. The molecule has an internal enantiomeric pair, which is analogous to a racemic mixture. Their individual effects on plane-polarized light cancel each other, and the compound is optically inactive. Note then that possession of chiral carbon atoms

meso compounds stereoisomers that are superimposable on their mirror images

Mirror

CO_2H / H—C—OH / H—C—OH / CO_2H
E

CO_2H / HO—C—H / HO—C—H / CO_2H
F

180°

CO_2H / H—C—OH / H—C—OH / CO_2H

Identical

Mirror

CO_2H / H—C—OH / HO—C—H / CO_2H
G

CO_2H / HO—C—H / H—C—OH / CO_2H
H

180°

CO_2H / HO—C—H / H—C—OH / CO_2H

Different

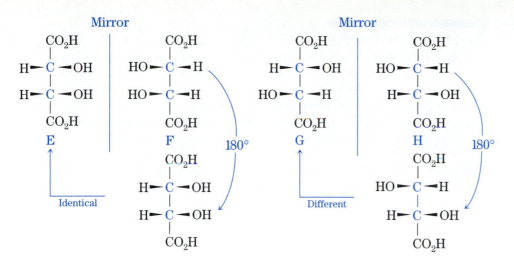

is not necessarily sufficient for optical activity; the compound must also be chiral in its overall structure.

Structures G and H are related as mirror images and are not superimposable even if rotated 180°. Thus G and H constitute an enantiomeric pair. There are two pairs of diastereomers: E and G, and E and H. Table 7.1 lists some properties of the various forms of tartaric acid, and Table 7.2 summarizes the definitions of terms used in describing this type of stereoisomerism.

Table 7.1 Properties of the Optical Isomers of Tartaric Acid

	Dextrorotatory Form	Levorotatory Form	Racemic Mixture	Meso Form
Rotation (α)	+12°	−12°	0°	0°
Melting point	168–170°C	168–170°C	206°C	140°C
Water-solubility 20°C, 100 ml H_2O	139 g	139 g	20.60 g	125 g
pK_A1 (acidity)	2.93	2.93	2.96	3.11
Density	1.7598	1.7598	1.697	1.666

Table 7.2 Terms Used to Describe Stereoisomers

Optically active, or chiral, compound A compound that is not superimposable on its mirror image. Such compounds rotate plane-polarized light.

Chiral carbon atom A carbon bonded to four different groups.

van't Hoff rule The maximum number of enantiomers a compound may have is 2^n; n represents the number of chiral carbon atoms.

Enantiomers Stereoisomers that are mirror images. Enantiomers have identical physical properties except for the rotation of plane-polarized light. One of the pair is levorotatory, and the other dextrorotatory, to equal extents.

Racemic mixture A mixture of equal parts of enantiomers. Racemic mixtures are optically inactive.

Diastereomers Stereoisomers that are not mirror images. All physical properties of diastereomers are usually different from one another.

Meso compound Stereoisomer that has more than one chiral center and that is superimposable on its mirror image. Meso compounds are optically inactive.

GETTING INVOLVED

✓ Structurally what are the differences among a pair of enantiomers, a pair of diastereomers, and a meso compound? How does each differ in physical properties?

✓ What is the difference between a chiral carbon atom and a chiral compound? Is a compound with chiral carbon atoms always chiral?

✓ Define the following: van't Hoff rule, optically active compound, and racemic mixture.

Problem 7.10

Draw the optical isomers of: **(a)** 2-bromo-3-chlorobutane, and **(b)** 2,3-dibromobutane. Identify pairs of enantiomers, pairs of diastereomers, and meso compounds. Use structures such as those in section 7.4.A and B.

See related problems 7.22–7.24 and 7.34–7.35.

7.5 Stereoisomerism in Cyclic Compounds

Cyclic compounds can exist as enantiomers as well as *cis-trans* isomers (section 2.8.E). Using 1,2-dibromocyclopropane as an illustration, we find that the *cis* isomer has its two bromines on the same side of the planar ring, while the *trans* isomer has one above and one below. Since 1,2-dibromocyclopropane has two chiral carbon atoms (the carbons bonded to bromines), it could have a maximum of four enantiomers (two pairs). Let us draw the mirror images of the *cis* and *trans* isomers and test for superimposability.

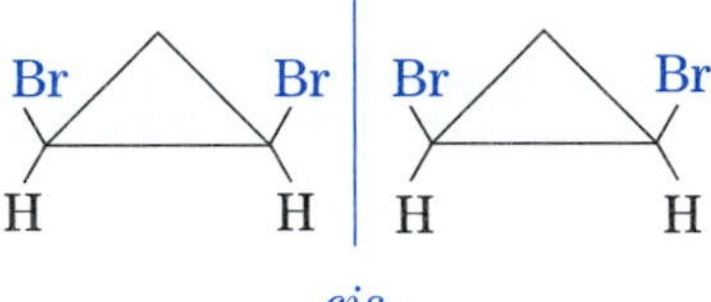

cis
Superimposable mirror images;
meso

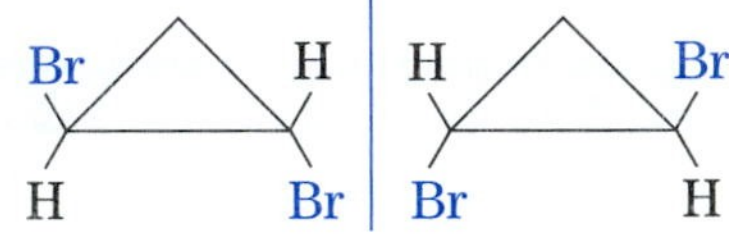

trans
Nonsuperimposable mirror images;
enantiomers

cis-1,2-Dibromocyclopropane is superimposable on its mirror image. Therefore, the molecule is not chiral (it is achiral), and it is an optically inactive meso structure. The *trans* isomer is not superimposable on its mirror image; it exists as an enantiomeric pair. The *cis* and *trans* isomers are related as diastereomers.

The compound 1-bromo-2-chlorocyclopropane has two dissimilar chiral carbon atoms and two pairs of enantiomers.

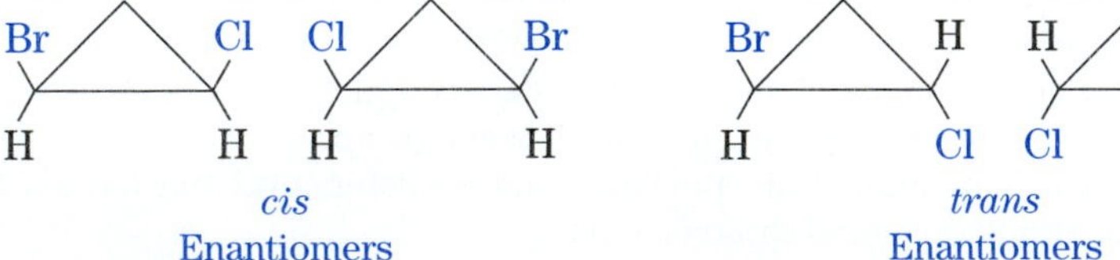

cis
Enantiomers

trans
Enantiomers

GETTING INVOLVED

Problem 7.11

Draw the optical isomers of: **(a)** 1,3-dibromocyclopentane and **(b)** 1-bromo-3-chlorocyclopentane. Identify pairs of enantiomers, pairs of diastereomers, and meso compounds.

See related problem 7.23.

Connections 7.1 www.prenhall.com/bailey

Stereoisomerism in the Biological World

Nowhere is the importance of stereoisomerism more evident than in the molecules of which living organisms are composed. The dietary carbohydrates, such as glucose, which we use as a source of food energy, are of a specific "family" of stereoisomers, the D family. The D refers to the right-handed orientation of the OH group on the chiral carbon (*) farthest from the functional carbonyl group. The L carbohydrates, that is, the mirror-image forms, or enantiomers, of the D molecules, do not exist to any great extent in nature.

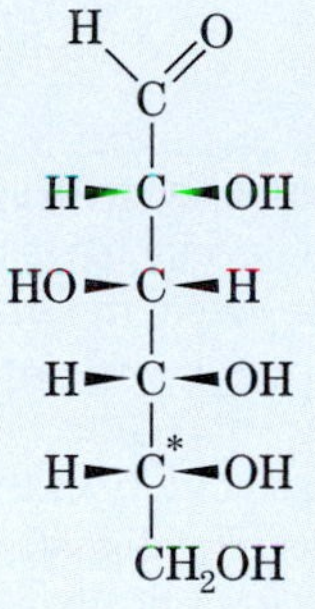

D-glucose 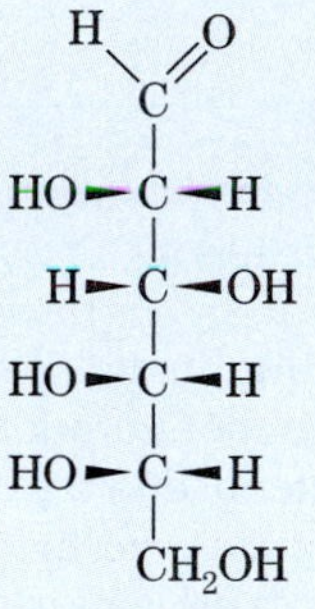 L-glucose

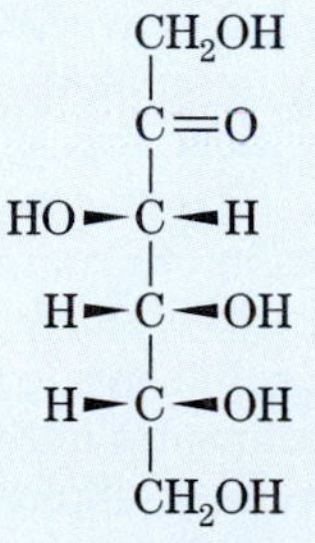

D-fructose 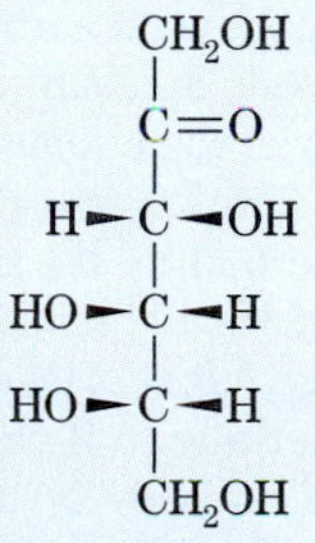 L-fructose

It is interesting to note that proteins, which are responsible for such varied physiological functions as catalysis, nutrient transport, and bone structure, are composed of L-amino acids. Their mirror images, the D-amino acids, only appear in lower life forms, such as in bacteria as components of their cell walls. The ages of fossilized proteins, and perhaps even of some living organisms, may be found by measuring the degree of conversion of L- to D-amino acids, which is a function of time.

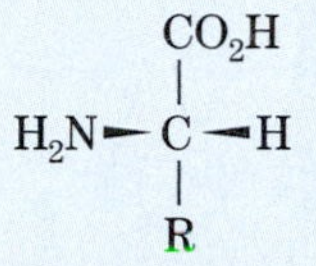

L-amino acid 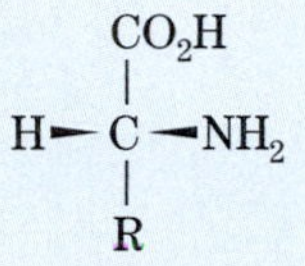 D-amino acid

A living organism uses stereoisomerism to increase its complexity and ensure its survival. The membranes of cells are composed of lipids and proteins, with carbohydrates covalently attached to the outside. These carbohydrates have multiple chiral centers and can thereby set up distinctive patterns that distinguish one type of tissue cell from another. In this way, heart tissue can be identified as different from kidney tissue, and circulating hormones and nutrients can be channeled to the correct endpoints. Our entire immune system depends on the identification of particular tissues as "self" rather than "nonself." The latter will be identified by their protein-carbohydrate markers and be destroyed.

The quest for artificial sweeteners has uncovered some interesting facts about the physiology of taste. First, not all sugars (carbohydrates) are sweet. Second, other types of compounds besides sugars can be sweet. Saccharin is 300 times sweeter than sucrose (table sugar). Aspartame, also a nonnutritive sweetener, is a dipeptide (protein), not a

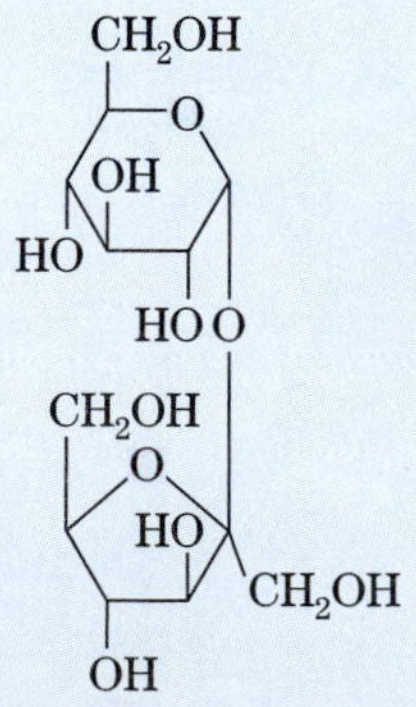

Sucrose 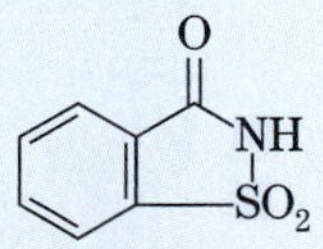 Saccharin 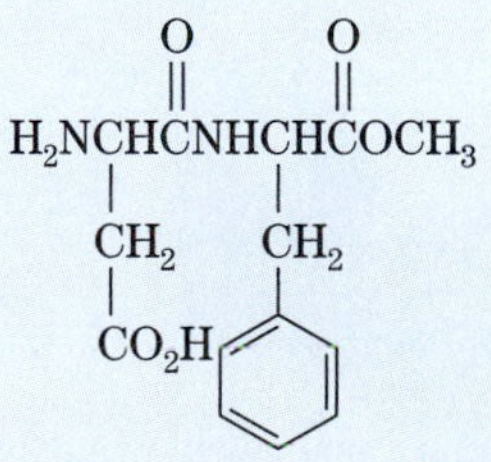 Aspartame

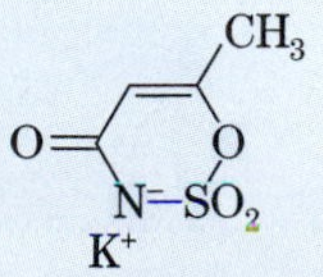

Acesulfame K 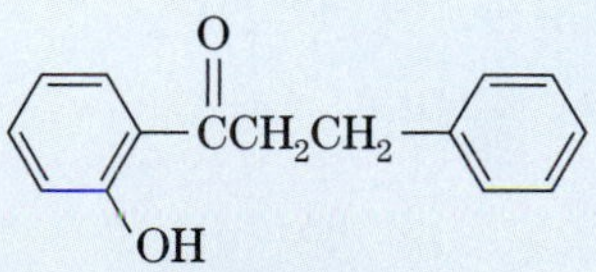 Dihydrochalcone

Connections 7.1 *(cont.)*

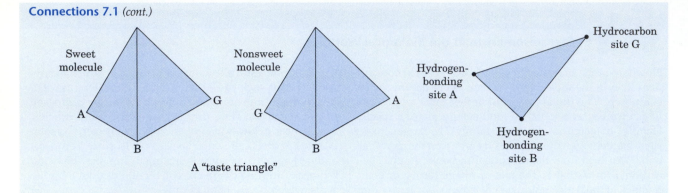

Sweet molecule
A
B
G

Nonsweet molecule
G
B
A

Hydrogen-bonding site A
Hydrogen-bonding site B
Hydrocarbon site G

A "taste triangle"

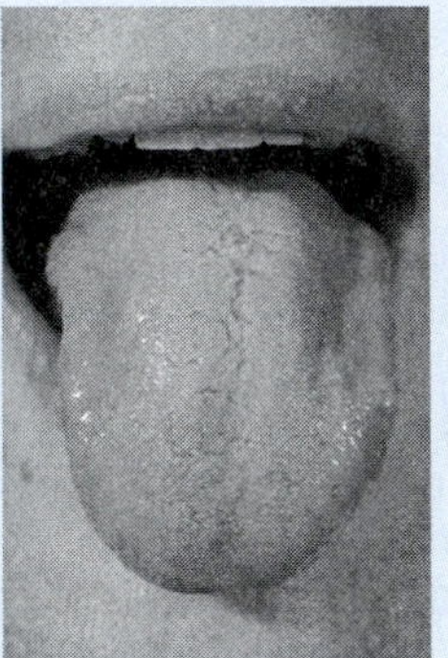

Taste buds on a tongue have taste triangles

carbohydrate. Dihydrochalcones are naturally sweet molecules found in the rind of citrus fruits, and Acesulfame K is a synthetic.

Taste is a sense that is difficult to research, since it has both objective and subjective elements and requires a sentient, intact subject who can communicate. Most of the study into taste therefore involves investigations of the structures of sweet molecules. It has been found that the most important factor is the arrangement of three groups on each molecule, two of which can hydrogen-bond to two points on a taste bud and the third being capable of nonpolar (hydrocarbon-type) interactions with a third taste-bud site. This sets up a "taste triangle," which is tied in with the configuration of groups on the sweet molecule.

Our ability to digest certain foods and not others depends on the configurations of the molecules of the potential food material and the optical selectivity of the enzymes that perform the digestive process. The digestive enzymes are themselves proteins composed of optically active amino acids. The carbohydrates starch and cellulose are both polymers of glucose. However, the linkage between the glucose monomer units in starch is referred to as an alpha (α) bond, axial at the glycosidic carbon (the carbon with two bonded oxygens), while that same linkage in cellulose is a beta (β) bond and is in the equatorial position. This subtle difference (α and β linkages are merely different configurations) can be discerned by the enzymes in our mouths and intestines, which will break the α bonds of starch but not the β bonds of cellulose. Meanwhile, ruminants (cows, sheep) harbor bacteria with β-cleaving enzymes in their guts to perform the vital conversion of grasses (cellulose) to nutritive glucose. This is the reason we can digest the toast, potatoes, or sweet rolls we have for breakfast but could not digest the morning newspaper.

Potatoes—starch

Cotton—cellulose

Connections 7.1 *(cont.)*

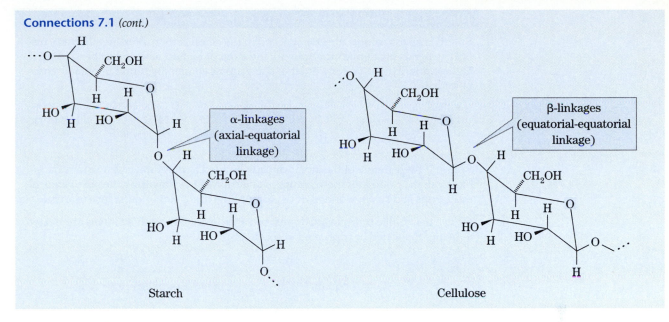

Starch

Cellulose

7.6 Specification of Configuration:

A. *R* and *S* Designations of Chiral Carbon Atoms

We have seen that a chiral carbon atom is bonded to four different groups and that the four groups can be arranged in two different ways that are related as mirror images. The specific arrangement of the groups characterizes a particular stereoisomer and is known as the *configuration*. This configuration can be described by actually drawing the compound. But how can we describe the configuration more conveniently?

A most effective method was developed by R. S. Cahn, C. Ingold, and V. Prelog (as was the E, Z method for geometric isomers in section 3.5B). It involves two steps:

Step 1: By a set of sequence rules, described in Table 7.3, the groups connected to the chiral carbon are assigned priorities.

Step 2: The molecule is then visualized such that the group of lowest priority is directed away from the observer. The remaining three groups are in a plane and project toward the observer. If the eye moves clockwise as it goes from the group of highest priority to the groups of second and third priority, the configuration is designated ***R*** (Latin, *rectus*, "right"). If it moves in a counterclockwise direction, the configuration is designated ***S*** (Latin, *sinister*, "left").

R, S
terms used to describe the configurations of chiral carbons

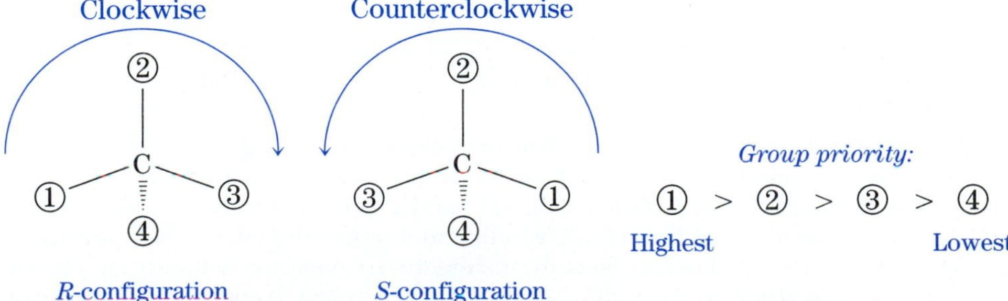

First, let us learn how to assign group priorities, and then we will apply this knowledge to determining *R* and *S* configurations.

Table 7.3 Sequence Rules for *R* and *S* Configurations

Rule 1: If all four atoms directly attached to the chiral carbon atom are different, priority depends on atomic number, with the atom of highest atomic number getting the highest priority. The priority order of some atoms commonly found in organic compounds is

High priority	I	>	Br	>	Cl	>	S	>	F	>	O	>	N	>	C	>	H	Low priority
	53		35		17		16		9		8		7		6		1	

Atomic numbers

Rule 2: If two or more of the atoms directly bonded to the chiral carbon atom are identical, the priority of these groups is determined by comparing the next atoms of the groups and so on, working outward until the first point of difference is found.

Rule 3: If a double or triple bond must be considered, the involved atoms are treated as being duplicated or triplicated, respectively.

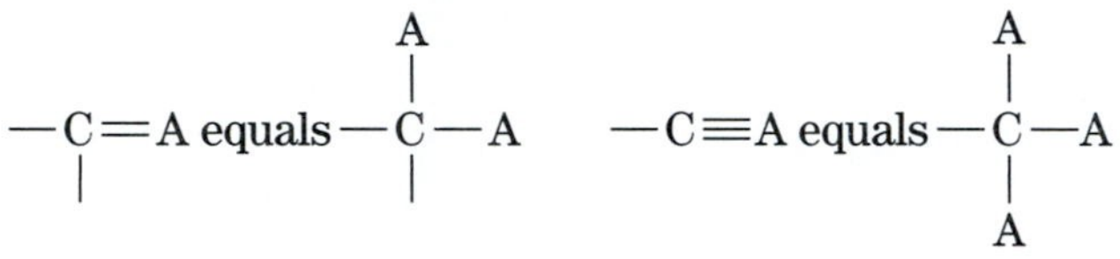

$$-\underset{|}{C}{=}A \text{ equals } -\overset{\overset{A}{|}}{\underset{|}{C}}-A \qquad -C{\equiv}A \text{ equals } -\overset{\overset{A}{|}}{\underset{\underset{A}{|}}{C}}-A$$

B. Determining Group Priorities

Priorities of groups attached to a chiral carbon are determined using the three rules in Table 7.3. These rules are illustrated by Examples 7.5, 7.6, and 7.7.

GETTING INVOLVED

✓ Describe the two steps for determining the configuration of a chiral carbon atom.

✓ How are the priorities of groups attached to a chiral carbon determined if **(a)** all the directly attached atoms are different; **(b)** two or more of the directly attached atoms are the same; **(c)** a group under consideration has a double or triple bond?

Example 7.5

Determine the priorities of the groups around the chiral carbon atom using Rules 1 and 2 of Table 7.3.

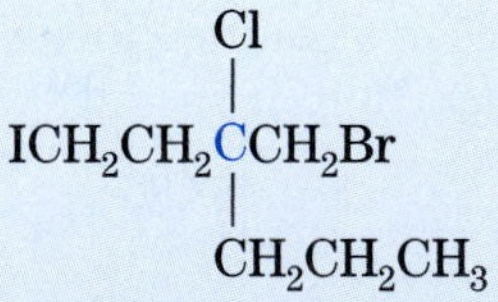

$$ICH_2CH_2\overset{\overset{Cl}{|}}{\underset{\underset{CH_2CH_2CH_3}{|}}{C}}CH_2Br$$

Priority sequence

$Cl > CH_2Br > CH_2CH_2I > CH_2CH_2CH_3$

Solution

Connected directly to the chiral carbon atom are a chlorine and three carbons. Chlorine has the highest atomic number and the highest priority. Connected, in each case, to the three carbons are 2 H's and a Br, 2 H's and a C, and 2 H's and a C. Bromine has the highest atomic number of C, H, and Br and thus CH_2Br is the highest priority of these three. The other two carbons are still identical. Connected to the second carbon of these groups are 2 H's and an I and 2 H's and a C. Iodine has the highest priority of these atoms (C, H, and I), so that $-CH_2CH_2I$ is next in the priority list and $-CH_2CH_2CH_3$ is last.

Example 7.6

Determine the priorities of the groups around the chiral carbon atom using Rules 1 and 2 of Table 7.3.

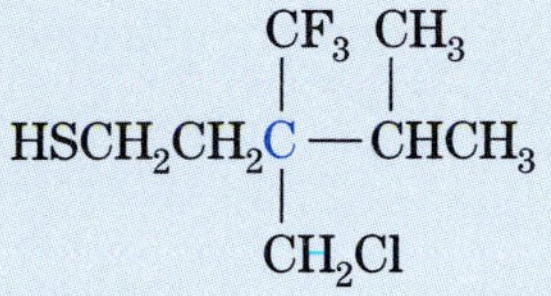

$$\mathrm{HSCH_2CH_2\overset{\overset{\large CF_3}{|}}{\underset{\underset{\large CH_2Cl}{|}}{C}}-\overset{\overset{\large CH_3}{|}}{C}HCH_3}$$

Priority sequence

$$\mathrm{CH_2Cl > CF_3 > CH_3\overset{\overset{\large CH_3}{|}}{C}H > CH_2CH_2SH}$$

Solution

The atoms directly bonded to the chiral carbon atom are all carbons, and it will be necessary to analyze the atoms bonded to these. Considering each individual carbon, we find that the bonded atoms are 3 F's; 2 C's and an H; 2 H's, and a Cl; 2 H's and a C. Of these bonded atoms (F, Cl, C, and H), chlorine has the highest atomic number and fluorine the next. Thus $-CH_2Cl$ has the highest priority, followed by CF_3; it makes no difference that 3 F's add up to more than 1 Cl and 2 H's. The remaining carbons both contain only carbon and hydrogen. However, 2 C's and 1 H (CH_3CHCH_3) take precedence over 1 C and 2 H's ($-CH_2CH_2SH$), making CH_3CHCH_3 next highest in priority, and leaving $-CH_2CH_2SH$ as the lowest-priority group.

Example 7.7

Determine the priorities of the groups around the chiral carbon atom using Rules 1, 2, and 3 of Table 7.3.

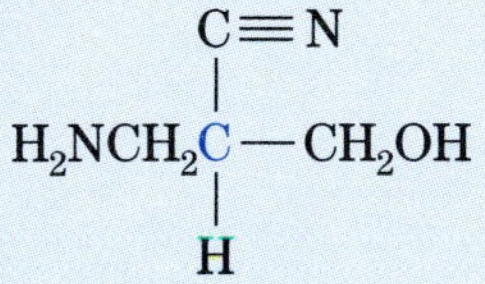

$$\mathrm{H_2NCH_2\overset{\overset{\large C\equiv N}{|}}{\underset{\underset{\large H}{|}}{C}}-CH_2OH}$$

Priority sequence

$$\mathrm{CH_2OH > C\equiv N > CH_2NH_2 > H}$$

Solution

Three carbons and a hydrogen are directly bonded to the chiral carbon atom. Hydrogen has the lowest priority. If we consider the three carbons, the bonded atoms are 3 N's (in $C\equiv N$ the nitrogens are triplicated), 2 H's and an O, and 2 H's and an N. Of these atoms (N, O, H), oxygen has the highest atomic number and $-CH_2OH$ the highest priority. Of the remaining two groups, 3 N's take precedence over 1 N and 2 H's, so that $-C\equiv N$ is next in priority, followed by $-CH_2NH_2$ and H.

Problem 7.12

Assume that the following sets of groups are attached to chiral carbon atoms. Arrange them in priority order according to the *R*, *S* sequence rules in Table 7.3.

(a) F, Cl, Br, I
(b) OCH_3, Br, H, CH_3
(c) $CH_2CH_2CH_3$, CH_2CH_2Br, CH_2OH, OH
(d) Cl, SH, CH_2OH, CO_2H

See related problem 7.25.

C. Determining *R* and *S* Configurations

Now, let us apply the two-step procedure for determining *R* and *S* configurations in a general way. First, we put the four groups bonded to the chiral carbon atom in order by priority (we shall use the numbers 1–4 to represent four groups here).

Highest priority ① > ② > ③ > ④ Lowest priority

Next, we put the groups of lowest priority behind the plane of the page. If groups 1, 2, and 3 are arranged in a clockwise fashion, the configuration is *R*. If they occur in a counterclockwise fashion, it is *S*.

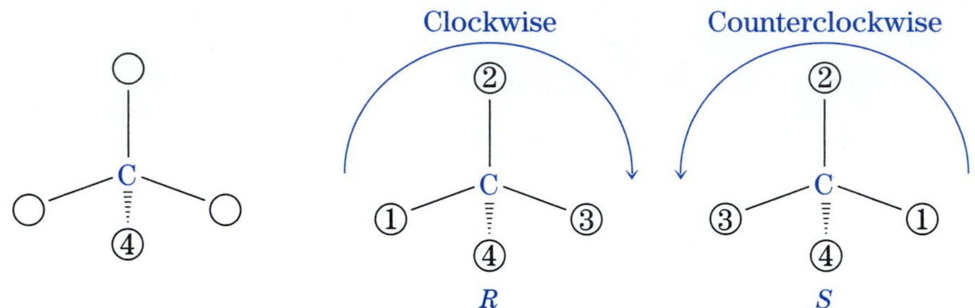

By drawing configurations in this manner, we are picturing the tetrahedral carbon as an inverted pyramid. The peak of the pyramid (tetrahedron) is behind the plane of the paper, and the three-cornered base is in the plane of the paper. To describe this molecule using the wedges and dashes representation for chiral carbon atoms introduced earlier in this chapter, we need only tilt the molecule a bit.

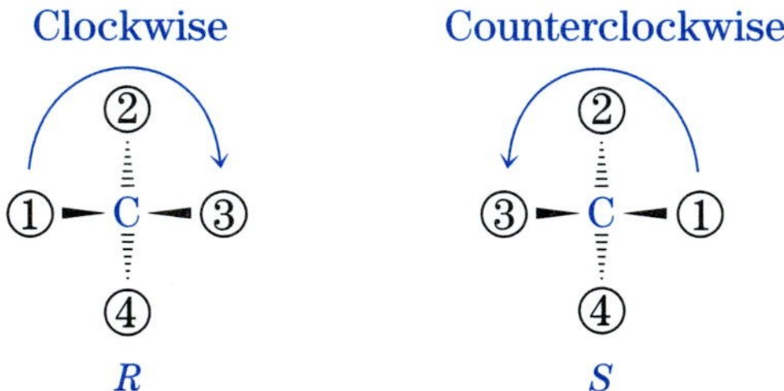

Now let us draw the *R* and *S* configurations of bromochlorofluoromethane. The priority sequence (Table 7.3) by atomic number is Br > Cl > F > H.

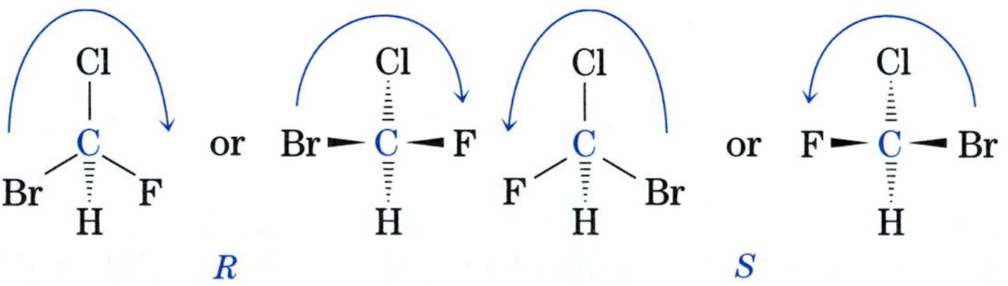

Suppose we wish to assign an *R* or *S* configuration to a structure that is not conveniently drawn so as to do this. For example, what is the configuration of the following molecule?

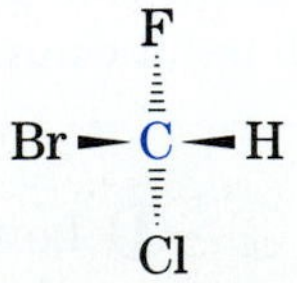

F
Br—C—H
Cl

For the configuration to be determined, the molecule must be positioned so that the lowest-priority group (H) is away from the observer.

F (top), Br—C—H, Cl (bottom) —Rotate 90°→ Br (top), Cl····C····F, H (bottom) —Flip over→ Br (top), F—C—Cl, H (bottom)

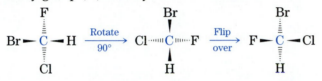

By tilting the molecule forward and drawing an arrow from highest to lowest priority (Br ⟶ Cl ⟶ F), we see that this is the *R* configuration.

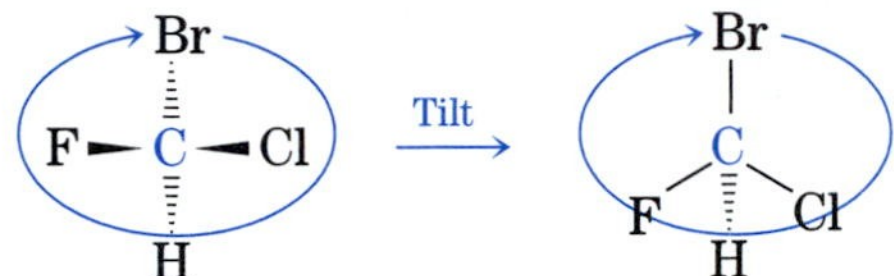

Clockwise
R configuration

The same result can be accomplished by interchanging groups bonded to the chiral carbon (see section 7.2.C). Remember that if two groups are switched, the configuration of the chiral atom is changed. If two more groups are interchanged, however, the molecule assumes its original configuration. Thus interchanges must be done in pairs to avoid changing the configuration. Using the above example, we switch the hydrogen and chlorine so as to get the lowest-priority group away from the observer. Then, to retain the original configuration, we interchange any two other groups, say the bromine and fluorine.

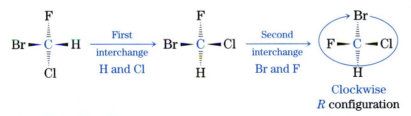

F (top), Br—C—H, Cl (bottom) —First interchange, H and Cl→ F (top), Br—C—Cl, H (bottom) —Second interchange, Br and F→ Br (top), F—C—Cl, H (bottom)

Clockwise
R configuration

GETTING INVOLVED

Example 7.8

Determine the *R* or *S* configuration of each of the following compounds:

A: CH_2Br (top), $Cl—C—CH_2CH_2I$, $CH_2CH_2CH_3$ (bottom)

B: CF_3 (top), $CH_3CH(CH_3)—C—CH_2CH_2SH$, CH_2Cl (bottom)

C: CH_2NH_2 (top), $H—C—CH_2OH$, $C{\equiv}N$ (bottom)

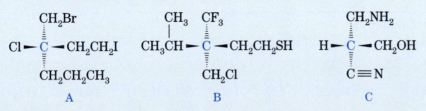

Solution

First, we determine the priority sequence of the four groups bonded to each chiral carbon. Priority sequences for each of these specific compounds are given in Examples 7.5, 7.6, and 7.7 and are shown with numbers here. Next, we interchange groups in pairs of interchanges so as to get the low-priority group away from the observer (this is already the case in the first example). Finally, we let our eyes proceed from priority group 1 to 2 to 3. If the eyes travel clockwise, the compound is *R*, and if counterclockwise, it is *S*.

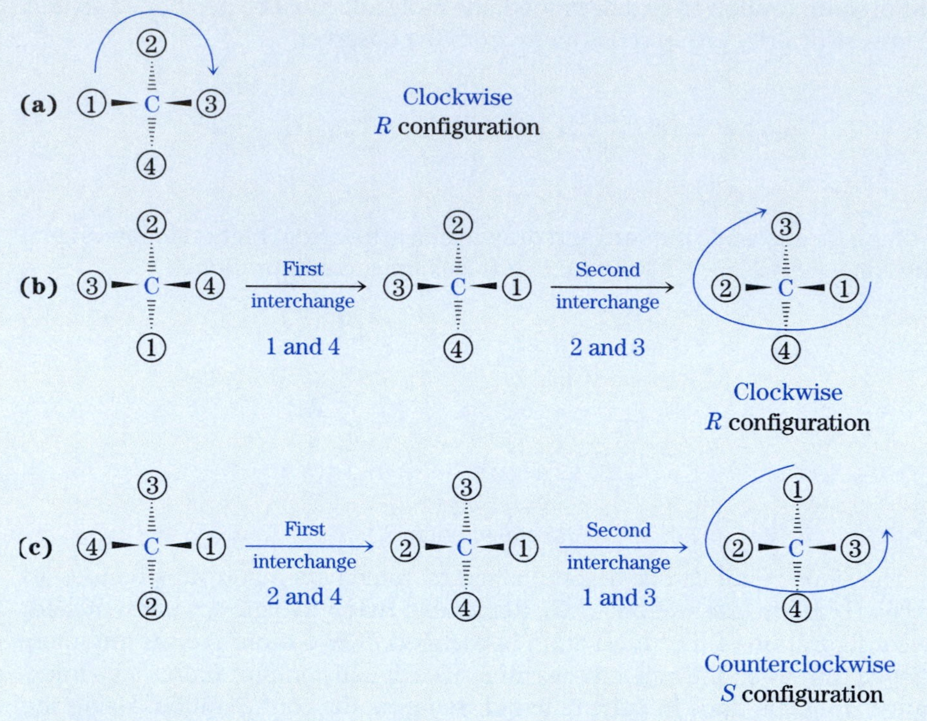

Example 7.9

Draw the *R* configuration of 2-bromobutane.

Solution

First, draw the compound and determine the priority sequence of groups attached to the chiral carbon. Then write a representational structure with wedges and dashes. Put the low-priority group on the bottom away from the observer (actually either bottom or top will accomplish projecting the group away from the observer). Arrange the other three groups clockwise from highest to lowest priority.

$CH_3CH_2CHCH_3$ (Br on C-2)

Priority sequence

$Br > CH_2CH_3 > CH_3 > H$

Clockwise

Br—C(CH_2CH_3 top, H bottom)—CH_3

The configuration is *R*

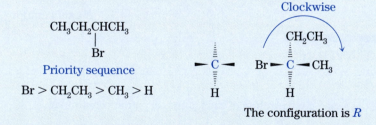

Problem 7.13

Specify configurations of the following compounds as *R* or *S*.

(a) Br—C—F, with CH_3 above and H below

(b) H_3C—C—CH_2Br, with CH_2CH(=O) above and NH_2 below

(c) HC≡C—C—CH_2COH(=O), with CH_2CH_2I above and $CH{=}CH_2$ below

(d) HO—C—COH(=O), with CH_3 above and CH_2CH_3 below

Problem 7.14

Draw and label the *R* and *S* configurations of 4-bromo-1-chloro-2-methylbutane.

See related problems 7.26–7.31, 7.36, 7.38.

7.7 Resolution of Enantiomers

Enantiomers, stereoisomers that are nonsuperimposable mirror images, have identical physical properties with the exception of the direction of rotation of plane-polarized light. Thus conventional separation methods, which rely on differences in solubilities, melting points, and boiling points, are ineffective.

Diastereomers, stereoisomers that are not mirror images, differ in all physical properties. A generally useful method for separation of enantiomers involves converting them into diastereomers, separating the diastereomers according to differences in melting point, boiling point, or solubility, and then reconverting the separated diastereomers to the original enantiomers. This method, called **resolution through diastereomers,** is illustrated in Figure 7.9.

resolution through diastereomers a method for separating enantiomers

In recent years, chromatography columns that are chiral have been developed for the separation of enantiomers. This method is a form of resolution through diastereomers. During the chromatographic process, enantiomers form transient diastereomeric complexes with properties that enable separation by the column before the complexes reconvert to the original enantiomers.

There are other methods for separating enantiomers, though most are not as practical as resolution through diastereomers. Louis Pasteur's separation of the sodium ammonium salt of tartaric acid relied on good fortune and visual acuity and is mainly of historical interest. The crystals with which he worked formed as mirror images, and Pasteur was able to recognize this and separate them with tweezers, like separating right- and left-handed gloves from a barrel of gloves. This method is called *mechanical resolution.* It is not generally practical, as most racemic mixtures do not readily form enantiomorphic crystals; even when such an event might occur, the separation by hand would be exceedingly tedious. Pasteur (1822–1895) is known primarily for his research on fermentation, the basis for microbiology; pasteurization, the process carried out on milk, is named for him.

Resolution can also be accomplished by biological means. Microorganisms produce enzymes that are themselves chiral and consequently react differently with each constituent of an enantiomeric pair. For example, Pasteur found that *Penicillium glaucum* selectively consumes the dextrorotatory form of tartaric acid.

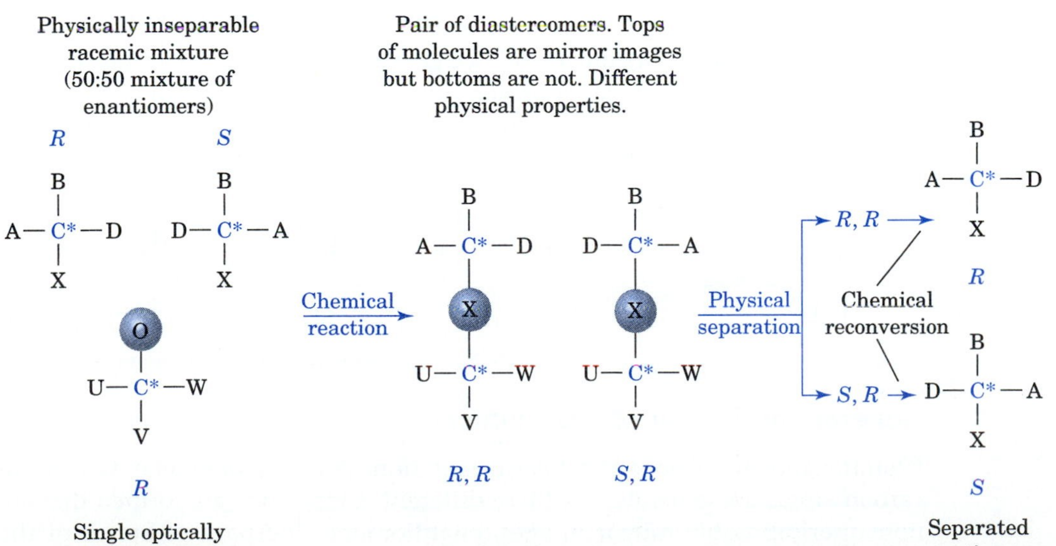

Figure 7.9 Resolution of enantiomers through the use of diastereomers. The group X of the enantiomeric pair reacts with the group O of the single enantiomer to form a pair of diastereomers. Asterisk (*) denotes a chiral carbon.

This method's disadvantage is that one enantiomeric form is metabolized and lost and the other is often isolated in poor yields.

GETTING INVOLVED

- ✓ Why is it not possible to separate enantiomers by conventional means such as distillation, crystallization, melting point, or solubility? Why is it helpful to convert enantiomers to diastereomers for eventual separation? Why can diastereomers be separated by conventional means?
- ✓ What are mechanical resolution and biological resolution of enantiomers?

7.8 Stereoisomerism and Chemical Reactions

Chiral carbon atoms can be generated in some chemical reactions and in others a chiral carbon atom may actually be the reaction site. In the next chapter, we will examine nucleophilic substitution and elimination reactions in which a chiral carbon atom can be the reaction site. In this section, however, we will look at reactions where chiral carbon atoms are generated using three reactions of alkenes we have already studied: catalytic hydrogenation (section 5.2B), potassium permanganate oxidation to form diols (section 5.7A), and addition of bromine (section 5.1B.2). All of these reactions proceed with a specific orientation. The first two involve syn or *cis* addition and the last one proceeds by anti or *trans* addition.

Generation of a Meso Compound

In reacting with cyclopentene, potassium permanganate introduces two OH groups in a *cis* configuration and generates two chiral carbon atoms. Since both the alkene and adding reagent are symmetrical and the addition is *cis*, whether the OH's add to the double bond from above or below, the mirror images formed are the same. The single product is a meso compound and is optically inactive.

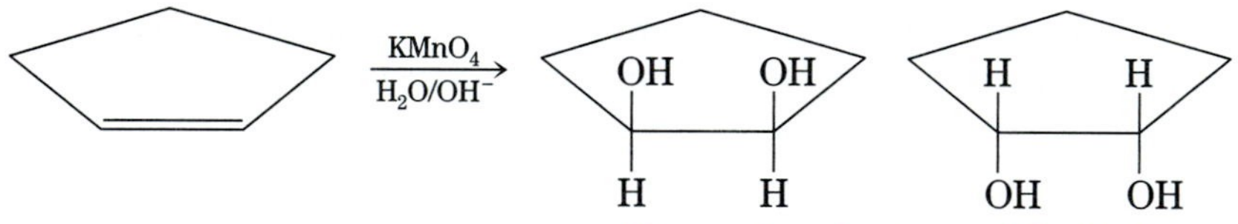

Meso: mirror images are identical

Hydrogenation of cyclopentene will not generate chiral carbon atoms, but hydrogenation of 1.2-dimethylcyclopentene will. Again, two chiral carbon atoms are formed, but since addition is *cis* and the alkene and adding reagent are both symmetrical, a meso compound results.

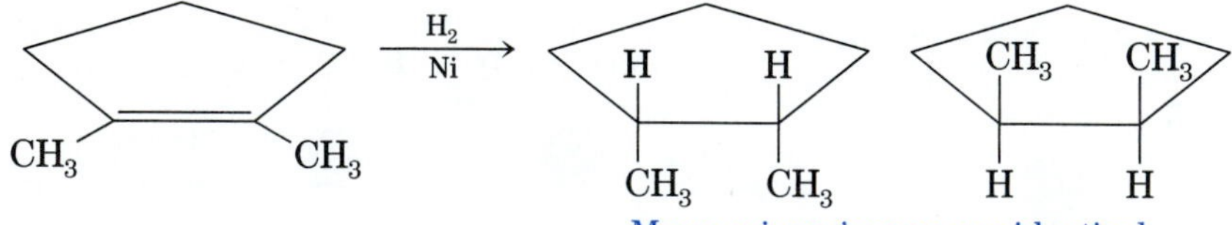

Meso: mirror images are identical

Generation of a Pair of Enantiomers

Bromine adds to alkenes in a *trans* orientation. With cyclopentene, two chiral carbon atoms are generated and two different compounds are formed that are nonsuperimposable mirror images, enantiomers. The path of attack of the bromines is identical for both and the two enantiomers are of equal stability. For these reasons, enantiomers always form in equal amounts.

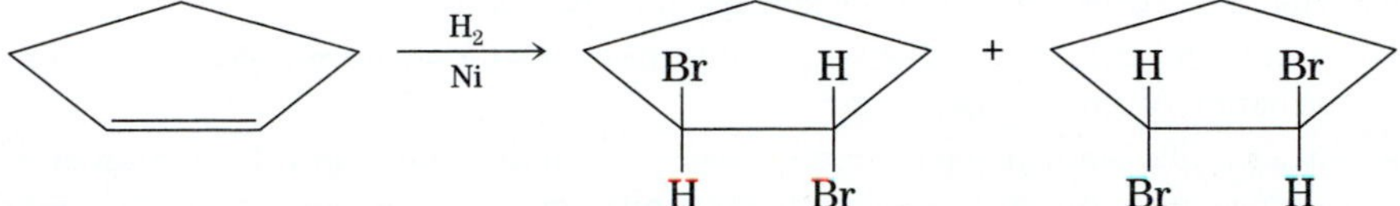

Enantiomers: nonsuperimposable mirror images; formed in equal amounts

For $KMnO_4$ hydroxylation or hydrogenation to form enantiomers, the starting alkene must be unsymmetrical. For example, hydrogenation of 1-ethyl-2-methyl-1-cyclopentene generates two chiral carbon atoms and a pair of enantiomers.

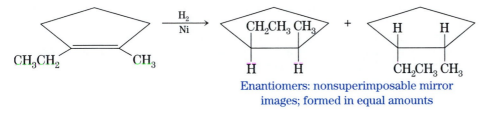

Enantiomers: nonsuperimposable mirror images; formed in equal amounts

Generation of a Pair of Diastereomers

Let's take a look at hydroxylation with potassium permanganate of optically active 3-methylcyclopentene. This is a pure enantiomer with one chiral carbon atom; two new chiral carbon atoms are generated by the reaction. Both the paths of attacks and product stabilities differ in this reaction. Since the methyl group is above the ring, introduction of the two OH groups by the permanganate ion will be less hindered from below. The product with the two OH's on the same side of the methyl group will have a different stability than the other where the groups are not in such close proximity. Because the reaction pathways are different and the stabilities of products are different, diastereomers always form in unequal amounts.

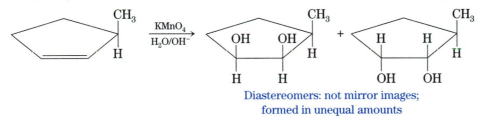

Diastereomers: not mirror images; formed in unequal amounts

Bromination of optically active 3-bromocyclopentene also generates two additional chiral carbon atoms and a pair of diastereomers. Note in this reaction, anti, or *trans*, addition occurs, and one of the diastereomers is a meso compound (the second product).

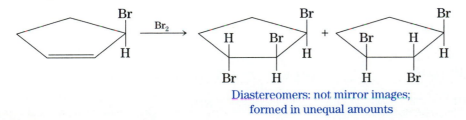

Diastereomers: not mirror images; formed in unequal amounts

GETTING INVOLVED

✓ What is the difference between syn and anti addition?

✓ In addition reactions of alkenes, what three characteristics are necessary to ensure formation of a meso compound?

✓ Structurally, what is the difference between enantiomers and diastereomers? In chemical reactions why are enantiomers always formed in equal amounts and diastereomers in unequal amounts (two reasons)?

✓ Using the examples given, describe why the reaction pathways are the same in the formation of enantiomers and different in the formation of diastereomers? Why do enantiomers have identical stabilities and diastereomers have unequal stabilities?

Problem 7.15

Write the reaction products of the following addition reactions of alkenes and describe the stereochemical results, i.e. what kinds of stereoisomers are formed. **(a)** 1,2-dimethylcyclopentene and $KMnO_4$; **(b)** methylcyclopentene and $KMnO_4$; **(c)** optically active 1,3-dimethylcyclopentene and H_2/Ni; **(d)** optically active 3-methylcyclopentene and Br_2; **(e)** 3,3-dimethylcyclopentene and Br_2.

Problem 7.16

When HBr adds to 1,2-dimethylcyclopentene, four optical isomers are formed. Draw the four products and identify pairs of enantiomers.

See related problems 7.32 and 7.35.

Problems

7.17 **Chiral Carbon Atoms:** Circle the chiral carbon atoms in each of the following molecules. What is the maximum number of enantiomers possible for each as predicted by the van't Hoff rule.

(a) [structure: cyclohexane ring with CH_3, OH, and $CH(CH_3)_2$ substituents]

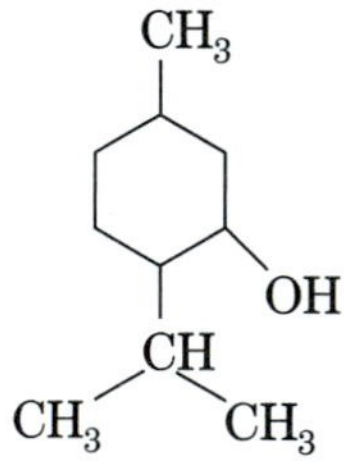

Menthol

(b) [structure: five-membered lactone ring with O, =O, HO, OH, and $CHCH_2OH$ bearing OH]

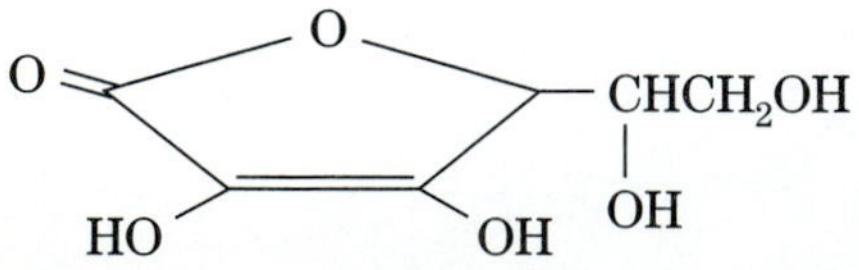

Vitamin C

(c) $NaO_2CCH_2CH_2CHCO_2H$ with NH_2 on CH

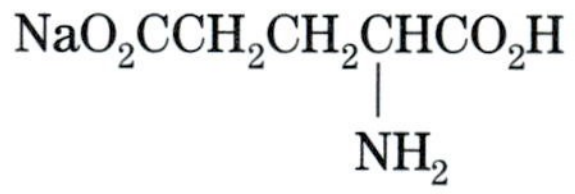

Monosodium glutamate

(d) [structure: steroid skeleton with HO, two H_3C groups, and side chain $CH(CH_3)CH_2CH_2CH_2CH(CH_3)CH_3$]

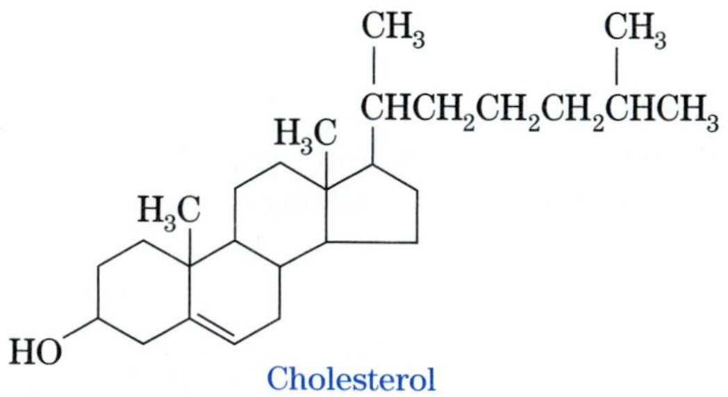

Cholesterol

(e) [structure: benzene ring attached to CH_2CHCH_3 with NH_2 on CH]

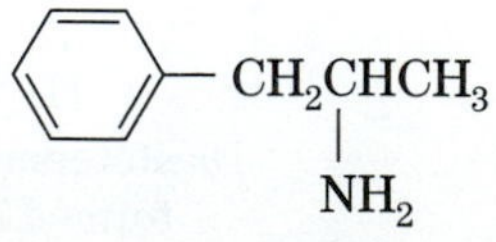

Amphetamine

(f) 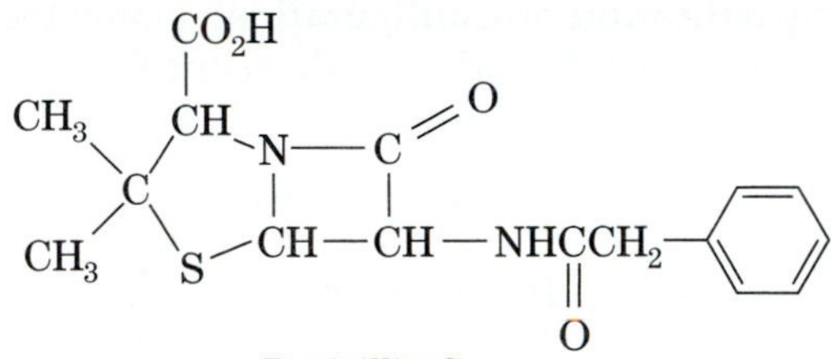

Penicillin G

(g) 

Nicotine

(h) 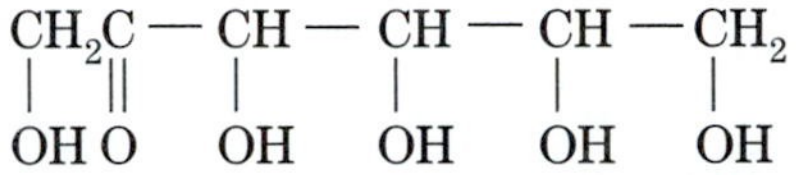

Fructose

7.18 **Chiral Carbon Atoms:** Draw the isomers described.

(a) the one ketone, formula $C_6H_{12}O$, with a chiral carbon

(b) the three aldehydes, formula $C_6H_{12}O$, with a chiral carbon

(c) the two isomers of $C_5H_{11}Br$ with a chiral carbon.

7.19 **Enantiomers:** Using wedges and dashes, draw the pair of enantiomers for each compound you found in Problem 7.18.

7.20 **Enantiomers and Diastereomers:** Draw three compounds of $C_6H_{13}Br$ so that two represent a pair of enantiomers and the third is a diastereomer of the others.

7.21 **Enantiomers:** Using wedges and dashes draw a pair of enantiomers for each of the following compounds:

(a) 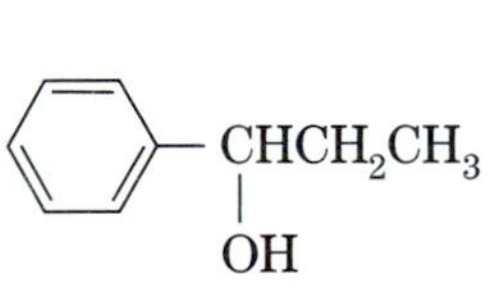 **(b)** 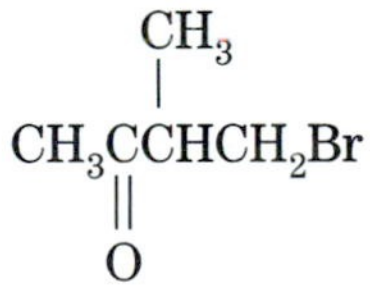

(c) 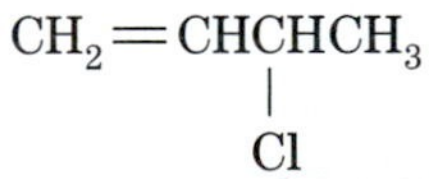

7.22 **Stereoisomers:** Draw the stereoisomers of the following compounds. Label pairs of enantiomers, pairs of diastereomers, and meso compounds.

(a) 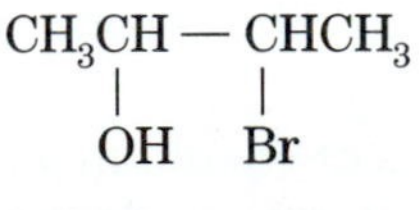 **(b)** 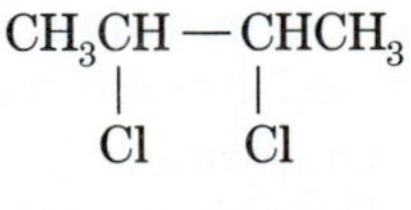

(c) 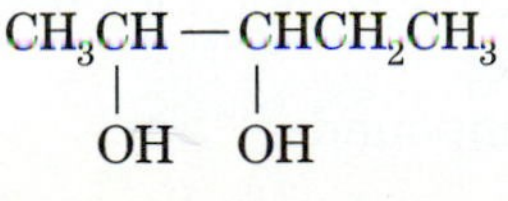 **(d)** 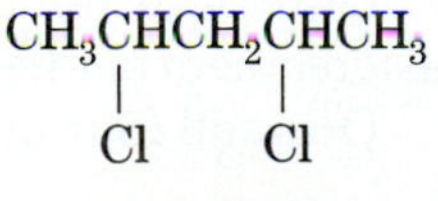

7.23 **Stereoisomerism in Cyclic Compounds:** Draw the optical isomers of the following compounds. Label pairs of enantiomers, pairs of diastereomers, and meso compounds.

(a) 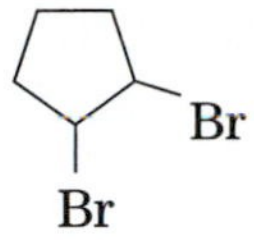 **(b)** 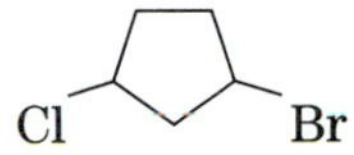

(c) 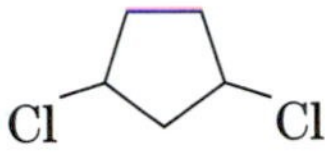

7.24 **Stereoisomers:** Draw the stereoisomers of the following compounds. Label pairs of enantiomers, pairs of diastereomers, and meso structures. Both structures have three chiral carbon atoms. In **(b),** carbon-3 does not appear chiral. However, there are chiral carbons on either side and since they may differ in configuration, C-3 is also chiral.

(a) 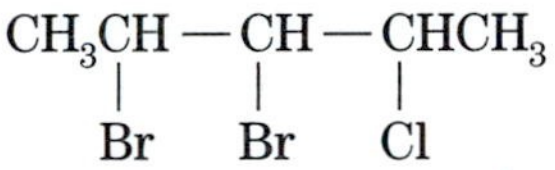

(b) 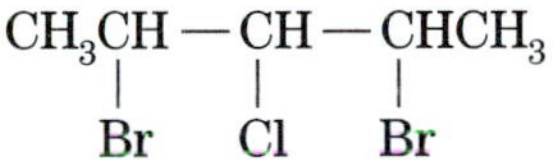

7.25 ***R, S* Configurations:** Assign priorities to the following sets of substituents:

(a) 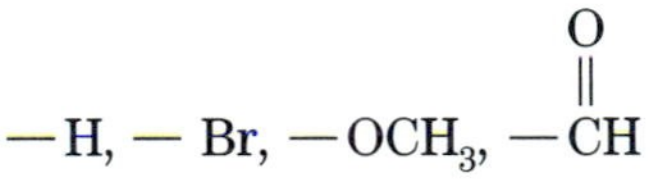

(b) $-CH_3$, $-CH_2CH_3$, $-CH(CH_3)_2$, $-C(CH_3)_3$

(c) $-F$, $-CH_2Cl$, $-CH_2CH_2I$, $-Br$

(d) $-CHCl_2$, $-CH_2Br$, $-I$, $-CH_3$

(e) $-NH_2$, $-C{\equiv}N$, $-OCH_3$, H

7.26 **Specification of Configuration:** Using the designations *R* or *S*, specify the configuration of each of the following:

(a) 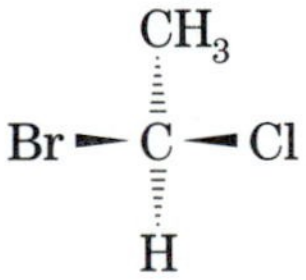 **(b)** 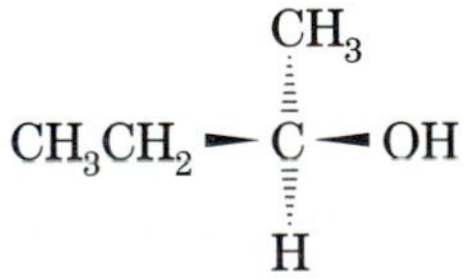

(c) 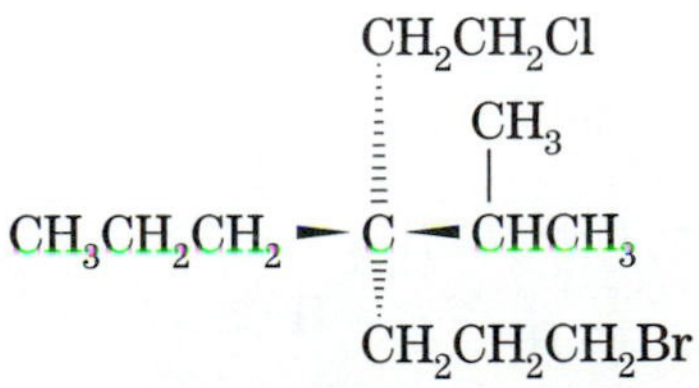

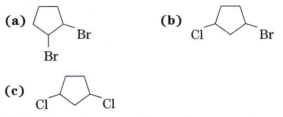

(d) 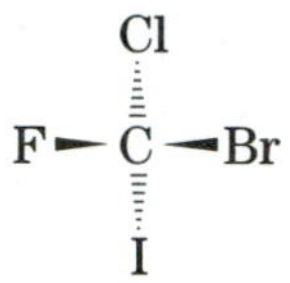 $F—C(Cl)(I)—Br$

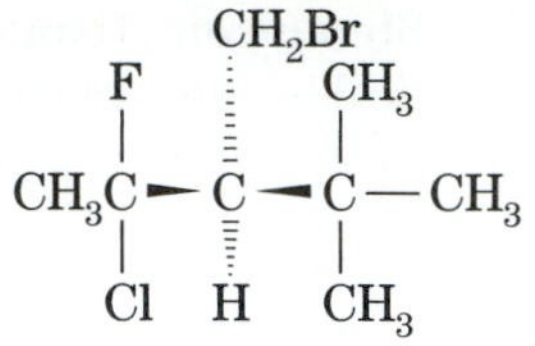

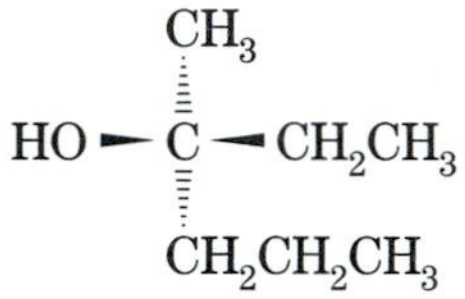

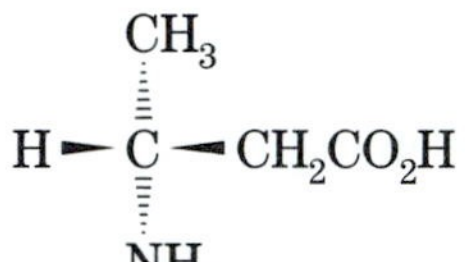

(e) $CH_3C(F)(Cl)—C(CH_2Br)(H)—C(CH_3)_2—CH_3$

(f) $HO—C(CH_3)(CH_2CH_2CH_3)—CH_2CH_3$

(g) $H—C(CH_3)(NH_2)—CH_2CO_2H$ (h) $H—C(Cl)(OCH_3)—(CH_2)_8CH_3$

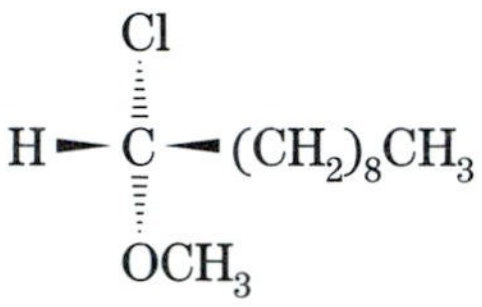

7.27 **Specification of Configuration:** Using the designations *Z* and *E*, specify the configurations of each of the following:

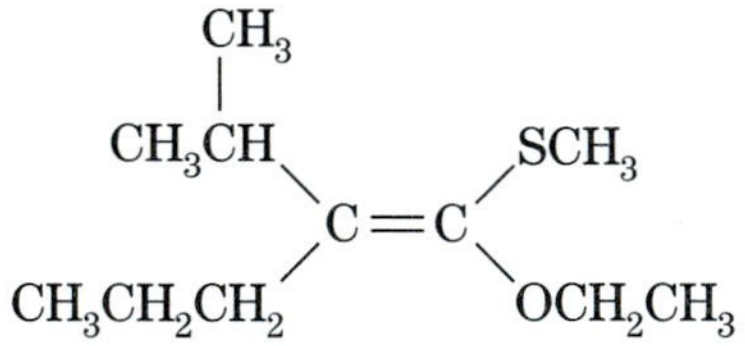

(a) 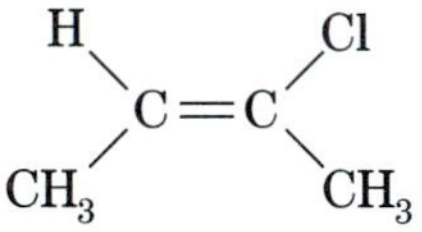 $(H)(CH_3)C{=}C(Cl)(CH_3)$

(b) $(CH_3CH(CH_3))(CH_3CH_2CH_2)C{=}C(SCH_3)(OCH_2CH_3)$

(c) $(HOCH_2)(CH_3)C{=}C(Br)(CH_2OH)$

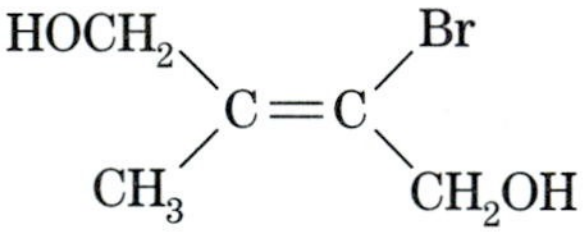

(d) 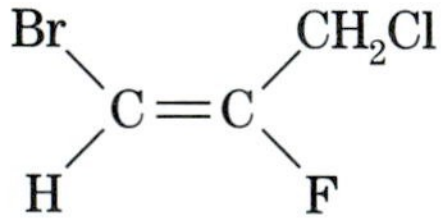 $(Br)(H)C{=}C(CH_2Cl)(F)$

(e) 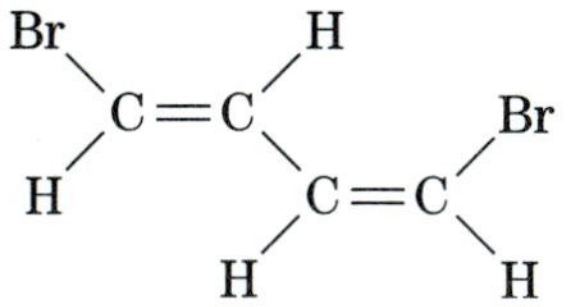 $(Br)(H)C{=}C(H)—C(H){=}C(Br)(H)$

7.28 **Specification of Configuration:** Using *R*, *S*, *Z*, and *E*, specify the configuration of the following molecule:

$CH_3—C(Br)(H)—C(H){=}C(CH_3)(H)$

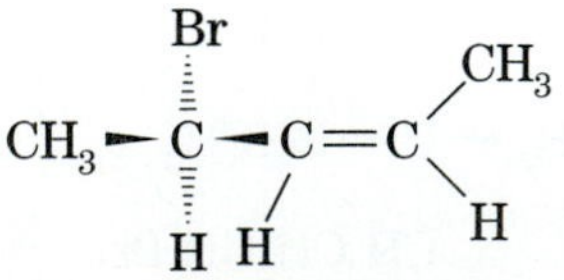

7.29 **Specification of Configuration:** Name the following compound, including in the name the *R*, *S* configurations of chiral carbons:

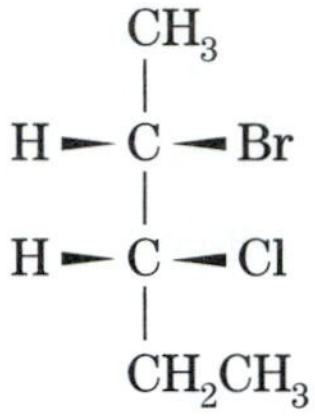

$CH_3—C(H)(Br)—C(H)(Cl)—CH_2CH_3$

7.30 **Specification of Configuration:** Draw the following molecules, clearly showing the stereochemistry:

(a) *S*-2-chlorobutane
(b) *S*-3-methylhexane
(c) *R*-1-bromo-2-methylbutane
(d) *R*-2,3-dimethylpentane

7.31 ***R*, *S* Configurations:** Is the following compound *R* or *S*? Draw the Newman projection of its enantiomer.

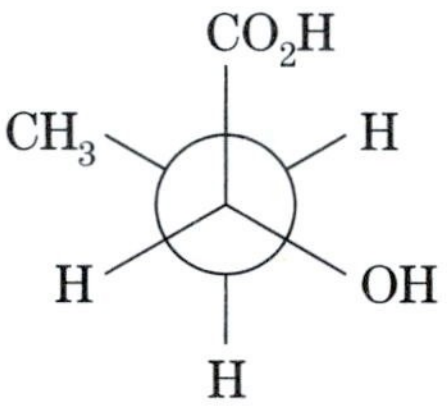

Newman projection — front carbon: CO_2H, H, OH; back carbon: CH_3, H, H

7.32 **Stereochemistry and Chemical Reactions:** Hydrogen adds to the following alkenes and produces two stereoisomers in each case. What are the stereoisomers? Are they formed in equal or unequal amounts? If formed in unequal amounts, give two reasons why and predict the likely predominant product.

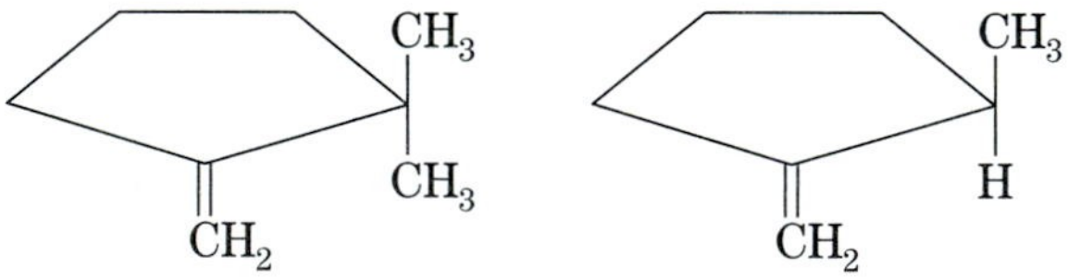

(cyclohexane ring with $=CH_2$ and adjacent carbon bearing CH_3, CH_3) (cyclohexane ring with $=CH_2$ and adjacent carbon bearing CH_3, H)

7.33 **Stereoisomers:** Draw the following:

(a) a chiral aldehyde
(b) a pair of enantiomeric alcohols
(c) a pair of diastereomeric carboxylic acids
(d) a pair of enantiomers with three chiral carbons
(e) two stereoisomers with four chiral carbons that are meso

7.34 **Stereoisomers:** Consider the compound 2,3,4,5,6,7-hexabromooctane.

(a) Draw one of the many pairs of enantiomers.
(b) Draw four optically active compounds that are diastereomers of one another.
(c) Draw all four meso compounds.

7.35 **Stereoisomers and Chemical Reactions:** Consider the following reaction equation for the hydrogenation of the carbohydrate galactose:

```
       H
       |
       C=O                              CH2OH
       |              H2                  |
   H — C — OH      -------->          H — C — OH
       |            catalyst              |
  HO — C — H                         HO — C — H
       |                                  |
  HO — C — H                         HO — C — H
       |                                  |
   H — C — OH                         H — C — OH
       |                                  |
      CH2OH                             CH2OH
```

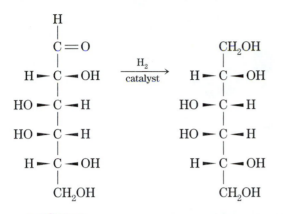

D-galactose

(a) Is galactose optically active? Is the reaction product optically active? Explain.

(b) Draw an optical isomer of galactose that would give an optically active hydrogenation product and one that would give an optically inactive product.

7.36 ***R, S* Configurations:** Draw:

(a) a Newman projection of (2*R*,3*S*)-2-bromo-3-chlorobutane looking down the $C_2—C_3$ bond

(b) *R*-1,1-dichloro-3-methyl cyclohexane

7.37 **Stereoisomers:** 4-Bromo-2-pentene exhibits both geometric isomerism and can exist as enantiomers. Using three-dimensional illustrations, draw the four stereoisomers.

7.38 ***R* and *S* Designations:** Problem 7.9 asks for structures of the four isomers of threonine. Indicate the configuration, using the *R* and *S* of each. Remember that each has two chiral carbons. There are four possibilities: 2*R*,3*R*; 2*R*,3*S*; 2*S*,3*R*; and 2*S*,3*S*.

7.39 **Enantiomers without Chiral Carbons:** Nitrogen and silicon can both act as chiral atoms and therefore exhibit optical isomerism. Draw an optically active compound with nitrogen and another with silicon in which the nitrogen and silicon are chiral.

7.40 **Enantiomers without Chiral Atoms:** For a compound to rotate plane-polarized light, it must be chiral overall. The presence of chiral atoms, however, is not a necessity for optical activity, just as the presence of chiral carbons does not guarantee optical activity (meso compounds, for example). The following compounds are chiral and rotate plane-polarized light even though neither possesses chiral atoms. Among the factors leading to chirality is the fact that one half of each molecule is perpendicular to the other half. Fully explain the geometry of each molecule. Draw a pair of enantiomers in each case, and show that they are not superimposable.

(a) [spiro compound: H, Cl on left carbon; Cl, H on right carbon]

(b) H(Cl)C=C=C(Cl)H

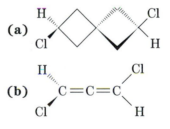

Activities with Molecular Models

1. Make a molecular model with one carbon and four different attached groups; this is a chiral carbon atom. Manipulate the two models to convince yourself that they are not superimposable. Make a similar model but with only three different groups. Show that these are superimposable.

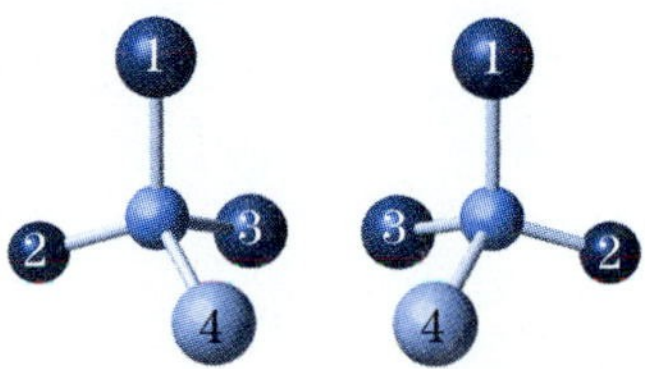

2. Make a molecular model of 1,2-dibromobutane. Put the model in the eclipsed conformation around carbons 2 and 3; this is a representation of the Fischer projection. A depiction is shown below for the meso compound. Exchange two groups on one of the carbon-bearing atoms and you will have one of the enantiomers. Make a mirror image to get the other. Notice in Fischer projections we have a totally eclipsed conformation that seems to curl toward the rear of the viewer.

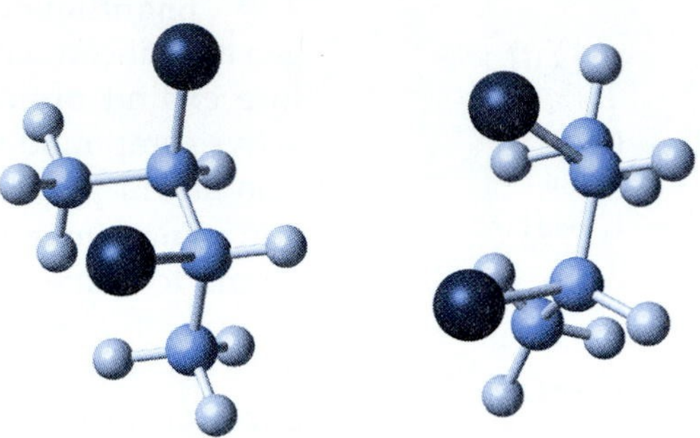

3. You may wish to use the molecular models with the chiral carbon atom from exercise 1 to assist you in visualizing R and S configurations.

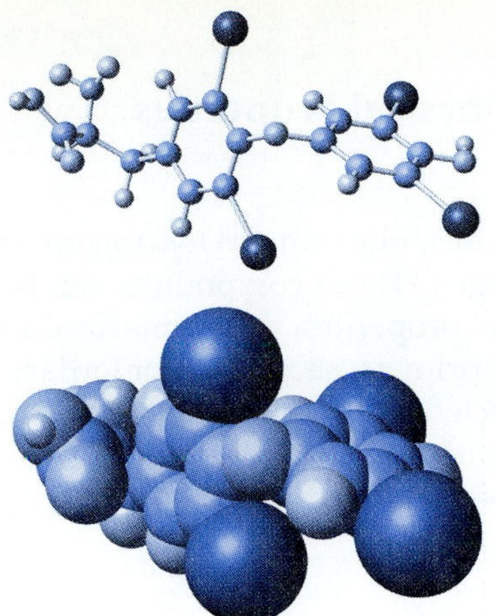

CHAPTER 8

Organic Halogen Compounds

8.1 Introduction

Organic halogen compounds are rarely found in nature but they do have a variety of commercial applications as insecticides, herbicides, dry-cleaning agents and degreasers, aerosol propellants and refrigerants such as the freons (CFC's), fire-extinguishing chemicals, and polymers. However, many chlorinated hydrocarbons have become suspect as carcinogens and are being replaced by other products.

For example, at one time over 2000 commercial products contained chloroform, the handkerchief anesthetic seen in many old movies; 80% of it was used as a flavoring in cough medicines. Because of evidence in laboratory animals of carcinogenic properties, chloroform was banned by the U.S. Food and Drug Administration in July, 1978. The well-known insecticide DDT was used during World War II to combat typhus epidemics in Europe and malaria epidemics in the South Pacific; the discoverer of its insecticidal properties, Paul Mueller, was awarded a Nobel Prize in 1948. Early in the 1970's, DDT and related organochlorine insecticides dieldrin, aldrin, chlordane, and heptachlor were banned by the Environmental Protection Agency, except for a few special uses.

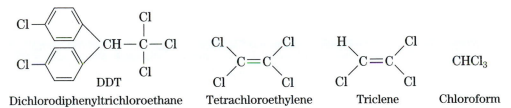

$(Cl{-}C_6H_4)_2CH{-}CCl_3$ DDT	$Cl_2C{=}CCl_2$	$HClC{=}CCl_2$	$CHCl_3$
Dichlorodiphenyltrichloroethane	Tetrachloroethylene	Triclene	Chloroform

One of the problems with DDT and other chlorinated hydrocarbon insecticides is that they are absorbed into the fatty tissue of living organisms and become concentrated as higher levels of the food chain are reached. An analogous property of lower molecular weight chlorinated hydrocarbons, their ability to dissolve fats, oils, and greases, however, make them excellent dry cleaning agents (among others, tetrachloroethylene and triclene). They are relatively nonflammable, have little or no structural effect on fabrics, and, because of their volatility, they are easily removed. They are hazardous air and ground water pollutants, however, and subject to regulation.

Organohalogen compounds also are used to produce important polymers such as polyvinyl chloride, Saran, Teflon, and Neoprene rubber (sections 5.3 and 5.6).

In this chapter, we will introduce two of the most thoroughly studied reactions in organic chemistry, nucleophilic substitution and elimination, characteristic reactions of alkyl halides.

8.2 Structure, Nomenclature, and Physical Properties

A. Structure and Properties

alkyl halide
alkane molecule in which a halogen has replaced a hydrogen

Alkyl halides are *organic halogen compounds* in which one or more hydrogens of a hydrocarbon have been replaced with a halogen. These compounds can be classified into groups that show similar chemical properties, as summarized in Table 8.1. **Alkyl halides** are further described as primary, secondary, or tertiary, depending on the number of alkyl groups connected to the halogenated carbon. If there is one carbon directly connected to the carbon bearing the halogen, the alkyl halide is primary; if there are two, it is secondary; and if three, it is tertiary.

```
     H              R              R
     |              |              |
 R — C — H      R — C — H      R — C — R
     |              |              |
     X              X              X
  Primary       Secondary       Tertiary
    1°             2°              3°
```

Like most classes of organic compounds, organic halides have boiling points that increase with molecular weight. Thus chloropropane boils at a higher temperature than chloroethane, which in turn has a higher boiling point than chloromethane. Since the atomic weights of halogens increase according to the order Cl < Br < I, the boiling points of particular alkyl halides increase as follows: R—Cl < R—Br < R—I. Generally, organic halides have densities greater than water and are insoluble in water. Table 8.2 tabulates the boiling points and densities of some representative compounds.

B. IUPAC Nomenclature

In IUPAC nomenclature, halogens are designated by the prefixes *fluoro-*, *chloro-*, *bromo-*, and *iodo-*. CH_3Cl is chloromethane; and $CH_3CHBrCH_2CH_3$, 2-bromobutane. Nomenclature of these compounds has been covered in section 2.6.D.

Table 8.1 Classes of Organic Halogen Compounds

Class	General Structure (X = halogen)	Examples
Alkyl halides	R—X	CH_3Cl, CH_3CH_2Br
Aryl halides	C_6H_5—X	C_6H_5—Br, CH_3—C_6H_4—I
Vinyl halides	>C=C< with X on C	CH_2=CHCl, CH_3CH=CHBr
Allylic halides	>C=C<—C(—)—X	CH_2=CHCH$_2$Br
Benzylic halides	C_6H_5—C(—)—X	C_6H_5—CH_2Cl

Table 8.2 Physical Properties of Some Organic Halogen Compounds

Organic Halogen Compound	Chloride bp, °C	Chloride Density, ~20°C	Bromide bp, °C	Bromide Density, ~20°C	Iodide bp, °C	Iodide Density, ~20°C
$CH_3—X$	–24	Gas	5	Gas	43	2.28
$CH_3CH_2—X$	12.5	Gas	38	1.44	72	1.93
CH_2X_2	40	1.34	99	2.49	180	3.33
CHX_3	61	1.49	151	2.89	Sublimes	4.01
CX_4	77	1.60	189.5	3.42	Sublimes	4.32
$CH_2{=}CHX$	–14	Gas	16	Gas	56	
$C_6H_5—X$	131	1.11	156	1.50	188	1.84

$CH_3CH(CH_3)CH_2CH(I)CH_2CH_3$ — 4-iodo-2-methylhexane

$H_2C{=}CHCH_2CHCl_2$ — 4, 4-dichloro-1-butene

(cyclopentane with CH_2CH_3 and Br) — 1-bromo-2-ethylcyclopentane

C. Common Nomenclature

A "salt-type" nomenclature is frequently used with alkyl halides in which the alkyl group's name precedes the name of the halide; thus CH_3Cl is chloromethane (IUPAC) or methyl chloride (common).

In addition, halogen derivatives of methane have nonsystematic names, which are often used.

CH_3Cl	CH_2Cl_2	$CHCl_3$	CCl_4
Methyl chloride	Methylene chloride	Chloroform	Carbon tetrachloride

GETTING INVOLVED

✓ What are the structural differences among alkyl, aryl, allyl, vinyl, and benzylic halides?

✓ How do boiling points of alkyl halides vary with molecular weight and with Cl, Br, I?

Problem 8.1

Name the following by the IUPAC system:

(a) $CH_3CH(Cl)CH_2CH_2CH_3$ **(b)** $BrCH_2CH{=}CHCH_2Br$ **(c)** F–C_6H_4–F (para) **(d)** CCl_3CHF_2

Problem 8.2

Draw the following compounds:
(a) carbon tetrabromide, **(b)** methylene bromide, **(c)** iodoform, **(d)** vinyl bromide, **(e)** *p*-nitrobenzyl chloride, **(f)** isopropyl iodide

See related problems 8.29–8.31.

8.3 Preparations of Organic Halogen Compounds

In previous chapters, we have seen several methods for preparing organic halogen compounds; these are summarized in this section. Halogenated alkanes are commonly prepared from alcohols, in a reaction we shall discuss in the next chapter.

A. Free-Radical Halogenation of Alkanes (section 4.4)

$$-\overset{|}{\underset{|}{C}}-H + X_2 \xrightarrow[\text{peroxides}]{\text{Light or}} -\overset{|}{\underset{|}{C}}-X + HX$$

$$X_2 = Cl_2, Br_2$$

B. Addition to Alkenes and Alkynes (section 5.1.A.1–2)

$$>C=C< + EA \longrightarrow -\overset{E}{\underset{|}{\overset{|}{C}}}-\overset{|}{\underset{A}{\underset{|}{C}}}-$$

$$-C\equiv C- + EA \longrightarrow \overset{E}{>}C=C\underset{A}{<} \xrightarrow{EA} -\overset{E}{\underset{E}{C}}-\overset{A}{\underset{A}{C}}-$$

$$EA = Cl_2, Br_2, HCl, HBr, HI$$

C. Electrophilic Aromatic Substitution (section 6.4)

$$C_6H_6 + X_2 \xrightarrow{FeX_3} C_6H_5-X + HX$$

$$X = Cl, Br$$

D. Conversion of Alcohols to Alkyl Halides (section 9.7.A–B)

$$-\overset{|}{\underset{|}{C}}-OH + \text{Reagent} \longrightarrow -\overset{|}{\underset{|}{C}}-X$$

$$\text{Reagent} = HCl, HBr, HI, SOCl_2, PCl_3, PBr_3$$

8.4 Nucleophilic Substitution

nucleophilic substitution substitution reaction in which a nucleophile replaces a leaving group such as a halide

Alkyl halides are important reagents in a wide variety of synthetic organic reactions. The halogen atom is more electronegative than carbon and withdraws electrons from it in a carbon-halogen bond. Under appropriate conditions, the halogen can be replaced in a **nucleophilic substitution** reaction. This is one of the simplest and most thoroughly studied reactions in organic chemistry. The reaction lends itself well to mechanistic studies and, as you will see in the next section, it has enormous synthetic utility. (You should review nucleophilicity and Lewis bases in section 4.2.C.)

nucleophile species with electron availability that donates electrons to an electrophile in a chemical reaction. Nucleophiles are Lewis bases

A. General Reaction

Nucleophilic substitution is widespread and varied. A common example is the reaction between an alkyl halide and a negative **nucleophile** (shown as the sodium salt). Because the carbon in the polar carbon-halogen bond is partially

Connections 8.1 www.prenhall.com/bailey

Drug Design

With the exception of the thyroid hormones (see Connections 8.2), halogenated compounds are seldom found in mammals. There are, however, many drugs with useful characteristics due at least in part to halogen substituents. An understanding of the effects of halogenation has been useful in the important area of drug design.

In order to be effective a drug must be designed to reach its site of action. In many cases this involves penetration of one or more membrane barriers between the site of application and the receptor location. Because the cell membrane is a lipid bilayer with a nonpolar interior, it tends to resist penetration by molecules that are not fat soluble; the more lipid soluble a molecule is, the better it will diffuse across the membrane. Such lipid solubility sometimes can be increased by halogen substituents.

The molecule cortisol is a corticosteroid hormone secreted by the adrenal cortex. Its biological functions are to help regulate carbohydrate and protein metabolism and salt balance, and to inhibit inflammation. Modifications to the structure change the biological actions of the resulting drug. They can enhance its absorptivity, slow its breakdown, and affect its potency. One such alteration, the substitution of a fluorine at C9, enhances corticosteroid activity. An increased potency means that less of a drug needs to be prescribed for a given effect.

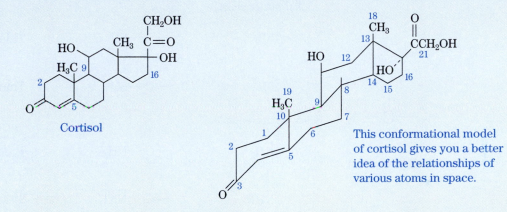

Cortisol

This conformational model of cortisol gives you a better idea of the relationships of various atoms in space.

Other types of alterations in the molecular structure can result in drugs that have very specific actions with fewer side reactions. Some predictions of structure and relative activity can be made by using information on existing drugs as well as using computer-aided drug analysis. Since most drugs have to bind at biomolecules known as receptors, there must be a correlation between the chemical and structural nature of the drug and that of its receptor. In addition, there are restrictions on a drug molecule's size and stereochemical orientation so that it can fit into the receptor's binding site effectively. Computer graphics and design are proving invaluable in utilizing this information in the search for better drug treatment.

Substituent	Position	Relative Antiinflammatory Potency	Salt-Retaining Potency	Duration of Action (half-life)
None		1	1	8–12 hours
F	9	10	125	8–12 hours
1,2-double bond				
F	6	10	0	12–36 hours
$-CH_3$	16			
1,2-double bond				
F	9	25	0	36–72 hours
$-CH_3$	16			

Table 8.3 Nucleophilic Substitution Reactions

Alkyl Halide	Nucleophile	Organic Product	Functional Group
	Oxygen nucleophiles		
	$^{-}:\ddot{O}H$	$-C-\ddot{O}H$	Alcohol
	$^{-}:\ddot{O}R$	$-C-\ddot{O}R$	Ether
	Sulfur nucleophiles		
	$^{-}:\ddot{S}H$	$-C-\ddot{S}H$	Thiol
	$^{-}:\ddot{S}R$	$-C-\ddot{S}R$	Thioether
$-C-X$ X = Cl, Br, I	*Nitrogen nucleophiles*		
	$^{-}:\ddot{N}H_2$	$-C-\ddot{N}H_2$	1° Amine
	$^{-}:\ddot{N}HR$	$-C-\ddot{N}HR$	2° Amine
	$^{-}:\ddot{N}R_2$	$-C-\ddot{N}R_2$	3° Amine
	Carbon nucleophiles		
	$^{-}:C\equiv N:$	$-C-C\equiv N:$	Nitrile
	$^{-}:C\equiv CR$	$-C-C\equiv CR$	Alkyne

positive, the negative nucleophile is attracted to it. The halide is replaced by the nucleophile; substitution results. Following is a general reaction equation for nucleophilic substitution. The reaction is summarized further in Table 8.3.

Nucleophilic Substitution:

$$-\overset{|}{\underset{|}{C}}{}^{\delta+}-X^{\delta-} + \overset{+}{Na}\overset{-}{Nu} \longrightarrow -\overset{|}{\underset{|}{C}}-Nu + \overset{+}{Na}\overset{-}{X}$$

$$X = Cl,\ Br,\ I,\ \text{and}\ \overline{Nu} = {}^{-}:\ddot{O}H,\ {}^{-}:\ddot{S}H,\ {}^{-}:\ddot{N}H_2,\ {}^{-}:CN$$
$$^{-}:\ddot{O}R,\ {}^{-}:\ddot{S}R,\ {}^{-}:\ddot{N}HR,\ \bar{:}C\equiv CR$$
$$^{-}:\ddot{N}R_2$$

Let's take a closer look at the reaction. The nucleophile, a Lewis base, has an unshared electron pair that is used to form the new carbon-nucleophile bond. As the halide departs, it retains the pair of electrons that composed the carbon-halogen bond. The halide is referred to as the *leaving group*. In nucleophilic substitution reactions, the leaving group is less nucleophilic than the nucleophile, and thus the reaction does not immediately reverse. In other words, the stronger nucleophile replaces the weaker nucleophile.

Nucleophilic substitution is a useful synthetic reaction, as illustrated by the Williamson synthesis of ethers. In the following example, the general anesthetic diethyl ether is prepared by a nucleophilic substitution reaction in which an ethoxide ion is the nucleophile ($CH_3CH_2O^-$) and chloride is the leaving group.

$$\overset{\delta+}{CH_3CH_2}\!:\!\overset{\delta-}{\ddot{\underset{..}{Cl}}}\!: + Na^+ :\!\overset{-}{\ddot{\underset{..}{O}}}CH_2CH_3 \longrightarrow CH_3CH_2\!:\!\ddot{\underset{..}{O}}CH_2CH_3 + Na^+ :\!\ddot{\underset{..}{Cl}}\!:^-$$

A similar process is used to prepare alkynes, with the salt of a terminal alkyne (section 5.8) used as the nucleophile:

$$CH_3CH_2C{\equiv}C\!:^- Na^+ + CH_3Br \longrightarrow CH_3CH_2C{\equiv}CCH_3 + NaBr$$

GETTING INVOLVED

- ✓ What is a nucleophile and how is it related to a Lewis base? Write structures for the negative nucleophiles (the oxygen, sulfur, nitrogen, and carbon nucleophiles) showing the nonbonding electron pairs that make them nucleophiles.
- ✓ Describe what happens in a nucleophilic substitution reaction. What role does the alkyl halide play? How are the nucleophile and leaving group different and how are they related? What bonds break and form?
- ✓ Write a general summary reaction equation for nucleophilic substitution that summarizes the halide leaving groups and the negative nucleophiles.

Example 8.1

Write equations for the nucleophilic substitution reactions between **(a)** 1-bromopropane and sodium hydrogen sulfide (NaSH) to form one of the compounds responsible for the odor of onions and **(b)** lithium dimethylamide ($LiN(CH_3)_2$) and methyl iodide to form one of the compounds responsible for the fishy odor of fish.

Solution

(a) Write structures for the two reactants. The negative nucleophile (^-SH) is associated with a cation and can be identified in this way. Now look for a carbon-halogen bond, the carbon-bromine bond. Bromide is the leaving group. Replace the bromide with SH; sodium bromide is the inorganic by-product.

$$CH_3CH_2CH_2Br + NaSH \longrightarrow CH_3CH_2CH_2SH + NaBr$$

(b) Replace the iodide ion with the dimethylamide ion. Lithium iodide is the inorganic by-product.

$$(CH_3)_2NLi + CH_3I \longrightarrow (CH_3)_2NCH_3 + LiI$$

Example 8.2

Write a reaction equation illustrating the preparation of 1-methoxypropane from an alkyl halide and a nucleophile.

Solution

Separate the molecule into two parts, an alkyl halide and a nucleophile, at the oxygen (the nucleophile). In this case there are two possibilities; the alkyl halide or the nucleophile can be a one-carbon or three-carbon fragment.

$$CH_3CH_2CH_2ONa + CH_3Cl \longrightarrow CH_3CH_2CH_2OCH_3 + NaCl$$

$$CH_3ONa + CH_3CH_2CH_2Cl \longrightarrow CH_3OCH_2CH_2CH_3 + NaCl$$

Problem 8.3

Test your comprehension of the nucleophilic substitution reaction by writing chemical equations illustrating the reaction of methyl iodide (CH_3I) with each of the following:

(a) $NaOH$ **(b)** $NaOCH_2CH_2CH_3$ **(c)** $NaSH$ **(d)** $NaSCH_3$
(e) $NaNH_2$ **(f)** $NaNHCH_2CH_3$ **(g)** $NaN(CH_3)_2$ **(h)** $NaCN$
(i) $NaC{\equiv}CCH_3$

Problem 8.4

Write nucleophilic substitution reactions for the preparations of the following compounds:

(a) $CH_3CH_2CH_2CN$ **(b)** $CH_3CH_2CH_2CH_2OH$ **(c)** CH_3SCH_3

See related problems 8.32 and 8.34–8.37.

B. Nucleophilic Substitution with Neutral Nucleophiles

Just as negative nucleophiles can replace the halogen of an alkyl halide, so can their neutral counterparts. For example, isopropyl alcohol can result from the reaction of 2-bromopropane with either $NaOH/H_2O$, in which OH^- is the nucleophile, or by heating with water/acetone, in which water is the neutral nucleophile.

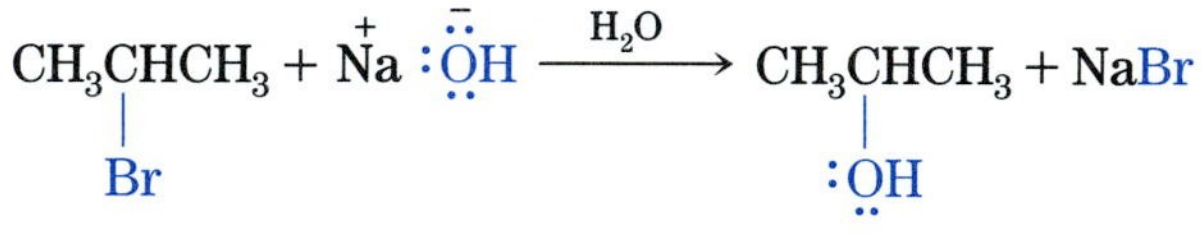

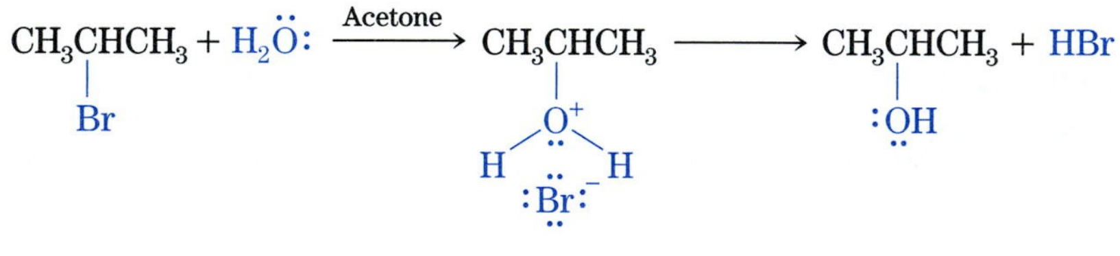

$$CH_3CH(Br)CH_3 + \overset{+}{Na}\,:\ddot{\underset{..}{O}}H^- \xrightarrow{H_2O} CH_3CH(:\ddot{O}H)CH_3 + NaBr$$

$$CH_3CH(Br)CH_3 + H_2\ddot{O}: \xrightarrow{\text{Acetone}} CH_3CH(\overset{+}{\ddot{O}}H_2)CH_3\ :\ddot{Br}:^- \longrightarrow CH_3CH(:\ddot{O}H)CH_3 + HBr$$

Using a neutral nucleophile, a charged intermediate is formed (the oxygen has three bonds) that readily loses hydrogen ion to produce a neutral product.

Similar reactions are possible with ammonia and its derivatives, the amines. For example, dimethylamine reacts with methyl chloride to produce trimethylammonium chloride, which becomes trimethylamine in the presence of a base [see Example 8.1(b) for comparison].

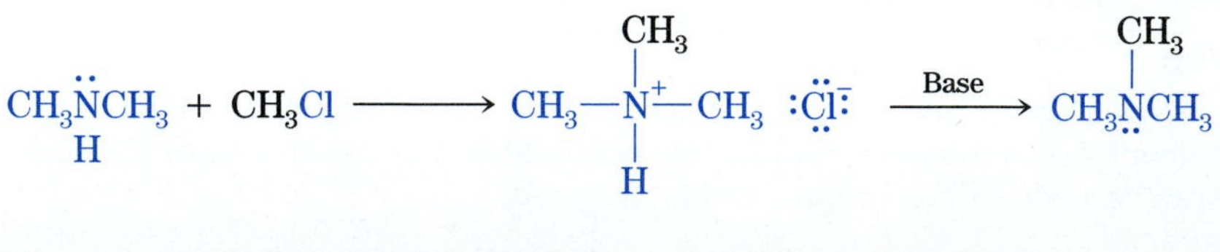

$$CH_3\ddot{N}(H)CH_3 + CH_3Cl \longrightarrow CH_3{-}\overset{+}{N}(CH_3)(H){-}CH_3\ :\ddot{Cl}:^- \xrightarrow{\text{Base}} CH_3\ddot{N}(CH_3)CH_3$$

If the ammonia derivative does not have a replaceable hydrogen, an ammonium salt is the result as in the preparation of the following molecule, an example of a cationic detergent used in shampoos.

$$CH_3(CH_2)_{14}CH_2\ddot{N}(CH_3)_2 + \overset{\delta+}{CH_3}-\overset{\delta-}{\ddot{Cl}}: \longrightarrow CH_3(CH_2)_{14}CH_2\overset{+}{N}(CH_3)_2CH_3\ :\ddot{Cl}:^-$$

Nucleophilic substitution reactions are common in biological chemistry, especially in reactions known as methylations. For example, adrenalin is formed from the methylation of the nitrogen (followed by loss of a hydrogen ion) by S-adenosylmethionine (a biological molecule composed of an amino acid, methione, and a nucleoside):

$$(HO)_2C_6H_3-CH(OH)CH_2\ddot{N}H_2 + CH_3:\overset{+}{\ddot{S}}(\text{Methione})-\text{Adenosine} \longrightarrow$$

$$(HO)_2C_6H_3-CH(OH)CH_2N(H)CH_3 + :\ddot{S}(\text{Methionine})-\text{Adenosine} + H^+$$

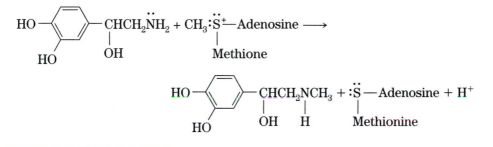

GETTING INVOLVED

✓ How are neutral nucleophiles similar to and different from negative nucleophiles? Write the structures for the neutral nucleophiles that are derivatives of water and ammonia; make sure you show the nonbonding electron pairs that make them nucleophiles.

✓ Why do neutral nucleophiles form a positive intermediate? Rationalize the formal charge. How is the neutral product eventually formed? Why does an amine with three alkyl groups form a cationic ammonium salt as the final product rather than a neutral organic compound?

Problem 8.5

Write the products resulting from the nucleophilic substitution reaction of 2-bromobutane with: **(a)** H_2O; **(b)** CH_3OH; **(c)** CH_3NHCH_3.

See related problem 8.33.

C. Introduction to Nucleophilic Substitution Reaction Mechanisms

How does nucleophilic substitution occur from a mechanistic standpoint? Basically, the reaction is simple—a halide ion is replaced by a nucleophile. If one thinks about the process logically, three ideas arise.

1. The nucleophile might enter and bond, followed by the halide leaving.
2. The nucleophile might attack and bond at exactly the same time the halide ion is leaving.
3. The halide ion might leave, followed by the attack and bonding of the nucleophile.

The first path requires that carbon accommodate five bonds, pentavalent carbon, and thus is not a realistic possibility. However, the other two ideas are sound, and both are common mechanisms for nucleophilic substitution. The second possibility is known as the S_N2 mechanism, and the third is referred to as the S_N1 mechanism. We will consider the S_N2 mechanism first.

D. The S_N2 Mechanism

$S_N2
substitution nucleophilic bimolecular; the one-step nucleophilic substitution mechanism

bimolecular
term that describes a reaction rate that depends on the concentration of two species

S_N2 stands for *substitution nucleophilic* ***bimolecular.*** This substitution mechanism is a one-step process with both the alkyl halide and the nucleophile involved simultaneously in the one step; hence the term bimolecular. In this mechanism, the formation of the carbon-nucleophile bond and the cleavage of the carbon-halogen bond occur simultaneously. The nucleophile enters as the halide ion leaves, attacking the carbon from the side opposite to that from which the halide departs. This is sterically favorable in that the nucleophile and halide do not hinder each other's movement.

The S_N2 Mechanism: One-Step Process

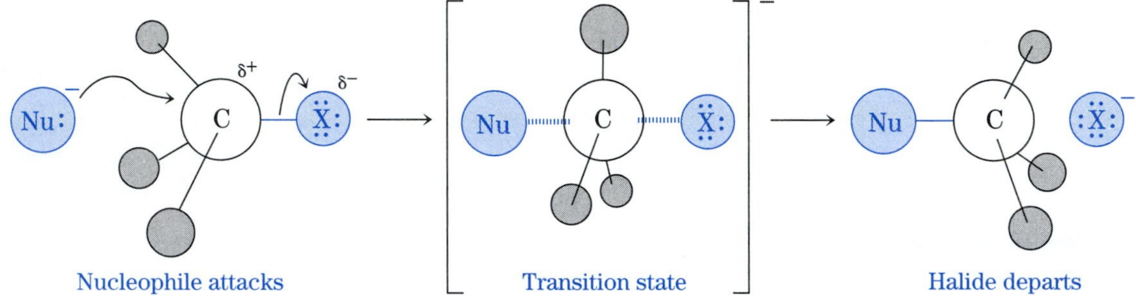

S_N2: Nucleophile with nonbonding electron pair attacks partially positive carbon from rear. Carbon-nucleophile bond forms simultaneously with cleavage of carbon-halogen bond in this one-step reaction that involves a five-centered transition state.

Although this appears to be two steps as written, the mechanism in reality is a concerted, one-step process. The transition state drawing is merely an illustration to conceptualize what is happening. A transition state is a dynamic process of change. Bonds are in the process of being broken and formed. Because of this, a transition state cannot be thought of as an isolateable intermediate.

GETTING INVOLVED

✓ What does the term S_N2 signify?

✓ Describe the S_N2 mechanism. Is it a one-step or two-step process? What does bimolecular mean? What is a transition state? Why does the nucleophile attack from the opposite side of departure of the halide?

Problem 8.6

Write the S_N2 mechanism for the reaction between bromoethane and sodium hydroxide, using illustrations as just shown.

Let us examine some observed characteristics of an S_N2 mechanism which offer evidence that it occurs.

1. ***Reaction Rates.*** The rate or speed at which a chemical reaction occurs depends on the concentrations of the reactants involved in the rate-determining step. An S_N2 reaction is a one-step process in which both of the reactants are involved. Thus the reaction rate of an S_N2 reaction should depend on the concentrations of both the alkyl halide and the nucleophile. This is actually observed and is expressed in the following reaction rate equation in which k is a rate constant (a proportionality constant specific to a particular reaction) and the brackets represent the concentrations of the reactants in moles per liter.

$$\text{Rate } S_N2 = k[\text{RX}][\text{Nu}^-] \qquad \text{General rate equation for } S_N2$$

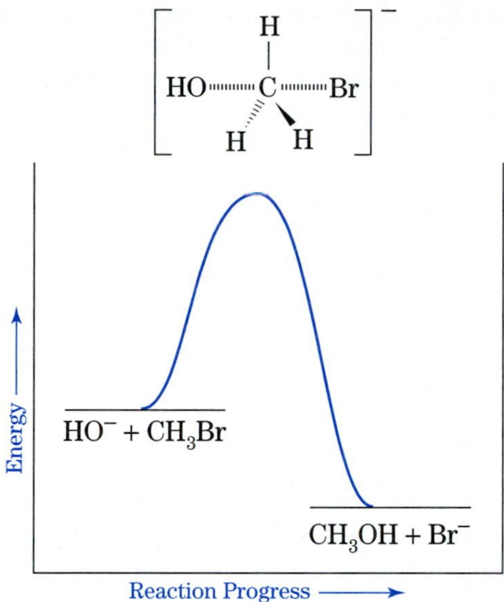

Figure 8.1 Potential energy diagram of an S_N1 reaction showing a one-step process with a single transition state.

Consider, for example, the reaction between bromomethane and the hydroxide nucleophile. The energy diagram (Figure 8.1) illustrates the S_N2 mechanism as one step, one transition state, and no intermediates.

$$HO^- + CH_3{-}Br \xrightarrow[\text{Transition state}]{[HO \cdots CH_3 \cdots Br]^-} CH_3OH + Br^-$$

$$\text{Rate} = k[CH_3Br][OH^-]$$

If one doubles the concentration of the hydroxide, the reaction rate doubles. With twice the concentration of hydroxide, there is twice the likelihood of the hydroxide ions attacking bromomethane. Likewise, doubling the concentration of bromomethane doubles the reaction rate. If both reactants are doubled, a quadrupling of the reaction rate is observed. If both concentrations are tripled, the reaction rate increases nine times.

GETTING INVOLVED

✓ Why does the rate of an S_N2 reaction depend on both the alkyl halide and nucleophile?

Problem 8.7

What would be the relative rates of an S_N2 reaction: **(a)** the concentration of bromomethane is tripled? **(b)** the concentration of hydroxide is quadrupled? **(c)** the concentration of bromomethane is doubled and that of hydroxide is tripled?

Problem 8.8

Write a rate equation for the S_N2 reaction described in problems 8.6 and 8.9.

2. ***Stereochemistry.*** The S_N2 reaction is a one-step process in which the nucleophile enters and bonds as the halide is leaving. Consequently, the nucleophile attacks from the rear to avoid interference with the path of the leaving halide. If the alkyl halide is optically active, inversion of configuration occurs; this is

actually observed by determining specific rotations using a polarimeter (section 7.3). This means that the optical activity is preserved but with opposite configuration to that of the reactant. Consider the following S_N2 reaction of optically active 2-bromobutane:

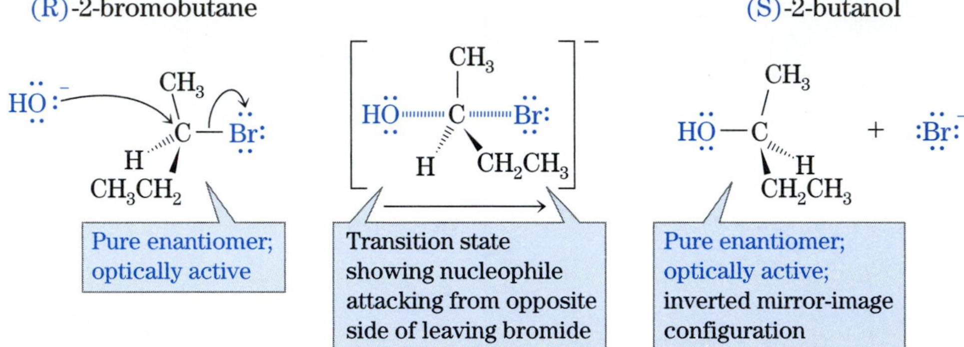

As the hydroxide bonds and the bromide leaves, the other three groups move from one side to the other, much like an umbrella blowing inside out in a strong wind. Inversion has occured! The product has the "mirror-image" configuration of the starting material.

Inversion of configuration in an S_N2 reaction can be shown simply using the wedge/dash representation we learned in Chapter 7 (see Example 8.3).

GETTING INVOLVED

✓ Why does the nucleophile attack from the rear? Why does inversion of configuration occur? What does this mean?

Example 8.3

Write an equation illustrating the reaction of (R)-2-chlorohexane with sodium ethoxide by an S_N2 mechanism.

Solution

Draw the *(R)*-2-chlorohexane, using wedge/dash structures. Replace the chlorine with ethoxy but on the opposite side to show inversion of configuration.

$$\underset{R}{Cl-\overset{CH_2CH_2CH_2CH_3}{\underset{H}{C}}-CH_3} + NaOCH_2CH_3 \longrightarrow \underset{S}{CH_3-\overset{CH_2CH_2CH_2CH_3}{\underset{H}{C}}-OCH_2CH_3} + NaCl$$

Problem 8.9

Using three-dimensional drawings as shown in this section, illustrate the S_N2 mechanism of the reaction between an enantiomer of 2-chloropentane and $NaSCH_3$.

Problem 8.10

Draw the products of the following S_N2 reactions, using wedge/dash projections as illustrated in the previous chapter. Remember, these undergo inversion of configuration.

$$\text{(a)}\ H_3C\text{—}\underset{H}{\overset{CH(CH_3)_2}{C}}\text{—}I + NaOH \longrightarrow \qquad \text{(b)}\ Br\text{—}\underset{CH_3}{\overset{CH_2CH_3}{C}}\text{—}H + NaOCH_3 \longrightarrow$$

Problem 8.11

Assign *R* and *S* configurations to the reactants and products in Problem 8.10.

See related problems 8.39, 8.42–8.43.

E. The S_N1 Mechanism

S_N1 means *substitution nucleophilic* ***unimolecular.*** In an S_N2 reaction mechanism, the nucleophile attacks the carbon-halogen bond and displaces the halogen in one step. The S_N1 reaction, on the other hand, is a two-step process. In the first step, the leaving group, the negative halide ion departs with the bonding pair of electrons and leaves a positive carbon, a carbocation. In the second step, the nucleophile enters with its nonbonding electron pair and bonds to the positive carbon.

S_N1
substitution nucleophilic unimolecular; the two-step nucleophilic substitution mechanism

unimolecular
term that describes a reaction rate that depends on the concentration of one species

The first, rate-determining step involves only one of the reacting species, the alkyl halide; thus the term *unimolecular.* In the second step, since the halide ion has left and no path of approach is restricted, the nucleophile can attack from either side of the carbocation (in contrast to the S_N2 mechanism).

S_N1 reactions commonly occur in neutral or acid conditions with neutral nucleophiles such as water, alcohols, or amines (section 8.4.B). In the following general mechanism, a neutral nucleophile bonds to the intermediate carbocation with subsequent loss of hydrogen ion to form the final neutral products.

The S_N1 Mechanism: A Two-Step Process

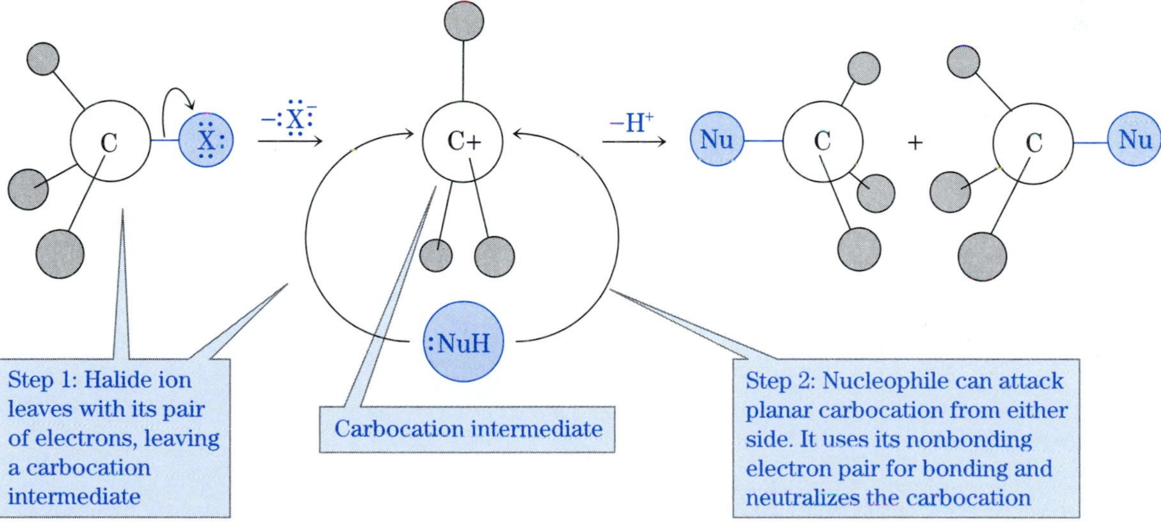
C
X:
−:X:⁻
C+
−H⁺
Nu
C
+
C
Nu
:NuH
Step 1: Halide ion leaves with its pair of electrons, leaving a carbocation intermediate
Carbocation intermediate
Step 2: Nucleophile can attack planar carbocation from either side. It uses its nonbonding electron pair for bonding and neutralizes the carbocation

Be sure you understand the difference between an intermediate and a transition state. An intermediate is the result of a transition but is a theoretically isolable entity. In the first step of the S_N1 mechanism, the carbon-halogen bond begins to break and finally completely severs. The process of breaking is the transition state; the result is the carbocation intermediate. In the second step, the carbon-nucleophile bond forms, a transition occurs, and the result is the final product. In a transition, bonds are changing by breaking or forming; in an intermediate or product, this change has come to conclusion and stopped.

The S_N1 mechanism of nucleophilic substitution using a neutral nucleophile is followed by a third step in which a hydrogen ion is lost. For example, in the reaction of 2-iodo-2-methylpropane with water, a carbocation forms that is neutralized by water to form on oxonium ion such as we have seen in hydration and dehydration mechanisms (sections 4.5.C and 5.1.B). Loss of a hydrogen ion leads to the final product. The first two steps characterize the S_N1 mechanism.

$$(CH_3)_3C{-}\ddot{\underset{\cdot\cdot}{I}}: \xrightarrow[\text{1st step } S_N1]{-:\ddot{I}:} (CH_3)_3C^+ \xrightarrow[\text{2nd step } S_N1]{H_2\ddot{O}:} (CH_3)_3C{-}\overset{H}{\overset{|+}{\underset{\cdot\cdot}{O}}}{-}H \xrightarrow[\text{Hydrogen ion loss}]{-H^+} (CH_3)_3C{-}\ddot{\underset{\cdot\cdot}{O}}H$$

Carbocation intermediate

Oxonium ion

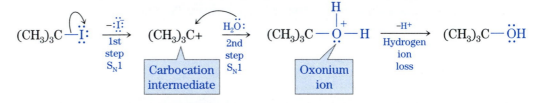

Figure 8.2 shows the energy diagram for this reaction with two transition states and one intermediate.

GETTING INVOLVED

- ✓ How do the terms S_N1 and S_N2 differ?
- ✓ Describe the S_N1 mechanism. Is it a one-step or two-step process? What does unimolecular mean? How does the carbocation intermediate form? How is the carbocation neutralized?
- ✓ How do S_N1 and S_N2 reactions differ in terms of number of steps, transition states versus intermediates, orientation of attack by the nucleophile, negative versus neutral nucleophiles, unimolecular versus bimolecular reactions?
- ✓ What is the difference between a transition state and intermediate? How many transition states and how many intermediates are in an S_N2 reaction and how many of each are in an S_N1 reaction?

Problem 8.12

Write the equation for the reaction between 2-chloro-2-methylbutane and ethanol to produce 2-ethoxy-2-methylbutane and HCl.

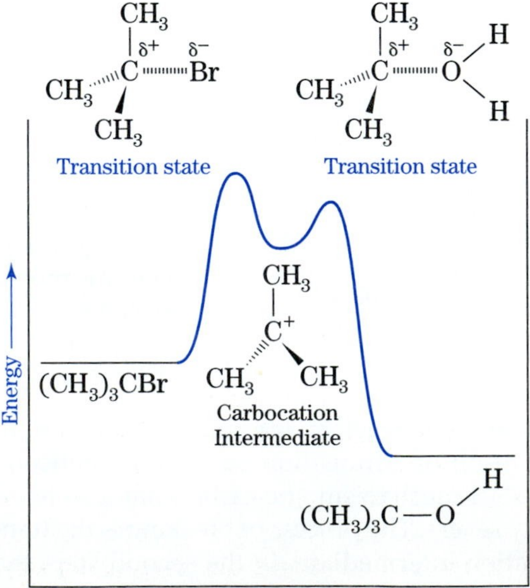

Figure 8.2 Potential energy diagram of an S_N2 reaction illustrating two steps, two transition states, and a carbocation intermediate.

Problem 8.13

Now write the S_N1 mechanism for the reaction described in Problem 8.12.

See related problem 8.39.

We will now examine reaction rates and stereochemistry for the S_N1 mechanism as we did for the S_N2 process.

1. ***Reaction Rates.*** The rate of a multistep chemical reaction depends on the rate of the slowest step. This concept is analogous to thousands of grains of sand falling in an hourglass. The rate depends entirely on how long it takes the individual grains of sand to reach and pass through the orifice. It is independent of the time required for them to fall from the orifice to the next chamber (the faster step). While a particle of sand is falling to the next chamber, others are making their way to the orifice; both processes are occurring at the same time and are not additive.

Hour glass

In an S_N1 mechanism, formation of the carbocation is the slow, rate-determining step. The carbocation is quickly neutralized in the second step. Thus the reaction rate is dependent only on the concentration of the alkyl halide, since it alone is involved in the rate-determining step; the nucleophile does not enter the picture until the second step. The reaction is unimolecular.

$$\text{Rate } S_N1 = k[\text{RX}] \qquad \text{General rate equation for } S_N1$$

Doubling or tripling the concentration of the alkyl halide doubles or triples the reaction rate. Doing the same to the nucleophile shows no effect. The potential energy diagram is illustrated in Figure 8.2.

GETTING INVOLVED

- ✓ What is the difference between unimolecular and bimolecular reactions and how do the terms describe the two mechanisms for nucleophilic substitution?
- ✓ Which is the slow step in an S_N1 reaction and why does the rate depend on it?
- ✓ How do the reaction rate equations for S_N1 and S_N2 reactions differ? How does changing the concentration of the alkyl halide or the nucleophile influence the rate of the two reactions?

Problem 8.14

Write the reaction rate equation for the S_N1 reaction described in Problem 8.13.

Problem 8.15

What effect will the following have on the rate of an S_N1 reaction? **(a)** Double the concentration of the nucleophile; **(b)** double the concentration of both the alkyl halide and the nucleophile; **(c)** quadruple the concentration of alkyl halide; **(d)** triple the concentration of the alkyl halide and double the concentration of the nucleophile.

2. ***Stereochemistry.*** In the first step of an S_N1 mechanism, the halide ion departs, leaving a planar carbocation that can be attacked from either side, equally. As a result, reaction of an optically active alkyl halide produces a pair of enantiomers in equal amounts, an optically inactive racemic mixture, as shown in the hydrolysis of *S*-3-bromo-2,3-dimethylpentane (remember there is an oxonium ion intermediate; this mechanism shows the final products after loss of hydrogen ion by the oxonium ion). Example 8.4 compares the stereochemistry of S_N2 and S_N1 reactions.

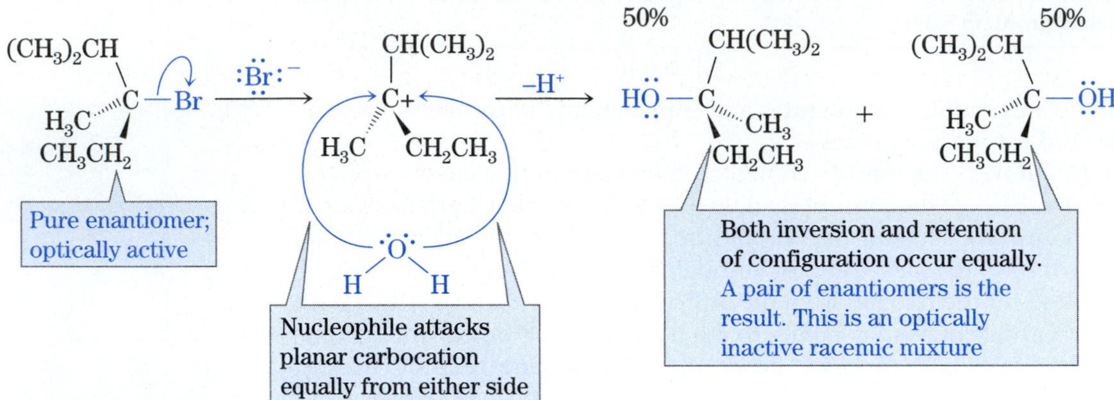

GETTING INVOLVED

- ✓ Why does an S_N2 reaction of an optically active halide give inversion of configuration and an S_N1, racemization?
- ✓ Why is the product of an S_N2 reaction of an optically halide still optically active whereas the product of an S_N1 reaction is optically inactive? In an S_N1 reaction, why are the enantiomers formed in equal amounts?

Example 8.4

Using optically active 2-bromobutane, write **(a)** an S_N2 mechanism for the reaction with $NaOCH_3$ and **(b)** an S_N1 mechanism for the reaction with CH_3OH. Show stereochemistry in each case.

Solution

(a) S_N2 Mechanism

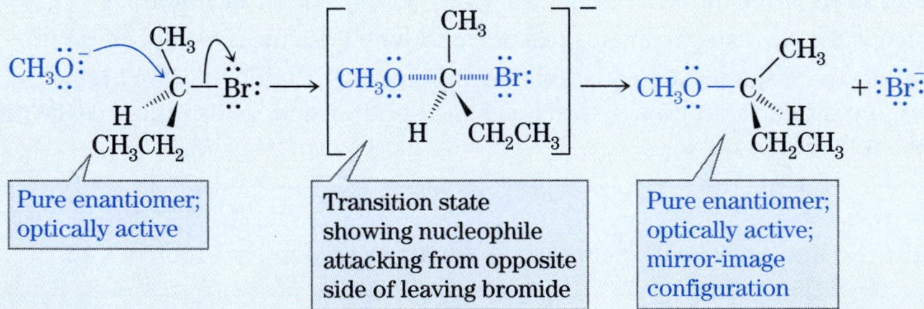

(b) S_N1 Mechanism

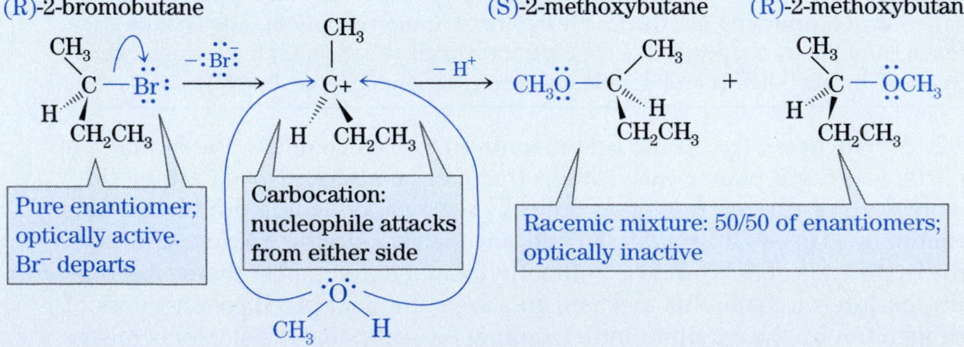

Problem 8.16

(a) Write an S_N1 mechanism for the reaction of optically active 3-methyl-3-chlorohexane with CH_3CH_2OH. Show stereochemistry clearly as in the example we just covered. **(b)** Write an S_N2 mechanism for the reaction between optically active 2-chlorohexane and $NaOCH_2CH_3$, showing stereochemistry.

Problem 8.17

Using wedges and dashes, write the products of the following S_N1 reactions. Remember that an enantiomeric pair results in this mechanism.

(a) H_3C—C(—I)(CH(CH$_3$)$_2$)(CH$_2$CH$_3$) + H_2O ⟶ **(b)** Br—C(—H)(CH$_2$CH$_3$)(C$_6$H$_5$) + CH_3OH ⟶

Problem 8.18

Assign *R* and *S* configurations to the reactants and products in Problem 8.17.

See related problems 8.39, 8.42, and 8.47.

F. Factors Influencing the Reaction Mechanism—S_N2 versus S_N1

We have seen that two mechanisms are possible for nucleophilic substitution. What determines which mechanism will be operative under specific conditions? The following are some of the factors that should be considered.

1. ***Carbocation Stability.*** An S_N1 reaction involves an intermediate carbocation; an S_N2 reaction does not. Alkyl halides that form relatively stable carbocations, will likely react by an S_N1 mechanism, whereas those that do not will react by an S_N2 mechanism, for which carbocation formation is unnecessary. We have previously explained the following order of carbocation stability (section 5.1.C.1):

Carbocation stability: R_3C^+ (3°) > R_2CH^+ (2°) > RCH_2^+ (1°) > CH_3^+

We should expect then that the propensity for an S_N1 mechanism to occur would increase as the reactant is changed from primary to secondary to tertiary halide, since tertiary halides will form relatively stable tertiary carbocations.

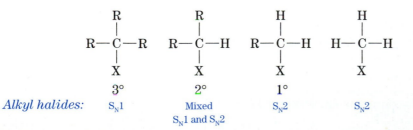

Alkyl halides:

R_3C—X	R_2CH—X	RCH_2—X	CH_3—X
3°	2°	1°	
S_N1	Mixed S_N1 and S_N2	S_N2	S_N2

GETTING INVOLVED

- ✓ How does carbocation stability influence S_N1 and S_N2 mechanisms?
- ✓ What is the order of carbocation stability?
- ✓ What alkyl halides most likely react by S_N1 and which by S_N2?

Example 8.5

Predict whether the isomers 1-bromobutane and 2-bromo-2-methylpropane would undergo nucleophilic substitution by S_N2 or S_N1 mechanisms.

Solution

1-Bromobutane forms an unstable 1° carbocation.

$$CH_3CH_2CH_2CH_2Br \xrightarrow{-Br^-} CH_3CH_2CH_2CH_2+ \qquad 1^\circ \text{ carbocation}$$

As a result, it reacts by an S_N2 mechanism, it does not involve carbocation intermediates.
2-Bromo-2-methylpropane forms a stable 3° carbocation and thus reacts by an S_N1 mechanism, which does involve a carbocation intermediate.

$$CH_3\overset{\overset{CH_3}{|}}{\underset{\underset{Br}{|}}{C}}CH_3 \xrightarrow{-Br^-} CH_3\overset{\overset{CH_3}{|}}{\underset{+}{C}}CH_3 \qquad 3^\circ \text{ carbocation}$$

Problem 8.19

There are four bromine derivatives of 2-methylbutane that are positional isomers of one another. Draw them and predict whether each would react by an S_N2 or S_N1 mechanism based on carbocation stability.

2. ***Steric Effects.*** In an S_N2 reaction, a nucleophile attacks a saturated carbon and pushes out a halide ion. For a brief period, five groups are coordinated around a single carbon, a relatively crowded condition. The bigger the groups around the carbon-halogen bond, the greater the difficulty the nucleophile has in reaching the carbon from the back side and displacing the halogen. In an S_N1 reaction, however, a tetravalent carbon loses a halide ion, forming a trivalent carbocation, a less crowded environment. Steric crowding is maximized in the S_N2 transition state and minimized in the S_N1 carbocation intermediate. Since alkyl groups are larger than hydrogen atoms, steric crowding increases in the direction from primary to tertiary alkyl halides, and the likelihood of an S_N1 reaction also increases.

Crowding increases, S_N1 increases ←

Alkyl halides: 3° 2° 1° CH_3X

→ Crowding decreases, S_N2 increases

This concept is illustrated in Figure 8.3.

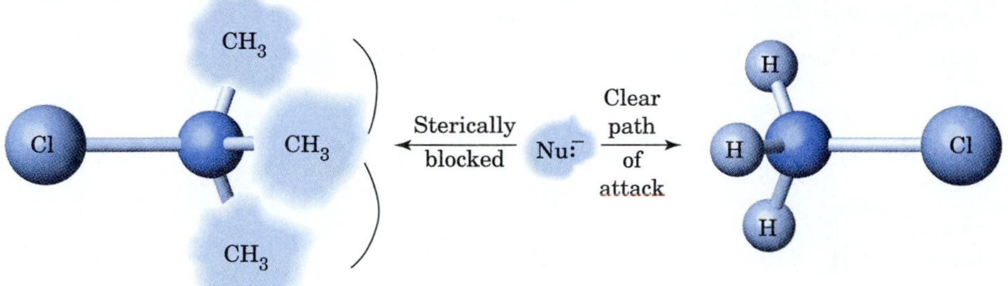

Figure 8.3 A nucleophile (Nu:⁻) can attack methyl chloride easily; tertiary butyl chloride is sterically very crowded, however, making nucleophilic attack by an S_N2 mechanism difficult.

GETTING INVOLVED

✓ Why does steric crowding increase in the S_N2 process and decrease in S_N1?

✓ Why is steric crowding greater in a 3° halide compared to a 1°?

Problem 8.20

Which compound do you predict would react faster by an S_N2 mechanism: 1-bromobutane or 2-bromobutane? Why?

See related problem 8.50.

3. ***Strength of Nucleophile.*** In an S_N2 reaction, the nucleophile physically displaces the halide and the reaction rate depends on its concentration. The rate of an S_N1 reaction, in contrast, is independent of the nucleophile. As a result, strong nucleophiles favor S_N2 processes.

Charged nucleophiles are stronger than their neutral counterparts. For example, HO^- is stronger than HOH, and RO^- is stronger than ROH.

More electronegative elements hold their electrons more tightly and are not as good nucleophiles as less electronegative atoms. Therefore, HS^- is stronger than HO^-, and H_3N is stronger than H_2O (oxygen is more electronegative than either sulfur or nitrogen).

GETTING INVOLVED

Problem 8.21

Select the stronger nucleophile from each of the following pairs: **(a)** CH_3O^- or CH_3OH; **(b)** H_2N^- or HO^-; **(c)** CH_3OH or CH_3NH_2; **(d)** HS^- or H_2O; **(e)** CH_3CH_2OH or CH_3CH_2SH.

4. ***Solvent.*** Polar solvents with unshared electron pairs such as water and alcohols can use these electrons to solvate carbocations, thereby promoting their formation and stabilization. In this way, they promote the S_N1 process that involves a carbocation intermediate.

GETTING INVOLVED

✓ Describe how carbocation stability, steric effects, strength of nucleophile, and solvent polarity each favor or disfavor S_N1 and S_N2 reactions.

Example 8.6

Would the following reactions proceed by an S_N1 or an S_N2 mechanism?

(a) $CH_3CH_2CH_2CH_2Cl + NaOH \longrightarrow CH_3CH_2CH_2CH_2OH + NaCl$

(b)
$$CH_3\underset{\displaystyle Cl}{\overset{\displaystyle CH_3}{\underset{|}{\overset{|}{C}}}}CH_3 + H_2O \longrightarrow CH_3\underset{\displaystyle OH}{\overset{\displaystyle CH_3}{\underset{|}{\overset{|}{C}}}}CH_3 + HCl$$

Solution

(a) The alkyl halide is primary and does not form a stable carbocation as would be required in an S_N1 mechanism. In addition, primary halides show relatively little steric hindrance to attack by a nucleophile. Hydroxide is a strong nucleophile. All factors favor the S_N2 mechanism.

(b) Tertiary halides from stable tertiary carbocations, the intermediate in the S_N1 mechanism. Tertiary halides are also relatively hindered, making attack by nucleophiles difficult. The nucleophile is neutral and not terribly strong. All factors favor the S_N1 mechanism.

Problem 8.22

Predict whether the following reactions would occur by S_N2 or S_N1 mechanisms: **(a)** 1-bromo-3-methylbutane and $NaOCH_3$; **(b)** 3-bromo-3-methylbutane and CH_3OH. Explain your prediction.

See related problem 8.38.

G. S_N1 and S_N2: Summary

Following are generalized S_N2 and S_N1 mechanisms for an optically active alkyl halide, with a summary of important characteristics. As you review nucleophilic substitution, it may be helpful to relate to the mechanisms with a simple analogy. Imagine that you are in a full classroom with a front and a back door and it is time to change classes. How can the students present be moved out and the new class moved in? The new class could come in and sit down and the other class could then get up and leave. This is operationally impossible, since available seating can accommodate only one group of students. For a similar reason, we earlier eliminated our first idea for a nucleophilic substitution mechanism—that in which the nucleophile entered and bonded and then the halide ion left; carbon cannot accommodate five bonds. However, if the new class entered from one door as the leaving class left through the other door, there would be a smooth transition. This is analogous to the S_N2 mechanism, in which the nucleophile enters at the same time the halide ion is leaving but from the opposite side. Finally, if the present class left the room first, then the new class could enter easily, using either door. Similarly, in an S_N1 mechanism, since the halide ion leaves completely in the first step, the incoming nucleophile can attack the carbocation freely from either side.

1. *Reaction.* Both S_N2 and S_N1 reactions are simple substitution reactions in which a leaving group, often a halogen, is replaced by a nucleophile.
2. *Mechanism.* An S_N2 reaction has a one-step mechanism that proceeds via a five-centered transition state; an S_N1 mechanism consists of two steps, with a carbocation intermediate.

S_N2
A one-step process

$Nu:^-$

Transition state of one-step reaction

Inversion of configuration; optically active

Optically active alkyl halide

S_N1
A two-step process

Step 1 :NuH Step 2 $-H^+$

Carbocation intermediate

Inversion and retention; optically inactive racemic mixture; enantiomeric pair

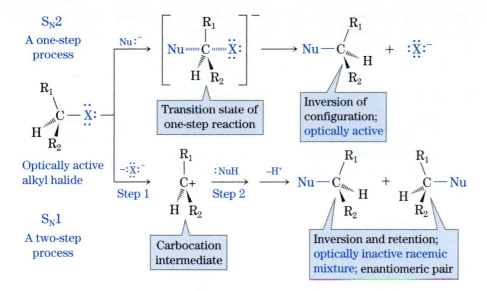

3. *Reaction Rates.* S_N2 reactions are bimolecular; the rate depends on the concentrations of both the alkyl halide and the nucleophile. S_N1 reactions are unimolecular; the rate depends only on the concentration of alkyl halide.
4. *Stereochemistry.* S_N2 reactions involving an optically active alkyl halide produce an optically active product but with an inverted configuration. S_N1 reactions proceed by racemization, giving approximately a 50/50 mixture of enantiomers; the product mixture is optically inactive.
5. *Structure and Reactivity.* S_N1 processes are favored by bulky alkyl halides that form stable carbocations; S_N2 processes are favored by just the opposite. Consequently, 3° halides usually react by an S_N1 mechanism, 1° by an S_N2, and 2° by either, depending on specific factors.
6. *Nucleophiles.* Strong nucleophiles favor S_N2 reactions.
7. *Solvent:* Polar solvents with unshared electron pairs such as water and alcohols favor S_N1 reactions.

8.5 Elimination Reactions of Alkyl Halides

We have just studied the reactions of alkyl halides with nucleophiles to produce substitution products. Earlier in section 4.5, however, we saw that alkyl halides can undergo elimination reactions in the presence of a base (nucleophile) such as potassium hydroxide. Hydroxide as a nucleophile can either displace halide ion to form an alcohol or effect elimination to form an alkene.

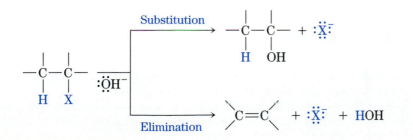

$:\ddot{O}H^-$

Substitution → $+ :\ddot{X}:^-$

Elimination → $C=C + :\ddot{X}:^- + HOH$

Competition between elimination and substitution is possible when any alkyl halide capable of undergoing elimination is treated with a nucleophile. Not only does elimination compete with nucleophilic substitution, but the elimination mechanisms, E_2 and E_1, are closely related to the substitution mechanisms.

A. The E_2 and E_1 Reaction Mechanisms

E_2
elimination bimolecular; the one-step elimination mechanism

The **E_2** mechanism is a concerted one-step process like the S_N2 mechanism. Instead of attacking the carbon to which the halogen is bonded, as in substitution, the nucleophile abstracts a hydrogen ion on an adjacent carbon. The halide leaves simultaneously, generating a double bond.

E_2 Mechanism: A One-Step Process

Nu:⁻ H C—C X → Nu:⁻ → [Nu H C---C X]⁻ → C=C + NuH + :X:⁻

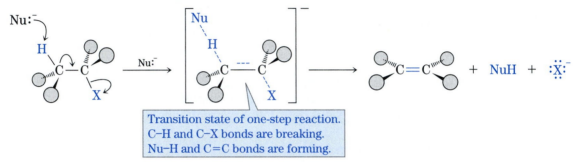

Transition state of one-step reaction. C–H and C–X bonds are breaking. Nu–H and C=C bonds are forming.

E_1
elimination unimolecular; the two-step elimination mechanism

The **E_1** reaction proceeds by a two-step mechanism. In the first step the carbon-halogen bond breaks and the halide leaves with the bonding pair of electrons; a carbocation results. The nucleophile abstracts a hydrogen ion from an adjacent carbon in the second step. The electrons from this carbon-hydrogen bond move toward the positively charged carbon and form the carbon-carbon double bond.

E_1 Mechanism: A Two-Step Process

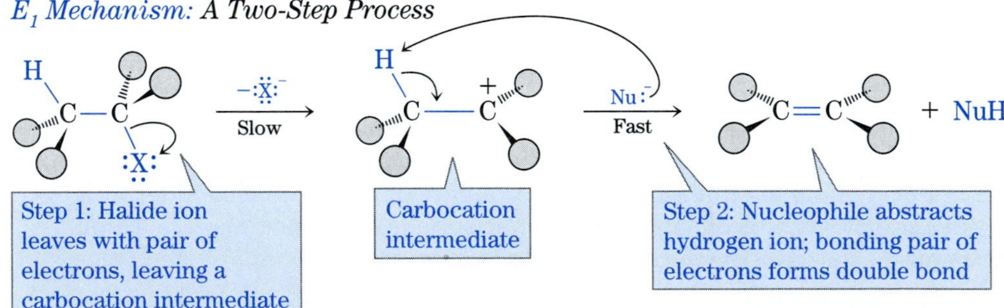
H
C—C
:X:
−:X:⁻
Slow
H
C—C+
Nu:⁻
Fast
C=C
+ NuH
Step 1: Halide ion leaves with pair of electrons, leaving a carbocation intermediate
Carbocation intermediate
Step 2: Nucleophile abstracts hydrogen ion; bonding pair of electrons forms double bond

GETTING INVOLVED

- ✓ First make sure you can still write dehydrohalogenation reactions and predict the predominant product using the Saytzeff Rule (see section 4.5).
- ✓ Spend a few minutes and describe E_2 and E_1 reactions to yourself. What bonds break in the starting material and what bonds form in the product? Which reaction

is one step and which is two steps? Which has a carbocation intermediate? How is it formed, how is it neutralized?

✓ Compare E_1 and E_2 reactions to S_N1 and S_N2 reactions in terms of **(a)** number of steps; **(b)** intermediates and transitions states; **(c)** anti elimination and backside attack in substitution; **(d)** function of the nucleophile; **(e)** elimination versus substitution products.

Example 8.7

Write E_2 and E_1 mechanisms for the reaction between 2-bromobutane and potassium hydroxide.

Solution

The reaction can give two products, with the more highly substituted product being predominant (section 4.5.A–B).

$$CH_3CH_2\underset{\displaystyle Br}{\underset{|}{C}}HCH_3 + KOH \longrightarrow \underset{\text{Predominant product}}{CH_3CH{=}CHCH_3} + CH_3CH_2CH{=}CH_2 + KBr + H_2O$$

E_2 mechanism: One step:

$$CH_3\underset{\displaystyle H}{\underset{|}{C}}H-\overset{\displaystyle Br}{\overset{|}{C}}HCH_3 + \bar{O}H \xrightarrow[\text{Transition state}]{\text{HO}\cdots\text{H}\cdots CH_3C\cdots CCH_3\ (\text{H, Br, H})} CH_3CH{=}CHCH_3$$

E_1 mechanism: Two steps:

$$CH_3\underset{\displaystyle H}{\underset{|}{C}}H-\underset{\displaystyle Br}{\underset{|}{C}}HCH_3 \xrightarrow{-Br^-} \underset{\text{Carbocation intermediate}}{CH_3\underset{\displaystyle H}{\underset{|}{C}}H-\overset{+}{C}HCH_3} \xrightarrow{^-OH} CH_3CH{=}CHCH_3 + H_2O$$

See Example 8.4 for the competing S_N2 and S_N1 processes.

Problem 8.23

Write the reaction of 2-bromopropane with potassium hydroxide by an E_1 and an E_2 mechanism.

See related problems 8.40–8.41.

B. Comparison of E_2 and E_1 Reactions

The different mechanisms of E_2 and E_1 reactions result in differences in their reaction rates and stereochemistry.

1. ***Reaction Rates.*** The E_2 reaction is a one-step process in which both the alkyl halide and nucleophile are involved. Thus the rate depends on both and the rate equation is identical to that of an S_N2 reaction (section 8.4.D.1).

Elimination Bimolecular $\quad Rate_{E_2} = k[RX][Nu^-]$

The E_1 reaction is two steps and the first is the slow, rate-determining step. Only the alkyl halide is involved in the first step and thus the reaction rate depends only on this species. The nucleophile is not involved until the second step and has no influence on the rate of reaction.

Elimination Unimolecular $\quad Rate_{E_1} = k[RX]$

GETTING INVOLVED

✓ Explain why an E_2 reaction is bimolecular and an E_1 reaction is unimolecular. What is the difference in the rate equations?

✓ Answer the question above for S_N2 and S_N1 reactions. You should find your answers essentially the same.

Problem 8.24

Describe the change in reaction rate for both E_2 and E_1 reactions for each of the following changes in concentrations: **(a)** double the concentration of the alkyl halide; **(b)** double the concentration of the nucleophile; **(c)** double the concentrations of both the alkyl halide and nucleophile; **(d)** triple the concentration of the alkyl halide and quadruple the concentration of the nucleophile.

Problem 8.25

Answer problem 8.24 for S_N2 and S_N1 reactions. Are your answers the same or different? Explain.

2. ***Stereochemistry.*** In elimination reactions, the groups being eliminated must be in an anti or syn relationship to one another, as illustrated by the following sawhorse and Newman projections. This is required because the developing p orbitals that will compose the new π bond must be parallel to each other so that overlap can occur.

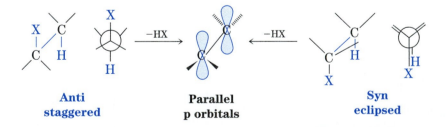

X C C H X H $-HX$ $-HX$ C C H X H X

Anti staggered **Parallel p orbitals** **Syn eclipsed**

The E_2 reaction occurs from the anti configuration. This allows maximum distance between the attacking nucleophile and the departing halide ion and thus is sterically favored. The situation is analogous to that in an S_N2 mechanism (section 8.4.C.2). This stereospecificity is very important, since anti and syn conformations can give two different geometric isomers in compounds where the reacting carbons are both chiral. Consider, for example, the dehydrobromination of 3*R*, 4*R* 3-bromo-4-methylhexane from both an anti and syn conformation.

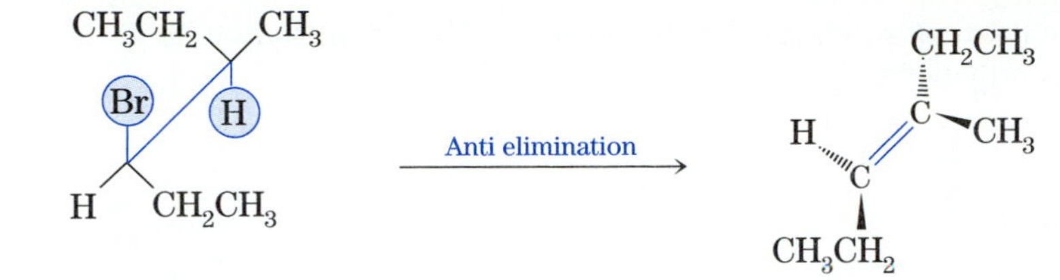

CH_3CH_2, CH_3, Br, H, H, CH_2CH_3 — Anti elimination → CH_2CH_3, H, C, C, CH_3, CH_3CH_2

Anti conformation — *trans* isomer

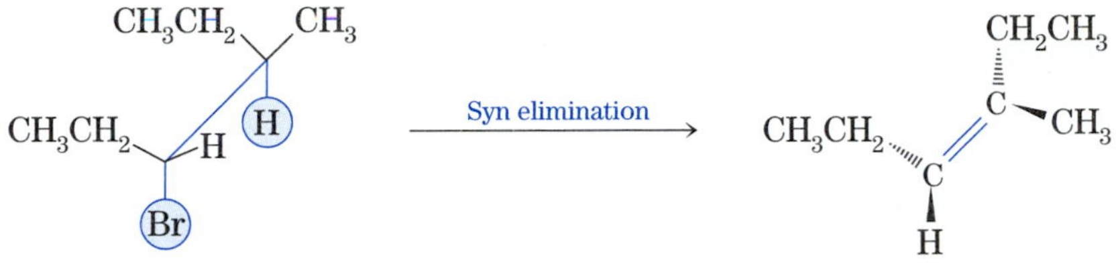

CH_3CH_2, CH_3, CH_3CH_2, H, H, Br — Syn elimination → CH_2CH_3, CH_3CH_2, C, C, CH_3, H

Syn conformation — *cis* isomer

As explained above, E_2 reactions proceed by anti elimination and consequently give only the *trans* isomer in the specific example described. However, E_1 reactions occur through a carbocation intermediate. The carbocation can form from either a syn or anti conformation and, once formed, shows free bond rotation, thus not maintaining any specific conformation. Therefore, E_1 reactions give products of both syn and anti elimination, in this case, both the *cis* and *trans* isomers.

E_2 Reactions $\xrightarrow[\text{elimination}]{\text{Anti}}$ *cis* or *trans* product

(3*R*, 4*R*) 3-bromo-4-methylhexane ⟶ *trans* product

E_1 Reactions $\xrightarrow[\text{elimination}]{\text{Both syn and anti}}$ *cis* and *trans* products

(3*R*, 4*R*) 3-bromo-4-methylhexane ⟶ *cis* and *trans* products

GETTING INVOLVED

- ✓ Why is either the syn or anti conformation of the departing halide and hydrogen necessary in E_2 and E_1 elimination? Why does anti elimination occur exclusively in E_2 reactions? Why is both syn and anti elimination possible in E_1 reactions?
- ✓ Why is elimination in compounds where the carbons bearing the departing halide and hydrogen are both chiral capable of producing geometric isomers? Why does E_2 elimination produce either *cis* or *trans*, whereas E_1 reactions can produce both *cis* and *trans?*

Problem 8.26

Take a look at the E_2 anti elimination of (3R, 4R) 3-bromo-4-methylhexane illustrated in this section; notice that the *trans* isomer formed. **(a)** You can make the sawhorse drawing 3S,4R by switching the hydrogen and ethyl group on carbon 3, why? What geometric isomer would be formed now? **(b)** You can now make the drawing 3S,4S by switching the ethyl and methyl groups on carbon 4 in addition to what you already did in part a. Is the product *cis* or *trans?* **(c)** What isomer would be formed from the 3R,4S configuration?

Problem 8.27

Write the geometric isomers that result from E_2 and E_1 elimination reactions of the following alkyl halide shown in the Newman projection. In the projection shown, the eliminating groups (H and Br) are neither in the syn nor anti conformation. To visualize the final products you should rotate the carbon-carbon bond to achieve these conformations.

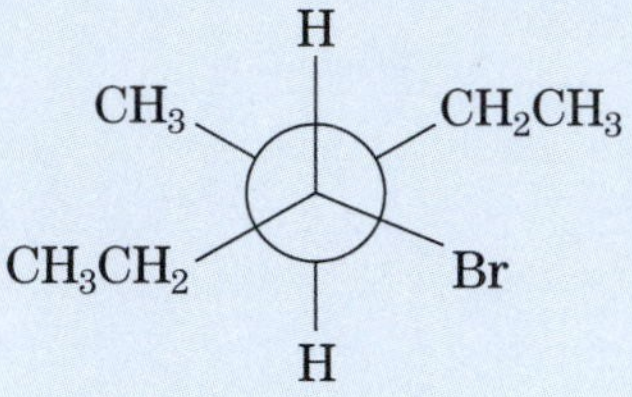

See related problems 8.44–8.46 and 8.48.

8.6 Substitution versus Elimination

When alkyl halides are treated with a nucleophile, they can undergo either substitution or elimination. The reaction mechanisms are strikingly similar and are summarized in Figure 8.4. Many factors influence the competition between substitution and elimination: solvent, nature of nucleophile, alkyl halide structure, and reaction conditions. Often the most important factor in determining which will occur is the stability of the alkene that could be formed by elimination should it predominate. Alkene stability increases with an increasing number of alkyl substituents (section 4.5.B). Since a tertiary halide is more highly substituted than is a primary halide, it will usually form a more highly substituted alkene. Thus, tertiary halides have a greater tendency toward elimination, since they form highly stable alkenes.

Alkyl Halides: Elimination increases ← 3° 2° 1° → Substitution increases

To illustrate this concept in a practical manner, let us consider the preparation of 2-methoxy-2-methylpropane by the Williamson synthesis of ethers. There

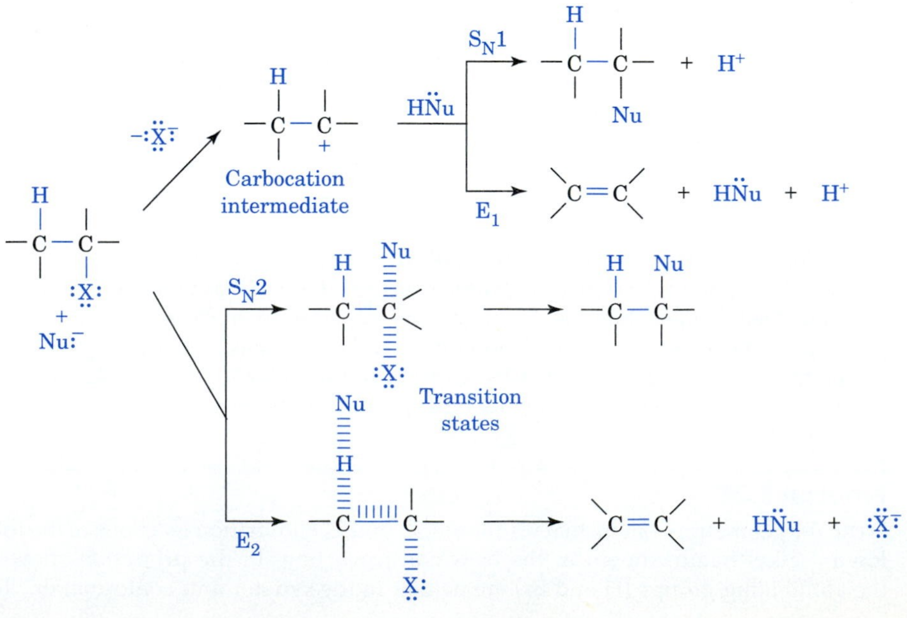

Figure 8.4 Competition between S_N1, S_N2, E_1, and E_2 reactions of alkyl halides with nucleophiles.

are two approaches. One is to combine tertiary butyl chloride with sodium methoxide.

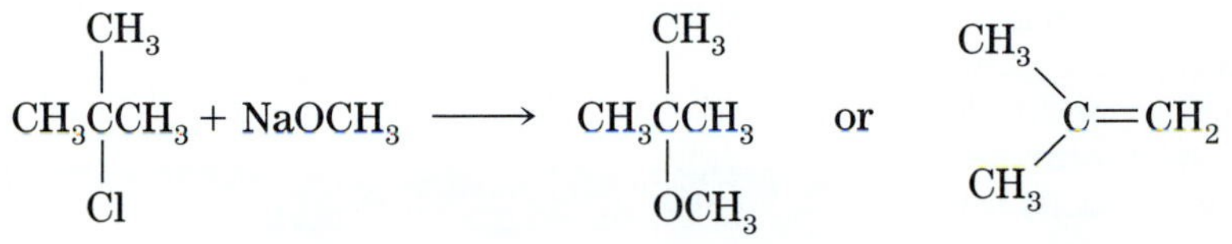

$$(CH_3)_3CCl + NaOCH_3 \longrightarrow (CH_3)_3COCH_3 \text{ or } (CH_3)_2C{=}CH_2$$

Substitution product

Elimination product (predominates)

Connections 8.2 www.prenhall.com/bailey

Thyroid Hormone

Most organisms do not contain halogenated hydrocarbons as functioning molecules. In fact, the introduction of such compounds can be highly toxic. However, there is one molecule that is highly halogenated and essential to the functioning of our bodies. That substance is thyroxine or thyroid hormone and a related material, triiodothyronine.

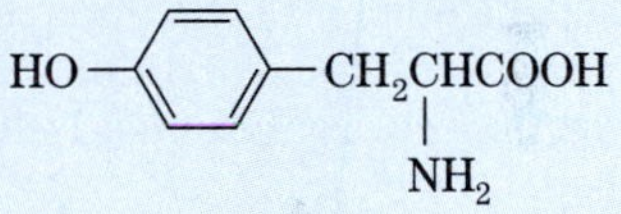

Tyrosine

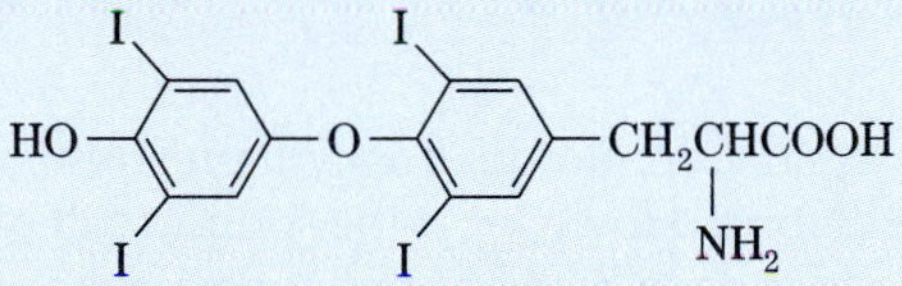

Thyroxine or T_4

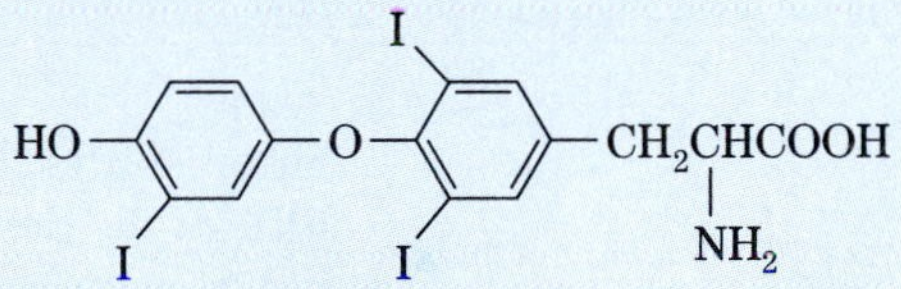

Triiodothyronine or T_3

These two hormones are produced in the thyroid gland, an endocrine organ, located in the front part of the neck. Their precursors are iodine and tyrosine, one of the amino acids comprising the proteins of the body. Dietary iodide is transported to, and absorbed by, the thyroid gland. Secretions from the area of the brain known as the hypothalamus stimulate the thyroid to iodinate the phenolic rings of tyrosine in a large carbohydrate-containing protein molecule called *thyroglobulin*. This reaction is catalyzed by an enzyme, iodoperoxidase. While still attached to the thyroglobulin, a diiodotyrosyl residue is coupled enzymatically to another diiodotyrosyl or to a monoiodotyrosyl species. The ether linkage is an unusual feature, for generally linkages to amino acids are peptide (amide) in nature.

Thyroglobulin protein chain

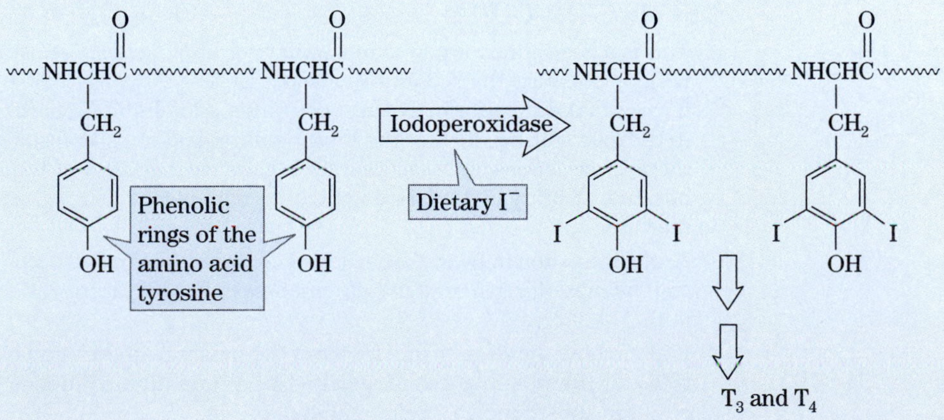

Phenolic rings of the amino acid tyrosine

Iodoperoxidase

Dietary I^-

T_3 and T_4

Connections 8.2 *(cont.)*

The thyroglobulin is then broken down into its component amino acids, releasing two to five thyroxine molecules per molecule of thyroglobulin. T_3 and T_4 are released by the thyroid into the bloodstream and carried by other protein carriers to target organs, which are responsible for metabolism. They then enter those organs and directly affect the synthesis of enzymes that control key metabolic reactions. (Although small amounts of triiodothyronine, the more potent hormone, are released by the thyroid, most is synthesized from thyroxine in the liver.)

A person may suffer from an overactive or underactive thyroid, known as hyperthyroidism and hypothyroidism, respectively. Symptoms of hyperthyroidism include nervousness, increased activity, rapid heartbeat, warm, moist skin, and insomnia. The signs of hypothyroidism include dry, coarse skin and hair, forgetfulness, and below-average body temperature. A newborn or baby lacking thyroid hormone suffers from deficiencies in both mental and physical development, a condition referred to as *cretinism,* Both hyper- and hypo- conditions may be detected by an enlargement of the thyroid gland itself—the formation of a goiter.

Therapy for thyroid excesses may include surgery, drugs, or treatment with radioactive iodine, while a lack of T_4 and T_3 may be treated with thyroid supplements.

There are many causes for thyroid malfunction, which range from congenital problems to environmental deficiencies. Diagnosis is important for the proper treatment to be prescribed.

Woman with goiter

Since *t*-butyl chloride is a tertiary halide, elimination will preferentially occur and little of the desired product will form. However, if the sodium salt of *t*-butyl alcohol is combined with methyl chloride, which has only one carbon, elimination is impossible, and the desired substitution product forms exclusively.

$$CH_3C(CH_3)_2\overset{-}{O}\overset{+}{Na} + CH_3Cl \longrightarrow CH_3C(CH_3)_2OCH_3 + NaCl$$

2-methoxy-2-methylpropane

GETTING INVOLVED

✓ This is a good time for you to integrate your knowledge of substitution and elimination. Compare the S_N2 and E_2 mechanisms and the S_N1 and E_1 mechanisms in terms of **(a)** difference between substitution and elimination; **(b)** number of steps in the mechanisms; **(c)** reaction intermediates and transition states; **(d)** molecularity of the reactions and reaction rates; **(e)** stereochemistry of the reactions; **(f)** carbocation stability in determining predominant mechanism; **(g)** steric effects on the mechanisms.

✓ Anti elimination in E_2 and inversion of configuration in S_N2 occur for a similar reason; likewise for syn and anti elimination in E_1 and racemization in S_N1. Explain both.

✓ Explain how the relationship between the structure of the alkyl halide and the stability of the possible alkene product of elimination influence the competition between substitution and elimination.

Problem 8.28

Write an equation showing the best way to prepare $(CH_3)_2CHOCH_2CH_3$ by the Williamson synthesis.

See related problem 8.51.

REACTION SUMMARY

A. Nucleophilic Substitution

Section 8.4; Examples 8.1–8.6; Table 8.3; Problems 8.3–8.22, 8.32–8.39, 8.41, 8.43, 8.47, 8.42–8.43.

Negative Nucleophiles

$$-\overset{|}{\underset{|}{C}}-X + Nu^- \longrightarrow -\overset{|}{\underset{|}{C}}-Nu + X^-$$

X = Cl, Br, I

Nu = ^-OH ^-SH $^-NH_2$ ^-CN

^-OR ^-SR ^-NHR $RC{\equiv}C^-$

$^-NR_2$

Neutral Nucleophiles

$$-\overset{|}{\underset{|}{C}}-X \quad HNu \longrightarrow -\overset{|}{\underset{|}{C}}-Nu + HX$$

X = Cl, Br, I

HNu = HOH HSH NH_3

HOR HSR RNH_2

R_2NH

B. Elimination Reactions: Dehydrohalogenation

Section 8.5 and review 4.5; Example 8.7; Problems 8.23–8.27, 8.40–8.41, 8.44–8.46, 8.48.

$$-\overset{|}{\underset{H}{C}}-\overset{|}{\underset{X}{C}}- \quad Nu^- \longrightarrow \;\rangle C{=}C\langle\; + HNu + X^-$$

Nu can be any of the negative or neutral nucleophiles listed in A, but we primarily used hydroxide ion (KOH in ethanol).

X = Cl, Br, I

Problems

8.29 **IUPAC Nomenclature:** Name the following compounds:

(a) $CH_3CH_2CHCH_2CH_3$ (Br on C-3)

(b) CH_3CH_2Cl

(c) $CH_3CH_2CH_2I$

(d) $(CH_3)_3CCH_2CH_2CBr_3$

(e) $CH_3CHCH_2CHCH_2CHCH_3$ (Cl, Cl, Cl on C-2, C-4, C-6)

(f) $CH_3(H)C{=}C(H)CH_2CHCHCH_3$ — CH_3 on C-4, Br on C-5 (cis: CH_3 and H on one carbon, H and $CH_2CHCHCH_3$ on the other)

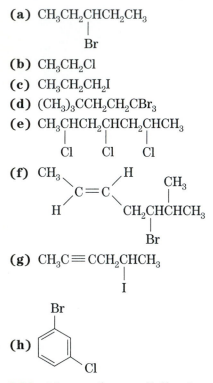

(g) $CH_3C{\equiv}CCH_2CHCH_3$ (I on C-5)

(h) Benzene ring with Br and Cl (meta)

8.30 **Nomenclature:** Following are some organohalogen compounds that are suspected of being dangerous to human health. Their names periodically appear in newspapers and magazines. Write structures for each.

(a) trichloromethane: chloroform; once in nonprescription cough medicines

(b) 1,2-dibromo-3-chloropropane: DBCP; an agricultural fumigant that diminished sperm count in chemical plant workers

(c) 1,2-dichloroethane: ethylene dichloride; used to make vinyl chloride from which PVC is made; may be a carcinogen

(d) tetrachloromethane: carbon tetrachloride; prolonged exposure can cause liver and kidney damage or failure

(e) tetrachloroethane and tetrachloroethylene: dry-cleaning agents; may be carcinogens

(f) dichlorodifluoromethane: a Freon that could lead to destruction of the ozone shield

8.31 **Common Nomenclature:** Draw the following compounds:

(a) methyl bromide
(b) methylene chloride
(c) bromoform
(d) carbon tetrafluoride
(e) allyl iodide
(f) vinyl chloride
(g) secondary butyl chloride
(h) isopropyl bromide

8.32 **Nucleophilic Substitution:** Complete the following reactions, showing the nucleophilic substitution products:

(a) CH_3CHCH_3 (Cl on C-2) $+ NaOH \longrightarrow$

(b) C_6H_5—$CH_2Br + NaCN \longrightarrow$

(c) $CH_3CH_2I + NaSH \longrightarrow$

(d) $CH_3CH_2CH_2Br + NaN(CH_3)_2 \longrightarrow$

(e) CH_3CHONa (CH_3 on C-2) $+ CH_3CH_2I \longrightarrow$

(f) CH_3CHCH_2Br (CH_3 on C-2) $+ CH_3SNa \longrightarrow$

(g) C_6H_5—$CH_2CH_2Cl + NaC{\equiv}CCH_3 \longrightarrow$

(h) $CH_3Cl + NaNH_2 \longrightarrow$

8.33 **Nucleophilic Substitution:** Complete the following reactions, showing the nucleophilic substitution products:

(a) $CH_3CCH_2CH_3$ (CH_3 and Cl on C-2) $+ CH_3OH \longrightarrow$

(b) C_6H_5—$CH_2Br + CH_3NHCH_3 \longrightarrow$

(c) $CH_3CH_2CH_2OH + CH_3CH$—C_6H_5 (Cl on CH) $\longrightarrow$

8.34 **Williamson Synthesis of Ethers:** Using alkyl halides and sodium alkoxides as starting materials, prepare the following ethers by the Williamson synthesis:

(a) $CH_3CH_2OCH_2CH_2CH_2CH_3$

(b) $CH_3CH_2OCH(CH_3)_2$

8.35 **Nucleophilic Substitution in Preparing Alkynes:** Prepare the following alkynes, using ethyne (acetylene) and alkyl halides as organic starting materials (see sections 5.8 and 8.4.A):

(a) $CH_3CH_2C{\equiv}CH$

(b) $CH_3C{\equiv}CCH_2CH_3$

(c) C_6H_5—$CH_2C{\equiv}CH$

8.36 **Nucleophilic Substitutions:** The following compounds are responsible for the odor and flavor of garlic. Write a nucleophilic substitution reaction showing the preparation of each.

(a) $CH_2{=}CHCH_2SH$

(b) $CH_2{=}CHCH_2SCH_2CH{=}CH_2$

8.37 **Synthesis:** Suggest an alkyl halide and a nucleophile from which the following could be prepared:

(a) $CH_3(CH_2)_8CH_2NH_2$

(b) $CH_3CH_2SCH_3$

(c) $CH_3CH_2CH_2CH_2OH$

8.38 **S_N1 and S_N2 Mechanisms:** Make a chart comparing an S_N1 and an S_N2 reaction of an alkyl halide and a nucleophile with regard to the following:

(a) rate expression

(b) reaction intermediates

(c) stereochemistry

(d) relative rates of reaction of 1°, 2°, and 3° halides

(e) effect of increasing concentration of nucleophile

(f) effect of increasing concentration of alkyl halide

(g) effect of an ionic or polar solvent

(h) effect of a nonpolar solvent

(i) effect of bulky groups around reaction center

(j) strength of nucleophile

8.39 **Nucleophilic Substitution Mechanisms:** Write S_N1 and S_N2 mechanisms for the following reaction. Clearly show stereochemistry.

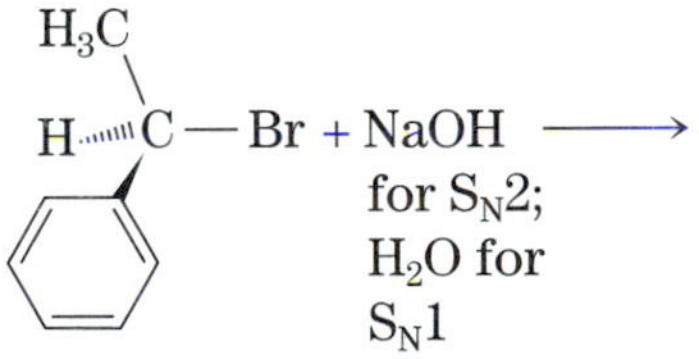

H_3C
H····C—Br + NaOH ⟶ for S_N2; H_2O for S_N1
(phenyl)

8.40 **Elimination Reactions:** Complete the following reactions, showing the major elimination products:

(a) 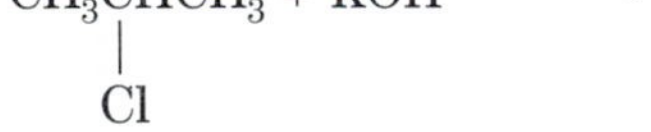 CH_3CHCH_3 (Cl on C-2) + KOH ⟶

(b) $CH_3CHCH_2CH_2CH_3$ (Br on C-2) + KOH ⟶

(c) $CH_3CHCHCH_2CH_3$ (CH_3 on C-2, I on C-3) + KOH ⟶

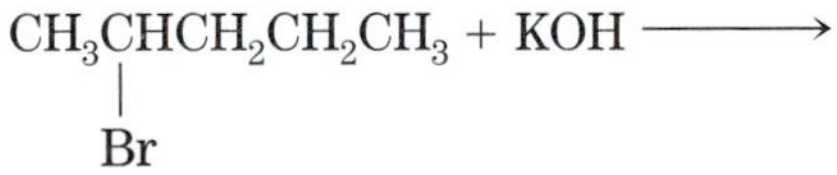

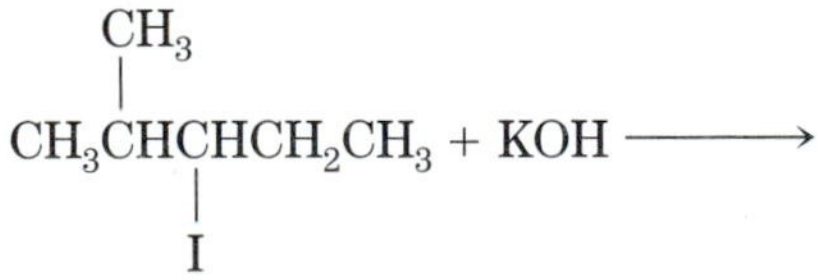

8.41 **Elimination Reaction Mechanisms:** Write an E_1 and an E_2 mechanism for the reaction of 2-bromo-2-methylpentane with potassium hydroxide (in aqueous alcohol solvent).

8.42 **S_N1 and S_N2 Stereochemistry:** Using dash/wedge projections, write reaction equations for the following, showing the stereochemistry of both the reactant and the product:

(a) *R*-2-bromopentane, CH_3SH, S_N1

(b) *S*-2-chlorobutane, $NaNH_2$, S_N2

(c) *S*-2-iodo-3-methylbutane, $NaOCH_2CH_3$, S_N2

(d) *R*-3-chloro-3-methylheptane, H_2O, S_N1

8.43 **S_N2 Stereochemistry:** Write the products of S_N2 reactions of the following optically active compounds with the reagents shown. Indicate whether the products are optically active or inactive.

(a) 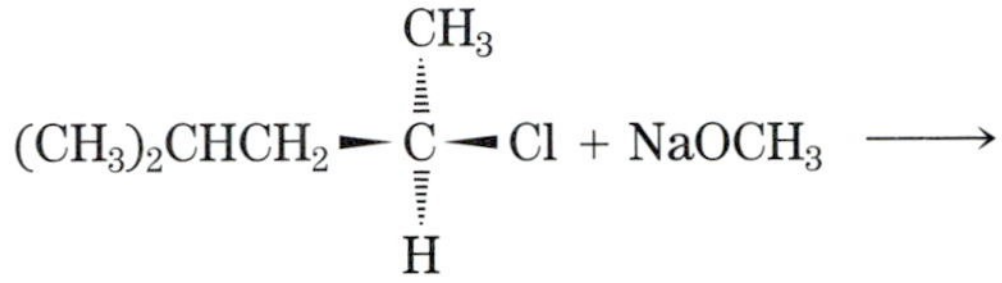 $(CH_3)_2CHCH_2$—C(CH_3)(H)—Cl + $NaOCH_3$ ⟶

(b) 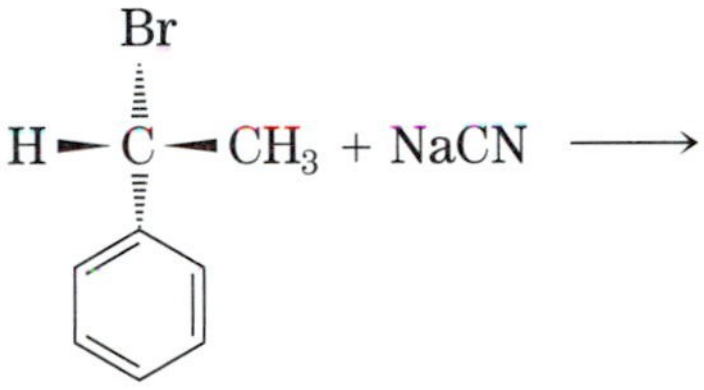 H—C(Br)(phenyl)—CH_3 + NaCN ⟶

8.44 **E_1 and E_2 Stereochemistry:** E_2 eliminations occur by anti elimination, in which the abstracted hydrogen and leaving halogen on the adjacent carbon are as far apart as possible (by C—C bond rotation). In E_1 reactions, because a carbocation is formed, such stereospecificity is not observed. Write the product or products of an E_2 elimination and an E_1 elimination of the following compound (shown as sawhorse diagram and Newman projection):

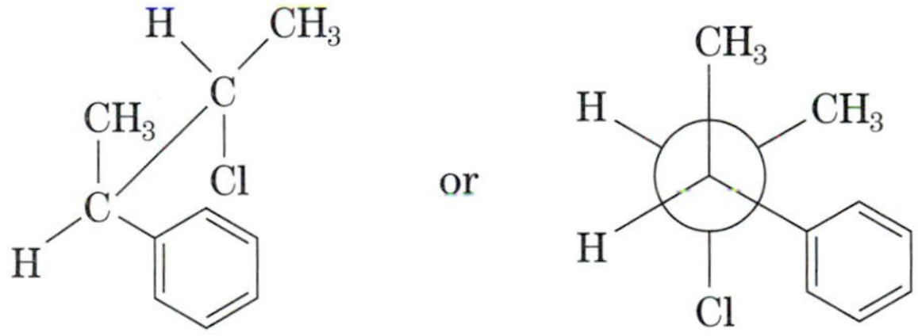

Sawhorse: front C: H, CH_3, Cl; back C: CH_3, H, phenyl — or — Newman: front: CH_3, H, H; back: CH_3, Cl, phenyl

8.45 **E_1 and E_2 Stereochemistry:** Write the products of E_2 and E_1 elimination reactions of (2*R*)(3*S*)2-chloro-2,3-diphenylbutane. What other configuration of this compound would yield the same geometric isomer? What two configurations would produce the other geometric isomer?

8.46 **E_1 and E_2 Stereochemistry:** If the dehydrochlorination reaction of *trans* 1-chloro-2-methylcyclopentane occurs by an E_2 mechanism, one would predict 3-methylcyclopentene as the product. If the mechanism is E_1, one would predict the normal Saytzeff rule product of 1-methylcyclopentene. Explain.

8.47 **Nucleophilic Substitution Reactions:** Predict the substitution product of the S_N2 reaction between *cis*

1-bromo-2-methylcyclopentane and $NaOCH_2CH_3$. What would be the products of the S_N1 reaction with CH_3CH_2OH?

8.48 **E_2 Elimination:** Write the chair form of 1-bromo-2-methylcyclohexane that, upon E_2 elimination, would yield **(a)** 1-methylcyclohexene; **(b)** 3-methylcyclohexene.

8.49 **Nucleophilic Substitution Reactions:** Draw the isomers of C_4H_9Cl and comment on their propensity toward S_N1 or S_N2 reaction mechanisms.

8.50 **Nucleophilic Substitution Reactions:** Draw the isomer of $C_5H_{11}Br$ that shows the least steric hindrance to an S_N2 reaction and the one that shows the most.

8.51 **Substitution versus Elimination:** Arrange the following isomeric alkyl halides in order of their likelihood of undergoing elimination in favor of nucleophilic substitution: (i) 1-bromopentane, (ii) 2-bromopentane, (iii) 2-bromo-2-methylpentane, and (iv) 2-bromo-2,3-dimethylpentane. Explain.

Activities with Molecular Models

1. Make a model of one of the enantiomers of 2-bromobutane. Make a model of the enantiomer that results from an S_N2 reaction in which the bromine is replaced by an OH. Make sure you have inversion of configuration. Look at the original enantiomer and visualize the OH coming in from the rear and displacing the bromine.

2. Now, using the 2-bromobutane enantiomer from exercise 1, make the models of the racemic mixture formed when the bromine is replaced by OH in an S_N1 reaction. Visualize the Br leaving first and the water attacking from either side of the carbocation to form the pair of enantiomers.

3. Make molecular models of the E_2 reactions described in section 8.5B.2. They may help you in understanding the stereochemistry.

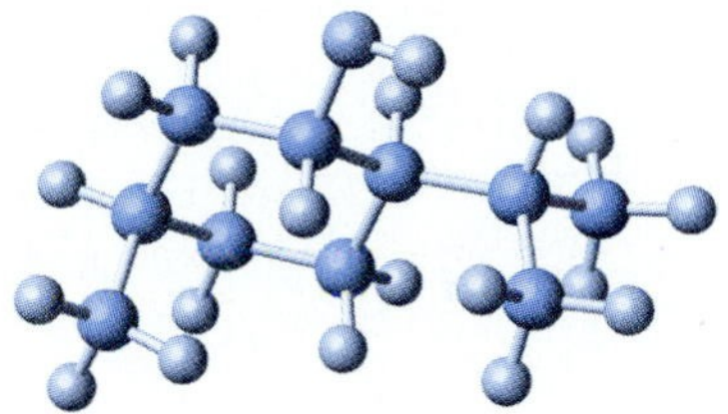

CHAPTER 9

Alcohols, Phenols, and Ethers

The organic compounds presented in this chapter are probably very familiar to you. Beverage alcohol, rubbing alcohol, antifreeze (ethylene glycol), the general anesthetic ether, and a class of medicinal compounds called phenols represent the types of substances we will study.

This chapter also begins a study of the functional groups commonly found in biological molecules. For example, carbohydrates are polyhydroxy aldehydes and ketones, and proteins are "polymers" of amino acids. Fats and oils are triesters of the alcohol glycerol; an ester is a derivative of an alcohol and carboxylic acid. Nucleic acids are composed of multifunctional amine bases and carbohydrate units.

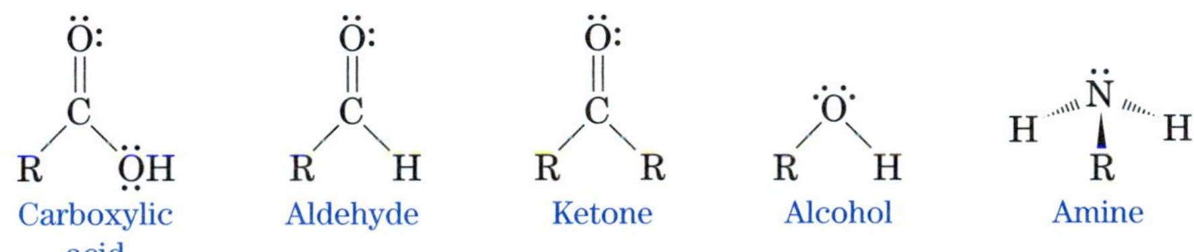

After studying the structure and chemistry of these functional groups and their derivatives, this knowledge will be applied to the major classes of biological molecules: carbohydrates, proteins, fats and oils, and nucleic acids.

9.1 Structure and Nomenclature

Alcohols, phenols, and **ethers** can be thought of as derivatives of water in which one or both hydrogens are replaced with hydrocarbon groups. Replacement of one hydrogen results in an alcohol, and replacement of both gives an ether. In phenols, one hydrogen of water is replaced by an aromatic ring.

alcohol
ROH, alkane in which a hydrogen is replaced with OH

phenol
ArOH, aromatic ring with bonded OH

ether
ROR, oxygen with two organic groups

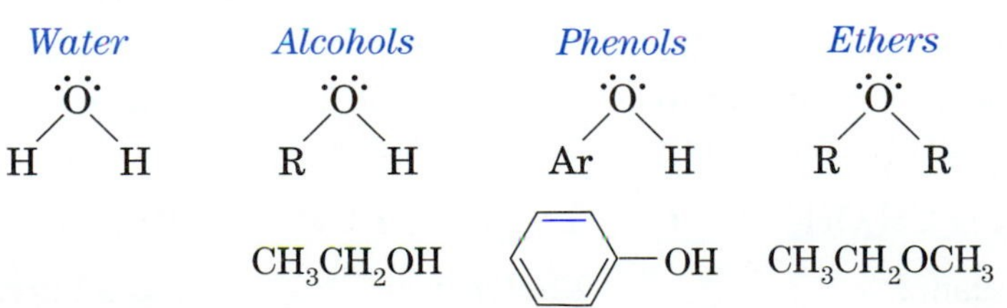

Alcohols are further classified as primary, secondary, and tertiary according to the number of alkyl groups directly bonded to the alcohol carbon.

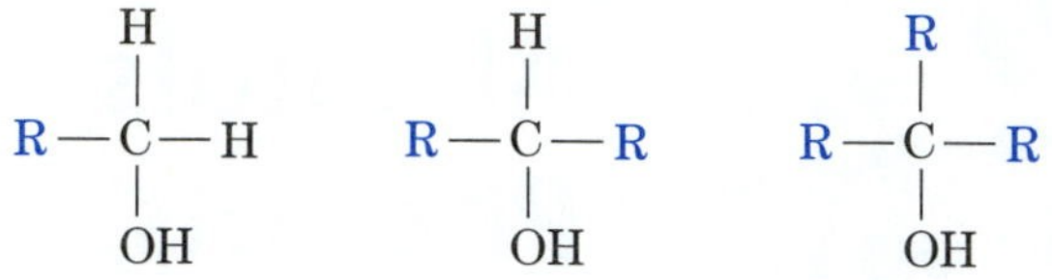

$$\underset{\substack{1° \\ \text{Primary}}}{R-\overset{\overset{H}{|}}{\underset{\underset{OH}{|}}{C}}-H} \qquad \underset{\substack{2° \\ \text{Secondary}}}{R-\overset{\overset{H}{|}}{\underset{\underset{OH}{|}}{C}}-R} \qquad \underset{\substack{3° \\ \text{Tertiary}}}{R-\overset{\overset{R}{|}}{\underset{\underset{OH}{|}}{C}}-R}$$

A. IUPAC Nomenclature of Alcohols

1. ***Saturated Alcohols.*** The names of most simple organic compounds are based on the name of the longest continuous chain of carbon atoms. To name alcohols, the *-e* of the parent hydrocarbon is replaced by *-ol,* as in the following examples:

CH_4	CH_3OH	CH_3CH_3	CH_3CH_2OH
Methane	Methanol	Ethane	Ethanol

When necessary, the position of the alcohol functional group is described by a number. The carbon chain is numbered to give the alcohol group the lowest possible number.

$CH_3CH_2CH_3$	$CH_3CH_2CH_2OH$	$CH_3\underset{\underset{OH}{|}}{C}HCH_3$
Propane	1-propanol	2-propanol

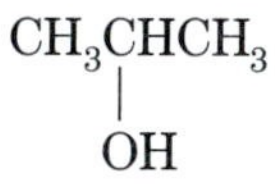

Compounds containing two or more alcohol groups are named as diols, triols, and so on, with a positional number assigned to each OH group.

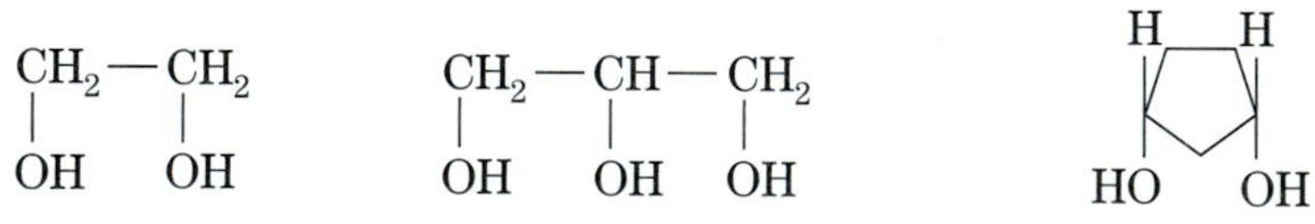

$\underset{\underset{OH}{|}}{CH_2}-\underset{\underset{OH}{|}}{CH_2}$	$\underset{\underset{OH}{|}}{CH_2}-\underset{\underset{OH}{|}}{CH}-\underset{\underset{OH}{|}}{CH_2}$	(cyclopentane ring with H, H and HO, OH on C-1 and C-3)
1,2-ethanediol	1,2,3-propanetriol	*cis*-1,3-cyclopentanediol

GETTING INVOLVED

Problem 9.1

Draw the four alcohols with the formula $C_4H_{10}O$ and label them as 1°, 2°, or 3°. Name the four alcohols.

See related problem 9.31–9.33.

2. ***Unsaturated Alcohols.*** To designate the double or triple bond of unsaturated alcohols, the *-an* of the parent hydrocarbon is changed to *-en* or *-yn,* respectively. The alcohol takes precedence in numbering the chain.

$CH_3CH_2CH_2CH_2OH$	$CH_3CH{=}CHCH_2OH$	$HC{\equiv}CCH_2CH_2OH$
1-Butanol	2-buten-1-ol	3-butyn-1-ol

The method for naming alcohols is as follows:

1. Use the Greek word for the number of carbons in the longest continuous chain (containing the alcohol group).
2. Follow this by the suffix *-an* if the chain is saturated, *-en* if it contains a carbon-carbon double bond, and *-yn* if it contains a carbon-carbon triple bond.
3. Next, add the suffix *-ol* to designate the alcohol function.
4. Number the carbon chain, giving the lowest possible number to the alcohol group (double bonds are next in precedence, followed by triple bonds, and finally by groups named by prefix). Incorporate the numbers indicating the positions of the various functional groups in the name.
5. Complete the name by naming all other groups by prefixes with their numerical positions.

GETTING INVOLVED

Example 9.1

Name the following compound:

$$\overset{6}{CH_2}{=}\overset{5}{C}\overset{4}{CH_2}\overset{3}{CH}\overset{2}{CH}(CH_3)\overset{1}{CH_3}$$

with Br on C5 and OH on C3.

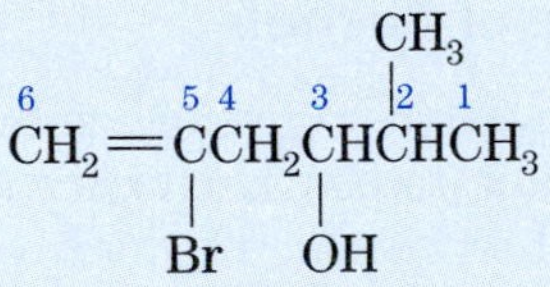

Solution

1. The longest chain is six carbons: hex.
2. There is a carbon-carbon double bond: hexen.
3. The alcohol function is designated by the suffix *-ol:* hexenol.
4. Number the chain to give the alcohol group the lowest possible number. Incorporate these numbers in the name: 5-hexen-3-ol. The first number, 5, refers to the position of the double bond; the second, 3, locates the alcohol group.
5. Name all other substituents (2-methyl and 5-bromo) with prefixes. The complete name is: 5-bromo-2-methyl-5-hexen-3-ol.

Problem 9.2

Name the following compounds by the IUPAC system of nomenclature:

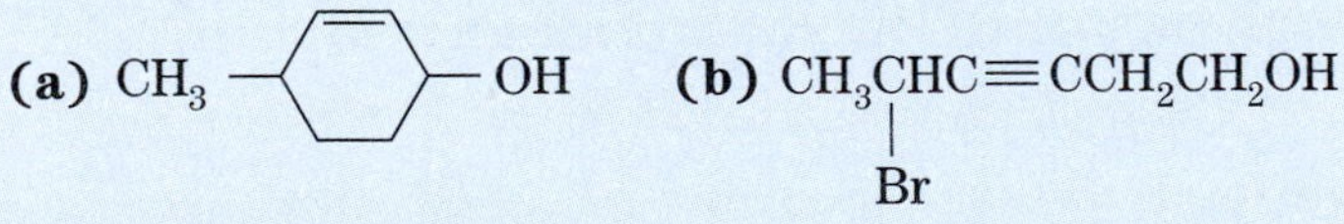

(a) CH_3—(cyclohexene ring)—OH **(b)** $CH_3CHC{\equiv}CCH_2CH_2OH$ (Br on C2)

See related problems 9.31–9.34.

B. IUPAC Nomenclature of Ethers

To name an ether, first find the longest continuous chain of carbon atoms. The substituents attached to this chain can be pictured as alkyl groups containing an oxygen. For this reason, they are referred to as *alkoxy groups.* Just as CH_3— is a methyl group, CH_3O— is a methoxy group.

CH_3CH_2—	Ethyl	CH_3CH_2O—	Ethoxy
$CH_3CH_2CH_2$—	Propyl	$CH_3CH_2CH_2O$—	Propoxy

These groups are named as prefixes and their positions designated by a number.

$CH_3OCH_2CH_2CH_3$ — 1-methoxypropane

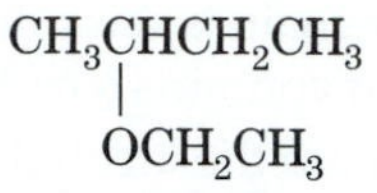

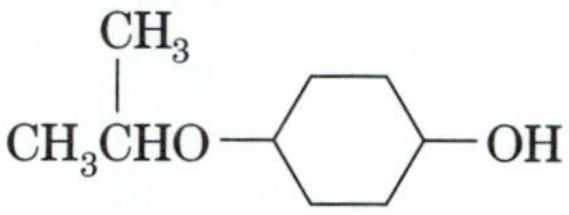

$CH_3CHCH_2CH_3$ with OCH_2CH_3 on C-2 — 2-ethoxybutane

$(CH_3)_2CHO$–cyclohexane–OH — 4-isopropoxy-1-cyclohexanol

GETTING INVOLVED

Problem 9.3

Name the following by the IUPAC system of nomenclature:

(a) $CH_3CH_2CH_2OCH_2(CH_2)_5CH_3$ (b) $(CH_3O)_2CH_2$
(c) $CH_3CH_2OCH_2CH_2OH$

See related problems 9.31, 9.35–9.36.

C. IUPAC Nomenclature of Phenols

Phenols are named according to the rules for a substituted benzene ring, except that the family name is phenol (rather than benzene, section 6.3.B). Numbering of the ring begins at the carbon bearing the hydroxyl. Common names such as phenol have been accepted into the IUPAC system (section 6.3.B).

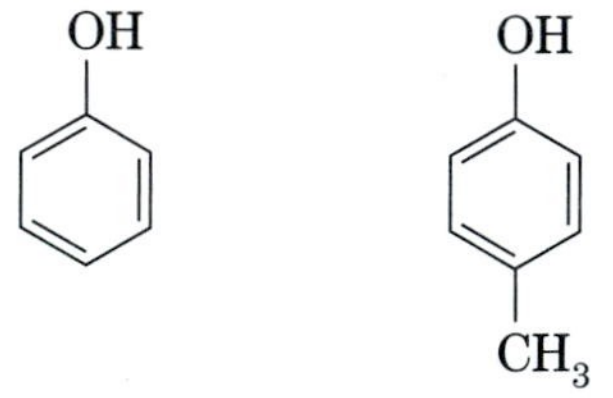

OH / CH3

Phenol 4-methylphenol

GETTING INVOLVED

Problem 9.4

Name the following by the IUPAC system of nomenclature:

(a) 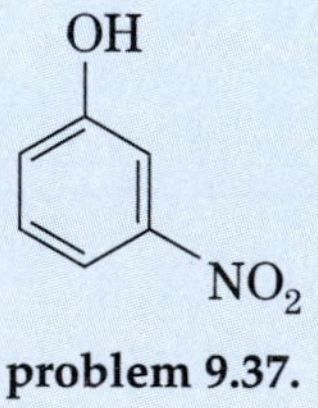 OH / NO2 (b) 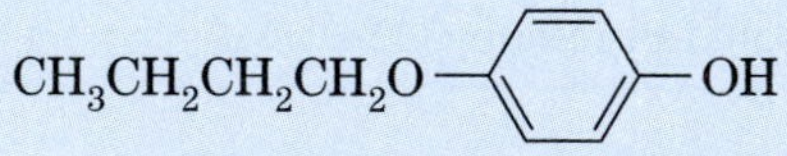 $CH_3CH_2CH_2CH_2O$–benzene–OH

See related problem 9.37.

D. Common Nomenclature of Alcohols and Ethers

Alcohols and ethers are frequently referred to by common names. In such terminology, the alkyl group or groups connected to the oxygen are named first in alphabetical order, followed by the class of compound, alcohol or ether.

CH_3CH_2OH	$CH_3CH(OH)CH_3$	$CH_3CH_2OCH_3$	$(CH_3)_2CHOCH(CH_3)_2$
Ethyl alcohol	Isopropyl alcohol	Ethyl methyl ether	Diisopropyl ether

GETTING INVOLVED

Problem 9.5

Draw the following compounds: **(a)** tertiary butyl alcohol, **(b)** pentyl alcohol, **(c)** diethyl ether, **(d)** ethyl cyclopentyl ether.

See related problem 9.38.

9.2 Physical Properties—Hydrogen-Bonding

Like other classes of organic compounds, the melting points and boiling points of alcohols and ethers generally increase with increasing molecular weight within a homologous series; this is illustrated in Table 9.1.

However, alcohols exhibit unusually high boiling points as illustrated by comparing alcohols and alkanes with approximately the same molecular weights (for example, methanol and ethane; ethanol and propane below).

	H_2O	CH_4	CH_3OH	CH_3CH_3	CH_3CH_2OH	$CH_3CH_2CH_3$
mol wt	18	16	32	30	46	44
bp °C	100	−164	65	−89	78.5	−42

Why is water's boiling point 264°C higher than methane's, methanol's 154°C higher than ethane's, and ethanol's 120°C higher than propane's? Methanol, in fact, has almost the same boiling point as hexane (mol wt = 86, bp = 69°C) even though hexane's molecular weight is 2.7 times greater. The answer is hydrogen bonding.

Table 9.1 Physical Properties of Alcohols, Phenols, and Ethers

Compound	Molecular Weight	Melting Point, °C	Boiling Point, °C
Alcohols			
CH_3OH	32	−94	65
CH_3CH_2OH	46	−117	78.5
$CH_3CH_2CH_2OH$	60	−127	97
$CH_3(CH_2)_3CH_2OH$	88	−79	137
$CH_3(CH_2)_5CH_2OH$	116	−34	176
Phenols			
C_6H_5OH (phenol)	94	43	182
Ethers			
CH_3OCH_3	46	−139	−23
$CH_3OCH_2CH_3$	60	—	11
$CH_3O(CH_2)_2CH_3$	74	—	39

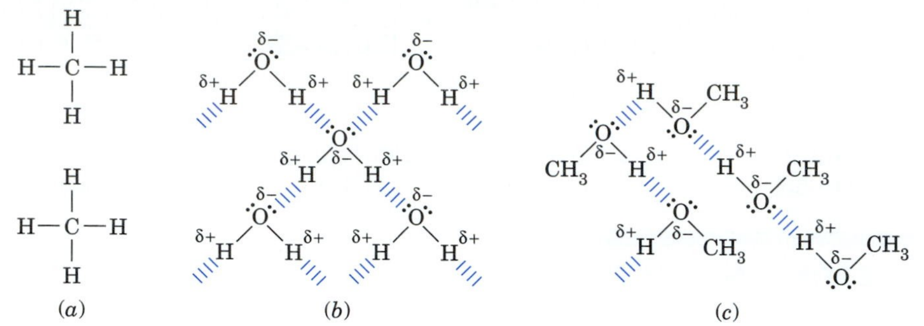

Figure 9.1 (a) Methane is a nonpolar compound with only weak intermolecular attractions. Consequently, it has a very low boiling point. (b) Water has strong attractions between molecules owing to its capacity to hydrogen-bond and thus has a relatively high boiling point. (c) Methanol and other alcohols can hydrogen-bond much like water and as a result have relatively high boiling points.

To convert a liquid to a gas, the forces of attraction between molecules in the liquid must be overcome. These intermolecular attractions are weak in alkanes, since alkanes are nonpolar molecules. However, the oxygen-hydrogen bond of alcohols is quite polar, and as a result dipole attractions exist between molecules, specifically between the hydrogen of one molecule and the oxygen (and its nonbonding electrons) of another. Furthermore, due to the minute size of hydrogen, close intermolecular association is possible, providing maximal attractions (see Figure 9.1). For these reasons, water, methanol, and ethanol have much higher boiling points than the alkanes of similar molecular weight. This phenomenon is called **hydrogen-bonding** and occurs in molecules where hydrogen is bonded to the strongly electronegative elements: nitrogen, oxygen, or fluorine.

hydrogen-bonding intermolecular attractions caused by hydrogen bonded to an electronegative element (O, N, F) being attracted to a nonbonding electron pair of another electronegative element

Since ethers have no oxygen-hydrogen bonds, hydrogen-bonding does not occur, resulting in considerably lower boiling points than isomeric alcohols of identical molecular weights (this is illustrated with the following pairs of compounds).

	CH_3CH_2OH	CH_3OCH_3	$CH_3CH_2CH_2OH$	$CH_3OCH_2CH_3$
mol wt	46	46	60	60
bp °C	78.5	−23	97	11

The strong intermolecular associations resulting from hydrogen-bonding can influence the viscosity (thickness) of a liquid as shown in the comparison of hexane—a gasoline component—and glycerol, a thick, syrupy liquid often used as a lubricant in laboratories (note that glycerol has three hydrogen bonding sites).

Hexane
mol wt = 86
bp = 69°C
$CH_3CH_2CH_2CH_2CH_2CH_3$

Glycerol
mol wt = 92
bp = 290°C
$CH_2CH\,CH_2$ (each carbon bearing OH: OH OH OH)

Since alcohols can hydrogen-bond with water (Figure 9.2), low-molecular-weight alcohols are water soluble. However, as the molecular weight of an alcohol increases, the proportion of it that is hydrocarbon increases. The alcohol becomes more like an alkane, less like water, and less soluble in water. Methanol, ethanol, and propanol are water soluble in all proportions, but solubility drops off significantly with butanol. Pentanol and hexanol are only slightly soluble, and heptanol and octanol are essentially insoluble in water.

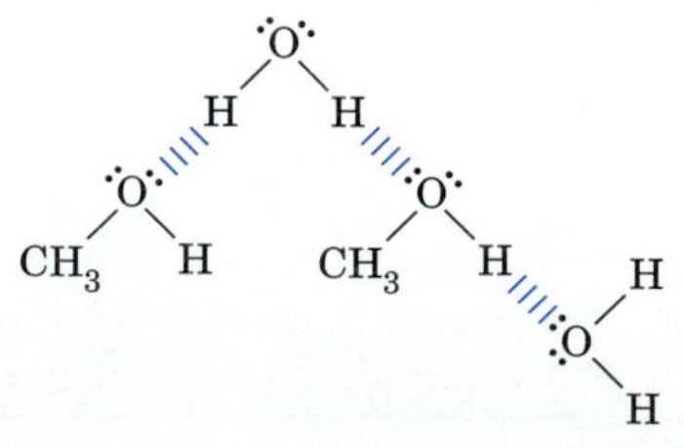

Figure 9.2 Solubility of methanol in water. Note the hydrogen-bonding between methanol and water.

Connections 9.1

www.prenhall.com/bailey

Methyl, Ethyl, and Isopropyl Alcohols

The three alcohols most commonly encountered in daily life are methanol, ethanol, and 2-propanol. All are precursors to other chemicals, have varied uses, and are produced in large quantities.

Methyl Alcohol, CH_3OH

Methyl alcohol or methanol, CH_3OH, is sometimes called wood alcohol; it was formerly produced by the destructive distillation of wood (in the absence of air to prevent ignition). In fact, the etymology of its name can be traced to this process. In Greek *methe* means wine and *hyle* means wood; methyl alcohol was the "wine of wood." Until the early part of the twentieth century, destructive distillation of wood was the source of methanol. Now, however, it is produced on a large scale by the reduction of carbon monoxide with hydrogen. Industrially, methanol is converted into formaldehyde or used to synthesize other chemicals. It is used as a solvent and as a clean-burning fuel, especially for racing cars.

Unlike beverage alcohol, methanol is toxic when ingested in small quantities. Blindness is a symptom of methanol poisoning, since it damages the optic nerve; death can also result.

Ethyl Alcohol, CH_3CH_2OH

Because of the way it is metabolized in the body, small amounts of ethanol may be ingested with little toxicity by most adults. Ethyl alcohol is commonly referred to as grain alcohol or beverage alcohol, because it can be produced by the fermentation of natural sugars and hydrolyzed starches found in grapes and grains. The fermentation process is used to produce alcoholic beverages, which can be divided into three categories—beers, wines, and spirits.

$$\underset{\text{Sucrose}}{C_{12}H_{22}O_{11}} + H_2O \xrightarrow{\text{Sucrase}} \underset{\text{Glucose}}{2C_6H_{12}O_6}$$

$$\underset{\text{Glucose}}{C_6H_{12}O_6} \xrightarrow[\text{Enzymes}]{\text{Yeast}} \underset{\text{Ethanol}}{2CH_3CH_2OH} + 2CO_2$$

> **Fermentation** is commonly thought of as the process for the natural production of alcohol. Yet fermentation is a much more universal process, encompassing a metabolic change caused by a living microorganism acting on organic materials. It is one of the oldest chemical processes used by humans. In addition to making alcoholic beverages, fermentation is responsible for the aging of meat and cheese and also the production of bread, foods (such as sauerkraut), animal feeds, drugs, antibiotics, hormones, and other materials. In 1857, Louis Pasteur proved that alcoholic fermentation is caused by living cells (yeast), and the ancient art of fermentation graduated from the realm of magic to the world of scientific understanding.

Beer is the fermentation product of barley and hops. Although most beers are 3.5%–5% alcohol, the alcohol content can vary from 2% to 12%.

Wines are fermented from the juice of grapes. Natural table wines and sparkling wines like champagne (highly carbonated) contain less than 14% alcohol. Fermentation ceases at this concentration because yeast cells die or stop reproducing due to the antiseptic action of alcohol. Wines such as sherry and aromatic wines such as vermouth are fortified with additional alcohol up to about 15% to 23%.

The distillation of fermented carbohydrate mashes produces spirits usually having 40%–50% alcohol. Congeners are also collected in the process, giving the beverages their characteristic flavors. Popular spirits and their predominant sources are whiskey, from corn and barley; rye whiskey, from rye grain, rum, from sugarcane or molasses; and gin and vodka, from grains.

The alcohol content of distilled spirits is expressed in terms of "proof," which originated from an old method of testing whiskey by pouring it on gunpowder. If the gunpowder could still be ignited, this was "proof" that the beverage did not contain too much water. Alcohol concentration in terms of proof is double the percentage of alcohol by volume; for example, a 100-proof vodka is 50% alcohol, 50% water by volume.

The production of fine alcoholic beverages is a time-honored art and science. However, inappropriate use of these beverages is a major health and safety problem in the United States. In most states, it is illegal to drive with a blood alcohol concentration of more than 0.08% to 0.10%. Individuals with 0.3% are visibly intoxicated; those with 0.4% are anesthetized and incapable of voluntary action; and a concentration of 0.5%–1% can lead to coma and death. Intoxication is the major factor in more than half of fatal traffic accidents, and alcoholism costs the country billions of dollars annually in lost productivity. Alcoholics have lowered life expectancy by 10 to 15 years due to liver degeneration and cardiovascular disease, especially if they smoke. Since it is an excellent organic solvent, ethanol readily crosses the blood-brain barrier and the placental membrane, endangering fetuses of pregnant women. Symptoms of fetal alcohol syndrome (FAS) include flattened facial features, smaller than normal brain size, learning disabilities, and retarded physical development. How much, if any, alcohol can be safely consumed by a pregnant woman without running the risk of having an FAS child is strongly debated.

Ethanol has many other important applications, including use as a solvent (vanilla and other extracts in your home are often ethanol solutions) and an antiseptic (mouthwashes are often 5%–30% alcohol). Ethyl alcohol produced for uses other than human consumption is denatured with methyl

Connections 9.1 *(cont.)*

and isopropyl alcohols and is not subject to beverages taxes. For commercial purposes, it is usually produced by the hydration of ethene.

Isopropyl Alcohol, $CH_3CHOHCH_3$

Isopropyl alcohol, the common rubbing alcohol sold in drugstores, is an even more effective antiseptic than ethyl alcohol. Isopropyl alcohol is oxidized industrially to produce acetone, an important solvent (and component of fingernail polish remover).

GETTING INVOLVED

✓ Why do alcohols have higher boiling points than alkanes of similar molecular weight. What is hydrogen bonding? What structural features are necessary for it to occur?

Problem 9.6

Explain the differences in boiling points between butane (mol wt = 58, bp = −0.5°C), 1-propanol (mol wt = 60, bp = 97°C), and 1,2-ethanediol (mol wt = 62, bp = 198°C).

Problem 9.7

Of the compounds shown below, the first is rose oil, the second is an isomer of the first, and the third is a gasoline component. Assign the boiling points 136°, 171°, and 221°C to the correct compounds. Explain.

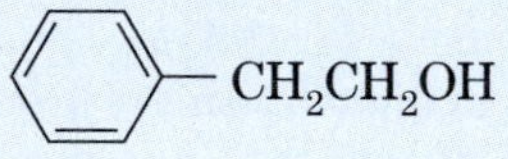

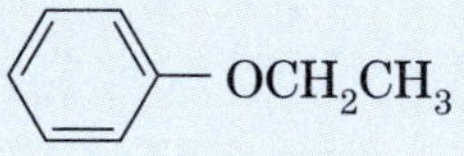

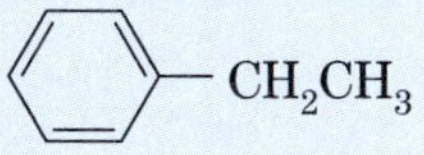

$C_6H_5-CH_2CH_2OH$ $\quad$ $C_6H_5-OCH_2CH_3$ $\quad$ $C_6H_5-CH_2CH_3$

Problem 9.8

Ethylene glycol ($HOCH_2CH_2OH$) is a good antifreeze because it has a high boiling point and is soluble in water in all proportions. These properties are due to hydrogen-bonding. Draw illustrations of ethylene glycol hydrogen-bonding with itself and with water in solution.

Problem 9.9

Which has the higher boiling point, butanoic acid (odor of rancid butter) or the isomer ethyl acetate (solvent in fingernail polish remover)? Explain.

$CH_3CH_2CH_2\overset{O}{\overset{\|}{C}}OH$ $\quad$ $CH_3\overset{O}{\overset{\|}{C}}OCH_2CH_3$

See related problems 9.40–9.43.

9.3 Uses of Alcohols, Ethers, and Phenols

A. Alcohols

Methyl, ethyl, and isopropyl alcohols are the most common of the simple alcohols. All are industrial chemicals and precursors for other industrial products. They are discussed in Connections 9.1.

CH_3OH	CH_3CH_2OH	$CH_3CH(OH)CH_3$
Methyl alcohol	Ethyl alcohol	Isopropyl alcohol

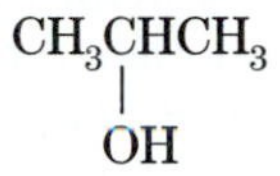

B. Polyhydric Alcohols

polyhydric alcohol alcohol with more than one hydroxy group

Polyhydric alcohols are alcohols with more than one hydroxy group per molecule. Two of the most important examples are ethylene glycol and glycerol.

$CH_2(OH)CH_2(OH)$	$CH_2(OH)CH(OH)CH_2(OH)$
Ethylene glycol (1,2-ethanediol)	Glycerol (1,2,3-propanetriol)

1. ***Ethylene Glycol.*** The principal commercial use of ethylene glycol is as an antifreeze in automobile radiators. Its unique properties make it especially suitable for this purpose. (See Problem 9.8). It has a high boiling point (198°C) and will not readily boil out of a hot radiator. It is soluble in water in all proportions, and it is noncorrosive. Other applications of ethylene glycol include its use as a hydraulic brake fluid and in the production of such polymers as Dacron.

2. ***Glycerol.*** Glycerol is a sweet, syrupy liquid obtained as a by-product of soap manufacture and through synthesis from propene. It is used commercially as a humectant to preserve moistness in tobacco, cosmetics, and the like. A *humectant* is an agent that attracts and retains moisture. Glycerol is particularly effective because of its capacity to hydrogen-bond with water. Another important application of glycerol occurs in the manufacture of polymers.

Glycerol can be converted into nitroglycerin by treatment with concentrated nitric and sulfuric acids (an example of inorganic ester preparation; see section 9.6.C).

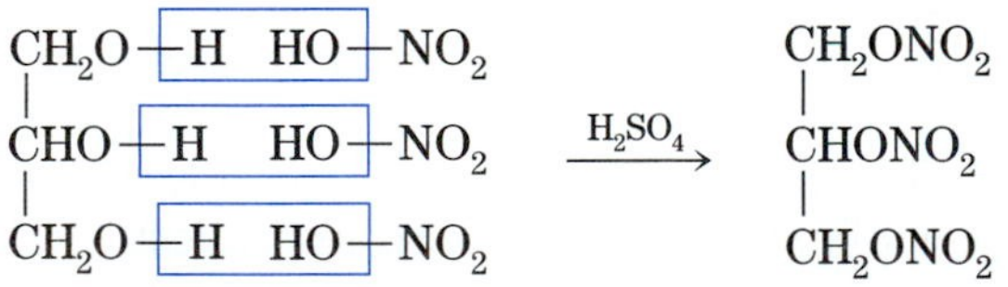

$$\begin{array}{l} CH_2O\text{-}H \;\; HO\text{-}NO_2 \\ | \\ CHO\text{-}H \;\; HO\text{-}NO_2 \\ | \\ CH_2O\text{-}H \;\; HO\text{-}NO_2 \end{array} \xrightarrow{H_2SO_4} \begin{array}{l} CH_2ONO_2 \\ | \\ CHONO_2 \\ | \\ CH_2ONO_2 \end{array}$$

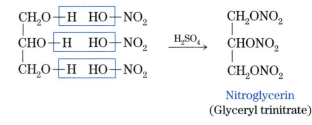

Nitroglycerin
(Glyceryl trinitrate)

The powerfully explosive character of nitroglycerin arises from its rapid conversion, sometimes merely on minor shock, from a liquid occupying a relatively small volume to a large volume of hot expanding gases. Four moles of nitroglycerin, occupying a volume of just over half a liter, decompose to 29 moles of hot gases (1 mole of a gas occupies 22.4 liters at STP) that probably expand to at least 10,000–20,000 times the original volume.

$$4\begin{array}{l}\text{CH}_2\text{ONO}_2\\ |\\ \text{CHONO}_2\\ |\\ \text{CH}_2\text{ONO}_2\end{array} \longrightarrow 6N_2 + O_2 + 12CO_2 + 10H_2O$$

Nobel Prize medal

Although nitroglycerin is a very powerful explosive, its shock sensitivity makes it extremely dangerous to use. In 1866, Alfred Nobel discovered that this undesirable property could be mitigated by mixing nitroglycerin with diatomaceous earth and sawdust. The resulting material, dynamite, made Nobel a very wealthy man. In his will, he specified that the income from the investment of his fortune be applied to the establishment of cash prizes in various disciplines. Nobel Prizes are awarded in physics, chemistry, physiology or medicine, literature, economics, and peace.

Many nitro compounds and nitrates are used as explosives. Most dynamites today are a mixture of ammonium nitrate and fuel oil. Nitroglycerin is also used medicinally for people with heart trouble to dilate blood vessels and arteries. Many nitrate and nitrite derivatives have vasodilatory properties.

C. Diethyl Ether, $CH_3CH_2OCH_2CH_3$

Diethyl ether, used since 1846 as a general anesthetic (a drug that acts on the brain and produces unconsciousness and insensitivity to pain), has been replaced by other anesthetics. Ethers play an important role as solvents for organic preparations and for the extraction of naturally occurring compounds.

D. Phenols

1. ***Medicinal Applications.*** In terms of its medicinal use, phenol has four properties worth noting:

1. ability to act as antiseptic, disinfectant
2. ability to act as a local anesthetic
3. skin irritancy
4. toxicity when ingested

OH

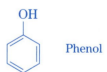

Phenol

Because of possible skin irritation and toxicity, phenol is found only in very small quantities in over-the-counter medications.

However, many related structures are much more effective for certain uses than is phenol itself. Because of their antiseptic and anesthetic activities, phenols are found in a variety of commercial products including soaps, deodorants, disinfectant sprays and ointments, first aid sprays, gargles, lozenges, and muscle rubs. Note that phenol units in the following examples:

HO OH

$(CH_2)_5CH_3$

Hexylresorcinol
Lozenges

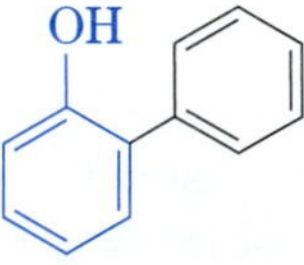

Ortho-phenylphenol
Disinfectant home sprays, gargles

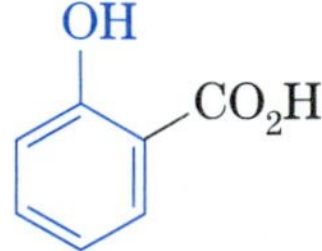

Salicylic acid
First aid disinfectants

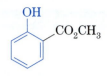

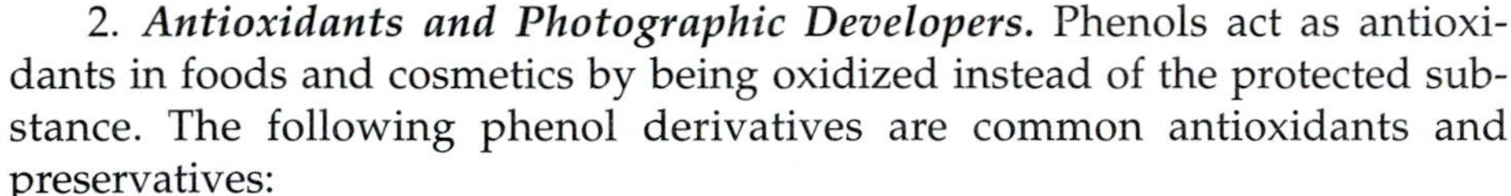

Methyl salicylate, oil of wintergreen
Flavoring, muscle soothers, mouthwashes, gargles

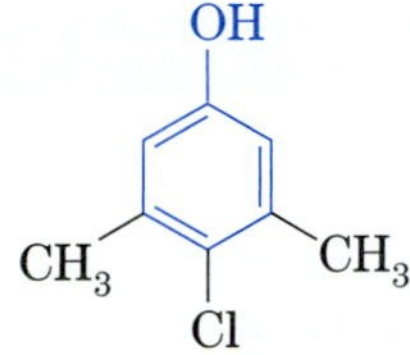

Parachlorometaxylenol, or chloroxylenol
Sunburn sprays; combats athlete's foot

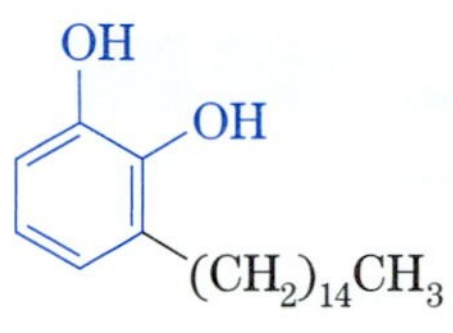

Pentadecylcatechol
Irritant in poison ivy and poison oak

Poison oak

2. ***Antioxidants and Photographic Developers.*** Phenols act as antioxidants in foods and cosmetics by being oxidized instead of the protected substance. The following phenol derivatives are common antioxidants and preservatives:

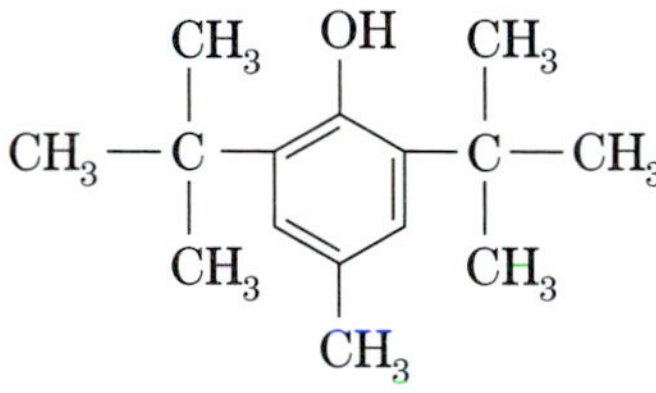

Butylated hydroxy toluene (BHT)

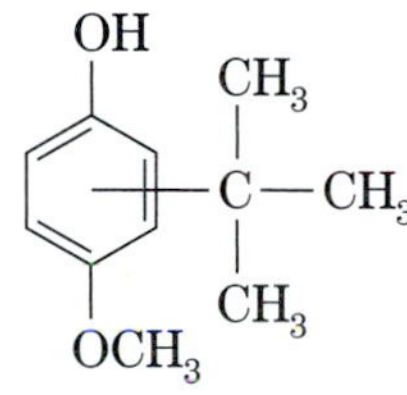

Butylated hydroxy anisole (BHA) (two isomers)

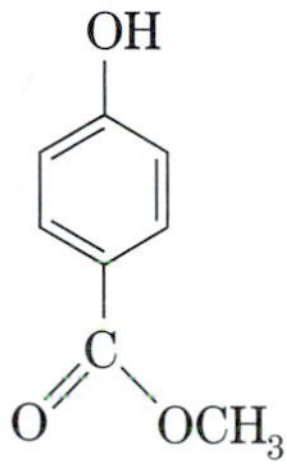

Methyl paraben

Positive image

Because of their ease of oxidation, the following phenols are good black-and-white photographic developers. During the development, the phenol is oxidized and the silver ion (AgBr dispersed in a gel) is reduced to metallic silver.

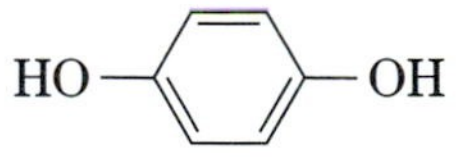

Hydroquinone

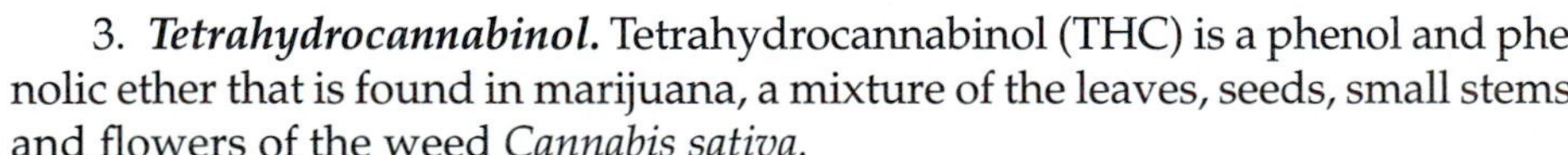

p-methylaminophenol

Negative image

3. ***Tetrahydrocannabinol.*** Tetrahydrocannabinol (THC) is a phenol and phenolic ether that is found in marijuana, a mixture of the leaves, seeds, small stems, and flowers of the weed *Cannabis sativa.*

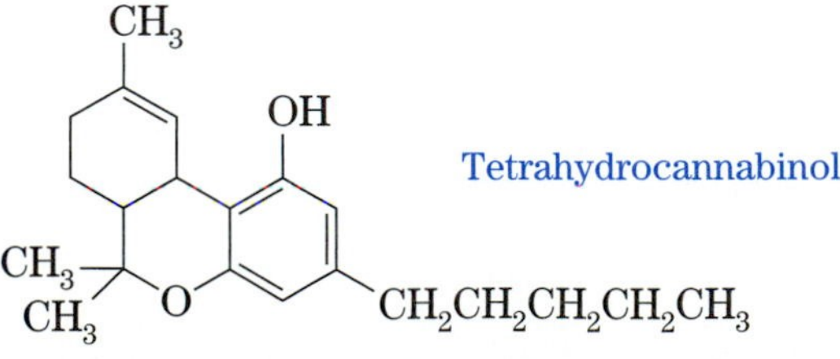

Tetrahydrocannabinol

Marijuana

Connections 9.2 www.prenhall.com/bailey

Neurotransmitters—The Heart of the Matter

Among the many biologically relevant phenol derivatives, the catechol neurotransmitters are some of the most valuable and interesting.

Norepinephrine: HO, HO–C_6H_3–$CHCH_2NH_2$ (with OH on the CH)

Dopamine: HO, HO–C_6H_3–$CH_2CH_2NH_2$

Catechol: HO, HO–C_6H_4

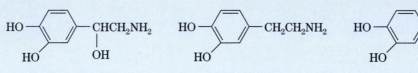

The nervous system runs on a series of physical and chemical reactions. Signals are carried from one nerve cell to another by simple chemical molecules known as *neurotransmitters.* Epinephrine (adrenalin), norepinephrine, dopamine, and acetylcholine are but four of the more than 20 known neurotransmitters. The first three substances are also called catecholamines because they are similar to catechol, or *o*-hydroxyphenol.

To illustrate the extent of current knowledge about the nervous system and the application of this knowledge, let us focus on one portion of the network located outside of the brain—the sympathetic nervous system (SNS). The SNS and the parasympathetic nervous system (PNS) stimulate almost every organ in the body in a complementary fashion. The PNS supplies the stimulation for normal physiological functions, while the SNS provides the necessary arousal for survival in the "cold, cruel world." The PNS is responsible for contraction of the pupils of the eyes, normal pulse and blood pressure, constriction of the bronchi, digestive enzyme-containing secretions in the mouth, and increased gastrointestinal activity. The SNS, on the other hand, in an effort to make the body alert and ready to respond to any outside threat, causes dilation of the pupils ("the better to see you with . . ."), increased pulse and blood pressure, and relaxation of the bronchi (oxygen is delivered to all tissues of the body to increase fuel burning), dry mouth, and decreased gastrointestinal motility.

The prime neurotransmitter in the SNS is norepinephrine. It is synthesized in an SNS nerve cell and, in response to a nerve impulse, is secreted into the space between two nerve cells, called the *synapse.* The neurotransmitter travels to the other side of the synapse and combines with a protein known as a *receptor* on the surface of the next nerve cell. This triggers the nerve impulse in that cell.

Molecules that are similar in structure to a natural neurotransmitter can either stimulate a nerve cell just like the natural chemical (in which case they are called *agonists*) or bind to the receptor without stimulation and block the access of the normal neurotransmitter (*antagonists*).

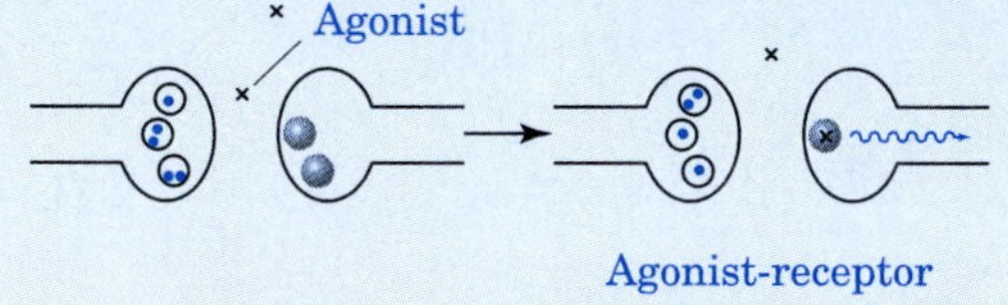

Agonist-receptor complex stimulates nerve transmission

Antagonist

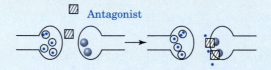

Antagonist-receptor complex causes no nerve stimulation and blocks normal stimulation

Amphetamines and decongestants are examples of SNS agonists. Although these drugs may be used for certain specific effects, such as in dieting and for nasal congestion, respectively, it is important to remember that they are similar to the natural neurotransmitters that generally affect the SNS and also the central nervous system, which is far

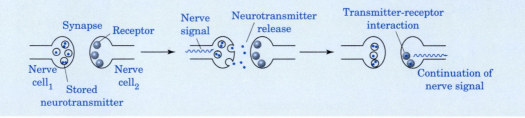

Synapse
Receptor
Nerve cell$_1$
Stored neurotransmitter
Nerve cell$_2$
Nerve signal
Neurotransmitter release
Transmitter-receptor interaction
Continuation of nerve signal

Connections 9.2 *(cont.)*

too complex to discuss in this space. Thus the warnings on the containers for many over-the-counter medications should be heeded, especially if the consumer has a preexisting condition such as high blood pressure, diabetes, or glaucoma.

Another aspect of drug use involves the design of neurotransmitter antagonists, as in the treatment of heart disease. Many types of neurotransmitter receptors exist, some of which are concentrated in specific tissues, such as heart tissue (β_1-receptors) and bronchial tissue (β_2-receptors). β_1-blockers have been designed to antagonize the nerve signals to the heart without having an effect of equal intensity on breathing.

Conversely, a drug that is specific for β_2-receptors could be used as an agonist to relieve asthma without worsening an existing cardiac condition.

The treatment of disease has reached a molecular level, leading to more potent and specific drugs with the possibility of living a longer and more enjoyable life.

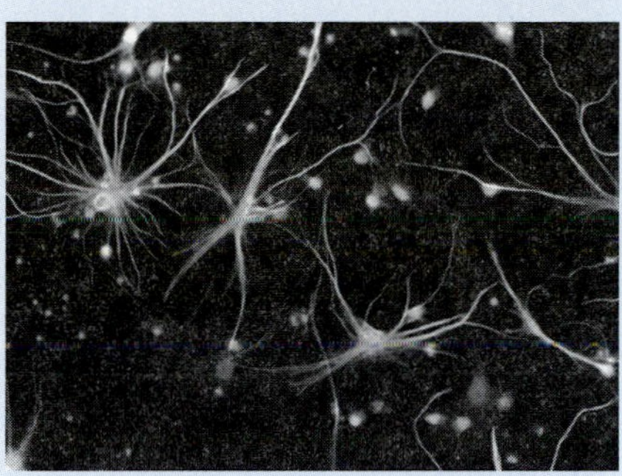

Brain section with fluorescent markers for neurotransmitters

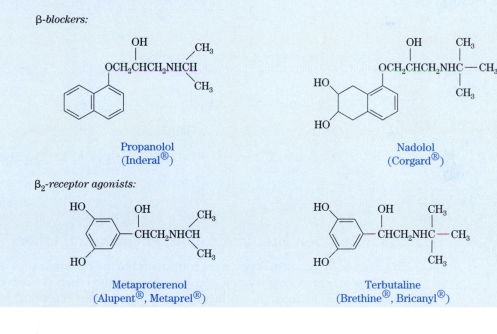

β-blockers:

OH, $OCH_2CHCH_2NHCH(CH_3)_2$ on naphthalene

Propanolol
(Inderal®)

HO, HO, OH, $OCH_2CHCH_2NHC(CH_3)_3$

Nadolol
(Corgard®)

β_2-receptor agonists:

HO, HO, OH, $CHCH_2NHCH(CH_3)_2$

Metaproterenol
(Alupent®, Metaprel®)

HO, HO, OH, $CHCH_2NHC(CH_3)_3$

Terbutaline
(Brethine®, Bricanyl®)

9.4 Preparation of Alcohols and Ethers

We have seen the preparation of alcohols by hydration of alkenes and nucleophilic substitution, and in Chapter 12 we will learn the Grignard synthesis of alcohols. Following is a summary of some common methods for synthesizing alcohols and ethers.

A. Hydration of Alkenes (sections 5.1.A.3, 5.1.B.3)

$$\rangle C{=}C\langle + H_2O \xrightarrow{H^+} -\underset{OH}{\overset{|}{C}}-\underset{H}{\overset{|}{C}}-$$

B. Nucleophilic Substitution (section 8.4)

$$-\overset{|}{\underset{|}{C}}-X + NaNu \longrightarrow -\overset{|}{\underset{|}{C}}-Nu + NaX$$

X = Cl, Br, I

Nu = OH for alcohol synthesis
= OR for ether synthesis (Williamson synthesis)

C. Reduction of Aldehydes and Ketones

1. *Catalytic Hydrogenation.* (section 11.5.D)

$$>C=O + H_2 \xrightarrow{Ni} -\overset{OH}{\underset{|}{\overset{|}{C}}}-H$$

2. *Reduction by Lithium Aluminum Hydride.* (section 11.5.E)

$$>C=O \xrightarrow{LiAlH_4} \xrightarrow[H^+]{H_2O,} -\overset{OH}{\underset{|}{\overset{|}{C}}}-H$$

3. *Grignard Synthesis of Alcohols.* (section 11.5.F)

$$R-X + Mg \xrightarrow{Ether} RMgX \quad \text{(Grignard reagent)}$$

$$>C=O + RMgX \longrightarrow -\overset{OMgX}{\underset{|}{\overset{|}{C}}}-R \xrightarrow[H^+]{H_2O,} -\overset{OH}{\underset{|}{\overset{|}{C}}}-R + MgXOH$$

9.5 Reaction Sites in Alcohols, Phenols, and Ethers

In Chapter 4 (section 4.2) you were introduced to the common sites of chemical reactions in organic compounds. Alcohol, phenols, and ethers possess two important structural features that influence their characteristic reactions: polar bonds and nonbonding electron pairs (Lewis base sites).

Each of these functional groups has two polar bonds. Alcohols and phenols have both a carbon-oxygen and oxygen-hydrogen polar bond, and ethers have two carbon-oxygen polar bonds. The oxygen takes on a partially negative charge in each of these bonds. In addition, the oxygen of these functional groups has two lone pairs of electrons, which further increase the electron availability at this site.

Reaction sites { *Polar bonds:* / *Unshared pairs of electrons:* }

$$\overset{\delta+}{-\overset{|}{\underset{|}{C}}}-\overset{\delta-}{\ddot{\underset{..}{O}}}-\overset{\delta+}{H}$$

Alcohols

$$\text{C}_6\text{H}_5\overset{\delta+}{-}\overset{\delta-}{\ddot{\underset{..}{O}}}-\overset{\delta+}{H}$$

Phenols

$$\overset{\delta+}{-\overset{|}{\underset{|}{C}}}-\overset{\delta-}{\ddot{\underset{..}{O}}}-\overset{\delta+}{\overset{|}{\underset{|}{C}}-}$$

Ethers

As would be expected, the reactions of these compounds occur at the polar bonds. Because of the charge separation and electron density on the oxygen, reagents are especially attracted to these bonds. In the following sections we will see reactions in which the O—H or C—O bond (or both) can be the reaction sites.

The unshared pairs of electrons on alcohols and ethers make these compounds Lewis bases (section 4.2.C). Just as water reacts with acids to form the hydronium ion (H_3O^+), alcohols and ethers form similar positive ions, called **oxonium ions,** in solutions of strong acids. In these reactions, the positive hydrogen ion, a Lewis acid, bonds to one of the lone pair of electrons on the oxygen. The oxygen becomes positive, since one of its lone pair of electrons is now being shared.

oxonium ion
ion formed by the bonding of a hydrogen ion to the oxygen of an alcohol or ether

Water: $H—\ddot{O}—H + H^+ \longrightarrow H—\overset{+}{O}(H)—H$

Alcohols: $R—\ddot{O}—H + H^+ \longrightarrow R—\overset{+}{O}(H)—H$

Ethers: $R—\ddot{O}—R + H^+ \longrightarrow R—\overset{+}{O}(H)—R$

The Lewis base character imparted by the lone pairs of electrons on oxygen is important in many of the reactions. The aromatic ring of phenol is responsible for another set of reactions for this class of compounds, electrophilic aromatic substitution, which we studied in Chapter 6.

GETTING INVOLVED

✓ What two types of reaction sites are common to alcohols, ethers, and phenols?
✓ What is an oxonium ion and how is one formed from alcohols or ethers?

Problem 9.10

Identify the possible reaction sites in the alcohol $H_2C{=}CHCH_2OH$, and write the product of the Lewis acid-base reaction of the alcohol with a hydrogen ion.

9.6 Reactions Involving the O — H Bond of Alcohols and Phenols

A. Relative Acidities of Alcohols and Phenols

1. *Acidity Constants:* By the **Bronsted-Lowry** definition, acids are hydrogen ion donors in chemical reactions and bases are hydrogen ion acceptors. Strengths of these acids are usually compared by measuring their degree of ionization in

Bronsted-Lowry acid
acid that is a hydrogen ion donor in a chemical reaction

strong acid
acid that is 100% ionized in water solution

water. **Strong acids,** such as HCl and HNO_3, are essentially 100% ionized in water.

$$HCl + H_2O \longrightarrow H_3O^+ + Cl^-$$ Strong acid; 100% ionized; no equilibrium

weak acid
acid that is only partially ionized in water solution

acidity constant
K_a, product of the concentrations of the ionized species of an acid divided by the concentration of the un-ionized form

However, most organic acids are **weak acids** and show only slight ionization in water. Because of this, a chemical equilibrium is established between the un-ionized and ionized forms. The equilibrium is described by an equilibrium expression and constant called an **acidity constant,** K_a.

$$HA + H_2O \rightleftharpoons H_3O^+ + A^-$$ Weak acid; partially ionized; equilibrium established

Un-ionized form   Ionized form

$$K_a = \frac{[H_3O^+][A^-]}{[HA]}$$ Acidity constant (brackets are concentrations in moles per liter)

Like any equilibrium constant, the acidity constant is defined as the product of the concentrations of the products (in moles per liter) divided by the concentrations of the reactants. Since water is usually present in large excess, its concentration remains essentially constant. Therefore, it is not included per se in the expression; as a constant, it becomes part of K_a.

The numerical value of the acidity constant describes the relative strengths of acids; the greater the degree of ionization in water, the stronger the acid. Solutions of highly ionized acids have greater concentrations of H_3O^+ and A^- (the numerator in the K_a expression) and lesser concentrations of the un-ionized form, HA (the denominator). Consequently, stronger acids will have numerically greater acidity constants.

***pK*$_a$**
negative logarithm of the acidity constant, K_a

Acid strengths also are often expressed by **pK_a,** which is defined as the negative logarithm of K_a.

$$pK_a = -\log K_a$$

pK_a is defined as the negative logarithm to allow the convenience of expressing most acidities with positive numbers. Because of this definition, however, numerically smaller pK_a's signify stronger acids and larger pK_a's, weaker acids. This is just the opposite of acidity constants; the relationship is illustrated in Table 9.2, which shows the relative acidities of several functional groups.

Typical acidity constants for phenols, carboxylic acids, and sulfonic acids—the most acidic of organic compounds—are given in Table 9.2, along with those of other classes of compounds for comparison purposes. Phenols have acidity constants of about 10^{-10} (this amounts to approximately 0.003% ionization for a 0.1 *M* solution), whereas carboxylic acids have constants around 10^{-5} (about 1% ionization in a 0.1 *M* solution). The difference between these acidity constants is five powers of ten; carboxylic acids are thus 100,000 times more acidic than phenols. Sulfonic acids have a structure similar to that of sulfuric acids and have acidities similar to those of strong mineral acids.

Table 9.2 Acidity Constants of Some Functional Groups

	Functional Group	Typical K_a	Typical pK_a	Example	K_a	pK_a
Weaker Acids	Alkanes	very small		CH_4	$\sim 10^{-50}$	~50
	Alkynes	10^{-25}	25	$HC{\equiv}CH$	$\sim 10^{-25}$	~25
	Alcohols	10^{-17}	17	$CH_3CHOHCH_3$	1.0×10^{-18}	18
				CH_3CH_2OH	1.3×10^{-16}	15.9
	Water			H_2O	1.8×10^{-16}	15.7
	Phenols	10^{-10}	10	C_6H_5OH	1.0×10^{-10}	10
				2-naphthol	3.1×10^{-10}	9.5
	Carboxylic acids	10^{-5}	5	CH_3CO_2H	1.8×10^{-5}	4.7
	Sulfonic acids	10^{0}	0	$C_6H_5SO_3H$	2×10^{-1}	0.7
				CH_3SO_3H	1.6×10^{1}	−1.2
Stronger Acids	Inorganic acids	large		HCl	10^{2}	−2
				H_2SO_4(first H^+)	10^{5}	−5

GETTING INVOLVED

✓ In terms of definitions, what is the difference between Bronsted-Lowry acids and bases? What is the difference between a strong acid and a weak acid?

✓ What is the definition of the acidity of constant K_a? Which denotes stronger and weaker acids, numerically higher or lower acidity constants? Explain in terms of the mathematical definition of the acidity constant.

✓ What is the definition of pK_a? Which denotes stronger and weaker acids, numerically higher or lower pK_a's? Is this the same or different from K_a? Explain.

Problem 9.11

Arrange the following from least to most acidic:
(a) pK_a's: 3.4, 6.2, and 11.8;
(b) K_a's: 3.4×10^{-3}, 9.8×10^{-12}, 6.7×10^{-5};
(c) $CH_3CH_2CH_3$, $CH_3CH_2CO_2H$, $CH_3CH_2CH_2OH$.
See related problem 9.46.

2. ***Conjugate Acids and Bases:*** The ion or molecule formed by loss of a proton from an acid is often referred to as the **conjugate base;** the species formed by the gain of a proton by a base is called a conjugate acid. Just as there are relative acidities among acids, there are relative basicities among conjugate bases. However, the relationship is an inverse one: the stronger the acid, the weaker the conjugate base; the weaker the acid, the stronger the conjugate base. This makes sense. A weak acid releases a hydrogen ion with great difficulty, because the conjugate base has such a strong attraction for the hydrogen ion; the conjugate base is strong. Strong acids, however, donate hydrogen ions freely; the conjugate base is weak and not attracted to hydrogen ions. Compare, for example, the relative acidities of hydrochloric acid (a strong acid) and water (an extremely weak acid).

conjugate base species formed by loss of a proton from an acid

conjugate acid species formed by gain of a proton by a base

Strong acid	$HCl + H_2O \longrightarrow$	$H_3O^+ + Cl^-$	Weak base
	Acid	Conjugate base	

Weak acid	$H_2O + H_2O \rightleftharpoons$	$H_3O^+ + OH^-$	Strong base
	Acid	Conjugate base	

The conjugate base of HCl is the chloride ion, a very weak base. Since the chloride ion is a poor base, it does not abstract H^+ from the hydronium ion and thus the reaction proceeds completely to the ionized form. However, the conjugate base of water is the hydroxide ion, which we consider to be a strong base. Since the hydroxide ion has such a strong attraction for hydrogen ions, the water molecule barely ionizes at all. Successful acid-base neutralization reactions occur between an acid and the conjugate base of a weaker acid. This is illustrated by the neutralization of hydrochloric acid with sodium hydroxide.

Acid-base neutralization reaction	HCl	+	NaOH	$\longrightarrow$	NaCl	+	H_2O
	Strong acid		Strong base ^-OH is conjugate base of weak acid, H_2O		Weak base Cl^- is conjugate base of strong acid, HCl		Weak acid

GETTING INVOLVED

✓ What are the similarities and differences among the terms acid, base, conjugate acid, and conjugate base?

✓ How are the strengths of an acid and its conjugate base related? How are the strengths of a base and its conjugate acid related? Explain this in terms of affinity or lack of affinity of these species for a hydrogen ion.

Example 9.2

Depending on the viewpoint one wishes to emphasize, an acid-base reaction can be described in terms of an acid and conjugate base or base and conjugate acid. Identify the acid, base, conjugate acid, and conjugate base in the neutralization reaction involving nitric acid and potassium hydroxide.

Solution

HNO_3	+	KOH	$\longrightarrow$	KNO_3	+	H_2O
Acid H^+ is the acid		Base ^-OH is the base		Conjugate base NO_3^- is the conjugate base		Conjugate acid

The anion (NO_3^-) associated with the acid (H^+) is the conjugate base. Since nitric acid is a strong acid, it follows that nitrate ion is a weak conjugate base. Hydroxide is the base; the neutralized base (water in this case) is the conjugate acid. Hydroxide is a strong base, and correspondingly, water is a weak conjugate acid.

Problem 9.12

Identify the acid, base, conjugate acid, and conjugate base for the following reaction in the direction written:

$$CH_3\ddot{\underset{..}{O}}H + H\ddot{\underset{..}{Cl}}: \longrightarrow CH_3\overset{H}{\underset{..}{\overset{..+}{O}}}H + :\ddot{\underset{..}{Cl}}:^-$$

Problem 9.13

Predict whether an acid-base neutralization would occur between the following pairs: **(a)** CH_4, CH_3CH_2ONa; **(b)** HCl, CH_3CO_2Na.

See related problems 9.44–9.46.

3. ***Acidity of Phenols.*** The characteristic property that differentiates phenols from alcohols is acidity. Phenols are weakly acidic (K_a for phenol is about 10^{-10}) and can be neutralized by sodium hydroxide. Alcohols with acidity constants of 10^{-16} to 10^{-19} are one million to one billion times less acidic than phenols and are not neutralized by sodium hydroxide. Note in Table 9.2 that phenols are more acidic than water and alcohols are less acidic. Thus the conjugate base of water, OH^-, can neutralize phenols but not alcohols.

Phenols: $C_6H_5{-}OH + NaOH \longrightarrow C_6H_5{-}ONa + H_2O$

Alcohols: $ROH + NaOH \longrightarrow$ No reaction

But how can this difference in acidity be explained? Both alcohols and phenols have polar O — H bonds. The phenol hydrogen is abstracted, however, because the resulting anion (phenoxide ion) is more stable than the alkoxide ion that would result from alcohol neutralization.

$$\underset{\text{Phenol}}{C_6H_5{-}\ddot{\underset{..}{O}}{:}H} \xrightarrow{^-:\ddot{\underset{..}{O}}:H} \underset{\text{Phenoxide ion}}{C_6H_5{-}\ddot{\underset{..}{O}}{:}^-} + H{:}\ddot{\underset{..}{O}}{:}H \qquad \underset{\text{Alcohol}}{R{-}\ddot{\underset{..}{O}}{:}H} \overset{^-:\ddot{\underset{..}{O}}:H}{\rightleftharpoons} \underset{\text{Alkoxide ion}}{R{-}\ddot{\underset{..}{O}}{:}^-} + H{:}\ddot{\underset{..}{O}}{:}H$$

In each case, a negative charge is left on the oxygen following abstraction of the hydrogen ion. But the phenoxide ion is capable of resonance (Figure 9.3), and some of the negative charge is dispersed throughout the benzene ring. The oxygen does not have to bear the entire brunt of the negative charge; the burden is shared and the phenoxide ion is stabilized. In contrast, no resonance is possible in the alkoxide ion, and the negative charge is concentrated on a single atom, the oxygen; this is a less stable condition.

There is other evidence supporting the dispersal of the negative charge of the phenoxide ion throughout the benzene ring, particularly at the *ortho* and *para* positions, as shown in Figure 9.3. For example, the nitro group (NO_2) is a powerful electron-withdrawing group and increases the acidity of phenol when placed on the benzene ring by further delocalizing the negative charge. This increase in acidity is only slight if the nitro is *meta*, but quite pronounced if it is *ortho* or *para*.

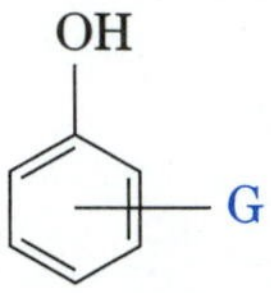

Acidity constants K_a

G =	K_a
Hydrogen	1.0×10^{-10}
Ortho nitro	6.8×10^{-8}
Meta nitro	5.3×10^{-9}
Para nitro	7×10^{-8}

Additional evidence of charge dispersal throughout the ring is seen in electrophilic aromatic substitution reactions of phenol. Recall the monobromination of benzene (section 6.4) requires pure bromine and iron or an iron halide catalyst (to generate the electrophile, Br^+). Bromination of phenol takes place much more readily. Delocalization of electrons in both phenol and phenoxide ion activates the *ortho* and *para* positions by making them partially negative (again see Figure 9.3) and consequently more attractive to the electrophile. In contrast to benzene, phenol tribrominates instantaneously when treated with a dilute water solution of bromine, without a catalyst.

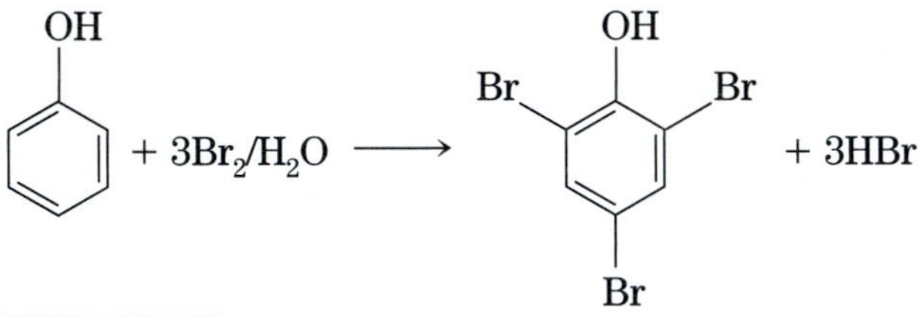

$$\text{phenol} + 3Br_2/H_2O \longrightarrow \text{2,4,6-tribromophenol} + 3HBr$$

GETTING INVOLVED

✓ Why are phenols more acidic than alcohols? How does resonance stabilize the phenoxide ion?

✓ In terms of resonance, explain the relative acidities of *ortho*, *meta*, and *para* nitrophenol? Why does phenol tribrominate so easily?

Problem 9.14

Write an equation illustrating the reaction between *o*-phenylphenol, an antiseptic in throat gargles and home disinfectant sprays, and sodium hydroxide.

See related problem 9.48.

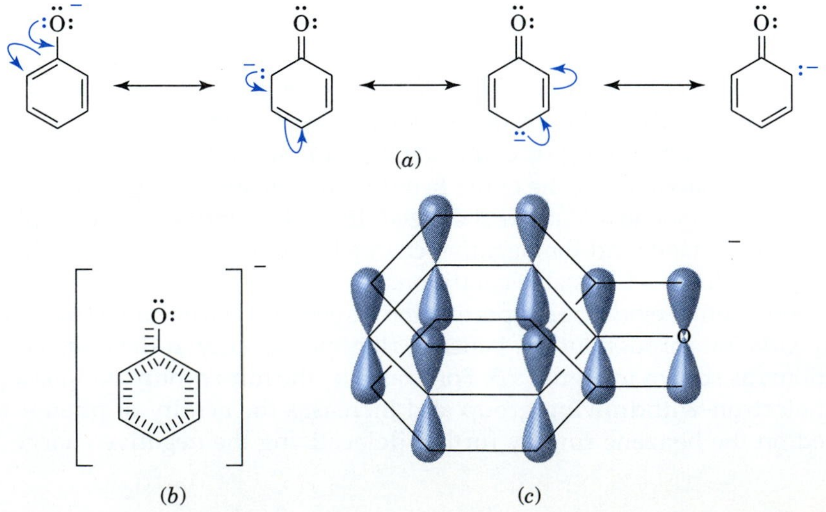

(a) (b) (c)

Figure 9.3 Resonance stabilization of the phenoxide ion. (a) Resonance forms. (b) Resonance hybrid. (c) Bonding picture.

B. Reaction of Alcohols with Sodium Metal: Reaction of the O — H Bond

Although alcohols are too weakly acidic to react appreciably with sodium hydroxide, they do react with active metals such as sodium. Thus sodium alkoxides can be prepared by using sodium metal. The reaction is not reversible as it would be with sodium hydroxide, since hydrogen gas evolves and escapes the system.

$$2ROH + 2Na \longrightarrow 2RONa + H_2\uparrow$$

The alkoxide ion, formed as a sodium salt (sodium alkoxide), is a strong base and nucleophile and is useful in organic synthesis, such as in the Williamson synthesis of ethers (section 8.6).

The reaction can be understood as an oxidation-reduction. Sodium releases its one outer-shell electron to the alcohol, forming a hydrogen atom and an alkoxide ion. The combination of two hydrogen atoms produces hydrogen gas.

$$\left.\begin{matrix} R\ddot{\underset{..}{O}}{:}H & \cdot Na \\ R\ddot{\underset{..}{O}}{:}H & \cdot Na \end{matrix}\right\} \longrightarrow \begin{matrix} R\ddot{\underset{..}{O}}{:}^- & Na^+ & H\cdot \\ R\ddot{\underset{..}{O}}{:}^- & Na^+ & H\cdot \end{matrix} \Big\} \longrightarrow H_2\uparrow$$

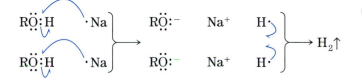

GETTING INVOLVED

Problem 9.15

Write equations illustrating the reaction (if any) of ethanol (beverage alcohol) with **(a)** sodium hydroxide and **(b)** sodium metal.

See related problem 9.49 a.

C. Formation of Esters: Reaction of the O — H Bond

Alcohols form esters with both inorganic and organic acids. This is a reaction that will be covered thoroughly in Chapter 13 (sections 13.3–13.6), so we will just introduce it here. We have already seen ester formation in the reaction of glycerol and nitric acid to form nitroglycerin. Other alcohols also form inorganic esters, as in the reaction of 1-dodecanol with sulfuric acid. An alkyl hydrogen sulfate results, which is used to prepare synthetic detergents.

$$CH_3(CH_2)_{10}CH_2OH + HOSO_3H \longrightarrow CH_3(CH_2)_{10}CH_2OSO_3H + H_2O$$

Organic esters are formed similarly by the reaction of alcohols with carboxylic acids, as in the following preparation of artificial rum flavoring:

$$CH_3CH_2OH + HO\overset{\overset{O}{\|}}{C}H \xrightarrow[\text{catalyst}]{H_2SO_4} CH_3CH_2O\overset{\overset{O}{\|}}{C}H + H_2O$$

GETTING INVOLVED

Problem 9.16

Write an equation illustrating the reaction between 1-butanol and nitrous acid (HONO) to produce butyl nitrite, a substance that dilates blood vessels and thus lowers blood pressure and is used illicitly for mood elevation.

See related problem 9.49 e.

Connections 9.3 www.prenhall.com/bailey

Insecticides and Nerve Gases

Alcohols can react with acids to form esters. Among the inorganic esters is a class of organophosphate esters. Phosphoric acid has three acidic groups, which can be derivatized or substituted.

Some organophosphates are toxic to the nerve cells of insects and animals. This makes them useful as insecticides, but some can also be agents of human destruction.

$$\mathrm{HO{-}\overset{O}{\overset{\|}{\underset{OH}{\underset{|}{P}}}}{-}OH + ROH \longrightarrow HO{-}\overset{O}{\overset{\|}{\underset{OH}{\underset{|}{P}}}}{-}OR \xrightarrow{R_1OH} HO{-}\overset{O}{\overset{\|}{\underset{OR_1}{\underset{|}{P}}}}{-}OR}$$

Organophosphate esters

Organophosphate insecticides are actually sulfur analogs of the esters shown above; that is, a double-bonded sulfur appears in place of the phosphorous-oxygen double bond. This decreases toxicity to humans and makes them safer to handle.

When the insecticide is sprayed on an organism, enzymes replace the sulfur with an oxygen, and the compound becomes toxic. The organophosphate can also undergo enzyme-catalyzed hydrolysis, that is, breakdown into its constituent acid and alcohols, which will detoxify the compound. In insects the toxifying reaction takes place faster and better than the detoxifying one. In humans the reverse situation exists. This makes the insecticide less harmful to higher animals than to insects.

Organophosphates interact with the parasympathetic portion of the nervous system, resulting in symptoms of salivation, diarrhea, indigestion, and vomiting, as well as slowed heartbeat and labored breathing. Chronic exposure to such materials can lead to poisoning. Field workers must be evaluated frequently as to the levels of pesticides in their bloodstreams.

Pesticides being sprayed

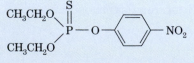

If the hydrolysis reaction is prevented, the potential risk is magnified hundreds, if not thousands, of times. Nerve gases are agents of chemical warfare which have rarely been used and are mainly a threat, albeit a persistent one. The four major nerve gases are shown.

Organophosphate insecticides

$$\mathrm{(CH_3CH_2O)_2P(=S){-}O{-}C_6H_4{-}NO_2}$$

Parathion

$$\mathrm{(CH_3O)_2P(=S){-}S\underset{\underset{CH_2\overset{O}{\overset{\|}{C}}OCH_2CH_3}{|}}{C}H\overset{O}{\overset{\|}{C}}OCH_2CH_3}$$

Malathion

Connections 9.3 *(cont.)*

Insects build up resistance to organophosphates by evolving enzymes that can detoxify them. Therefore, other types of insecticides, such as the carbamates (esters of phenols), have been developed that also impede the parasympathetic nervous system and are resilient to insect resistance until their use also stimulates insect adaptation.

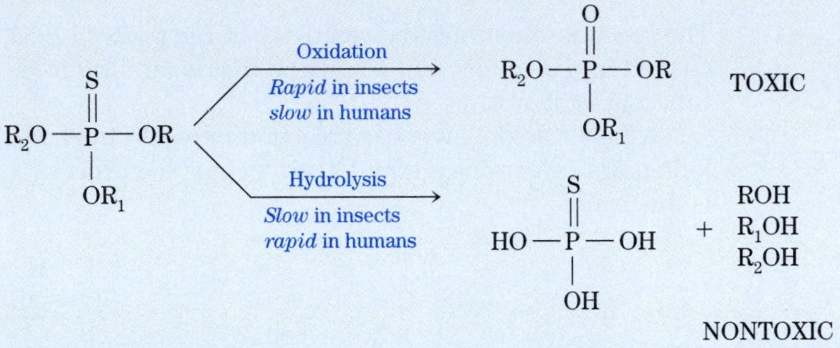

$R_2O-P(=S)(OR_1)-OR$ —Oxidation (*Rapid* in insects, *slow* in humans)→ $R_2O-P(=O)(OR_1)-OR$ TOXIC

$R_2O-P(=S)(OR_1)-OR$ —Hydrolysis (*Slow* in insects, *rapid* in humans)→ $HO-P(=S)(OH)-OH$ + ROH, R_1OH, R_2OH NONTOXIC

Nerve gases

$(CH_3)_2CHO(CH_3)P(=O)-F$
Sarin

$CH_3(CH_3CH_2O)P(=O)-SCH_2CH_2N(CH(CH_3)_2)_2$
VX

$(CH_3)_2N(CH_3CH_2O)P(=O)-CN$
Tabun

$(CH_3)_3CCH(CH_3)O(CH_3)P(=O)-F$
Soman

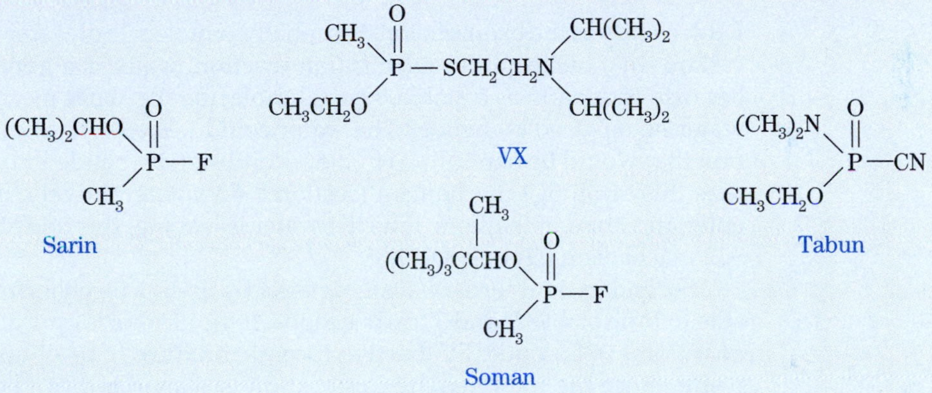

Carbamate insecticides

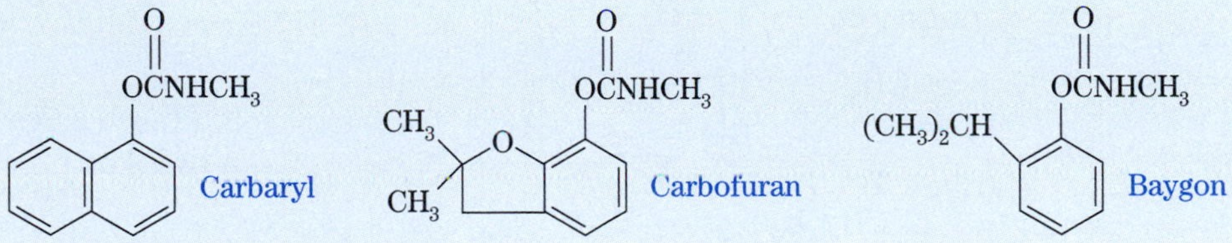

Carbaryl ($OC(=O)NHCH_3$ on naphthalene)

Carbofuran ($OC(=O)NHCH_3$; CH_3, CH_3, O)

Baygon ($OC(=O)NHCH_3$; $(CH_3)_2CH$)

9.7 Reactions of Alcohols and Ethers with Hydrogen Halides: Reaction of the C — O Bond by Nucleophilic Substitution

A. Reactions of Alcohols with Hydrogen Halides: S_N1 and S_N2 Mechanisms

Alcohols react with hydrogen halides to produce alkyl halides by a nucleophilic substitution reaction.

$$-\overset{|}{\underset{|}{C}}-OH + HX \longrightarrow -\overset{|}{\underset{|}{C}}-X + H_2O$$

where
HX = HCl, HBr, HI

For example, 2-butanol reacts readily with hydrogen bromide to form 2-bromobutane.

$$CH_3CH(OH)CH_2CH_3 + HBr \longrightarrow CH_3CH(Br)CH_2CH_3 + H_2O$$

The reaction illustrates the reactivity of the polar C—O bond, the Lewis base character of alcohols, and a reaction that is familiar to us—nucleophilic substitution (Chapter 8).

A Lewis acid–Lewis base reaction between the hydrogen ion of HX and a lone pair of electrons on the alcohol begins the process; a positive oxonium ion is the result.

$$-\overset{|}{\underset{|}{C}}-\ddot{O}H + H^+ \rightleftharpoons -\overset{|}{\underset{|}{C}}-\overset{+}{\ddot{O}}H_2$$

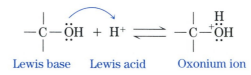

Lewis base Lewis acid Oxonium ion

Look closely at the oxonium ion. Essentially, a molecule of water is bonded to the carbon. In a nucleophilic substitution reaction, water is a good leaving group because it departs as a stable, neutral molecule. By what mechanisms can the water be replaced by halide? The water could leave first, producing a carbocation that would be neutralized in the second step by halide ion. You may recognize this as an S_N1 mechanism (section 8.4.E). Alternatively, the halide could enter and bond at the same time the water is leaving; this one-step process is an S_N2 mechanism (section 8.4.D).

Secondary and tertiary alcohols react by an S_N1 mechanism, since they are able to form stable 2° and 3° carbocations. If the alcohol is optically active (a pure enantiomer), an optically inactive racemic mixture (pair of enantiomers) will result, since the intermediate carbocation is planar and can be attacked from either side. This S_N1 mechanism is illustrated by the reaction of optically active 2-butanol with HBr. Nucleophilic substitution occurs after formation of the oxonium ion.

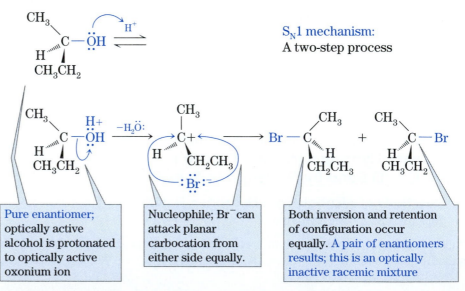

S_N1 mechanism:
A two-step process

$$CH_3(H)(CH_3CH_2)C-\ddot{O}H \xrightarrow{H^+} CH_3(H)(CH_3CH_2)C-\overset{+}{\ddot{O}}H_2 \xrightarrow{-H_2\ddot{O}:} CH_3(H)(CH_3CH_2)C^+ \xrightarrow{:\ddot{Br}:^-} Br-C(CH_3)(H)(CH_2CH_3) + (CH_3)(H)(CH_3CH_2)C-Br$$

Pure enantiomer; optically active alcohol is protonated to optically active oxonium ion

Nucleophile; Br^- can attack planar carbocation from either side equally.

Both inversion and retention of configuration occur equally. A pair of enantiomers results; this is an optically inactive racemic mixture

Primary alcohols can only form unstable primary carbocations and thus react with hydrogen halides only through an S_N2 mechanism in which a carbocation is unnecessary. The one-step displacement process occurs following the formation of an oxonium ion. We can illustrate this mechanism using 1-butanol and HBr.

S_N2 mechanism:
A one-step process
Primary alcohol protonated to form primary oxonium ion; oxonium ion is attacked by bromide
Transition state showing bromide displacing water molecule from the opposite side to form the final product

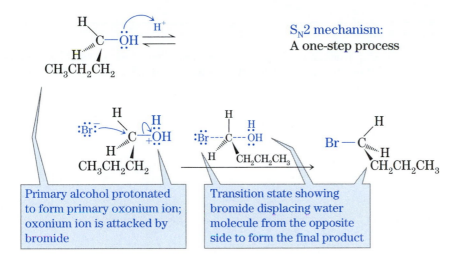

These mechanisms are illustrated in Figure 9.4b (see pg. 278).

Tertiary, secondary, and primary alcohols react with HCl (in $ZnCl_2$ solution) at different rates and this difference is the basis of the historical Lucas test, once used for distinguishing among low-molecular-weight alcohols. Although the alcohols are soluble in the Lucas reagent, the alkyl halide formed is not and the solution becomes cloudy. The reaction rate varies with the stability of the potential carbocation intermediate formed in the S_N1 mechanism. Tertiary alcohols react almost instantaneously at room temperature, and secondary alcohols react in 5 to 15 minutes when heated; both react by S_N1 mechanisms. Primary alcohols, which react by an S_N2 mechanism, are the slowest, requiring several hours of reaction time, even with heat.

GETTING INVOLVED

- ✓ Describe the reaction of alcohols with hydrogen halides. What bonds break in the reactants, what bonds form in the products? How is the reaction initiated? Describe the first step as a Lewis acid–Lewis base reaction.
- ✓ What is the difference in the S_N1 and S_N2 mechanisms for the reaction of alcohols with hydrogen halides in terms of number of steps, reaction intermediates, and stereochemistry?
- ✓ Why do tertiary and secondary alcohols react by an S_N1 mechanism and primary by S_N2? Why do tertiary halides react faster than secondary?
- ✓ Describe the Lucas test and how and why it works.

Problem 9.17

Write equations for the following reaction of alcohols with hydrogen halides: **(a)** 2-methyl-2-butanol and HCl; **(b)** 1-pentanol and HBr (heat); **(c)** 2-pentanol and HCl.

Problem 9.18

Arrange the reactions in problem 9.17 from fastest to slowest. In each case predict whether the substitution mechanism is S_N1 or S_N2.

Problem 9.19

There are four alcohols with the molecular formula $C_5H_{12}O$ and with only four carbons in the longest chain. Draw them and predict their relative reaction times with the Lucas reagent.

Problem 9.20

Write step-by-step reaction mechanisms for the following. In each case show the formation of the oxonium ion first and then illustrate the S_N1 or S_N2 mechanism.
(a) ethanol and HCl by an S_N2 process

(b) $C_6H_5(H)(CH_3)C{-}OH + HBr \longrightarrow S_N1$ mechanism

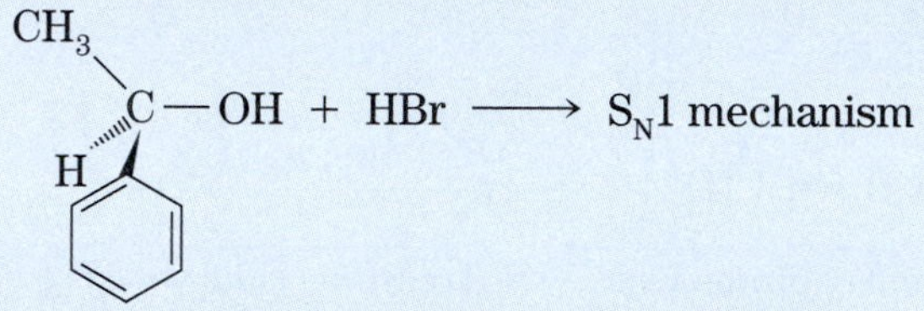

See related problems 9.49 c, 9.50, 9.54, 9.56–9.58.

B. Methods for Converting Alcohols to Alkyl Halides: Reaction of the C—O Bond

There are other synthetic methods for converting alcohols to alkyl halides. Thionyl chloride is a convenient reagent for making alkyl chlorides since the byproducts, HCl and SO_2, are both gases and exit the reaction mixture, leaving a rather pure product. Phosphorus trihalides are also effective reagents for synthesizing alkyl halides from alcohols. The following examples illustrate these important synthetic reactions.

Thionyl chloride

$$ROH + SOCl_2 \longrightarrow RCl + HCl + SO_2$$

$$CH_3CH_2CH_2CH_2CH_2OH + SOCl_2 \longrightarrow CH_3CH_2CH_2CH_2CH_2Cl + HCl + SO_2$$

Phosphorus trihalides

$$3ROH + PX_3 \longrightarrow 3RX + P(OH)_3 \quad X = Cl, Br$$

$$3CH_3CH_2CH(CH_3)CH_2OH + PBr_3 \longrightarrow 3CH_3CH_2CH(CH_3)CH_2Br + P(OH)_3$$

GETTING INVOLVED

Problem 9.21

Write equations showing the reaction of ethanol with **(a)** thionyl chloride and **(b)** phosphorus tribromide.

Problem 9.22

Show three ways for preparing 1-chlorobutane from butanol.
See related problem 9.50.

C. Reactions of Ethers with Hydrogen Halides: S_N1 and S_N2 Mechanisms

The reaction of ethers with hydrogen halides is very similar to that of alcohols. Whereas an alcohol reacts with a hydrogen halide to form an alkyl halide and water, an ether forms an alkyl halide and an alcohol.

Alcohol: $ROH + HX \longrightarrow RX + HOH$

Ether: $ROR + HX \longrightarrow RX + ROH$

HX = HCl, HBr, HI

For example, consider the reaction of 2-methoxypropane with HBr.

$$CH_3CH(CH_3)OCH_3 + HBr \longrightarrow CH_3CH(CH_3)Br + CH_3OH$$

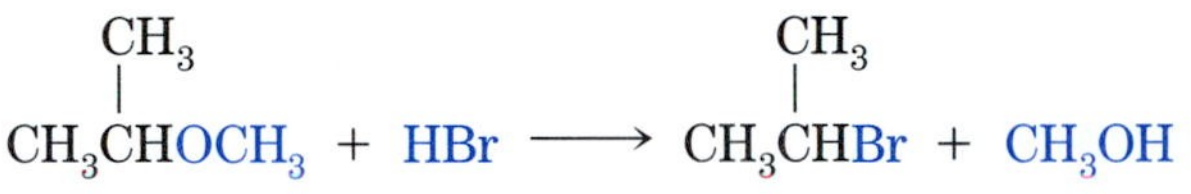

Of course, the alcohol formed in the reaction of ethers with HX can react further, if additional reagent is present, to form another molecule of alkyl halide and water, as illustrated by the reaction of alcohols above. Thus in the presence of at least two mole-equivalents of hydrogen halide, the reaction of ethers can be generalized as follows:

Ethers with two moles HX $ROR + 2HX \longrightarrow 2RX + H_2O$ HX = HCl, HBr, HI

This reaction illustrates the reactivity of the polar carbon-oxygen bond and the Lewis base character of ethers. Let us examine the mechanism of the reaction of 2-methoxypropane with HBr illustrated above. A Lewis acid–Lewis base reaction between the hydrogen ion of HX and a nonbonding electron pair on the ether begins the reaction; an oxonium ion results. Since the ether involves a secondary carbon, an S_N1 mechanism is operative.

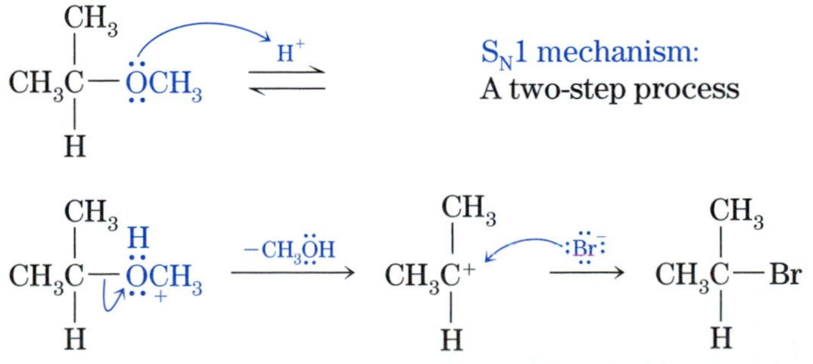

$$CH_3C(CH_3)(H)-\ddot{O}CH_3 \underset{}{\overset{H^+}{\rightleftharpoons}}$$

S_N1 mechanism:
A two-step process

$$CH_3C(CH_3)(H)-\overset{+}{O}(H)CH_3 \xrightarrow{-CH_3\ddot{O}H} CH_3C^+(CH_3)(H) \xrightarrow{:\ddot{Br}:^-} CH_3C(CH_3)(H)-Br$$

Oxonium ion — Carbocation — Alkyl bromide product

Cleavage of ethers occurs by an S_N1 process at secondary and tertiary carbons, since these carbons can form stable carbocations. However, at primary carbons, the S_N2 process is dominant, since the very unstable primary carbocations are not a part of this mechanism, as illustrated by the following example.

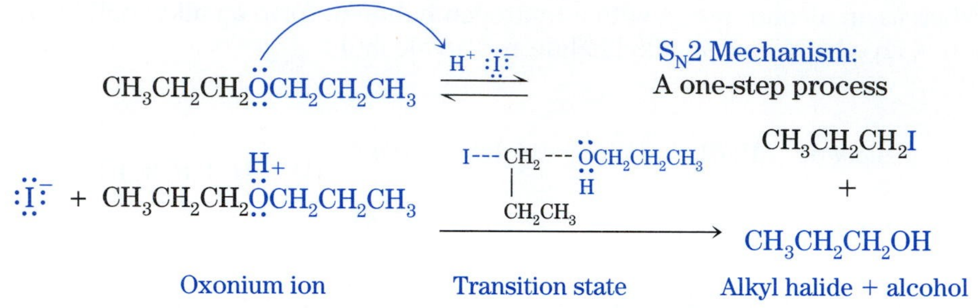

S_N2 Mechanism: A one-step process

$$CH_3CH_2CH_2\ddot{O}CH_2CH_2CH_3 \xrightleftharpoons{H^+\ :\ddot{I}:^-}$$

$$:\ddot{I}:^- + CH_3CH_2CH_2\overset{H+}{\ddot{O}}CH_2CH_2CH_3 \longrightarrow [I{-}{-}{-}CH_2(CH_2CH_3){-}{-}{-}\ddot{O}(H)CH_2CH_2CH_3] \longrightarrow CH_3CH_2CH_2I + CH_3CH_2CH_2OH$$

Oxonium ion — Transition state — Alkyl halide + alcohol

These nucleophilic substitution mechanisms for the reactions of alcohols and ethers with hydrogen halides are compared and summarized in Figure 9.4.

GETTING INVOLVED

- ✓ Why do alcohols react with only one equivalent of HX to produce one equivalent of alkyl halide and ethers react with two equivalents of HX to produce two equivalents of alkyl halide?
- ✓ What determines whether an ether or an alcohol reacts with hydrogen halide by an S_N1 or S_N2 mechanism? Why is an oxonium ion formed prior to the nucleophilic substitution in both cases? What are the leaving groups in the reaction of ether and the reaction of alcohols with HX?

Problem 9.23

Write an equation illustrating the reaction of 2-methoxy-2-methylpropane with **(a)** one mole of hydrogen bromide; **(b)** two moles of HBr.

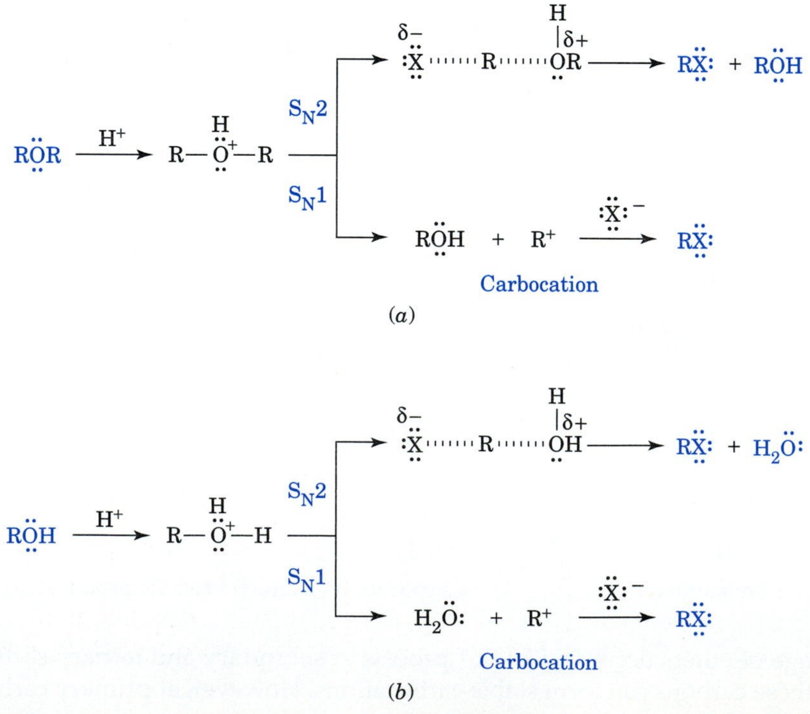

(a)

$$R\ddot{O}R \xrightarrow{H^+} R{-}\overset{H}{\ddot{O}^+}{-}R$$

S_N2: $\overset{\delta-}{:\ddot{X}}\cdots R\cdots \overset{H\,|\,\delta+}{\ddot{O}R} \longrightarrow R\ddot{X}: + R\ddot{O}H$

S_N1: $R\ddot{O}H + R^+ \xrightarrow{:\ddot{X}:^-} R\ddot{X}:$ (Carbocation)

(b)

$$R\ddot{O}H \xrightarrow{H^+} R{-}\overset{H}{\ddot{O}^+}{-}H$$

S_N2: $\overset{\delta-}{:\ddot{X}}\cdots R\cdots \overset{H\,|\,\delta+}{\ddot{O}H} \longrightarrow R\ddot{X}: + H_2\ddot{O}:$

S_N1: $H_2\ddot{O}: + R^+ \xrightarrow{:\ddot{X}:^-} R\ddot{X}:$ (Carbocation)

Figure 9.4 Comparison of mechanisms of the reactions of (a) ethers and (b) alcohols with hydrogen halides.

Problem 9.24

Write the mechanism for the reactions described in Problem 9.23 by writing **(a)** S_N1 mechanism with one mole of HBr; and **(b)** the S_N2 mechanism for the reaction of the remaining alcohol with HBr.

Problem 9.25

Write the product of the following reaction:

(tetrahydrofuran) + 2HI ⟶

See related problems 9.50, 9.56–9.58.

9.8 Dehydration of Alcohols by E_1 Elimination: Reaction of the C—O Bond

Alcohols can be dehydrated with acid to form alkenes (section 4.5).

$$-\overset{|}{\underset{\text{H}}{\overset{}{C}}}-\overset{|}{\underset{\text{OH}}{C}}- \xrightarrow{H_2SO_4} \;>C=C< \; + H_2O$$

Once again, the reaction illustrates the reactivity of the carbon-oxygen bond and the Lewis base character of alcohols. The mechanism is similar to that of the S_N1 reaction of alcohols with hydrogen halides; it differs only in the final step.

The reaction begins with a Lewis acid–Lewis base reaction between a hydrogen ion from sulfuric acid and a nonbonding electron pair on the alcohol: an oxonium ion results. The oxonium ion loses a water molecule to form a carbocation. In the final step, the carbocation is neutralized by elimination of a hydrogen ion with the resultant formation of a carbon-carbon double bond. Note that the acid serves as a catalyst; it provides a hydrogen ion in the first step, which is returned in the last step. This is an E_1-type mechanism and has been introduced previously in sections 4.5.C and 8.5, and is merely the reverse of hydration of alkenes (sections 5.1 A.3 and B.3).

E_1 Mechanism for Dehydration of Alcohols

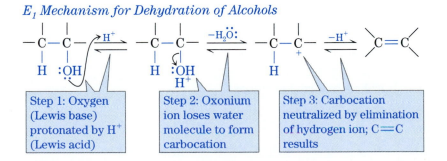

$$-\underset{\text{H}}{C}-\underset{:\ddot{O}H}{C}- \underset{}{\overset{H^+}{\rightleftharpoons}} -\underset{\text{H}}{C}-\underset{:\overset{+}{O}H_2}{C}- \overset{-H_2\ddot{O}:}{\rightleftharpoons} -\underset{\text{H}}{C}-\overset{+}{C}- \overset{-H^+}{\rightleftharpoons} >C=C<$$

Step 1: Oxygen (Lewis base) protonated by H^+ (Lewis acid)

Step 2: Oxonium ion loses water molecule to form carbocation

Step 3: Carbocation neutralized by elimination of hydrogen ion; C=C results

The S_N1 reaction of alcohols with hydrogen halides is identical except in the neutralization of the carbocation. In that mechanism, a negative halide ion neutralizes the carbocation by forming a carbon-halogen bond (see Figures 9.4b and 8.2).

You will recall from section 4.5.B that in dehydrations where more than one alkene is possible, the most substituted one predominates, as illustrated in the following example:

$$CH_3CH(OH)CH(CH_3)CH_3 \xrightarrow{H_2SO_4} CH_3C(CH_3){=}CHCH_3 + CH_3CH(CH_3)CH{=}CH_2$$

Predominant product (trisubstituted alkene) — Minor product (monosubstituted alkene)

GETTING INVOLVED

✓ Describe how the dehydration of alcohols by an E_1 mechanism, and the S_N1 reactions of both alcohols and ethers with hydrogen halides are alike in the first two steps and how they differ in the neutralization step.

Problem 9.26

Write in a way to illustrate similarities and differences for **(a)** the E_1 mechanism for the dehydration of 2-propanol; **(b)** the S_N1 mechanism for the reaction of 2-propanol with HBr; and **(c)** the S_N1 mechanism for the reaction of 2-methoxypropane with HBr.

Problem 9.27

Write reaction equations illustrating the dehydration of the following alcohols with sulfuric acid: **(a)** 2-propanol; **(b)** 2-methyl-2-butanol; and **(c)** 3-methyl-2-pentanol.

See related problems 9.53, 9.59.

9.9 Oxidation of Alcohols: Reaction of the C—O and O—H Bonds

oxidation
removal of hydrogen from carbon-oxygen single bond or insertion of oxygen in a molecule

An important reaction of alcohols is their oxidation to carbonyl compounds, compounds with a carbon-oxygen double bond. Primary alcohols generally oxidize to carboxylic acids via an aldehyde intermediate; aldehydes are very easily oxidized and form carboxylic acids quickly under most conditions. **Oxidation** involves initial loss of hydrogen from the carbon-oxygen bond of the alcohol, followed by insertion of oxygen in the remaining carbon-hydrogen bond.

$$\underset{1^\circ \text{ alcohol}}{RCH_2OH} \xrightarrow{\text{Oxidation}} \underset{\text{Aldehyde}}{RCH{=}O} \xrightarrow{\text{Further oxidation}} \underset{\text{Carboxylic acid}}{RC({=}O)OH}$$

Secondary alcohols oxidize to ketones.

$$\underset{2^\circ \text{ alcohol}}{RCH(OH)R} \xrightarrow{\text{Oxidation}} \underset{\text{Ketone}}{RC({=}O)R}$$

Tertiary alcohols do not have a carbon-hydrogen bond on the alcohol carbon and usually do not oxidize under mild conditions.

Connections 9.4

www.prenhall.com/bailey

Measuring Blood Alcohol

The familiar Breathalyzer test for estimating alcohol in the blood, whether administered in the field by having the subject breathe into a plastic bag or more precisely in a lab, is based on the oxidation of ethyl alcohol to acetic acid using potassium dichromate.

Potassium dichromate is a very vivid reddish-orange substance. As the breath being tested is passed through a dichromate reagent, the reddish-orange color changes to green as the ethyl alcohol is oxidized and Cr^{6+} of the orange red dichromate is reduced to the green Cr^{3+}. The dichromate reagent, absorbed on a silica gel solid base, is packed into a tube through which the test breath is passed. As the breath passes along the tube the oxidation-reduction occurs; the alcohol content of the breath is estimated by the length of the tube that changes from orange to green during the test. An estimate of blood alcohol is calculated from this measurement.

$$\underset{\text{Ethyl alcohol}}{CH_3CH_2OH} + \underset{\text{Dichromate (orange/red)}}{Cr_2O_7^{2-}} \xrightarrow{H^+/H_2O} \underset{\text{Acetic acid}}{CH_3C(=O)OH} + \underset{\text{Chromium ion green}}{Cr^{3+}}$$

Chromium trioxide and sodium dichromate are common oxidizing agents which oxidize primary alcohols to carboxylic acids and secondary alcohols to ketones, as illustrated by the following reactions:

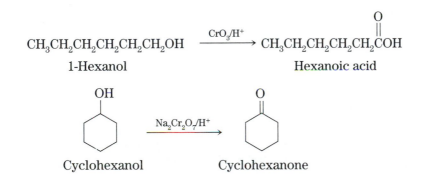

$$\underset{\text{1-Hexanol}}{CH_3CH_2CH_2CH_2CH_2CH_2OH} \xrightarrow{CrO_3/H^+} \underset{\text{Hexanoic acid}}{CH_3CH_2CH_2CH_2CH_2\overset{O}{\overset{\|}{C}}OH}$$

$$\underset{\text{Cyclohexanol}}{C_6H_{11}OH} \xrightarrow{Na_2Cr_2O_7/H^+} \underset{\text{Cyclohexanone}}{C_6H_{10}O}$$

Under certain conditions, it is possible to stop the oxidation of a primary alcohol at the aldehyde stage. Pyridinium chlorochromate (PCC, $C_5H_6NCrO_3Cl$) is an effective reagent in this regard.

$$\underset{\text{1-Hexanol}}{CH_3CH_2CH_2CH_2CH_2CH_2OH} \xrightarrow{PCC} \underset{\text{Hexanal}}{CH_3CH_2CH_2CH_2CH_2\overset{O}{\overset{\|}{C}}H}$$

GETTING INVOLVED

Problem 9.28

Write reaction equations illustrating the oxidation of the four isomeric alcohols with the formula $C_4H_{10}O$, using chromium trioxide.

Connections 9.5

www.prenhall.com/bailey

Methanol and Ethylene Glycol Poisoning

Although most of us can assimilate and metabolize small amounts of ethanol, consumption of other alcohols can be extremely toxic. To begin with, all alcohols are nervous system depressants. They impair the transmission of nerve signals, ultimately leading to a block of respiration. This in itself is part of their toxicity. In addition, biological catalysts called enzymes, especially those found in the liver, oxidize alcohols to aldehydes, ketones, and carboxylic acids, many of which are toxic by virtue of their chemical reactivity and/or solubility characteristics.

Methanol is readily oxidized to methanal (formaldehyde), then to methanoic acid (formic acid). Formic acid is a relatively strong organic acid and can upset the acid-base balance of the body, causing a condition called *metabolic acidosis.* The oxidation products also attack the optic nerve, leading to blindness.

$$CH_3OH \longrightarrow HCH(=O) \longrightarrow HCOH(=O)$$

Methanol → Methanal (formaldehyde) → Methanoic acid (formic acid)

These reactions are catalyzed by liver enzymes

Ethylene glycol, antifreeze, is converted to oxalic acid (ethanedioic acid), which forms insoluble "stones" with circulating calcium ions. Oxalic acid can be found naturally in rhubarb leaves and in spinach. This is the reason that nonoxalic-containing rhubarb stems are used for cooking. Even spinach, though high in vitamins and minerals, can be a problem if consumed in excess.

$$HOCH_2-CH_2OH \rightsquigarrow HOC(=O)-COH(=O)$$

Ethylene glycol → Oxalic acid

Why, then, can we consume limited amounts of ethanol? Ethanol is as much a respiratory depressant as any other alcohol. Many deaths have occurred with rapid drinking of large quantities. When consumed in moderate amounts, ethanol is converted to ethanal (acetaldehyde) and then to ethanoic or acetic acid. Acetic acid can enter the metabolic Krebs cycle to act as a fuel source for the body or, if not needed, can be converted to fat. However, the caloric content is said to be "empty" because it contains no vitamins, minerals, proteins, carbohydrates, or lipids. An alcoholic may consume up to 75% of his or her daily caloric needs through ethanol alone, leaving the body deficient in other essential nutrients.

$$CH_3CH_2OH \longrightarrow CH_3CH(=O) \longrightarrow CH_3COH(=O) \dashrightarrow \text{Krebs cycle}$$

Ethanol → Acetaldehyde → Acetic acid

If a person has consumed methanol (often a contaminant of homemade liquor or "moonshine") or ethylene glycol, primary medical treatment deals with the acidosis produced. Next, ethanol is given because it competes with the methanol or ethylene glycol for the same oxidizing enzymes. Ethanol is actually a better substrate for the enzymes and will preferentially tie up enzyme molecules. This allows time for the kidneys to filter out and excrete the other water-soluble alcohols.

Problem 9.29

Write reaction equations illustrating the preparations of each of the following compounds from alcohols:

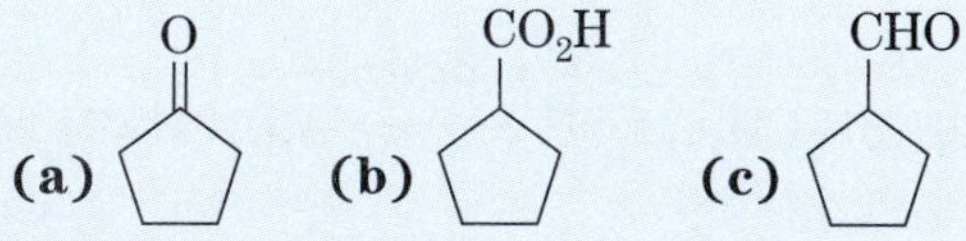

(a) cyclopentanone (C=O on cyclopentane ring) (b) cyclopentane-CO_2H (c) cyclopentane-CHO

See related problems 9.49 d, 9.51, 9.64.

9.10 Epoxides

epoxide (or **oxirane**)
three-membered ring cyclic ether

Three-membered cyclic ethers are called **epoxides** or **oxiranes.** The simplest and most commercially important epoxide is ethylene oxide, which is used in the petrochemical industry as an intermediate in the production of antifreeze, synthetic fibers, resins, paints, adhesives, films, cosmetics, and synthetic detergents. Ethylene oxide is prepared from ethylene.

$$\text{Ethylene} \quad CH_2{=}CH_2 + O_2 \xrightarrow[\text{Heat, pressure}]{Ag} \underset{\diagdown\,O\,\diagup}{CH_2{-}CH_2} \quad \text{Ethylene oxide}$$

They are prepared from alkenes more generally using a peroxycarboxylic acid.

$$\text{Cyclohexene} + RCOOH\ (\text{with } C{=}O) \longrightarrow \text{cyclohexene oxide} + RCOH\ (\text{with } C{=}O)$$

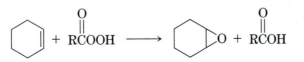

A. Reactions of Ethylene Oxide

The importance of ethylene oxide as an industrial chemical lies in its propensity toward ring-opening reactions. Like cyclopropane (section 2.8.A), ethylene oxide suffers from acute angle strain because of the distortion of normal bond angles from 109° to approximately 60°. This strain is relieved by cleavage of a polar carbon-oxygen bond under either acidic or basic conditions. For example, more than half of the ethylene oxide produced commercially is hydrolyzed to ethylene glycol, which is used as antifreeze, in brake fluids, and in the manufacture of polyester fibers. The reaction mechanism is very similar to that of acid cleavage of ethers to alkyl halides (section 9.7.C). First, the epoxide oxygen is protonated. Nucleophilic attack of a water molecule and loss of a hydrogen ion follow.

$$\underset{\diagdown\,\ddot{O}\,\diagup}{CH_2{-}CH_2} \xrightarrow{H^+} \underset{\diagdown\,\underset{H^+}{\ddot{O}}\,\diagup}{CH_2{-}CH_2} \xrightarrow{H_2\ddot{O}:} \underset{H\ddot{O}:\quad :\overset{+}{\ddot{O}}H_2}{CH_2{-}CH_2} \xrightarrow{-H^+} \underset{OH\quad OH}{CH_2{-}CH_2}$$

The use of alcohols instead of water to effect the ring opening produces ether-alcohol compounds commercially known as cellosolves, such as methyl cellosolve, which is added to jet fuels to prevent formation of ice crystals.

$$CH_3OH + \underset{\diagdown\,O\,\diagup}{CH_2{-}CH_2} \xrightarrow{H^+} CH_3OCH_2CH_2OH$$

Methyl cellosolve
(2-methoxyethanol)

Reaction of ethylene oxide with ammonia produces ethanolamine, which is used to remove hydrogen sulfide and carbon dioxide from natural gas.

$$NH_3 + \underset{\diagdown\,O\,\diagup}{CH_2{-}CH_2} \longrightarrow H_2NCH_2CH_2OH$$

Ethanolamine

GETTING INVOLVED

✓ Why does the ring structure of epoxides make them so reactive?

Example 9.3

Write the product of the following reaction and propose a reaction mechanism:

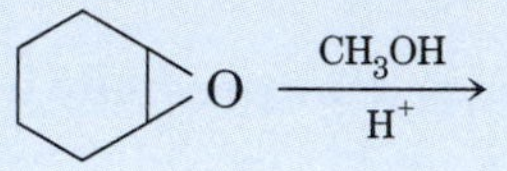

$$\text{cyclohexene oxide} \xrightarrow[H^+]{CH_3OH}$$

Solution

The polar O-H bond of methanol cleaves the strained epoxide ring to form 2-methoxy-cyclohexanol by the following reaction mechanism:

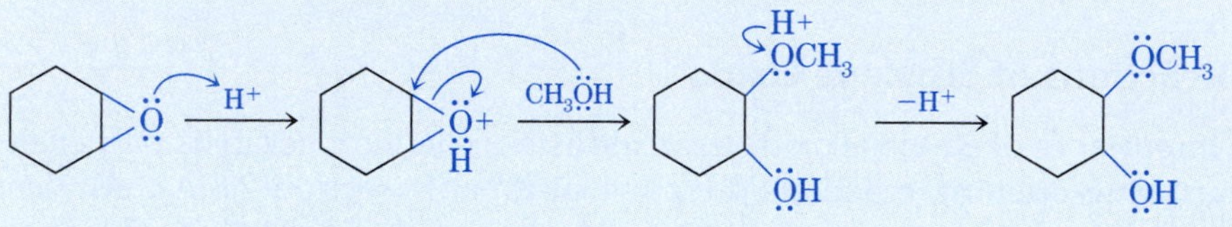

H+ CH$_3$ÖH −H+ ÖCH$_3$ ÖH

2-methoxycyclohexanol

Problem 9.30

Predict the product of the reaction of the following epoxide with one mole of each of the following reagents:

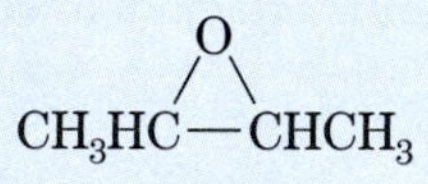

$CH_3HC{-}CHCH_3$ (epoxide, O bridging)

(a) HBr; **(b)** H_2O/H^+; **(c)** CH_3CH_2OH/H^+.

See related problems 9.55 and 9.61.

B. Epoxy Resins

Epoxy resins are polymers with tremendous adhesive properties and are used to bind glass, porcelain, metal, and wood. They are manufactured from epichlorohydrin and bisphenol A.

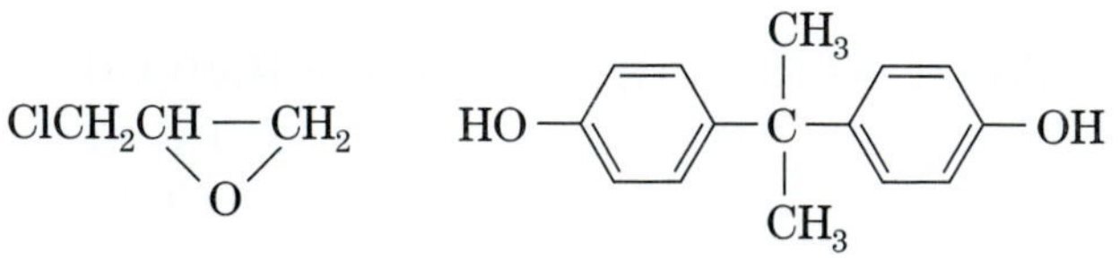

$ClCH_2CH{-}CH_2$ (with O bridging) HO–C$_6$H$_4$–C(CH_3)$_2$–C$_6$H$_4$–OH

Epichlorohydrin Bisphenol A

The somewhat involved process includes chemistry already studied: acidity of phenols, ring opening of epoxides, and nucleophilic substitution of alkyl halides. The structure of an epoxy resin follows. Treatment of the developing polymer with a triamine causes crosslinking between polymer chains and gives the resin added strength.

$$\underset{\diagdown O \diagup}{CH_2{-}CHCH_2} \sim\!\!\left(O{-}C_6H_4{-}\underset{CH_3}{\overset{CH_3}{C}}{-}C_6H_4{-}OCH_2\underset{OH}{CHCH_2}\sim\right)_n O{-}C_6H_4{-}\underset{CH_3}{\overset{CH_3}{C}}{-}C_6H_4{-}OCH_2\underset{\diagdown O \diagup}{CH{-}CH_2}$$

9.11 Sulfur Analogues of Alcohols and Ethers

Since sulfur is directly below oxygen in group VI of the periodic table, there are sulfur counterparts of alcohols and ethers. The sulfur analogues of alcohols are called mercaptans, **thiols,** or alkyl hydrogen sulfides, and the sulfur analogues of ethers are thioethers or **sulfides.**

thiol
RSH, alkane in which a hydrogen has been replaced by SH

sulfide
RSR, sulfur with two bonded alkyl groups

$CH_3CH_2CH_2CH_2SH$	$CH_3SCH_2CH_2CH_3$
Butanethiol	Methyl propyl sulfide

Thiols and sulfides are especially noted for their strong, often unpleasant odors, as is illustrated by the following examples.

H_2S	CH_3SH, CH_3CH_2SH	$CH_3CH_2CH_2SH$
Hydrogen sulfide (rotten eggs)	Methanethiol ethanethiol (added to natural gas to provide a warning odor)	Propanethiol (from fresh onions)

Skunk

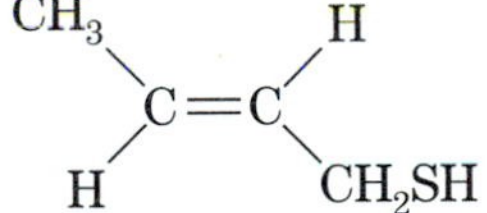

$$\overset{CH_3\quad\quad H}{\underset{H\quad\quad CH_2SH}{C{=}C}}$$

$$\overset{CH_3}{\underset{|}{}}CH_3CHCH_2CH_2SH$$

$$\overset{CH_3\quad\quad H}{\underset{H\quad\quad CH_2SSCH_3}{C{=}C}}$$

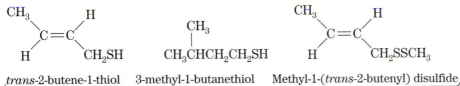

trans-2-butene-1-thiol   3-methyl-1-butanethiol   Methyl-1-(*trans*-2-butenyl) disulfide

(Main constituents of the scent of skunks)

$ClCH_2CH_2SCH_2CH_2Cl$	$CH_2{=}CHCH_2SH$, $(CH_2{=}CHCH_2)_2S$
2-chloroethyl sulfide (mustard gas used in chemical warfare)	Allyl mercaptan, allyl sulfide (responsible for the flavor and odor of garlic)

Garlic

This amino acid cysteine is a thiol, and cystine, another amino acid, is a disulfide. They can be interconverted by oxidation and reduction.

$$2HO\overset{O}{\overset{\|}{C}}\underset{NH_2}{\underset{|}{C}}HCH_2SH \underset{\text{Reduction}}{\overset{\text{Oxidation}}{\rightleftharpoons}} HO\overset{O}{\overset{\|}{C}}\underset{NH_2}{\underset{|}{C}}HCH_2S{-}SCH_2\underset{NH_2}{\underset{|}{C}}H\overset{O}{\overset{\|}{C}}OH$$

Cysteine   Cystine

The disulfide unit in cystine is important in determining the shapes of protein molecules. The cleavage and recombination of the disulfide units of cystine in hair is the basis of hair permanent waves.

Enzymes possessing the thiol group react with heavy metal ions such as those of mercury and lead. This can precipitate or deactivate the enzyme and is partly the basis of mercury and lead poisoning.

GETTING INVOLVED

✓ What is the difference between thiols and sulfides?

REACTION SUMMARY

A. Preparations of Alcohols

See summary in section 9.4; Problem 9.62.

B. Reaction of Phenols with Base

Section 9.6.A.2; Problems 9.14, 9.47–9.48.

$$ArOH + NaOH \longrightarrow ArONa + H_2O \qquad Ar = \text{aromatic ring like benzene}$$

C. Reaction of Alcohols with Sodium Metal

Section 9.6.B; Problems 9.15, 9.49(a).

$$2ROH + 2Na \longrightarrow 2RONa + H_2$$

D. Formation of Esters

Section 9.6.C; Problems 9.16 and 9.49(e).

$$ROH + HO\overset{\overset{\displaystyle O}{\|}}{C}R \xrightarrow{H^+} RO\overset{\overset{\displaystyle O}{\|}}{C}R + H_2O$$

E. Reactions of Alcohols with Hydrogen Halides

Section 9.7.A; Problems 9.17–9.20, 9.49(c), 9.50, 9.54, 9.56–9.58.

$$ROH + HX \longrightarrow RX + H_2O \qquad HX = HCl, HBr, HI$$

F. Reactions of Alcohols with $SOCl_2$ and PX_3

Section 9.7.B; Problems 9.21–9.22, 9.50.

$$ROH + SOCl_2 \longrightarrow RCl + SO_2 + HCl$$

$$ROH + PX_3 \longrightarrow RX + P(OH)_3 \qquad Px_3 = PCl_3, PBr_3$$

G. Reactions of Ethers with Hydrogen Halides

Section 9.7.C; Problems 9.23–9.25, 9.50, 9.56–9.58.

$$ROR + 1HX \longrightarrow RX + ROH$$

$$ROR + 2HX \longrightarrow 2RX + H_2O \quad HX = HCl, HBr, HI$$

H. Dehydration of Alcohols

Section 9.8; Problems 9.26–9.27; 9.53, 9.59, 9.64.

$$\underset{\text{H}\quad\text{OH}}{-\overset{|}{\underset{|}{C}}-\overset{|}{\underset{|}{C}}-} \xrightarrow{H_2SO_4} \;\rangle C{=}C\langle \; + H_2O$$

Orientation of Elimination: most substituted (with alkyl groups) alkene is formed predominantly.

I. Oxidation of Alcohols

Section 9.9; Problems 9.28, 9.49(d), 9.51, 9.64.

1° $RCH_2OH \xrightarrow{PCC} R\overset{O}{\overset{\|}{C}}H$ Aldehyde

1° $RCH_2OH \xrightarrow{CrO_3/H^+} R\overset{O}{\overset{\|}{C}}OH$ Carboxylic acid

2° $R\overset{OH}{\overset{|}{C}}HR \xrightarrow[CrO_3/H^+]{PCC\ or} R\overset{O}{\overset{\|}{C}}R$ Ketone

3° $R\overset{OH}{\overset{|}{\underset{\underset{R}{|}}{C}}}R \longrightarrow$ No reaction under these conditions

J. Reactions of Epoxides

Section 9.10; Example 9.3; Problems 9.31, 9.55, 9.61.

$$-\overset{|}{C}\overset{O}{\frown}\overset{|}{C}- + HA \longrightarrow -\overset{|}{\underset{\underset{A}{|}}{C}}-\overset{|}{\underset{\underset{OH}{|}}{C}}-$$

HA = H_2O, ROH, R_2NH, HX(X = Cl, Br, I)

Problems

9.31 **Isomerism and Nomenclature:** For the molecular formula $C_5H_{12}O$:

(a) draw all alcohols
(b) classify the alcohols as 1°, 2°, or 3°
(c) name the alcohols by the IUPAC system
(d) draw all ethers
(e) name the ethers by the IUPAC system

9.32 **IUPAC Nomenclature of Alcohols:** Name the following compounds by the IUPAC system of nomenclature:

(a) $CH_3(CH_2)_7CH_2OH$

(b) $CH_3(CH_2)_3\underset{\underset{OH}{|}}{C}HCH_3$

(c) $CH_3\overset{\overset{CH_3}{|}}{\underset{\underset{OH}{|}}{C}}CH_2CH_3$

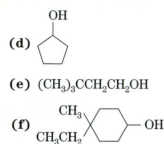

(d) cyclopentanol structure (OH on cyclopentane)

(e) $(CH_3)_3CCH_2CH_2OH$

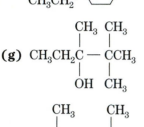

(f) cyclohexane ring bearing CH_3 and CH_3CH_2 on one carbon and OH on the opposite carbon

(g) $CH_3CH_2C(CH_3)(OH)-C(CH_3)_2CH_3$

(h) $CH_3C(CH_3)(OH)CH_2CH_2CH(CH_3)CH_3$

9.33 **IUPAC Nomenclature of Alcohols:** Name the following compounds by the IUPAC system of nomenclature:

(a) $CH_3CH(Br)-CH(Br)-CH(OH)CH_2CH_3$

(b) $CH_3CH(CH_2CH_3)CH_2CH(OH)CH_2CH_3$

(c) $HO(CH_2)_5OH$

(d) cyclohexane with OH, HO, OH at alternating positions (1,3,5)

9.34 **IUPAC Nomenclature of Unsaturated Alcohols:** Name the following compounds by the IUPAC system of nomenclature:

(a) $CH_3CH(OH)CH{=}CH_2$

(b) $CH_3CH_2CH(CH_2CH_3)C{\equiv}CCH_2OH$

(c) $HOCH_2CH{=}CHCH{=}CHCH_2OH$

(d) cyclopentene with OH

(e) benzene ring $-CH_2CH_2OH$

(f) $HC{\equiv}CCH(OH)CH{=}CHCH_3$

9.35 **IUPAC Nomenclature of Ethers:** Name the following compounds by the IUPAC system of nomenclature:

(a) $CH_3OCH_2CH_3$ (b) $CH_3CH_2OCH_2CH_3$

(c) $CH_3OCH_2(CH_2)_4CH_2OCH_2CH_3$

(d) cyclopentane $-OCH_2CH_2CH_3$ (e) cyclopropane with OCH_3

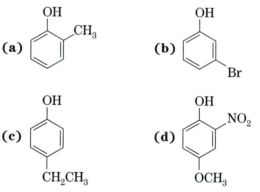

9.36 **IUPAC Nomenclature of Ethers:** Name the following compounds by the IUPAC system of nomenclature:

(a) $(CH_3O)_4C$

(b) $CH_3OCH_2CH_2CH_2OH$

(d) $CH_3CH_2OCH{=}CHCH_3$

(d) $CH_3OCH_2CH{=}CHCH_2OH$

9.37 **IUPAC Nomenclature of Phenols:** Name the following compounds by the IUPAC system of nomenclature:

(a) benzene ring with OH and ortho CH_3 (b) benzene ring with OH and meta Br

(c) benzene ring with OH and para CH_2CH_3 (d) benzene ring with OH, ortho NO_2, and para OCH_3

9.38 **Common Nomenclature:** Draw structures for the following compounds:

(a) secondary butyl alcohol
(b) neopentyl alcohol
(c) ethyl isopropyl ether
(d) cyclohexyl methyl ether
(e) allyl alcohol
(f) phenyl vinyl ether

9.39 **IUPAC Nomenclature:** Draw structures for the following compounds:

(a) 1,2,4-cyclopentanetriol
(b) 2-hexanethiol
(c) ethyl propyl disulfide
(d) ethyl propyl sulfide
(e) *p*-methoxyphenol
(f) 2-ethyl-4-isopropylcyclohexanol
(g) 2-methoxybutanol

9.40 **Physical Properties:** For each of the following sets of compounds, arrange the members in order of increasing boiling point:

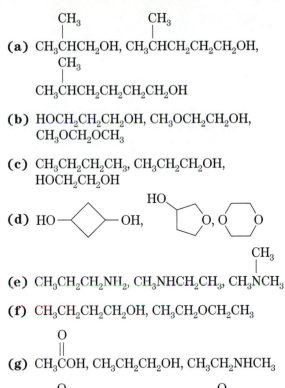

(a) $CH_3CH(CH_3)CH_2OH$, $CH_3CH(CH_3)CH_2CH_2CH_2OH$, $CH_3CH(CH_3)CH_2CH_2CH_2CH_2OH$

(b) $HOCH_2CH_2CH_2OH$, $CH_3OCH_2CH_2OH$, $CH_3OCH_2OCH_3$

(c) $CH_3CH_2CH_2CH_3$, $CH_3CH_2CH_2OH$, $HOCH_2CH_2OH$

(d) HO—(cyclobutane)—OH, HO-(tetrahydrofuran ring with O), O(1,4-dioxane ring)O

(e) $CH_3CH_2CH_2NH_2$, $CH_3NHCH_2CH_3$, $CH_3N(CH_3)CH_3$

(f) $CH_3CH_2CH_2CH_2OH$, $CH_3CH_2OCH_2CH_3$

(g) $CH_3\overset{O}{\overset{\|}{C}}OH$, $CH_3CH_2CH_2OH$, $CH_3CH_2NHCH_3$

(h) $CH_3\overset{O}{\overset{\|}{C}}CH_3$, $CH_3CH_2CH_2CH_3$, $CH_3\overset{O}{\overset{\|}{C}}OH$

(i) $HOCH_2CH_2\overset{O}{\overset{\|}{C}}H$, $CH_3CH_2\overset{O}{\overset{\|}{C}}OH$, $CH_3\overset{O}{\overset{\|}{C}}OCH_3$

(j) CH_4, CH_3Cl, CH_2Cl_2, $CHCl_3$, CCl_4, CBr_4

9.41 **Physical Properties:** For each of the following pairs of compounds, the *ortho* isomer has the lower boiling point. Hydrogen-bonding between the two groups, intramolecular in the *ortho* isomer and intermolecular in the *para,* is responsible for the difference in boiling point in each case. Draw the compounds and show how hydrogen bonding affects the boiling points.

(a) *ortho* and *para* nitrophenol (boiling points: 216°C, 279°C)

(b) *ortho* and *meta* hydroxybenzaldehyde (boiling points: 197°C, 240°C)

(c) *ortho* and *para* methoxyphenol (boiling points: 205°C, 243°C)

9.42 **Water-Solubility:** Sucrose ($C_{12}H_{22}O_{11}$), table sugar, dissolves to the extent of 200 g per 100 ml of water. How can one account for this tremendous solubility?

9.43 **Water Solubility:** Arrange the following compounds in order of increasing water solubility:

(a) ethanol, pentanol, hexanol

(b) pentane, heptanol, propanol

(e) hexane, hexanol, 1,2-ethanediol

(d) pentane, ethoxyethane, butanol

9.44 **Acidity:** Predict whether an acid-base neutralization reaction would occur between the following species. Consult Table 9.2.

(a) CH_3CH_3, NaOH **(b)** $CH_3CH_2SO_3H$, CH_3CO_2Na

(c) H_2SO_4, NaOH **(d)** CH_3OH, CH_3Na

(e) CH_3ONa, CH_3CO_2H **(f)** CH_3SO_3H, NaCl

(g) C_6H_5—OH, $NaOCH_3$

(h) C_6H_5—ONa, CH_3CH_2OH

9.45 **Acid-Base Neutralization:** For those combinations in problem 9.44 that react, write the acid-base neutralization equation.

9.46 **Acidity Constants:** Use the compounds i–iv for the comparisons that follow.

(i) $CH_3CH_2CH_2OH$ **(ii)** $CH_3CH_2CH_3$

(iii) $CH_3CH_2CO_2H$ **(iv)** H_3C—C_6H_4—OH

(a) Match the following approximate K_a's with the compounds: 10^{-49}, 10^{-11}, 10^{-5}, 10^{-16}.

(b) Match the following approximate pK_a's with the compounds: 11, 16, 5, 49.

(c) Arrange the compounds in order of acidity from least to most acidic.

9.47 **Acidity of Phenols:** Arrange the following phenols in order of increasing acidity. Be sure to determine whether the attached groups are electron-releasing or electron-withdrawing.

(a)

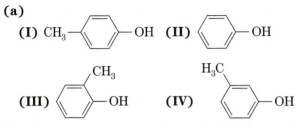

(I) CH_3—C_6H_4—OH (para) **(II)** C_6H_5—OH

(III) ortho-CH_3-C_6H_4—OH **(IV)** meta-H_3C-C_6H_4—OH

(b) Same structures as above, but replace —CH_3 with $CH_3\overset{O}{\overset{\|}{C}}$—

(c) (i) meta nitrophenol, (ii) para methylphenol, (iii) 2,4-dinitrophenol, (iv) phenol.

9.48 **Acidity of Phenols:** Write reaction equations showing the neutralization of the following phenols with sodium hydroxide:

(a) para methylphenol;
(b) ortho nitrophenol.

9.49 **Reactions of Alcohols:** Write equations illustrating the reaction of

(I) $CH_3CH_2CH_2CH_2OH$ **(II)** $CH_3CHCH_2CH_3$ (OH on C-2)

CH_3
|
(III) CH_3CCH_3
|
OH

with the following reagents:
(a) Na
(b) H_2SO_4 (dehydration)
(c) $HCl/ZnCl_2$
(d) CrO_3/H^+
(e) HNO_3

9.50 **Reactions of Alcohols to Form Alkyl Halides:** Write an equation showing the reaction between members of each of the following pairs of substances:
(a) CH_3CHCH_3 (OH on C-2), HBr
(b) CH_3CCH_3 (with CH_3 and OH on the central C), HI
(c) CH_3CHCH_3 (OH on C-2), $HCl/ZnCl_2$
(d) $CH_3CH_2CH_2OH$, $SOCl_2$
(e) $CH_3CHCH_2CH_3$ (OH on C-2), PBr_3

9.51 **Oxidation of Alcohols:** Write the products of the following oxidations:
(a) CH_3CH_2OH, CrO_3/H^+
(b) $CH_3CHCHCH_3$ (CH_3 on C-2, OH on C-3), CrO_3/H^+
(c) $CH_3(CH_2)_{10}CH_2OH$, PCC

9.52 **Reactions of Ethers:** Write an equation showing the reaction between members of each of the following pairs of reactants:
(a) $CH_3OCH_2CHCH_3$ (CH_3 on CH), 2HBr
(b) $CH_3CH_2OCH_2CH_3$, HCl
(c) $CH_3CHOCH_2CH_3$ (CH_3 on CH), 2HI
(d) O (tetrahydrofuran ring) + 2HBr

9.53 **Dehydration of Alcohols:** Write reaction equations for the dehydration of the following alcohols, using sulfuric acid. Show the predominant product when more than one elimination product is possible.
(a) $CH_3CHCHCH_3$ (CH_3 on C-2, OH on C-3)
(b) $CH_3CH—CCH_2CH_2CH_3$ (CH_3 on CH; CH_3 and OH on C)
(c) H_3C, OH, CH_3, CH_3 (substituted cyclohexane)

9.54 **Reaction of Alcohols with Hydrogen Halides:** Write the products of reaction of each of the alcohols in problem 9.53 with HBr.

9.55 **Reactions of Epoxides:** Write equations for the reaction of ethylene oxide with:

(a) H_2O/H^+; **(b)** CH_3CH_2OH/H^+;
(c) CH_3NH (CH_3 on N); **(d)** HBr;
(e) 2HBr

9.56 **Reaction Mechanisms:** Predict whether the following isomers would react with hydrogen halides by an S_N1 mechanism, an S_N2 mechanism, or both. Assume excess hydrogen halide.

(a) $CH_3CH_2CH_2CH_2OH$ **(b)** $CH_3OCH_2CH_2CH_3$
(c) $CH_3CH_2CHCH_3$ (OH on C-2) **(d)** CH_3OCHCH_3 (CH_3 on CH)
(e) $CH_3CH_2OCH_2CH_3$

9.57 **Nucleophilic Substitution Mechanisms:** Write mechanisms for the following reactions: **(a)** 1-propanol and HBr by S_N2; **(b)** 2-propanol and HBr by S_N1; **(c)** 1-methoxybutane with 2 HCl by S_N2 in both displacements; **(d)** 2-methoxybutane with 2HCl by S_N1 in one displacement and S_N2 in the other.

9.58 **Nucleophilic Substitution Mechanisms:** Write S_N1 mechanisms for the following reactions. Clearly show stereochemistry.

(a) reaction of optically active 3-methyl-3-hexanol with HCl;

(b) optically active 2-methoxybutane with HBr.

9.59 **Dehydration Mechanisms:** Write step-by-step reaction mechanisms for dehydration of the following alcohols with sulfuric acid:

(a) 2-methyl-2-butanol;

(b) cyclohexanol.

9.60 **Qualitative Analysis:** Suggest and explain a chemical method (preferably a simple test-tube reaction) for distinguishing between the members of the following sets of compounds. Tell what you would do and see.

(a) p-ethylphenol and 4-ethylcyclohexanol

(b) 1-butanol, 2-butanol, and 2-methyl-2-propanol

(c) 2-methyl-2-propanol and 2-butanol

9.61 **Epoxide Chemistry:** Ammonia reacts with three molecules of ethylene oxide to form triethanolamine, used as an intermediate in the manufacture of detergents, waxes, polishes, herbicides, toiletries, and cement additives. Write a structure for triethanolamine and rationalize its formation.

9.62 **Preparations of Alcohols:** Write reaction equations illustrating the hydration of the following alkenes with water and sulfuric acid to produce alcohols:

(a) 2-butene;

(b) 2-methyl-2-pentene;

(c) 1-hexene.

9.63 **Williamson Synthesis of Ethers:** There are two ways to prepare ethoxypropane by the Williamson synthesis. Write a reaction equation for each.

9.64 **Synthesis Using Alcohols:** From which alcohols and oxidizing agents can the following compounds be prepared?

(a) $CH_3CH_2CH_2\overset{\overset{\large O}{\|}}{C}CH_3$ **(b)** $CH_3CH_2CH_2CH_2\overset{\overset{\large O}{\|}}{C}H$

(c) $CH_3CH_2CH_2CH_2\overset{\overset{\large O}{\|}}{C}OH$

9.65 **Synthesis Using Alcohols:** From which alcohols and reagents can the following compounds be prepared? **(a)** methylcyclohexene; **(b)** 1-bromo-hexane; **(c)** 1-methoxyhexane.

Activities with Molecular Models

1. Make molecular models of C_2H_6O, one alcohol and one ether.

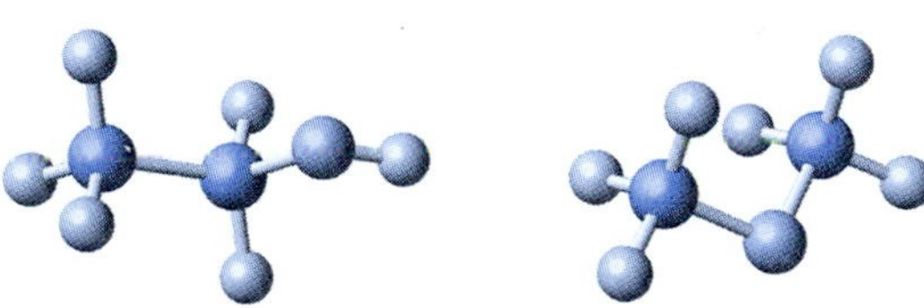

2. Make molecular models of the isomers of C_3H_8O, two alcohols and one ether. How many nonbonding electron pairs reside on the oxygen? What is the hybridization and geometric orientation of the carbons and the oxygen?

3. Make a model of one of the seven isomers of $C_4H_{10}O$ and then convert it into the other alcohols (four total) and ethers (three total). Draw each structure. Identify skeletal, positional, and functional isomers.

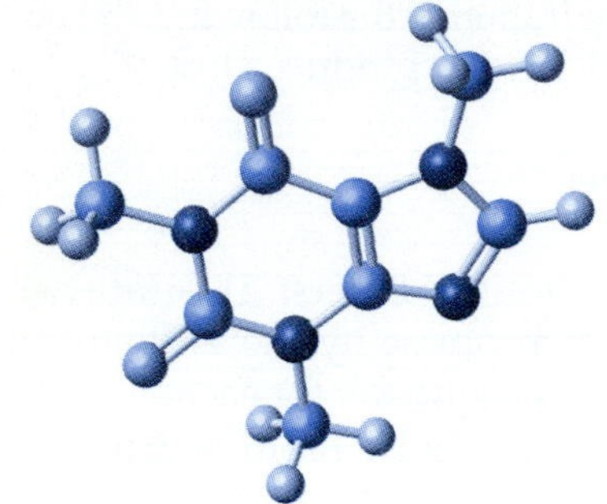

CHAPTER 10

Amines

Amines are nitrogen-containing compounds that can be described as derivatives of the inorganic compound ammonia, NH_3. Organic nitrogen compounds are found in all living organisms in varied forms including amino acids and proteins, genetic material (DNA, RNA), hormones, vitamins, and neurotransmitters. The main metabolic end-product through which nitrogen is excreted from the body is urea, $H_2N\overset{O}{\overset{\|}{C}}NH_2$, found in urine. The unpleasant odors we associate with organic decomposition arise from nitrogen-containing waste products, as illustrated in the following examples.

$H_2NCH_2CH_2CH_2CH_2NH_2$ Putrescine
$H_2NCH_2CH_2CH_2CH_2CH_2NH_2$ Cadaverine
} Bacterial decomposition products of protein

CH_3NH_2 Methylamine
$(CH_3)_2NH$ Dimethylamine
$(CH_3)_3N$ Trimethylamine
} "Fishy" odor of old fish; found in herring brine

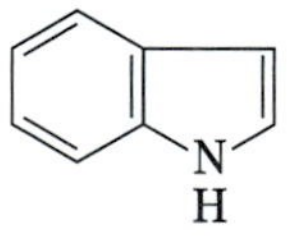

Indole 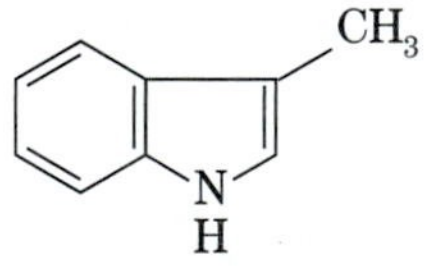   Skatole

Formed during putrefaction of protein; contribute to odor of feces

10.1 Structure of Amines

amine
derivative of ammonia in which one or more hydrogens are replaced by organic groups

Amines are derivatives of ammonia in which one or more hydrogens have been replaced by organic groups. If one hydrogen is replaced, a primary amine results; if two hydrogens are replaced, the amine is secondary; and if all three hydrogens are replaced, the amine is tertiary. This is illustrated by the following alkyl amines.

NH_3	$CH_3CH(CH_3)NH_2$	$CH_3CH_2N(H)CH_3$	$CH_3N(CH_3)CH_3$
	Primary amine	Secondary amine	Tertiary amine
Ammonia	Isopropylamine	Ethylmethylamine	Trimethylamine

Note that 1°, 2°, and 3° refer to the degree of substitution on the nitrogen, not the carbon, as we saw with alcohols in the previous chapter.

Should one or more of the organic groups be aromatic, the compound is an **arylamine.**

arylamine
derivative of ammonia in which at least one hydrogen is replaced by an aromatic ring

C_6H_5–NH_2 Aniline C_6H_5–$NHCH_3$ N-methylaniline

The nitrogen of amines has three bonded atoms and a nonbonding electron pair; as a result it is tetrahedral and sp^3-hybridized with bond angles very close to 109°. All bonds are σ bonds.

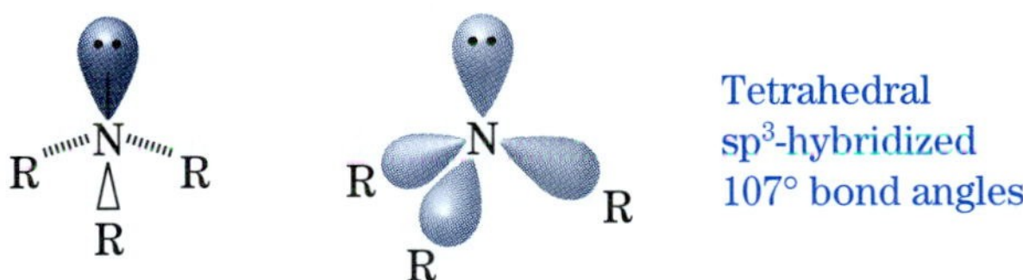

GETTING INVOLVED

✓ How do primary, secondary, and tertiary amines differ structurally?

✓ Why is the nitrogen in amines tetrahedral? What is the hybridization?

Problem 10.1

For the formula $C_4H_{11}N$, there are four primary, three secondary, and one tertiary amines. Draw those in each classification.

10.2 Nomenclature of Amines

A. Alkyl and Aromatic Amines

Simple amines can be acceptably named by common nomenclature as alkyl amines in which -amine is added to the name of the organic group.

CH_3NH_2	$(CH_3)_2NH$	$CH_3CH_2NH_2$	$(CH_3CH_2)_3N$
Methylamine	Dimethylamine	Ethylamine	Triethylamine

In systematic nomenclature, the suffix -amine is added to the name of the longest continuous carbon chain possessing the functional group (the e-ending in the alkane name is changed to amine). The simplest aromatic amine is named *aniline.*

CH_3NH_2	$CH_3CH_2CH_2NH_2$	$CH_3CH(NH_2)CH_2CH_3$	$C_6H_5NH_2$
Methanamine (Methylamine)	1-Propanamine (Propylamine)	2-butanamine (Isobutylamine)	Aniline

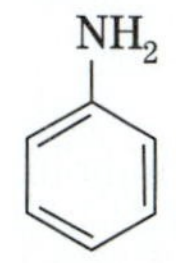

To name a substituted amine, both the name and location of the substituent must be identified. Substituents on a carbon chain are located by a number, whereas those on a nitrogen are identified by a capital N. These two principles are illustrated in the following examples:

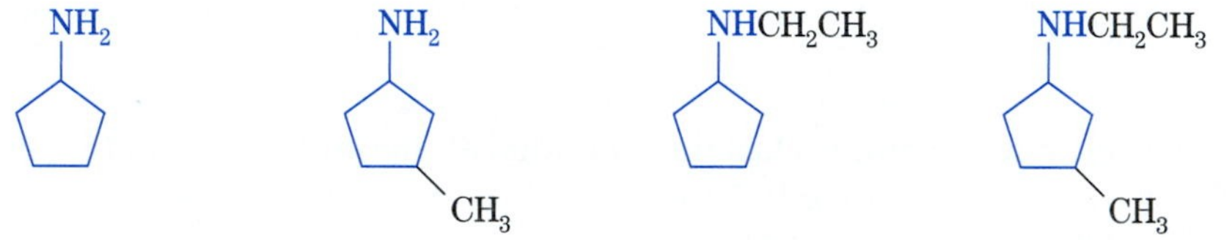

Cyclopentanamine

3-methyl-cyclopentanamine

N-ethyl-cyclopentanamine

N-ethyl-3-methyl-cyclopentanamine

N-substituents are named individually, in alphabetical order if they are different; the prefix *di-* is used if they are the same.

GETTING INVOLVED

Example 10.1

Name the following substituted amines:

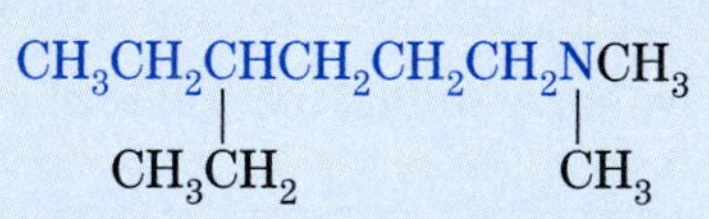

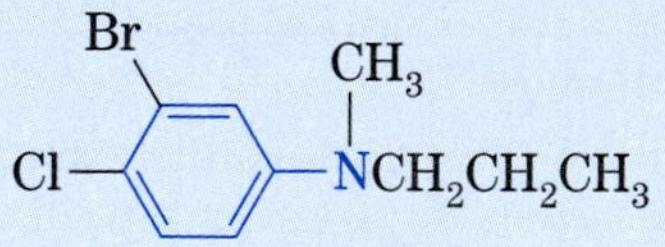

$CH_3CH_2CH(CH_2CH_3)CH_2CH_2CH_2N(CH_3)CH_3$

Br, Cl-substituted benzene ring with $-N(CH_3)CH_2CH_2CH_3$

Solution

The first step is to determine the base names of these compounds without substituents. In your mind, replace all substituents with hydrogen and name the simple compounds that result.

$CH_3CH_2CH_2CH_2CH_2CH_2NH_2$

1-Hexanamine

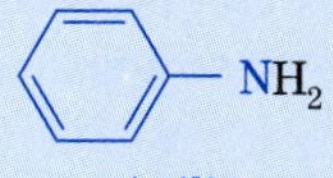

$C_6H_5NH_2$

Aniline

Now, add the name and location of each substituent.

4-ethyl-N,N-dimethyl-1-hexanamine

3-bromo-4-chloro-N-methyl-N-propylaniline

Problem 10.2

Name the following amines:

(a) $CH_3(CH_2)_7CH_2NH_2$

(b) 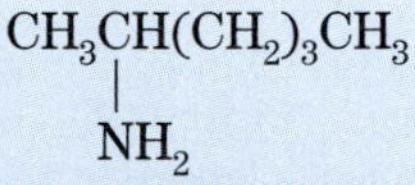 $CH_3CH(NH_2)(CH_2)_3CH_3$ — $CH_3CH(CH_2)_3CH_3$ with NH_2 on C-2

(c) $CH_3CH_2CHCH_2CH_3$ with $N(CH_3)_2$ on C-3

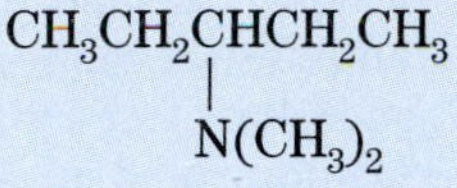

(d) 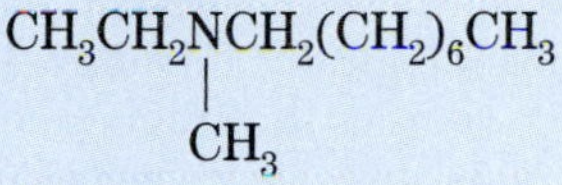 $CH_3CH_2NCH_2(CH_2)_6CH_3$ with CH_3 on N

(e) 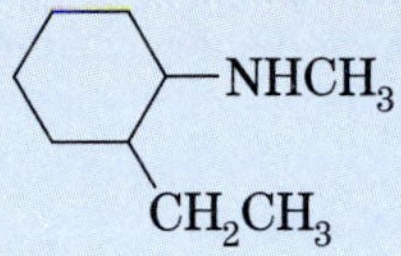 Cyclohexane ring bearing —$NHCH_3$ and, on the adjacent carbon, CH_2CH_3

(f) 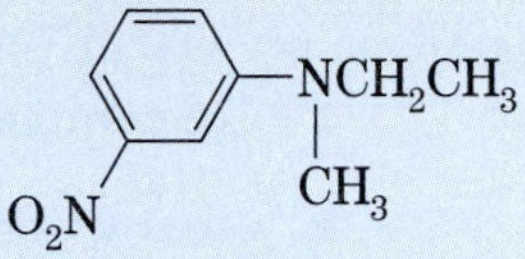 Benzene ring bearing —NCH_2CH_3 (with CH_3 on N) and, meta to it, O_2N

Problem 10.3

Give systematic names for the isomers drawn in Problem 10.1.

See related problems 10.25–10.26, 10.28.

B. Unsaturated Amines

Unsaturated amines are named in a systematic manner as illustrated in Example 10.2.

GETTING INVOLVED

Example 10.2

Name the following compound:

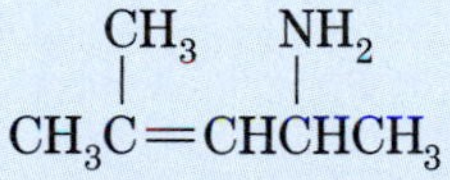

$$CH_3\overset{\overset{\large CH_3}{|}}{C}{=}CH\overset{\overset{\large NH_2}{|}}{C}HCH_3$$

1. Name the longest chain of carbons: pent.
2. Name carbon-carbon double and triple bonds with suffixes: penten.
3. Name the amine group with a suffix: pentenamine.
4. Number the carbon chain giving priority to the amine group since it was named with a suffix. Complete the base of the name: 3-penten-2-amine.
5. Name and locate all other groups with prefixes. The complete name is: 4-methyl-3-penten-2-amine.

Problem 10.4

Name the following unsaturated amines:

(a) $CH_3CH_2CH{=}CHCH_2NH_2$

(b) $(CH_3)_2CHC{\equiv}CCH_2CH_2NH_2$

(c) $H_2NCH_2CH{=}CHCH{=}CHCH_2NH_2$

(d) 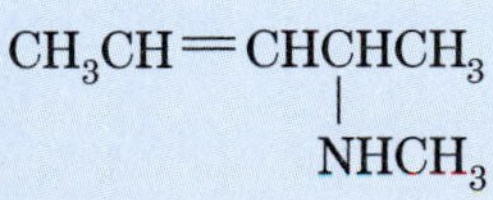 $CH_3CH{=}CH\underset{\underset{\large NHCH_3}{|}}{C}HCH_3$

See related problem 10.27.

Connections 10.1 www.prenhall.com/bailey

Nasal Decongestants, Diet Pills, and Stimulants

All of us have had an adrenalin rush; adrenalin (epinephrine) is released by the human adrenal gland in times of stress, fear, or excitement. The structure of adrenalin, a secondary amine, follows, along with some physiological effects that you might recognize.

Adrenalin belongs to a group of compounds sometimes referred to as *phenylalkylamines;* these compounds have a benzene ring, an alkyl group, and an amine group. A number of these bases are found in the herb *ma huang,* which has been used medicinally in China for more than 5000 years.

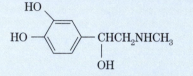

Adrenalin or epinephrine, a phenylalkylamine

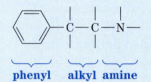

phenyl alkyl amine

The physiological effects we experience with adrenalin are common, in varying degrees, to other phenylalkylamines. For example, peyote, used in the religious rituals of Indian tribes in Mexico and legally in religious ceremonies of the Native American Church in the United States, is a Mexican cactus that produces the hallucinogenic drug mescaline. Amphetamine (also called dexedrine, benzedrine, dexxies, bennies, uppers, pep pills) was introduced in 1932 as a nasal decongestant; it was used in World War II to keep front-line troops alert. Today, it is found in some prescription diet pills. However, psychological dependence on amphetamines can occur, and withdrawal can lead to fatigue and depression. Ritalin, a somewhat more complex phenylalkylamine, is used to assist children and adults in coping with diagnosed attention deficit disorder (ADD).

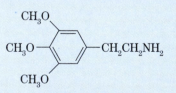

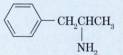

1. **Accelerated heartbeat**
2. **Contraction of blood vessels; increased blood pressure**
3. **Relaxation of bronchi and mucous membranes: runny nose, clear nasal passages**
4. **Restriction of digestive secretions: decreased appetite**
5. **Excitement, alertness**
6. **Energy: release of glucose from glycogen storage**

Mescaline

Amphetamine

Ritalin

Many over-the-counter nasal decongestants, both topical and oral, contain phenylalkylamines, most commonly ephedrine, phenylephrine, and phenylpropanolamine hydrochloride.

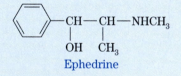

Ephedrine

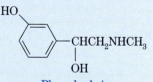

Phenylephrine, neosynephrine

Phenylpropanolamine hydrochloride

These drugs function by contracting the arterioles within the nasal mucous membranes, thereby restricting blood flow to this area. Swelling is reduced, nasal passages are opened, and the ventilation and drainage of sinuses are possible. However, prolonged use of decongestants, especially topical sprays, can result in restricted nutrient flow

Connections 10.1 *(cont.)*

to the area and in reduced waste removal from the sinuses, leaving the affected tissues swollen and susceptible to infection. Long-duration nasal decongestants contain compounds like xylometazoline hydrochloride, a compound that is structurally related to phenylalkylamines.

Chlorpheniramine

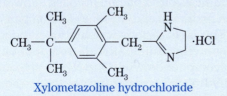

Xylometazoline hydrochloride

Phenylpropanolamine hydrochloride is also used in diet pills, often in doses similar to oral nasal decongestants. The appetite suppressant effect of phenylalkylamines is at work here, but other physiological actions can lead to side effects. As a consequence, oral decongestants and diet pills containing phenylpropanolamine hydrochloride often have printed cautions to people with heart or blood pressure problems (effects 1 and 2) and diabetes (effect 6).

Many oral nasal decongestants and allergy preparations contain antihistamines. When the body begins to experience an allergic reaction such as to pollen (hay fever), insect stings, and many other irritants, histamine is produced. Most symptoms of allergies are caused by histamine. Antihistamines reduce or eliminate the effects of histamines; some common nonprescription ones follow.

Some antihistamines are used as sleeping pills and sedatives, and many that treat allergies can cause drowsiness. Fexofenadine (Allegra®) and loratadine (Claritin®) are common prescription antihistamines that do not cause drowsiness because they do not penetrate the blood-brain barrier.

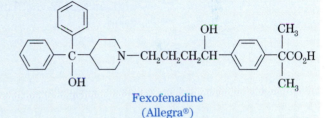

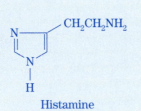

Histamine

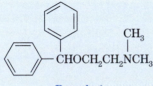

Benadryl

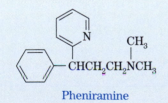

Pheniramine

Fexofenadine (Allegra®)

Loratadine (Claritin®)

Peyote cacti, a source of mescaline

10.3 Physical Properties of Amines

hydrogen-bonding intermolecular attractions caused by hydrogen bonded to electronegative element (O, N, F) being attracted to a nonbonding electron pair of another electronegative element

Like other classes of organic compounds, amines have melting points and boiling points that generally increase with molecular weight, as illustrated in Table 10.1. However, the magnitude of the boiling points and the fact that lower molecular weight amines are water-soluble can be attributed to their ability to participate in **hydrogen-bonding** (Figure 10.1). The electronegativity difference between nitrogen and hydrogen produces a partial negative charge on nitrogen and a partial positive charge on hydrogen. This polarity, the minute size of hydrogen, and the presence of a nonbonding electron pair on nitrogen allow the strong intermolecular attraction between hydrogen and the lone electron pair on nitrogen of a second amine molecule. This is characteristic of hydrogen-bonding and is illustrated in Figure 10.1 using methylamine. (Recall from section 9.2 that hydrogen-bonding is possible in compounds with O—H, N—H, and F—H bonds.) In a water solution, the hydrogen on water is attracted to the nitrogen lone pair, and the hydrogen bonded to nitrogen on methylamine is attracted to the oxygen lone pairs.

Methylamine, because it can hydrogen-bond, has a considerably higher boiling point than does ethane, a nonpolar alkane. But its boiling point is not as high as that of methanol, since the N—H bond is not as polar as the O—H bond.

	CH_3CH_3	CH_3NH_2	CH_3OH
mol wt	30	31	32
bp	−89°C	−6°C	65°C

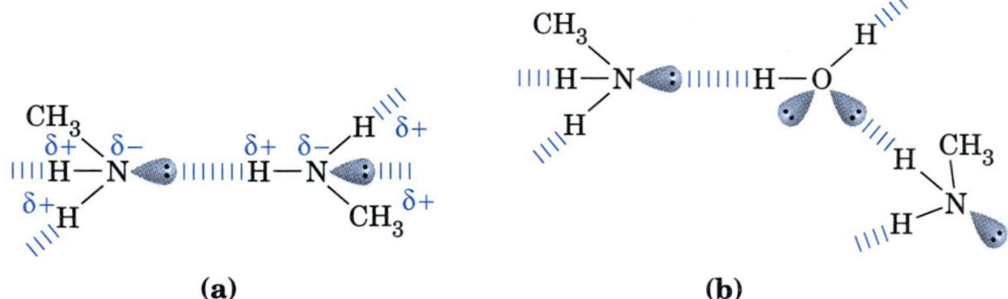

Figure 10.1 Hydrogen-bonding in methylamine. (a) Pure liquid. (b) Water solution.

Table 10.1 Physical Properties of Amines

Structure	Molecular Weight	Melting Point, °C	Boiling Point, °C
CH_3NH_2	31	− 94	− 6
$CH_3CH_2NH_2$	45	− 81	17
$CH_3CH_2CH_2NH_2$	59	− 83	48
$(CH_3CH_2)_2NH$	73	− 48	56
$(CH_3)_3N$	59	−117	3
$C_6H_5-NH_2$	93	− 6	184

Connections 10.2 www.prenhall.com/bailey

Local Anesthetics and Cocaine

We can all appreciate the spray that relieves the pain of a severe sunburn or the injection that numbs the mouth for dental work. These are local anesthetics, a class of compounds that cause a loss of sensation to the area to which they are applied. The most common over-the-counter formulations include benzocaine (Anbesol®, Lanacane®, Solarcaine®), xylocaine (Lidocaine®), and tetracaine (Cetacaine®). Throat lozenges and sprays as well as Ayds® diet candy also contain benzocaine. The anesthetic Novacain® is actually procaine. Medications taken for coughing may also contain local anesthetics. All of these compounds are amines.

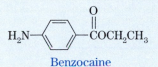

$H_2N-C_6H_4-COCH_2CH_3$ (C=O)

Benzocaine

$CH_2(CH_2)_3HN-C_6H_4-COCH_2N(CH_3)_2$ (C=O)

Tetracaine

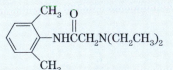

2,6-$(CH_3)_2C_6H_3-NHCCH_2N(CH_2CH_3)_2$ (C=O)

Xylocaine

$H_2N-C_6H_4-COCH_2CH_2N(CH_2CH_3)_2$ (C=O)

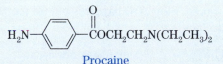

Procaine

Probably the most infamous local anesthetic today is cocaine. Used as an aide to nasal surgery, this compound is also abused for its effects on the central nervous system: euphoria, assertiveness, alertness, and general stimulation. It can be smoked, inhaled, injected, or rubbed on the gums.

Cocaine is isolated from the leaves of *Erythroxylon coca,* which grows at high elevations in the Andes mountains of Bolivia, Columbia, and Peru. The oval plant leaves can be harvested four to five times per year. Natives of South America mix the leaves with ashes, packing the mixture between cheek and gums. This procedure causes very slow absorption of the active compound and is stimulating though not usually euphoric. It is meant to aid in the adaptation to high altitudes and hard, servile labor.

The isolation of pure cocaine can be performed under acidic conditions, which produce a salt form. Extraction with a nonpolar solvent like diethyl ether allows the substance to be easily volatilized (called "free-basing"). Free-basing with bicarbonate added leads to a solid, rocklike form which, when burned, produces a popping sound due to the liberation of CO_2 from the bicarbonate. This is "crack" cocaine. Such purified forms can be quickly addicting. Excessive use results in hypertension, delirium, increased body temperature, seizures, and respiratory failure. There also exists the risk of excessive cardiac stimulation, which can lead to sudden death even upon one use.

CH_3, N, $COCH_3$ (C=O), $OC-C_6H_5$ (C=O)

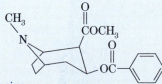

Cocaine

Cocaine was so venerated by the Incas that only priests and aristocrats were allowed to use it. The invading Spanish brought it back to Europe where it was cultivated in the 1800s.

Native South American with coca leaves

As the ability to hydrogen-bond decreases, so does the boiling point, as is illustrated by the following compounds of similar molecular weight:

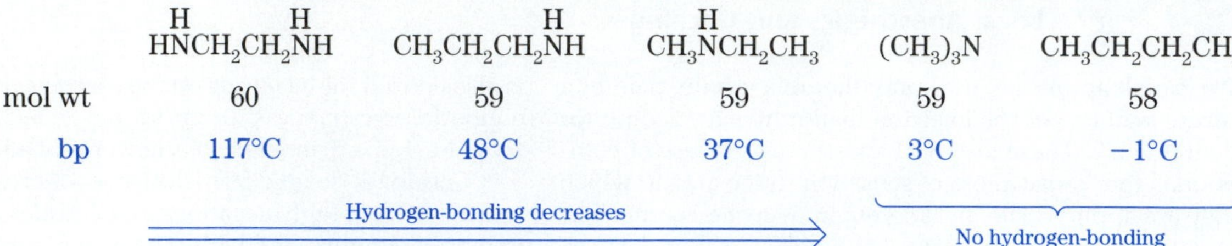

GETTING INVOLVED

✓ Why do the boiling points of amines generally increase with molecular weight? Why do primary amines have higher boiling points than alkanes of comparable molecular weight? Why do alcohols of comparable molecular weight have boiling points higher than those of primary amines? Why do boiling points of isomeric amines generally decrease in the order primary, secondary, tertiary?

✓ What are the structural characteristics that allow hydrogen bonding in amines?

Problem 10.5

Arrange the following compounds in order of increasing boiling point: Explain your answers.

(a) (i) $CH_3CH_2CH_2CH_2NH_2$ (ii) $CH_3CH_2N(CH_3)_2$ (iii) $CH_3CH_2CH_2NHCH_3$

(b) (i) CH_3CH_2OH (ii) $CH_3CH_2NH_2$ (iii) $CH_3CH_2CH_3$ (iv) CH_3NHCH_3

See related problems 10.29–10.31.

10.4 Basicity of Amines

A. Salt Formation

Basicity and the ability to react with acids are the most characteristic properties of amines. The presence of a nonbonding electron pair on the nitrogen makes amines Lewis bases, and, like ammonia, they can share this pair of electrons with hydrogen ions from strong mineral acids.

$$NH_3 + HNO_3 \longrightarrow NH_4^+NO_3^-$$

Ammonium nitrate, an important fertilizer

$$CH_3NH_2 + HCl \longrightarrow CH_3NH_3^+Cl^-$$

Methyl ammonium chloride

$$CH_3NH_2\!: + H^+ :\ddot{Cl}:^- \longrightarrow CH_3\overset{+}{N}H_2\!:\!H + :\ddot{Cl}:^-$$

GETTING INVOLVED

✓ Why are amines basic?

Example 10.3

Write an equation illustrating the reaction of N-methylbutanamine with HCl.

Solution

The lone pair of electrons on nitrogen bonds to the hydrogen ion of HCl, forming an ammonium salt.

$$CH_3CH_2CH_2CH_2\overset{H}{\underset{\cdot\cdot}{N}}CH_3 + HCl \longrightarrow CH_3CH_2CH_2CH_2\overset{H}{\underset{H}{\overset{+}{N}}}CH_3\ Cl^-$$

Problem 10.6

Write reaction equations for the following acid-base reactions: **(a)** propylamine/HBr; **(b)** dimethylamine/HNO_3; **(c)** triethylamine/HCl.

Problem 10.7

Ammonium sulfate and ammonium phosphate are also important fertilizers that can be prepared from ammonia and sulfuric and phosphoric acids, respectively. Write the reaction equations.

B. Expressing Relative Basicities of Amines: The Basicity Constant

Amines are weak bases. When they are dissolved in water, an equilibrium is established in which the water donates a hydrogen ion to the amine. The extent to which this reaction occurs is a measure of the amine's basicity. This is expressed by an equilibrium constant called the **basicity constant,** K_b, or its negative logarithm, **pK_b**. The basicity constant is conceptually analogous to the acidity constant presented in the previous chapter, section 9.6.A. Remember, water is in excess; its concentration therefore is considered constant, and it does not appear in the K_b expression. The basicity of amines is often expressed using K_a and pK_a; this is described in section 10.4D.

basicity constant

pK$_b$
the negative logarithm of the basicity constant

$$R_3N\!: + H_2O \rightleftharpoons R_3NH^+ + OH^-$$

$$K_b = \frac{[R_3NH^+]\,[OH^-]}{[R_3N]} \qquad pK_b = -\log K_b$$

$$K_a \times K_b = 10^{-14} \qquad pK_a + pK_b = 14$$

Since the protonated amine is in the numerator of the equilibrium expression, larger K_b's signify greater basicity. The opposite is true of pK_b's, because they are defined as the negative logarithm of K_b; the smaller the pK_b, the stronger the base.

10^{-14} Small K_b		Large K_b 10^0
14 Large pK_b		Small pK_b 0
Weak bases	Increasing basicity →	Strong bases

Table 10.2 describes the relative basicities of some selected amines in terms if K_b and pK_b.

GETTING INVOLVED

✓ How do the mathematical definitions of K_b and pK_b differ?

✓ How do K_b's and pK_b's vary numerically with increasing or decreasing basicities of amines?

Table 10.2 Basicities of Selected Amines

Amine	Structure	K_b[1]	pK_b[1]	pK_a[1]
Ammonia	NH_3	1.79×10^{-5}	4.74	9.26
Primary Amines				
Methylamine	CH_3NH_2	4.42×10^{-4}	3.35	10.65
Ethylamine	$CH_3CH_2NH_2$	4.37×10^{-4}	3.36	10.64
Secondary Amines				
Dimethylamine	$(CH_3)_2NH$	5.29×10^{-4}	3.28	10.72
Diethylamine	$(CH_3CH_2)_2NH$	9.80×10^{-4}	3.01	10.99
Tertiary Amines				
Trimethylamine	$(CH_3)_3N$	5.49×10^{-5}	4.36	9.74
Triethylamine	$(CH_3CH_2)_3N$	5.71×10^{-4}	3.25	10.75
Aromatic Amines				
Aniline	$C_6H_5-NH_2$	4.00×10^{-10}	9.40	4.60
p-Methylaniline	$H_3C-C_6H_4-NH_2$	1.20×10^{-9}	8.92	5.08
p-Nitroaniline	$O_2N-C_6H_4-NH_2$	1.00×10^{-13}	13.00	1.00
Amides				
Acetamide	$CH_3\overset{O}{\overset{\|}{C}}NH_2$	3.10×10^{-15}	14.51	- 0.51

[1]These are basicity constants for the amine; high K_b's and low pK_b's signify high basicities. See section 10.4.B.

[2]These are pK_a's for the ammonium salt of the amine; high pK_a's signify high basicities for the amines (low acidities for the ammonium salts). See section 10.4.D.

Problem 10.8

Arrange the following K_b's and pK_b's from least to most basic: **(a)** K_b's: 5.6×10^{-5}, 9.1×10^{-10}, 3.6×10^{-4}; **(b)** pK_b's: 3.2, 9.1, 4.3.

Problem 10.9

Using K_b's and pK_b's in Table 10.2, determine which amine in each of the following pairs is more basic: **(a)** ethylamine or diethylamine; **(b)** methylamine or triethylamine; **(c)** triethylamine or aniline; **(d)** *p*-methylaniline or *p*-nitroaniline.

See related problem 10.32.

C. Relationship of Structure and Basicity in Amines

Although the K_b's of amines fall within a fairly narrow range, we can make several observations from Table 10.2 concerning the relationship of structure to amine basicity.

1. *Electron-releasing groups increase basicity; alkyl amines are more basic than ammonia.* Alkyl amines are generally more basic than ammonia, since the electron-releasing alkyl groups increase electron density around the nitrogen, thereby increasing the availability of the lone pair of electrons. They also stabilize the positive charge in the ammonium ion that results from reaction of the amine with hydrogen ion. From Table 10.2 we can see pK_b's mostly between 3 and 4 for

primary, secondary, and tertiary amines, compared to 4.74 for ammonia; alkyl amines are about ten times more basic.

$$\begin{matrix} & H & & \\ H & N: & \xrightarrow{H^+} & \\ & H & & \end{matrix} \begin{matrix} & H_+ & \\ H & N:H & \\ & H & \end{matrix} \qquad \begin{matrix} & & H \\ H_3C & \longrightarrow & N: \\ & & H \end{matrix} \quad \xrightarrow{H^+} \quad \begin{matrix} & & H_+ \\ H_3C & \longrightarrow & N:H \\ & & H \end{matrix}$$

Ammonia; no electron-releasing groups

Electron release increases lone pair availability

Electron release stabilizes positive charge

2. *Electron-withdrawing groups decrease basicity; amides are much less basic than ammonia.* Amides are 10 billion times less basic than ammonia (pK_b around 15 compared to around 5 for ammonia in Table 10.2) and do not form ammonium salts with mineral acids. The carbon-oxygen double bond of an amide is a strong electron-withdrawing group and delocalizes the nonbonding electron pair by resonance, making it unavailable for reaction with the hydrogen ion.

$$\begin{matrix} :\ddot{O} \\ \| \\ CH_3C - \ddot{N}H_2 \end{matrix}$$

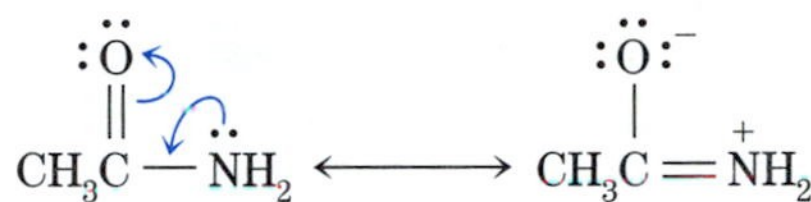

Polar C=O bond withdraws electrons

Nonbonding electrons delocalized by resonance

GETTING INVOLVED

✓ Explain why electron-releasing groups increase basicity of amines; consider both the amine and ammonium salt. What is an example of an electron-releasing group?

✓ Explain why electron-withdrawing groups decrease basicity. What is an example of an electron-withdrawing group?

Problem 10.10

Arrange the following compounds in order of increasing basicity. Explain your order.

(a) (i) $CH_3CH_2CH_2NH_2$ (ii) NH_3 (iii) $CH_3CH_2NHCH_3$

(b) (i) NH_3 (ii) $CH_3CH_2CH_2\overset{O}{\overset{\|}{C}}NH_2$ (iii) $CH_3\overset{O}{\overset{\|}{C}}NH\overset{O}{\overset{\|}{C}}CH_3$

(c) (i) NH_3 (ii) $CH_3CH_2NH_2$ (iii) $CH_3\overset{O}{\overset{\|}{C}}NH_2$ (iv) CH_3NHCH_3 (v) $CH_3\overset{O}{\overset{\|}{C}}NH\overset{O}{\overset{\|}{C}}CH_3$

(d) (i) $CH_3CH_2CH_2NH_2$ (ii) $CH_3CH_2\overset{O}{\overset{\|}{C}}NH_2$ (iii) $CH_3\overset{O}{\overset{\|}{C}}CH_2NH_2$ (iv) $H\overset{O}{\overset{\|}{C}}CH_2CH_2NH_2$ (v) NH_3

3. *Aromatic amines are considerably less basic than alkyl amines.* This dramatic decrease in basicity is due to resonance of nitrogen's lone pair of electrons with the benzene ring, as shown in Figure 10.2. The resonance not only decreases the availability of nitrogen's lone pair but also adds stability to aromatic amines, making them less reactive in acid-base reactions. Note in Table 10.2 that the simplest alkyl amine, methylamine, has a pK_b of 3.35, whereas the simplest aromatic amine, aniline, has a pK_b of 9.40; this represents six powers of ten (one million times) less basicity for aniline.

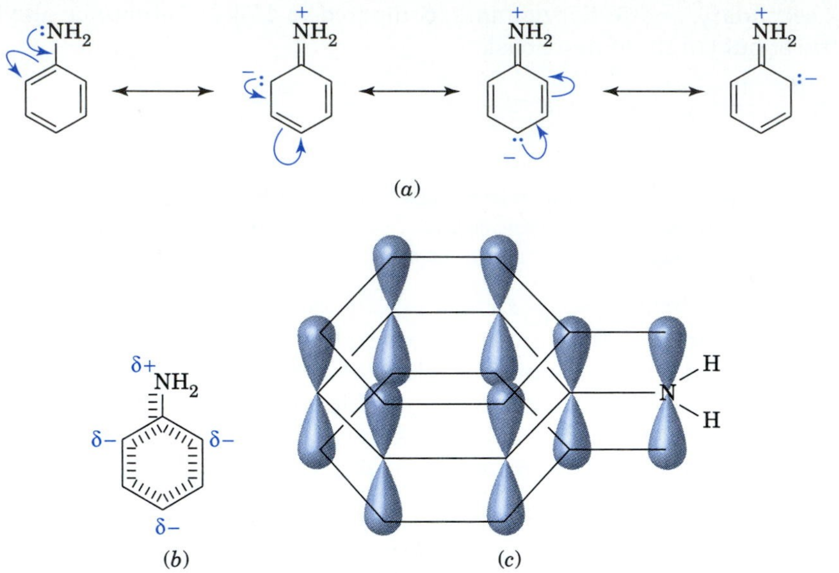

(a)

(b)

(c)

Figure 10.2 Resonance in aniline and aromatic amines. (a) Resonance forms. (b) Resonance hybrid. (c) π bonding picture

This is analogous to the increased acidity of phenols compared to alcohols as presented in section 9.6A.3. Aniline is less basic than alkylamines because its nonbonding pair is drawn into the ring by resonance. Phenol is more acidic than alcohols because a nonbonding electron pair of the phenoxide ion is drawn into the ring thus stabilizing the ion by resonance.

4. *The presence of electron-releasing groups on aromatic amines increases basicity, whereas electron-withdrawing groups decrease basicity.* Electron-releasing groups tend to increase electron availability (*p*-methylaniline is more basic than aniline, Table 10.2). Withdrawing groups, however, further pull the nonbonding pair of electrons on nitrogen to the ring, making it less available for reaction with acids (*p*-nitroaniline is almost 4000 times less basic than aniline). Since the resonance shown in Figure 10.2 has special significance at ortho and para positions, the placement of groups at these positions has a larger effect on basicity than groups at meta positions. This is illustrated with nitroanilines. Compare this concept with the acidity of nitrophenols discussed in section 9.6.A.2; also compare Figure 10.2 with Figure 9.3.

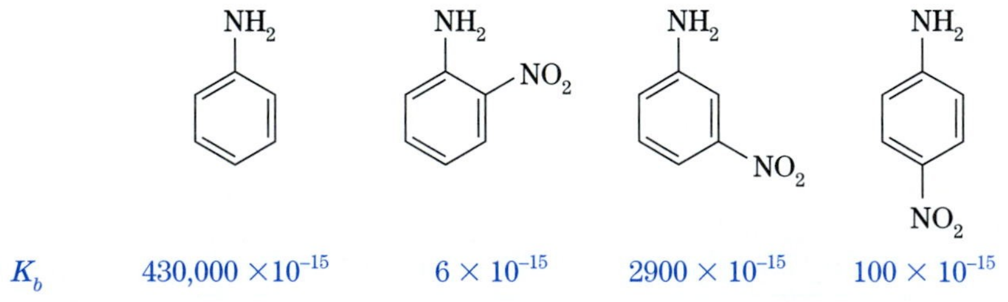

	aniline	o-nitroaniline	m-nitroaniline	p-nitroaniline
K_b	$430{,}000 \times 10^{-15}$	6×10^{-15}	2900×10^{-15}	100×10^{-15}

GETTING INVOLVED

✓ Why are aromatic amines less basic than alkyl amines?

✓ How do electron-releasing groups and electron-withdrawing groups on the aromatic ring affect basicity? Why are the influences of these groups stronger at ortho and para positions relative to meta positions?

Problem 10.11

Arrange the following compounds in order of increasing basicity. Explain your order.

(a) 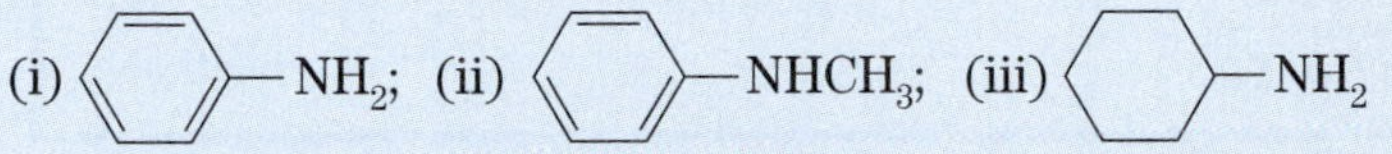 (i) $C_6H_5-NH_2$; (ii) $C_6H_5-NHCH_3$; (iii) $C_6H_{11}-NH_2$

(b) 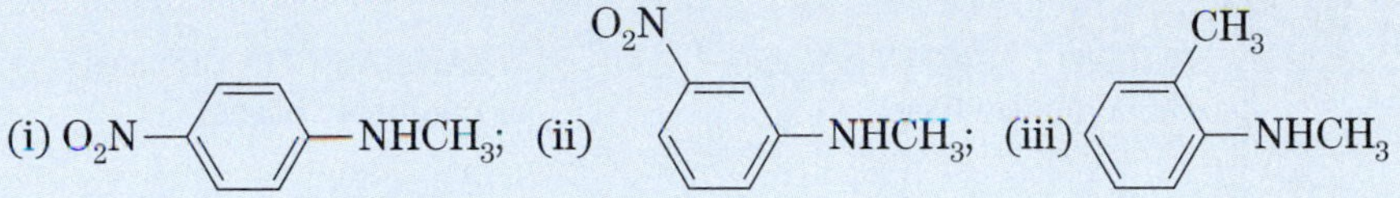 (i) $O_2N-C_6H_4-NHCH_3$ (para); (ii) $C_6H_4(NO_2)-NHCH_3$ (meta, O_2N); (iii) $C_6H_4(CH_3)-NHCH_3$ (meta, CH_3)

(c) (i) $CH_3\overset{O}{\overset{\|}{C}}-C_6H_4-NH_2$; (ii) $CH_3\overset{O}{\overset{\|}{C}}-C_6H_3(NO_2)-NH_2$; (iii) $CH_3-C_6H_4-NH_2$

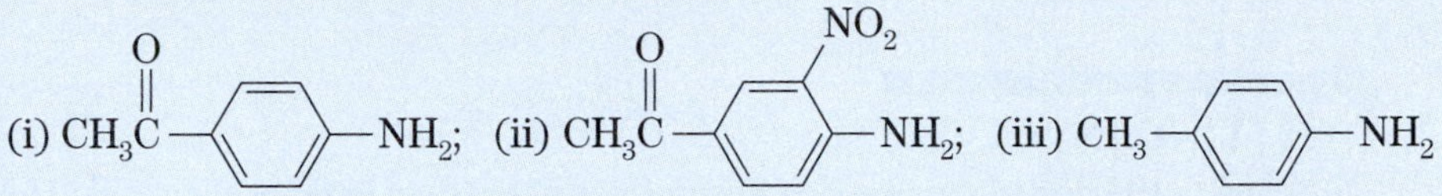

(d) 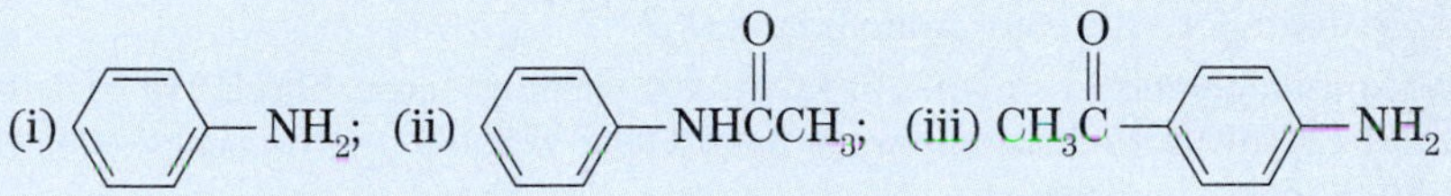 (i) $C_6H_5-NH_2$; (ii) $C_6H_5-NH\overset{O}{\overset{\|}{C}}CH_3$; (iii) $CH_3\overset{O}{\overset{\|}{C}}-C_6H_4-NH_2$

See related problems 10.34–10.35.

D. Expressing Basicity with Acidity Constants

For convenience and consistency, relative basicities are sometimes expressed by using **acidity constants, K_a's** and **pK_a's**, which were described in section 9.6.A.1. This is common when amino acids are described, and we will use this expression in Chapter 17. To express basicity of amines as acidity constants, we must write the equilibrium showing the ammonium salt, the conjugate acid, ionizing in water. The acidity constant is defined as shown.

acidity constant
K_a, product of the concentrations of the ionized form of an acid divided by the concentration of the un-ionized form

pK_a
the negative logarithm of the acidity constant

$$R_3NH^+ + H_2O \rightleftharpoons H_3O^+ + R_3N$$

$$K_a = \frac{[H_3O^+][R_3N]}{[R_3NH^+]} \qquad pK_a = -\log K_a$$

Because the concentrations of the products, including H_3O^+, appear in the numerator, a high K_a signifies a comparatively high degree of ionization of the ammonium salt, R_3NH^+. If the ammonium salt is highly ionized, the amine must have little affinity for the hydrogen ion and therefore is a comparatively weak conjugate base (section 9.6.A.1). Thus, the higher the K_a, the weaker the amine is as a base. Since pK_a is the negative logarithm of K_a, low pK_a's mean low basicity (or high acidity of the salt). On the other end of the scale, low K_a's or high pK_a's mean high basicity. If the ammonium salt is poorly ionized, its acidity is low (low K_a) and its conjugate base, the amine, is a relatively strong base that holds tightly to the bonded hydrogen ion. The general meaning of the K_a and pK_a scales is summarized in the following diagram.

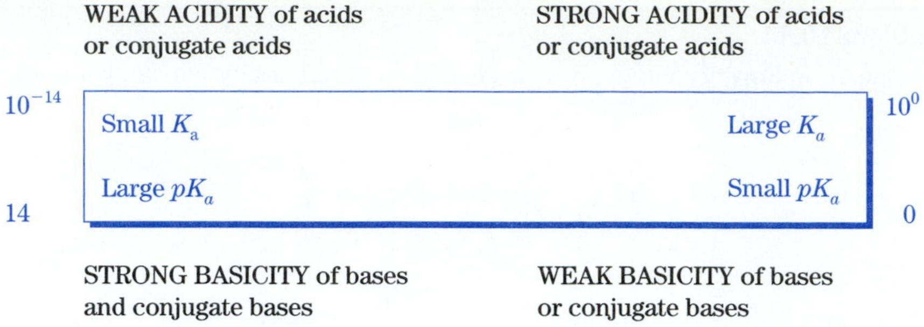

WEAK ACIDITY of acids or conjugate acids		STRONG ACIDITY of acids or conjugate acids	
10^{-14}	Small K_a	Large K_a	10^0
14	Large pK_a	Small pK_a	0
STRONG BASICITY of bases and conjugate bases		WEAK BASICITY of bases or conjugate bases	

GETTING INVOLVED

✓ How are K_a and pK_a mathematically defined in terms of basicity of amines? How are these definitions different from K_b and pK_b?

✓ What do high and low values of K_a and pK_a mean in terms of relative basicities of amines? How does this compare to K_b and pK_b? What do high and low values of K_a and pK_a mean in terms of acidity of acids?

Problem 10.12

Assume that the following K_a's or pK_a's are being used to describe the relative basicity of a group of amines. Arrange from least basic to most basic.

(a) K_a's: 9.9×10^{-10}, 8.3×10^{-10}, 2.3×10^{-11}

(b) pK_a's: 5.25, 10.74, 9.81

Problem 10.13

Assume the K_a's and pK_a's in problem 10.12 are being used to describe the relative acidities of a group of carboxylic acids. Arrange them from least acidic to most acidic.

See related problem 10.33.

alkylation introduction of an alkyl group into a molecule

10.5 Amines as Nucleophiles: Alkylation by Nucleophilic Substitution

Most of the reactions of amines are due to the presence of a nonbonding pair of electrons on the nitrogen. We have already seen that amines are Lewis bases and react with acids because of this nonbonding pair (section 10.4). Amines are also effective nucleophiles and will react with alkyl halides.

Consider, for example, the reaction of methylamine with methyl chloride. The nonbonding pair of electrons on nitrogen is attracted to the positive carbon of the polar carbon-chlorine bond. A new bond forms between the nitrogen and carbon using this pair of electrons, and the chloride is displaced by an S_N2 mechanism (section 8.4.D).

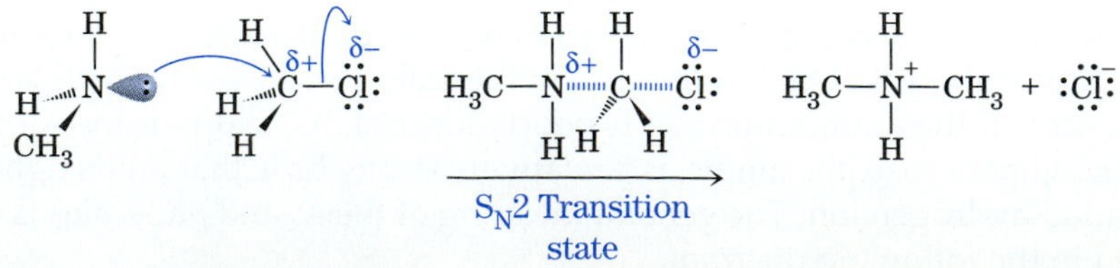

$$CH_3NH_2 + CH_3^{\delta+}-Cl^{\delta-} \longrightarrow [H_3C-NH_2\cdots CH_3\cdots Cl]^{\ddagger} \longrightarrow H_3C-\overset{+}{N}H_2-CH_3 + :Cl:^-$$

S_N2 Transition state

Dimethylammonium chloride, an amine salt, results. But the reaction does not stop here. A hydrogen ion on this salt can be transferred to unreacted methylamine (it is basic), which, especially early in the reaction, is in high concentration.

$$(CH_3)_2NH_2^+Cl^- + CH_3NH_2 \longrightarrow (CH_3)_2NH + CH_3NH_3^+Cl^-$$

This frees dimethylamine to react with methyl chloride, in the same way methylamine does, by an S_N2 reaction.

$$(CH_3)_2\ddot{N}H + CH_3Cl \longrightarrow (CH_3)_3NH^+Cl^-$$

The salt formed can also be neutralized by methylamine. The resulting trimethylamine can react with yet another molecule of methyl chloride (again S_N2) to form what is called a *quaternary ammonium salt*, a positive nitrogen with four bonded alkyl groups. Because there are no more replaceable hydrogens on the nitrogen, the reaction stops here.

$$(CH_3)_3NH^+\ Cl^- + CH_3\ddot{N}H_2 \longrightarrow (CH_3)_3N\colon + CH_3NH_3^+\ Cl^-$$

$$(CH_3)_3N\colon + CH_3Cl \longrightarrow (CH_3)_4N^+\ Cl^-$$

A quaternary ammonium salt

Quaternary ammonium salts are easily prepared by this method from ammonia or an amine (a weak base such as CO_3^{2-} is used to free the amine from the salt in each step). For example, ammonia can be completely alkylated by this reaction; each step is nucleophilic substitution, S_N2.

$$:NH_3 \xrightarrow[2)\ CO_3^{2-}]{1)\ RX} R\ddot{N}H_2 \xrightarrow[2)\ CO_3^{2-}]{1)\ RX} R_2\ddot{N}H \xrightarrow[2)\ CO_3^{2-}]{1)\ RX} R_3\ddot{N} \xrightarrow{RX} R_4N^+\ X^-$$

However, any of the intermediate amines are difficult to isolate in reasonable yields because they tend to become further alkylated very easily. Thus this alkylation reaction is seldom a practical one for synthesis of 1°, 2°, or 3° amines.

GETTING INVOLVED

- ✓ Describe the alkylation of amines and explain why it is a nucleophilic substitution reaction.
- ✓ Why does the alkylation of ammonia proceed through four alkylations? What is a quaternary ammonium salt and why does the alkylation stop here? What is the function of sodium carbonate in the complete alkylation of an amine?

Connections 10.3

www.prenhall.com/bailey

Acetylcholine and Neuromuscular Blockade

Acetylcholine is a biologically important quaternary ammonium salt. This major neurotransmitter has two significant functional groups, the quaternary salt at one end and an acetyl (ester) group at the other. As a neurotransmitter, it is responsible for signals to the autonomic functions of the body such as digestion and for stimulation of muscles, both voluntary and involuntary. Acetylcholine is also essential to the brain, especially in the areas of orientation, learning, and memory.

As a neurotransmitter it is a chemical link between nerve cells, or neurons, being synthesized and released from one cell and traveling across the synapse to the next. After stimulating the second cell it must be broken down so that the cell can rest for subsequent stimulation. An enzyme called *acetylcholineesterase* does this job of breakdown. Excessive stimulation by acetylcholine will lead to diarrhea, vomiting, constriction of the pupils of the eyes, slowed heartbeat, and eventual collapse of the respiratory system.

There are natural and synthetic compounds that bear a chemical resemblance to the neurotransmitter and will block its action specifically at muscle cells. This will result in paralysis, which could be beneficial for the hunter capturing prey or the physician having to put a patient under the control of a respirator or preparing him or her for major surgery.

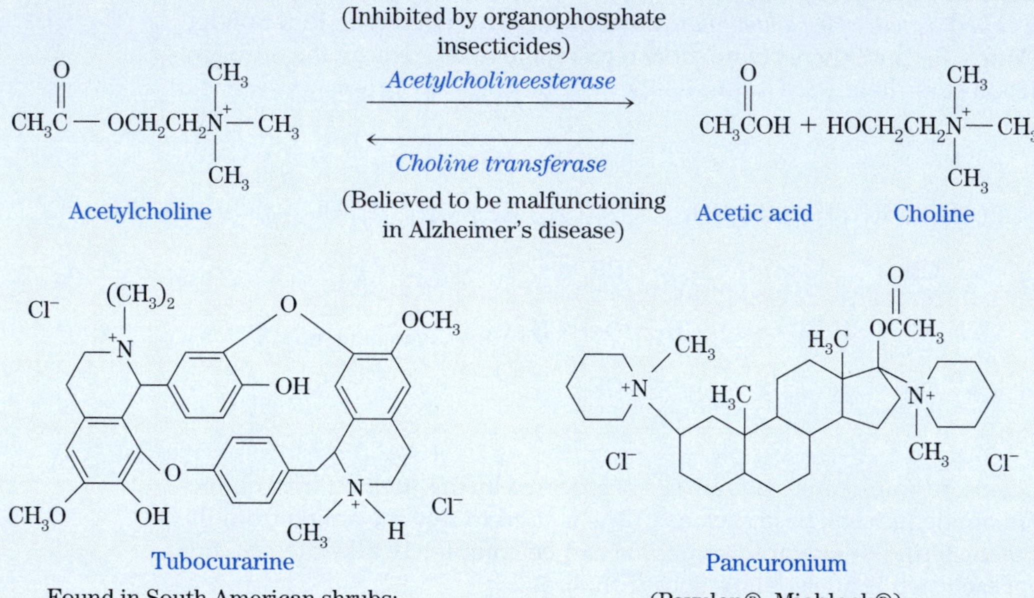

Tubocurarine

Found in South American shrubs; used to paralyze prey

Pancuronium

(Pavulon®, Mioblock®) used for surgical paralysis

Example 10.4

Write reaction equations illustrating the reaction between **(a)** aniline and three moles of methylbromide (in Na_2CO_3) and **(b)** trimethylamine and ethyl chloride.

Solution

(a) $$C_6H_5\ddot{N}H_2 + 3CH_3Br \xrightarrow{Na_2CO_3} C_6H_5\overset{+}{N}(CH_3)_3\ Br^-$$

(b) $$(CH_3)_3N\!: + CH_3CH_2Cl \longrightarrow (CH_3)_3\overset{+}{N}CH_2CH_3\ Cl^-$$

Problem 10.14

Write the products of the following alkylation reactions:

(a) $(CH_3)_3N$, $C_6H_5CH_2Br$ **(b)** $CH_3CH_2NH_2$, $3CH_3I$, Na_2CO_3

(c) pyrrolidine (N–H), $2CH_3CH_2Cl$, Na_2CO_3 **(d)** NH_3, $4CH_3I$, Na_2CO_3

See related problems 10.36, 10.42.

10.6 Preparations of Amines by Reduction Reactions

A. Reduction of Aromatic Nitro Compounds

Aromatic nitro compounds can be **reduced** to primary aromatic amines with hydrogen and a metal catalyst (such as platinum) or with iron or tin in acid solution.

reduction introduction of hydrogen into a molecule, often resulting in the loss of oxygen or conversion of double bonds to single bonds

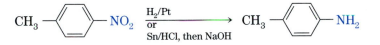

$$CH_3-C_6H_4-NO_2 \xrightarrow[\text{Sn/HCl, then NaOH}]{H_2/Pt \text{ or}} CH_3-C_6H_4-NH_2$$

Combining this method of reduction with the electrophilic aromatic substitution reactions we covered in Chapter 6 (section 6.4) offers pathways to a number of aromatic amines, as shown in Example 10.5.

GETTING INVOLVED

Example 10.5

Devise a synthesis for meta chloroaniline from benzene.

Solution

Since the nitro group is meta-directing, it should be introduced first (nitration is accomplished by using nitric and sulfuric acids). Chlorination ($Cl_2/FeCl_3$) followed by reduction gives the desired compound.

$$C_6H_6 \xrightarrow[H_2SO_4]{HNO_3} C_6H_5NO_2 \xrightarrow[FeCl_3]{Cl_2} m\text{-}ClC_6H_4NO_2 \xrightarrow[2)\ NaOH]{1)\ Sn/HCl} m\text{-}ClC_6H_4NH_2$$

Problem 10.15

From what nitro compound could the following amines be produced by reduction? **(a)** *p*-methylaniline; **(b)** *o*-bromoaniline.

Problem 10.16

Using electrophilic aromatic substitution reactions and reduction of nitro groups, propose syntheses for the following: **(a)** *p*-chloroaniline from benzene; **(b)** 3-bromo-4-methylaniline from toluene.

See related problems 10.37–10.38.

B. Reduction of Nitriles

nitrile
functional group of a carbon-nitrogen triple bond

Addition is the characteristic reaction of most multiple bonds. Addition of two moles of hydrogen to the carbon-nitrogen triple bond of **nitriles** produces primary amines. We saw in Chapter 8 that nitriles can be produced by nucleophilic substitution from an alkyl halide using NaCN.

$$R-C\equiv N + 2H_2 \xrightarrow{Ni} R-CH_2NH_2$$

GETTING INVOLVED

Problem 10.17

Complete the following reaction sequence: 1-chloropentane plus NaCN followed by $2H_2/Ni$.

See related problem 10.39.

Connections 10.4 www.prenhall.com/bailey

Sulfa Drugs

Amines are basic and nucleophilic and they react with aromatic sulfonyl chlorides to produce aromatic sulfonamides. The polar S—Cl bond is attacked by nucleophilic substitution. This reaction is the basis for the preparation of sulfa drugs.

$$C_6H_5-\overset{O}{\underset{O}{\overset{\|}{\underset{\|}{S^{\delta+}}}}}-Cl^{\delta-} + :\overset{R}{\underset{H}{N}}-R \longrightarrow C_6H_5-\overset{O}{\underset{O}{\overset{\|}{\underset{\|}{S}}}}-\overset{R}{N}-R + HCl$$

A sulfonyl chloride   A sulfonamide

Research on sulfa drugs began in 1935 when a physician, Gerhard Domagk, gave his young daughter an oral dose of a sulfonamide dye in a desperate attempt to save her from death from a streptococcal infection. The discovery that sulfonamides retard bacterial growth led to the synthesis and testing of over 5000 sulfonamides, particularly those related to sulfanilamide, during the ensuing dozen years.

Sulfa drugs do not kill bacteria but only inhibit their growth. This limits the infection to a small colony which can be destroyed by natural body mechanisms. To reproduce, some bacteria require a chemical, *p*-aminobenzoic acid (PABA, used as a sunscreen in some antisunburn creams).

Sulfa drugs chemically resemble *p*-aminobenzoic acid, and the bacteria mistakenly absorb the sulfa drug instead of the needed material, and stop growing. Sulfa drugs are effective only on bacteria requiring *p*-aminobenzoic acid for growth. The drugs are used to treat a variety of bacterial infections,

$$H_2N-C_6H_4-SO_2NH_2 \qquad H_2N-C_6H_4-CO_2H$$

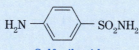

Sulfanilamide   *p*-aminobenzoic acid

including respiratory, gastrointestinal, and urinary tract infections, gonorrhea, and some eye and skin infections. Some common examples of sulfa drugs follow.

Although sulfa drugs are still used in veterinary medicine, their use with humans has declined with the advent of antibiotics.

$$NH_2-C_6H_4-SO_2NH-C_4H_3N_2 \qquad NH_2-C_6H_4-SO_2NH-\overset{O}{\overset{\|}{C}}CH_3 \qquad NH_2-C_6H_4-SO_2NH-C_3HNO-CH_3$$

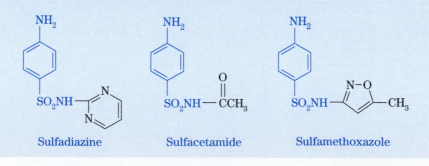

Sulfadiazine   Sulfacetamide   Sulfamethoxazole

C. Reduction of Amides

Amide reduction with lithium aluminum hydride can be used to prepare primary, secondary, and tertiary amines.

amide
functional group in which a trivalent nitrogen is bonded to a carbon-oxygen double bond

$$\mathrm{R\overset{\overset{\large O}{\|}}{C}NH_2} \xrightarrow[\text{2) } H_2O]{\text{1) LiAlH}_4} \mathrm{RCH_2NH_2}$$

We shall see in chapter 13 (sections 13.2–13.7) that amides can be produced from carboxylic acid and their derivatives.

GETTING INVOLVED

Example 10.6

From what amide could N-isopropylbutanamine be prepared by reduction with lithium aluminum hydride?

Solution

In amide reduction the C=O is reduced to a CH_2. There is only one CH_2 on the nitrogen, so it must be the C=O.

$$\mathrm{CH_3CH_2CH_2\overset{\overset{\large O}{\|}}{C}{-}NH{-}CH(CH_3)_2} \xrightarrow[\text{2) } H_2O]{\text{1) LiAlH}_4} \mathrm{CH_3CH_2CH_2CH_2{-}NH{-}CH(CH_3)_2}$$

Notice in problem 10.18 there are two ways to prepare the amine from an amide.

Problem 10.18

Show two ways that N-ethylpentanamine can be produced by $LiAlH_4$ reduction of an amide.

See related problems 10.40–10.41.

10.7 Aromatic Diazonium Salts

A. Preparation

Primary aromatic amines react with nitrous acid to form an interesting and synthetically useful compound called a **diazonium salt.** Since nitrous acid, HONO, is an unstable substance, it must be generated in the reaction mixture from sodium nitrite under acid conditions. The diazonium salt is also unstable, and the reaction must be performed in a cold solution.

diazonium salt
compound in which a molecule of nitrogen is bonded to an aromatic ring

$$\underset{\text{Primary aromatic amine}}{\mathrm{C_6H_5{-}NH_2}} \xrightarrow[\text{HCl, 0°C}]{\text{NaNO}_2} \underset{\text{Aromatic diazonium salt}}{\mathrm{C_6H_5{-}\overset{+}{N}{\equiv}N\ \ Cl^-}}$$

Although aromatic diazonium salts are stable in cold solutions, they are dangerously explosive if isolated. Consequently, they are used for immediate reaction in solution. Alkyl diazonium salts are considerably less stable and cannot be utilized in the same manner as their aromatic counterparts.

Since diazonium salts are important for their synthetic utility, it is important to recognize that they can be derived from nitro groups that are introduced onto benzene rings by electrophilic aromatic substitution (section 6.4). Nitro groups can be reduced to primary amines, NH_2 groups, with iron and hydrochloric acid.

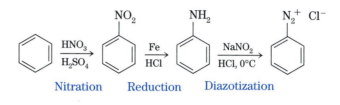

NO_2 NH_2 N_2^+ Cl^-

$\xrightarrow[H_2SO_4]{HNO_3}$ $\xrightarrow[HCl]{Fe}$ $\xrightarrow[HCl,\ 0°C]{NaNO_2}$

Nitration Reduction Diazotization

Diazonium salts undergo two general types of reactions: replacement reactions, in which nitrogen is evolved, and coupling reactions, with the retention of nitrogen. The following resonance forms are useful when these reaction types are considered.

Resonance forms: $\overset{+}{N}\equiv N:$ $\longleftrightarrow$ $\ddot{N}=\ddot{N}^+$

Important in replacement reactions

Important in coupling reactions

GETTING INVOLVED

✓ What is an aromatic diazonium salt? Draw the two resonance forms. How are diazonium salts prepared?

Problem 10.19

(a) Write a reaction equation showing the preparation of a diazonium salt from *p*-methylaniline. **(b)** Write a reaction sequence showing the preparation of the diazonium salt in part a from toluene.

B. Replacement Reactions

You have probably noticed that a diazonium salt is basically a benzene ring with a bonded nitrogen molecule. Nitrogen is a very stable species and can easily leave the diazonium salt as a gas, N_2. For this reason, diazonium salts undergo replacement reactions readily in which nitrogen can be replaced with a variety of groups, some of which are difficult to introduce on an aromatic ring in any other way. Figure 10.3 summarizes these replacement reactions, and Example 10.8 illustrates a synthetic application in conjunction with electrophilic aromatic substitution reactions (see section 6.4).

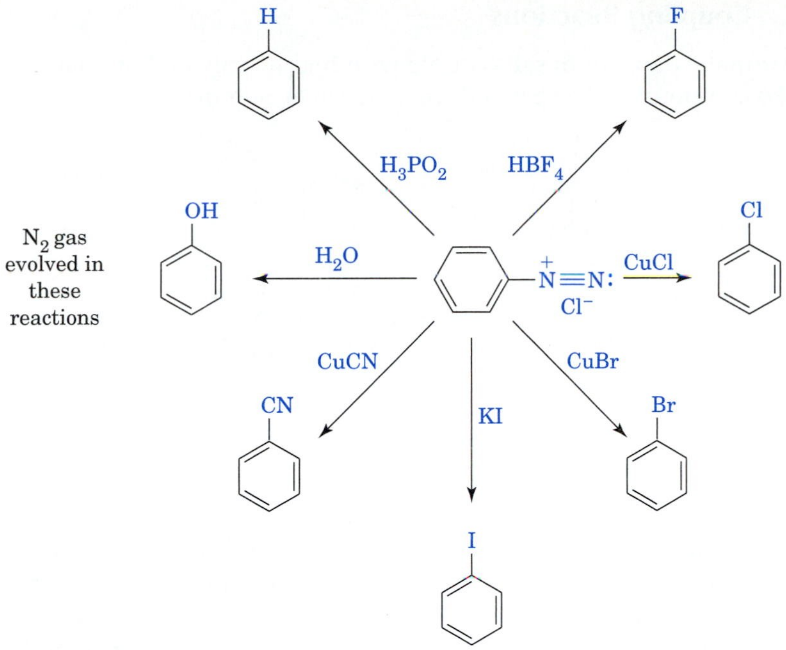

H
F
H_3PO_2
HBF_4
OH
Cl
N_2 gas evolved in these reactions
H_2O
$\overset{+}{N}{\equiv}N{:}$ Cl^-
CuCl
CuCN
CuBr
CN
KI
Br
I

Figure 10.3 Replacement reactions of aromatic diazonium salts.

GETTING INVOLVED

✓ Describe what happens in a replacement reaction. Be sure you know the reagents for introducing the various groups by replacement. Also, review electrophilic aromatic substitution.

Example 10.8

Show the synthesis of meta bromoiodobenzene from benzene. Use a diazonium salt replacement reaction for introducing the iodine.

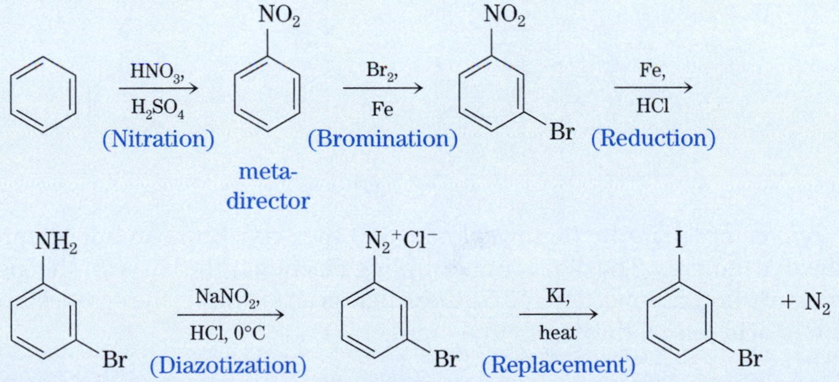

NO_2
NO_2
HNO_3, H_2SO_4
(Nitration)
meta-director
Br_2, Fe
(Bromination)
Br
Fe, HCl
(Reduction)
NH_2
Br
$NaNO_2$, HCl, 0°C
(Diazotization)
$N_2^+Cl^-$
Br
KI, heat
(Replacement)
I
Br
$+ N_2$

Problem 10.20

Draw the diazonium salt formed from the reaction of *p*-chloroaniline with $NaNO_2/HCl$. Write the products of the reactions of this salt with **(a)** HBF_4; **(b)** CuCl; **(c)** CuBr; **(d)** KI; **(e)** CuCN; **(f)** H_2O; **(g)** H_3PO_2.

Problem 10.21

Show how the following compounds can be prepared: **(a)** phenol from aniline; **(b)** *p*-fluorotoluene from *p*-methylaniline; **(c)** benzonitrile (cyanobenzene) from nitrobenzene; **(d)** *m*-dichlorobenzene from nitrobenzene.

See related problems 10.43–10.44.

C. Coupling Reactions

Aromatic diazonium salts couple with highly activated aromatic rings to form azo compounds. The general reaction can be summarized as follows.

:N$^+$ Cl$^-$ ‖ N: (Diazonium salt) + G-substituted benzene ⟶ Azo compound

G = electron-releasing groups: —OH,—NH_2, —NHR, —NR_2

Diazonium salt

Azo compound

Note that the reaction involves aromatic substitution and nitrogen is retained in the product. The reaction mechanism is electrophilic aromatic substitution. The positive nitrogen of the diazonium salt, a weak electrophile, is attracted to the π cloud of the activated ring. It usually bonds to the para position (which is less crowded, since the group G is an ortho, para director), forming a carbocation. Loss of a hydrogen ion re-forms the aromatic ring.

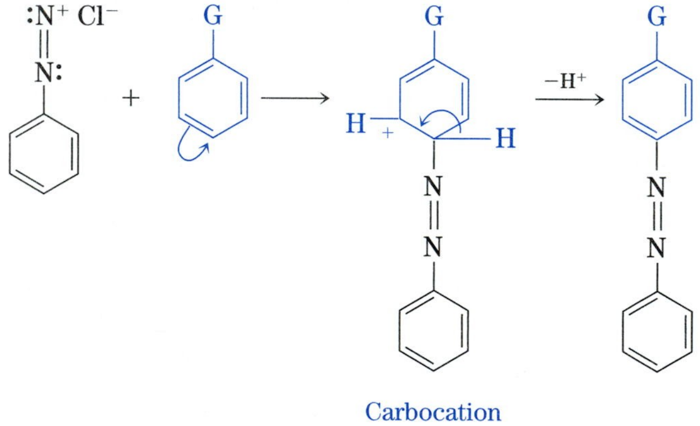

$-H^+$

Carbocation

Azo compounds are highly colored, and they constitute an important part of the dye industry. The diazonium coupling reaction is the basis for the ingrain dyeing method (Connections 10.5). Consider, as an example, the synthesis of the dye and acid-base indicator methyl orange.

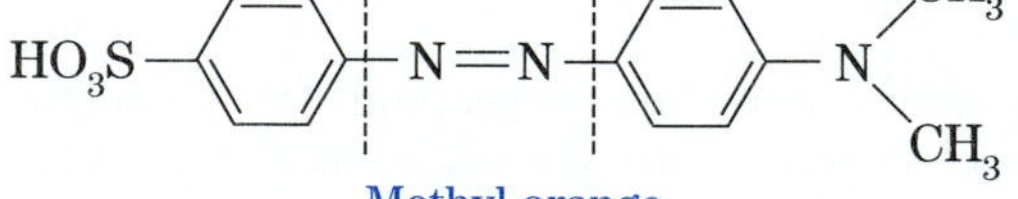

HO_3S—C_6H_4—N=N—C_6H_4—N(CH_3)$_2$

Methyl orange

One might consider making this by coupling a diazonium salt to benzenesulfonic acid or to N,N-dimethylaniline (see dashed lines of structure). The sulfonic acid group is deactivating toward electrophilic substitution, and attempting to couple *p*-dimethylaminobenzene diazonium chloride to benezenesulfonic acid would be unproductive. The dimethylamino group is strongly activating, however, and the following sequence can produce methyl orange in excellent yield:

$$\text{NH}_2\text{–C}_6\text{H}_4\text{–SO}_3\text{H} \xrightarrow[\text{HCl, 0°C}]{\text{NaNO}_2,} \text{Cl}^-\,\text{N}^+{\equiv}\text{N–C}_6\text{H}_4\text{–SO}_3\text{H} \xrightarrow[\text{Coupling}]{\text{C}_6\text{H}_5\text{N(CH}_3)_2} (\text{CH}_3)_2\text{N–C}_6\text{H}_4\text{–N{=}N–C}_6\text{H}_4\text{–SO}_3\text{H}$$

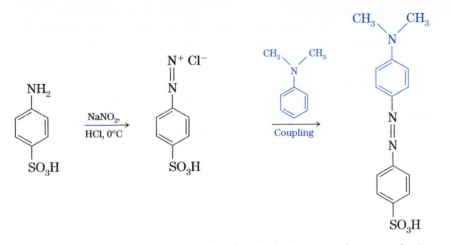

To apply the dye to a fabric, the fabric is immersed in a solution of N,N-dimethylaniline and then in a solution of the diazonium salt. The two reactants meet deep in the fabric, and the dye is synthesized.

Connections 10.5 www.prenhall.com/bailey

Dyes and Dyeing

Compounds that absorb one or more wavelengths of visible light appear colored to the human eye. White light possesses all wavelengths of visible light. When a beam of white light strikes a colored surface, certain wavelengths are absorbed and others are reflected; we see what is reflected. For example, if an object absorbs wavelengths in the blue-green region, the object will appear red, because this color constitutes the remaining wavelengths. Conversely, if red light is absorbed, the object will appear blue-green.

What structural features cause an organic molecule to appear colored? Basically there are two: (1) the compound usually has a chromophore group (color-bearing group); and (2) there is an extensive network of alternating single and double bonds (conjugation) of which the chromophore is a part. Following are a few examples of chromophore groups; dyes are often classified according to the chromophore group present.

Chromophore Groups

Nitro	Azo	Dicarbonyl
$—NO_2$	$—N{=}N—$	$—\overset{O}{\overset{\|}{C}}—\left(\overset{\|}{C}{=}\overset{\|}{C}\right)_n—\overset{O}{\overset{\|}{C}}—$ (n = 0 or greater)

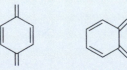

p-quinoid   *o*-quinoid

For a compound to be a dye, it must not only show color, it must also be able to adhere to a fabric. Auxochrome groups are acidic or basic groups that can cause a dye to bind to a fabric by ionic attractions and hydrogen-bonding. For example, wool and silk are proteinaceous materials composed of amino acids (amine group and carboxylic acid group) linked by amide bonds. These acidic and basic groups in proteins often exist in their salt forms, which are charged. Charged groups on a dye are attracted to groups of opposite charge on a fabric, and binding results. The abundant opportunities for hydrogen-bonding between auxochrome groups and wool and silk fibers also allow dyes to adhere to these fabrics.

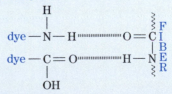

Auxochrome Groups

Basic $—\ddot{N}H_2$, $—\ddot{N}HR$, $—\ddot{N}R_2$

Acidic $—CO_2H$, $—SO_3H$, $—OH$

Ionic attractions

$$\text{fiber}—\overset{O}{\overset{\|}{C}}O^- \cdots\cdots H_3N^+—\text{dye}$$

$$\text{fiber}—\overset{+}{N}H_3 \cdots\cdots \overset{-}{O}_3S—\text{dye}$$

Hydrogen-bonding

$$\text{dye}—\overset{H}{\overset{|}{N}}—H\cdots\cdots O{=}C \text{ (FIBER)}$$

$$\text{dye}—\underset{OH}{\underset{|}{C}}{=}O\cdots\cdots H—N \text{ (FIBER)}$$

Connections 10.5 *(cont.)*

Following are structures of typical dyes. Look closely and identify chromophore and auxochrome groups and the extensive networks of conjugated double and single bonds. These are the three essential structural features of dyes.

Dyes can be classified according to the method of application to a fabric. Direct dyes are applied by immersing the fabric in a water solution of the dye; adherence is due to acid-base interactions and hydrogen-bonding. Disperse dyes are insoluble in water but "soluble" in the fabric. They are applied to modern fabrics such as nylon and polyesters as finely milled particles colloidally dispersed in water. Mordant dyes are applied to a fabric treated with a mordant (Latin *mordere*, to bite), which can bind both to the fabric and the dye; in this method a dye with little affinity to a fabric but great affinity to a mordant can be applied. Reactive dyes actually form covalent bonds with the fiber. Basic groups on the fiber attack carbon-chlorine bonds on the dye via nucleophilic substitution to form the bond. Ingrain dyes are water-insoluble dyes that are synthesized right in the fabric from water-soluble reactants. Usually the first of the reactants applied has an auxochrome that allows it to bind to the fabric; the second reactant reacts with the first to form the dye. Vat dyes are compounds that are water soluble in a reduced colorless form. After application, however, oxidation (sometimes merely in the air) converts the dye to the water-insoluble colored form.

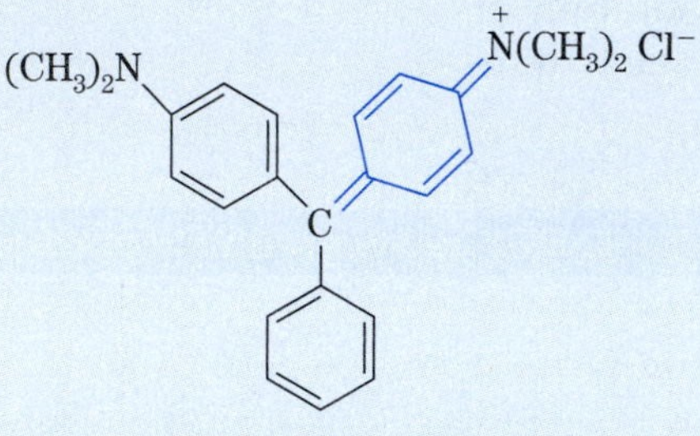

$(CH_3)_2N$, $\overset{+}{N}(CH_3)_2\ Cl^-$, C

Malachite green

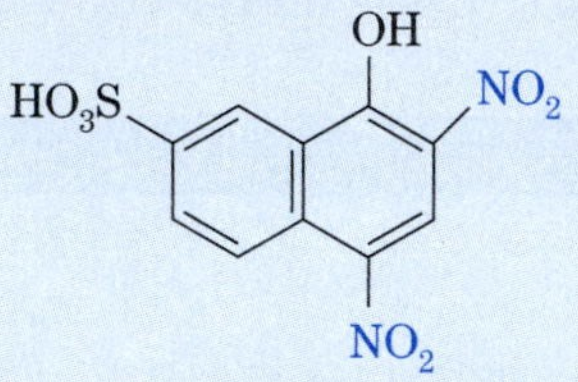

OH, HO_3S, NO_2, NO_2

Naphthol yellow S

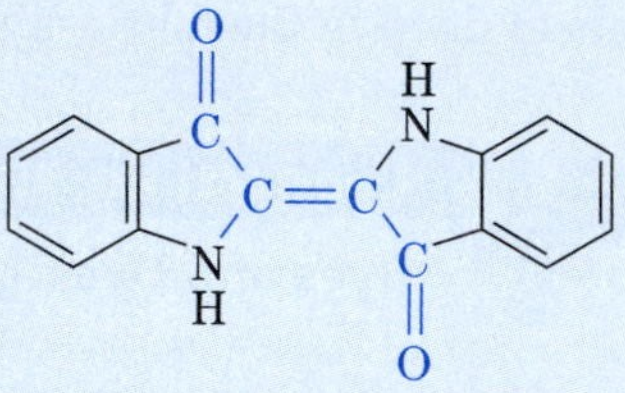

O, C, H, N, C=C, N, H, C, O

Indigo

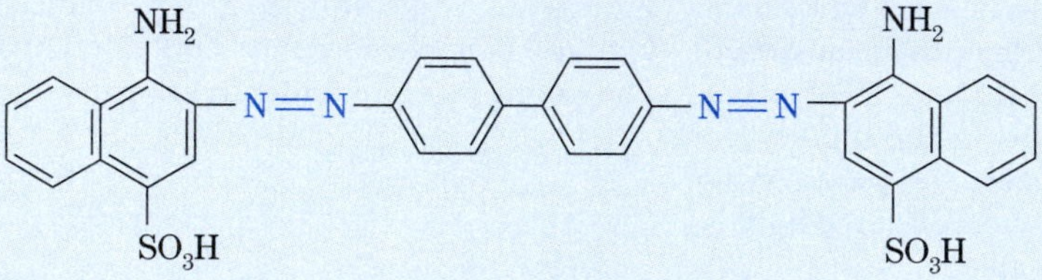

NH_2, $-N{=}N-$, $-N{=}N-$, NH_2, SO_3H, SO_3H

Congo red

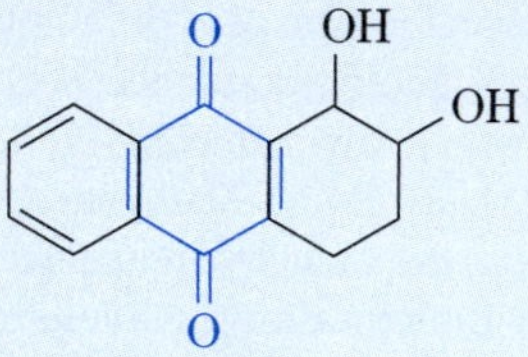

O, OH, OH, O

Alizarin

GETTING INVOLVED

✓ Describe the mechanism of the coupling reaction as electrophilic aromatic substitution. What type of group must be on the aromatic ring for substitution to occur?

Problem 10.22

Synthesize the following compound by a coupling reaction, starting with available compounds:

$$H_2N-C_6H_4-N{=}N-C_6H_4-NO_2$$

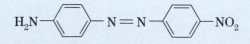

See related problem 10.45.

10.8 Heterocyclic Amines

Many important amines are members of the large class of heterocyclic compounds. **Heterocycles** are cyclic compounds in which one or more of the ring atoms is not carbon. Heterocyclic amines, compounds in which at least one of the ring atoms is nitrogen, are especially interesting compounds, since many have a biological source or application. Many of these naturally occurring compounds are in a class called **alkaloids,** which are loosely defined as plant-produced nitrogenous bases that have a physiological effect on humans.

heterocycle
cyclic compound where at least one ring atom is not carbon

alkaloids
plant-produced nitrogenous bases that have physiological effects on humans

A. Structure and Basicity of Heterocyclic Amines

Like other amines, heterocyclic amines are basic, but their basicities can vary dramatically depending on structure and the availability of nitrogen's nonbonding electron pair. For example, compare piperidine, a nonaromatic amine, and pyridine, an aromatic amine.

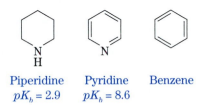

N H N

Piperidine $pK_b = 2.9$ Pyridine $pK_b = 8.6$ Benzene

Pyridine has a nitrogen in place of one C—H in benzene. Consequently, like benzene, pyridine is aromatic. Each is cyclic and planar, has a p orbital on each ring atom, and has six p electrons; all of these are defining features of aromaticity in benzene (section 6.2.C). Pyridine is about 500,000 times less basic than piperidine. Because it is aromatic, pyridine is unusually stable and not as susceptible to reaction, including reaction with acid.

A similar but more dramatic comparison exists between pyrrolidine, a nonaromatic five-membered heterocyclic amine, and pyrrole, its aromatic counterpart. Pyrrole is almost 100 billion times less basic than pyrrolidine!

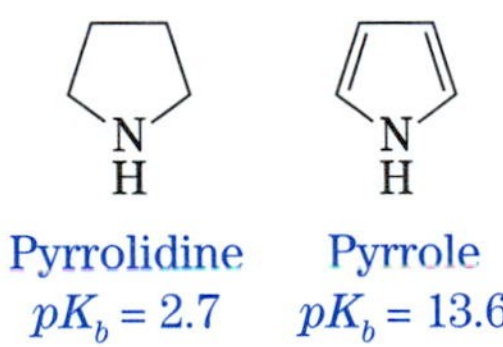

Pyrrolidine $pK_b = 2.7$ Pyrrole $pK_b = 13.6$

Pyrrole is cyclic, planar, and has a p orbital on each ring atom, and has six π electrons, the same structural characteristics as benzene, even though pyrrole has only a five-membered ring. The nonbonding electron pair on nitrogen exists in a p orbital to complete the aromatic sextet, four π electrons from the two double bonds and two from nitrogen, and allow aromaticity. The nitrogen is sp^2-hybridized so the lone pair can reside in the overlapping p orbitals (Figure 10.4). Unlike pyridine, pyrrole's nonbonding electron pair is directly involved in the

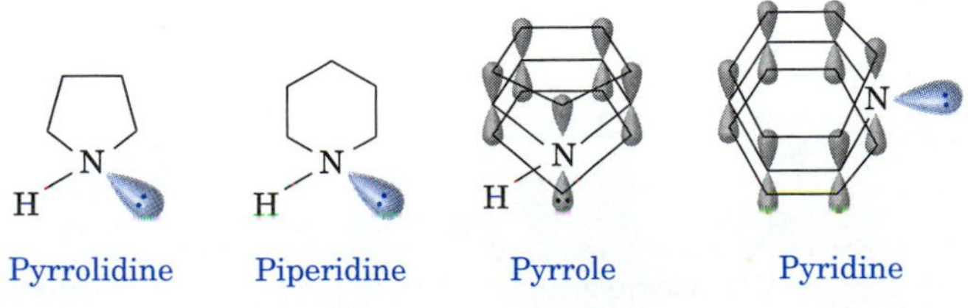

Figure 10.4 π bonding pictures.

aromatic structure and, as a result, is much less available for reaction with acids. In fact, for the lone pair to react with a hydrogen ion and form a salt requires disruption of the aromatic π cloud, which would destroy the compound's aromaticity and stability. This is not the case with pyridine as its lone pair is not part of the aromatic sextet; the sextet is provided by the three double bonds. Piperidine and pyrrolidine have relative strong basicities; in effect, they are secondary alkyl amines and have corresponding basicity constants (see Table 10.2). The electronic structures of these four heterocyclic amines are depicted in Figure 10.4.

GETTING INVOLVED

- ✓ What are the four structural features necessary in a compound for aromaticity?
- ✓ Why is pyridine less basic than piperidine? Why is pyrrole less basic than pyrrolidine? Why is pyrrole less basic than pyridine?
- ✓ Describe structurally why pyridine and pyrrole are aromatic. Indicate whether the lone pair of electrons is part of the aromatic sextet in each case.

Problem 10.23

Quinoline and indole (see section 10.8.B for structures) are both aromatic heterocyclic amines. Explain why they are aromatic. Indicate in each case whether or not the nitrogen lone pair is part of the aromatic π electron system.

See related problems 10.46–10.47.

Atropa belladonna

B. Naturally Occurring Heterocyclic Amines: Alkaloids

Alkaloids can be classified to some extent by the heterocyclic ring systems found in their structures. For example, you should be able to identify the pyrrolidine, pyrrole, piperidine, and pyridine ring systems in the following alkaloids. Coniine, a piperidine, is the principal alkaloid in hemlock, the poison used to execute the Greek philosopher Socrates around 400 B.C. Also in the piperidine class are piperine, which occurs in black pepper, and lobeline from the seeds of Indian tobacco and the basic ingredient in some nonprescription deterrents to cigarette-smoking. Cuscohygrine is a pyrrolidine alkaloid found in deadly nightshade (*Atropa belladonna*) and Peruvian coca shrub; pyrrolidine itself is found in wild carrots. The most familiar pyridine-pyrrolidine alkaloid is nicotine, the principal alkaloid component of tobacco (4%–6% in leaves) and one of the most toxic alkaloids known; it is fatal to all forms of animal life (by respiratory paralysis) and is used as an agricultural insecticide.

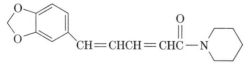

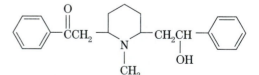

Piperine — $CH{=}CHCH{=}CHC(=O){-}N$ (piperidine ring), attached to a benzodioxole ring

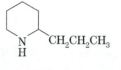

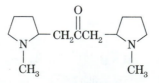

Lobeline — $C_6H_5C(=O)CH_2$–(N-CH_3 piperidine)–$CH_2CH(OH)C_6H_5$

Coniine — piperidine (N–H) with $CH_2CH_2CH_3$

Cuscohygrine — (N-CH_3 pyrrolidine)–$CH_2C(=O)CH_2$–(N-CH_3 pyrrolidine)

Nicotine — pyridine ring attached to N-CH_3 pyrrolidine

There are four pyrrole units (shown in several resonance forms) in heme (of hemoglobin, the oxygen transport system, and red pigment of blood) and chlorophyll, a green material that is responsible for photosynthesis in green plants.

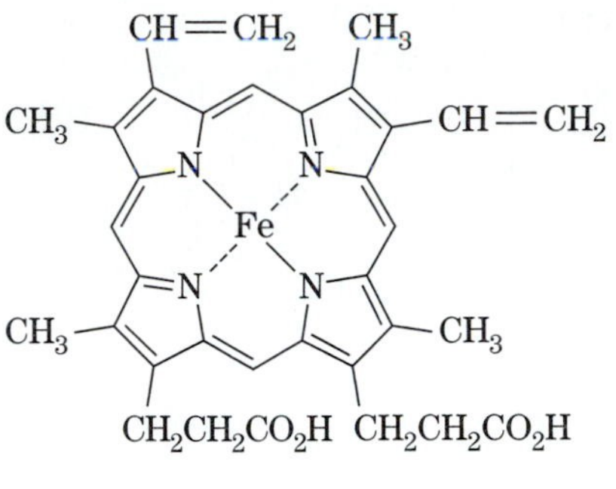

Heme

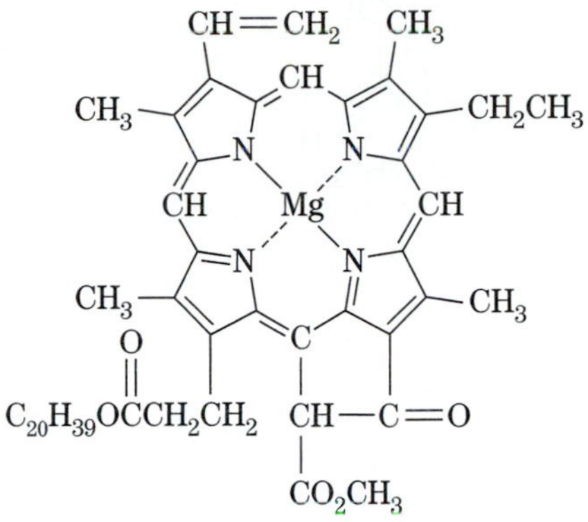

Chlorophyll a

Other classes of alkaloids possess the fused-ring heterocycles quinoline, isoquinoline, indole, and purine.

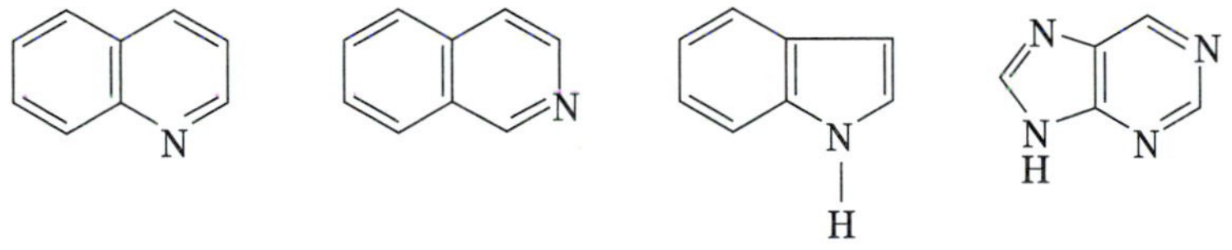

Quinoline Isoquinoline Indole Purine

Quinine, found in tonic water, is the most important of several alkaloids in the bark of the cinchona tree, which is native to the eastern slopes of the Andes; the bark was used by the Jesuits around 1600 for antimalarial preparations, and quinine was one of the first antimalarial drugs. Two isoquinoline-type rings can be seen in tubocurarine chloride, an exceedingly potent poison that has been used on arrows and blowdarts by African and South American tribes.

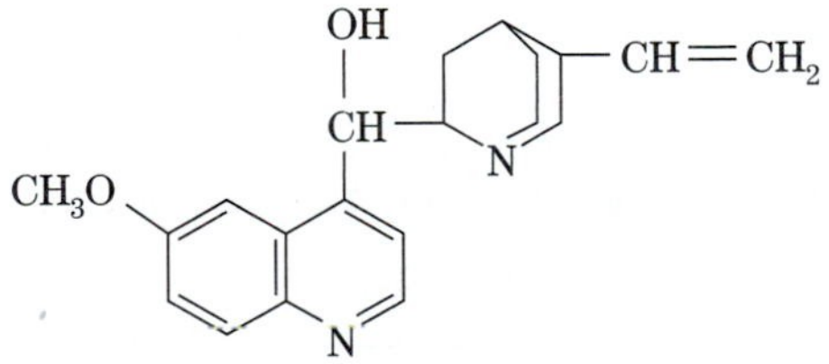

Quinine

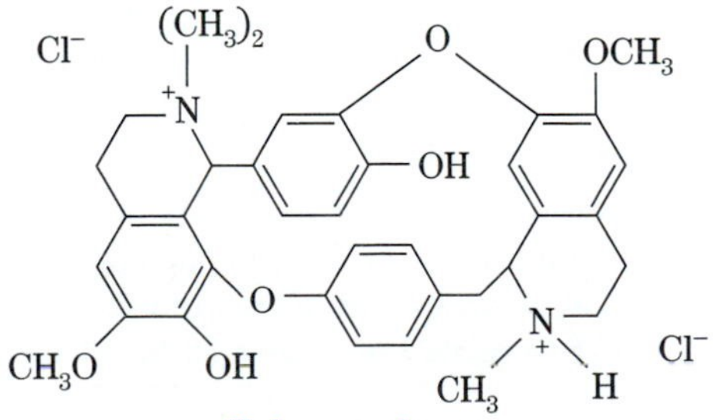

Tubocurarine

Found in South American shrubs; used to paralyze prey

Quite a few isoquinoline alkaloids can be isolated from the opium poppy, including the opium alkaloids morphine and codeine; heroin is a synthetic derivative. All of these compounds have a pain-relieving effect and generate a feeling of well-being. Unfortunately, all three are addictive in various degrees. A structurally similar synthetic compound, dextromethorphan, controls coughing like codeine by working on the cough control center in the medulla; it is as effective as codeine but nonaddictive and is used in nonprescription cough medicines. An example of a purine alkaloid, caffeine (a stimulant in coffee and tea), is also shown below.

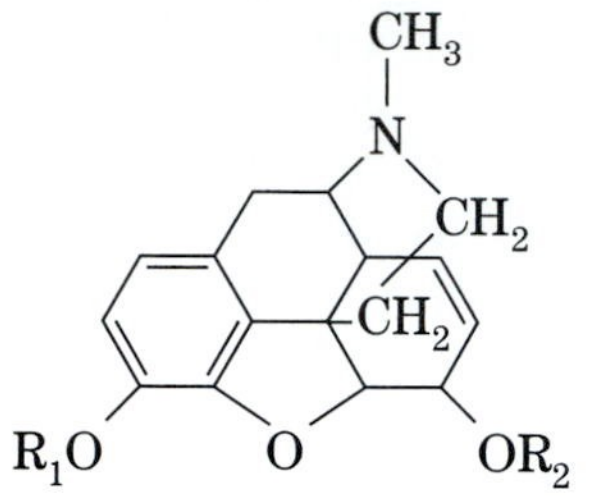

$R_1 = R_2 =$ —H for morphine
$R_1 =$ —CH_3, $R_2 =$ —H for codeine
$R_1 = R_2 =$ —$\overset{O}{\|}{C}CH_3$ for heroin

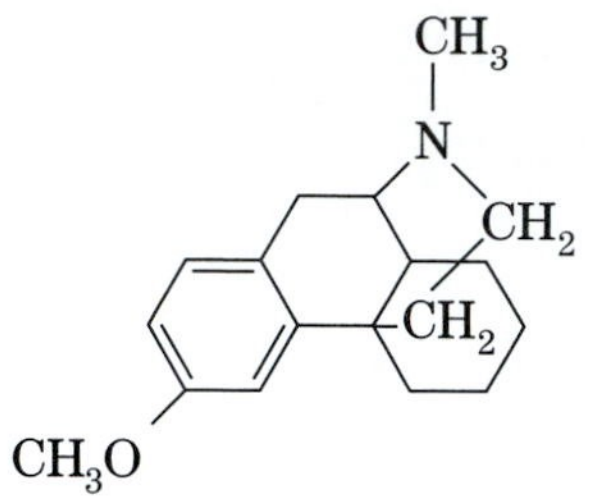

Dextromethorphan

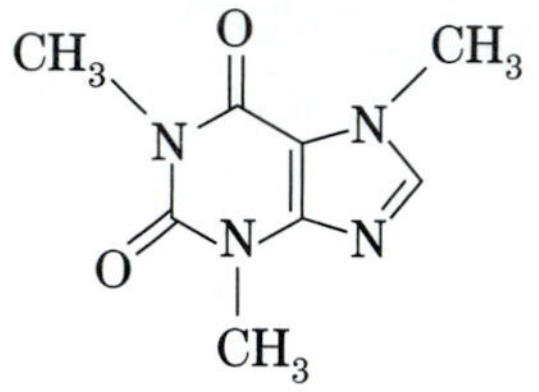

Caffeine

Lysergic acid diethylamide (the hallucinogenic drug LSD) and strychnine, a rodenticide, are examples of indole alkaloids.

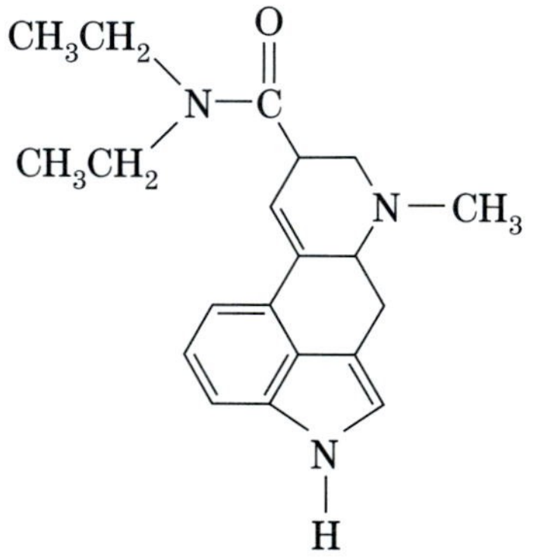

Lysergic acid diethylamide (LSD)

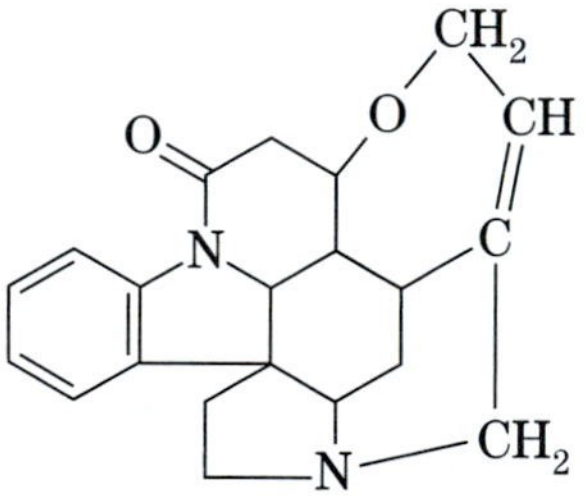

Strychnine

GETTING INVOLVED

✓ Examine each of the structures in section 10.8B. Identify pyridine, piperidine, pyrrole, pyrrolidine, quinoline, isoquinoline, indole, and purine structural units.

Problem 10.24

Examine the structures of isoquinoline and purine and determine whether these compounds are aromatic. For each of the nitrogens, indicate whether or not the lone pair of electrons is part of an aromatic sextet.

See related problems 10.46–10.47.

REACTION SUMMARY

1. Reaction of Amines with Acids to Form Ammonium Salts

Section 10.4.A; Example 10.3; Problems 10.6–10.7, 10.42(a)–(b).

$$R_3N + HA \longrightarrow R_3NH^+ \quad A^-$$

2. Preparation of Amines

A. Alkylation of Amines by Nucleophilic Substitution

Section 10.5; Example 10.4; Problems 10.14, 10.35, 10.42(c).

$$NH_3 + RX \longrightarrow RNH_2$$
$$RNH_2 + RX \longrightarrow R_2NH$$
$$R_2NH + RX \longrightarrow R_3N$$
$$R_3N + RX \longrightarrow R_4N^+X^-$$

B. Reduction of Aromatic Nitro Compounds

Section 10.6.A; Example 10.5; Problems 10.15–10.16, 10.37–10.38.

$$ArNO_2 \xrightarrow[\text{Sn/HCl}]{\text{H}_2\text{/Pt or}} ArNH_2$$

C. Reduction of Nitriles

Section 10.6.B; Problems 10.17, 10.39.

$$RC{\equiv}N + 2H_2 \xrightarrow{\text{Ni}} RCH_2NH_2$$

D. Reduction of Amides

Section 10.6.C.3; Example 10.6; Problems 10.18, 10.40–10.41.

$$R\overset{\overset{\displaystyle O}{\|}}{C}NR_2 \xrightarrow[\text{2) H}_2\text{O}]{\text{1) LiAlH}_4} RCH_2NR_2$$

3. Diazonium Salts

A. Preparation

Section 10.7.A. Problem 10.20.

$$ArNH_2 \xrightarrow[\text{HCl, 0°C}]{\text{NaNO}_2} ArN_2{}^+Cl^-$$

B. Replacement Reactions

Section 10.7.B; Example 10.8; Problems 10.20–10.21, 10.43–10.44. (See Reaction Summary in Figure 10.3.)

C. Coupling Reactions

Section 10.7.C; Problems 10.22, 10.45.

$$C_6H_5{-}N{=}N^+\,Cl^- + C_6H_5{-}G \longrightarrow C_6H_5{-}N{=}N{-}C_6H_4{-}G$$

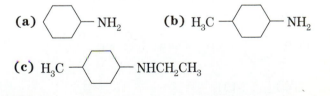

$$G = OH,\ NR_2$$

Problems

10.25 **IUPAC Nomenclature:** Name the following compounds:

(a) $CH_3(CH_2)_5CH_2NH_2$

(b) $CH_3CH_2CHCH_3$ (with NH_2 on the CH carbon)

(c) $H_2N(CH_2)_8NH_2$

10.26 **IUPAC Nomenclature:** Name the following compounds:

(a) cyclohexyl–NH_2

(b) H_3C–cyclohexyl–NH_2 (1,4-substituted)

(c) H_3C–cyclohexyl–$NHCH_2CH_3$ (1,4-substituted)

(d) $H_3C-C_6H_4-NH_2$ (para)

(e) $H_3C-C_6H_4-NHCH_2CH_3$ (para)

(f) $C_6H_5-N(CH_3)CH_2CH_3$

(g) $CH_3CH(Br)CH_2N(CH_2CH_3)_2$

(h) $CH_3CH(N(CH_2CH_2CH_3)_2)CH_2CH_2CH_2CH_2C(CH_3)_2CH_3$

10.27 **IUPAC Nomenclature:** Name the following compounds:

(a) $CH_3CH_2C{\equiv}CCH_2CH_2CH_2NH_2$

(b) $CH_3CH_2CH_2CH{=}CHCH_2NH_2$

(c) $CH_3CH{=}CHCH{=}CHCH_2NHCH_2CH_3$

(d) $CH_3C{\equiv}CCH_2N(CH_3)_2$

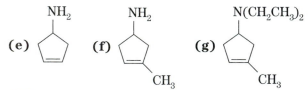

(e) cyclopentene ring bearing NH_2 **(f)** cyclopentene ring bearing NH_2 and CH_3 **(g)** cyclopentene ring bearing $N(CH_2CH_3)_2$ and CH_3

10.28 **Nomenclature:** Draw the following compounds:
(a) cycloheptanamine
(b) ethylpropylamine
(c) tributylamine
(d) ethylisopropylmethylamine
(e) N,N-dimethylaniline
(f) 2,4,6-trichloroaniline
(g) N-ethyl-1-heptanamine
(h) N-ethyl-N-methyl-3-propylcyclopentanamine

10.29 **Physical Properties:** Arrange the following in order of increasing boiling point:
(a) (i) methanamine, (ii) propanamine, (iii) heptanamine, (iv) decanamine
(b) (i) ethanamine, (ii) ethanol, (iii) ethane
(c) (i) propylamine, (ii) ethylmethylamine, (iii) trimethylamine

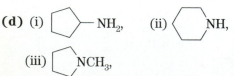

(d) (i) cyclopentyl$-NH_2$, (ii) piperidine ring NH, (iii) pyrrolidine ring NCH_3,

10.30 **Physical Properties:** Explain the following boiling points: methylamine, −6°C; dimethylamine, 7°C; trimethylamine, 3°C.

10.31 **Physical Properties:** Explain the boiling points of these compounds with similar molecular weights: pentane, 36°C; butylamine, 78°C; diethylamine, 56°C; 1-butanol, 117°C.

10.32 **Basicity Constants**
(a) Arrange the following basicity constants from least basic to most basic amines: 10^{-3}, 10^{-10}, 10^{-5}.
(b) Convert the basicity constants in part (a) to pK_b's.
(c) Arrange the following pK_b's from least basic to most basic for amines: 6, 11, 3.
(d) Convert the pK_b's in part (c) to K_b's.

10.33 **Acidity Constants**
(a) Arrange the following acidity constants from least acidic to most acidic for acids: 10^{-3}, 10^{-12}, 10^{-8}.
(b) Arrange the acidity constants in part (a) so that they express the basicities of a group of amines from least basic to most basic.
(c) Convert the acidity constants in part (a) to pK_a's.
(d) Arrange the following pK_a's from least acidic to most acidic for acids: 13, 4, 9.
(e) Arrange the pK_a's in part (d) so that they express the basicities of a group of amines from least basic to most basic.
(f) Convert the pK_a's in part (d) to K_a's.
(g) Write an equilibrium reaction equation for the reaction of methanamine with water.
(h) Write an equilibrium reaction equation for the ionization of methylammonium ion in water.

10.34 **Basicity of Amines:** Select the more basic amine from each of the following pairs. Explain your selection.
(a) ammonia or propylamine
(b) ethylamine or diethylamine
(c) aniline or cyclohexylamine
(d) aniline or N-methylaniline
(e) aniline or N-phenylaniline
(f) aniline or *p*-chloroaniline
(g) 2,4-dinitroaniline or *p*-nitroaniline
(h) propanamine or 2-chloropropanamine
(i) 2-chloropropanamine or 3-chloropropanamine

10.35 **Acidity and Basicity of Phenol and Aniline:** Aniline is much less basic than methylamine for essentially the same reasons that phenol is much more acidic than methyl alcohol. Both aniline and phenol are extremely reactive with respect to electrophilic aromatic substitution. Explain these observations. See sections 10.4.C.3 and 9.6.A.2 for assistance.

10.36 **Alkylation of Amines:** Write reaction equations for the exhaustive alkylation of the following amines to form the quaternary ammonium salt with the alkyl halide specified:

(a) hexanamine and methyl bromide
(b) N-methylpropanamine and ethyl chloride
(c) N,N-diethylaniline and methyl iodide
(d) trimethylamine and bromooctane

10.37 **Reduction of Nitro Compounds:** Write equations for the following reactions:

(a) *p*-ethylnitrobenzene and Sn/HCl
(b) *m*-chloronitrobenzene and H_2/Pt

10.38 **Reduction of Nitro Compounds:** Offer syntheses for the following aromatic amines:

(a) *p*-bromoaniline from *p*-bromonitrobenzene
(b) *m*-bromoaniline from benzene
(c) *p*-methylaniline from benzene

10.39 **Reduction of Nitriles:** Write the reaction sequence for the preparation of 1-heptanamine from 1-bromohexane via a nitrile.

10.40 **Reduction of Amides:** Write reaction equations showing the preparation of the following amines by reduction of amides:

(a) 1-hexanamine
(b) N-propylbutanamine (two ways)

10.41 **Reductions to Form Amines:** 1,4-hexandiamine, a precursor in nylon production, can be synthesized by either of the following schemes. Write reaction sequences describing the syntheses.

(a) reduction of the corresponding diamide
(b) treatment of 1,4-dichloro-2-butene with 2NaCN followed by hydrogenation with H_2/Ni (five moles H_2 consumed)

10.42 **Reactions of Amines:** Write equations showing reactions of (i) propylamine, (ii) ethylmethylamine, and (iii) trimethylamine with each of the following:

(a) HCl **(b)** H_2SO_4
(c) excess $CH_3Br(Na_2CO_3)$

10.43 **Reactions of Diazonium Salts:** Write the product of the reaction between *p*-methylaniline and $NaNO_2$/HCl at 0°C. Then show the reaction of this product with each of the following reagents:

(a) CuCl **(b)** CuBr
(c) KI **(d)** CuCN
(e) H_2O **(f)** HBF_4
(g) H_3PO_2 **(h)** phenol
(i) N,N-dimethylaniline

10.44 **Syntheses Using Diazonium Salts:** Synthesize the following compounds using electrophilic aromatic substitution (section 6.4) and diazonium salt replacement reactions:

(a) *p*-bromotoluene from *p*-methylaniline
(b) *m*-iodobromobenzene from *m*-bromoaniline
(c) bromobenzene from nitrobenzene
(d) *m*-chlorofluorobenzene from nitrobenzene
(e) *m*-bromophenol from nitrobenzene
(f) 3-chloro-4-bromophenol from bromobenzene
(g) cyanobenzene from benzene
(h) *p*-bromoiodobenzene from benzene
(i) *m*-chlorofluorobenzene from benzene

10.45 **Diazonium Salts—Coupling Reactions:** With chemical equations, show how the following dyes can be synthesized by the diazonium coupling reaction. Start with stable available compounds. (These are the same reactions that would be used to apply these dyes by the ingrain dyeing method.)

(a) 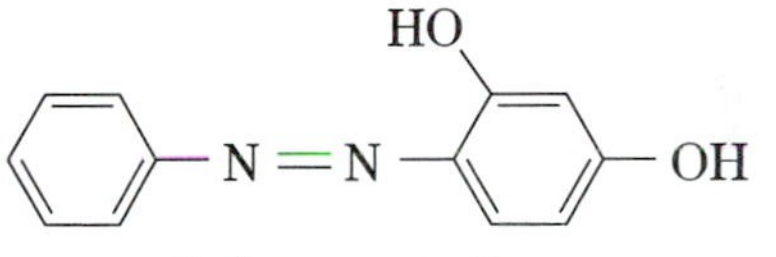 HO, N=N, OH

Sudan orange G

(b) 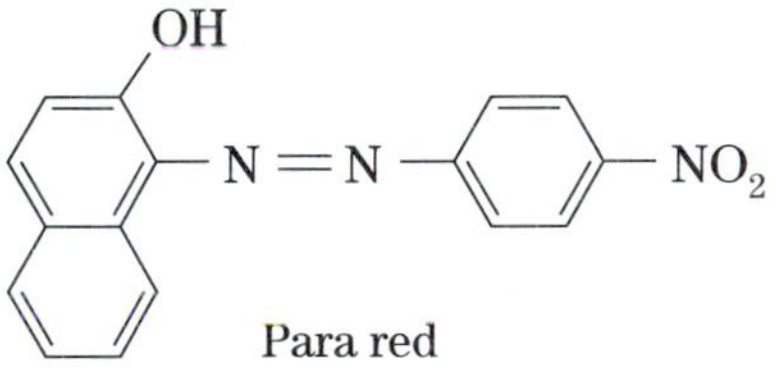 OH, N=N, NO_2

Para red

10.46 **Basicity of Heterocyclic Amines:** Both pyrrole and imidazole are aromatic. However, imidazole is four million times more basic. Explain this difference in terms of π bonding patterns and availability of the unshared electron pairs.

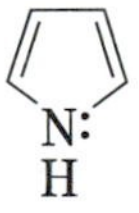

N: H Pyrrole and N: N: H Imidazole

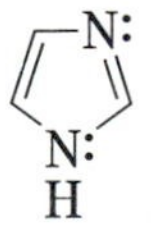

10.47 **Aromaticity of Heterocyclic Compounds:** Furan and thiophene are both aromatic heterocycles. Explain. Be sure to describe the role of both nonbonding electron pairs on the oxygen and the sulfur.

Furan  O  S Thiophene

10.48 **Dyes** For all the dyes in Connections 10.5, identify the chromophore and auxochrome groups.

Activities with Molecular Models

1. Make models of a primary and a secondary amine of C_2H_7N.

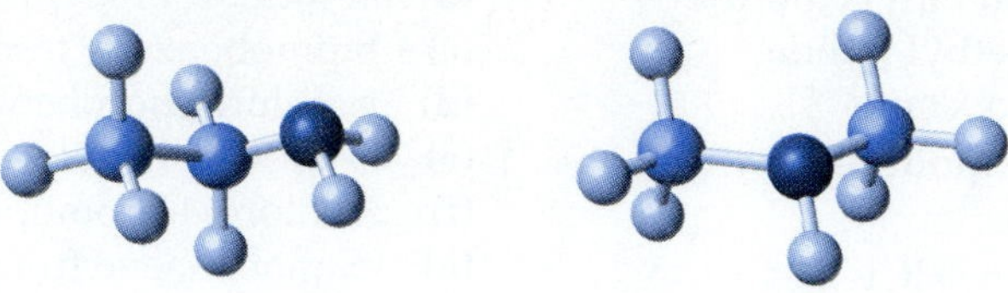

2. Make models of the four isomers of C_3H_9N. What is the hybridization of each carbon and the nitrogen? How many nonbonding electron pairs are on the nitrogen? Identify primary, secondary, and tertiary amines.

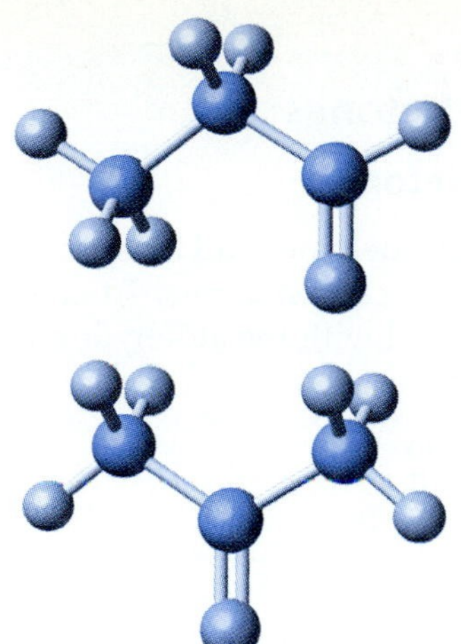

CHAPTER 11

Aldehydes and Ketones

11.1 Structure of Aldehydes and Ketones

Aldehydes and **ketones** are structurally very similar; both have a carbon-oxygen double bond called a **carbonyl** group. They differ in that aldehydes have at least one hydrogen atom bonded to the carbonyl group, whereas in ketones the carbonyl is bonded to two carbons. The biological preservative, formaldehyde, is the simplest aldehyde; benzaldehyde, oil of bitter almond, is the simplest aromatic aldehyde. Acetone is the simplest ketone; it is an important industrial solvent and a principal ingredient in some fingernail polish removers. Acetophenone is the simplest aromatic ketone; it is used in perfumery.

aldehyde
functional group in which at least one H is bonded to a carbonyl

ketone
functional group in which two organic substituents are bonded to a carbonyl

carbonyl
the carbon-oxygen double bond, C=O

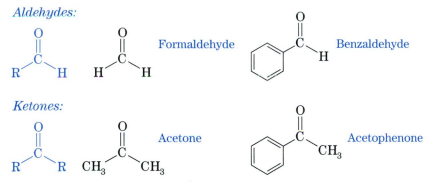

Aldehydes:

$R{-}C({=}O){-}H$ $\quad$ $H{-}C({=}O){-}H$ Formaldehyde $\quad$ $C_6H_5{-}C({=}O){-}H$ Benzaldehyde

Ketones:

$R{-}C({=}O){-}R$ $\quad$ $CH_3{-}C({=}O){-}CH_3$ Acetone $\quad$ $C_6H_5{-}C({=}O){-}CH_3$ Acetophenone

Aldehydes and ketones are quite prevalent in nature. They occur as natural fragrances and flavorings. In addition, carbonyl groups and their derivatives are the main structural features of carbohydrates and appear in other natural compounds, including dyes, vitamins, and hormones.

The carbonyl group is exceedingly important in organic chemistry. In addition to aldehydes and ketones, it is found in carboxylic acids and carboxylic acid derivatives, compounds that we will consider in the next two chapters.

$R{-}C({=}O){-}OH$	$R{-}C({=}O){-}Cl$	$R{-}C({=}O){-}O{-}C({=}O){-}R$	$R{-}C({=}O){-}OR$	$R{-}C({=}O){-}NH_2$
Carboxylic acid	Acid chloride	Acid anhydride	Ester	Amide

GETTING INVOLVED

- ✓ What is the difference structurally among aldehydes, ketones, and carboxylic acids?
- ✓ What is a carbonyl group?

11.2 Nomenclature of Aldehydes and Ketones

A. IUPAC Nomenclature of Aldehydes and Ketones

Like the names of other organic compounds, those of aldehydes and ketones are based on the name of the longest continuous chain of carbon atoms. To name aldehydes, the *-e* of the parent hydrocarbon is replaced with the aldehyde suffix *-al*, and to name ketones, by the ketone suffix *-one*.

$$\underset{\text{Propane}}{CH_3CH_2CH_3} \qquad \underset{\text{Propanal}}{CH_3CH_2\overset{\overset{O}{\|}}{C}H} \qquad \underset{\text{Propanone}}{CH_3\overset{\overset{O}{\|}}{C}CH_3}$$

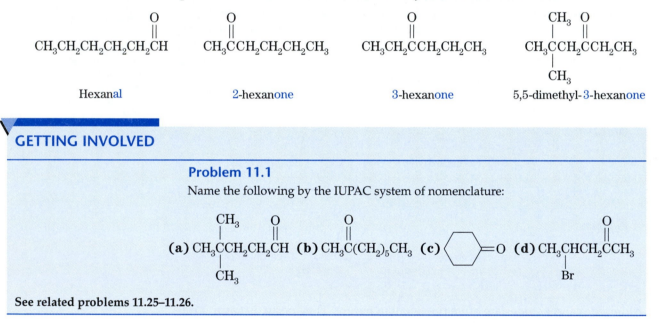

The aldehyde function is at the end of the chain and numbering starts there. It is usually not necessary to express the position in the name. For ketones, the chain is numbered so as to give the ketone group the lowest possible number. The position of the ketone function is usually indicated in the name.

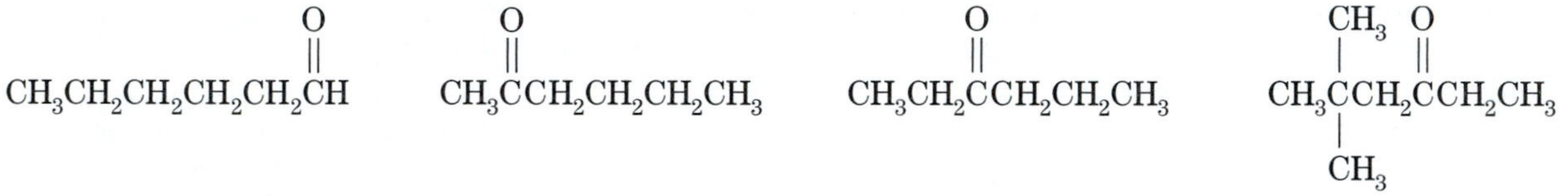

$$\underset{\text{Hexanal}}{CH_3CH_2CH_2CH_2CH_2\overset{\overset{O}{\|}}{C}H} \qquad \underset{\text{2-hexanone}}{CH_3\overset{\overset{O}{\|}}{C}CH_2CH_2CH_2CH_3} \qquad \underset{\text{3-hexanone}}{CH_3CH_2\overset{\overset{O}{\|}}{C}CH_2CH_2CH_3} \qquad \underset{\text{5,5-dimethyl-3-hexanone}}{CH_3\underset{\underset{CH_3}{|}}{\overset{\overset{CH_3}{|}}{C}}CH_2\overset{\overset{O}{\|}}{C}CH_2CH_3}$$

GETTING INVOLVED

Problem 11.1

Name the following by the IUPAC system of nomenclature:

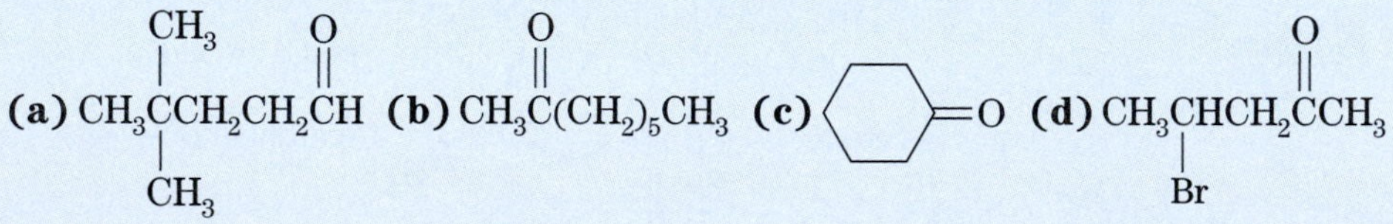

$$\textbf{(a)}\ CH_3\underset{\underset{CH_3}{|}}{\overset{\overset{CH_3}{|}}{C}}CH_2CH_2\overset{\overset{O}{\|}}{C}H \quad \textbf{(b)}\ CH_3\overset{\overset{O}{\|}}{C}(CH_2)_5CH_3 \quad \textbf{(c)}\ \text{cyclohexane}{=}O \quad \textbf{(d)}\ CH_3\underset{\underset{Br}{|}}{C}HCH_2\overset{\overset{O}{\|}}{C}CH_3$$

See related problems 11.25–11.26.

B. Polyfunctional Aldehydes and Ketones

Compounds with two aldehyde or two ketone groups are named *dials* and *diones*, respectively. But what about a compound that possesses both an aldehyde and a ketone group, or maybe an alcohol or amine group as well? In these cases, one group is named using the normal suffix, and the rest are named by prefixes. The group highest in the following table takes the suffix, and the chain is numbered to give it the lowest possible number.

	Functional group	Suffix	Prefix
Priority for use of suffix ⇑	Aldehyde	*al*	*oxo*
	Ketone	*one*	*oxo*
	Alcohol	*ol*	*hydroxy*
	Amine	*amine*	*amino*

The following examples apply these principles:

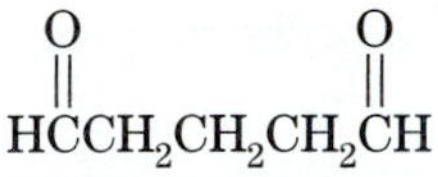

$\overset{O}{\overset{\Vert}{H C}}CH_2CH_2CH_2\overset{O}{\overset{\Vert}{C}}H$	$CH_3\overset{O}{\overset{\Vert}{C}}CH_2\overset{O}{\overset{\Vert}{C}}CH_3$	$CH_3\overset{O}{\overset{\Vert}{C}}CH_2\overset{O}{\overset{\Vert}{C}}CH_2\overset{O}{\overset{\Vert}{C}}CH_3$
1,5-pentandial	2,4-pentandione	2,4,6-heptantrione
$CH_3\overset{O}{\overset{\Vert}{C}}CH_2CH_2\overset{O}{\overset{\Vert}{C}}H$	$CH_3\overset{OH}{\overset{\vert}{C}}HCH_2\overset{O}{\overset{\Vert}{C}}CH_3$	$CH_3\overset{NH_2}{\overset{\vert}{C}}HCH_2\overset{O}{\overset{\Vert}{C}}CH_2\overset{O}{\overset{\Vert}{C}}H$
4-oxopentanal	4-hydroxy-2-pentanone	5-amino-3-oxohexanal

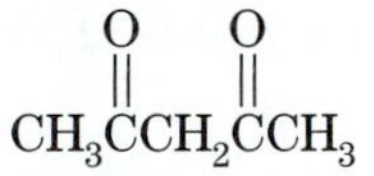

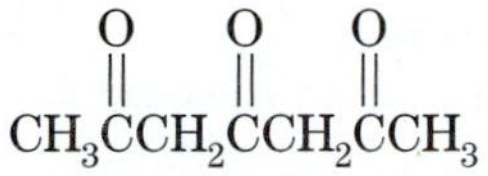

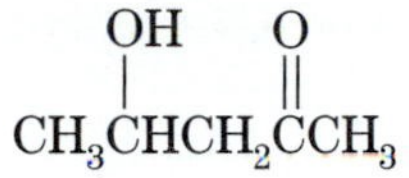

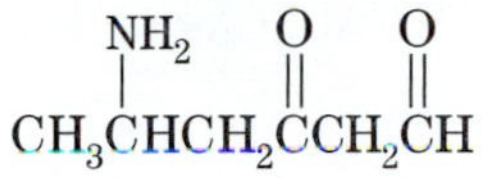

GETTING INVOLVED

Problem 11.2

Name the following by the IUPAC system of nomenclature:

(a)

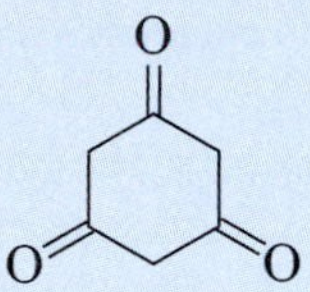

(b) 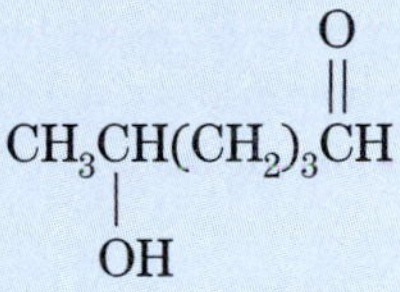 $CH_3\underset{OH}{\underset{\vert}{C}}H(CH_2)_3\overset{O}{\overset{\Vert}{C}}H$

(c) $CH_3\underset{Br}{\underset{\vert}{\overset{CH_3}{\overset{\vert}{C}}}}CH_2\underset{O}{\underset{\Vert}{C}}CH_2\underset{OH}{\underset{\vert}{C}}HCH_2\overset{O}{\overset{\Vert}{C}}H$

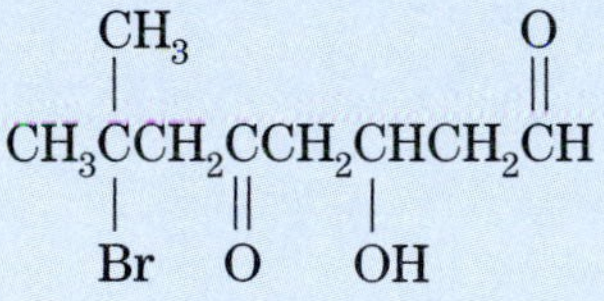

(d) 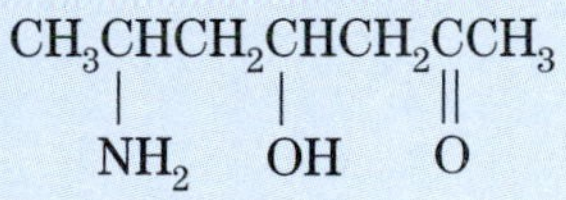 $CH_3\underset{NH_2}{\underset{\vert}{C}}HCH_2\underset{OH}{\underset{\vert}{C}}HCH_2\underset{O}{\underset{\Vert}{C}}CH_3$

See related problem 11.27.

C. Unsaturated and Polyfunctional Aldehydes and Ketones

Let us apply the nomenclature we have learned to compounds containing several important structural features. The following procedure is useful in naming more complex molecules:

1. Determine and name the longest continuous chain of carbon atoms containing the functional group of highest priority.
2. Follow the root name with the suffix *-an* if all carbon-carbon bonds are single bonds, *-en* if the chain contains a carbon-carbon double bond, and *-yn* if it contains a carbon-carbon triple bond.
3. Name the most important functional group (aldehyde > ketone > alcohol > amine) with the appropriate suffix.
4. Number the carbon chain, giving the lowest possible number to the functional group named by the suffix (next in precedence are carbon-carbon multiple bonds, with carbon-carbon double bonds taking precedence over triple bonds when otherwise the direction of the numbering makes no difference.) Complete the suffix by assigning numbers to the most important functional group and carbon-carbon multiple bonds.
5. Name all other groups with prefixes (in alphabetical order) and assign them the appropriate numbers.

GETTING INVOLVED

✓ Take a minute to make sure you understand and can describe the procedure for naming aldehydes and ketones.

Example 11.1

Give the IUPAC name for citral (lemon flavor and odor):

$$\underset{8}{CH_3}\overset{CH_3}{\underset{7}{C}}{=}\underset{6}{C}H\underset{5}{C}H_2\underset{4}{C}H_2\overset{CH_3}{\underset{3}{C}}{=}\underset{2}{C}H\overset{O}{\underset{1}{C}}H$$

Solution

1. There are eight carbons in the longest chain: oct.
2. There are two carbon-carbon double bonds: octadien.
3. The aldehyde is the only functional group and is named with a suffix: octadienal.
4. The chain is numbered from the aldehyde group. The double bonds at the second and sixth carbons are identified. Since the aldehyde is at carbon-1, it is not necessary to give it a number: 2,6-octadienal.
5. The methyls are named with prefixes and assigned numbers. The complete name is

3,7-dimethyl¦2,6¦oct¦a¦dien¦al

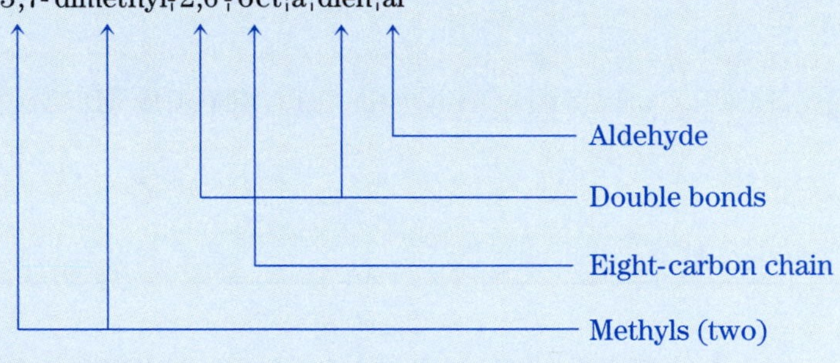

Example 11.2

Name this compound:

$$\underset{1}{C}H_3\overset{O}{\underset{2}{C}}\underset{3}{C}{\equiv}\underset{4}{C}\underset{5}{C}H_2OH$$

Solution

1. Five-carbon chain: pent.
2. One carbon-carbon triple bond: pentyn.
3. Ketone higher in list than alcohol; gets suffix: pentynone.
4. Chain numbered to give lowest number to ketone: 3-pentyn-2-one.
5. Alcohol is named with a prefix; the final name is

5-hydroxy-3-pentyn-2-one

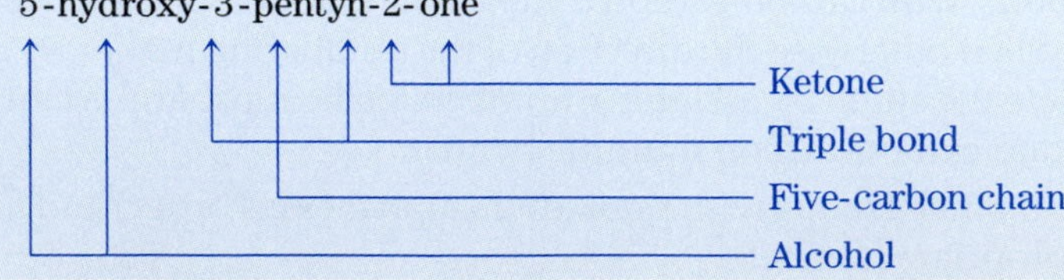

Problem 11.3

Name the following by the IUPAC system of nomenclature:

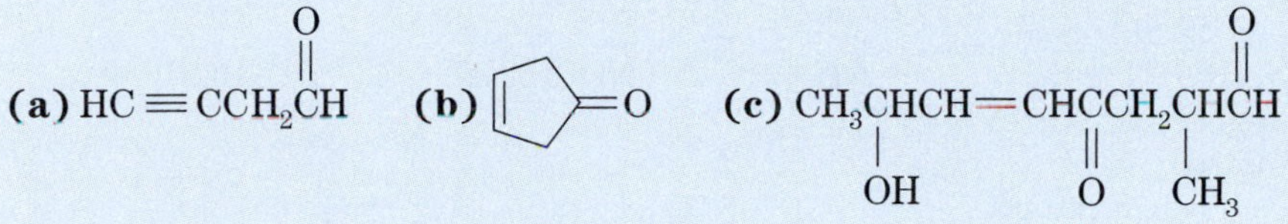

(a) $HC{\equiv}CCH_2\overset{O}{\overset{\|}{C}}H$ (b) cyclopentenone (2-cyclopenten-1-one ring, =O) (c) $CH_3\underset{OH}{\underset{|}{C}}HCH{=}CH\underset{O}{\underset{\|}{C}}CH_2\underset{CH_3}{\underset{|}{C}}H\overset{O}{\overset{\|}{C}}H$

See related problems 11.28–11.29.

D. Common Nomenclature

The use of trivial names for aldehydes, particularly simple ones, is very prevalent. As we will see in Chapter 12, the common names of aldehydes are related to those of carboxylic acids.

$H\overset{O}{\overset{\|}{C}}H$	$CH_3\overset{O}{\overset{\|}{C}}H$	$CH_3CH_2\overset{O}{\overset{\|}{C}}H$	$CH_3CH_2CH_2\overset{O}{\overset{\|}{C}}H$
Formaldehyde	Acetaldehyde	Propionaldehyde	Butyraldehyde

The simplest ketone, 2-propanone, is commonly called *acetone*. The common names of other ketones are derived by naming the alkyl groups attached to the carbonyl carbon.

$CH_3\overset{O}{\overset{\|}{C}}CH_3$	$CH_3\overset{O}{\overset{\|}{C}}CH_2CH_3$	$CH_3CH_2\overset{O}{\overset{\|}{C}}CH_2CH_3$
Acetone	Methyl ethyl ketone	Diethyl ketone

GETTING INVOLVED

Problem 11.4

Write structures for the following compounds: **(a)** isobutyraldehyde; **(b)** 2-chloropropionaldehyde; **(c)** methyl propyl ketone; **(d)** methyl phenyl ketone.

See related problem 11.30.

11.3 Some Preparations of Aldehydes and Ketones

Following is a summary of preparations of aldehydes and ketones covered in other parts of the book.

A. Hydration of Alkynes (section 5.2.D)

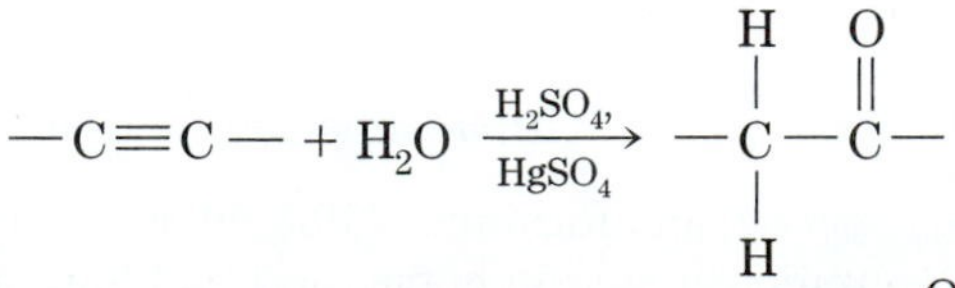

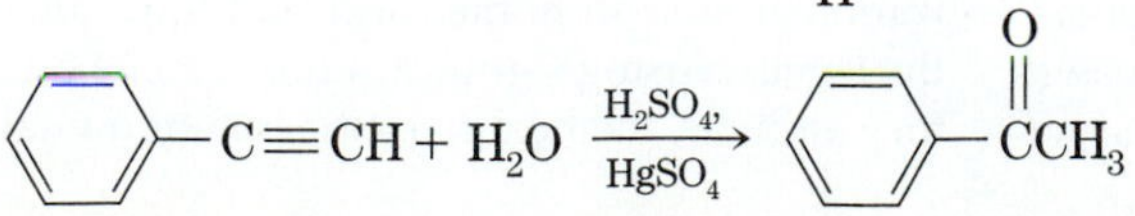

$$-C{\equiv}C- + H_2O \xrightarrow[HgSO_4]{H_2SO_4,} -\underset{H}{\overset{H}{\overset{|}{\underset{|}{C}}}}-\overset{O}{\overset{\|}{C}}-$$

$$C_6H_5-C{\equiv}CH + H_2O \xrightarrow[HgSO_4]{H_2SO_4,} C_6H_5-\overset{O}{\overset{\|}{C}}CH_3$$

Connections 11.1 www.prenhall.com/bailey

Formaldehyde and Synthetic Polymers

Formaldehyde, the simplest aldehyde, is a starting material for a number of polymers. Because its carbonyl group is highly polarized and unhindered, it is highly reactive. In fact, it can self-polymerize if stored undiluted.

The first manufactured polymer of commercial importance was developed by Leo Baekeland in 1907. Baekeland was already a successful and recognized chemist (while in his thirties, he had invented Velox®, the first photographic paper that could be exposed by artificial light) when he embarked on the investigation that led to the development of Bakelite, the first synthetic polymer. While searching for an artificial shellac to replace that processed from the Indian female lac insect, Baekeland came across an 1871 article by Adolph von Baeyer describing a hornlike, insoluble material resulting from the heating of a mixture of phenol and formaldehyde. Von Baeyer had difficulty in isolating the material, but Baekeland, recognizing its potential value, was able to control the reaction. The result was a thermosetting plastic he called Bakelite®.

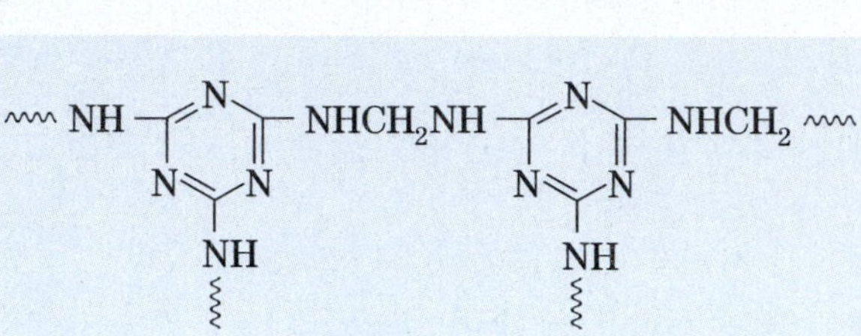

~~~ NH–(triazine)–NHCH$_2$NH–(triazine)–NHCH$_2$ ~~~ (each triazine bearing NH ~~~)

Melmac®

HCH(=O) + phenol (OH) ⟶ Bakelite network (phenol rings linked by CH$_2$ bridges)

Formaldehyde Phenol

Bakelite®

Over the past century, Bakelite and related phenolic resins have been widely used in molded products such as the handles on cooking and electrical utensils, electrical plates and switches, and some appliances and business machines. Bakelite's resistance to heat, electricity, and organic solvents gives it great versatility.

Formaldehyde forms polymers with itself [Delrin® and Celcon®, $(CH_2O)_n$] that are sturdy enough to be used in gears, bearings, pump parts, and instrument housings. When polymerized with melamine, formaldehyde forms the familiar material used for plastic dinnerware, Melmac®. Urea-formaldehyde polymers are employed as adhesives in plywood and are used to make foam insulation, carpeting, textiles, paper products, and furniture.

Produced from the vapor-phase air oxidation of methanol, formaldehyde is a flammable, colorless gas with a suffocating odor, intensely irritating to the mucous membranes. The Environmental Protection Agency classifies it as a carcinogen because it causes cancer in laboratory animals, especially nasal tumors in rats.

One of the largest single uses of formaldehyde resins is as a bonding adhesive in plywood and particleboard. It has become a significant indoor air pollutant, being slowly emitted from particleboard, plywood, insulating foam, carpeting, and furniture. Although even the highest concentrations found in homes and buildings are still far below the levels causing significant numbers of cancers in laboratory animals, the health issue remains a vital concern.

~~~ NC(=O)NCH$_2$NC(=O)N ~~~ (with N–CH$_2$ ~~~ branches)

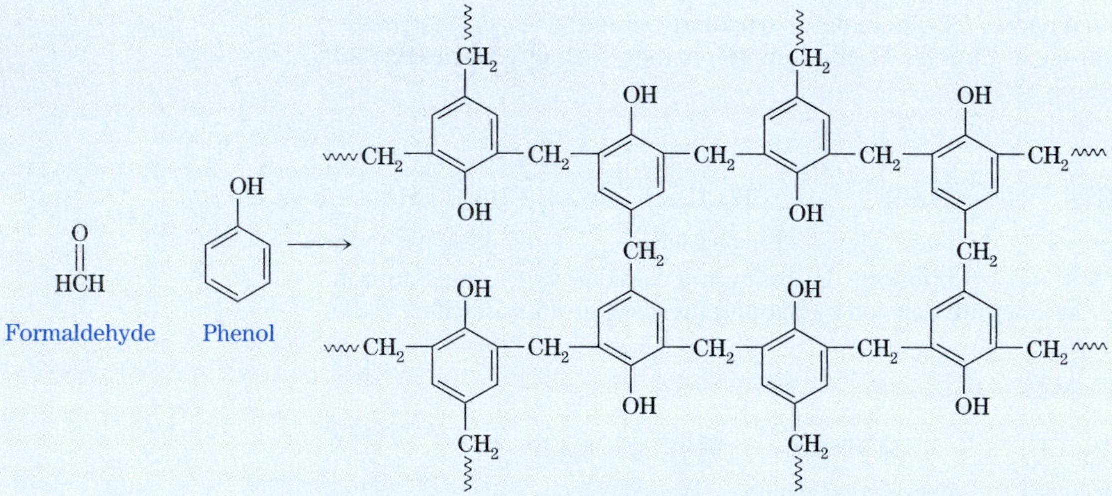

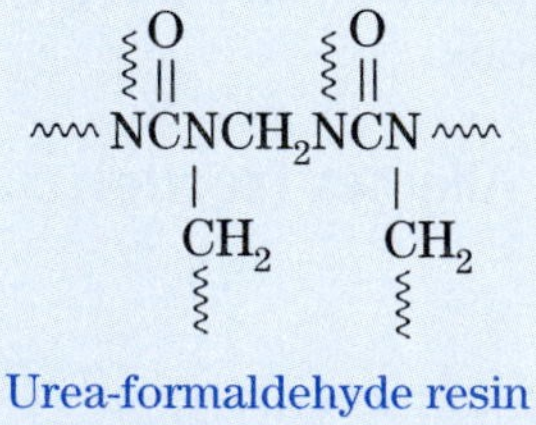

Urea-formaldehyde resin

B. Ozonolysis of Alkenes (section 5.7.B)

$$\mathrm{>C{=}C<} \xrightarrow{O_3} \xrightarrow[H_2O]{Zn,} \mathrm{>C{=}O + O{=}C<}$$

$$\text{1-methylcyclohexene} \xrightarrow{O_3} \xrightarrow[H_2O]{Zn,} \mathrm{H\overset{O}{\overset{\|}{C}}(CH_2)_4\overset{O}{\overset{\|}{C}}CH_3}$$

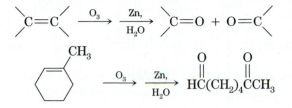

C. Friedel-Crafts Reaction (section 6.4.A and 6.4.B.2)

$$\text{benzene} + \mathrm{R\overset{O}{\overset{\|}{C}}Cl} \xrightarrow{AlCl_3} \mathrm{C_6H_5{-}\overset{O}{\overset{\|}{C}}R + HCl}$$

$$\text{benzene} + \mathrm{CH_3\overset{O}{\overset{\|}{C}}Cl} \xrightarrow{AlCl_3} \mathrm{C_6H_5{-}\overset{O}{\overset{\|}{C}}CH_3}$$

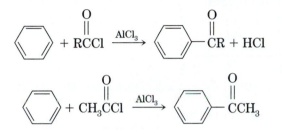

D. Oxidation of Alcohols (section 9.9)

PCC for aldehydes or ketones; $Na_2Cr_2O_7$ or CrO_3 for ketones

$$\mathrm{-\overset{OH}{\overset{|}{\underset{|}{C}}}-H} \xrightarrow[\text{agent}]{\text{Oxidizing}} \mathrm{>C{=}O}$$

$$\mathrm{CH_3(CH_2)_5CH_2OH} \xrightarrow{PCC} \mathrm{CH_3(CH_2)_5\overset{O}{\overset{\|}{C}}H}$$

$$\text{cyclohexyl}\mathrm{{-}OH} \xrightarrow{CrO_3} \text{cyclohexanone } (\mathrm{{=}O})$$

11.4 Oxidation of Aldehydes: Tollens' "Silver Mirror" Test

Aldehydes and ketones are structurally similar and consequently show similar chemical properties. They do differ significantly in one chemical property—susceptibility to oxidation. Aldehydes are easily oxidized under mild conditions; ketones are not. The susceptibility to oxidation of aldehydes is due to the hydrogen on the carbonyl carbon, which is lost during oxidation.

Aldehydes: $\mathrm{R\overset{O}{\overset{\|}{C}}H} \xrightarrow[\text{oxidation}]{\text{Mild}} \mathrm{R\overset{O}{\overset{\|}{C}}OH}$ (carboxylic acid)

Ketones: $\mathrm{R\overset{O}{\overset{\|}{C}}R} \xrightarrow[\text{oxidation}]{\text{Mild}}$ no reaction

Aldehydes can be distinguished from ketones by using Tollens' reagent, which is a solution of silver nitrate in ammonium hydroxide [actually $Ag(NH_3)_2OH$].

As the aldehyde is oxidized to the salt of a carboxylic acid, silver ion (Ag^+) is reduced to metallic silver. Ketones don't react.

$$\mathrm{RCH(=O)} + 2Ag(NH_3)_2^+ + 3OH^- \longrightarrow \mathrm{RCO^-(=O)} + 2Ag\downarrow + 4NH_3 + 2H_2O$$

If the reaction is allowed to proceed slowly in a clean test tube, metallic silver is deposited on the glass walls, creating a smooth reflective surface; hence the name silver mirror test.

GETTING INVOLVED

✓ What is produced when an aldehyde is oxidized? What happens to a ketone under mild oxidizing conditions?

Problem 11.5

Write equations showing the reactions of each of the isomers propanal and 2-propanone with Tollens' reagent.

See related problem 11.33(a).

11.5 Addition Reactions of Aldehydes and Ketones

A. General Considerations

1. ***Reactivity of the Carbonyl Group.*** The carbonyl group of aldehydes and ketones is very reactive for the following reasons: (1) the carbon-oxygen double bond is electron rich because of the presence of the π bond; (2) the carbon-oxygen double bond is polar; (3) the oxygen has two nonbonding electron pairs; and (4) the carbonyl group has a flat open structure that makes it accessible to other reagents. These structural features are illustrated in Figure 11.1.

Aldehydes are generally more reactive than ketones, since they have only one alkyl group (and a small hydrogen), whereas ketones have two alkyl groups that can hinder the approach of reacting species. In addition, many reactions of carbonyl groups depend on the positive character of the carbonyl carbon. As electron-releasing groups, alkyl substituents diminish the partially positive charge; this also decreases the reactivity of ketones.

Because of their polarity, carbonyl groups attract both electrophilic and nucleophilic reagents. Electrophilic (electron-seeking) reagents are electron deficient and thus are attracted to the partially negative carbonyl oxygen and its nonbonding electron pairs. In contrast, nucleophilic reagents are electron rich and seek positive centers; they are attracted to the partially positive carbon.

Nucleophiles and bases are attracted to the carbonyl carbon

$$\delta+ \; C{=}\ddot{O}\!: \; \delta-$$

Electrophiles and acids are attracted to the carbonyl oxygen

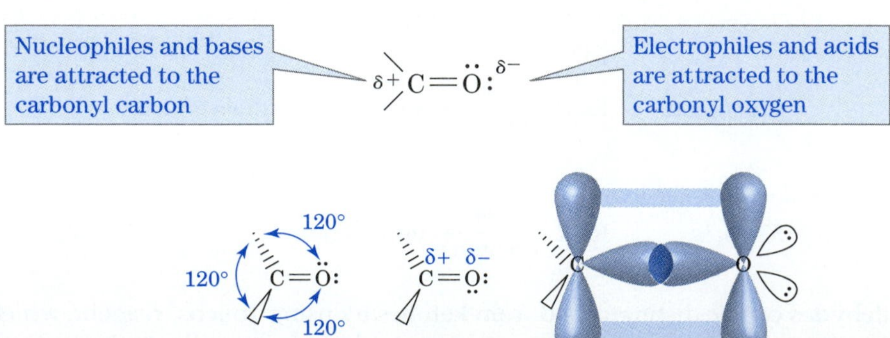

Figure 11.1 Structure of the carbonyl group.

A number of nucleophiles react with aldehydes and ketones, including hydride ($H:^-$), carbanions ($R_3C:^-$), water ($H_2\ddot{O}:$), alcohols ($R\ddot{O}H$), and amines ($R\ddot{N}H_2$). All of these species are Lewis bases with nonbonding electron pairs and are attracted to the carbonyl carbon.

GETTING INVOLVED

✓ Describe the four structural features that make aldehydes and ketones especially reactive. Why are ketones less reactive than aldehydes?

2. ***General Reaction.*** Addition is the characteristic chemical reaction of most compounds possessing a multiple bond. For example, we saw in Section 5.1 that alkenes add a variety of reagents, such as hydrogen, halogens, hydrogen halides, and water, as summarized in the following equation:

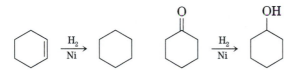

Alkenes: $$\text{>C=C<} + EA \longrightarrow -\overset{E}{\underset{|}{\overset{|}{C}}}-\overset{|}{\underset{A}{\underset{|}{C}}}- \qquad EA = H_2, X_2, HX, H_2O$$

Aldehydes and ketones possess a carbon-oxygen double bond, and, as we might expect, addition is their most characteristic chemical reaction. For example, alkenes as well as aldehydes and ketones add hydrogen in the presence of a metal catalyst.

$$\text{cyclohexene} \xrightarrow[Ni]{H_2} \text{cyclohexane} \qquad \text{cyclohexanone} \xrightarrow[Ni]{H_2} \text{cyclohexanol}$$

Unlike the double bond in alkenes, the carbonyl group has a permanent polarity. Consequently, unsymmetrical reagents (H—Nu) always add so the positive portion bonds to the negative oxygen and the negative portion to the positive carbon.

Aldehydes and Ketones: $$\text{>}\overset{:\ddot{O}^{\delta-}}{\overset{\|}{C}}{}^{\delta+}\text{<} + \overset{\delta+}{H}-\overset{\delta-}{Nu} \longrightarrow -\overset{O-H}{\overset{|}{\underset{|}{C}}}-Nu$$

Although aldehydes and ketones add a variety of reagents, the reactions are generally not as simple as those of alkenes. This is because the product of straight addition is frequently unstable and either exists in equilibrium with the starting materials or reacts further to form a more stable substance. For example, the product of addition of water or hydrogen halide to a carbonyl compound usually comprises only a small portion of the equilibrium mixture between it and the starting materials. Even when it is formed in significant amounts, it can seldom be isolated from the reaction mixture.

$$\text{>}\overset{O}{\overset{\|}{C}}\text{<} + H_2O \underset{\text{Elimination}}{\overset{\text{Addition}}{\rightleftharpoons}} -\overset{OH}{\overset{|}{\underset{|}{C}}}-OH \qquad \text{>}\overset{O}{\overset{\|}{C}}\text{<} + HX \underset{\text{Elimination}}{\overset{\text{Addition}}{\rightleftharpoons}} -\overset{OH}{\overset{|}{\underset{|}{C}}}-X$$

Compounds of this type, in which a carbon possesses an —OH or —NH group and one or more —OH, —OR, —NH_2, or —X (halogen) groups, are usually

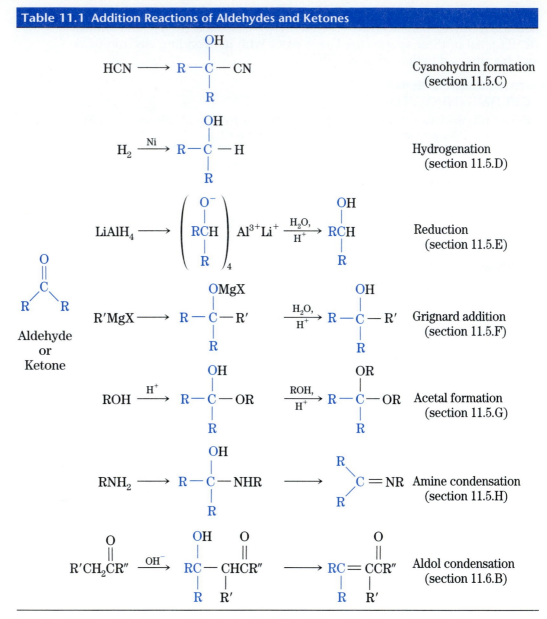

Table 11.1 Addition Reactions of Aldehydes and Ketones

Starting compound (Aldehyde or Ketone): $R_2C{=}O$

Reagent	Addition product	Final product	Reaction
HCN	$R_2C(OH)CN$		Cyanohydrin formation (section 11.5.C)
H_2, Ni	$R_2C(OH)H$		Hydrogenation (section 11.5.D)
$LiAlH_4$	$(RCH(R)O^-)_4\ Al^{3+}Li^+$	$\xrightarrow{H_2O,\ H^+}$ $RCH(OH)R$	Reduction (section 11.5.E)
$R'MgX$	$R_2C(OMgX)R'$	$\xrightarrow{H_2O,\ H^+}$ $R_2C(OH)R'$	Grignard addition (section 11.5.F)
ROH, H^+	$R_2C(OH)OR$	$\xrightarrow{ROH,\ H^+}$ $R_2C(OR)OR$	Acetal formation (section 11.5.G)
RNH_2	$R_2C(OH)NHR$	$\longrightarrow$ $R_2C{=}NR$	Amine condensation (section 11.5.H)
$R'CH_2CR''$ (C=O), OH^-	$RC(OH)(R)-CH(R')CR''(=O)$	$\longrightarrow$ $RC(R){=}C(R')CR''(=O)$	Aldol condensation (section 11.6.B)

Note: The first step of all of these reactions is simple addition to the carbon-oxygen double bond.

unstable and readily undergo elimination. Exceptions exist; two are illustrated in Example 11.3 and Problem 11.6.

Table 11.1 summarizes the addition reactions of aldehydes and ketones. If you take a look at these, you will notice that in each case the first step of the reaction is simple addition.

GETTING INVOLVED

✓ The common reaction of aldehydes and ketones is the same as that for alkenes. What is it and why is it the same? How does this reaction differ for aldehydes and ketones compared to alkenes in orientation of addition and stability of products?

Example 11.3

As a rule, the products of the addition of water to an aldehyde or ketone are a minor part of the equilibrium mixture and cannot be isolated from the solution. An exception is a substance commonly called chloral hydrate (an anesthetic used in veterinary medicine also called "knock-out drops"), which is formed from the addition of water to trichloroethanal. Write an equation for this reaction.

Solution

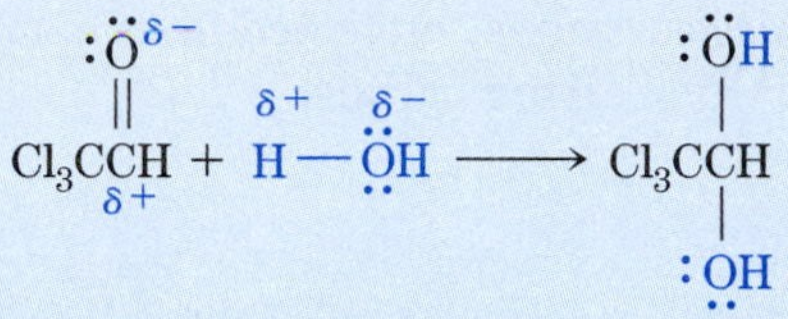

$$\underset{\delta+}{Cl_3C\overset{\overset{:\ddot{O}^{\delta-}}{\|}}{C}H} + \overset{\delta+}{H}-\overset{\delta-}{\ddot{O}H} \longrightarrow Cl_3C\overset{\overset{:\ddot{O}H}{|}}{\underset{\underset{:\ddot{O}H}{|}}{C}}H$$

The product is a 1,1 diol called a **gem diol**

gem diol carbon with two bonded OH groups

Notice that the partially positive hydrogen of the water molecule adds to the aldehyde's electron-rich oxygen, and the partially negative oxygen of water (OH) adds to the partially positive carbonyl carbon. This aldehyde is especially reactive because of the electron-withdrawing effect on the chlorines. The carbonyl group is made less stable and thus more reactive.

Problem 11.6

Another exception to the general rule that water does not form stable addition products with aldehydes and ketones is the reaction of cyclopropanone with water. The trigonal carbon in the reactant becomes tetrahedral in the product, thus reducing angle strain in the three-membered ring. Write an equation for this reaction.

B. Mechanisms of Nucleophilic Addition Reactions of Aldehydes and Ketones

Most addition reactions of aldehydes and ketones are nucleophilic additions, since the group adding to the carbonyl carbon is almost always a nucleophile (see Table 11.1). Recall that the carbonyl group has a constant polarity. The partially positive carbonyl carbon attracts nucleophiles or Lewis bases, and the carbonyl oxygen, which is partially negative and possesses two nonbonding electron pairs, attracts electrophiles or Lewis acids. Depending on the reaction conditions, these nucleophilic addition reactions can be either base-initiated or acid-initiated.

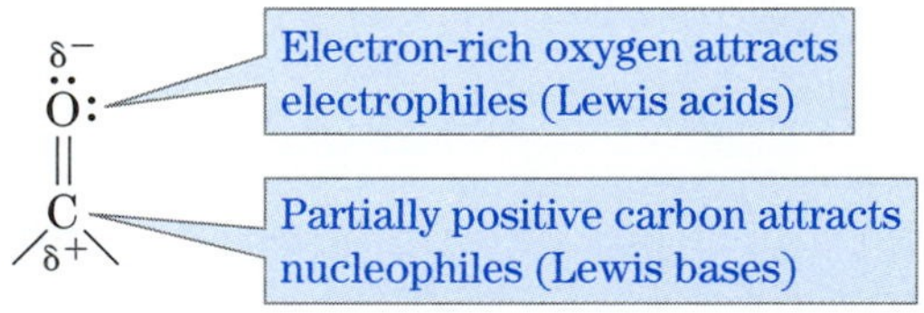

$\delta-$:Ö: — Electron-rich oxygen attracts electrophiles (Lewis acids)

C $\delta+$ — Partially positive carbon attracts nucleophiles (Lewis bases)

In base-initiated nucleophilic addition, the nucleophile attacks the carbonyl carbon first and provides both electrons for the new carbon-nucleophile bond. The π electrons of the carbonyl are displaced to the oxygen, forming an anion. Abstraction of a hydrogen ion from HNu by the negative oxygen (or from neutralization with acid) completes the addition process.

Base-Initiated Addition

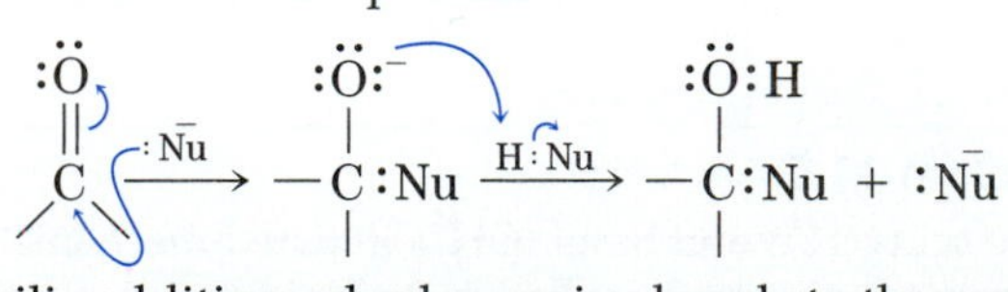

$$\overset{\overset{:\ddot{O}}{\|}}{C} \xrightarrow{:\bar{N}u} -\overset{\overset{:\ddot{O}:^-}{|}}{\underset{|}{C}}:Nu \xrightarrow{H:Nu} -\overset{\overset{:\ddot{O}:H}{|}}{\underset{|}{C}}:Nu + :\bar{N}u$$

In acid-initiated nucleophilic addition, a hydrogen ion bonds to the partially negative carbonyl oxygen; a carbocation results. The formation of a carbocation enhances the attraction of the nucleophile to the carbonyl carbon. Note that the reaction is acid-catalyzed: a hydrogen ion initiates the process and is returned in the final step.

Acid-Initiated Addition

$$\underset{\diagup\;\diagdown}{\overset{:\ddot{O}:}{\overset{\|}{C}}} \xrightarrow{H^+} \underset{\diagup\;\diagdown}{\overset{:\ddot{O}:H}{\overset{|}{C^+}}} \xrightarrow{H:\ddot{N}u} -\underset{|}{\overset{:\ddot{O}:H}{\overset{|}{C}}}:\underset{\ddot{}\;H}{\overset{+}{N}u} \xrightarrow{-H^+} -\underset{|}{\overset{:\ddot{O}:H}{\overset{|}{C}}}:\ddot{N}u$$

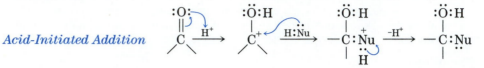

In these addition reactions the carbonyl carbon is converted from a trigonal, sp^2-hybridized atom to a tetrahedral, sp^3-hybridized carbon.

GETTING INVOLVED

✓ What is nucleophilic addition as applied to aldehydes and ketones? What is the difference between acid-initiated and base-initiated nucleophilic addition? Describe each in words before doing Problem 11.7.

Example 11.4

Write mechanisms for the acid- and base-initiated addition of water to trichloroethanal, as shown in Example 11.3.

Solution

Acid-initiated: H^+ adds first, forming a carbocation that is neutralized by a water molecule. Release of H^+ in the final step releases the product.

$$Cl_3C\overset{:\ddot{O}}{\overset{\|}{C}}H \xrightarrow{H^+} Cl_3C\underset{+}{\overset{:\ddot{O}H}{\overset{|}{C}}}H \xrightarrow{H_2\ddot{O}:} Cl_3C\underset{\underset{H}{:\overset{+}{O}H}}{\overset{:\ddot{O}H}{\overset{|}{C}}}H \xrightarrow{-H^+} Cl_3C\underset{:\ddot{O}H}{\overset{:\ddot{O}H}{\overset{|}{C}}}H$$

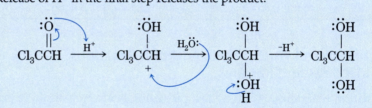

Base-initiated: Two mechanisms are possible. If the solution is slightly basic, hydroxide adds first and a hydrogen ion is abstracted by the resulting anion from a water molecule.

$$Cl_3C\overset{:\ddot{O}}{\overset{\|}{C}}H \xrightarrow{:\ddot{O}H^-} Cl_3C\underset{:\ddot{O}H}{\overset{:\ddot{O}:^-}{\overset{|}{C}}}H \xrightarrow{H_2\ddot{O}:} Cl_3C\underset{:\ddot{O}H}{\overset{:\ddot{O}H}{\overset{|}{C}}}H + :\ddot{O}H^-$$

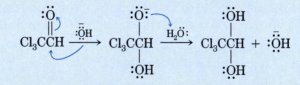

Water itself can act as the nucleophile and provide the hydrogen ion needed to complete the addition process.

$$Cl_3C\overset{:\ddot{O}}{\overset{\|}{C}}H \xrightarrow{H_2\ddot{O}:} Cl_3C\underset{\underset{H^+}{:\ddot{O}H}}{\overset{:\ddot{O}:^-}{\overset{|}{C}}}H \xrightarrow{H_2\ddot{O}:} Cl_3C\underset{:\ddot{O}H}{\overset{:\ddot{O}H}{\overset{|}{C}}}H + :\ddot{O}H^- + H^+$$

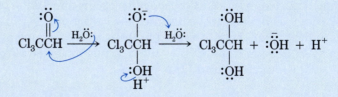

Problem 11.7

Write acid- (H^+) and base- (OH^-) initiated mechanisms for the addition of water to cyclopropanone, as described in Problem 11.6.

C. Addition of Hydrogen Cyanide

1. ***General Reaction.*** Hydrogen cyanide adds to aldehydes and ketones to form a class of compounds known as **cyanohydrins** or *hydroxy* **nitriles.** In adding, the hydrogen ion bonds to the negative oxygen, and the negative cyanide bonds to the positive carbonyl carbon.

cyanohydrin
carbon with both an OH and CN bonded

nitrile
compound with a carbon-nitrogen triple bond

$$\overset{O^{\delta-}}{\overset{\|}{C^{\delta+}}} + \overset{\delta+\delta-}{HCN} \longrightarrow -\overset{OH}{\overset{|}{\underset{|}{C}}}-CN$$

Because the cyanide group is easily hydrolyzed to a carboxylic acid, cyanohydrins are useful intermediates in organic synthesis. The reaction is useful in preparing biological molecules such as hydroxy acids and carbohydrates. A variation of the reaction can lead to amino acids. For example, the hydroxy acid lactic acid (one enantiomer is found in sore muscles, the other in sour milk) can be prepared from ethanal by addition of HCN followed by hydrolysis.

$$CH_3\overset{O}{\overset{\|}{C}}H + HCN \xrightarrow[\text{HCN}]{\text{Addition of}} CH_3\overset{OH}{\overset{|}{\underset{H}{\underset{|}{C}}}}-CN \xrightarrow[\text{(Hydrolysis of nitrile)}]{H_2O/H^+} CH_3\overset{OH}{\overset{|}{\underset{H}{\underset{|}{C}}}}-\underset{O}{\underset{\|}{C}}OH$$

Since hydrogen cyanide is extremely toxic (cyanide ion binds to blood hemoglobin and respiratory cytochromes more strongly than oxygen), this reaction must be performed carefully in a fume hood.

2. ***Reaction Mechanism.*** Since hydrogen cyanide is a weak acid and poor nucleophile, its addition to aldehydes and ketones is often performed by mixing the aldehyde or ketone with a sodium cyanide solution followed by neutralization with acid. Under these conditions, the cyanide ion is the nucleophile and adds first.

$$\overset{:\ddot{O}}{\overset{\|}{C}} \leftarrow :\bar{C}{\equiv}N: \longrightarrow -\overset{:\ddot{O}:^-}{\overset{|}{\underset{|}{C}}}:C{\equiv}N: \xrightarrow{\text{HCN}} -\overset{:\ddot{O}:H}{\overset{|}{\underset{|}{C}}}:C{\equiv}N: + :\bar{C}{\equiv}N:$$

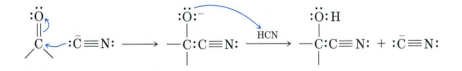

GETTING INVOLVED

✓ What are the two species that add in hydrogen cyanide addition to aldehydes and ketones?

✓ The reaction is usually base-initiated, why? What is the nucleophile? How is the intermediate anion neutralized?

Problem 11.8

(a) Write reaction equations illustrating the addition of HCN to 2-butanone and to benzaldehyde. **(b)** Write a reaction mechanism for the addition of HCN to ethanal.

See related problems 11.33(b) and 11.46(b).

D. Reduction to Alcohols: Catalytic Hydrogenation

Addition of hydrogen to aldehydes and ketones, catalytically and under pressure, results in the formation of primary and secondary alcohols, respectively.

$$\text{\textit{Aldehydes:}}\quad \mathrm{R\overset{\overset{\displaystyle O}{\|}}{C}H} + \mathrm{H_2} \xrightarrow[\text{Pressure}]{\text{Ni,}} \mathrm{R\underset{\underset{\displaystyle H}{|}}{\overset{\overset{\displaystyle OH}{|}}{C}}H} \quad 1^\circ \text{ alcohol}$$

$$\text{\textit{Ketones:}}\quad \mathrm{R\overset{\overset{\displaystyle O}{\|}}{C}R} + \mathrm{H_2} \xrightarrow[\text{Pressure}]{\text{Ni,}} \mathrm{R\underset{\underset{\displaystyle H}{|}}{\overset{\overset{\displaystyle OH}{|}}{C}}R} \quad 2^\circ \text{ alcohol}$$

The reaction mechanism is analogous to that of addition of hydrogen to alkenes (section 5.1.A.4). It does not involve nucleophilic addition as do other reactions of carbonyl compounds.

GETTING INVOLVED

✓ What is produced by hydrogenation of an aldehyde? of a ketone? Why cannot 3° alcohols be made by hydrogenation of aldehydes or ketones?

Problem 11.9

Write reaction equations illustrating the conversion of the isomers propanal and 2-propanone to primary and secondary alcohols, respectively, by hydrogenation.

See related problems 11.33(c) and 11.43.

E. Reduction to Alcohols with Sodium Borohydride and Lithium Aluminum Hydride

1. ***General Reaction.*** A second and often more convenient method for the reduction of aldehydes and ketones to alcohols involves the use of metal hydrides such as sodium borohydride ($NaBH_4$) or lithium aluminum hydride ($LiAlH_4$). The procedure involves treating a carbonyl compound with sodium borohydride followed by hydrolysis in water or dilute acid.

As with catalytic hydrogenation, primary alcohols can be prepared from aldehydes and secondary alcohols from ketones.

$$\text{\textit{Aldehydes:}}\quad \mathrm{R\overset{\overset{\displaystyle O}{\|}}{C}H} \xrightarrow[\substack{\text{or}\\ \mathrm{LiAlH_4}}]{\mathrm{NaBH_4}} \mathrm{R\overset{\overset{\displaystyle OH}{|}}{C}H_2} \quad 1^\circ \text{ alcohols}$$

$$\text{\textit{Ketones:}}\quad \mathrm{R\overset{\overset{\displaystyle O}{\|}}{C}R} \xrightarrow[\substack{\text{or}\\ \mathrm{LiAlH_4}}]{\mathrm{NaBH_4}} \mathrm{R\overset{\overset{\displaystyle OH}{|}}{C}HR} \quad 2^\circ \text{ alcohols}$$

2. ***Reaction Mechanism.*** When this reaction is examined closely, it proves to be an example of nucleophilic addition. Sodium borohydride has a boron in the 3+ oxidation state with four bonded hydride ions (negative hydrogen ions, $H:^-$). Being negative, the hydride ions are attracted to the positive carbonyl carbon and provide the electrons for a new carbon-hydrogen bond (remember, reduction involves the formation of new carbon-hydrogen bonds). The reaction is base initiated; hydride ion ($H:^-$) is the nucleophile. One mole of sodium borohydride actually reduces four moles of carbonyl and produces 4 moles of alkoxide ion. Treatment with water neutralizes the alkoxides to alcohols.

$$Na^+ \left[H-\overset{H}{\underset{H}{B}}-H \right]^- + R_2C{=}\ddot{O}: \xrightarrow{-BH_3} R-\overset{:\ddot{O}:^-}{\underset{R}{C}}-H \xrightarrow{H_3O^+} R-\overset{OH}{\underset{R}{C}}-H + H_2O$$

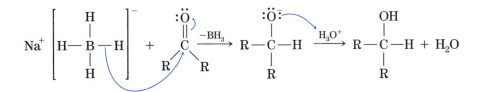

GETTING INVOLVED

- ✓ Overall in the $NaBH_4$ reduction of aldehydes and ketones, what are the two atoms that end up bonded to the carbon and oxygen? Why do aldehydes yield primary alcohols and ketones, secondary alcohols?
- ✓ What is the nucleophile in the nucleophilic addition mechanism? Is the reaction base-initiated or acid-initiated? What results from the nucleophilic attack and how is it neutralized?
- ✓ Why can four moles of aldehyde or ketone be reduced by one mole of $NaBH_4$?

Example 11.5

Write a reaction equation showing the preparation of 2-pentanol by the reduction of a ketone with $NaBH_4$.

Solution

$$4CH_3\underset{\underset{O}{\|}}{C}CH_2CH_2CH_3 + NaBH_4 \longrightarrow \xrightarrow[H^+]{H_2O} 4CH_3\underset{\underset{OH}{|}}{C}HCH_2CH_2CH_3$$

Problem 11.10

(a) Write reaction equations showing the reductions of propanal and 2-propanone to primary and secondary alcohols, respectively, with $NaBH_4$. **(b)** Write a detailed reaction mechanism for the reduction of ethanal with $NaBH_4$.

See related problems 11.33(d), 11.43, and 11.46(c).

3. ***Biological Reductions.*** The processes of metabolism include several series of oxidations and reductions. During the fermentation of sugars by certain strains of yeast, acetaldehyde is reduced to ethanol with the help of an enzyme (biological catalyst) and its cofactor NADH (the reduced form of nicotinamide ade-

nine dinucleotide). The reduction probably involves a transfer of a hydride ion from the NADH to the carbonyl of the acetaldehyde.

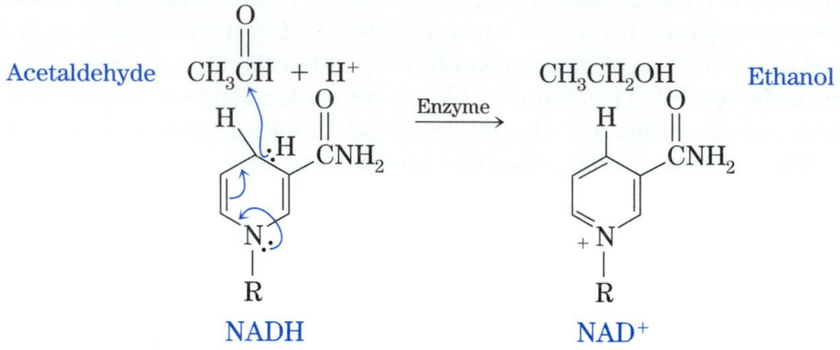

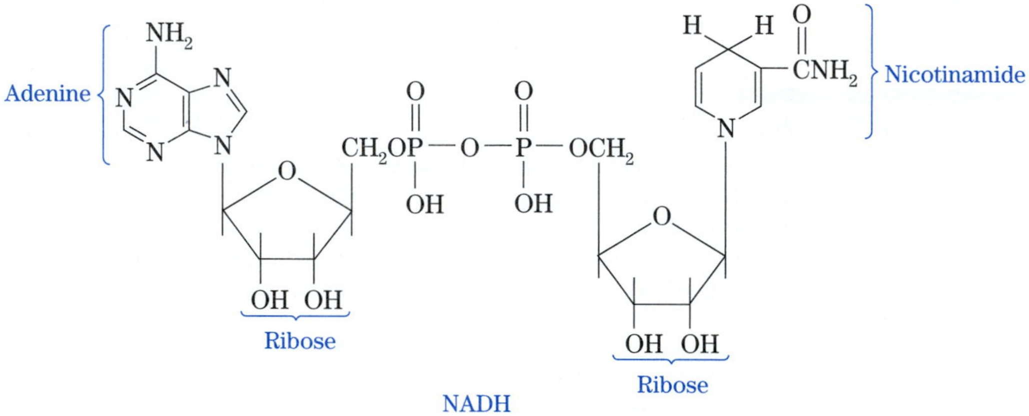

Acetaldehyde $CH_3CHO + H^+$ + NADH $\xrightarrow{\text{Enzyme}}$ CH_3CH_2OH Ethanol + NAD^+

NADH (Adenine, Ribose, Nicotinamide)

F. Grignard Addition—Preparation of Alcohols

Grignard reagent
the reagent RMgX developed by Nobel laureate Victor Grignard

1. ***General Reaction.*** One of the most versatile preparations of alcohols was developed by the French chemist Victor Grignard (1871–1935). His efforts won him a Nobel Prize in 1912. The **Grignard reagent** is prepared by treating an alkyl or aryl halide with magnesium metal in dry ether. The magnesium metal reacts slowly, forming a solution of the Grignard reagent.

$$RX + Mg \xrightarrow{\text{Ether}} \underset{\text{Grignard reagent}}{RMgX} \quad (\overset{\delta-}{R}-\overset{\delta+}{MgX})$$

R = alkyl or aryl group
X = Cl, Br, I

In reacting with aldehydes and ketones, the Grignard reagent adds to the carbon-oxygen bond, with the negative alkyl group attacking the carbonyl carbon and the positive magnesium going to the negative oxygen. The resulting alkoxide is then neutralized to an alcohol. Primary, secondary, and tertiary alcohols can be prepared in this fashion. Reaction with formaldehyde results in primary alcohols. With any other aldehyde, secondary alcohols are formed. Ketones are used to synthesize tertiary alcohols.

Formaldehyde:

$$\mathrm{H\overset{O}{\overset{\|}{C}}H + RMgX \longrightarrow H\overset{OMgX}{\overset{|}{\underset{R}{\underset{|}{C}}}}H \xrightarrow[H^+]{H_2O,} H\overset{OH}{\overset{|}{\underset{R}{\underset{|}{C}}}}H} \quad 1^\circ \text{ alcohols}$$

Aldehydes:

$$\mathrm{R'\overset{O}{\overset{\|}{C}}H + RMgX \longrightarrow R'\overset{OMgX}{\overset{|}{\underset{R}{\underset{|}{C}}}}H \xrightarrow[H^+]{H_2O,} R'\overset{OH}{\overset{|}{\underset{R}{\underset{|}{C}}}}H} \quad 2^\circ \text{ alcohols}$$

Ketones:

$$\mathrm{R'\overset{O}{\overset{\|}{C}}R'' + RMgX \longrightarrow R'\overset{OMgX}{\overset{|}{\underset{R}{\underset{|}{C}}}}R'' \xrightarrow[H^+]{H_2O,} R'\overset{OH}{\overset{|}{\underset{R}{\underset{|}{C}}}}R''} \quad 3^\circ \text{ alcohols}$$

2. ***Reaction Mechanism.*** The reaction of Grignard reagents with carbonyl compounds is an example of base-initiated nucleophilic addition. Due to the electropositive nature of magnesium, the carbon-magnesium bond is very polar, with the carbon having carbanion character.

$$\overset{\delta^-}{\mathrm{R}} - \overset{\delta^+}{\mathrm{MgX}}$$

When a Grignard reagent is mixed with an aldehyde or a ketone, the negative hydrocarbon group quickly attacks the positive carbonyl carbon, providing the two electrons needed for the new carbon-carbon bond. The π electrons are displaced to the oxygen, forming the alcohol salt that is then neutralized to an alcohol with water and acid.

$$\mathrm{\overset{\delta-}{:\ddot{O}}{=}\overset{\delta+}{C}\langle \;\; \overset{\delta-}{R}-\overset{\delta+}{MgX} \longrightarrow -\overset{:\ddot{O}:^- \;\; ^+MgX}{\overset{|}{\underset{|}{C}}}{:}R \xrightarrow[H^+]{H_2O} -\overset{:\ddot{O}:H}{\overset{|}{\underset{|}{C}}}{:}R + MgXOH}$$

Note that the hydrocarbon portion of a Grignard reagent essentially acts as a carbanion. It is for this reason that Grignard reactions must be performed in scrupulously dry ether. Even traces of moisture can neutralize the reagent.

$$\mathrm{-\overset{|}{\underset{|}{\overset{\delta-}{C}}}-\overset{\delta+}{MgX} + \overset{\delta+}{H}-\overset{\delta-}{OH} \longrightarrow -\overset{|}{\underset{|}{C}}{:}H + MgXOH}$$

GETTING INVOLVED

- ✓ From what is a Grignard reagent prepared?
- ✓ What is the difference in polarity of the carbon-halogen bond of an alkyl halide and the carbon-magnesium bond of a Grignard reagent? Why is the polarity of the Grignard reagent so important in its ability to react with aldehydes and ketones?
- ✓ In the Grignard preparation of alcohols, an alkyl group and a hydrogen overall add to the carbon-oxygen double bond of aldehydes and ketones. What is the source of the alkyl group and the hydrogen? Formaldehyde, other aldehydes, and ketones give primary, secondary, and tertiary alcohols, respectively. Explain this.
- ✓ Is the Grignard addition to a carbonyl acid-initiated or base-initiated? What is the nucleophile? What is the intermediate of nucleophilic addition and how is it neutralized?

Problem 11.11

(a) Write a reaction equation illustrating the preparation of a Grignard reagent from chloromethane. **(b)** Write reaction equations showing the preparation of alcohols using the Grignard reagent methyl magnesium chloride and each of the following carbonyl compounds: formaldehyde (methanal), propanal, and 2-propanone.

Problem 11.12

Write a reaction mechanism for the reaction of methyl magnesium chloride with formaldehyde, followed by hydrolysis.

See related problems 11.33(e–f) and 11.46(a).

3. ***Grignard Synthesis of Alcohols.*** How can a Grignard synthesis be planned? First, one must recognize that during the reaction the Grignard reagent always provides one alkyl group to the final alcohol product, and the others, if any, must come from the carbonyl compound. So to make a primary alcohol, one chooses formaldehyde as the carbonyl compound because it possesses no alkyl groups—the one alkyl group is provided by the Grignard reagent. For a secondary alcohol, an aldehyde provides one alkyl group and the Grignard reagent the other. In synthesizing a tertiary alcohol, the ketone provides two alkyl groups and the Grignard one. If all three alkyl groups are different, there are three possible Grignard syntheses. Equations illustrating these concepts are in section 11.5.F.1.

Let us illustrate with a specific problem, the synthesis of the secondary alcohol 1-phenyl-1-propanol.

GETTING INVOLVED

✓ Why is there only one way to prepare a primary alcohol by the Grignard synthesis using formaldehyde, but a maximum of two ways to make a secondary alcohol from an aldehyde, and three for making a tertiary alcohol from a ketone?

Example 11.6

Synthesize

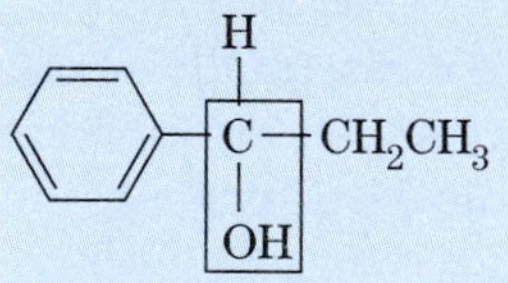

H
C₆H₅—C—CH_2CH_3
OH

Solution

First, identify the alcohol function (boxed in the formula as shown) and then realize that one of the attached alkyl groups comes from a carbonyl compound (an aldehyde) and the other from a Grignard reagent as shown. There are two ways to make this secondary alcohol by the Grignard synthesis:

From CH_3CH_2MgBr

From C₆H₅—MgBr | From $CH_3CH_2CH{=}O$ (propanal)

From C₆H₅—CH=O

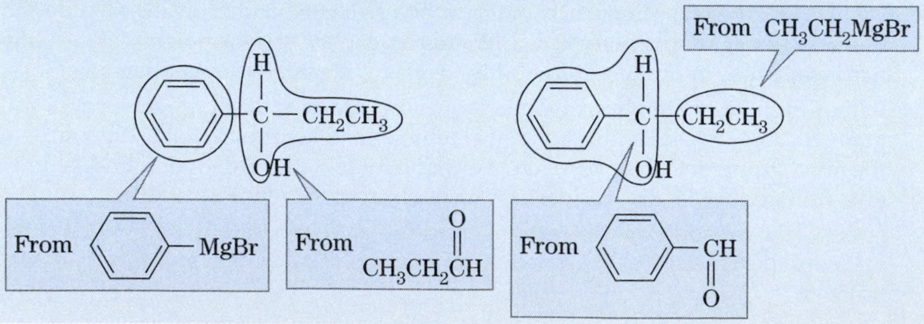

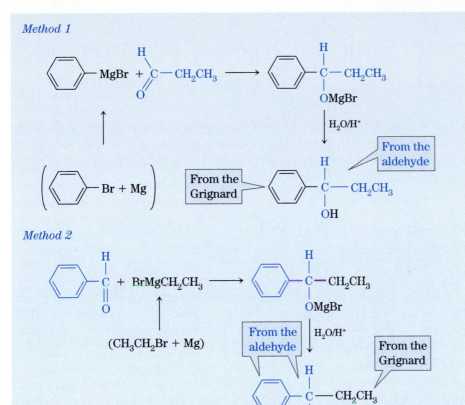

Method 1

$C_6H_5MgBr + H(C{=}O)CH_2CH_3 \longrightarrow C_6H_5CH(OMgBr)CH_2CH_3 \xrightarrow{H_2O/H^+} C_6H_5CH(OH)CH_2CH_3$

($C_6H_5Br + Mg$)

From the Grignard

From the aldehyde

Method 2

$C_6H_5CH{=}O + BrMgCH_2CH_3 \longrightarrow C_6H_5CH(OMgBr)CH_2CH_3 \xrightarrow{H_2O/H^+} C_6H_5CH(OH)CH_2CH_3$

($CH_3CH_2Br + Mg$)

From the aldehyde

From the Grignard

Problem 11.13

Considering the Grignard synthesis of alcohols from aldehydes and ketones and using the molecular formula $C_7H_{16}O$, draw: **(a)** a secondary alcohol that can be synthesized in only one way; **(b)** a secondary alcohol that can be synthesized in two ways; **(c)** a tertiary alcohol that can be synthesized in only one way; **(d)** a tertiary alcohol that can be synthesized in two ways; **(e)** a tertiary alcohol that can be synthesized in three ways.

Problem 11.14

Using the Grignard synthesis of alcohols, show three methods for preparing the tertiary alcohol 3-methyl-3-hexanol.

See related problems 11.34 and 11.43.

G. Alcohol Addition—Acetal Formation

1. *General Reaction.* Alcohols add to aldehydes and ketones to form **hemiacetals,** which can condense with a second molecule of alcohol to produce **acetals.** A hemiacetal is a compound that has an —OR (ether) and one —OH attached to the same carbon; an acetal has two —OR groups attached to the same carbon (diether). The first step of hemiacetal formation, simply involves addition of the polar O—H group of the alcohol to the polar C=O of the aldehyde or ketone. The acetal is formed by intermolecular dehydration ($-H_2O$) between the hemiacetal and a second molecule of alcohol.

hemiacetal
carbon bonded to both an OH and an OR group

acetal
carbon bonded to two OR groups

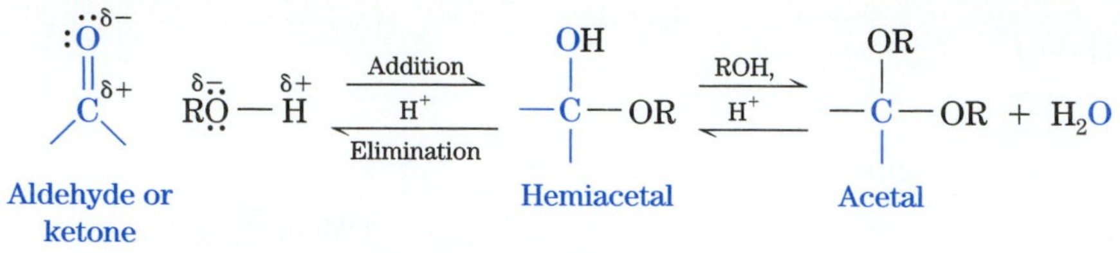

Aldehyde or ketone

Hemiacetal

Acetal

The hemiacetal is in equilibrium with the starting carbonyl compound, but the acetal can be isolated in a stable state if the water by-product is removed during its formation. The following example shows the formation of the methyl hemiacetal and the dimethyl acetal of cyclopentanone:

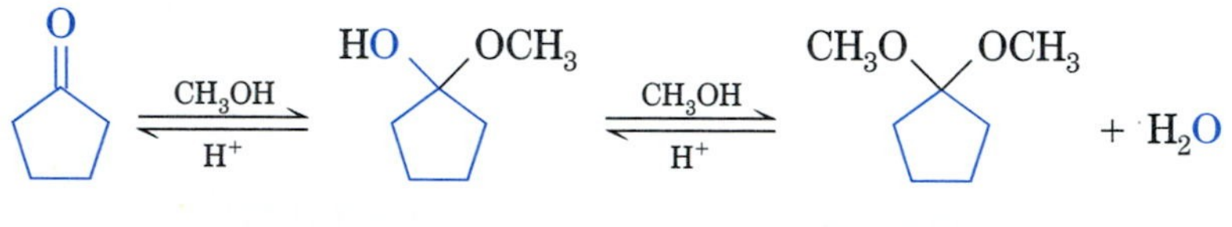

Carbohydrates usually exist in hemiacetal or acetal forms. The most prevalent example is glucose. Glucose in the open-chain form possesses both aldehyde and alcohol functions. In nature, glucose exists predominantly in a cyclic hemiacetal form, which arises by addition of the alcohol function on carbon-5 to the carbonyl group.

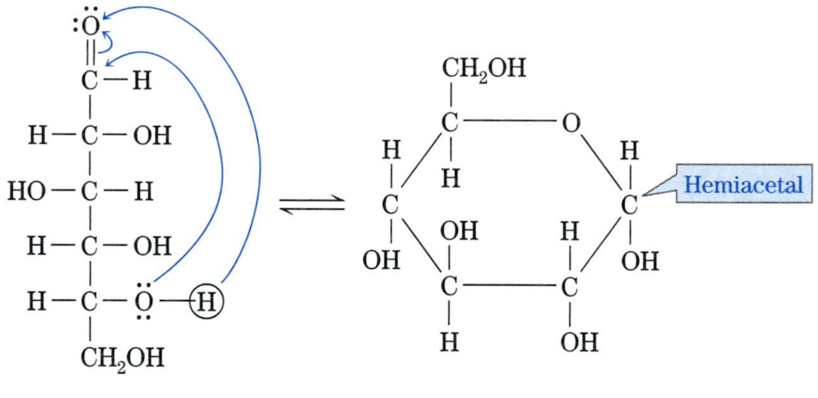

Hemiacetal

Glucose

2. ***Reaction Mechanism.*** Hemiacetal formation occurs by an acid-initiated nucleophilic addition mechanism. The alcohol molecule (ROH) is the nucleophile and the whole process is in equilibrium. The individual steps are described below.

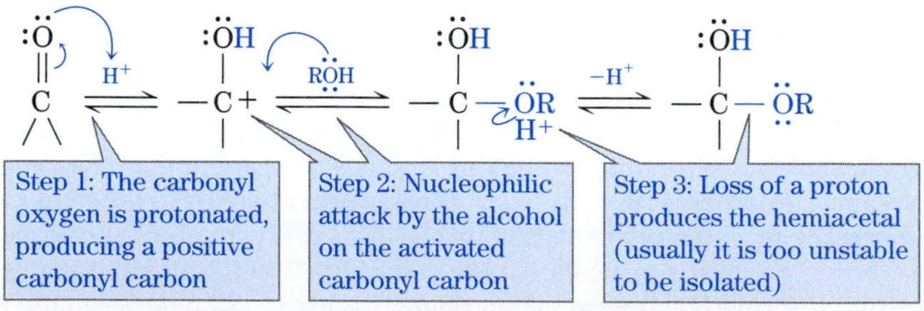

Step 1: The carbonyl oxygen is protonated, producing a positive carbonyl carbon

Step 2: Nucleophilic attack by the alcohol on the activated carbonyl carbon

Step 3: Loss of a proton produces the hemiacetal (usually it is too unstable to be isolated)

Reaction with a second mole of alcohol yields an acetal, essentially by an intermolecular dehydration.

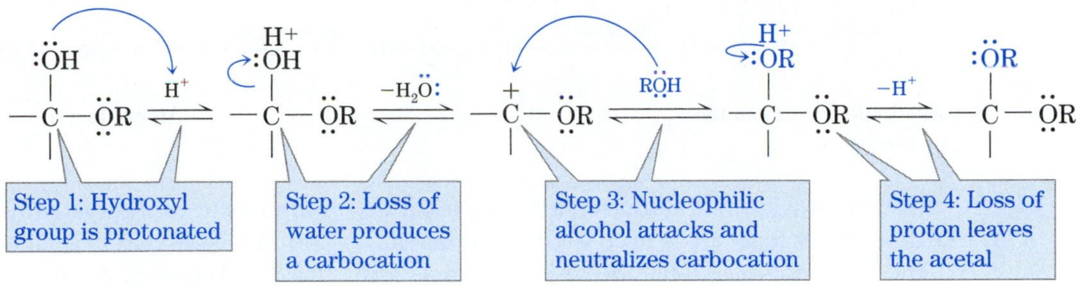

Step 1: Hydroxyl group is protonated

Step 2: Loss of water produces a carbocation

Step 3: Nucleophilic alcohol attacks and neutralizes carbocation

Step 4: Loss of proton leaves the acetal

GETTING INVOLVED

✓ What is the difference between a hemiacetal and an acetal? From what are each made?

✓ Is hemiacetal formation an acid- or base-initiated nucleophilic addition? What is the nucleophile? What are the two atoms/groups that end up on the carbon and oxygen of the carbonyl?

✓ Describe the mechanisms of formation of hemiacetals and acetals.

Problem 11.15

Write structures for the **(a)** hemiacetal and **(b)** acetal that are formed in the reaction of benzaldehyde with ethanol.

Problem 11.16

Write mechanisms for the formation of the two compounds in Problem 11.15.

Problem 11.17

From what ketone and alcohol is the following acetal prepared?

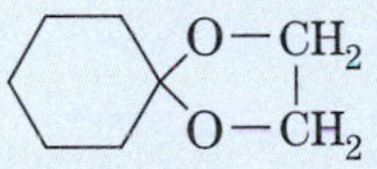

Problem 11.18

Write a step-by-step mechanism for the hydrolysis of the acetal in Problem 11.17 to the original ketone and alcohol.

See related problems 11.33(i–j), 11.41–11.42, and 11.46(f).

H. Addition of Amines

Primary amines (RNH_2) are effective Lewis bases and add to aldehydes and ketones in a way analogous to the way alcohols add to form hemiacetals. The nucleophilic addition reaction involves attack of nitrogen's nonbonding electron pair on the partially positive carbonyl carbon and addition of a hydrogen from the amine to the oxygen of the double bond. The initial addition product is not

imine compound with a carbon-nitrogen double bond

stable, and a molecule of water is eliminated between the carbon and nitrogen to form a double bond. The product is called an **imine.**

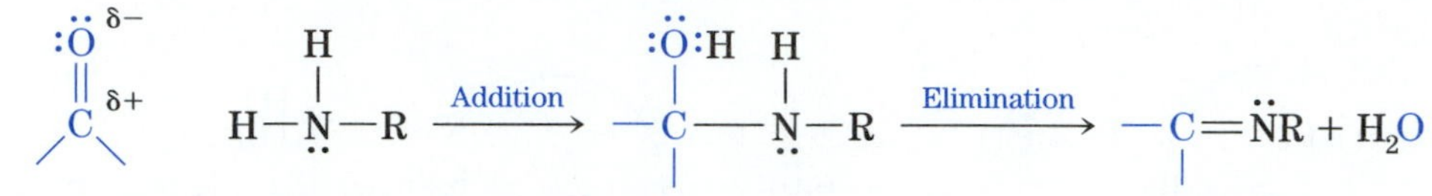

Aldehyde or ketone + 1° amine —Addition→ —Elimination→ An imine $+ H_2O$

Aldehydes and ketones react with a variety of amines to form crystalline derivatives that can be used to characterize the compound. Note that in each of the following reactions, the carbonyl reacts with an $-NH_2$ group. A carbon-nitrogen double bond forms in place of the original carbon-oxygen double bond.

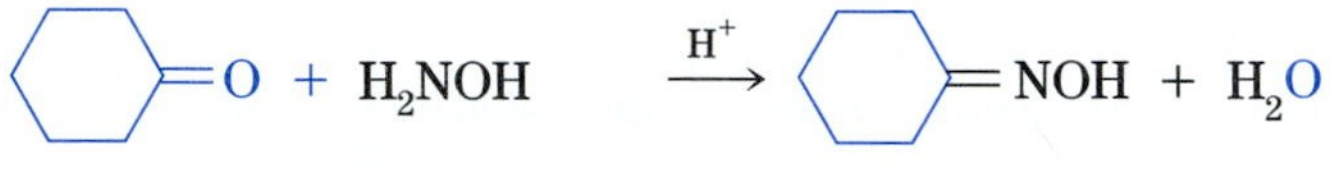

Cyclohexanone $+ H_2NOH$ (Hydroxylamine) $\xrightarrow{H^+}$ cyclohexanone oxime $=NOH + H_2O$ (An oxime)

$CH_3CH_2CHO + H_2NNH-C_6H_3(NO_2)_2$ (2,4-dinitrophenylhydrazine) $\longrightarrow CH_3CH_2CH{=}NNH-C_6H_3(NO_2)_2 + H_2O$ (A 2,4-dinitrophenylhydrazone)

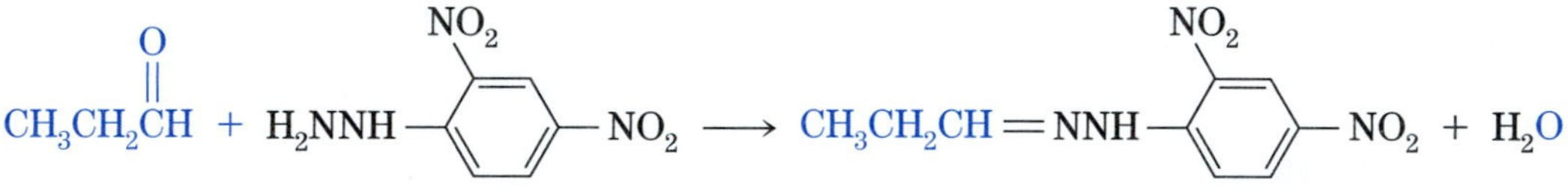

Imine formation is important biochemically since many enzymes use an $-NH_2$ group of an amino acid to react with and bind a carbonyl substrate to the enzyme. In the rods of the eye, for example, 11-*cis*-retinal combines with a large protein molecule, opsin, through an imine function to form rhodopsin, which is operative in converting light impulses into nerve impulses (see Connections 3.3, Chapter 3).

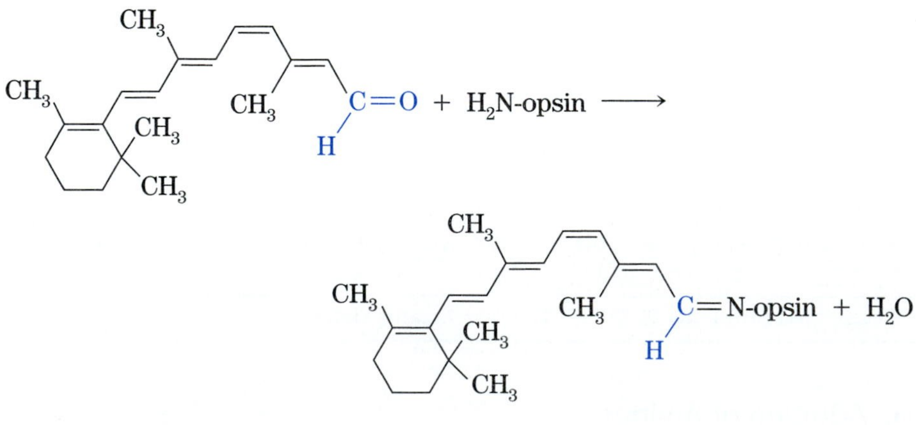

11-*cis*-retinal $C{=}O + H_2N\text{-opsin} \longrightarrow C{=}N\text{-opsin} + H_2O$

GETTING INVOLVED

✓ Describe the reaction of a primary amine with an aldehyde or ketone. What is the nucleophile and why? What adds to the carbonyl? What happens after the addition?

Problem 11.19

Write equations showing the reactions between the following substances: **(a)** benzaldehyde and 2,4-dinitrophenylhydrazine; **(b)** 2-pentanone and hydroxylamine.

See related problem 11.33(g–h).

Table 11.1 summarizes the addition reactions of aldehydes and ketones.

11.6 Reactions Involving α-Hydrogens

A. Acidity of α-Hydrogens

α-hydrogen
hydrogen on a carbon connected to a carbonyl group

Hydrogens on a carbon directly attached to a carbonyl group are referred to as **α-hydrogens** (*alpha*-hydrogens).

$$\underset{\epsilon}{C}-\underset{\delta}{C}-\underset{\gamma}{C}-\underset{\beta}{C}-\underset{\alpha}{C}-\overset{O}{\overset{\|}{C}}-$$

Although most carbon-hydrogen bonds are nonpolar and quite unreactive, the carbon-hydrogen bonds adjacent to a carbonyl group are polar and important reaction sites. The pK_a of a typical α-hydrogen of the aldehyde or ketone is around 20. This makes it much more acidic than alkane carbon-hydrogen bonds (pK_a around 50) but much less acidic than alcohols (pK_a around 17) or carboxylic acids (pK_a around 5). The carbon-oxygen double bond of the aldehyde or ketone is a strong electron-withdrawing group and polarizes the adjacent carbon-hydrogen bonds, making α-hydrogens weakly acidic. Strong bases such as sodium hydroxide can abstract α-hydrogens, forming an equilibrium with the corresponding carbanion.

$$-\overset{\delta-}{\underset{\underset{\delta+H}{|}}{\overset{|}{C}}}-\underset{\delta+}{\overset{\overset{O^{\delta-}}{\|}}{C}}- + \bar{O}H \rightleftharpoons -\underset{\ddot{}^{-}}{\overset{|}{C}}-\overset{\overset{O}{\|}}{C}- + H_2O$$

Once formed, the carbanion is resonance-stabilized (Figure 11.2; see section 5.5 for discussion of resonance in reactive intermediates). The negative charge is not concentrated on one atom but delocalized between the α-carbon and the carbonyl oxygen. This charge dispersal stabilizes the carbanion. The greater the number of atoms involved in delocalizing the negative charge, the greater is the stability of the carbanion.

Of particular importance is the resonance form in which the negative charge resides on the oxygen atom. The electronegative oxygen is more able to accommodate a negative charge than is carbon; thus the resonance hybrid is more like this resonance form than the one with the negative charge on the carbon. As a result this anion is more like an alkoxide ion than a carbanion, even though it was formed by abstraction of a hydrogen from an α-carbon.

α-Hydrogens are acidic, then, because the carbon-hydrogen bond is polarized by the adjacent carbonyl function and the carbanion resulting from hydrogen abstraction is resonance-stabilized. The carbanion is usually referred to as an

Base

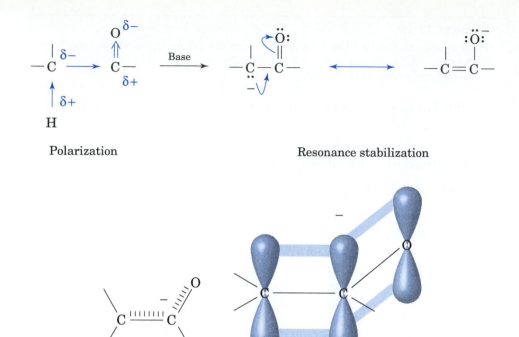

Polarization

Resonance stabilization

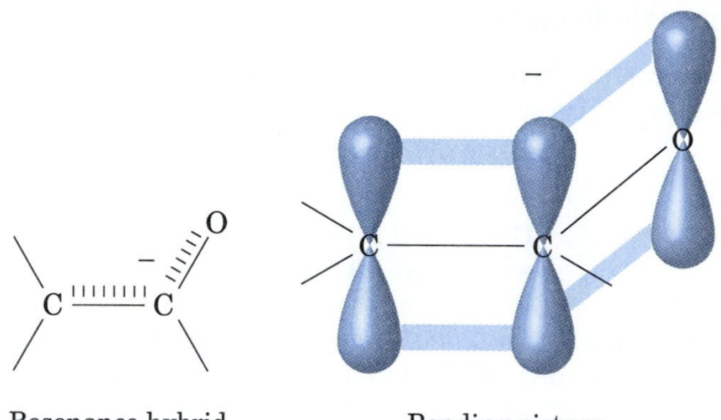

Figure 11.2 α-Hydrogens are acidic due to polarization of the carbon-hydrogen bond by the carbonyl group and resonance stabilization of the carbanion (enolate ion).

enolate resonance-stabilized carbanion resulting from abstraction of an α-hydrogen

enol compound with OH bonded to a carbon-carbon double bond

enolate ion because it is the anion of the **enol** formed when the carbanion is neutralized by acid.

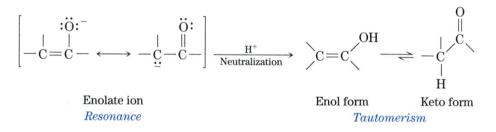

H^+

Neutralization

Enolate ion

Resonance

Enol form

Keto form

Tautomerism

tautomers two easily interconvertible structural isomers

tautomerism an equilibrium between two structural isomers

Aldehydes and ketones with α-hydrogens are in equilibrium with the corresponding enol form in which there is an O—H group on a carbon-carbon double bond. The interconversion between the aldehyde or ketone and the enol form is called *keto-enol tautomerism.* **Tautomers** are isomers that are interconvertible by the simple movement of electrons and atoms, and **tautomerism** is essentially just a special type of isomerization. Tautomerism differs from resonance in that the former involves the movement of both electrons and atoms (a hydrogen in this case), whereas the latter is represented by varying the positions of electrons only. Tautomers are actual species in equilibrium, whereas resonance structures are theoretical constructs used to describe the resonance hybrid—they do not in fact exist.

Although they exist in equilibrium with the free aldehyde or ketone form, enols are relatively unstable and the equilibrium usually favors the keto form. For example, the enol form of cyclohexanone comprises less than 0.02% of the equilibrium mixture. In some cases, the enol form is exceptionally stable and is the predominant if not the exclusive member of the equilibrium mixture. The enol form of 2,4-cyclohexadienone is an aromatic compound (phenol), and the stability of the benzene ring causes the enol to be favored, exclusively.

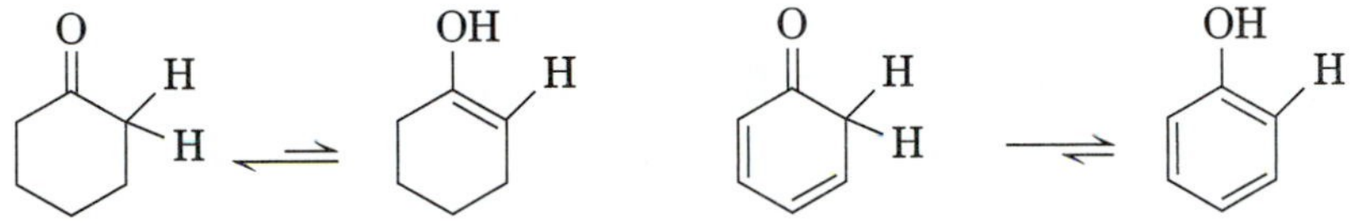

99.98% Keto form cyclohexanone ⇌ 0.02% Enol form

Keto form 2,4-cyclohexadienone ⇌ Preferred enol form

Evidence of the activity of α-hydrogens can be found by mixing cyclohexanone in a weakly basic solution of D_2O (water in which the hydrogens have been replaced by the isotope deuterium). The α-hydrogens are slowly abstracted by the base, and the resulting carbanions neutralized by D_2O. Eventually all of the α-hydrogens are replaced by deuterium; none of the other hydrogens is affected.

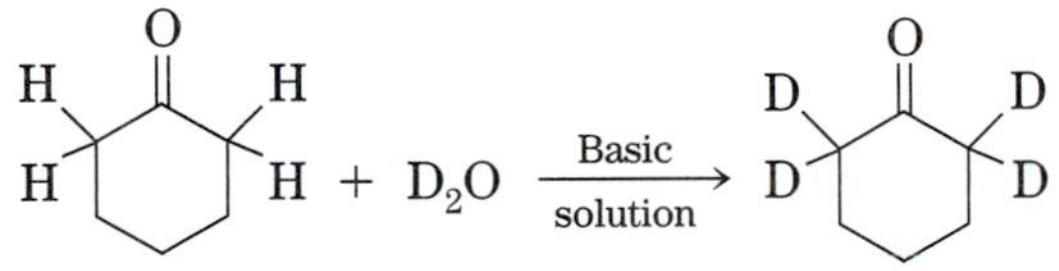

$$\text{cyclohexanone} + D_2O \xrightarrow[\text{solution}]{\text{Basic}} \text{2,2,6,6-tetradeuteriocyclohexanone}$$

GETTING INVOLVED

- ✓ What is an alpha hydrogen? What are the two reasons it is acidic? How does its acidity compare to alkane, alcohol, and carboxylic acid hydrogens?
- ✓ Why is the carbanion produced from abstraction of an alpha hydrogen often called an enolate ion? Can you draw the resonance forms of such a carbanion?
- ✓ What is tautomerism and how does it differ from resonance? What is keto-enol tautomerism? Which usually prevails, the keto or enol form?
- ✓ Why would an aldehyde or ketone with alpha hydrogens exchange them with deuterium in a D_2O/NaOD solution? Why would none of the other hydrogens exchange?

Problem 11.20

For propanal, $CH_3CH_2\overset{\overset{\large O}{\|}}{C}H$ **(a)** write the enol form; **(b)** write the anion formed upon treatment with base (show both resonance forms); **(c)** write the product of reaction with D_2O/NaOD.

See related problems 11.33(k), 11.39, and 11.47.

B. The Aldol Condensation

The **aldol condensation** is an important example of a reaction that depends on the acidity of α-hydrogens. In the mechanism, one aldehyde or ketone molecule adds to the carbon-oxygen double bond of another by base-initiated nucleophilic addition.

aldol condensation base-catalyzed reaction between two aldehyde or ketone molecules to form a product with both alcohol and carbonyl groups

1. ***General Reaction.*** Under conditions of basic catalysis, an aldehyde or ketone α-carbon, made negative by abstraction of a hydrogen, bonds to the partially positive carbonyl carbon of a second molecule. Since the product formed

when an aldehyde is subjected to these conditions has both an aldehyde and an alcohol function, the reaction is called the *aldol condensation.*

$$2RCH_2\overset{\overset{\displaystyle O}{\|}}{C}H \xrightarrow{OH^-} RCH_2\overset{\overset{\displaystyle OH}{|}}{C}H\underset{\underset{\displaystyle R}{|}}{C}H\overset{\overset{\displaystyle O}{\|}}{C}H$$

An aldol

$$\left(RCH_2\underset{\underset{\displaystyle H}{|}}{\overset{\overset{\displaystyle \delta-\,:\ddot{O}}{\|}}{C^{\delta+}}} \quad {}^{\delta-}\underset{\underset{\displaystyle R}{|}}{\overset{\overset{\displaystyle \delta+\,H}{|}}{C}}\overset{\overset{\displaystyle \ddot{O}:}{\|}}{C}H \xrightarrow[\text{Addition}]{:\ddot{O}H^-} RCH_2\underset{\underset{\displaystyle H}{|}}{\overset{\overset{\displaystyle :\ddot{O}H}{|}}{C}}-\underset{\underset{\displaystyle R}{|}}{C}H\overset{\overset{\displaystyle \ddot{O}:}{\|}}{C}H \right)$$

Aldols are easily dehydrated, because the resulting double bond is conjugated with the carbonyl group, which creates an extended system of overlapping p orbitals that is a resonance-stabilized structure.

$$RCH_2\overset{\overset{\displaystyle \boxed{OH}}{|}}{C}H-\underset{\underset{\displaystyle R}{|}}{\overset{\overset{\displaystyle \boxed{H}}{|}}{C}}-\overset{\overset{\displaystyle O}{\|}}{C}H \xrightarrow{\text{Dehydration}} RCH_2CH=\underset{\underset{\displaystyle R}{|}}{C}\overset{\overset{\displaystyle O}{\|}}{C}H$$

In some cases, the aldol dehydrates spontaneously on formation or during acid neutralization of the reaction mixture, and its isolation becomes impossible. In summary, the aldol condensation involves addition of an α-carbon of one aldehyde or ketone to the carbonyl group of a second molecule. The resulting aldol can sometimes be isolated, but it often dehydrates. The overall process resembles the reaction of aldehydes and ketones with amines (section 11.5.G and Table 11.1).

GETTING INVOLVED

✓ In terms of what happens (not mechanism) describe the aldol condensation. What adds to the carbon oxygen double bond? What results when an aldol dehydrates?

Example 11.7

Write reaction equations showing the aldol condensation of **(a)** ethanal and **(b)** propanal.

Solution

$$\textbf{(a)}\ CH_3\underset{\underset{\displaystyle H}{|}}{\overset{\overset{\displaystyle \delta-\,:\ddot{O}}{\|}}{C^{\delta+}}} \quad {}^{\delta-}\overset{\overset{\displaystyle \delta+\,H}{|}}{C}H_2\overset{\overset{\displaystyle \ddot{O}:}{\|}}{C}H \xrightarrow{OH^-} CH_3\overset{\overset{\displaystyle OH}{|}}{C}HCH_2\overset{\overset{\displaystyle O}{\|}}{C}H \xrightarrow{H^+} CH_3CH=CH\overset{\overset{\displaystyle O}{\|}}{C}H$$

$$\textbf{(b)}\ CH_3CH_2\underset{\underset{\displaystyle H}{|}}{\overset{\overset{\displaystyle \delta-\,:\ddot{O}}{\|}}{C^{\delta+}}} \quad {}^{\delta-}\underset{\underset{\displaystyle CH_3}{|}}{\overset{\overset{\displaystyle \delta+\,H}{|}}{C}}H\overset{\overset{\displaystyle O}{\|}}{C}H \xrightarrow{OH^-} CH_3CH_2\overset{\overset{\displaystyle OH}{|}}{C}H\underset{\underset{\displaystyle CH_3}{|}}{C}H\overset{\overset{\displaystyle O}{\|}}{C}H \xrightarrow{H^+} CH_3CH_2CH=\underset{\underset{\displaystyle CH_3}{|}}{C}\overset{\overset{\displaystyle O}{\|}}{C}H$$

Problem 11.21

Write the product of the aldol condensation of butanal.
See related problems 11.36 and 11.38(a–b).

2. ***Mechanism of the Aldol Condensation.*** The aldol condensation depends on the acidity of α-hydrogens in aldehydes and ketones. Let us consider the base-catalyzed condensation of acetaldehyde. To initiate the reaction, a hydroxide base abstracts an α-hydrogen, generating the resonance-stabilized enolate anion.

Step 1:

$$HC(=O)CH_2{-}H + {:}\ddot{O}H^- \longrightarrow \left[HC(=\ddot{O}{:})\ddot{C}H_2^- \longleftrightarrow H(^-{:}\ddot{O}{:})C{=}CH_2 \right] + H_2O$$

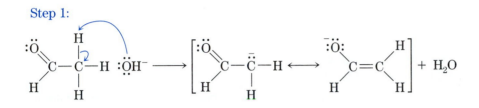

Once the carbanion is formed, it attacks the positive carbonyl carbon of another acetaldehyde molecule by a nucleophilic addition mechanism in an effort to neutralize itself. As the carbanion bonds, the π electrons of the carbonyl group are transferred completely to the oxygen, forming an alkoxide ion.

Step 2:

$$CH_3\overset{\delta+}{C}H{=}\overset{\delta-}{O} + {:}\bar{C}H_2CH{=}O \longrightarrow CH_3CH(O^-)CH_2CH{=}O$$

The alkoxide ion is neutralized by a water molecule, and the catalyst is regenerated in the process.

Step 3:

$$CH_3CH(\ddot{O}{:}^-)CH_2CH{=}O + H{-}\ddot{O}{-}H \longrightarrow CH_3CH(\ddot{O}H)CH_2CH{=}O + {:}\ddot{O}H^-$$

GETTING INVOLVED

- ✓ Describe in words the three steps of the mechanism of the aldol condensation.
- ✓ Why can the alpha hydrogen be abstracted in step 1? How is the resulting carbanion stabilized?
- ✓ Is step 2 acid- or base-initiated? What is the nucleophile? What is the result of step 2 and how is the species neutralized in step 3?

Problem 11.22

Write the mechanism illustrating the aldol condensation of butanal.
See related problem 11.46(e).

crossed aldol condensation aldol condensation between two different aldehydes or ketones

3. ***Crossed Aldol Condensations.*** Aldol condensations between two different carbonyl compounds can be performed successfully as long as one of the reactants has no α-hydrogens. For example, by mixing benzaldehyde (which has no α-hydrogens) and a base and slowly adding acetaldehyde a drop at a time (to prevent its condensing with itself), one can synthesize cinnamaldehyde, the primary component of cinnamon oil.

$$C_6H_5CHO + CH_3CHO \xrightarrow{OH^-} C_6H_5CH{=}CHCHO + H_2O$$

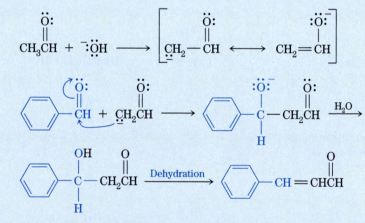

GETTING INVOLVED

✓ What is a crossed aldol condensation? Why is it best that one of the reactants have no alpha hydrogens? Why is this compound mixed with the base and the other added to the solution slowly?

Example 11.8

Write a mechanism for the aldol condensation between benzaldehyde and ethanal.

Solution

$$CH_3CHO + {}^-OH \longrightarrow \left[{}^-CH_2{-}CHO \longleftrightarrow CH_2{=}CH{-}O^- \right]$$

$$C_6H_5CHO + {}^-CH_2CHO \longrightarrow C_6H_5CH(O^-){-}CH_2CHO \xrightarrow{H_2O}$$

$$C_6H_5CH(OH){-}CH_2CHO \xrightarrow{\text{Dehydration}} C_6H_5CH{=}CHCHO$$

Dehydration usually occurs at this point.

Problem 11.23

Write the product of the crossed aldol condensation between p-chlorobenzaldehyde and butanal.

Problem 11.24

Write a mechanism for the reaction in problem 11.23.

See related problems 11.37(c) and 11.38(c).

4. ***Aldol Additions in Nature.*** One step in the synthesis of glucose by higher organisms, involves a crossed aldol addition (without subsequent dehydration)

between the monophosphates of glyceraldehyde and dihydroxyacetone to produce fructose 1,6-diphosphate. The reaction is catalyzed by the enzyme aldolase.

$$\begin{array}{c} CH_2OPO_3H^- \\ | \\ C{=}O \\ | \\ HO{-}C{-}H \\ | \\ H \end{array} + \begin{array}{c} H{-}C{=}O \\ | \\ H{-}C{-}OH \\ | \\ CH_2OPO_3H^- \end{array} \xrightarrow{\text{Aldolase}} \begin{array}{c} CH_2OPO_3H^- \\ | \\ C{=}O \\ | \\ HO{-}C{-}H \\ | \\ H{-}C{-}OH \\ | \\ H{-}C{-}OH \\ | \\ CH_2OPO_3H^- \end{array}$$

α-hydrogen

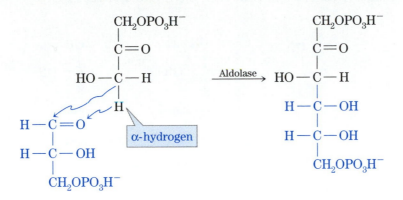

REACTION SUMMARY

1. Preparations of Aldehydes and Ketones

See summary in section 11.3; Problems 11.31–11.32.

2. Addition of HCN to Aldehydes and Ketones

Section 11.5.C; Problems 11.8, 11.33(b), 11.46(b).

$$>C{=}O + HCN\ (NaCN, H^+) \longrightarrow -\overset{OH}{\underset{|}{\overset{|}{C}}}-CN$$

3. Reduction of Aldehydes and Ketones

Section 11.5.D–E; Example 11.5; Problems 11.9–11.10, 11.33(c–d), 11.43, 11.46(c).

Catalytic hydrogenation

$$>C{=}O + H_2 \xrightarrow{\text{Ni}} -\overset{OH}{\underset{|}{\overset{|}{C}}}-H$$

Lithium aluminum hydride

$$>C{=}O \xrightarrow[\text{or } NaBH_4]{LiAlH_4} \xrightarrow{H_2O/H^+} -\overset{OH}{\underset{|}{\overset{|}{C}}}-H$$

4. Grignard Reagent with Aldehydes and Ketones

Section 11.5.F.1–3; Example 11.6; Problems 11.11–11.14, 11.33(e–f), 11.34, 11.43, 11.46(a).

$$>C{=}O \xrightarrow{RMgX} \xrightarrow{H_2O/H^+} -\overset{OH}{\underset{|}{\overset{|}{C}}}-R$$

5. Hemiacetal and Acetal Formation

Section 11.5.G; Problems 11.16–11.18, 11.33(i–j), 11.41–11.42, 11.46(f).

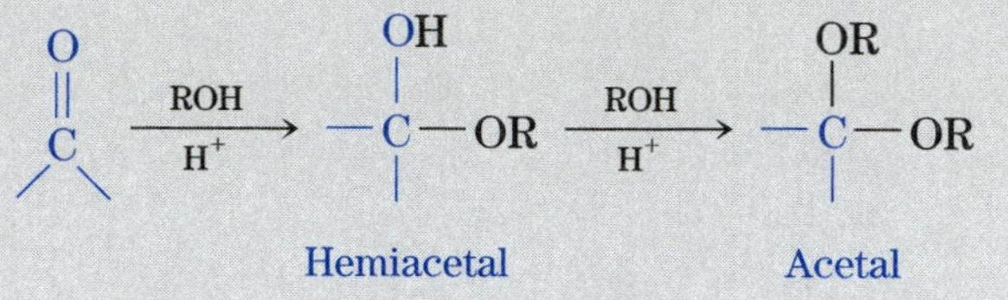

$$\begin{matrix} O \\ \| \\ C \\ / \ \backslash \end{matrix} \xrightarrow[H^+]{ROH} \underset{\text{Hemiacetal}}{-\overset{OH}{\overset{|}{\underset{|}{C}}}-OR} \xrightarrow[H^+]{ROH} \underset{\text{Acetal}}{-\overset{OR}{\overset{|}{\underset{|}{C}}}-OR}$$

6. Reaction of 1° Amines with Aldehydes and Ketones

Section 11.5.H; Problems 11.19, 11.33(g–h), 11.46(d).

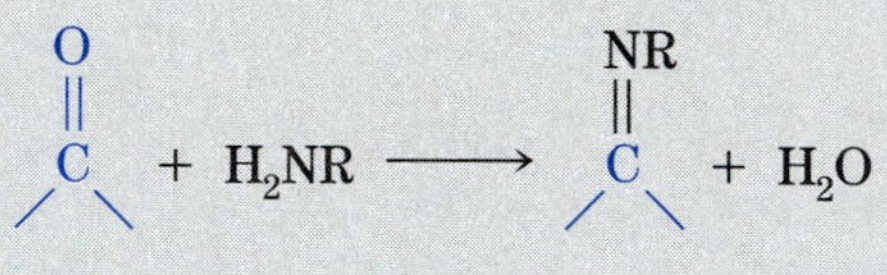

$$\begin{matrix} O \\ \| \\ C \\ / \ \backslash \end{matrix} + H_2NR \longrightarrow \begin{matrix} NR \\ \| \\ C \\ / \ \backslash \end{matrix} + H_2O$$

7. Aldol Condensation

Section 11.6.B; Examples 11.7–11.8; Problems 11.21–11.24, 11.36–11.38, 11.46(e), 11.48–11.49.

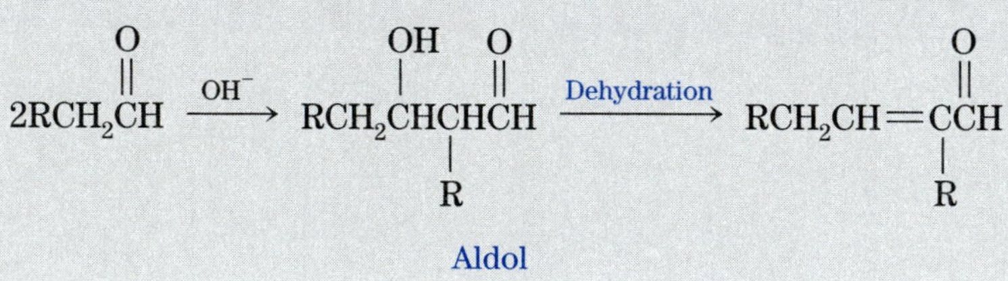

$$2RCH_2\overset{O}{\overset{\|}{C}}H \xrightarrow{OH^-} \underset{\text{Aldol}}{RCH_2\overset{OH}{\overset{|}{C}}H\underset{R}{\underset{|}{C}}H\overset{O}{\overset{\|}{C}}H} \xrightarrow{\text{Dehydration}} RCH_2CH{=}\underset{R}{\underset{|}{C}}\overset{O}{\overset{\|}{C}}H$$

Problems

11.25 **IUPAC Nomenclature of Aldehydes:** Name the following by the IUPAC system of nomenclature:

(a) 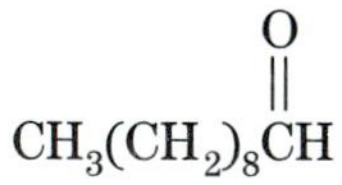 $CH_3(CH_2)_8\overset{O}{\overset{\|}{C}}H$

(b) $(CH_3)_2CHCH_2CH_2\overset{O}{\overset{\|}{C}}H$

(c) $CH_3CH_2\overset{CH_3CH_2}{\overset{|}{C}}HCH_2\overset{CH_3}{\overset{|}{C}}HCH_2\overset{O}{\overset{\|}{C}}H$

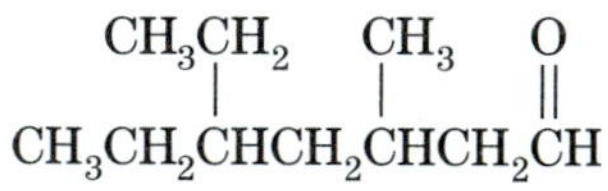

(d) 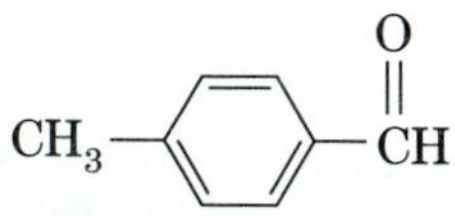 $CH_3-C_6H_4-\overset{O}{\overset{\|}{C}}H$ (para-substituted benzene ring)

(e) 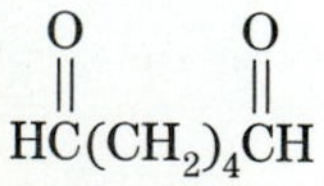 $H\overset{O}{\overset{\|}{C}}(CH_2)_4\overset{O}{\overset{\|}{C}}H$

11.26 **IUPAC Nomenclature of Ketones:** Name the following by the IUPAC system of nomenclature:

(a) 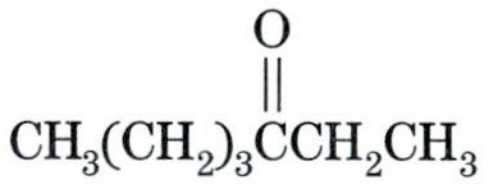 $CH_3(CH_2)_3\overset{O}{\overset{\|}{C}}CH_2CH_3$

(b) $CH_3\overset{CH_3}{\overset{|}{C}}HCH_2\overset{O}{\overset{\|}{C}}CH_2CH_2CH_3$

(c) CH_3-(cyclohexane ring)$=O$ (4-position)

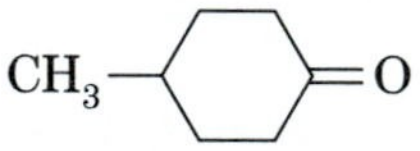

(d) 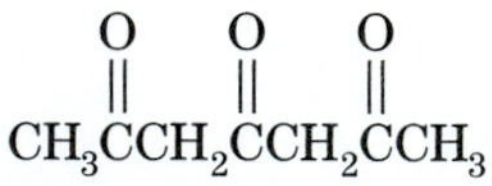 $CH_3\overset{O}{\overset{\|}{C}}CH_2\overset{O}{\overset{\|}{C}}CH_2\overset{O}{\overset{\|}{C}}CH_3$

(e) 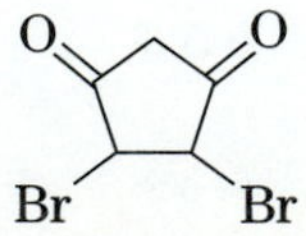 cyclopentane ring with two adjacent C=O groups' neighbors: O, O at positions 1 and 3; Br, Br at positions 4 and 5

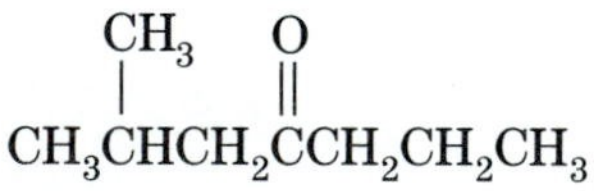

11.27 **IUPAC Nomenclature of Aldehydes and Ketones:** Name the following by the IUPAC system of nomenclature:

(a) $CH_3CH_2\overset{\overset{O}{\|}}{C}H$

(b) $CH_3CH_2\overset{\overset{O}{\|}}{C}CH_2CH_3$

(c) $CH_3\overset{\overset{OH}{|}}{C}HCH_2\overset{\overset{OH}{|}}{C}HCH_2\overset{\overset{O}{\|}}{C}H$

(d) $CH_3\overset{\overset{NH_2}{|}}{C}HCH_2\overset{\overset{O}{\|}}{C}CH_3$

(e) $CH_3\overset{\overset{OH}{|}}{C}HCH_2\overset{\overset{O}{\|}}{C}CH_2\overset{\overset{O}{\|}}{C}H$

(f) $CH_3\underset{\underset{O}{\|}}{C}CH_2\underset{\underset{OH}{|}}{C}HCH_2\underset{\underset{O}{\|}}{C}CH_2CH_3$

(g) $CH_3\underset{\underset{CH_3}{|}}{C}HCH_2\underset{\underset{NH_2}{|}}{C}HCH_2\underset{\underset{O}{\|}}{C}H$

(h) $HO-$(cyclohexane ring)$=O$

11.28 **IUPAC Nomenclature of Aldehydes and Ketones:** Name the following compounds by the IUPAC system of nomenclature.

(a) $CH_2{=}CHCH_2\overset{\overset{O}{\|}}{C}H$

(b) $HC{\equiv}C\overset{\overset{O}{\|}}{C}CH_2OH$

(c) $CH_3CH{=}CH{-}CH{=}CH\overset{\overset{O}{\|}}{C}H$

(d) $CH_2{=}CH\underset{\underset{NH_2}{|}}{C}HCH{=}CH\underset{\underset{O}{\|}}{C}CH_2\underset{\underset{O}{\|}}{C}H$

(e) $CH_3\underset{\underset{O}{\|}}{C}CH{=}CH\underset{\underset{O}{\|}}{C}CH_2CH_3$

(f) $HC{\equiv}C\underset{\underset{O}{\|}}{C}CH{=}CH\underset{\underset{O}{\|}}{C}H$

11.29 **IUPAC Nomenclature:** Draw structures for the following compounds:

(a) 3-heptanone
(b) octanal
(c) 5-oxohexanal
(d) 3,7-dihydroxy-5-oxoheptanal
(e) 3-cyclopentenone
(f) 1,1,1,5,5,5-hexabromo-2,4-pentandione
(g) 4-oxo-7-bromo-7-ethyl-9-hydroxy-2,5-nonadiynal
(h) *m*-methylbenzaldehyde
(i) 1-phenyl-2-butanone

11.30 **Common Nomenclature:** Draw structures for the following compounds:

(a) butyl ethyl ketone
(b) acetone
(c) formaldehyde
(d) chloroacetaldehyde
(e) dipropyl ketone
(f) diphenyl ketone

11.31 **Preparations of Aldehydes and Ketones:** Complete the following reactions, which illustrate some preparations of aldehydes and ketones:

(a) $CH_3C{\equiv}CCH_3 + H_2O \xrightarrow[HgSO_4]{H_2SO_4,}$

(b) $CH_3\overset{\overset{CH_3}{|}}{C}{=}CHCH_3 \xrightarrow{O_3} \xrightarrow[H_2O]{Zn,}$

(c) (benzene) $+ CH_3CH_2CH_2\overset{\overset{O}{\|}}{C}Cl \xrightarrow{AlCl_3}$

(d) $CH_3CH_2\overset{\overset{OH}{|}}{C}HCH_2CH_3 \xrightarrow{Na_2Cr_2O_7}$

11.32 **Preparations of Aldehydes and Ketones:** Write reaction equations illustrating the preparations of the following compounds:

(a) phenyl butyl ketone by a Friedel-Crafts reaction
(b) 2-butanone by ozonolysis of an alkene
(c) 2-pentanone from an alkyne
(d) 3-methylcyclopentanone by oxidation of an alcohol
(e) decanal by oxidation of an alcohol

11.33 **Reactions of Aldehydes and Ketones:** Write equations illustrating the reaction (if any) of the following two compounds with each of the reagents listed:

(I) Benzaldehyde, $C_6H_5\overset{O}{\overset{\|}{C}}H$

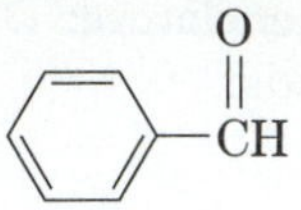

(II) Acetophenone, $C_6H_5\overset{O}{\overset{\|}{C}}CH_3$

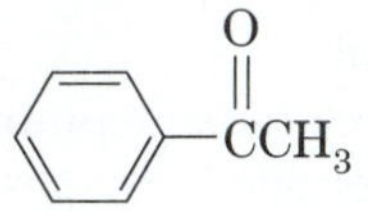

(a) Tollens' reagent
(b) HCN in basic solution
(c) 1 mole H_2/Ni
(d) $NaBH_4$, then H_2O, H^+
(e) CH_3MgCl, then H_2O, H^+
(f) 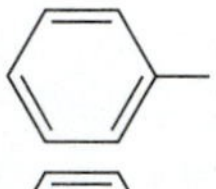 C_6H_5–MgBr, then H_2O, H^+
(g) 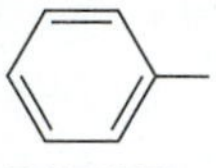 C_6H_5–$NHNH_2$
(h) H_2NOH
(i) 1 mole CH_3OH/H^+
(j) 2 moles CH_3OH/H^+
(k) D_2O, NaOD

11.34 **Grignard Synthesis of Alcohols:** Prepare the following alcohols in all the possible ways using the Grignard synthesis:

(a) $CH_3CH_2CH_2CH_2OH$ **(b)** $CH_3\underset{OH}{\underset{|}{C}}HCH_2CH_3$

(c) C_6H_5–$\underset{OH}{\underset{|}{\overset{CH_2CH_3}{\overset{|}{C}}}}CH_2CH_2CH_3$

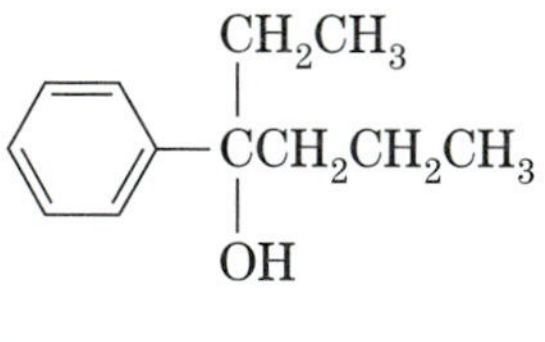

11.35 **Grignard Reaction:** Complete the following reactions showing the major organic products:

(a) $CH_3CH_2\overset{O}{\overset{\|}{C}}H + C_6H_5\text{–}MgCl \longrightarrow \xrightarrow[H^+]{H_2O,}$

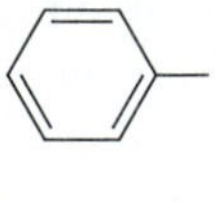

(b) 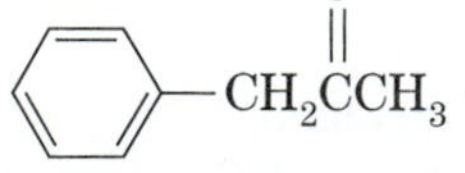 $C_6H_5\text{–}CH_2\overset{O}{\overset{\|}{C}}CH_3 + CH_3MgBr \longrightarrow \xrightarrow[H^+]{H_2O,}$

(c) $CH_3\overset{O}{\overset{\|}{C}}CH_2CH_3 + C_6H_5\text{–}MgBr \longrightarrow \xrightarrow[H^+]{H_2O,}$

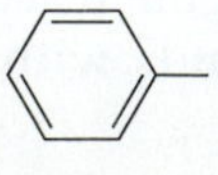

11.36 **Aldol Condensation:** Write reaction equations illustrating the aldol condensation of the following compounds using sodium hydroxide as the base. Show the aldol initially formed and the unsaturated aldehyde or ketone produced by dehydration.

(a) $CH_3CH_2CH_2CH_2\overset{O}{\overset{\|}{C}}H$

(b) $CH_3\overset{CH_3}{\overset{|}{C}}HCH_2\overset{O}{\overset{\|}{C}}H$ **(c)** 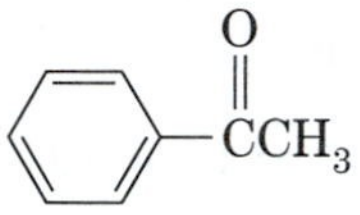 $C_6H_5\text{–}\overset{O}{\overset{\|}{C}}CH_3$

11.37 **Crossed Aldol Condensation:** Write equations showing the aldol condensation of benzaldehyde and acetophenone (methyl phenyl ketone). Show both the initial aldol and the dehydration product.

11.38 **Aldol Condensation:** Show how the following compounds could be prepared by the aldol condensation:

(a) $CH_3(CH_2)_4CH{=}\underset{CH_3(CH_2)_3}{\underset{|}{\overset{O}{\overset{\|}{C}}}}H$

(b) 2-cyclopentylidenecyclopentanone (cyclopentanone ring with C=O, double-bonded at the adjacent carbon to a cyclopentane ring)

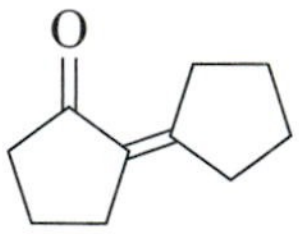

(c) 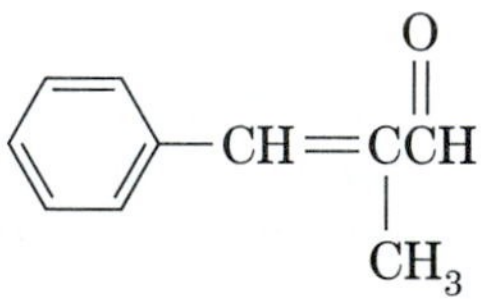 $C_6H_5\text{–}CH{=}\underset{CH_3}{\underset{|}{\overset{O}{\overset{\|}{C}}}}H$

11.39 **Enolate Ions:** Write the carbanion formed during the aldol condensations in problem 11.36. Draw the resonance structures.

11.40 **Keto-Enol Tautomerism:** Draw the enols that would result initially from acidification of the ions in problem 11.39.

11.41 **Acetal Formation:** Write the acetal or ketal that would result from reaction of the following compounds:

(a) propanal and 2 moles ethanol
(b) propanone and 2 moles methanol
(c) cyclohexanone and 1 mole 1,2-ethanediol

11.42 **Acetal Formation:** When heated with methanol under acidic conditions, 4-hydroxypentanal gives a cyclic acetal in which only one mole of methanol is consumed. Draw the starting material and the acetal.

11.43 **Preparation of Alcohols:** Show in as many ways as possible how 2-pentanol could be prepared from carbonyl compounds by the Grignard synthesis and by reduction with H_2/Ni and $NaBH_4$.

11.44 **Keto-Enol Tautomerism:** Draw the keto and enol forms of the following molecules:

(a) 3-pentanone
(b) cyclopentanone
(c) ethanal

11.45 **Tautomerism:** The enamine with the formula $CH_2{=}CH{-}NHCH_3$ is part of a tautomeric mixture in which the other tautomer is the more stable component. Draw the other tautomer.

11.46 **Reaction Mechanisms:** Write (1) products and (2) reaction mechanisms for the reaction of propanal with the following reagents:

(a) CH_3MgCl, then H_2O, H^+
(b) NaCN, then H^+
(c) $NaBH_4$, then H_2O
(d) H_2NOH/H^+
(e) NaOH (aldol condensation)
(f) $2CH_3OH/H^+$

11.47 **Acidity of α-Hydrogens:** There are three distinct types of hydrogens in the following molecule. Arrange them in order of increasing acidity. Explain your order.

$$CH_3CH_2\overset{\overset{\displaystyle O}{\|}}{C}CH_2\overset{\overset{\displaystyle O}{\|}}{C}CH_2CH_3$$

11.48 **Aldol-Type Condensations:** Aldehydes and ketones can engage in aldol-type condensations with other molecules that have acidic hydrogens. Show the products of the base-catalyzed aldol condensation of benzaldehyde with the following compounds:

(a) CH_3NO_2 **(b)** CH_3CN

(c) $CH_3O\overset{\overset{\displaystyle O}{\|}}{C}CH_2\overset{\overset{\displaystyle O}{\|}}{C}OCH_3$ ($NaOCH_3$ base)

11.49 **Reaction Mechanisms of Aldol-Type Condensations:** Write reaction mechanisms for the following aldol reactions in base (NaOH); do not do a mechanism for the dehydration.

(a) propanone
(b) benzaldehyde and propanal

11.50 **Organic Qualitative Analysis:** Describe how you could chemically distinguish between the following compounds. Tell what you would do and see.

(a) propanal and propanone
(b) 2-propanol and propanone
(c) 1-butanol, butanal, and butanone

11.51 **Carbohydrate Chemistry:** Below is the structure of lactose (5% of human milk and cow's milk). Identify any acetal or hemiacetal linkages.

OH, H, CH_2OH, HO, H, HO, H, OH, O, H, H, H, H, HO, O, H, OH, O, CH_2OH, H, H

Lactose

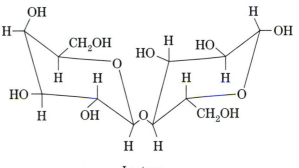

Activities with Molecular Models

1. Make models of the aldehyde and ketone with the formula C_3H_6O.

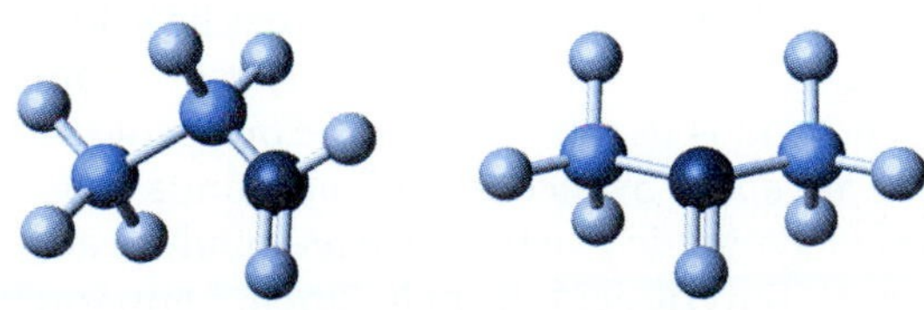

2. Make models of the three isomers of C_4H_8O. Identify aldehydes and ketones. How many nonbonding electron pairs are on the oxygen of each model? What are the hybridizations of the carbons and the oxygen?

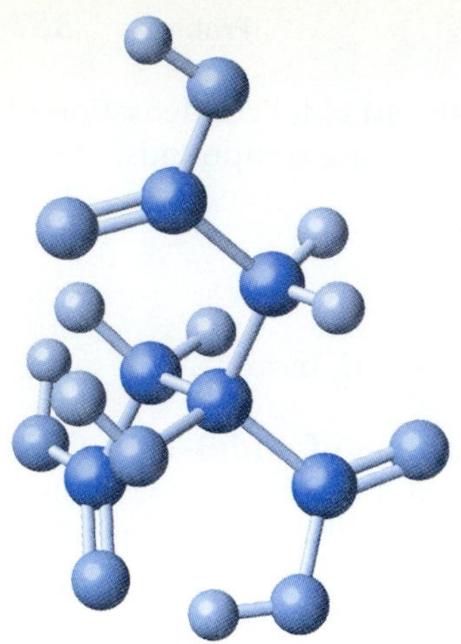

CHAPTER 12
Carboxylic Acids

12.1 Structure of Carboxylic Acids

carboxylic acid functional group in which OH is attached to carbon-oxygen double bond

Carboxylic acids are structurally characterized by the carboxyl group, which is commonly represented in three ways:

$$-\overset{\displaystyle O}{\overset{\|}{C}}OH \qquad -CO_2H \qquad -COOH$$

Because of the following structural features, the carboxylic acid group is very reactive.

1. There are three polar bonds—the carbon-oxygen double and single bonds and the oxygen-hydrogen bond.
2. The electrons in the π bond of the carbonyl group (C=O) are susceptible to attack.
3. The carbonyl oxygen is electron rich because of bond polarity and two unshared electron pairs.

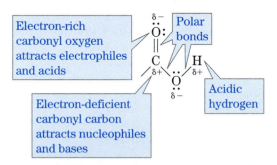

Like inorganic acids, carboxylic acids often have an unpleasant, acrid odor and sour taste. The simplest carboxylic acid, formic acid, is a dangerously caustic liquid with an irritating odor; it is a component of the sting of some ants. Acetic acid is responsible for the pungent taste and odor of vinegar (most vinegars are about 5% acetic acid) and finds extensive use in the industrial production of synthetic plastics such as cellulose acetate (acetate rayon) and polyvinyl acetate. Butyric acid (from *butyrum,* Latin for "butter") contributes to the strong odor of rancid butter and other fats. Lactic acid is formed when milk sours and as muscles tire. It is also a product of bacterial degradation of sucrose by microorganisms in the plaque on teeth.

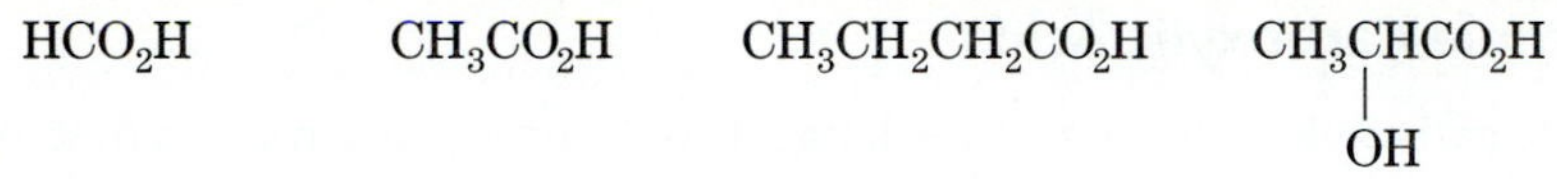

HCO_2H	CH_3CO_2H	$CH_3CH_2CH_2CO_2H$	$CH_3CH(OH)CO_2H$
Formic acid	Acetic acid	Butyric acid	Lactic acid

Caproic, caprylic, and capric acids (from *caper,* Latin for "goat") are present in the skin secretions of goats.

$CH_3(CH_2)_4CO_2H$	$CH_3(CH_2)_6CO_2H$	$CH_3(CH_2)_8CO_2H$
Caproic acid	Caprylic acid	Capric acid

The sour, biting taste of many citrus fruits is due to citric acid (6%–7% in lemon juice). Tartaric acid and its salts are found in grapes and tartar sauce.

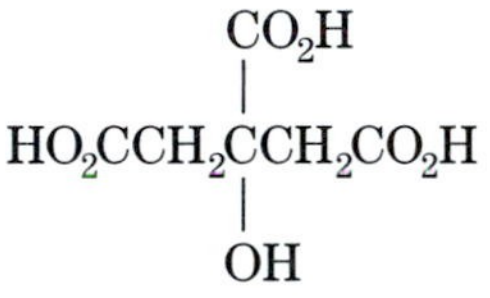

$HO_2CCH_2C(OH)(CO_2H)CH_2CO_2H$	$HO_2CCH(OH)—CH(OH)CO_2H$
Citric acid	Tartaric acid

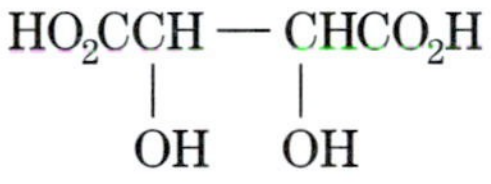

Oleic acid is a precursor in the biological synthesis of fats and oils and is the primary fatty acid component of lard, butter, and olive oil. Cholic acid is a component of intestinal bile in vertebrates that allows the emulsification of ingested fats and oils.

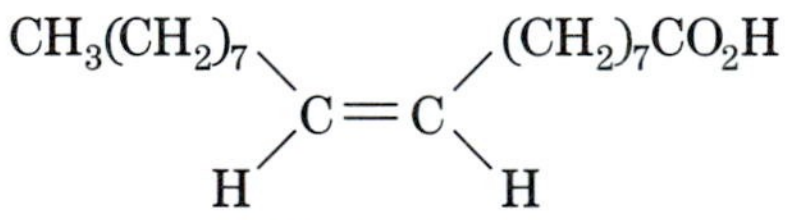

$CH_3(CH_2)_7CH=CH(CH_2)_7CO_2H$ (H, H cis)

Oleic acid

HO, CH_3, CH_3, $CHCH_2CH_2CO_2H$ (CH_3), HO, OH

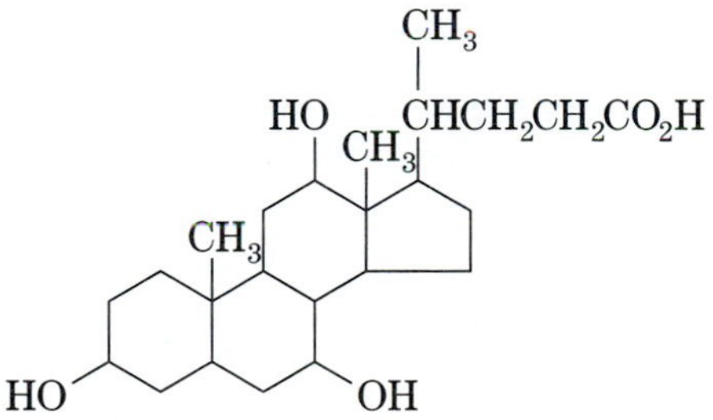

Cholic acid

GETTING INVOLVED

✓ Write the general structure of a carboxylic acid and identify the sites that make it reactive; include polar bonds, π-bonds, sites for electrophilic and nucleophilic attack, Lewis acid and base sites, and the acidic hydrogen.

12.2 Nomenclature of Carboxylic Acids

Carboxylic acids are the last of the major functional groups to be presented in this text. With this in mind, we will integrate and summarize the nomenclature previously covered as we consider the nomenclature of carboxylic acids.

A. Simple Carboxylic Acids

To name carboxylic acids, name the longest continuous carbon chain that includes the acid group and replace the final *-e* with the suffix *-oic* and the word *acid.*

CH_3CH_3	$CH_3C(=O)OH$	$CH_3(CH_2)_4CH_3$	$CH_3(CH_2)_4C(=O)OH$
Ethane	Ethanoic acid	Hexane	hexanoic acid

If the acid group is directly attached to a ring, the suffix *carboxylic acid* is used in naming. An aromatic acid in which the acid group is attached to a benzene ring is called *benzoic acid.* In numbering substituted carboxylic acids, start with the carboxyl carbon.

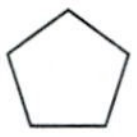

Cyclopentane 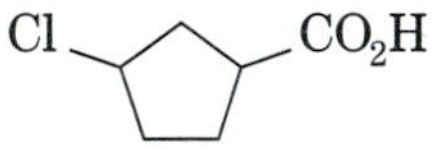   3-chlorocyclopentanecarboxylic acid 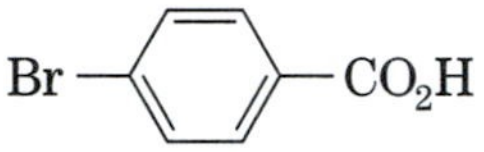   *p*-bromobenzoic acid

GETTING INVOLVED

Problem 12.1

Name the following compounds by the IUPAC system:

(a) $CH_3(CH_2)_5CO_2H$ **(b)** $Br_3CCH_2CH_2CO_2H$ **(c)** $HO_2CCH_2CH_2CH_2CO_2H$

(d) 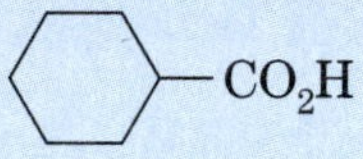 (cyclohexane)—CO_2H **(e)** 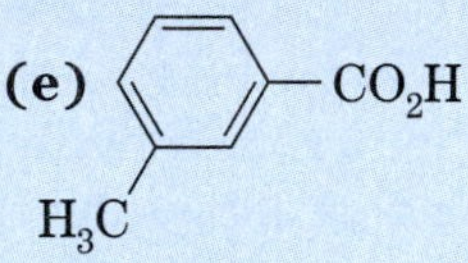 (benzene with H_3C)—CO_2H **(f)** 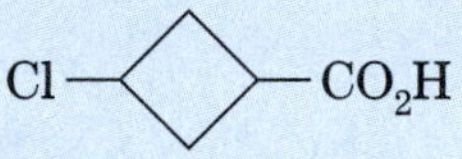 Cl—(cyclobutane)—CO_2H

See related problems 12.18–12.20.

B. Polyfunctional Carboxylic Acids

Recall that carbon-carbon double and triple bonds in a carbon chain are indicated by the suffixes *-ene* and *-yne,* respectively. The carboxylic acid group is almost always named with a suffix; any other functional groups present (aldehyde, ketone, alcohol, or amine) are named with prefixes (see Table 12.1).

$CH_3CH{=}CHCO_2H$	$CH_3CH(NH_2)CO_2H$	$CH_3CH_2C(=O)CH_2CO_2H$
2-butenoic acid	2-aminopropanoic acid	3-oxopentanoic acid

GETTING INVOLVED

Problem 12.2

Name the following by the IUPAC system:

(a) $CH_3CH(OH)CO_2H$ **(b)** 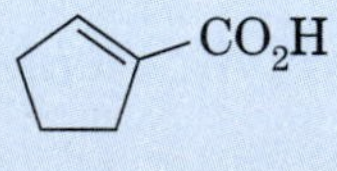 (cyclopentene)—CO_2H **(c)** $HO_2CCH_2C{\equiv}CCH_2CO_2H$

See related problem 12.21.

Table 12.1

Functional Group	General Formula	Suffix	Prefix
Carboxylic acid	$-CO_2H$	*oic acid*	*carboxy*
Aldehyde	$-\overset{O}{\overset{\parallel}{C}}H$	*al*	*oxo*
Ketone	$-\overset{O}{\overset{\parallel}{C}}-$	*one*	*oxo*
Alcohol	$-OH$	*ol*	*hydroxy*
Amine	$-NH_2$	*amine*	*amino*

C. General Procedure for Naming Organic Compounds

The following systematic procedure is useful for naming most of the types of organic compounds presented in this text. The appendix of the book has a comprehensive summary of organic nomenclature.

1. Name the longest chain of carbon atoms containing the highest priority functional group.
2. If all carbon-carbon bonds are single bonds, retain the *-an* suffix. For carbon-carbon double bonds, use the suffix *-en,* and for carbon-carbon triple bonds, *-yn.*
3. Name the highest priority functional group with a suffix and other groups with prefixes. The order of precedence for selecting the group named by a suffix is the order shown in Table 12.1.
4. Number the carbon chain, giving preference to groups in the following order: (a) functional groups named by a suffix, (b) carbon-carbon multiple bonds (carbon-carbon double bonds take precedence over triple bonds when there is a choice in determining lowest numbers), and (c) groups named with prefixes. Identify the positions of groups named by suffixes with numbers.
5. Name all other groups with prefixes, and number them.

GETTING INVOLVED

✓ Do you know and understand the five steps for naming a polyfunctional organic compound?

✓ Can you draw and recognize the groups in Table 12.1? Do you know the suffix and prefix designations for each and when to use them?

Example 12.1

Name the following compound by the IUPAC system:

$$\overset{5}{C}H_3\overset{4}{\underset{\underset{O}{\parallel}}{C}}\overset{3}{C}{\equiv}\overset{2}{C}\overset{1}{C}O_2H$$

1. There are five carbons in the longest chain: pent.
2. The triple bond takes a *-yn* suffix: pentyn.
3. The carboxylic acid is highest in Table 12.1 and takes the *-oic acid* suffix: pentynoic acid.

4. Number the carbon chain from the carboxylic acid, the group highest in Table 12.1. Locate the triple bond: 2-pentynoic acid.
5. Name the ketone group with the prefix *oxo-* and indicate its position. The complete name is 4-oxo-2-pentynoic acid.

Example 12.2

Name the following compound by the IUPAC system.

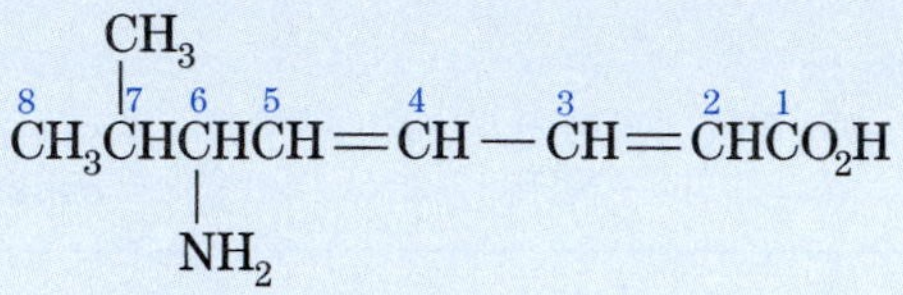

$$\overset{8}{CH_3}\overset{7}{\underset{|}{\overset{|}{C}}}H(CH_3)\overset{6}{\underset{|}{C}}H(NH_2)\overset{5}{C}H=\overset{4}{C}H-\overset{3}{C}H=\overset{2}{C}H\overset{1}{C}O_2H$$

1. Eight carbons are in the longest chain: oct.
2. Two double bonds are indicated by *-diene:* octadien.
3. The carboxylic acid group is higher than the amine in Table 12.1 and takes the suffix: octadienoic acid.
4. Numbering is from the group highest in Table 12.1. Identify locations of the double bonds; the carboxylic acid group is understood to be on carbon-1: 2,4-octadienoic acid.
5. The amine and CH_3 groups are named with prefixes. The complete name is 6-amino-7-methyl-2,4-octadienoic acid.

Problem 12.3

Name the following compounds by the IUPAC system:

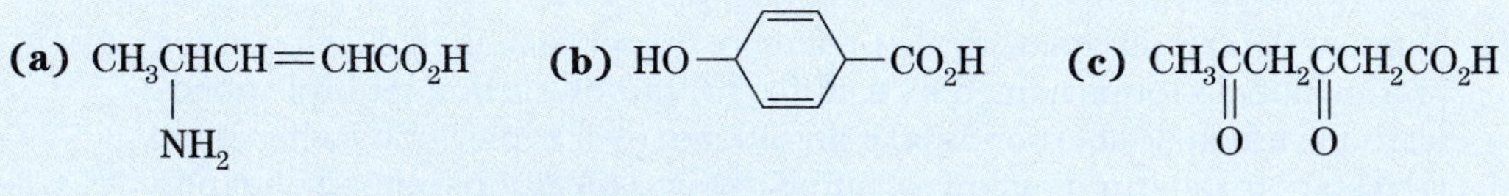

(a) $CH_3CH(NH_2)CH=CHCO_2H$ **(b)** HO–(1,4-cyclohexadiene ring)–CO_2H **(c)** $CH_3C(=O)CH_2C(=O)CH_2CO_2H$

See related problems 12.22 and 12.24.

D. Common Names of Carboxylic Acids

Carboxylic acids have long been referred to by common names which often describe familiar sources or properties of these compounds. Some of these are summarized in Table 12.2; we saw others in section 12.1.

12.3 Physical Properties of Carboxylic Acids

hydrogen-bonding intermolecular attractions are caused by hydrogen bonded to an electronegative element (O, N, F) being attracted to a lone pair of electrons of another electronegative element

The most striking and important physical characteristic of carboxylic acids stem from their ability to **hydrogen-bond.** You will recall that hydrogen-bonding is common in organic compounds that possess either O—H or N—H bonds (section 9.2). Hydrogen-bonding is a form of strong intermolecular attraction among molecules and causes these substances to have higher boiling points than compounds of similar molecular weight that are incapable of hydrogen-bonding. Since water is capable of hydrogen-bonding with carboxylic acids, lower-molecular-weight acids are water-soluble. These properties are illustrated in Table 12.3. Notice that the two compounds capable of hydrogen-bonding (the alcohol and carboxylic acid) have much higher boiling points than the other two. But also note that the carboxylic acid has a significantly higher boiling point than

Table 12.2 Physical Properties of Carboxylic Acids

Structure	Common Name	Derivation of Name	Melting Point, °C	Boiling Point, °C	Water Solubility, g/100 g H_2O	Acidity Constant K_a	pK_a
HCO_2H	Formic acid	L. *formica*, "ant"	8	101	∞	1.77×10^{-4}	3.75
CH_3CO_2H	Acetic acid	L. *acetum*, "vinegar"	17	118	∞	1.76×10^{-5}	4.75
$CH_3CH_2CO_2H$	Propionic acid	Gr. *proto*, "first"; *pion*, "fat"	−22	141	∞	1.34×10^{-5}	4.87
$CH_3(CH_2)_2CO_2H$	Butyric acid	L. *butyrum*, "butter"	−8	164	∞	1.54×10^{-5}	4.81
$CH_3(CH_2)_3CO_2H$	Valeric acid	L. *valere*, "to be strong" (valerian root)	−35	187	4.97	1.51×10^{-5}	4.82
$CH_3(CH_2)_4CO_2H$	Caproic acid	L. *caper*, "goat"	−3	205	1.08	1.43×10^{-5}	4.84
$CH_3(CH_2)_5CO_2H$	Enanthic acid	Gr. *oinánth(ē)*, the vine blossom	−8	223	0.24	1.42×10^{-5}	4.85
$CH_3(CH_2)_6CO_2H$	Caprylic acid	L. *caper*, "goat"	17	240	0.07	1.28×10^{-5}	4.89
$CH_3(CH_2)_7CO_2H$	Pelargonic acid	Pelargonium plant	13	253	0.03	1.09×10^{-5}	4.96
$CH_3(CH_2)_8CO_2H$	Capric acid	L. *caper*, "goat"	31	270	0.02	1.43×10^{-5}	4.84
$CH_3(CH_2)_{10}CO_2H$	Lauric acid	Laurel	44	—	0.006	—	

Table 12.3 Comparison of Carboxylic Acids and Other Compounds in Physical Properties

	Mol Wt	Boiling Point, °C	Water Solubility, g/100 g
$CH_3CH_2CH_2CH_2CH_3$	72	36	0.04
$CH_3CH_2CH_2CH_2OH$	74	117	7.4
$CH_3CH_2C(=O)OH$	74	141	∞
$CH_3C(=O)OCH_3$	74	57	32

the alcohol. This is due to the greater polarity of the O—H bond in carboxylic acids (C=O is electron-withdrawing and further polarizes the O—H) and the ability of the carboxylic acids to hydrogen-bond in two places. This hydrogen-bonding is so strong that some carboxylic acids exist as **dimers** (two molecules) even in the vapor phase.

dimer
two structural units

$$R-C(=O\cdots H-O)(-O-H\cdots O=)C-R$$

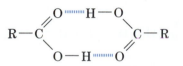

As in other classes of organic compounds, the boiling points of carboxylic acids increase with molecular weight (section 2.9). This steady trend is evident in Table 12.2, which summarizes the physical properties of a homologous series of carboxylic acids. Notice also that the proportion of nonpolar hydrocarbon (water-insoluble) to polar carboxylic acid group (water soluble) in the molecule increases with molecular weight. Thus water solubility decreases.

GETTING INVOLVED

- ✓ What is hydrogen-bonding and what types of compounds are capable of it?
- ✓ Why are carboxylic acids capable of hydrogen-bonding? How do they form dimers? Why does hydrogen-bonding increase boiling points of carboxylic acids relative to compounds that cannot hydrogen-bond? Why do carboxylic acids have higher boiling points than alcohols of similar molecular weight?
- ✓ How does hydrogen-bonding influence the solubility of carboxylic acids in water?
- ✓ Why do higher molecular weight carboxylic acids have lower water solubility than those with smaller molecular weights?

Problem 12.4

Arrange the following compounds in order of increasing boiling point.

$$HO\overset{O}{\overset{\|}{C}}(CH_2)_3\overset{O}{\overset{\|}{C}}OH \qquad CH_3O\overset{O}{\overset{\|}{C}}CH_2CH_2\overset{O}{\overset{\|}{C}}OH \qquad CH_3O\overset{O}{\overset{\|}{C}}CH_2\overset{O}{\overset{\|}{C}}OCH_3$$

Problem 12.5

Match these acids—ethanoic acid, pentanoic acid, and decanoic acid—with the following water solubilities: 3.7 g/100 g, 0.2 g/100 g, soluble in all proportions. Explain your answer.

See related problems 12.27–12.28.

12.4 Acidity of Carboxylic Acids

A. Reactions of Acids with Base: Salt Formation

Of the principal classes of organic compounds, only carboxylic acids and phenols are significantly acidic. (Refer back to Table 9.2 for acidity constants for various organic compounds.) This acidity can be detected by reaction of carboxylic acids and phenols with base, and, in fact, these neutralization reactions are qualitative tests in organic analysis. For example, both carboxylic acids and phenols are neutralized by sodium hydroxide; the acidic hydrogen ion of the hydroxyl group is abstracted by the hydroxide base. Alcohols (pK_a's around 17 or 18) are up to 100 million times less acidic than phenols (pK_a's around 10) and generally are not neutralized by sodium hydroxide.

Carboxylic acids: $$R\overset{O}{\overset{\|}{C}}—O—H + NaOH \longrightarrow R\overset{O}{\overset{\|}{C}}—O^- \; Na^+ + H_2O$$

Phenols: $$C_6H_5—O—H + NaOH \longrightarrow C_6H_5—O^- \; Na^+ + H_2O$$

Alcohols: $$R—O—H + NaOH \longrightarrow \text{no reaction}$$

Although phenols are definitely acidic compared to alcohols, they are considerably less acidic than carboxylic acids (pK_a's around 5). Both phenols and carboxylic acids are neutralized by the stronger base sodium hydroxide, but only carboxylic acids react with the weaker base sodium bicarbonate (this is the basis of the familiar reaction between vinegar and baking soda in which carbon dioxide effervesces).

Carboxylic acids: $RC(=O)-O-H + NaHCO_3 \longrightarrow RC(=O)-O^-Na^+ + H_2O + CO_2\uparrow$

Alcohols or phenols: $R-O-H$ or $C_6H_5-O-H + NaHCO_3 \longrightarrow$ no reaction

GETTING INVOLVED

✓ What is the order of acidity of alcohols, phenols, and carboxylic acids? How do they differ in their reactions with sodium hydroxide and sodium bicarbonate?

Problem 12.6

Write balanced equations showing the preparations of the food preservatives **(a)** sodium benzoate, **(b)** calcium propionate, **(c)** potassium sorbate, and **(d)** monosodium glutamate from the corresponding carboxylic acids. The structures of these compounds are shown in Connections 12.1.

Problem 12.7

If you have two test tubes, one containing an aqueous solution of phenol and the other containing an aqueous solution of propanoic acid, how would you chemically determine which is which? What simple reagent would you use and what would you see?

See related problem 12.30.

B. Explanation for the Acidity of Carboxylic Acids

Why are carboxylic acids so much more acidic than alcohols when both have O—H groups? The answer lies in the increased polarity of the O—H bond due to the carbonyl group in carboxylic acids, and in the stability of the anion formed upon ionization or neutralization. In the alkoxide ion, formed from neutralization of alcohols, the negative charge is localized on one atom, the oxygen. In contrast, the negative charge is delocalized by resonance in the carboxylate ion and thus it is considerably more stable.

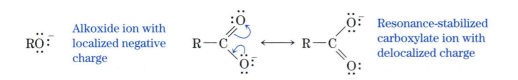

$R\ddot{O}:^-$ Alkoxide ion with localized negative charge

$R-C(=\ddot{O}:)-\ddot{O}:^- \longleftrightarrow R-C(-\ddot{O}:^-)=\ddot{O}:$ Resonance-stabilized carboxylate ion with delocalized charge

In Figure 12.1 you can see that the lone pair of electrons responsible for the negative charge exists in a p orbital that overlaps with the p orbitals of the adjacent π bond spreading the charge throughout the three-atom system. As an electronegative atom, oxygen is especially able to accommodate a negative charge, and there are two of them doing this in the carboxylate ion, compared to only one in the alkoxide ion.

X-ray studies support this resonance stabilization hypothesis. In un-ionized formic acid, the carbon-oxygen double bond is shorter (1.23Å) than the carbon-oxygen single bond (1.36Å). However, in sodium formate, the two carbon-oxygen

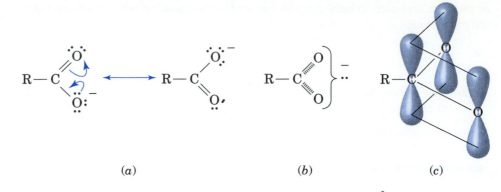

Figure 12.1 Resonance stabilization of carboxylate anion. (a) Resonance forms. (b) Resonance hybrid. (c) Bonding picture.

(*a*) (*b*) (*c*)

bond lengths are equivalent and intermediate in length (1.27Å) between a single and a double bond.

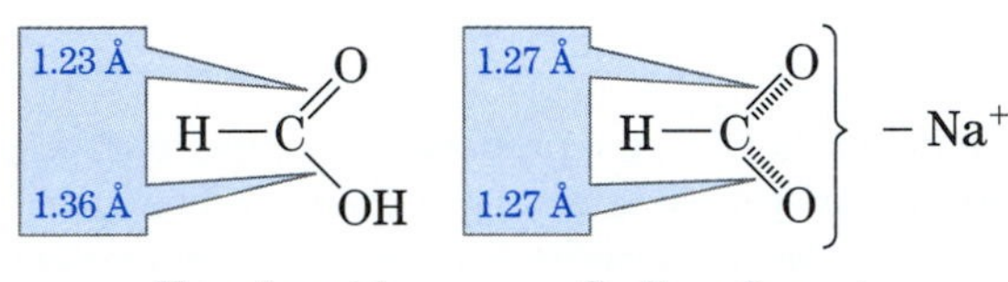

In the ionized form of phenol, the negative charge is delocalized over the oxygen and aromatic ring (section 9.6.A.2 and Figure 9.3) but there is no second oxygen to help accommodate the negative charge. Because of these structural differences, carboxylic acids have acidity constants (section 9.6 and Table 9.2) of about 10^{-5}, phenols of around 10^{-10}, and alcohols in the range of 10^{-18}. Comparing acidity constants, we find that carboxylic acids are 100,000 times more acidic than phenols and ten trillion times more acidic than alcohols.

GETTING INVOLVED

✓ What are the two reasons that carboxylic acids are more acidic than alcohols? Why are they more acidic than phenols?

✓ Why are the two carbon-oxygen bonds in un-ionized carboxylic acids of different lengths (which is shorter, why?), but of the same length in ionized acids? Can you draw the resonance forms for a carboxylate ion?

Problem 12.8

Arrange the following compounds in order of increasing acidity: nitric acid; butanoic acid; butanol; butane; phenol.

C. Structure and Relative Acidities of Carboxylic Acids

acidity constant
K_a product of the concentrations of the ionized form of an acid divided by the concentration of the un-ionized form

pK$_a$
the negative logarithm of K_a

Carboxylic acids are weak acids and only partially ionized in water; an equilibrium is established. The extent of ionization is described by the **acidity constant,** K_a or by ***pK$_a$***, which is the negative logarithm of the acidity constant.

$$\mathrm{RCOOH} + \mathrm{H_2O} \rightleftharpoons \mathrm{RCO_2^-} + \mathrm{H_3O^+} \qquad K_a = \frac{[\mathrm{RCO_2^-}][\mathrm{H_3O^+}]}{[\mathrm{RCO_2H}]} \qquad pK_a = -\log K_a$$

You will recall from our introduction to acidity in the chapter on alcohols (section 9.6 and Table 9.2) that the larger the K_a and the smaller the pK_a, the greater the acidity of an acid. This is because the concentrations of the ionized species of the acid are in the numerator of the acidity constant equation and the extent of ionization is the measure of acidity.

The presence of various substituents on a carboxylic acid molecule can measurably affect the acidity. Carboxylic acids are acidic because the negative charge of the carboxylate ion is delocalized by resonance. Any group that can enhance this effect, that is, further diminish the effect of the negative charge, will increase acidity. In general, electron-withdrawing groups increase acidity because they diminish the intensity of the negative charge and in doing so stabilize the carboxylate anion. Electron-releasing groups, however, decrease acidity because they intensify the negative charge, thereby destabilizing the carboxylate anion.

Electron-withdrawing and releasing groups

In the following example, the methyl group of ethanoic acid is electron-releasing and intensifies the negative charge on the carboxylate anion; ethanoic acid is thus less acidic than methanoic acid. However, replacing one of the hydrogens on the methyl group with the strongly electron-withdrawing nitro group reverses this effect and dramatically increases acidity.

	HCO_2H	CH_3CO_2H	$O_2N—CH_2CO_2H$
K_a	17.7×10^{-5}	1.75×10^{-5}	2100×10^{-5}
pK_a	3.75	4.76	1.68

Strength of electron-withdrawing groups

The electron-withdrawing strength of a group determines the magnitude of its effect on acidity. The electronegativity (and thus the electron-attracting capability) of halogens is of the order F > Cl > Br > I. This trend is exemplified in the haloacetic acids.

		FCH_2CO_2H	$ClCH_2CO_2H$	$BrCH_2CO_2H$	ICH_2CO_2H	CH_3CO_2H	
Most acidic	K_a	260×10^{-5}	136×10^{-5}	125×10^{-5}	67×10^{-5}	1.76×10^{-5}	Least acidic
	pK_a	2.59	2.87	2.90	3.17	4.75	

Number of electron-withdrawing groups

As the number of electron-withdrawing substituents increases, so does acidity.

		CH_3CO_2H	$ClCH_2CO_2H$	Cl_2CHCO_2H	Cl_3CCO_2H	
Least acidic	K_a	1.76×10^{-5}	136×10^{-5}	$5{,}530 \times 10^{-5}$	$23{,}200 \times 10^{-5}$	Most acidic
	pK_a	4.75	2.87	1.26	0.63	

Proximity of electron-withdrawing groups

The proximity of the electron-withdrawing group is also important in considering acidity. The nearer the group to the carboxyl, the greater the effect.

		$CH_3CH_2CHCO_2H$ (Cl on C-2)	$CH_3CHCH_2CO_2H$ (Cl on C-3)	$CH_2CH_2CH_2CO_2H$ (Cl on C-4)	$CH_3CH_2CH_2CO_2H$	
Most acidic	K_a	139×10^{-5}	8.9×10^{-5}	3.0×10^{-5}	1.5×10^{-5}	Least acidic
	pK_a	2.86	4.05	4.52	4.82	

Aromatic carboxylic acids

With some modification, these same principles apply to aromatic carboxylic acids. Electron-withdrawing groups enhance acidity. Because their effect is largely due to resonance, they have their greatest impact if positioned ortho or para to the acid group.

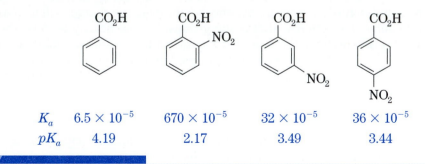

	CO_2H (benzoic acid)	CO_2H, ortho-NO_2	CO_2H, meta-NO_2	CO_2H, para-NO_2
K_a	6.5×10^{-5}	670×10^{-5}	32×10^{-5}	36×10^{-5}
pK_a	4.19	2.17	3.49	3.44

GETTING INVOLVED

✓ What are K_a and pK_a? For each of these, what denotes higher acidities, large or small values?

✓ Why do electron-releasing groups generally decrease acidity of carboxylic acids whereas electron-withdrawing groups increase acidity? How does strength, number, and proximity of electron-withdrawing groups influence acidity?

✓ Why does an electron-withdrawing group ortho or para to the acid group of benzoic acid increase acidity more than a meta group?

Example 12.3

Arrange the following compounds from least acidic to most acidic.

(i) CH_3CHFCO_2H **(ii)** $CH_2FCHFCO_2H$ **(iii)** $CH_3CHClCO_2H$ **(iv)** $CH_2ClCH_2CO_2H$

Strength, proximity, and number of electron-withdrawing groups are involved in this problem. iv is the least acidic because Cl is less electronegative than F, there is only one Cl, and the Cl is as far away as possible. iii is next; the Cl is closer than in iv. Next is i; the F is in the same place as the Cl in iii, but F is more electronegative. ii is most acidic; there are two F's.

Problem 12.9

Arrange the following K_a's and pK_a's in order of increasing acidity:

(a) K_a's of (i) 5.9×10^{-2}, (ii) 3×10^{-5}, (iii) 6.9×10^{-5}, (iv) 1.5×10^{-3}, (v) 9.3×10^{-4}

(b) pK_a's of (i) 1.17, (ii) 2.86, (iii) 4.41, (iv) 3.52

Problem 12.10

Arrange the following carboxylic acids in order of increasing acidity:

(a) (i) F_3CCO_2H, (ii) Br_3CCO_2H, (iii) I_3CCO_2H, (iv) Cl_3CCO_2H

(b) (i) $Cl_2CHCH_2CO_2H$, (ii) $CH_3CCl_2CO_2H$, (iii) $ClCH_2CHClCO_2H$, (iv) $ClCH_2CH_2CO_2H$

(c) (i) $CH_3CH_2CO_2H$, (ii) HCO_2H, (iii) $HO_2C{-}CO_2H$

(d) ortho, meta, and para chlorobenzoic acids.

See related problem 12.29.

Connections 12.1 www.prenhall.com/bailey

Food Preservatives

Salts of carboxylic acids, or sometimes the acids themselves, are added to a wide variety of processed foods as food preservatives. They act to retard food spoilage by inhibiting or preventing growth of bacteria and fungi and other microorganisms. Some common food preservatives are shown below. You may especially recognize calcium propionate, which is often added to breads to prevent molding; sodium benzoate, which is a common additive to unrefrigerated bottled citrus juices and drinks; and monosodium glutamate, which is also used as a flavor enhancer.

Since carboxylic acid salts prevent bacterial growth in foods, it is not surprising that some find other, related applications. Calcium and zinc undecylates, for example, are components of some foot and baby powders, where they retard bacterial and fungal growth. Soaps are the sodium salts of long-chain fatty acids derived from fats and oils.

$C_6H_5-CO_2^-Na^+$

Sodium benzoate

$(CH_3CH_2CO_2^-)_2Ca^{2+}$

Calcium propionate

$CH_3CH{=}CH-CH{=}CHCO_2^-K^+$

Potassium sorbate

$HO_2CCH(NH_2)CH_2CH_2CO_2^-Na^+$

Monosodium glutamate

$(CH_3(CH_2)_9CO_2^-)_2Zn^{2+}$

Zinc undecylate

$R\sim\sim\sim C(=O)O^-Na^+$

A soap (R = 12–18 CH_2)

Phenols, such as butylated hydroxytoluene (BHT), are also used as food preservatives, since many are effective antioxidants and, in some cases, also act as antimicrobial agents (see section 9.3.D).

D. Nomenclature of the Salts of Carboxylic Acids

We saw in section A that a base can abstract a proton from a carboxylic acid, leaving a carboxylate ion. A **salt** of a carboxylic acid is the carboxylate ion plus the cation from the base. The salts of many inorganic acids are named by changing the suffix *-ic acid* to *-ate* and prefixing the name with the name of the cation that replaced the acidic hydrogen.

salt
ionic compound composed of cation from a base and anion from neutralized acid

HNO_3	Nitr*ic acid*	H_2SO_4	Sulfur*ic acid*
$NaNO_3$	*Sodium* nitr*ate*	$(NH_4)_2SO_4$	*Ammonium* sulf*ate*

Salts of carboxylic acids are named in the same way. First, name the parent acid. If you are looking at the salt form, imagine the cation as a hydrogen to aid in visualizing the acid. Then to name the salt, change the *-ic acid* of the parent acid to *-ate* and precede this with the name of the cation.

$CH_3CH_2C(=O)OH$ Propano*ic acid* $C_6H_5-C(=O)OH$ Benzo*ic acid*

$CH_3CH_2C(=O)O^-Na^+$ *Sodium* propano*ate* $C_6H_5-C(=O)O^-NH_4^+$ *Ammonium* benzo*ate*

GETTING INVOLVED

Problem 12.11

Name the following carboxylic acid salts:

(a) $CH_3\overset{O}{\overset{\|}{C}}ONa$ **(b)** $(CH_3CH_2\overset{O}{\overset{\|}{C}}O)_2Ca$ **(c)** $CH_3CH{=}CH\overset{O}{\overset{\|}{C}}OK$ **(d)** $Br-C_6H_4-\overset{O}{\overset{\|}{C}}ONH_4$

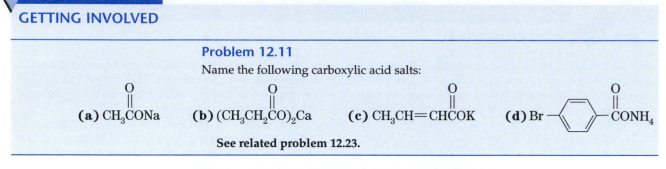

See related problem 12.23.

12.5 Preparations of Carboxylic Acids

Some of the reactions of the functional groups we have previously covered are useful in the synthesis of carboxylic acids. We can classify these simply into two types: preparations in which the acid results from groups already in the starting material and those in which a carbon is added in the synthesis.

A. Oxidation of Alkylbenzenes (section 6.5)

Primary and secondary alkyl side chains on an aromatic ring can be oxidized to carboxylic acid groups with potassium permanganate; given enough reagent, multiple groups can be converted. Of course, the oxidation can be combined with electrophilic aromatic substitution reactions to produce substituted benzoic acids as illustrated by the synthesis of *p*-chlorobenzoic acid.

$$CH_3-C_6H_4-CH_3 \xrightarrow[OH^-]{KMnO_4} HO_2C-C_6H_4-CO_2H$$

$$C_6H_5-CH_3 \xrightarrow[FeCl_3]{Cl_2} Cl-C_6H_4-CH_3 \xrightarrow[OH^-]{KMnO_4} Cl-C_6H_4-CO_2H$$

It is useful to note that the alkyl group is an ortho-para director and the resulting acid group is a meta director. If we wished to synthesize *m*-chlorobenzoic acid, the oxidation would be performed first and then the chlorine would be introduced.

GETTING INVOLVED

Problem 12.12

Write the product of $KMnO_4$ oxidation of: **(a)** propylbenzene; **(b)** 1,3,5-trimethylbenzene.

Problem 12.13

Propose syntheses for **(a)** both *p*-nitrobenzoic acid and *m*-nitrobenzoic acid from benzene; **(b)** 2-bromo-4-nitrobenzoic acid from toluene.

B. Oxidation of Primary Alcohols (section 9.9)

Primary alcohols are oxidized to carboxylic acids (via the corresponding aldehyde) with reagents such as CrO_3 or $Na_2Cr_2O_7$ under acid conditions.

$$\underset{\text{1-octanol}}{CH_3CH_2CH_2CH_2CH_2CH_2CH_2CH_2OH} \xrightarrow[H^+]{CrO_3} \underset{\text{Octanoic acid}}{CH_3CH_2CH_2CH_2CH_2CH_2CH_2\overset{O}{\overset{\|}{C}}OH}$$

GETTING INVOLVED

Problem 12.14

Write a reaction equation illustrating the preparation of 4,4-dimethylpentanoic acid by oxidation of the corresponding alcohol.

C. Hydrolysis of Nitriles (sections 8.4.A and 11.5.C)

Nitriles can be hydrolyzed under acidic or basic conditions to carboxylic acids. They can be prepared by nucleophilic substitution or by addition of HCN to an aldehyde or ketone. In these examples, an additional carbon is introduced into the molecule.

$$C_6H_5CH_2Br \xrightarrow{NaCN} C_6H_5CH_2CN \xrightarrow{H_2O/H^+} C_6H_5CH_2CO_2H$$

$$C_6H_5CH{=}O \xrightarrow[(NaCN,\ H^+)]{HCN} C_6H_5CH(OH)CN \xrightarrow{H_2O/H^+} C_6H_5CH(OH)CO_2H$$

GETTING INVOLVED

Problem 12.15

Show the synthesis of hexanoic acid from 1-bromopentane using nitrile hydrolysis.

Problem 12.16

Propose a synthesis for 2-methyl-2-hydroxyhexanoic acid from an aldehyde or ketone with six carbons.

D. Carbonation of Grignard Reagents

Carboxylic acids can be synthesized from alkyl or aryl halides by converting the halide to the corresponding Grignard reagent followed by treatment with carbon dioxide. The product has one more carbon than the starting material.

$$C_6H_5CH_2Br \xrightarrow[Ether]{Mg} C_6H_5CH_2MgBr \xrightarrow[2)\ H_2O/H^+]{1)\ CO_2} C_6H_5CH_2CO_2H$$

The Grignard reagent adds to CO_2 by nucleophilic addition. The resulting salt is neutralized by aqueous acid.

$$C_6H_5\overset{\delta-}{C}H_2\text{—}\overset{\delta+}{MgBr} + O^{\delta-}{=}C^{\delta+}{=}O^{\delta-} \longrightarrow C_6H_5CH_2C(=O)O^- \overset{+}{[MgBr]} \xrightarrow[H_2O]{HCl} C_6H_5CH_2C(=O)OH + MgBrCl$$

GETTING INVOLVED

Problem 12.17

Show a synthesis of hexanoic acid from 1-bromopentane by a carbonation of a Grignard reagent.

See related problems 12.25–12.26.

REACTION SUMMARY

1. Formation of Carboxylic Acid Salts

Section 12.4.A; Problems 12.6–12.7, 12.30.

$$\mathrm{RC(=O)OH} + \mathrm{M^+OH^-} \longrightarrow \mathrm{RC(=O)O^-M^+} + \mathrm{H_2O}$$

2. Preparations of Carboxylic Acids

Section 12.5; Problems 12.12–12.17, 12.25–12.26.

A. Oxidation of Alkylbenzenes

$$\mathrm{C_6H_5{-}R} \xrightarrow{\mathrm{KMnO_4}} \mathrm{C_6H_5{-}CO_2H}$$

B. Oxidation of Primary Alcohols

$$\mathrm{RCH_2OH} \xrightarrow{\mathrm{CrO_3/H^+}} \mathrm{RCO_2H}$$

C. Hydrolysis of Nitriles

$$\mathrm{RC{\equiv}N} \xrightarrow{\mathrm{H_2O/H^+}} \mathrm{RCO_2H}$$

D. Carbonation of Grignard Reagents

$$\mathrm{RX} \xrightarrow[\text{Ether}]{\mathrm{Mg}} \mathrm{RMgX} \xrightarrow[\text{2) } \mathrm{H_2O/H^+}]{\text{1) } \mathrm{CO_2}} \mathrm{RCO_2H}$$

Problems

12.18 **Nomenclature of Carboxylic Acids:** Name the following compounds by the IUPAC system of nomenclature:

(a) $CH_3(CH_2)_7CO_2H$

(b) $CH_3CH_2CH_2CH_2CO_2H$

(c) $CH_3CH(CH_3)CH_2CH_2CO_2H$

(d) $CH_3CH(CH_3)CH_2CH(CH_2CH_3)CH_2CO_2H$

(e) $HO_2CCH_2CH_2CO_2H$

(f) Cl_3CCO_2H

12.19 **Nomenclature of Carboxylic Acids:** Name the following compounds by the IUPAC system:

(a) 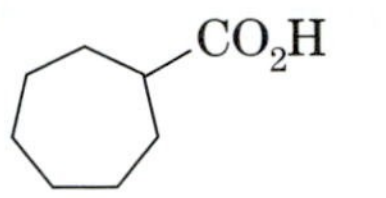 CO_2H (on a cycloheptane ring) **(b)** 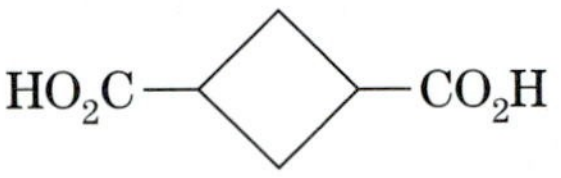 HO_2C—(cyclobutane)—CO_2H

(c) 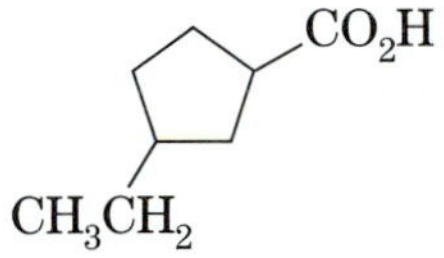 CO_2H, CH_3CH_2 (on a cyclopentane ring)

12.20 **Nomenclature of Carboxylic Acids:** Name the following compounds by the IUPAC system:

(a) 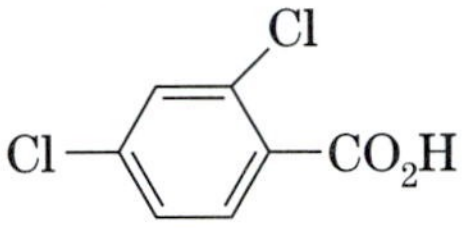 Cl—(benzene ring, Cl)—CO_2H **(b)** 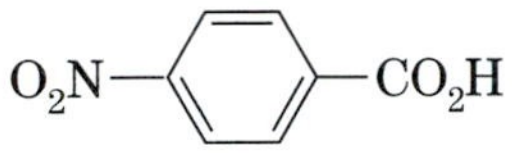 O_2N—(benzene ring)—CO_2H

(c) 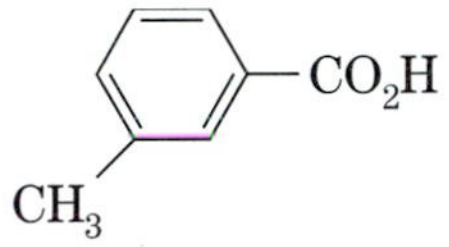 (benzene ring)—CO_2H, CH_3

12.21 **Nomenclature of Polyfunctional Carboxylic Acids:** Name the following compounds by the IUPAC system of nomenclature:

(a) $CH_3CH_2CH_2CH{=}CHCO_2H$

(b) $CH_3\underset{OH}{\underset{|}{C}H}CH_2CH_2CH_2CO_2H$

(c) 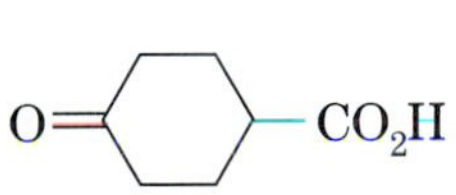 $O{=}$(cyclohexane ring)—CO_2H

(d) 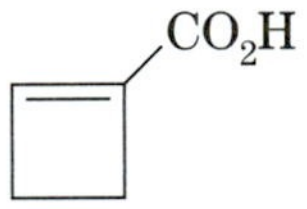 (cyclobutene ring)—CO_2H

(e) $HO_2CCH{=}CHCO_2H$

(f) $CH_3\underset{O}{\underset{\|}{C}}CH_2\underset{O}{\underset{\|}{C}}CH_2CO_2H$

(g) $CH_3CH{=}CH{-}CH{=}CHCO_2H$

(h) $CH_3\underset{O}{\underset{\|}{C}}C{\equiv}CCO_2H$

(i) $H_2NCH_2CH{=}CHCO_2H$

12.22 **Nomenclature of Organic Compounds:** Name the following compounds by the IUPAC system of nomenclature:

(a) $CH_3CH{=}CH\overset{O}{\overset{\|}{C}}H$

(b) $H_2NCH_2CH{=}CH{-}CH{=}CHCH_2OH$

(c) $CH_3\overset{O}{\overset{\|}{C}}\underset{OH}{\underset{|}{C}H}\overset{O}{\overset{\|}{C}}CH_3$

(d) $CH_3\overset{O}{\overset{\|}{C}}C{\equiv}C\overset{O}{\overset{\|}{C}}CH_3$

(e) 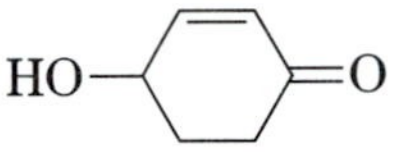 HO—(cyclohexene ring)$=O$

(f) 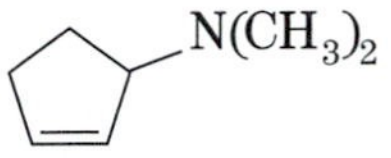 (cyclopentene ring)—$N(CH_3)_2$

(g) $CH_3CH_2CH_2\underset{OH}{\underset{|}{C}H}CH{=}CHCO_2H$

(h) $CH_3CH{=}CH\underset{O}{\underset{\|}{C}}C{\equiv}C\underset{OH}{\underset{|}{C}H}CH_2Br$

12.23 **Nomenclature of Carboxylic Acid Salts:** Name the following by the IUPAC system of nomenclature:

(a) $CH_3CH_2CH_2CO_2Na$

(b) $(CH_3CO_2)_2Ca$

(c) $Br_3CCH_2CH_2CO_2K$

(d) 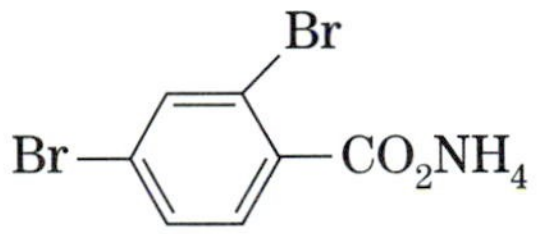 Br—(benzene ring, Br)—CO_2NH_4

(e) 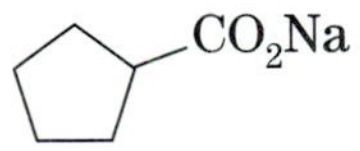 (cyclopentane ring)—CO_2Na

(f) $CH_3\underset{O}{\underset{\|}{C}}CH{=}CHCO_2Na$

12.24 **IUPAC Nomenclature:** Draw the following compounds: **(a)** 3-methylbutanoic acid; **(b)** 5-bromo-3-hexynoic acid; **(c)** 4-oxopentanoic acid; **(d)** 1,3,5,7-cyclooctatetraene carboxylic acid; **(e)** 5-hydroxy-2,4-hexandione

12.25 **Preparations of Carboxylic Acids:** Write reaction equations illustrating the preparation of benzoic acid by the following methods:

(a) oxidation of alkylbenzenes
(b) oxidation of primary alcohols
(c) hydrolysis of nitriles
(d) carbonation of Grignard reagents

12.26 **Preparations of Carboxylic Acids:** Offer syntheses for the following compounds:

(a) 1-butanol to pentanoic acid
(b) toluene to *m*-bromobenzoic acid
(c) 2-chloroheptane to 2-methylheptanoic acid
(d) pentanal to 2-hydroxyhexanoic acid
(e) 1-heptanol to heptanoic acid

12.27 **Physical Properties:** Arrange the following compounds in order of increasing boiling point. Explain your answer.

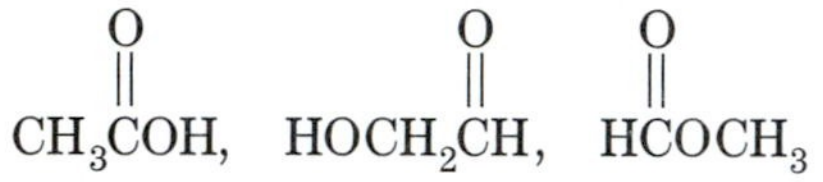

$CH_3\overset{O}{\overset{\|}{C}}OH$, $HOCH_2\overset{O}{\overset{\|}{C}}H$, $H\overset{O}{\overset{\|}{C}}OCH_3$

12.28 **Physical Properties:** Although they have similar molecular weights, chloroethane has a boiling point of 12° C and ethanoic acid's boiling point is 118°C. Bromoethane, with a molecular weight almost double these, boils at 38°C. Explain these boiling temperatures.

12.29 **Acidity:** Arrange each of the following groups of compounds in order of increasing acidity:

(a) $CH_3\underset{\underset{F}{|}}{C}HCO_2H$, $CH_3\underset{\underset{Br}{|}}{C}HCO_2H$, $CH_3\underset{\underset{Cl}{|}}{C}HCO_2H$, $CH_3CH_2CO_2H$

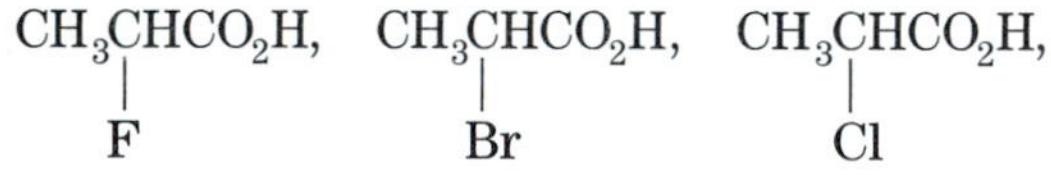

(b) $CH_3CH_2CH_2\overset{O}{\overset{\|}{C}}CO_2H$, $CH_3CH_2\overset{O}{\overset{\|}{C}}CH_2CO_2H$, $CH_3\overset{O}{\overset{\|}{C}}CH_2CH_2CO_2H$

(c) $CH_3\underset{\underset{Br}{|}}{C}H\underset{\underset{Br}{|}}{C}HCO_2H$, $\underset{\underset{Br}{|}}{C}H_2\underset{\underset{Br}{|}}{C}HCH_2CO_2H$, $CH_3CH_2\overset{\overset{Br}{|}}{\underset{\underset{Br}{|}}{C}}-CO_2H$, $\underset{\underset{Br}{|}}{C}H_2CH_2\underset{\underset{Br}{|}}{C}HCO_2H$

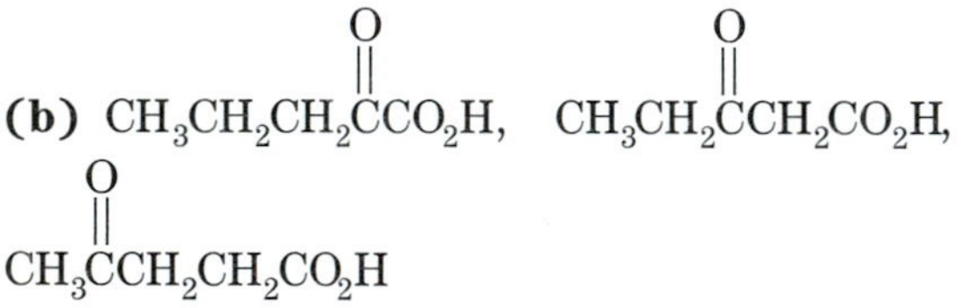

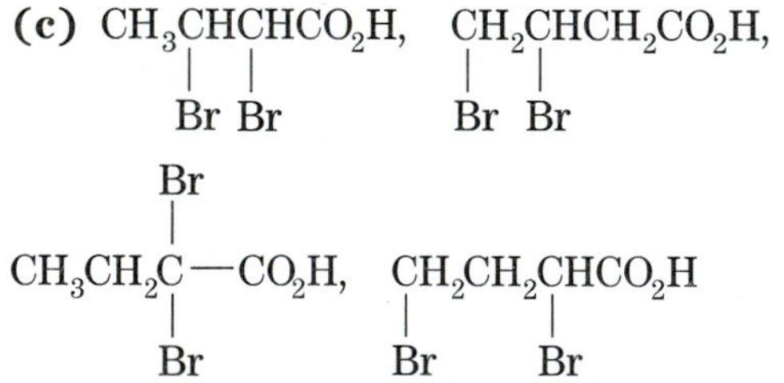

(d) CO_2H / Cl; CO_2H / Cl / Cl; CO_2H / Cl / Cl; CO_2H / Cl

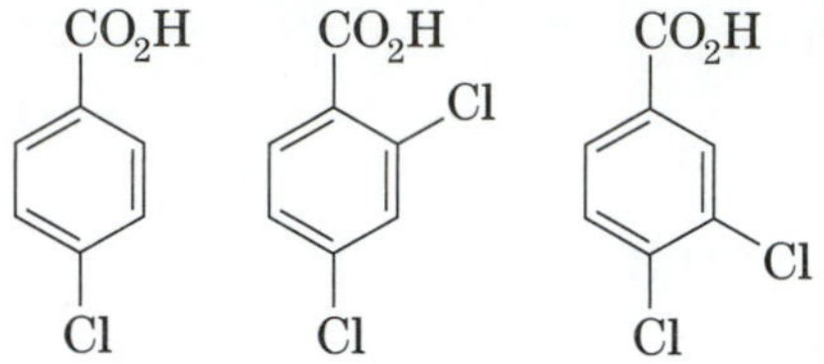

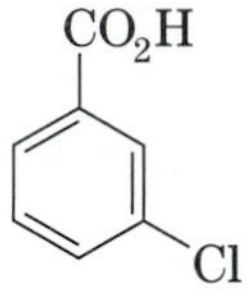

(e) HO_2CCO_2H, $HO_2CCH_2CO_2H$, $HO_2CCH_2CH_2CO_2H$

(f) CH_3CH_2OH, CH_3CO_2H, CH_3CH_3, C_6H_5—OH

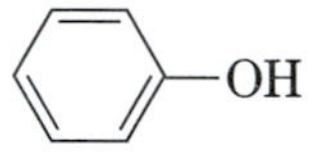

12.30 **Neutralization Reactions of Carboxylic Acids:** Write products for the reactions between the following pairs of reactants:

(a) $CH_3(CH_2)_5CO_2H/NaOH$

(b) C_6H_5—CH_2CO_2H/KOH

(c) $HO_2C(CH_2)_3CO_2H/Ca(OH)_2$

(d) CH_3CO_2H/NH_4OH

Activities with Molecular Models

1. Make a molecular model of (a) formic acid found in the sting of ants and (b) acetic acid that makes up 5% of most vinegars. Make molecular models of each. How many nonbonding electron pairs reside on each oxygen? What is the hybridization and geometrical orientation of each carbon and each oxygen? What are the bond angles around these atoms?
2. There are two carboxylic acids with the molecular formula of $C_4H_8O_2$. Make molecular models of each.

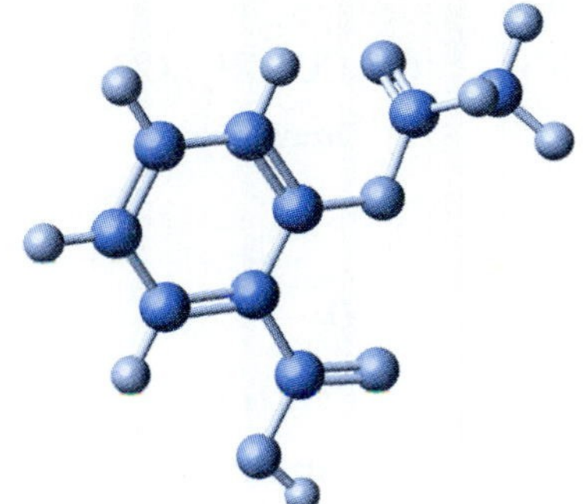

CHAPTER 13

Derivatives of Carboxylic Acids

13.1 Structure and Nomenclature of Carboxylic Acid Derivatives

A. Structure

Carboxylic acids and their derivatives can be expressed as variations of a single formula in which an electronegative atom—oxygen, nitrogen, or halogen—is bonded to an *acyl group.*

$$\mathrm{R{-}C({=}O){-}L}$$

L = Cl (acid chloride); $\mathrm{OC({=}O)R}$ (acid anhydride); OH (carboxylic acid); OR (ester); NH_2, NHR, or NR_2 (amide).

$$\mathrm{R{-}C({=}O){-}}$$ acyl group

acyl group

$\mathrm{RC({=}O){-}}$

carboxylic acid
functional group in which OH, hydroxy, is attached to an acyl group

ester
functional group in which OR, alkoxy, is attached to an acyl group

amide
functional group in which NH_2, NHR, or NR_2 is attached to an acyl group

acid chloride
functional group in which Cl, chloride, is attached to an acyl group

acid anhydride
functional group in which RCO_2 of one acid molecule is bonded to the acyl group of another

General structures for each of the types of derivatives follow.

$\mathrm{R{-}C({=}O){-}OH}$ **Carboxylic acid**

$\mathrm{R{-}C({=}O){-}OR}$ **Ester**

$\mathrm{R{-}C({=}O){-}NH_2}$ **Amide**

$\mathrm{R{-}C({=}O){-}Cl}$ **Acid chloride**

$\mathrm{R{-}C({=}O){-}O{-}C({=}O){-}R}$ **Acid anhydride**

Carboxylic acids, esters, and amides are abundant in nature. We have already seen familiar examples of carboxylic acids such as acetic, lactic, and citric acids (section 12.1). Proteins are amides, polyamides to be exact, as they are large molecules composed of amino acids connected by amide linkages. Ester linkages are found in fats, oils, and natural waxes. Many simple esters have a pleasant odor and, in combination with other compounds, are responsible for the taste and fragrance of fruits and flowers.

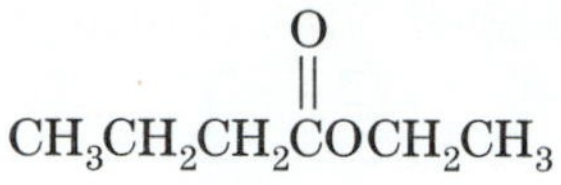

$CH_3CH_2CH_2COCH_2CH_3$ (C=O)
Pineapple odor

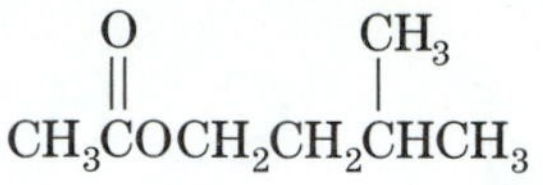

$CH_3COCH_2CH_2CH(CH_3)CH_3$ (C=O)
Banana odor

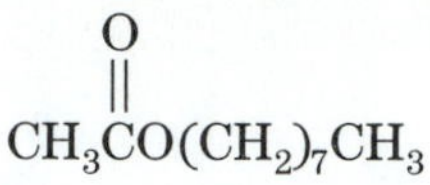

$CH_3CO(CH_2)_7CH_3$ (C=O)
Orange odor

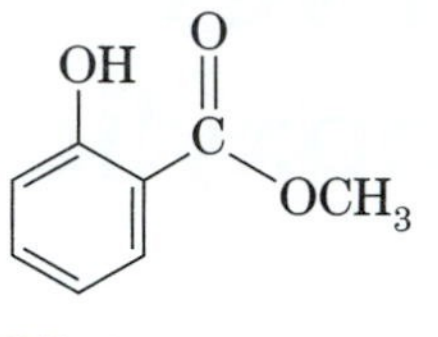

$o\text{-}HOC_6H_4COOCH_3$
Oil of wintergreen

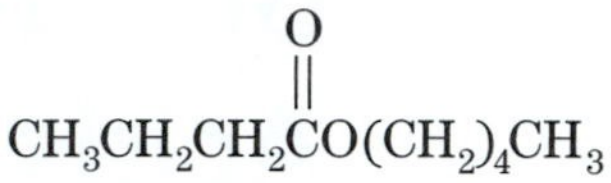

$CH_3CH_2CH_2CO(CH_2)_4CH_3$ (C=O)
Apricot odor

$HCOCH_2CH_3$ (C=O)
Artificial rum flavor

Acid chlorides and acid anhydrides are very reactive compounds and, for this reason, are not found in nature. They are very useful laboratory chemicals for organic synthesis. The term *acid anhydride* results from picturing the structure as two carboxylic acid molecules minus one molecule of water.

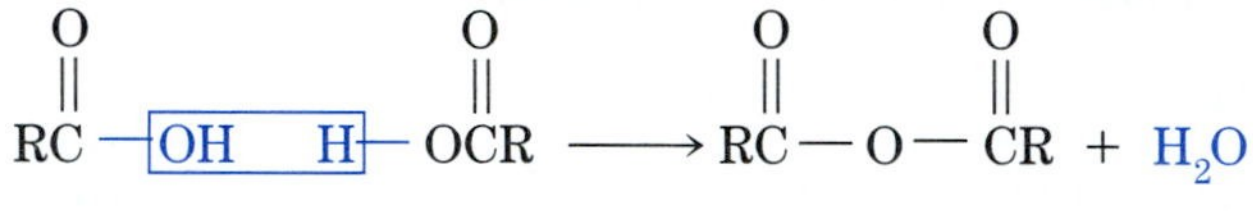

$$RC(=O){-}\boxed{OH\quad H}{-}OC(=O)R \longrightarrow RC(=O){-}O{-}C(=O)R + H_2O$$

Carboxylic acids and their derivatives engage in a variety of chemical reactions. Their chemistry is a result of the reactive sites summarized below.

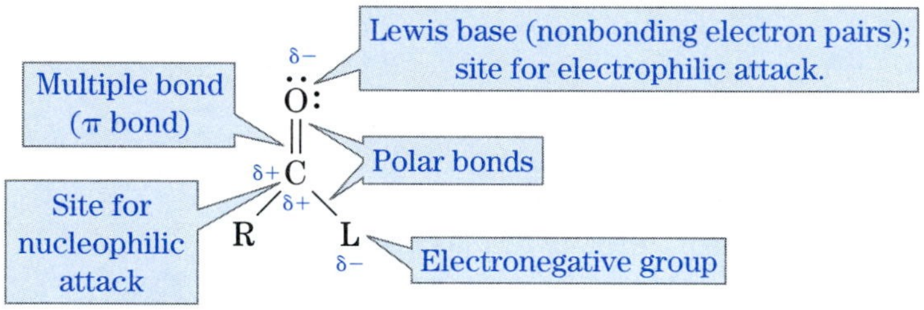

Lewis base (nonbonding electron pairs); site for electrophilic attack.
Multiple bond (π bond)
Polar bonds
Site for nucleophilic attack
Electronegative group

$$\overset{\delta-}{\ddot{O}}{:}=\overset{\delta+}{C}(R){-}\overset{\delta-}{L}$$

GETTING INVOLVED

✓ What are the structural differences among the five acid derivatives described in this section?

✓ Describe the reaction sites in a carboxylic acid derivative.

Problem 13.1

Write the structures for the compounds described:
(a) the six acids and esters with the formula $C_4H_8O_2$; **(b)** the four amides with the formula C_3H_7NO; **(c)** the two acid chlorides with the formula C_4H_7OCl; **(d)** the three anhydrides that are derivatives of ethanoic and propanoic acids.

B. Nomenclature of Carboxylic Acid Derivatives

Carboxylic acid derivatives are named by modifying the ending on the name of the parent acid. The following derivatives of propanoic acid and benzoic acid serve to illustrate this.

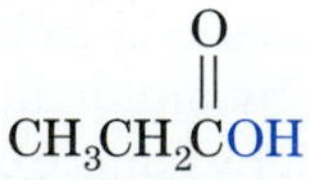

CH_3CH_2COH (C=O)
Propano*ic acid*

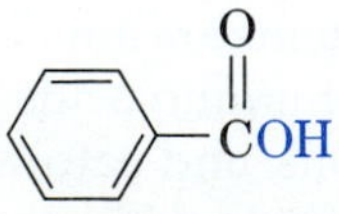

$C_6H_5{-}COH$ (C=O)
Benzo*ic acid*

1. *Acid Chlorides.* Acid chlorides are named by changing *-ic acid* to *-yl chloride.*

$$CH_3CH_2\overset{\overset{\displaystyle O}{\|}}{C}Cl \qquad C_6H_5\text{—}\overset{\overset{\displaystyle O}{\|}}{C}Cl$$

Propano*yl chloride* Benzo*yl chloride*

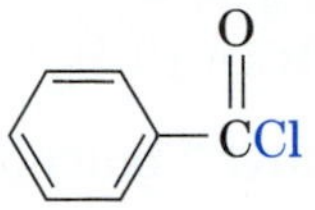

GETTING INVOLVED

Problem 13.2

Name the following acid chlorides:

(a) $CH_3(CH_2)_3\overset{\overset{\displaystyle O}{\|}}{C}Cl$ **(b)** $CH_2{=}CH\overset{\overset{\displaystyle O}{\|}}{C}Cl$ **(c)** 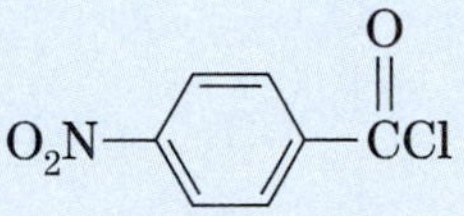 $O_2N\text{—}C_6H_4\text{—}\overset{\overset{\displaystyle O}{\|}}{C}Cl$

See related problem 13.28.

2. *Acid Anhydrides.* Acid anhydrides are named by changing the word *acid* of the parent acids to *anhydride.*

$$CH_3CH_2\overset{\overset{\displaystyle O}{\|}}{C}O\overset{\overset{\displaystyle O}{\|}}{C}CH_2CH_3 \qquad C_6H_5\text{—}\overset{\overset{\displaystyle O}{\|}}{C}O\overset{\overset{\displaystyle O}{\|}}{C}\text{—}C_6H_5 \qquad C_6H_5\text{—}\overset{\overset{\displaystyle O}{\|}}{C}O\overset{\overset{\displaystyle O}{\|}}{C}CH_2CH_3$$

Propanoic *anhydride* Benzoic *anhydride* 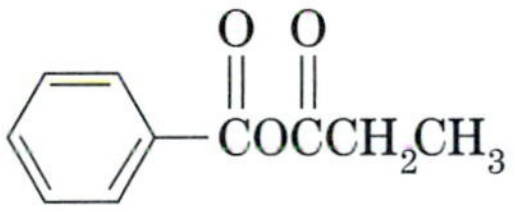 Benzoic propanoic *anhydride*

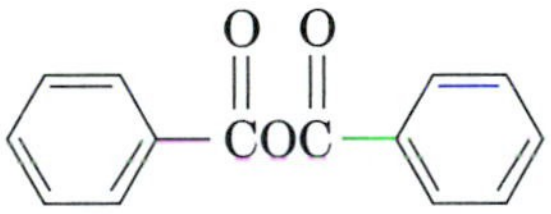

GETTING INVOLVED

Problem 13.3

Name the following acid anhydrides:

(a) $CH_3\overset{\overset{\displaystyle O}{\|}}{C}O\overset{\overset{\displaystyle O}{\|}}{C}CH_3$ **(b)** $CH_3(CH_2)_3\overset{\overset{\displaystyle O}{\|}}{C}O\overset{\overset{\displaystyle O}{\|}}{C}(CH_2)_3CH_3$ **(c)** $CH_3\overset{\overset{\displaystyle O}{\|}}{C}O\overset{\overset{\displaystyle O}{\|}}{C}(CH_2)_3CH_3$

See related problem 13.29.

3. *Esters.* Esters are named in the same way we named salts of carboxylic acids (section 12.4.D). Change the *-ic acid* to *-ate* and precede the name with the *alkyl group* attached to the ester oxygen.

$$CH_3CH_2\overset{\overset{\displaystyle O}{\|}}{C}OCH_2CH_3 \qquad C_6H_5\text{—}\overset{\overset{\displaystyle O}{\|}}{C}OCH_2CH_3$$

Ethyl propano*ate* *Ethyl* benzo*ate*

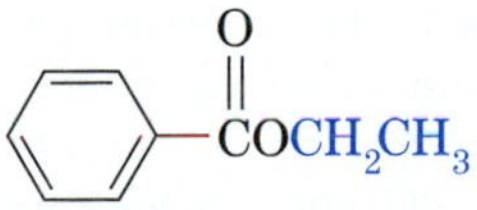

To name more complex esters and salts, mentally replace the cation or organic group with a hydrogen and name the parent acid. Then make the necessary changes to name the salt or ester. For example, let us name the following ester:

$$CH_3C(CH_3){=}CHC({=}O)OCH(CH_3)CH_3 \qquad CH_3C(CH_3){=}CHC({=}O)OH$$

The parent acid is 3-methyl-2-butenoic *acid*, and the ester is *isopropyl* 3-methyl-2-butenoate.

GETTING INVOLVED

Problem 13.4

Name the following esters:

(a) $CH_3C(=O)OCH_3$ (b) $CH_2{=}CHC(=O)OCH_2CH_3$ (c) 3-Cl-C_6H_4–$C(=O)OCH(CH_3)CH_3$

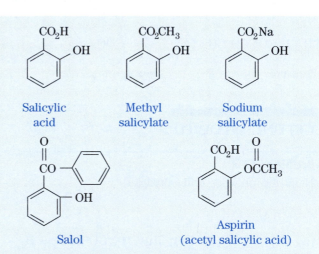

See related problems 13.27, 13.30, and 13.55.

Connections 13.1

www.prenhall.com/bailey

Aspirin and Other Analgesics

Analgesics (pain relievers) are among the most important medicinal applications of carboxylic acid derivatives. One of the oldest analgesics, a drug that is amazing for its continuing and varied usefulness, is aspirin, acetyl salicylic acid, the salicylate ester of acetic acid (see section 13.4.B).

As an antipyretic, aspirin reduces fever but does not lower normal body temperatures. Its analgesic properties are effective against pains accompanying colds, flu, nervous tension, rheumatism, and arthritis. Recent evidence suggests that continuous small doses over long periods could decrease the chances of heart problems and increase the chances of surviving a heart attack should one occur.

The name *aspirin* comes from that of a willow, *Salix spirea.* Jesuit missionaries in the Middle Ages used the bark of this tree for medicinal purposes. In the seventeenth century, it was found that extracts of willow bark had fever-reducing properties. In 1826 the active principle, salicylic acid, was isolated. By 1852 salicylic acid had been independently synthesized, and by 1874 relatively large-scale production had made it available as a medicine.

Salicylic acid is a bifunctional molecule (acid and phenol) from which many familiar substances are derived. Salicylic acid itself is used as a disinfectant in some first aid sprays and ointments, and its methyl ester, methyl salicylate (oil of wintergreen), is used in topical rubs for sore muscles. Although salicylic acid is an effective antipyretic, it causes severe stomach irritation in some people, and for this reason the search for a pain reliever continued in the late 1800s. It was hypothesized that the neutralized acid would cause less gastric irritation; so in 1875 sodium salicylate was introduced. Unfortunately, it did not prove to be any better.

Salicylic acid (CO_2H, OH) — Methyl salicylate (CO_2CH_3, OH) — Sodium salicylate (CO_2Na, OH)

Salol ($C(=O)O$–C_6H_5, OH) — Aspirin (acetyl salicylic acid) (CO_2H, $OC(=O)CH_3$)

Salol, a phenol ester of salicylic acid, was introduced in 1886, and its use did lead to greatly decreased incidence of gastric distress. In the small intestine, it hydrolyzes to sodium salicylate, which had previously been used as a pain reliever. The simultaneous liberation of phenol led to the danger of phenol poisoning, however.

Toward the end of the nineteenth century, Felix Hofmann, who worked for the Bayer Company, investigated other derivatives of salicylic acid and tested acetyl salicylic acid on his father, who suffered from arthritis. This and other tests revealed its excellent medicinal properties and a decreased frequency of gastric irritation. Acetyl salicylic acid, aspirin, was marketed in 1899 by the Bayer Company.

Unfortunately, even aspirin causes stomach distress in some individuals and minor, usually clinically unimportant, gastric or intestinal bleeding. Other products have been

Connections 13.1 *(cont.)*

introduced that do not have these unpleasant side effects. The most familiar of these is acetaminophen. It and phenacetin (both are amides and derivatives of *p*-aminophenol) are essentially equivalent to aspirin in their antipyretic and analgesic properties, but unlike aspirin, neither has a significant effect on inflamed joints caused by rheumatoid arthritis. Phenacetin has been implicated in kidney damage, and though it was once a popular ingredient in APC (aspirin and phenacetin and caffeine) tablets, its use has been largely discontinued. Ibuprofen in low-strength doses is a relative newcomer to the nonprescription pain reliever market, although it was available as a prescription drug for some time.

Combination pain relievers are preparations in which aspirin is combined with other pain relievers, stomach antacids, or both. Acetaminophen, salicylamide, and caffeine are commonly found along with aspirin in these products. Salicylamide is much less effective than aspirin and too weak and unreliable to be generally useful as a pain reliever alone. The rationale for adding caffeine is still not completely clear.

Antacids are added to pain relievers to raise gastric pH and thus minimize stomach upset (the extent of this effect is controversial) and to accelerate tablet dissolution. Antacids found in pain relievers or over-the-counter antacid preparations include $NaHCO_3$ (baking soda, bicarbonate of soda), $CaCO_3$ (calcium carbonate), $Mg(OH)_2$ (milk of magnesia), $Al(OH)_3$ (aluminum hydroxide), $NaAl(OH)_2CO_3$ (dihydroxyaluminum sodium carbonate), and $Mg_2Si_3O_8$ (magnesium trisilicate).

$C(=O)NH_2$, OH — Salicylamide

$NHC(=O)CH_3$, OCH_2CH_3 — Phenacetin

$NHC(=O)CH_3$, OH — Acetaminophen

CH_3, O, CH_3, N, N, N, N, O, CH_3 — Caffeine

CO_2H, $CHCH_3$, $CH_2CH(CH_3)_2$ — Ibuprofen

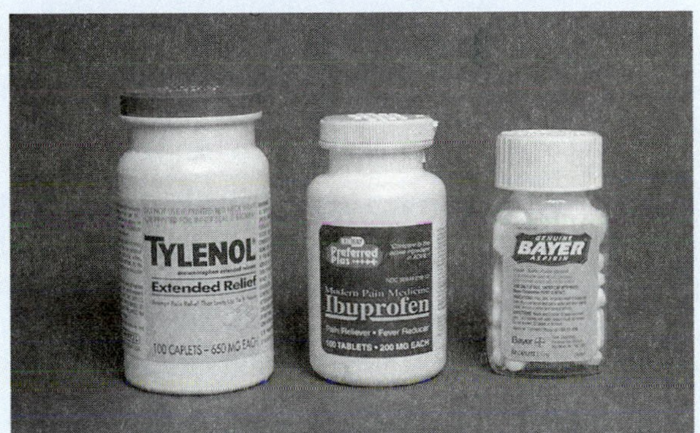

Acetaminophen, ibuprofen, and aspirin

4. *Amides.* Amides are named by changing *-oic acid* to *-amide.*

$$CH_3CH_2\overset{O}{\overset{\|}{C}}NH_2 \qquad C_6H_5\!-\!\overset{O}{\overset{\|}{C}}NH_2$$

Propan*amide* Benz*amide*

Substituted amides are named merely by locating the position of any substituents. For example, the following amide is a derivative of *p*-nitrobenzoic acid and is named as shown:

$$O_2N\!-\!C_6H_4\!-\!\overset{O}{\overset{\|}{C}}OH \qquad O_2N\!-\!C_6H_4\!-\!\overset{O}{\overset{\|}{C}}\!-\!\underset{}{\overset{CH_3}{\overset{|}{N}}}\!-\!CH_2CH_3$$

p-nitrobenzoic acid N-ethyl-N-methyl-*p*-nitrobenzamide

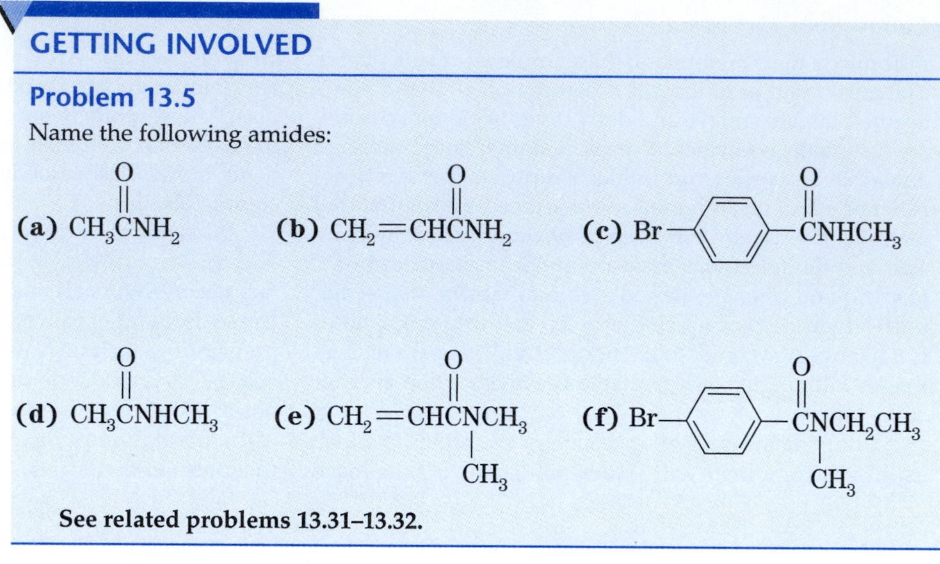

13.2 Nucleophilic Acyl Substitution Reactions

A. The Reaction

nucleophilic acyl substitution nucleophilic substitution in which an atom or group attached to an acyl group, RC=O, is replaced

Nucleophilic acyl substitution reactions are similar in some respects to the nucleophilic alkyl substitution reactions we studied in Chapter 8 (section 8.4). In the latter, a negative or neutral nucleophile replaced a leaving group, usually halide ion in the cases we studied, to produce the final product. The reaction proceeded by an S_N1 or S_N2 mechanism depending on the structure of the alkyl halide and reaction conditions.

Nucleophilic Alkyl Substitution

$$-\overset{|}{\underset{|}{C}}-X \;+\; \begin{matrix}Nu:^- \\ \text{or} \\ HNu:\end{matrix} \;\longrightarrow\; -\overset{|}{\underset{|}{C}}-Nu \;+\; \begin{matrix}X^- \\ \text{or} \\ HX\end{matrix}$$

Alkyl halide
X = Cl, Br, I

Nu = OH, OR, SH, SR, NH_2, NHR, NR_2, CN, and others

In nucleophilic acyl substitution a nucleophile, either negative or neutral, also replaces a leaving group to form the substitution product. Because of the structures of the effective nucleophiles and leaving groups, the reaction usually involves the conversion of one acid derivative into another, in most cases, one that is less reactive.

Nucleophilic Acyl Substitution

$$\underset{\text{Usually a more reactive acid derivative}}{R\overset{O}{\overset{\|}{C}}L} \;+\; \begin{matrix}Nu:^- \\ \text{or} \\ H{-}Nu:\end{matrix} \;\longrightarrow\; \underset{\text{Usually a less reactive acid derivative}}{R\overset{O}{\overset{\|}{C}}Nu} \;+\; \begin{matrix}L:^- \\ \text{or} \\ H{-}L\end{matrix}$$

$$L = Cl,\ \overset{O}{\overset{\|}{OCR}},\ OH,\ OR,\ NR_2 \qquad H{-}Nu = \overset{O}{\overset{\|}{HOCR}},\ H{-}OH,\ H{-}OR,\ H{-}NR_2$$

Order of Reactivity: Most reactive $R\overset{O}{\overset{\|}{C}}Cl > R\overset{O}{\overset{\|}{C}}O\overset{O}{\overset{\|}{C}}R > R\overset{O}{\overset{\|}{C}}OH, R\overset{O}{\overset{\|}{C}}OR > R\overset{O}{\overset{\|}{C}}NR_2$ Least reactive

Acid chloride > Acid anhydride > Acid, Ester > Amide

Nucleophilic acyl substitution reactions are summarized in Figure 13.1. You will notice that if an acid derivative reacts with water, a carboxylic acid is produced; with an alcohol, an ester is formed; and with an amine, an amide results.

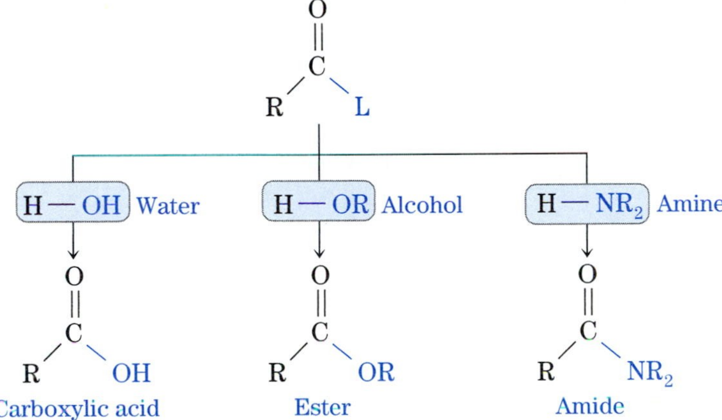

GETTING INVOLVED

- ✓ How are nucleophilic alkyl substitution and nucleophilic acyl substitution similar and different?
- ✓ What are the possible leaving groups in nucleophilic acyl substitution? What acid derivative is formed when an acid derivative reacts with water? alcohols? amines?
- ✓ In nucleophilic acyl substitution, acid derivatives are interconverted. What does this mean (take a preliminary look at Figure 13.1 and then come back to it after you have studied all of the reactions)?

B. The Reaction Mechanism

Nucleophilic acyl substitution reactions can involve either a negative or neutral nucleophile; the nucleophile is strongly attracted to the partially positive carbonyl carbon. Let's look at the reaction mechanism using a negative nucleophile. This process can be compared to the S_N2 mechanism of nucleophilic alkyl substitution. Recall that in the S_N2 mechanism, the nucleophile-carbon bond forms at the same time that the carbon-halide bond is breaking; the nucleophile enters and the halide leaves in one step (section 8.4.D).

S_N2 Mechanism: One Step

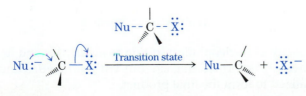

$$Nu:^- + C{-}\ddot{X}: \xrightarrow{\text{Transition state: } Nu{-}{-}C{-}{-}\ddot{X}:} Nu{-}C + :\ddot{X}:^-$$

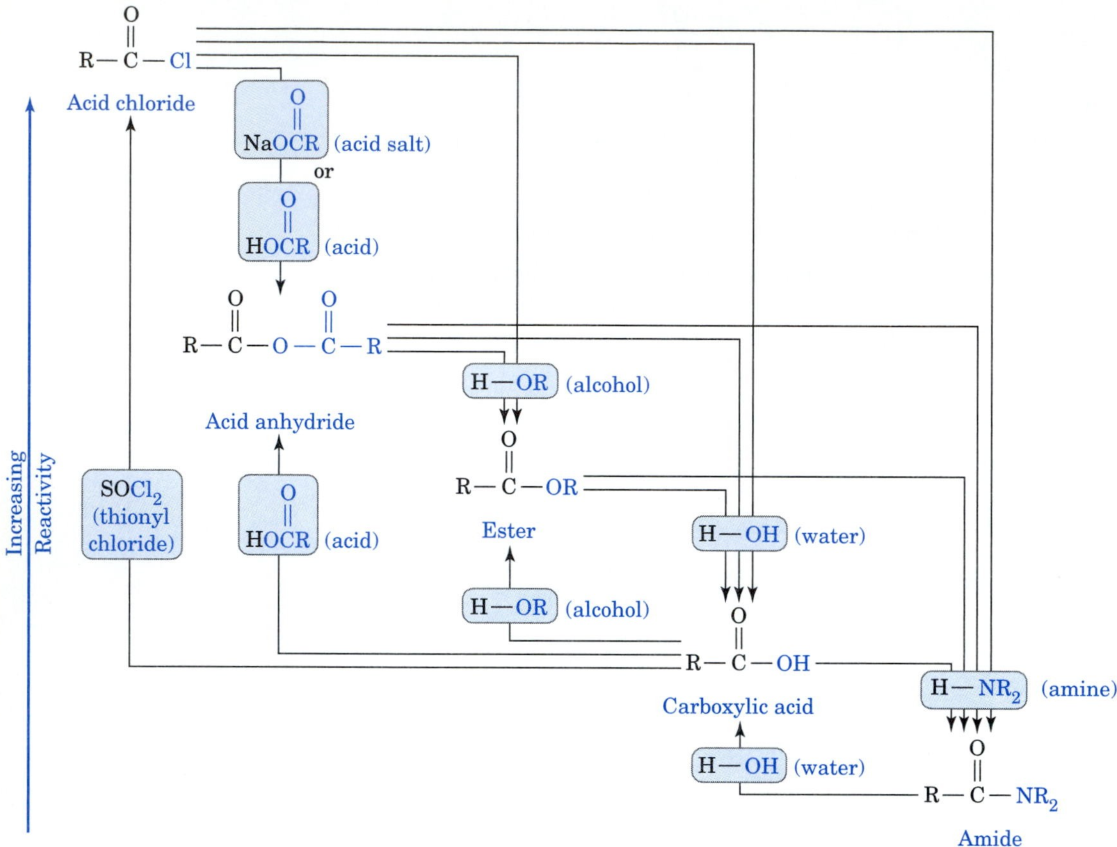

Figure 13.1 *Interconversions of Acid Derivatives:* Only several reactants are shown, not reaction conditions. Esters and acids are of similar reactivity. Converting a less reactive derivative into one that is more reactive is more difficult than converting a more reactive derivative into a less reactive one.

In nucleophilic acyl substitution, the nucleophile is attracted to the partially positive carbonyl carbon as it is to the alkyl carbon in the S_N2 process. Unlike the S_N2 mechanism, however, the nucleophile actually bonds by an addition reaction similar to the nucleophilic addition mechanism we studied with aldehydes and ketones (section 11.5.B). The leaving group then departs; the mechanism is two steps instead of one.

Nucleophilic Acyl Substitution: Two Steps
Negative Nucleophile

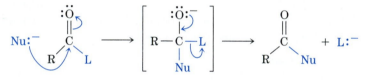

Nu:⁻ + R—C(=O)—L ⟶ [R—C(—O:⁻)(—L)—Nu] ⟶ R—C(=O)—Nu + L:⁻

Tetrahedral intermediate

Note the similarity of nucleophilic addition to aldehydes and ketones. A tetrahedral intermediate is formed in each. But with aldehydes and ketones, it is merely neutralized to form the final product.

Nucleophilic Addition to Aldehydes and Ketones: Two Steps

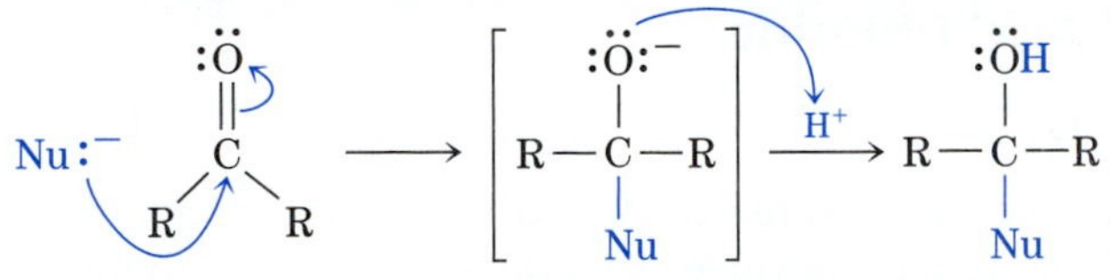

Tetrahedral intermediate

A tetrahedral intermediate is formed in nucleophilic acyl substitution using a neutral nucleophile. Proton transfers accompany the departure of the leaving group.

Nucleophilic Acyl Substitution: Two Steps
Neutral Nucleophile

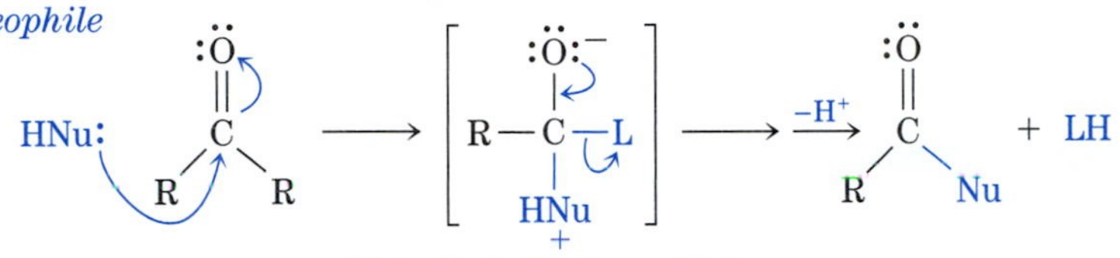

Tetrahedral intermediate

We shall also see that with some of the less reactive acid derivatives, acid catalysis promotes the reaction by forming a positive species which more strongly attracts nucleophiles. A tetrahedral intermediate appears in the mechanism.

Nucleophilic Acyl Substitution
Acid Catalysis, Neutral Nucleophile

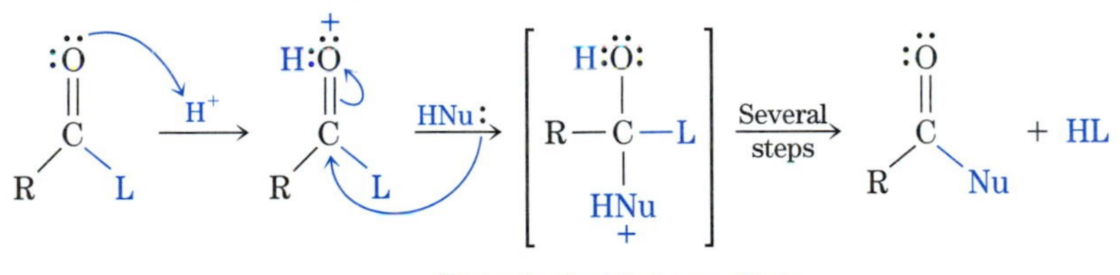

Tetrahedral intermediate

In the following sections, we will examine the specific reactions and reaction mechanisms for each of the common acid derivatives.

GETTING INVOLVED

✓ Do you see that the nucleophilic acyl substitution mechanism can be initiated by a negative nucleophile, a neutral nucleophile, or by acid catalysis with a neutral nucleophile? Do you see that all three possibilities proceed through a tetrahedral intermediate? Do you see that in each case one acid derivative is converted into another?

✓ Take a moment to compare nucleophilic acyl substitution to nucleophilic addition to aldehydes and ketones. Remember these reactions can be base initiated with a negative or neutral nucleophile, or acid initiated with a neutral nucleophile, and that all proceed through a tetrahedral intermediate similar to nucleophilic acyl substitution.

✓ Finally compare to nucleophilic alkyl substitution; remember S_N2 reactions occur with both negative and neutral nucleophiles. What is the difference between nucleophilic alkyl substitution and nucleophilic acyl substitution?

13.3 Nucleophilic Acyl Substitution Reactions of Acid Chlorides

A. Synthesis

Acid chlorides are the most reactive of carboxylic acid derivatives. Because of this, it is relatively simple to produce other derivatives from them, but special methods are required for their synthesis. One such method involves the reaction of carboxylic acids with thionyl chloride. Although a more reactive substance (an acid chloride) is produced from a less reactive one (a carboxylic acid), the reaction is not reversible since the by-products are gases, which escape the reaction solution.

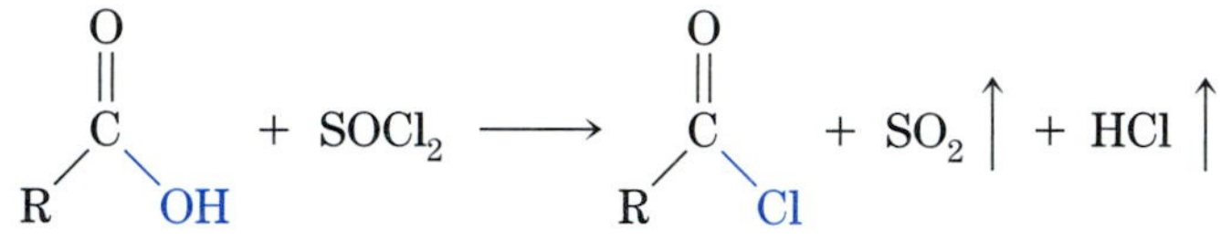

$$RCOOH + SOCl_2 \longrightarrow RCOCl + SO_2\uparrow + HCl\uparrow$$

GETTING INVOLVED

Problem 13.6
Write an equation showing the preparation of benzoyl chloride from benzoic acid.

B. Reactions

As the most reactive of the carboxylic acid derivatives, acid chlorides are useful in the synthesis of the other derivatives. Reaction with the sodium salt of a carboxylic acid is the preferred method for producing acid anhydrides. The other reactions are common to most carboxylic acid derivatives. Reaction with an alcohol produces an ester; with water, a carboxylic acid; and with ammonia or an amine, an amide. HCl is the inorganic by-product in all of these reactions.

Reactions of Acid Chlorides

$RCOCl$

- Acid salt: $NaOCR$ (O over C) → $RCOOCR$ (O over each C) — Acid anhydride + NaCl
- Alcohol: $H—OR$ → $RCOOR$ — Ester + HCl
- Water: $H—OH$ → $RCOOH$ — Acid + HCl
- Amine: $H—NR_2$ → $RCONR_2$ — Amide + HCl

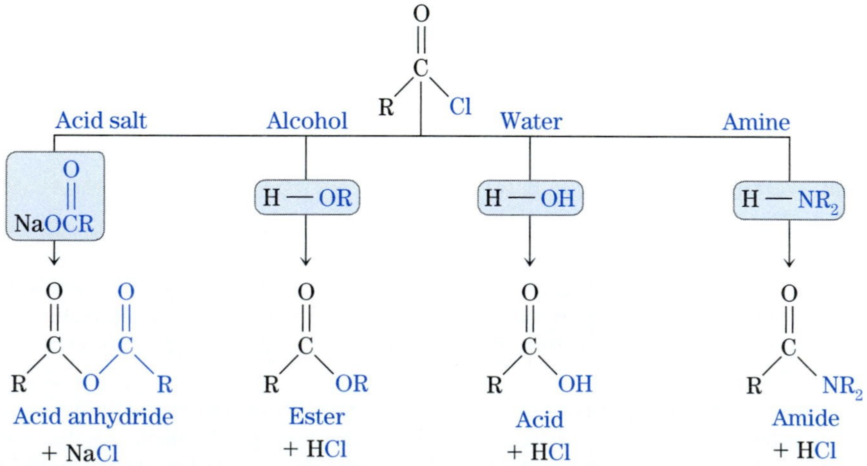

Let us take a look at some specific examples. Remember, an oxygen or nitrogen with a lone pair of electrons is the nucleophile and replaces the chloride. The polar O—H bond of water, an alcohol, or the N—H bond of an amine cleaves in the reaction; HCl is the by-product. Reaction of butanoyl chloride with ethanol produces the ester ethyl butanoate, an ester that has the odor of pineapple.

$$CH_3CH_2CH_2\overset{O}{\overset{\|}{C}}-Cl + H-OCH_2CH_3 \longrightarrow CH_3CH_2CH_2\overset{O}{\overset{\|}{C}}-OCH_2CH_3 + HCl$$

Acetyl chloride and water form acetic acid, the acid found in vinegar.

$$CH_3\overset{O}{\overset{\|}{C}}-Cl + H-OH \longrightarrow CH_3\overset{O}{\overset{\|}{C}}-OH + HCl$$

Acetyl chloride and *p*-hydroxyaniline produce the nonprescription pain reliever acetaminophen, an amide.

$$CH_3\overset{O}{\overset{\|}{C}}-Cl + H-\overset{H}{\overset{|}{N}}-C_6H_4-OH \longrightarrow CH_3\overset{O}{\overset{\|}{C}}-\overset{H}{\overset{|}{N}}-C_6H_4-OH + HCl$$

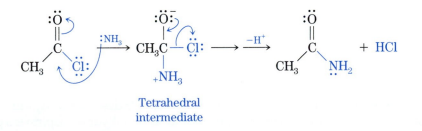

GETTING INVOLVED

✓ Be sure you see that an acid chloride can be converted into all the other acid derivatives. Why? What derivative is formed when an acid chloride reacts with each of the following: an acid or acid salt; water; an alcohol; an amine? What is the inorganic by-product?

Problem 13.7

Write equations showing the reactions of ethanoyl chloride with the following: **(a)** H_2O; **(b)** CH_3CH_2OH; **(c)** CH_3CO_2Na; **(d)** NH_3; **(e)** CH_3NH_2; **(f)** $CH_3CH_2NHCH_2CH_3$.
See related problem 13.34.

C. Nucleophilic Acyl Substitution Mechanism

Consider the reaction of the Lewis base ammonia with ethanoyl chloride. As a strong nucleophile, the ammonia attacks the partially positive carbonyl carbon and bonds, using its lone pair of electrons. Elimination of HCl produces the amide product.

$$CH_3\overset{:\ddot{O}}{\overset{\|}{C}}-\ddot{\underset{..}{Cl}}: \xrightarrow{:NH_3} CH_3\overset{:\ddot{O}:^-}{\overset{|}{\underset{\underset{+NH_3}{|}}{C}}}-\ddot{\underset{..}{Cl}}: \longrightarrow \xrightarrow{-H^+} CH_3\overset{:\ddot{O}}{\overset{\|}{C}}-\underset{..}{NH_2} + HCl$$

Tetrahedral intermediate

GETTING INVOLVED

✓ Note in the mechanism the neutral nucleophile and tetrahedral intermediate.

Problem 13.8

Show the mechanism for the reaction of water with ethanoyl chloride.
See related problem 13.44(a).

13.4 Nucleophilic Acyl Substitution Reactions of Acid Anhydrides

A. Synthesis of Acid Anhydrides

Acid anhydrides are best prepared by the reaction of an acid chloride and a salt of a carboxylic acid in a nucleophilic acyl substitution reaction as shown in section 13.3.B.

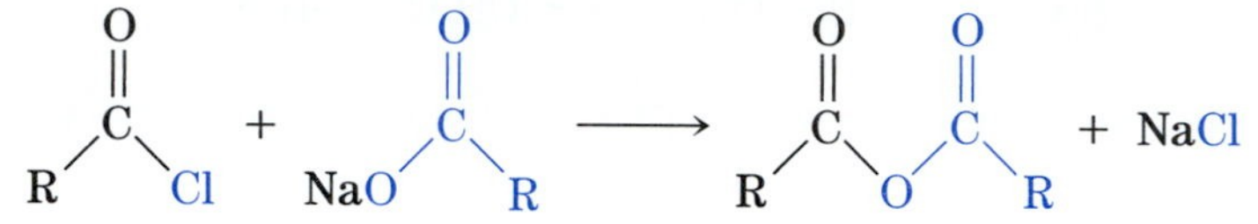

$$RC(=O)Cl + NaOC(=O)R \longrightarrow RC(=O)OC(=O)R + NaCl$$

GETTING INVOLVED

Problem 13.9

Show two ways to make ethanoic propanoic anhydride.

B. Reactions of Acid Anhydrides

Acid anhydrides, like acid chlorides, react with water to form carboxylic acids, with alcohols to form esters, and with ammonia or amines to form amides. They differ from acid chlorides in these reactions only in the by-product, which is a carboxylic acid instead of hydrogen chloride.

Reactions of Acid Anhydrides

$$RC(=O)OC(=O)R$$

Alcohol: $H{-}OR \rightarrow RC(=O)OR$ (Ester) and $RC(=O)OH$

Water: $H{-}OH \rightarrow RC(=O)OH$ (Acid) and $RC(=O)OH$

Amine: $H{-}NR_2 \rightarrow RC(=O)NR_2$ (Amide) and $RC(=O)OH$

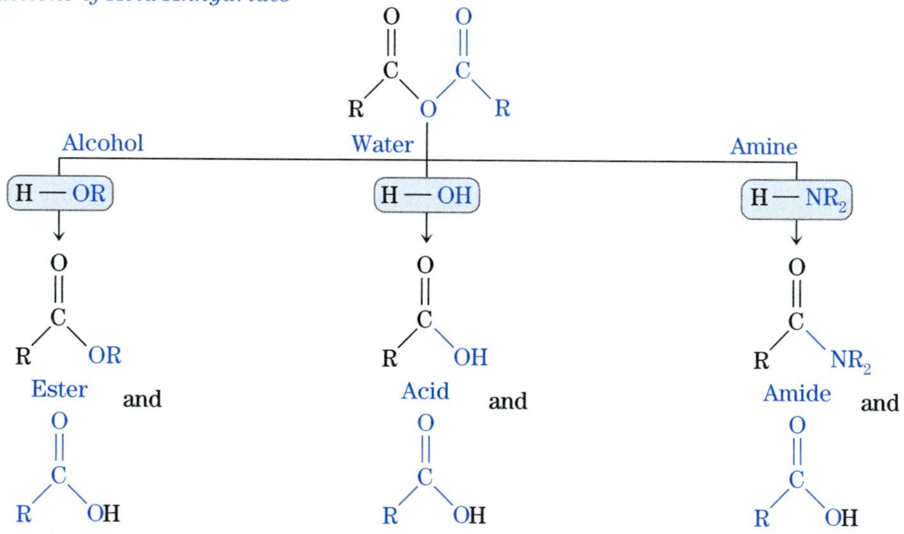

Some specific examples of the reactions of acid anhydrides follow. Aspirin is the result of the reaction below between a phenol and acetic anhydride; aspirin is a phenolic ester. Acetic anhydride and *p*-aminophenol react to produce acetaminophen.

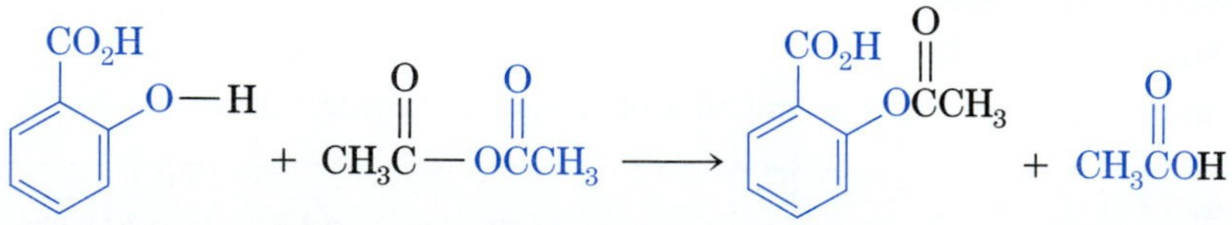

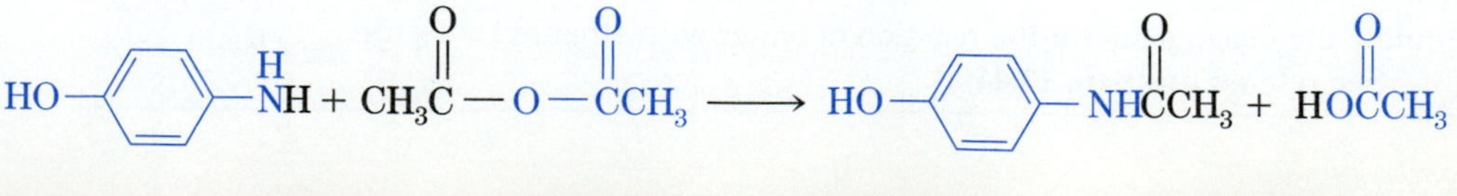

$$o\text{-}HO_2C{-}C_6H_4{-}O{-}H + CH_3C(=O){-}OC(=O)CH_3 \longrightarrow o\text{-}HO_2C{-}C_6H_4{-}OC(=O)CH_3 + CH_3C(=O)OH$$

$$HO{-}C_6H_4{-}NH(H) + CH_3C(=O){-}O{-}C(=O)CH_3 \longrightarrow HO{-}C_6H_4{-}NHC(=O)CH_3 + HOC(=O)CH_3$$

GETTING INVOLVED

✓ Compare the products of acid anhydrides with alcohols, water and amines with those of acid chlorides. Also compare the by-products. Look at both charts.

Problem 13.10

Write equations describing the reaction of ethanoic anhydride ($CH_3\overset{O}{\overset{\|}{C}}O\overset{O}{\overset{\|}{C}}CH_3$) with **(a)** H_2O; **(b)** CH_3CH_2OH; **(c)** NH_3; **(d)** CH_3NH_2.

See related problem 13.35.

C. Nucleophilic Acyl Substitution Mechanism

The nucleophilic acyl substitution reaction mechanism involves a Lewis base (alcohol, water, ammonia, or amine) attacking and adding to the partially positive carbonyl carbon, using a lone pair of electrons. A tetrahedral intermediate results, which quickly eliminates a molecule of carboxylic acid to form the final product. This mechanism is illustrated using ethanoic anhydride and methanol.

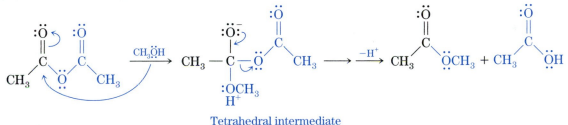

$$CH_3COCOCH_3 \xrightarrow{CH_3\ddot{O}H} CH_3-\underset{\underset{H^+}{:OCH_3}}{\overset{:\ddot{O}:^-}{C}}-\ddot{O}-\overset{:\ddot{O}}{\overset{\|}{C}}CH_3 \longrightarrow \xrightarrow{-H^+} CH_3\overset{:\ddot{O}}{\overset{\|}{C}}\ddot{O}CH_3 + CH_3\overset{:\ddot{O}}{\overset{\|}{C}}\ddot{O}H$$

Tetrahedral intermediate

GETTING INVOLVED

✓ Notice again the mechanism involves attack by a neutral nucleophile to form a tetrahedral intermediate.

Problem 13.11

Write the nucleophilic acyl substitution mechanism for the reaction between ethanoic anhydride and ammonia.

See related problem 13.44(b).

13.5 Nucleophilic Acyl Substitution Reactions of Carboxylic Acids

The preparations of carboxylic acids are summarized in section 12.5. In addition to these, carboxylic acids can be produced from any acid derivative—acid chloride, acid anhydride, ester, or amide—by treatment with water (hydrolysis).

A. Reactions of Carboxylic Acids

Carboxylic acids react with alcohols to produce esters and with ammonia and amines to produce amides. With thionyl chloride they can be used to prepare the more reactive acid derivatives, acid chlorides (section 13.3.A). Since the leaving group is OH in these reactions, water is the by-product in most.

Reactions of Carboxylic Acids

Thionyl chloride | Alcohol | Amine

$$RCOOH \xrightarrow{SOCl_2} RCOCl \text{ (Acid chloride)} + SO_2 + HCl$$

$$RCOOH \xrightarrow{H-OR} RCOOR \text{ (Ester)} + H_2O$$

$$RCOOH \xrightarrow{H-NR_2} RCONR_2 \text{ (Amide)} + H_2O$$

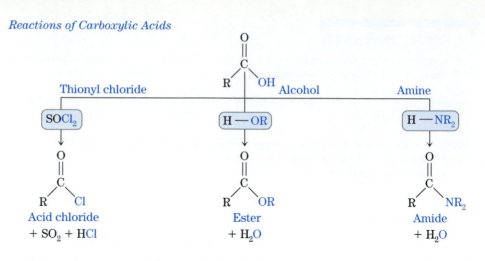

An example of esterification is the reaction of ethanoic acid, which is responsible for the smell of vinegar, and 3-methylbutanol to produce an ester with the aroma of bananas. Because of the diminished reactivity of carboxylic acids compared to acid chlorides and anhydrides, an acid, such as sulfuric acid, is needed to catalyze this reaction.

$$CH_3C(=O)-OH + H-OCH_2CH_2CH(CH_3)CH_3 \xrightarrow{H^+} CH_3C(=O)-OCH_2CH_2CH(CH_3)CH_3 + H_2O$$

Salicylic acid reacts with ammonia, forming the pain reliever salicylamide.

$$HO-C_6H_4-C(=O)-OH + H-NH_2 \xrightarrow[H^+]{Heat} HO-C_6H_4-C(=O)-NH_2 + H_2O$$

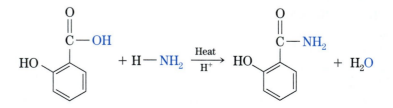

GETTING INVOLVED

Problem 13.12

Write equations showing the reactions of propanoic acid with **(a)** thionyl chloride; **(b)** methanol; **(c)** ammonia; **(d)** methanamine.

Problem 13.13

From what carboxylic acid and amine could the following amide be prepared?

$$C_6H_5-C(=O)N(CH_3)_2$$

See related problem 13.36.

B. Nucleophilic Acyl Substitution Mechanism

The most important step in any nucleophilic acyl substitution reaction is the attack of the nucleophile on the carbonyl carbon. Carboxylic acids are less reactive than acid chlorides and anhydrides, and thus the esterification reaction must be catalyzed with strong acid. One role of the catalyst is to make the carbonyl carbon more attractive to the nucleophile; in the first step of the mechanism, protonation results in a positively charged intermediate and that activates the carbonyl carbon to attack by the nucleophile. Following is the acid-catalyzed esterification of acetic acid with ethyl alcohol, which produces ethyl acetate, a component of fingernail polish remover.

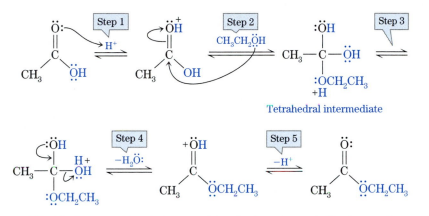

Step 1 H^+ — Step 2 $CH_3CH_2\ddot{O}H$ — Step 3

Tetrahedral intermediate

Step 4 $-H_2\ddot{O}:$ — Step 5 $-H^+$

Step 1: The reaction is initiated by the bonding of a hydrogen ion to the partially negative oxygen of the carbonyl group.

Step 2: The Lewis base ethanol, the nucleophile, is attracted to the partially positive acyl carbon. A tetrahedral intermediate results.

Step 3: A simple hydrogen ion transfer occurs.

Step 4: A water molecule is lost. The leaving group is gone.

Step 5: Loss of a hydrogen ion in the last step results in the ester.

Basically, the first two steps involve the addition of ethanol, and the last two steps involve the elimination of water to form the ester. Hydrogen ion is truly a catalyst—it initiates the reaction in the first step and is returned in the last. Note that each step in the mechanism is reversible; the reverse of this process is the mechanism for the acid-catalyzed **hydrolysis** of an ester.

hydrolysis
cleavage of a bond by water

GETTING INVOLVED

✓ Look at the esterification mechanism and make sure you understand that it is an acid catalyzed nucleophilic acyl substitution with the normal tetrahedral intermediate. Why is acid catalysis necessary with ester formation from acids?

Problem 13.14

Write a step-by-step mechanism for the reaction between benzoic acid and methanol with an acid catalyst.

See related problems 13.44(e) and 13.60.

13.6 Nucleophilic Acyl Substitution Reactions of Esters

Esters can be prepared readily from any acid derivative except amides (see Figure 13.1). In each case the acid derivative is mixed with the appropriate alcohol. The reaction conditions may vary depending on the reactivity of the acid derivative and, of course, the by-product depends on which derivative is employed.

A. Reactions of Esters

transesterification conversion of one ester into another by replacing the OR group

Esters can be converted to carboxylic acids with water, to other esters by treatment with an alcohol (a process called **transesterification**), and to amides upon reaction with amines. In each case the by-product is a molecule of alcohol corresponding to the structure of the leaving group.

Reactions of Esters

$$RC(=O)OR$$

Water	Alcohol	Amine
$H—OH$	$H—OR'$	$H—NR_2$
↓	↓	↓
$RC(=O)OH$	$RC(=O)OR'$	$RC(=O)NR_2$
Carboxylic acid + ROH	Ester + ROH	Amide + ROH

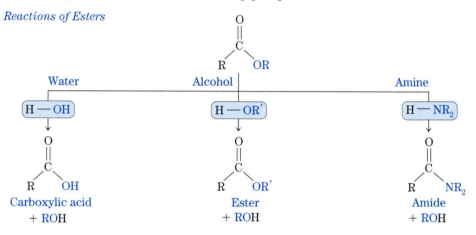

Since carboxylic acids and esters have similar reactivities, their interconversion is an equilibrium process. For example, consider the acid-catalyzed hydrolysis of ethyl acetate, fingernail polish remover.

$$CH_3\overset{O}{\overset{\|}{C}}—OCH_2CH_3 + H—OH \overset{H^+}{\rightleftharpoons} CH_3\overset{O}{\overset{\|}{C}}—OH + HOCH_2CH_3$$

You may have noticed that we used the reverse of this reaction as an example of esterification in the previous section. The direction of the reaction depends on the reaction conditions. To effect hydrolysis as shown, we use an excess of water to force the equilibrium to the right; to do this, water could be the solvent. To shift the equilibrium to the left, we use an excess of either acetic acid or ethanol; using ethanol as the solvent will cause the shift. Similar considerations also apply to the transesterification reaction.

Amides are formed when an ester is treated with ammonia or an amine. This is not a preferred method for the synthesis of amides, as they are more easily produced from acid chlorides or acid anhydrides.

$$CH_3\overset{O}{\overset{\|}{C}}OCH_2CH_3 + NH_3 \longrightarrow CH_3\overset{O}{\overset{\|}{C}}NH_2 + CH_3CH_2OH$$

GETTING INVOLVED

✓ Compare the reactions of acid chlorides, acid anhydrides, acids, and esters with water, alcohols, and amines. Compare the acid derivative formed in each case and the by-products.

Problem 13.15

Write equations indicating the reaction of ethyl ethanoate ($CH_3\overset{O}{\overset{\|}{C}}OCH_2CH_3$) with **(a)** H_2O/H^+; **(b)** CH_3OH/H^+; **(c)** NH_3; **(d)** $CH_3CH_2NHCH_2CH_3$.

See related problems 13.37 and 13.50–13.51.

B. Nucleophilic Acyl Substitution Mechanism

As an example of this mechanism, let us consider the base-catalyzed hydrolysis of ethyl acetate. This type of reaction is called *saponification* because when it is applied to fats and oils (ester of long-chain, "fatty" acids), soap is produced (section 15.4.C). The reaction is not reversible, since the salt of a carboxylic acid results. The negative carboxylate ion does not attract nucleophiles.

$$CH_3\overset{O}{\overset{\|}{C}}OCH_2CH_3 + NaOH/H_2O \longrightarrow CH_3\overset{O}{\overset{\|}{C}}O^-Na^+ + CH_3CH_2OH$$

The mechanism is initiated by attack of the hydroxide ion, a negative nucleophile, to produce an unstable tetrahedral intermediate. Ethoxide departs, making the substitution complete. The carboxylic acid that results is quickly neutralized by the very basic ethoxide ion.

$$CH_3\overset{:\ddot{O}}{\overset{\|}{C}}OCH_2CH_3 \xrightarrow{:\ddot{O}H^-} CH_3\overset{:\ddot{O}:^-}{\underset{OH}{C}}-\ddot{O}CH_2CH_3 \longrightarrow CH_3\overset{\ddot{O}:}{\overset{\|}{C}}OH + CH_3CH_2\ddot{O}:^- \longrightarrow CH_3CH_2\ddot{O}H + CH_3\overset{:\ddot{O}}{\overset{\|}{C}}\ddot{O}:^-$$

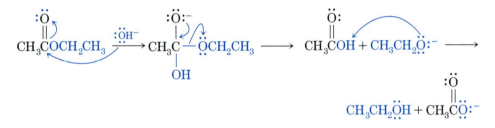

GETTING INVOLVED

✓ Do you see that this mechanism is initiated by a negative nucleophile and that a tetrahedral intermediate results? Why is the reaction irreversible?

Problem 13.16

Write reaction mechanisms illustrating the hydrolysis of methyl benzoate under **(a)** acidic conditions (see section 13.5B for help; this mechanism is the reverse) and **(b)** basic conditions.

See related problems 13.44(c–d).

C. Synthesis of Esters by Nucleophilic Acyl Substitution

How does one determine what materials to use in the synthesis of a particular ester, such as ethyl benzoate? First, focus your attention on the carbon-oxygen double bond, as this is common to all acid derivatives. The singly bonded oxygen came from an alcohol; mentally break the bond and place a hydrogen on the oxygen. The rest of the molecule comes from the carboxylic acid; mentally put an OH on it.

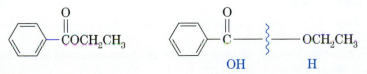

$$C_6H_5\overset{O}{\overset{\|}{C}}OCH_2CH_3 \qquad C_6H_5\overset{O}{\overset{\|}{C}}\underset{OH}{} \;\wr\; \underset{H}{O}CH_2CH_3$$

The ester can be prepared from benzoic acid and ethanol under acid conditions. Or, if you wished to use an acid chloride, it could be prepared from benzoyl chloride and ethanol.

$$C_6H_5\text{—}\overset{\overset{\large O}{\|}}{C}\text{—}OH + CH_3CH_2OH \xrightarrow{H^+} C_6H_5\text{—}\overset{\overset{\large O}{\|}}{C}\text{—}OCH_2CH_3 + H_2O$$

$$C_6H_5\text{—}\overset{\overset{\large O}{\|}}{C}\text{—}Cl + CH_3CH_2OH \longrightarrow C_6H_5\text{—}\overset{\overset{\large O}{\|}}{C}\text{—}OCH_2CH_3 + HCl$$

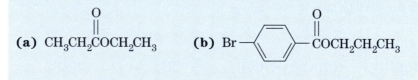

GETTING INVOLVED

Problem 13.17

Write the structure of the carboxylic acid and alcohol from which each of the following esters can be produced:

(a) $CH_3CH_2\overset{\overset{\large O}{\|}}{C}OCH_2CH_3$ **(b)** $Br\text{—}C_6H_4\text{—}\overset{\overset{\large O}{\|}}{C}OCH_2CH_2CH_3$

See related problems 13.40–13.42 and 13.55.

13.7 Nucleophilic Acyl Substitution Reactions of Amides

Amides are the least reactive of the acid derivatives and consequently do not engage in reactions as extensively as other acid derivatives do. Amides can be hydrolyzed to acids, however, under either acidic or basic conditions, with prolonged heating.

$$R\overset{\overset{\large O}{\|}}{C}\text{—}NH_2 + H\text{—}OH \xrightarrow{H^+} R\overset{\overset{\large O}{\|}}{C}OH + NH_4^+$$

$$R\overset{\overset{\large O}{\|}}{C}\text{—}NH_2 + H\text{—}OH \xrightarrow{OH^-} R\overset{\overset{\large O}{\|}}{C}O^- + NH_3$$

Acid hydrolysis gives an acid and an ammonium salt (since ammonia is basic), whereas in base, free ammonia and the acid salt result.

As the least reactive acid derivative, amides can be prepared from each of the others as illustrated in Example 13.1. Notice that in each of these reactions the same product is formed, but the by-products differ depending on the acid derivative used.

GETTING INVOLVED

Example 13.1

Write reaction equations showing the preparation of ethanamide from an acid chloride, an acid anhydride, and an ester.

Solution

$$CH_3\overset{\overset{\displaystyle O}{\|}}{C}Cl + NH_3 \longrightarrow CH_3\overset{\overset{\displaystyle O}{\|}}{C}NH_2 + HCl$$

$$CH_3\overset{\overset{\displaystyle O}{\|}}{C}O\overset{\overset{\displaystyle O}{\|}}{C}CH_3 + NH_3 \longrightarrow CH_3\overset{\overset{\displaystyle O}{\|}}{C}NH_2 + CH_3\overset{\overset{\displaystyle O}{\|}}{C}OH$$

$$CH_3\overset{\overset{\displaystyle O}{\|}}{C}OCH_3 + NH_3 \longrightarrow CH_3\overset{\overset{\displaystyle O}{\|}}{C}NH_2 + CH_3OH$$

Problem 13.18

Write reaction equations for the acidic and basic hydrolysis of ethanamide.

Problem 13.19

Write equations illustrating the preparation of $CH_3CH_2\overset{\overset{\displaystyle O}{\|}}{C}\underset{\underset{\displaystyle CH_3}{|}}{N}CH_3$ from the three types of acid derivatives shown in Example 13.1.

See related problems 13.38–13.39, 13.44f and 13.49.

13.8 Polyamides and Polyesters

A variety of familiar **polyamide** and **polyester** polymers are prepared by amidification and esterification reactions such as those presented in this chapter. An alcohol will react with an acid to produce an ester, and an amine with an acid to form an amide. But imagine the results if the acid were a dicarboxylic acid and the alcohol or amine were a dialcohol or diamine. Both ends of a dicarboxylic acid molecule can react with diamine molecules, and each end of a diamine can react with a diacid. The result would be a repetitive amidification producing a gigantic polymer.

polyamide
polymer (large molecule) in which the repeating structural units are connected by amide linkages

polyester
polymer (large molecule) in which the repeating structural units are connected by ester linkages

For example, Nylon 66 is produced from adipic acid (a dicarboxylic acid) and hexamethylene diamine (a diamine). Both ends of both molecules react repeatedly to produce a long polymer in stepwise growth.

$$H-\overset{H}{N}(CH_2)_6\overset{H}{N}-H \; HO-\overset{\overset{\displaystyle O}{\|}}{C}(CH_2)_4\overset{\overset{\displaystyle O}{\|}}{C}-OH \; H-\overset{H}{N}(CH_2)_6\overset{H}{N}-H \; HO-\overset{\overset{\displaystyle O}{\|}}{C}(CH_2)_4\overset{\overset{\displaystyle O}{\|}}{C}-OH \longrightarrow$$

Hexamethylene diamine — Adipic acid — Amide formation

$$\sim\left(\overset{H}{N}(CH_2)_6\overset{H}{N}-\overset{\overset{\displaystyle O}{\|}}{C}(CH_2)_4\overset{\overset{\displaystyle O}{\|}}{C}\right)_n\sim$$

Amide linkage

Nylon 66 (a polyamide)

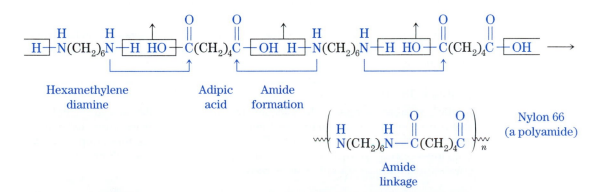

This polymer, named from the fact that both reactants have six carbons, is one of the most important synthetic fibers, being used in, among other things, clothing, sails, parachutes, fishing line, brushes, combs, gears, carpets, and bearings.

There are a variety of nylons that vary only in the number of carbons in the starting diamine and diacid. For example, Nylon 6-10 is formed from a reaction identical to that of Nylon 66 except that the diacid has ten carbons instead of six (the diamine still has six). Nylon 4-6 is produced from a four-carbon diamine and a six-carbon diacid.

Nylon 6 is formed from caprolactam (a six-carbon molecule that is an internal cyclic amide). When heated, the ring opens, and the resulting species forms amide bonds repeatedly along a long chain to produce a polyamide.

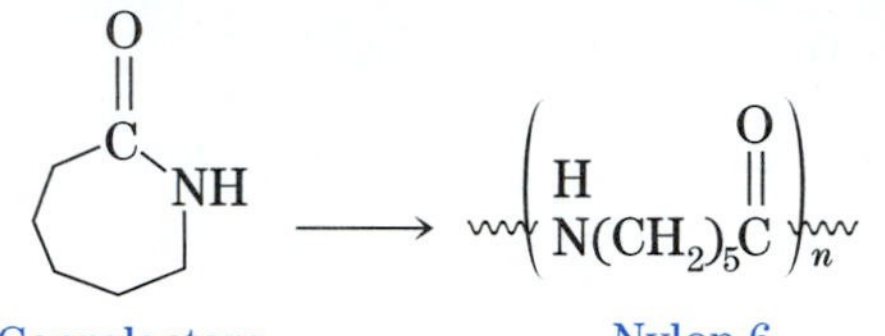

$$\text{Caprolactam} \longrightarrow \left(\overset{\text{H}}{\text{N}}(CH_2)_5\overset{\text{O}}{\overset{\|}{\text{C}}} \right)_n \ \text{Nylon 6}$$

Nylon 6 and Nylon 66 are the most heavily used nylons for fiber manufacture.

The formation of polyesters is theoretically analogous to that of polyamides. A diester is condensed with a diol. Both ends of both molecules can react continuously to form ester linkages by a transesterification process. Textile fibers known as Dacron® and transparent films marketed as Mylar® are polyesters produced form the dimethyl ester of terephthalic acid and ethylene glycol.

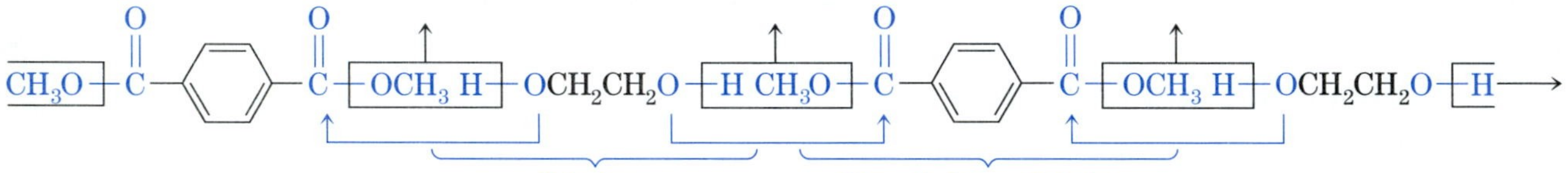

$$CH_3O-\overset{O}{\overset{\|}{C}}-C_6H_4-\overset{O}{\overset{\|}{C}}-OCH_3\ \ H-OCH_2CH_2O-H\ \ CH_3O-\overset{O}{\overset{\|}{C}}-C_6H_4-\overset{O}{\overset{\|}{C}}-OCH_3\ \ H-OCH_2CH_2O-H \longrightarrow$$

Ethylene glycol — Dimethyl terephthalate

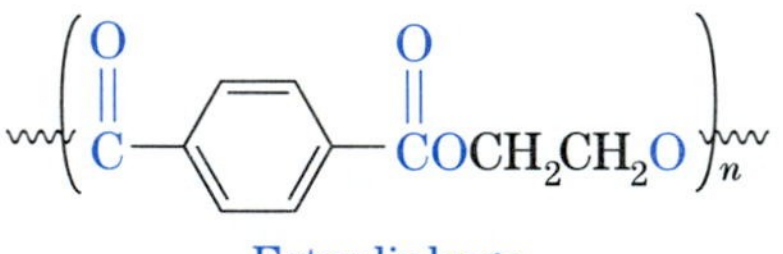

$$\left(\overset{O}{\overset{\|}{C}}-C_6H_4-\overset{O}{\overset{\|}{C}}OCH_2CH_2O\right)_n$$

Ester linkage
Dacron® or Mylar®

GETTING INVOLVED

✓ What are a polyamide and a polyester? Do you see that the formation of these polymers involves nucleophilic acyl substitution reactions we have studied? What are the reactions?

✓ Why do the compounds used to form condensation polymers have to be bifunctional? What should a dicarboxylic acid be reacted with to produce a polyamide? a polyester?

Problem 13.20

Kevlar®, an aromatic polyamide called an aramid, is an exceptionally strong polymer that is used for cord in radial tires and in bulletproof vests. Its *meta*-oriented equivalent, Nomex®, is used in flame-resistant clothing (for firefighters, for example) and for both internal and external parts in aircraft, spacecraft, and boats. From what diacid chloride and diamine could Kevlar be made?

$$\text{Kevlar}\quad \left(HN-C_6H_4-NH-\overset{O}{\overset{\|}{C}}-C_6H_4-\overset{O}{\overset{\|}{C}} \right)_n$$

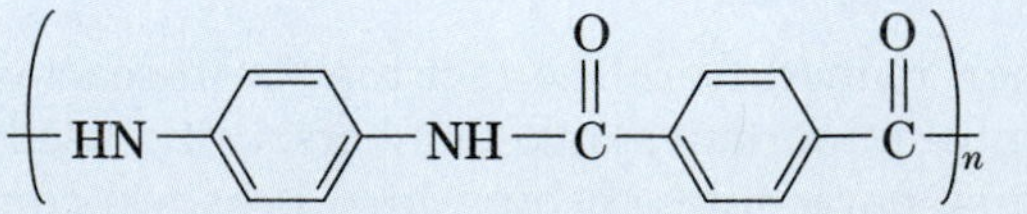

Problem 13.21

Kodel® polyester is made from the following diacid and dialcohol. Write a structure for Kodel.

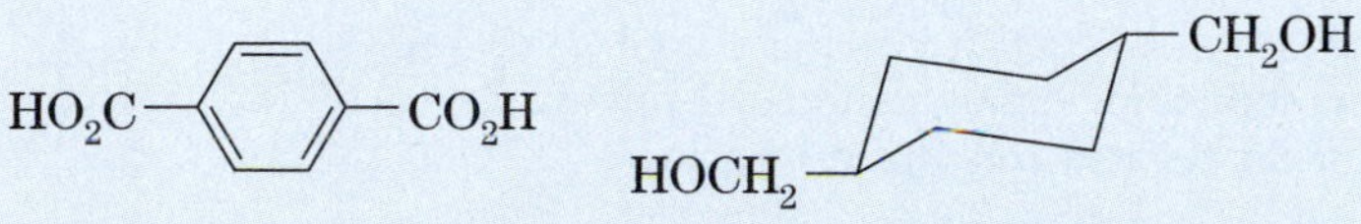

$$HO_2C-C_6H_4-CO_2H \qquad HOCH_2-C_6H_{10}-CH_2OH$$

See related problems 13.52–13.53.

13.9 Nucleophilic Addition Reactions of Carboxylic Acid Derivatives

A. Reduction with Lithium Aluminum Hydride

All of the acid derivatives, except amides, can be reduced to primary alcohols using lithium aluminum hydride.

$$RC(=O)-L \xrightarrow{LiAlH_4} \xrightarrow{H_2O/H^+} RCH_2OH + LH \qquad L = Cl, OC(=O)R, OR, OH$$

The first part of this reaction is a type of nucleophilic acyl substitution. A hydride ion from $LiAlH_4$ attacks the carbonyl carbon, forming a tetrahedral intermediate that expels the leaving group. The resulting aldehyde is quickly reduced to the alcohol by a nucleophilic addition reaction, as described in the chapter on aldehydes and ketones, section 11.5.E.

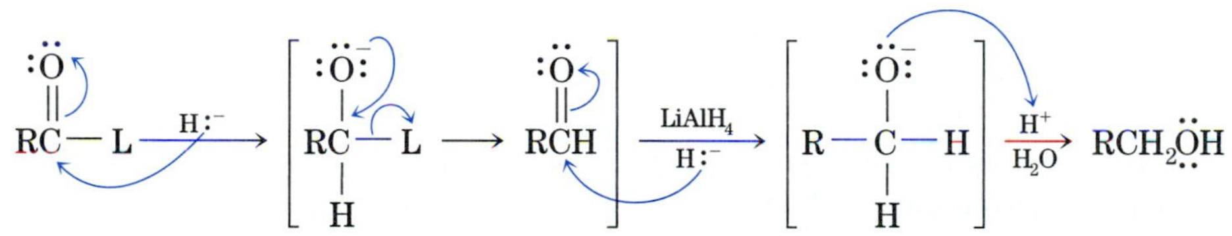

$$RC(=O)-L \xrightarrow{H:^-} \left[RC(O^-)(H)-L \longrightarrow RCH(=O) \right] \xrightarrow[H:^-]{LiAlH_4} \left[R-C(O^-)(H)-H \right] \xrightarrow[H_2O]{H^+} RCH_2OH$$

Acid derivative — Tetrahedral and aldehyde intermediates — Tetrahedral intermediate — 1° alcohol

Following is an example of this reaction using an ester; the 2-phenylethanol produced has the odor of roses and is found in a number of essential oils from plants.

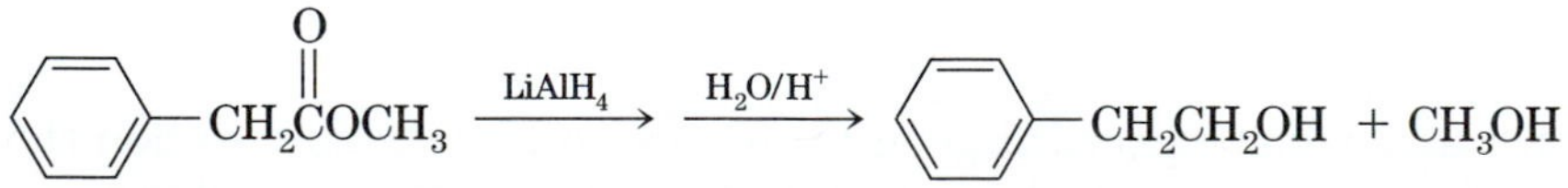

$$C_6H_5-CH_2C(=O)OCH_3 \xrightarrow{LiAlH_4} \xrightarrow{H_2O/H^+} C_6H_5-CH_2CH_2OH + CH_3OH$$

Lithium aluminum hydride reduction of amides produces amines.

$$RC(=O)NR_2 \xrightarrow{LiAlH_4} \xrightarrow{H_2O} RCH_2NR_2 \qquad R = H, \text{alkyl, or aryl}$$

GETTING INVOLVED

✓ What is the product of the $LiAlH_4$ reduction of acid chlorides, acid anhydrides, carboxylic acids, esters, and amides?

✓ Describe the mechanism of $LiAlH_4$ reduction of acid derivatives to primary alcohols. Do you see that tetrahedral intermediates are formed twice, once as a result of nucleophilic acyl substitution and finally from nucleophilic addition?

Problem 13.22

Write equations showing the preparation of 2-phenylethanol by $LiAlH_4$ reduction of an **(a)** acid chloride, **(b)** acid anhydride, **(c)** acid.

Problem 13.23

Write the products of the reaction of the following amides with lithium aluminum hydride:

(a) $CH_3C(=O)NH_2$ **(b)** $CH_3CH_2CH_2C(=O)NHCH_3$ **(c)** $C_6H_5C(=O)N(CH_2CH_3)_2$

See related problems 13.45 and 13.47(a).

B. Reaction with Grignard Reagents

Esters and acid chlorides react with Grignard reagents (section 13.9.B) to produce tertiary alcohols. The following example using an ester illustrates this reaction.

$$C_6H_5C(=O)OCH_2CH_3 + 2CH_3MgBr \longrightarrow \xrightarrow[H^+]{H_2O} C_6H_5C(OH)(CH_3)CH_3 + CH_3CH_2OH$$

The reaction mechanism clearly shows the relationship between the nucleophilic substitution mechanism of the acid derivative and the nucleophilic addition mechanism of aldehydes and ketones. The first mole of Grignard reagent, a strong nucleophile, adds to the carbonyl group of the ester to produce an unstable tetrahedral intermediate. Subsequent elimination of the ethoxy group (the leaving group of the acid derivative) generates a ketone.

$$C_6H_5\overset{\delta+}{C}(=\overset{\delta-}{O})OCH_2CH_3 \xrightarrow{\overset{\delta-}{C}H_3\overset{\delta+}{Mg}\overset{\delta-}{Br}} C_6H_5C(O^-\ \overset{+}{Mg}Br)(CH_3)OCH_2CH_3 \longrightarrow C_6H_5C(=O)CH_3 + CH_3CH_2OMgBr$$

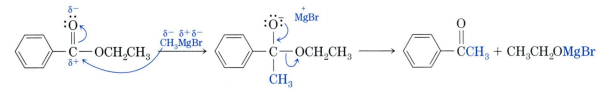

Ketones react with Grignard reagents to produce tertiary alcohols (section 11.5.F). Again, the nucleophilic alkyl group of the Grignard is attracted to the carbonyl carbon; addition occurs, and the salt of the alcohol forms. This salt is neutralized to the alcohol, the final product of the reaction.

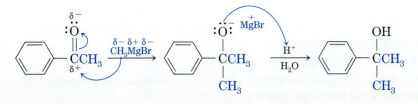

$$C_6H_5\overset{\delta+}{C}(=\overset{\delta-}{O})CH_3 \xrightarrow{\overset{\delta-}{C}H_3\overset{\delta+}{Mg}\overset{\delta-}{Br}} C_6H_5C(O^-\ \overset{+}{Mg}Br)(CH_3)CH_3 \xrightarrow[H_2O]{H^+} C_6H_5C(OH)(CH_3)CH_3$$

GETTING INVOLVED

- ✓ Why are tertiary alcohols produced from the reaction of esters with Grignard reagents? Why are two moles of the Grignard reagent required?
- ✓ In the mechanism of the reaction of esters with Grignard reagents, do you see that tetrahedral intermediates are formed twice, once as a result of nucleophilic acyl substitution and finally from nucleophilic addition?

Problem 13.24

For the following reaction: **(a)** write the product and **(b)** write the structures of the tetrahedral intermediate of nucleophilic acyl substitution, the intermediate ketone, and the tetrahedral intermediate of nucleophilic addition to the ketone.

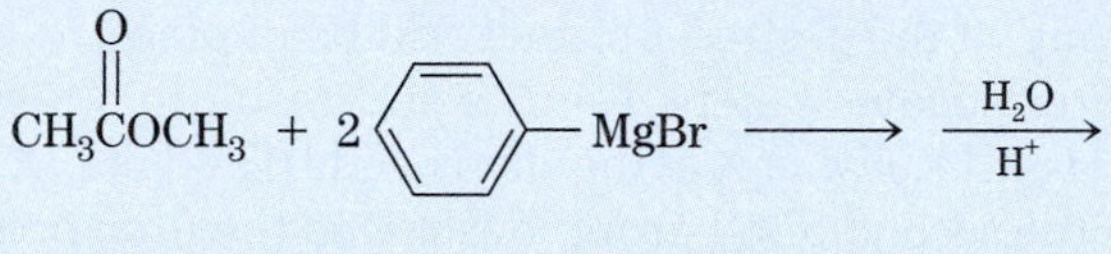

$$CH_3\overset{O}{\overset{\|}{C}}OCH_3 + 2\ C_6H_5\text{—MgBr} \longrightarrow \xrightarrow[H^+]{H_2O}$$

See related problems 13.46, 13.47(b), and 13.48.

13.10 Reactions of Acid Derivatives Involving Carbanions

A. Malonic Ester Synthesis

The **malonic ester synthesis** is useful in preparing substituted acetic acids and their derivatives. Follow the steps in Figure 13.2 as this synthetic procedure is discussed for the preparation of 2-ethyl-5-methylhexanoic acid, a disubstituted acetic acid.

malonic ester synthesis a method for preparing disubstituted acetic acids (at the α carbon)

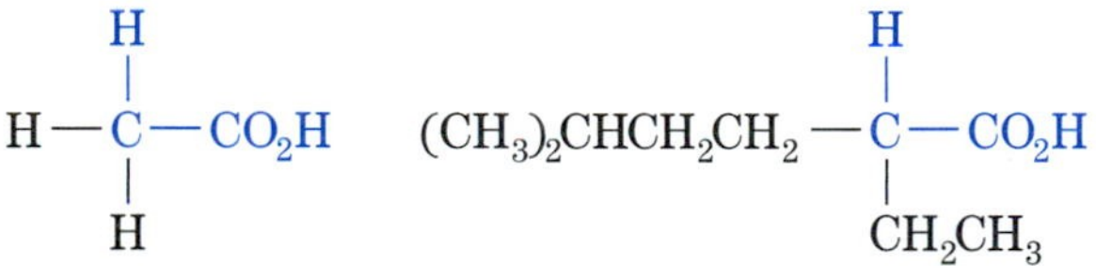

$$H\text{—}\underset{H}{\overset{H}{C}}\text{—}CO_2H \qquad (CH_3)_2CHCH_2CH_2\text{—}\underset{CH_2CH_3}{\overset{H}{C}}\text{—}CO_2H$$

Acetic acid — A disubstituted acetic acid

Step 1: Malonic ester is acidic, and its α-hydrogens can be extracted by base because of the polarization of the carbon-hydrogen bonds by the adjacent car-

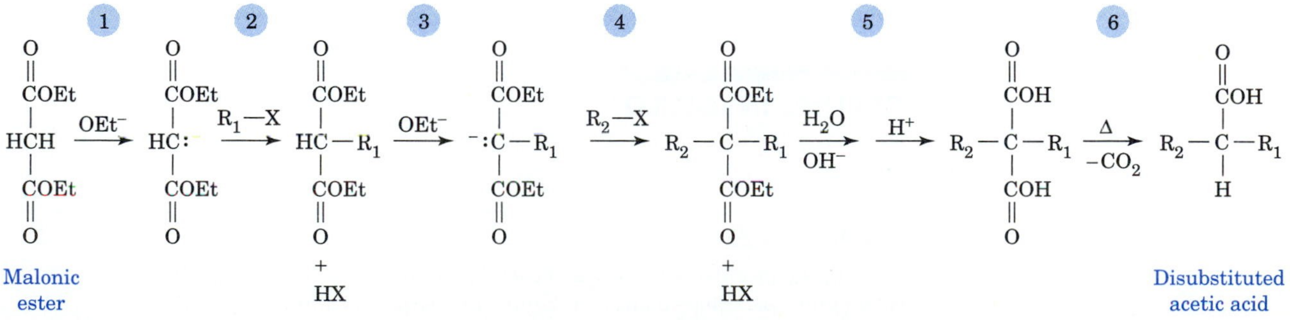

$$\underset{\text{Malonic ester}}{HC H(COEt)_2} \xrightarrow[\text{(1)}]{OEt^-} \ddot{H C}(COEt)_2 \xrightarrow[\text{(2)}]{R_1\text{—}X} HC(R_1)(COEt)_2 + HX \xrightarrow[\text{(3)}]{OEt^-} {}^-\!:C(R_1)(COEt)_2 \xrightarrow[\text{(4)}]{R_2\text{—}X} R_2\text{—}C(R_1)(COEt)_2 + HX \xrightarrow[OH^-]{H_2O} \xrightarrow[\text{(5)}]{H^+} R_2\text{—}C(R_1)(COH)_2 \xrightarrow[-CO_2]{\Delta\ \text{(6)}} \underset{\text{Disubstituted acetic acid}}{R_2\text{—}\underset{H}{C}(R_1)\text{—}\overset{O}{\overset{\|}{C}}OH}$$

Figure 13.2 Malonic ester synthesis of disubstituted acetic acids. Numbers mark the successive steps as described in the text. Et stands for CH_2CH_3.

bonyl groups and because of the resonance stabilization of the resulting carbanion (see section 11.6.A for an explanation of the acidity of α-hydrogens).

$$\mathrm{EtO{-}\overset{\overset{\displaystyle :\ddot{O}}{\|}}{C}\cdots\underset{\underset{\displaystyle H}{|}}{\overset{-}{C}}\cdots\overset{\overset{\displaystyle \ddot{O}:}{\|}}{C}{-}OEt}$$

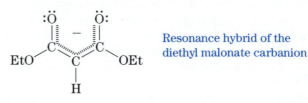

Resonance hybrid of the diethyl malonate carbanion

Step 2: When the carbanion is treated with an alkyl halide, the halide is displaced by nucleophilic substitution. If the alkyl halide is CH_3CH_2X (R_1 = CH_2CH_3), the ethyl group of the desired product will be in place.

Steps 3–4: These are repeats of steps 1 and 2 and result in a disubstituted malonic ester. Use of $(CH_3)_2CHCH_2CH_2X$ as the alkyl halide (R = $(CH_3)_2CHCH_2CH_2$ in step 4) will provide the second alkyl group desired in the final product.

Step 5: The diester is hydrolyzed to a dicarboxylic acid.

Step 6: Dicarboxylic acids in which the two acid groups are separated by one carbon atom decarboxylate (lose CO_2) when heated. The final product is the desired disubstituted acetic acid. If the disubstituted malonic acid or ester is isolated before decarboxylation, it can be used for the preparation of barbiturates (see Connections 13.2).

$$\mathrm{CH_2(COOEt)_2 \xrightarrow{\text{Steps 1–4}} CH_3CH_2C(COOEt)_2CH_2CH_2CH(CH_3)_2 \xrightarrow{\text{Step 5}}}$$

$$\mathrm{CH_3CH_2C(COOH)_2CH_2CH_2CH(CH_3)_2 \xrightarrow[-CO_2]{\text{Step 6,}} CH_3CH_2CH(COOH)CH_2CH_2CH(CH_3)_2}$$

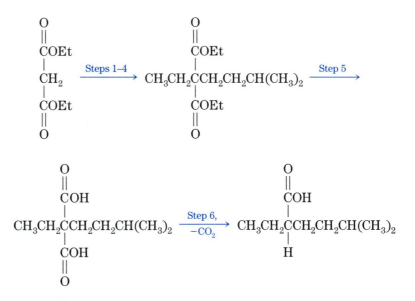

GETTING INVOLVED

✓ Explain the steps of the malonic ester synthesis.

✓ Why is the carbanion formed in step 1 stable?

Problem 13.25

Prepare the monosubstituted acetic acid hexanoic acid (caproic acid) by the malonic ester synthesis. (Steps 3 and 4 in Figure 13.2 would be eliminated.)

See related problem 13.54.

Connections 13.2 www.prenhall.com/bailey

Barbiturates

Barbiturates are made by condensing urea or thiourea with malonic esters and substituted malonic esters. The reaction is a condensation between an ester and amine (urea is actually an amide) to form two new amide linkages.

$$R_1R_2C(C(=O)OCH_2CH_3)_2 + (HNH)_2C{=}O \longrightarrow$$

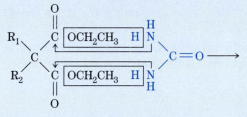

A disubstituted malonic ester + urea

$$R_1R_2C(C(=O)NH)_2C{=}O + 2CH_3CH_2OH$$

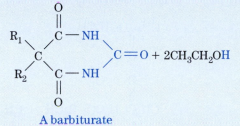

A barbiturate

Barbiturate Structures:

$R_1 = CH_3CH_2—$, $R_2 = (CH_3)_2CHCH_2CH_2—$ } Amytal, amobarbital

$R_1 = —CH_2CH_3$, $R_2 = CH_3CHCH_2CH_2CH_3$ (branched) } Nembutal, pentobarbital

$R_1 = —C_6H_5$ (phenyl), $R_2 = —CH_2CH_3$ } Phenobarbital, luminal

$R_1 = CH_2{=}CHCH_2—$, $R_2 = CH_3CHCH_2CH_2CH_3$ (branched) } Seconal, secobarbital

Barbiturates depress activity in the central nervous system and are useful as hypnotics and sedatives in both human and veterinary medicine. For example, in human medicine they are used (by prescription) as sleeping pills, to control blood pressure, to combat epileptic seizures, and to control colic in young babies. Since they depress a wide range of other biological functions, such as oxygen intake and heart activity, they must be used cautiously. When a barbiturate is applied as a general anesthetic, as sodium pentothal is, the effective dose is as much as 50%–75% of the lethal dose.

Persons taking barbiturates as sleeping pills must be particularly careful to avoid drinking alcohol while under the influence of a barbiturate (or vice versa) because of the synergistic effect; that is, the combined effect of barbiturate and alcohol is greater than the expected sum of the two.

$$(CH_3CH_2)(CH_3CH_2CH_2CH(CH_3))C(C(=O)\ddot{N}^{-}\,{}^{+}Na)(C(=O)NH)C{=}S$$

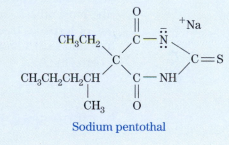

Sodium pentothal

B. Claisen Condensation

The **Claisen condensation** is a carbanion-type reaction in which an ester is converted to a β-keto ester. Consider, for example, the condensation of ethyl acetate to ethyl 3-oxobutanoate:

Claisen condensation a method for making β keto esters from esters with α hydrogens

$$2CH_3\overset{O}{\overset{\|}{C}}OCH_2CH_3 \xrightarrow[CH_3CH_2OH]{NaOCH_2CH_3,} CH_3\overset{O}{\overset{\|}{C}}CH_2\overset{O}{\overset{\|}{C}}OCH_2CH_3 + CH_3CH_2OH$$

The reaction mechanism involves initial abstraction of an α-hydrogen from an ester molecule by ethoxide ion. The α-hydrogens of the ester are acidic due to polarization of the carbon-hydrogen bond by the carbonyl group and resonance stabilization of the resulting carbanion (see section 11.6.A).

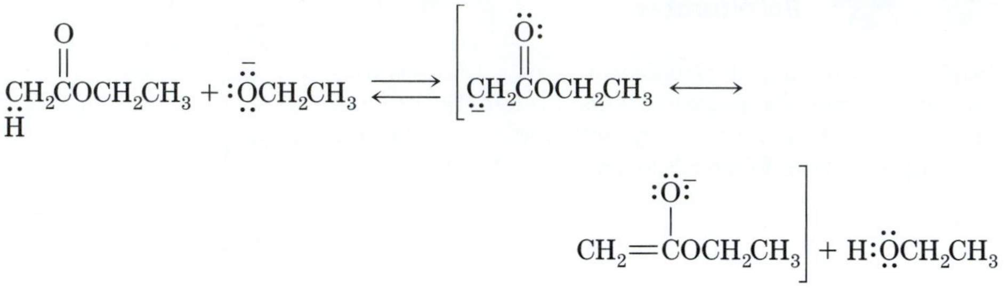

$$\underset{\ddot{H}}{CH_2}\overset{O}{\overset{\|}{C}}OCH_2CH_3 + :\ddot{\bar{O}}CH_2CH_3 \rightleftharpoons \left[\ddot{\bar{C}}H_2\overset{\ddot{O}:}{\overset{\|}{C}}OCH_2CH_3 \longleftrightarrow CH_2{=}\overset{:\ddot{O}:^-}{\overset{|}{C}}OCH_2CH_3 \right] + H:\ddot{O}CH_2CH_3$$

The carbanion attacks the partially positive carbonyl carbon of an ester molecule and displaces an ethoxide ion, by nucleophilic acyl substitution.

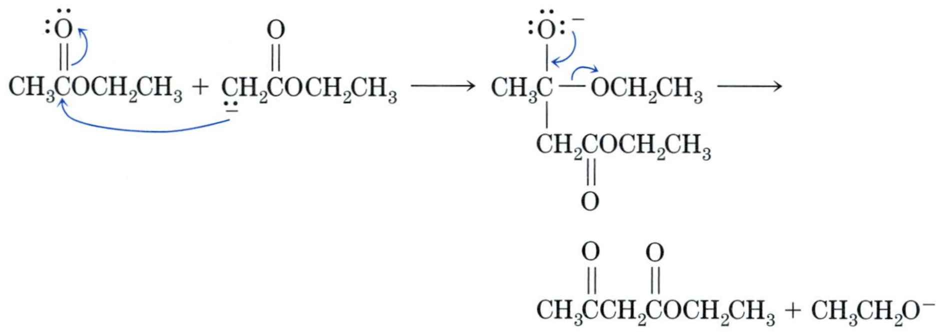

$$CH_3\overset{:\ddot{O}}{\overset{\|}{C}}OCH_2CH_3 + \ddot{\bar{C}}H_2\overset{O}{\overset{\|}{C}}OCH_2CH_3 \longrightarrow CH_3\overset{:\ddot{O}:^-}{\overset{|}{C}}(CH_2\overset{O}{\overset{\|}{C}}OCH_2CH_3){-}OCH_2CH_3 \longrightarrow CH_3\overset{O}{\overset{\|}{C}}CH_2\overset{O}{\overset{\|}{C}}OCH_2CH_3 + CH_3CH_2O^-$$

Claisen-type condensations occur in some biological systems. For example, the biosynthesis of acetoacetyl coenzyme A, an intermediate in the biosynthesis of terpenes (see Connections 5.4), steroids (section 15.6.D), and fatty acids (section 15.3), involves the Claisen-type condensation of two molecules of acetyl coenzyme A.

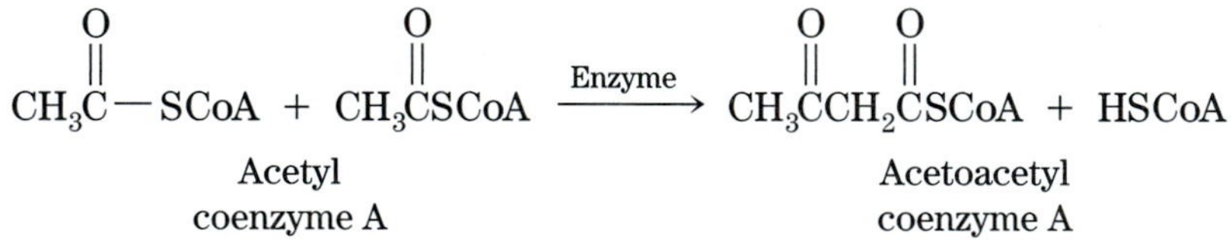

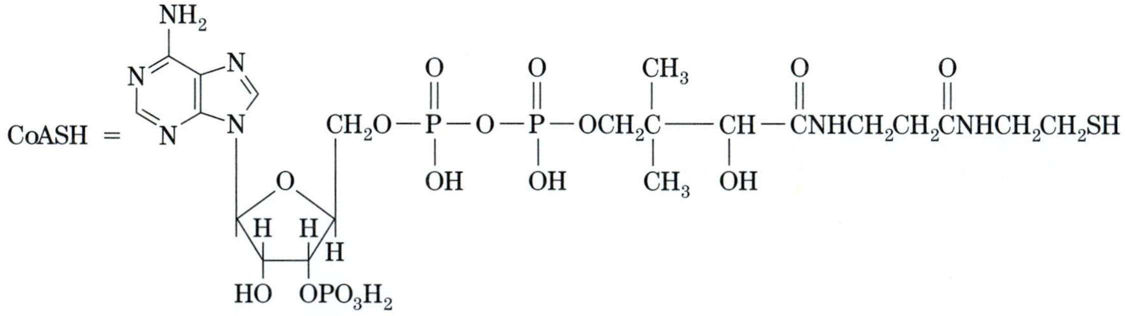

$$CH_3\overset{O}{\overset{\|}{C}}{-}SCoA + CH_3\overset{O}{\overset{\|}{C}}SCoA \xrightarrow{\text{Enzyme}} CH_3\overset{O}{\overset{\|}{C}}CH_2\overset{O}{\overset{\|}{C}}SCoA + HSCoA$$

Acetyl coenzyme A — Acetoacetyl coenzyme A

CoASH = (adenine: NH_2, N, N, N, N) ribose (O; H H H; HO OPO_3H_2) $-CH_2O-\overset{O}{\overset{\|}{P}}(OH)-O-\overset{O}{\overset{\|}{P}}(OH)-OCH_2C(CH_3)_2-CH(OH)-\overset{O}{\overset{\|}{C}}NHCH_2CH_2\overset{O}{\overset{\|}{C}}NHCH_2CH_2SH$

GETTING INVOLVED

- ✓ Why is it possible to form a stable carbanion? Do you see the α-hydrogens?
- ✓ Do you see that this reaction is nucleophilic acyl substitution with the tetrahedral intermediate?

Problem 13.26

Write the product of the Claisen condensation, using ethyl propanoate and sodium ethoxide.

See related problems 13.56–13.57.

REACTION SUMMARY

1. Reactions of Carboxylic Acids and Derivatives

See Figure 13.1 for a complete summary.

A. Reactions of Acid Chlorides

Sections 13.3, 13.9; Problems 13.7–13.8; 13.22; 13.33(a), 13.34, 13.44(a), 13.45(a), 13.46(a).

With acid salt to form acid anhydrides

$$\mathrm{RC(=O)-Cl + NaOC(=O)R \longrightarrow RC(=O)-OC(=O)R + NaCl}$$

With water to form carboxylic acids

$$\mathrm{RC(=O)-Cl + H-OH \longrightarrow RC(=O)-OH + HCl}$$

With alcohols to form esters

$$\mathrm{RC(=O)-Cl + H-OR \longrightarrow RC(=O)-OR + HCl}$$

With amines to form amides

$$\mathrm{RC(=O)-Cl + H-NR_2 \longrightarrow RC(=O)-NR_2 + HCl}$$

With lithium aluminum hydride to form 1° alcohols

$$\mathrm{RC(=O)-Cl \xrightarrow{LiAlH_4} \xrightarrow{H_2O/H^+} RCH_2OH}$$

With Grignard reagents to form 3° alcohols

$$\mathrm{RC(=O)-Cl \xrightarrow{R'MgX} \xrightarrow{H_2O/H^+} RC(OH)(R')R'}$$

B. Reactions of Acid Anhydrides

Sections 13.4, 13.9.A; Problems 13.10–13.11, 13.22–13.23; 13.33(b), 13.35, 13.44(b).

With water to form carboxylic acids

$$\mathrm{RC(=O)-OC(=O)R + H-OH \longrightarrow RC(=O)-OH + RC(=O)OH}$$

With alcohols to form esters

$$\mathrm{RC(=O)-OC(=O)R + H-OR \longrightarrow RC(=O)-OR + RC(=O)OH}$$

With amines to form amides

$$\mathrm{RC(=O)-OC(=O)R + H-NR_2 \longrightarrow RC(=O)-NR_2 + RC(=O)OH}$$

With lithium aluminum hydride to form 1° alcohols

$$\mathrm{RC(=O)OC(=O)R \xrightarrow{LiAlH_4} \xrightarrow{H_2O/H^+} RCH_2OH}$$

C. Reactions of Carboxylic Acids

Sections 13.5, 13.9.A; Problems 13.12–13.14, 13.22–13.23; 13.36, 13.44(e), 13.60. Also section 12.4.A; Problems 12.6.

With thionyl chloride to form acid chlorides

$$\mathrm{RC(=O)OH + SOCl_2 \longrightarrow RC(=O)Cl + SO_2 + HCl}$$

With alcohols to form esters

$$\mathrm{RC(=O)OH + HOR \xrightarrow{H^+} RC(=O)OR + H_2O}$$

With amines to form amides

$$\mathrm{RC(=O)OH + HNR_2 \xrightarrow{Heat} RC(=O)NR_2 + H_2O}$$

With lithium aluminum hydride to form 1° alcohols

$$\mathrm{RC(=O)OH \xrightarrow{LiAlH_4} \xrightarrow[H^+]{H_2O} RCH_2OH}$$

D. Reactions of Esters

Sections 13.6, 13.9; Problems 13.15–13.17; 13.23–13.24; 13.33(c), 13.37, 13.40–13.41, 13.44(c)–(d), 13.46–13.48, 13.50–13.51, 13.55(b).

With water to form acids

$$\mathrm{RC(=O)-OR + HOH \xrightarrow{H^+} RC(=O)-OH + ROH}$$ (acid salt if basic conditions)

With alcohols to form another ester

$$\mathrm{RC(=O)-OR + HOR' \xrightarrow{H^+} RC(=O)-OR' + ROH}$$

With amines to form amides

$$\mathrm{RC(=O)—OR + HNR_2 \xrightarrow{H^+} RC(=O)—NR_2 + ROH}$$

With lithium aluminum hydride to form 1° alcohols

$$\mathrm{RC(=O)OR \xrightarrow{LiAlH_4} \xrightarrow{H_2O/H^+} RCH_2OH}$$

With Grignard reagents to form 3° alcohols

$$\mathrm{RC(=O)OR \xrightarrow{R'MgX} \xrightarrow{H_2O/H^+} RC(OH)(R')R'}$$

E. Reactions of Amides

Sections 13.6, 13.9.A; Example 13.1; Problems 13.18–13.19, 13.33(d), 13.38–13.39, 13.44(f), 13.45(d–e), 13.49.

Reaction with water to form acids

$$\mathrm{RC(=O)—NR_2 + H_2O \xrightarrow[OH^-]{H^+ or} RC(=O)OH + HNR_2}$$

Reaction with lithium aluminum hydride to form amines

$$\mathrm{RC(=O)NR_2 \xrightarrow{LiAlH_4} \xrightarrow{H_2O} RCH_2NR_2}$$

2. Malonic Ester Synthesis

Section 13.10.A; Problems 13.25, 13.54. See Figure 13.2 for a summary.

3. Claisen Condensation

Section 13.10.B; Problems 13.26, 13.56–13.57.

$$\mathrm{2RCH_2C(=O)OR \xrightarrow[ROH]{NaOR} RCH_2C(=O)CH(R)C(=O)OR + ROH}$$

Problems

13.27 **Nomenclature of Carboxylic Acids:** Name the following carboxylic acids by the IUPAC system:

(a) $CH_3CH_2CH_2CO_2H$

(b) $(CH_3)_2CH(CH_2)_5CO_2H$

(c) $CH_3CH_2CH{=}CHCO_2H$

(d) $CH_3C(=O)CH_2CH_2CO_2H$

(e) $H_2NCH_2CO_2H$

(f) $CH_3CH(OH)CH{=}CHCH{=}CHCO_2H$

(g) 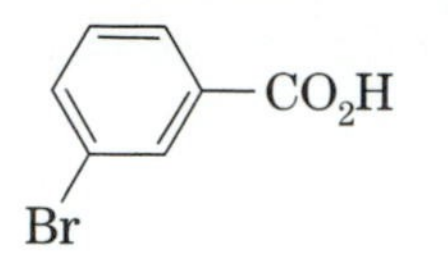 CO_2H, Br

(h) 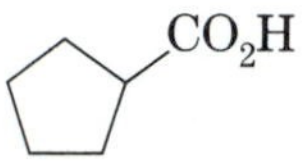 CO_2H

(i) $HO_2CCH_2CH_2CH_2CH_2CO_2H$

13.28 **Nomenclature of Acid Chlorides:** Name the following by the IUPAC system of nomenclature:

(a) $CH_3CH_2CH_2\overset{O}{\overset{\|}{C}}Cl$ **(b)** $CH_3CH{=}CH\overset{O}{\overset{\|}{C}}Cl$

(c) $CH_3CH_2\overset{O}{\overset{\|}{C}}CH_2\overset{O}{\overset{\|}{C}}Cl$ **(d)** 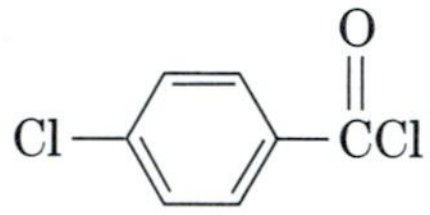 $Cl-C_6H_4-\overset{O}{\overset{\|}{C}}Cl$

13.29 **Nomenclature of Acid Anhydrides:** Name the following compounds by the IUPAC system of nomenclature:

(a) $CH_3CH_2CH_2\overset{O}{\overset{\|}{C}}O\overset{O}{\overset{\|}{C}}CH_2CH_2CH_3$

(b) $CH_3CH_2\overset{O}{\overset{\|}{C}}O\overset{O}{\overset{\|}{C}}CH_2CH_3$

(c) $CH_3CH_2CH_2\overset{O}{\overset{\|}{C}}O\overset{O}{\overset{\|}{C}}CH_2CH_3$

13.30 **Nomenclature of Esters:** Name the following esters by the IUPAC system of nomenclature:

(a) $CH_3CH_2CH_2CH_2\overset{O}{\overset{\|}{C}}OCH_3$

(b) $CH_3CH_2CH_2\overset{O}{\overset{\|}{C}}OCH_2CH_3$

(c) $CH_3CH_2\overset{O}{\overset{\|}{C}}OCH_2CH_2CH_3$

(d) $CH_3\overset{O}{\overset{\|}{C}}OCH_2CH_2CH_2CH_3$

(e) $H\overset{O}{\overset{\|}{C}}OCH_2CH_2CH_2CH_2CH_3$

(f) $CH_3CH_2\overset{O}{\overset{\|}{C}}OCH(CH_3)_2$

(g) 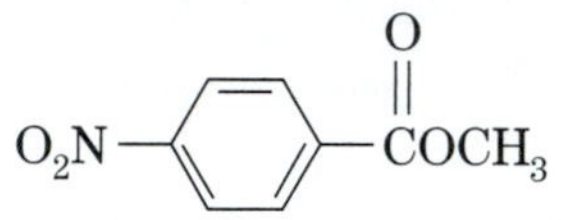 $O_2N-C_6H_4-\overset{O}{\overset{\|}{C}}OCH_3$

(h) $CH_3CH{=}CHCH{=}CH\overset{O}{\overset{\|}{C}}OCH_2CH_2CH_2CH_3$

13.31 **Nomenclature of Amides:** Name the following compounds by the IUPAC system of nomenclature:

(a) $CH_3CH_2CH_2CH_2\overset{O}{\overset{\|}{C}}NH_2$

(b) $CH_3CH_2CH_2\overset{O}{\overset{\|}{C}}NHCH_3$

(c) $CH_3CH_2\overset{O}{\overset{\|}{C}}NHCH_2CH_3$

(d) $CH_3CH_2\overset{O}{\overset{\|}{C}}N(CH_3)_2$

(e) $CH_3\overset{O}{\overset{\|}{C}}-\underset{}{N}(CH_3)CH_2CH_3$

(f) 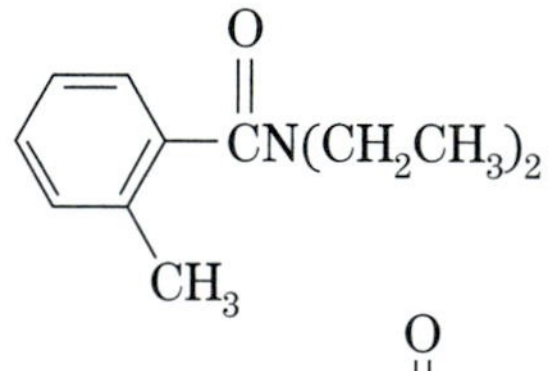 $\overset{O}{\overset{\|}{C}}N(CH_2CH_3)_2$, CH_3

(g) $CH_3\underset{OH}{\underset{|}{C}}HCH{=}CH\overset{O}{\overset{\|}{C}}NHCH_2CH_2CH_3$

13.32 **Nomenclature of Carboxylic Acid Derivatives:** Draw structures for the following compounds: **(a)** butanoic acid (in rancid butter); **(b)** *p*-aminobenzoic acid (a sunscreen); **(c)** ethyl *p*-aminobenzoate (the local anesthetic benzocaine); **(d)** pentyl butanoate (apricot odor); **(e)** potassium 2,4-hexadienoate (the food preservative, potassium sorbate); **(f)** *o*-hydroxybenzamide (a pain reliever); **(g)** N,N-dimethylmethanamide (the solvent DMF); **(h)** N,N-diethyl *m*-methylbenzamide (insect repellent); **(i)** butanoic hexanoic anhydride; **(j)** butanoic anhydride; **(k)** *m*-chlorobenzoyl chloride.

13.33 **Reactions of Acid Derivatives:** Write structures for the **(a)** acid chloride, **(b)** acid anhydride, **(c)** ester, and **(d)** amide of propanoic acid. Show the products and by-products of the reactions of these compounds with water.

13.34 **Reactions of Acid Chlorides:** Write the products and inorganic by-product of the reaction of benzoyl chloride with each of the following compounds:

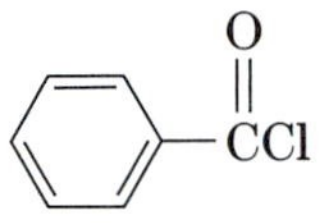

C_6H_5–C(=O)Cl

(a) $CH_3CH_2CO_2Na$ **(b)** H_2O **(c)** CH_3OH

(d) NH_3 **(e)** CH_3NH_2 **(f)** $CH_3NHCH_2CH_3$

13.35 **Reactions of Acid Anhydrides:** Write the products and by-product of the reaction of propanoic anhydride with the following:

(a) H_2O **(b)** $CH_3CHOHCH_3$

(c) NH_3 **(d)**  (cyclic) NH

13.36 **Reactions of Carboxylic Acids:** Write the products and inorganic by-products of the reaction of benzoic acid with the following reagents:

(a) $CH_3CH_2CH_2OH/H^+$ **(b)** NH_3
(c) CH_3NHCH_3 **(d)** $SOCl_2$

13.37 **Reactions of Esters:** Write the products and by-products of the reaction of methyl propanoate ($CH_3CH_2CO_2CH_3$) with the following reagents:

(a) H_2O/OH^- **(b)** NH_3
(c) CH_3NH_2 **(d)** $CH_3CH_2CH_2CH_2CH_2OH/H^+$

13.38 **Reactions of Amides:** Write reaction equations showing the reaction of the following amides with water (hydrolysis):

(a) 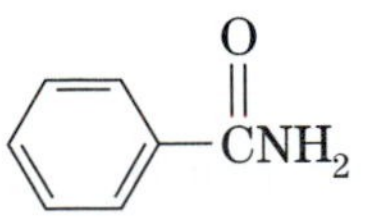 C_6H_5–C(=O)NH_2 **(b)** CH_3CH_2C(=O)$N(CH_3)_2$

13.39 **Preparations of Amides:** Write equations showing how each of the amides in problem 13.38 can be prepared from the following acid derivatives: **(a)** acid chloride; **(b)** acid anhydride; **(c)** carboxylic acid; **(d)** ester.

13.40 **Preparations of Esters:** Write equations showing how the ester shown in problem 13.37 can be prepared from the following acid derivatives: **(a)** acid chloride; **(b)** acid anhydride; **(c)** carboxylic acid.

13.41 **Preparations of Esters:** For each of the esters shown in problem 13.30, write the structure of the acid and alcohol from which they could be prepared.

13.42 **Lactones:** Upon heating with acid, 4-hydroxybutanoic acid eliminates water and forms a cyclic ester called a lactone. Write the structure of the lactone.

13.43 **Reactions of Diacids:** In a common experiment in full-year organic chemistry labs, students synthesize the following compound. When they take the melting point of the product, many are intrigued to observe moisture forming in the capillary tube. A chemical reaction is occurring; write the structure of the probable product.

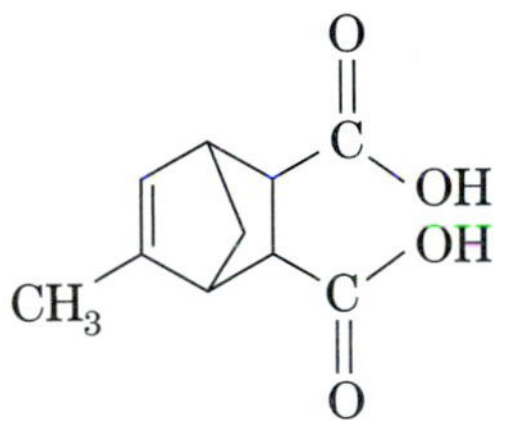

CH_3 (bicyclic ring) with two C(=O)OH groups

13.44 **Nucleophilic Acyl Substitution Mechanisms:** Write step-by-step mechanisms for the following nucleophilic acyl substitution reactions:

(a) CH_3C(=O)Cl + CH_3OH

(b) CH_3C(=O)OC(=O)CH_3 + CH_3CH_2OH

(c) CH_3C(=O)OCH_3 + NH_3

(d) 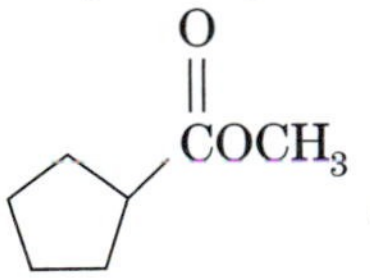 (cyclopentyl)–C(=O)OCH_3 + $NaOH/H_2O$

(e) CH_3CH_2C(=O)OH + CH_3CH_2OH/H^+

(f) 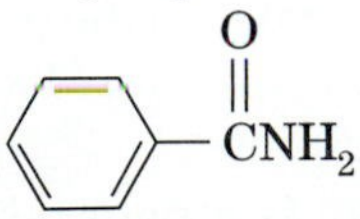 C_6H_5–C(=O)NH_2 + $NaOH/H_2O$

13.45 Reactions with $LiAlH_4$: Write the products of the reactions of the compounds shown with lithium aluminum hydride followed by neutralization with acid:

(a) 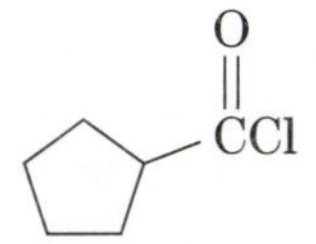 cyclopentyl–$\overset{O}{\overset{\|}{C}}Cl$

(b) 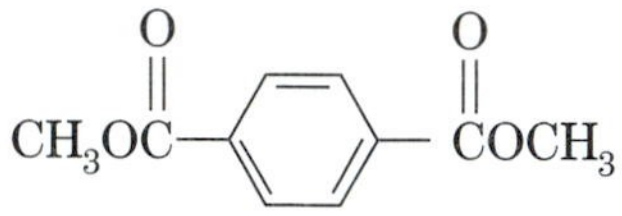 $CH_3O\overset{O}{\overset{\|}{C}}$–$C_6H_4$–$\overset{O}{\overset{\|}{C}}OCH_3$

(c) $CH_3(CH_2)_{12}CO_2CH_3$

(d) $CH_3CH_2CH_2CH_2\overset{O}{\overset{\|}{C}}NH_2$

(e) 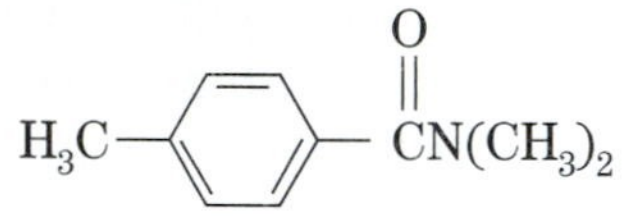 H_3C–C_6H_4–$\overset{O}{\overset{\|}{C}}N(CH_3)_2$

13.46 Acid Derivatives and Grignard Reagents: Write the products of the reaction between each of the acid derivatives shown and Grignard reagent shown followed by neutralization:

(a) $CH_3CH_2CH_2\overset{O}{\overset{\|}{C}}Cl$ and $2CH_3CH_2MgBr$

(b) 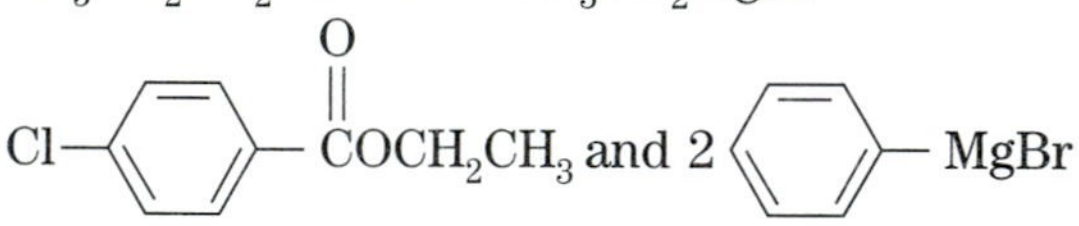 Cl–C_6H_4–$\overset{O}{\overset{\|}{C}}OCH_2CH_3$ and 2 C_6H_5–MgBr

(c) γ-butyrolactone (five-membered ring, O, C=O) and 2 C_6H_5–MgBr

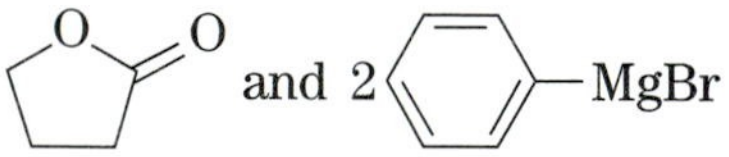

13.47 Nucleophilic Acyl Addition Mechanisms: Write step-by-step mechanisms for the following reactions: **(a)** $CH_3CO_2CH_3$ and $LiAlH_4$ followed by neutralization; **(b)** $CH_3CO_2CH_3$ and CH_3MgBr followed by neutralization.

13.48 Grignard Synthesis of Alcohols: Starting with an organic halogen compound and an ester, provide a synthesis for each of the following compounds: **(a)** 1,1-diphenyl-1-butanol; **(b)** 3-butyl-1-phenyl-3-heptanol.

13.49 Hydrolysis of Urea: Areas in which there is an accumulation of urine, such as a cat box, develop an ammonia-like odor. Is there a chemical basis for this odor? Explain and illustrate with a chemical equation.

13.50 Decomposition of Aspirin: On prolonged standing, aspirin tablets sometimes take on the odor of vinegar. Is there a chemical basis for this odor? Explain and illustrate with a chemical equation.

13.51 Hydrolysis of Salol: Salol is a pain reliever that was introduced in 1886. Although it had some positive aspects, the possibility of phenol poison was a concern. Explain and illustrate how phenol could be formed in the basic environment of the small intestine. Also show in your explanation the formation of sodium salicylate, the actual active chemical.

13.52 Condensation Polymers: Write structures for the following condensation polymers:

(a) Nylon 6-10 formed from $H_2N(CH_2)_6NH_2$ and $HO_2C(CH_2)_8CO_2H$

(b) Nylon 4-6 formed from $H_2N(CH_2)_4NH_2$ and $HO_2C(CH_2)_4CO_2H$

(c) polycarbonate plastics from

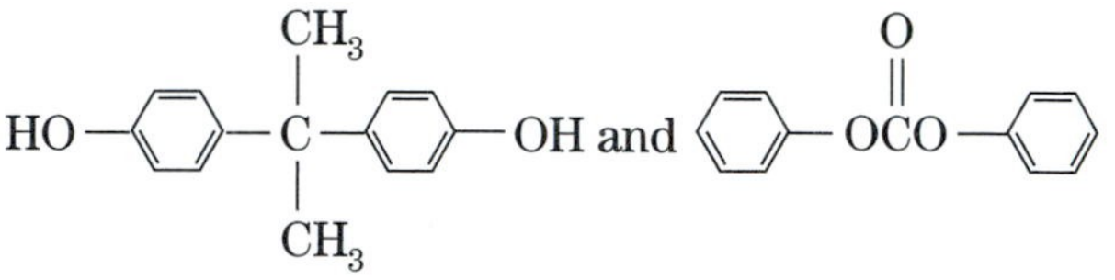

HO–C_6H_4–$C(CH_3)_2$–C_6H_4–OH and C_6H_5–$O\overset{O}{\overset{\|}{C}}O$–$C_6H_5$

13.53 Polyurethanes: Following is the structure of polyurethane, a polymer used in elastic fibers and semirigid construction foams. What type of condensation polymer does it appear to be (polyester, polyamide, etc.)?

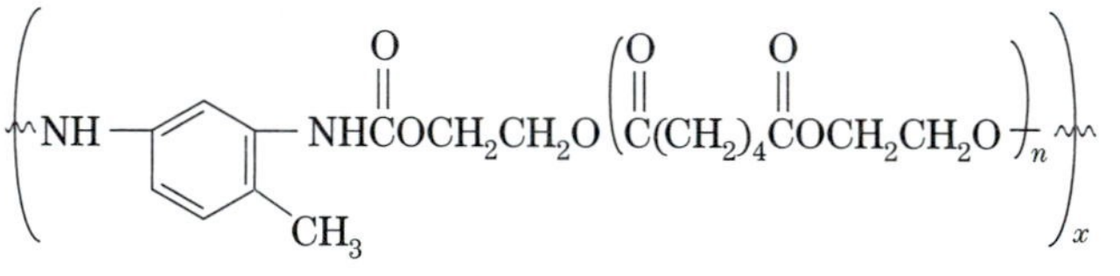

$$\left(\sim NH-C_6H_3(CH_3)-NH\overset{O}{\overset{\|}{C}}OCH_2CH_2O\left(\overset{O}{\overset{\|}{C}}(CH_2)_4\overset{O}{\overset{\|}{C}}OCH_2CH_2O\right)_n\sim\right)_x$$

A polyurethane

13.54 Malonic Ester Synthesis: Prepare the following compounds from diethyl malonate, using the malonic ester synthesis:

(a) butanoic acid

(b) 2-methylbutanoic acid

13.55 Familiar Esters:

(a) Name the esters in section 13.1.A.

(b) Write reaction equations showing the preparation of these esters from an acid and an alcohol.

13.56 Claisen Condensation: Write a step-by-step reaction mechanism for the Claisen condensation of methyl ethanoate ($CH_3CO_2CH_3$) with sodium methoxide as the catalyst.

13.57 Claisen Condensation: Write reaction equations for the following Claisen condensations:

(a) $CH_3\overset{O}{\overset{\|}{C}}OCH_3$, $NaOCH_3$

(b) $CH_3CH_2\overset{O}{\overset{\|}{C}}OCH_3$, $NaOCH_3$

(c) 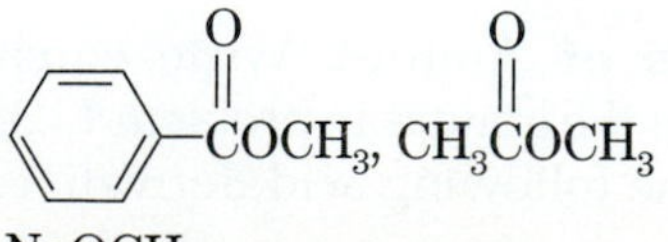 C_6H_5–$\overset{O}{\overset{\|}{C}}OCH_3$, $CH_3\overset{O}{\overset{\|}{C}}OCH_3$, $NaOCH_3$

13.58 **Proteins:** Proteins are large molecules composed of many amino acid units connected by amide bonds. Write the structure of a protein composed of the amino acids glycine, phenylalanine, and proline. See Table 17.1 for structures of amino acids.

13.59 **Preparation of Medicinal Compounds:** Write reaction equations showing the syntheses of the following materials from the indicated starting materials:

(a) sodium salicylate from salicylic acid
(b) phenacetin from *p*-ethoxyaniline
(c) acetominophen from *p*-aminophenol
(d) benzocaine from *p*-aminobenzoic acid
(e) methyl salicylate from salicylic acid
(f) salicylamide from salicylic acid
(g) aspirin from salicylic acid
(h) phenobarbital from urea and a substituted diethylmalonate

13.60 **Reaction Mechanisms—Condensation Reactions:** One of the important experiments used to elucidate the mechanism of acid-catalyzed esterification was to show whether the oxygen in the ester (the oxygen of —OR) came from the original acid or the alcohol. When ordinary benzoic acid is allowed to react with isotopically enriched methanol, $CH_3O^{18}H$, the methyl benzoate produced contains the labeled oxygen. Using words and reaction equations, show how this experiment answers the question.

Activities with Molecular Models

1. Make molecular models of the simplest acid, ester, amide, and acid chloride with only two carbons.

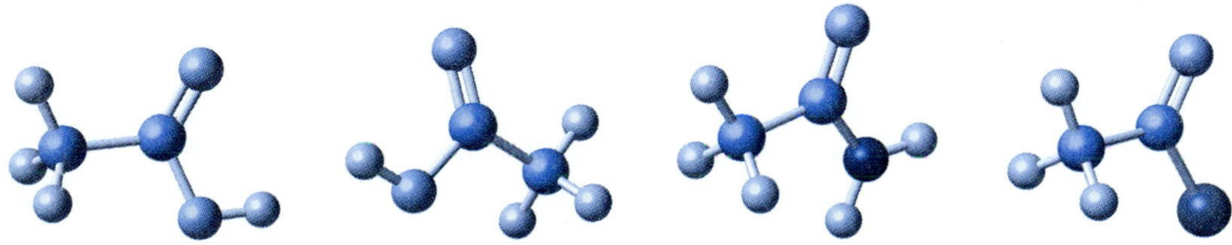

2. Make models of the four esters with the formula $C_4H_8O_2$.
3. Make models of the three amides with the formula C_3H_7NO.
4. Make models of the two acid anhydrides with the formula $C_4H_6O_3$.
5. Make models of the two acid chlorides with the formula C_4H_7OCl.

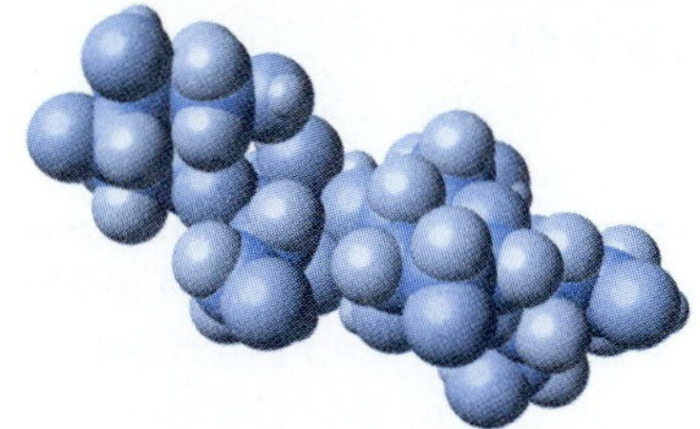

CHAPTER 14
Carbohydrates

14.1 Chemical Nature of Carbohydrates—Polyhydroxy Aldehydes and Ketones

Carbohydrates comprise one of the four major classes of biologically active organic molecules, or biomolecules. (The other classes are lipids, proteins, and nucleic acids.) Simple and complex carbohydrates are the main source of metabolic energy for all the organism's activities, from locomotion to the building of other molecules.

carbohydrate
a polyhydroxy—aldehyde or ketone; the polymers and derivatives of such compounds

The general formula for many common **carbohydrates**—including glucose and fructose ($C_6H_{12}O_6$) and sucrose and lactose ($C_{12}H_{22}O_{11}$)—is $C_n(H_2O)_m$, which would seem to support the old concept that these compounds are hydrates of carbon. However, their true chemical structures are those of polyhydroxy (more than one hydroxy, or —OH groups) aldehydes and ketones. The term *carbohydrate* can also refer to derivatives and polymers of polyhydroxy aldehydes and ketones. Carbohydrates contain a carbonyl group as an aldehyde or ketone as well as more than one alcohol group. The two simplest carbohydrate molecules illustrate this.

O=CH–CHOH–CH_2OH	Glyceraldehyde An aldose A triose An aldotriose	CH_2OH–C=O–CH_2OH	Dihydroxyacetone A ketose A triose A ketotriose

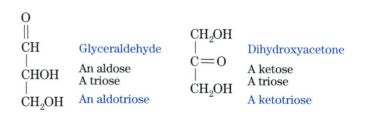

Note that these two compounds have the same molecular formula, $C_3H_6O_3$, and both contain a carbonyl group—one as an aldehyde and the other as a ketone—as well as two alcohol (hydroxy) groups. In addition, they are functional isomers of each other.

14.2 Nomenclature of Carbohydrates

In chemical and biochemical discussions carbohydrates are more frequently referred to as *saccharides,* from the Greek word for something sweet. This term is a misnomer, as many, if not most, saccharides are not sweet. However, this terminology does allow us to talk conveniently about individual carbohydrate units, or **monosaccharides,** as well as their polymer, **oligosaccharides** (two to ten units) and **polysaccharides** (more than ten units).

The molecules shown before as well as other carbohydrates can be named according to IUPAC rules of nomenclature. Glyceraldehyde is 2,3-dihydroxypropanal, while dihydroxyacetone would be 1,3-dihydroxypropanone. There are also general ways of naming monosaccharides that, using the suffix *-ose* to indicate a carbohydrate, can specify the carbonyl functional group (**aldose** or **ketose**), the number of carbon atoms (*tri-, tetr-, pent-*), or both (*aldohexose, ketopentose*). In addition, each monosaccharide has its own individual name, which is dependent upon the overall structure of the molecule, as we shall soon see.

monosaccharide
a single carbohydrate unit

oligosaccharide
a polymer of two to ten saccharide units

polysaccharide
a polymer with more than ten saccharide units

aldose
a polyhydroxy aldehyde

ketose
a polyhydroxy ketone

14.3 Structures of Monosaccharides

A. D, L-Aldoses: Open Chain Structures

Glyceraldehyde, the simplest aldose, has one chiral carbon atom and therefore two (2^1) optical isomers (Chapter 7). These are enantiomers, or mirror images.

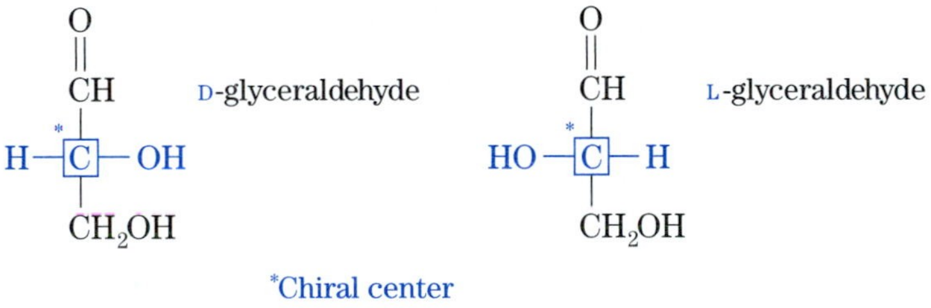

The L- and D- designations refer to the physical placement of the —OH group on the left and right side of the chiral carbon atom, respectively, and have no intended correlation with the direction of the rotation of plane-polarized light.

The structures of monosaccharides with more than three carbons can be drawn by introducing a CHOH group into glyceraldehyde between the carbonyl group and the chiral carbon atom. Notice that the new CHOH will be another chiral atom with the possibility of two orientations, one with the —OH group on the right and one with the —OH group on the left. L- and D- glyceraldehydes will each give rise to two aldotetroses, for a total of four stereoisomers (two chiral centers, 2^2 stereoisomers).

D-sugar
the most common carbohydrates; D- refers to the right-hand orientation of the —OH group on the chiral carbon atom farthest from the carbonyl group

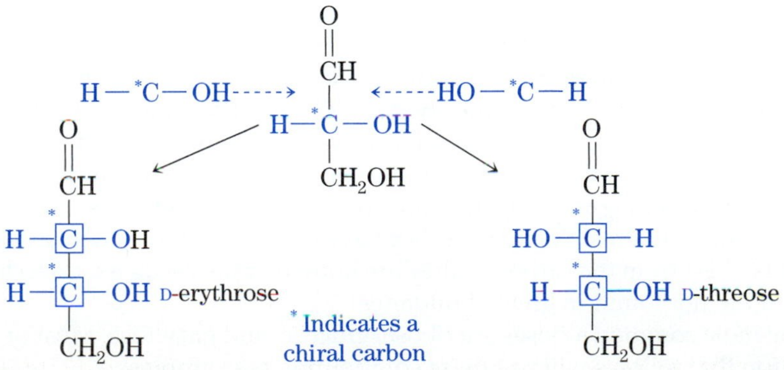

Connections 14.1 www.prenhall.com/bailey

Diabetes

More than 15 million people in the United States currently suffer from the effects of a condition known as *diabetes mellitus.* It may take several forms, all of which result in faulty metabolism of glucose, our primary energy source. Normally, when we ingest food our bodies trigger the release of the endocrine hormone insulin from the beta cells of the pancreas. Insulin, in turn, facilitates the entrance and metabolism of glucose in our cells. If insulin is absent or malfunctions, glucose circulating in the blood increases in concentration. This condition is known as **hyperglycemia.** The kidneys are responsible for cleaning the blood of unnecessary materials and usually reabsorb normal quantities of glucose for metabolism. However, if the blood glucose concentration exceeds the *renal threshold* of 162–180 mg/100 mL (deciliter) in a person who is fasting, glucose will "spill" into the urine and can be detected there by simple tests.

The tragic outcome of this glucose excess is starvation, because the glucose cannot be used by the brain, muscles, and other organs. In addition, all the tissues of the body are being bathed in a concentrated sugar solution that can modify the proteins of the body and severely upset metabolism. Some of the consequences of unchecked diabetes include atherosclerosis (narrowing of the blood vessels), blindness (retinopathy), kidney failure, and coma, all of which could lead to premature death.

The treatment of diabetes depends upon the type and severity of the condition. Those who produce little or no insulin usually due to the destruction of pancreatic beta cells, have Type 1 (previously called insulin-dependent diabetes mellitus, IDDM) and require daily injections of insulin. Type 2, (noninsulin-dependent diabetes mellitus, NIDDM) in which insulin is produced but is not effective, frequently affects those who are overweight or genetically predisposed. Type 1 diabetes may arise from destruction of the beta cells of the pancreas due to the body's reaction to a virus. Types 1 and 2 may also have genetic or drug-induced causes. Often found in older people who are obese, type 2 is frequently controlled by diet, exercise, and/or the administration of oral drugs such as tolbutamide or glipizide (sulfonylurea class), Metformin® (biguanide class), or miglitol (an antiglucosidase).

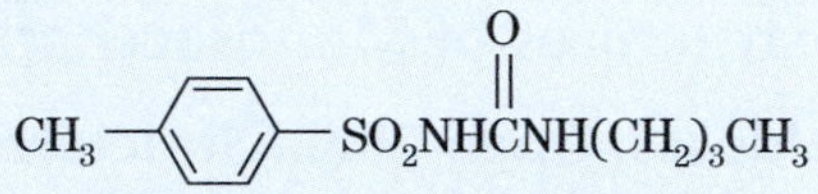

CH_3–C_6H_4–$SO_2NHC(=O)NH(CH_2)_3CH_3$

Generic drug name: tolbutamide
Trade (proprietary) name: Orinase®

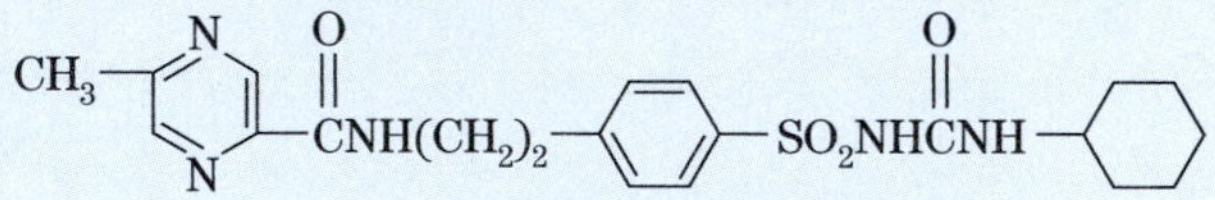

Generic drug name: glipizide
Trade name: Glucotrol®

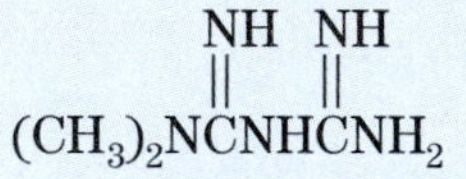

$(CH_3)_2NC(=NH)NHC(=NH)NH_2$

metformin (Diabex®)

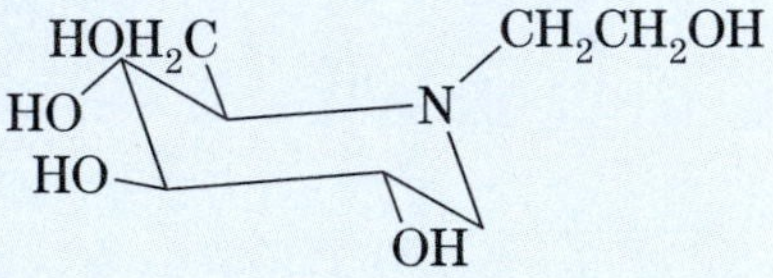

miglitol

Research into the causes and treatments for diabetes is very active. Biotechnology has produced human insulin, which has fewer side effects than animal types used in the past. The quality and length of life is being increased for millions every day.

Both of these compounds are D-sugars because the chiral carbon farthest from the carbonyl is derived from D-glyceraldehyde. They are related to each other as diastereomers. Recall that diastereomers (section 7.5.B) are distinct chemical entities having different physical properties. More specifically, since D-erythrose and D-threose are different at only one chiral carbon, they are referred to as **epimers.**

epimer
one of two diastereomers that differ in the orientation of groups at only one carbon

There are two chiral carbon atoms in the aldotetroses, which means 2^2 or 4 stereoisomers are possible. Only two are shown. (What are the other two?) Each of these tetroses can be extended to two aldopentoses, and so on. Most of the common naturally occurring monosaccharides have been derived from D-glyceraldehyde. Figure 14.1 illustrates the aldotroses, pentoses, and hexoses derived from D-glyceraldehyde. Since they have in common the D-orientation on the carbon farthest from the carbonyl, they are known as the D-sugars. L-saccharides exist but are not found in great abundance.

The most common aldoses are ribose, glucose, and galactose. Most of the information that follows will use these compounds as examples.

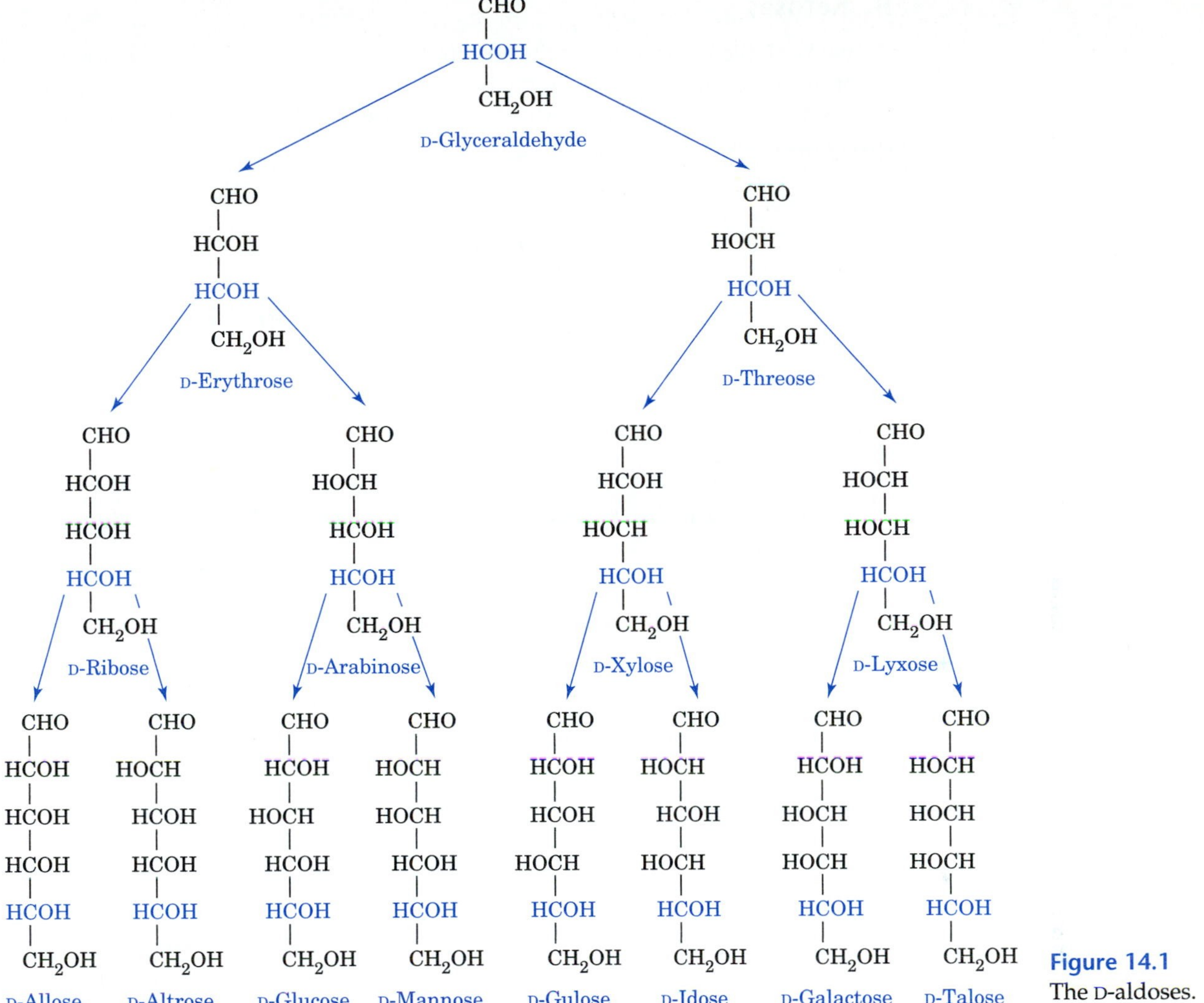

Figure 14.1
The D-aldoses.

GETTING INVOLVED

- ✓ What suffix in a molecular name usually indicates that the structure is a carbohydrate?
- ✓ What functional groups are present in carbohydrate molecules?
- ✓ Draw the structure of an aldopentose; a ketopentose.
- ✓ Now draw the enantiomers of the aldopentose and ketopentose you just made. How are these four structures alike and how are they different?

Problem 14.1

What is the general term that could be used to describe allose? xylose?

Problem 14.2

Using Figure 14.1 find an epimer and two diastereomers for D-glucose.

Problem 14.3

Draw the structures of the aldoheptoses that would arise if you inserted a new chiral carbon atom at position-2 of D-galactose.

B. Ketoses

As yet we have not discussed the structures of ketoses. A family of ketotetroses, ketopentoses, and ketohexoses can be drawn up in the same way as the aldoses, that is, by inserting a chiral carbon atom following the carbonyl group in dihydroxy acetone.

CH_2OH — $C{=}O$ — CH_2OH

H—*C—OH → ← HO—*C—H

CH_2OH — $C{=}O$ — H—*C—OH — CH_2OH D-erythrulose

CH_2OH — $C{=}O$ — HO—*C—H — CH_2OH L-erythrulose

*Indicates a chiral carbon atom

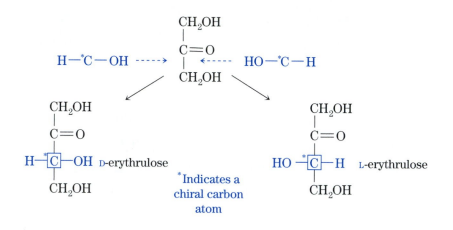

The most common ketose is the ketohexose fructose (fruit sugar). Notice that it is a functional isomer of glucose.

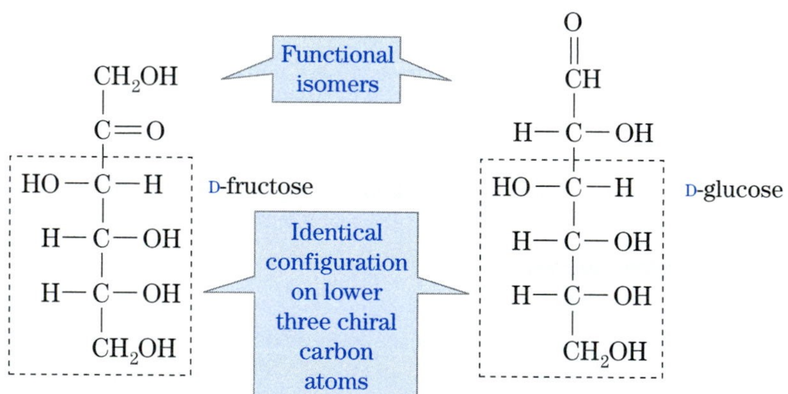

GETTING INVOLVED

✓ To draw the family of ketopentoses derived from D-erythrulose, where do you place the new chiral carbon atom?

✓ How does the placement of a new chiral carbon in ketoses compare to that for building the aldoses?

✓ Identify the chiral carbon atoms in D-fructose and D-glucose.

Problem 14.4

Draw the family of ketopentoses derived from D-erythrulose. Which of these are epimers?

See related problem 14.18.

C. Fischer Projections

The way in which the previous structures are drawn is called a *Fischer projection;* the most highly oxidized carbon is at the top and the rest of the chain is drawn below with the groups attached to the chiral carbons projected to the left and right in order to simulate the tetrahedral bonding angles of a saturated carbon atom (see section 7.4).

GETTING INVOLVED

✓ Draw a Fischer projection for each of the following:

$CH_3CH(OH)CO_2H$ $HOCH_2CH(OH)CHO$ $CH_3COCH(OH)CH_3$

D. Cyclic Structures—Hemiacetal Formation

In section 11.5.G we saw that the carbonyl group of an aldehyde is polar and can react with a polar alcohol group to form an alcohol-ether known as a **hemiacetal.** The reaction could proceed one step further with another mole of alcohol to form a diether known as an **acetal.** Note that a new chiral carbon atom is formed in the reaction.

hemiacetal
the alcohol-ether product of the reaction between an aldehyde and one mole of an alcohol

acetal
the diether product of the reaction between an aldehyde and two moles of alcohol

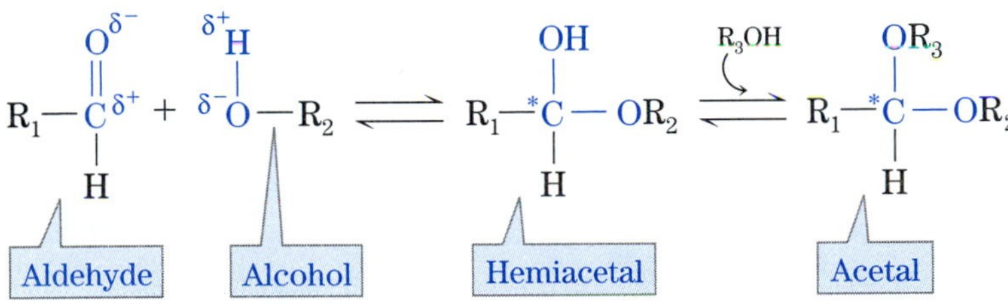

*Indicates a chiral carbon center

GETTING INVOLVED

✓ What are the similarities and differences between an acetal and a ketal?

✓ How are hemiacetals formed? acetals?

✓ What groups are attached to the new chiral carbon atom in a hemiacetal? in an acetal?

1. ***Ring-Forming Reaction.*** The tetrahedral nature of carbon and the length of the chain in aldopentoses and aldohexoses allow the carbonyl in a monosaccharide molecule to come into close proximity with an alcohol group on the same molecule. The hemiacetal that results from an intramolecular reaction will make a ring, a cyclic hemiacetal. As we saw earlier in section 2.8 on the cyclic structures, five- and six-membered rings are stable. The same holds true for the cyclic hemiacetals with the one difference that one member of the ring is an oxygen.

hemiketal
the alcohol-ether product of the reaction between a ketone and one mole of alcohol

ketal
the diether product of the reaction between ketone and two moles of alcohol

The most prevalent cyclic hemiacetal structure for glucose is a six-membered ring. To form this structure, the carbonyl carbon is the first ring member and the hydroxy group oxygen on the fifth carbon adds to the carbonyl to produce the six-membered ring hemiacetal.

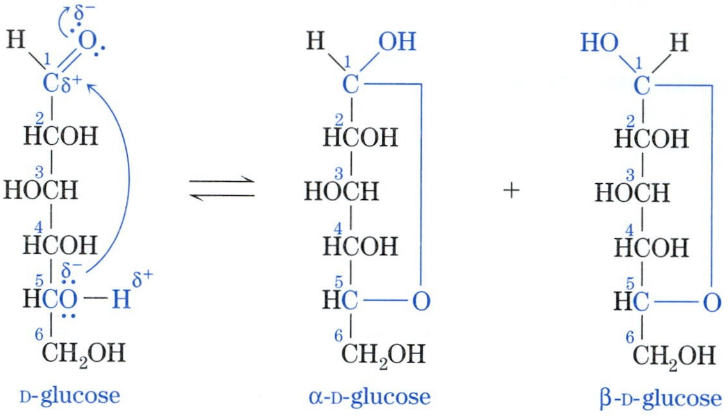

D-glucose ⇌ α-D-glucose + β-D-glucose

anomer
one of two optical isomers formed at the new chiral carbon produced when an aldehyde forms a hemiacetal or a ketone forms a ketal

α-anomer
the cyclic hemiacetal that has the —OH group on the new chiral carbon below the ring; on the right in a Fischer projection

β-anomer
the cyclic hemiacetal form that has the —OH group on the new chiral carbon above the ring

Notice that a new chiral center has been formed. The symbol **α** or **β** is used to indicate whether the hemiacetal alcohol group is on the right or left, respectively, in the Fischer projection. Specifically, these two new optical isomers (diastereomers) are called **anomers.**

In order to make this connection, the glucose molecule must bend back on itself to form the cyclic structure. The carbonyl carbon is the first member of the ring, while an alcohol oxygen is the last member.

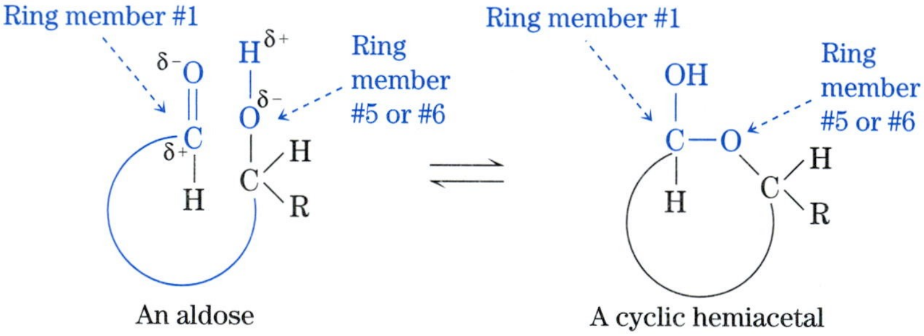

Ring member #1
Ring member #5 or #6

An aldose ⇌ A cyclic hemiacetal

GETTING INVOLVED

✓ How are anomers and epimers alike? How are they different?
✓ How are diastereomers related to anomers and epimers?

Example 14.1

Draw the five- and six-membered cyclic hemiacetal forms of D-ribose.

Solution

First draw the Fischer projection D-ribose. Number the members of the proposed ring starting with the carbonyl carbon as #1 and the alcohol oxygen as the last member, #5 or #6, of the ring.

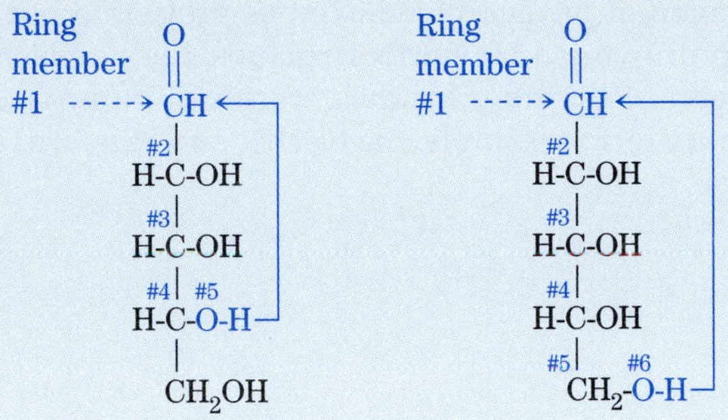

Ring member #1 → CH (=O)
#2 H-C-OH
#3 H-C-OH
#4 H-C-O-H #5
CH_2OH

Ring member #1 → CH (=O)
#2 H-C-OH
#3 H-C-OH
#4 H-C-OH
#5 CH_2-O-H #6

Next, perform the conversion to the hemiacetal.

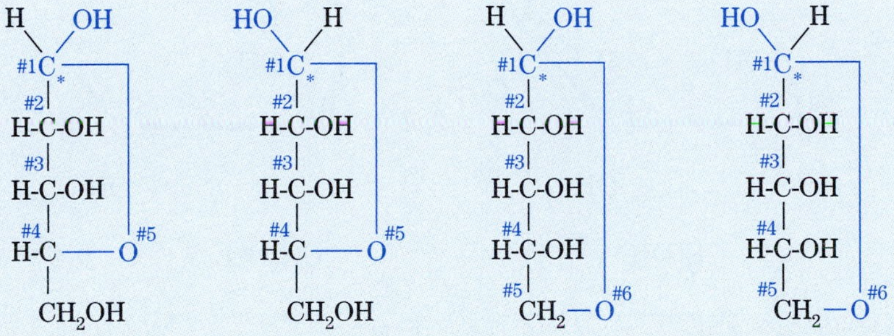

H OH
#1C*
#2 H-C-OH
#3 H-C-OH
#4 H-C—O #5
CH_2OH
α

HO H
#1C*
#2 H-C-OH
#3 H-C-OH
#4 H-C—O #5
CH_2OH
β

H OH
#1C*
#2 H-C-OH
#3 H-C-OH
#4 H-C-OH
#5 CH_2—O #6
α

HO H
#1C*
#2 H-C-OH
#3 H-C-OH
#4 H-C-OH
#5 CH_2—O #6
β

*Indicates the new chiral carbon atom

Problem 14.5

Among the four compounds drawn below, which are anomers, which are epimers, and which are diastereomers? Are there any structures which are none of these? Which ones?

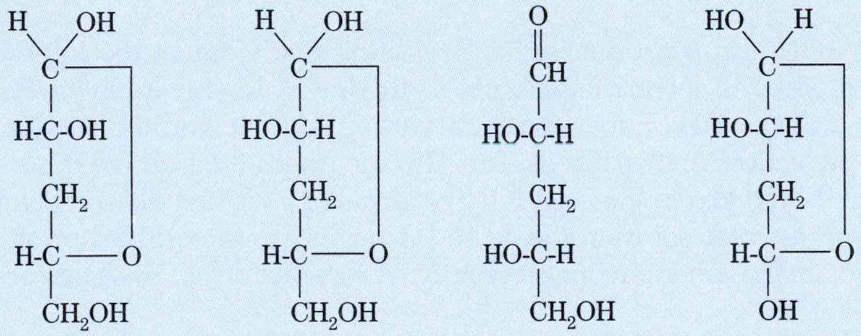

H OH
C
H-C-OH
CH_2
H-C—O
CH_2OH

H OH
C
HO-C-H
CH_2
H-C—O
CH_2OH

O
CH
HO-C-H
CH_2
HO-C-H
CH_2OH

HO H
C
HO-C-H
CH_2
H-C—O
OH

Problem 14.6

Draw the six-membered cyclic hemiacetal forms of D-mannose and D-glucose. Designate the α- and β-anomers.

See related problems 14.19, 14.22, 14.24.

These Fischer projections obviously do not adequately represent the correct bond lengths and atom orientations of the cyclic structures. There are other ways to represent cyclic monosaccharides that may seem more familiar to you.

2. ***Structural Representations: Haworth Structures.*** The cyclic form of a monosaccharide is most frequently represented by using a Haworth formula in

Haworth structure two-dimensional five- or six-membered ring representation of the cyclic form of a monosaccharide; —OH groups that appear on the right in a Fischer projection are drawn down (below the plane of the ring) in a Haworth structure and those —OH groups on the left in a Fischer projection are drawn up

which a planar pentagon or hexagon is viewed as projecting out of the paper toward the reader. In drawing a **Haworth structure,** the hemiacetal carbon is placed on the right end of the ring; for the D-series of monosaccharides, the —CH_2OH is up. Compare, for example, the Fischer and Haworth structures for the α- and β-D-glucose:

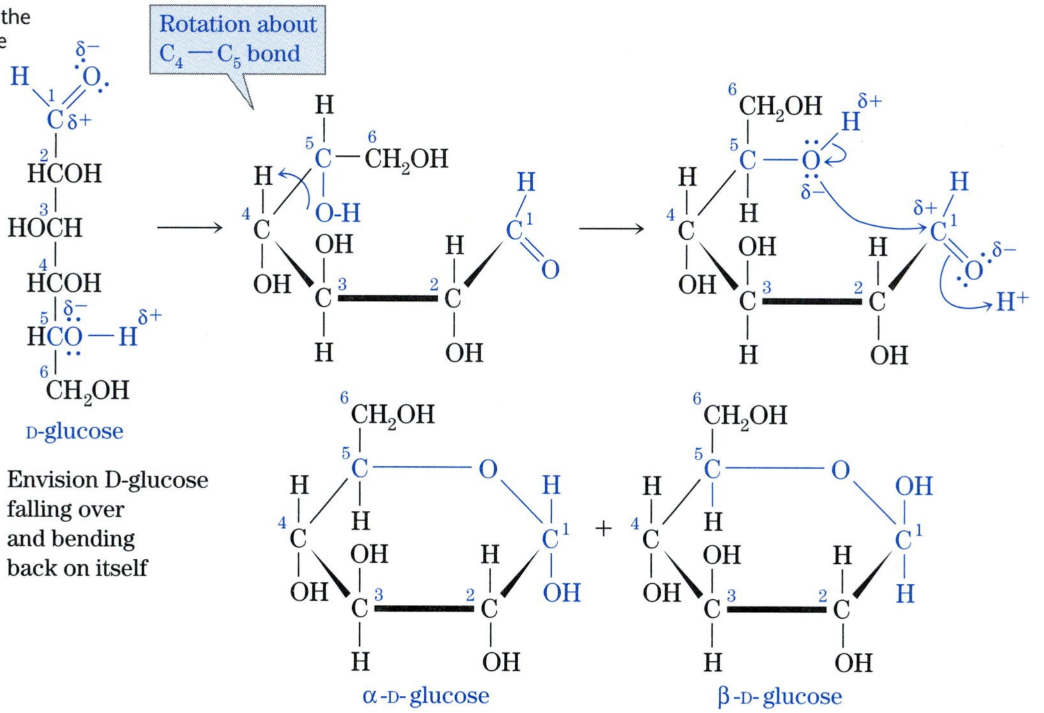

Rotation about C_4—C_5 bond

D-glucose

Envision D-glucose falling over and bending back on itself

α-D-glucose + β-D-glucose

The —OH groups in the Fischer projection that were on the right (except that on C-6, which is not on a chiral carbon) are drawn downward in the Haworth structure. If the —OH on the new chiral carbon is up, it would be on the left in a Fischer projection, that is, it is β-. This also means that it is on the same side of the ring as the oxygen in the ring. If that group were written down, it would be the α-form. In water solution the α- and β-forms are in equilibrium with the open chain and so can interconvert readily. The cyclic forms, however, predominate in solution.

$$\alpha\text{-form} \rightleftarrows \begin{matrix}\text{Open chain}\\ \text{aldehyde or}\\ \text{ketone}\end{matrix} \leftrightarrows \beta\text{-form}$$

Haworth structures can also be drawn as are other cyclic forms, that is, without showing the carbon atoms and using the intersection of bonds to indicate the position of each carbon. The hydrogen atoms may also be assumed, since the cyclic forms have saturated carbons. Because it is the alcohol and hemiacetal groups that are reactive, the figures representing saccharides commonly use a line to indicate an —OH group. Please note that this is used only in carbohydrate chemistry.

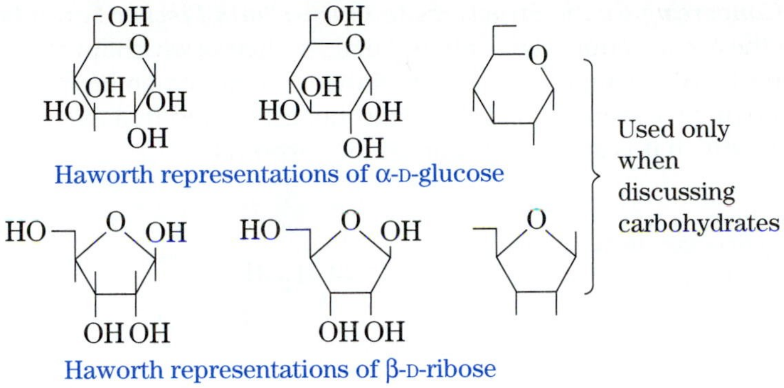

Haworth representations of α-D-glucose

Haworth representations of β-D-ribose

Five-membered rings are called **furanose** forms because they resemble the heterocyclic compound furan, while six-membered rings are termed **pyranose** forms because of their resemblance to pyran.

furanose five-membered ring form of a monosaccharide

pyranose six-membered ring form of a monosaccharide

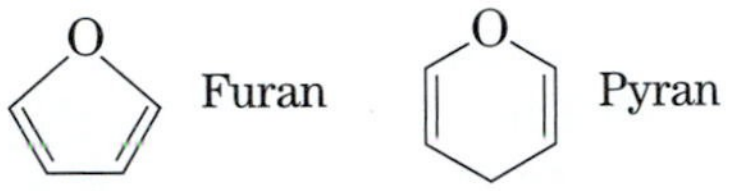

GETTING INVOLVED

- ✓ Draw the Haworth structures for the three forms of D-glucose, α-, β-, and open chain, as an equilibrium.
- ✓ Besides being anomers, are the α- and β- forms of a monosaccharide also epimers, diastereomers, and/or enantiomers?
- ✓ How is a furanose version of a monosaccharide the same and how is it different from a pyranose form?
- ✓ Can aldotetroses form both furanose and pyranose rings?

Problem 14.7

Draw the Haworth structures for the five-membered and six-membered cyclic hemiacetal forms of D-arabinose and D-xylose.

3. ***Structural Representations: Conformational Structures.*** Recall that cyclohexane could also be drawn in its **conformational structure,** that is, as a boat and a chair form. This also can be done with cyclic monosaccharides. The chair form is the stable form for sugars as it is for cyclohexane. Placement of the —OH groups will again be up and down, but this time in relation to axial and equatorial positions around the ring. Recall that axial and equatorial alternate as to which is above the ring as you go around it.

conformational structure relating to carbohydrates, this is the chair form of the cyclic hemiacetal or hemiketal

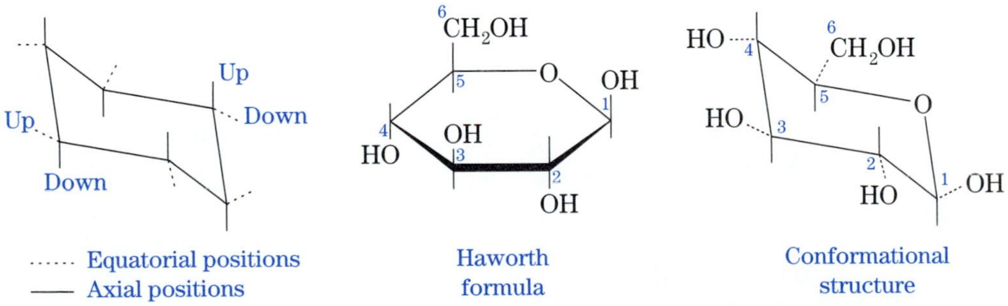

Representations of β-D-glucose

It is of interest to note that β-D-glucose is probably the most abundant form of carbon in the biosphere. As you can see from the equatorial placement of all of the substituent —OH groups, it is a very stable structure.

4. ***Converting Cyclic Structures to Open-Chain Fischer Formulas.*** You can convert the cyclic hemiacetals or hemiketals to their open-chain aldose or ketose structures by reversing the reaction that made them, remembering that the oxygen in the ring was contributed by the alcohol group and that the —OH-bearing carbon to which it is attached is part of the carbonyl.

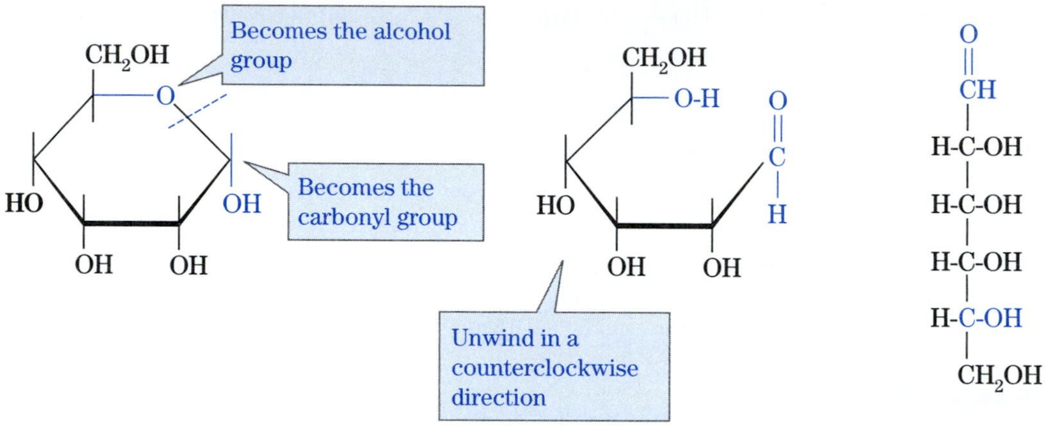

5. ***Cyclic Hemiketals and Ketals.*** Ketoses form ring structures in exactly the same way as do aldoses. The difference will lie in the groups attached to the new chiral carbon center.

GETTING INVOLVED

- ✓ Can the right and left —OH groups in the Fischer projection of a monosaccharide be correlated directly to up and down positions in the conformational form? Explain the correlation.
- ✓ Compare the cyclic forms of α-D-glucose and α-D-fructose, both in the pyranose Haworth forms. What is the major distinguishing characteristic of an aldose as compared to its corresponding ketose?

Example 14.2

Draw the Haworth and conformational structures for the α-anomer of the six-membered cyclic hemiacetal form of D-talose.

Solution

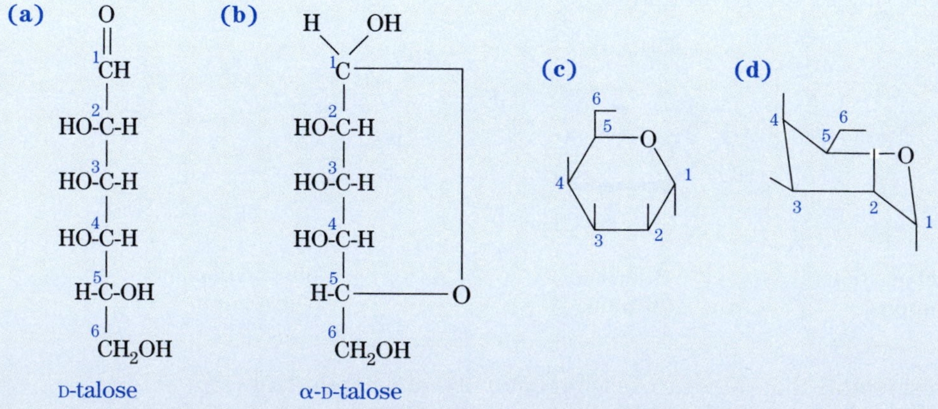

(a) Begin by drawing the open-chain Fischer projection of D-talose.

(b) Form the hemiacetal linkage for a six-membered ring. Since this is an aldose, the first carbon, the carbonyl carbon, is the first member of the ring. For the six-membered ring the last atom will be the oxygen on carbon-5. Be sure to draw the α-form.

(c) Convert to the Haworth structure. The three remaining —OH groups on chiral carbons are on the left in the Fischer projection and will be drawn up in the Haworth form.

(d) Draw the corresponding conformational structure.

Example 14.3

Draw the Haworth structures for the β-isomer of the five- and six-membered cyclic Haworth structures of D-fructose.

Solution

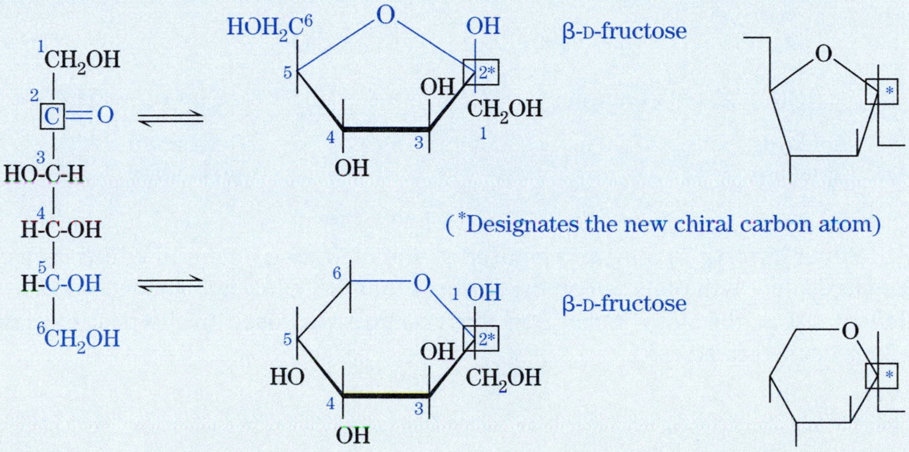

$^{1}CH_2OH$, ^{2}C=O, ^{3}HO-C-H, ^{4}H-C-OH, ^{5}H-C-OH, $^{6}CH_2OH$ ⇌ HOH_2C^6 (5), O, OH, 2*, OH, CH_2OH (1), 4, 3, OH — β-D-fructose

(*Designates the new chiral carbon atom)

⇌ 6, O, 1 OH, 5, 2*, OH, HO, CH_2OH, 4, 3, OH — β-D-fructose

Problem 14.8

Draw the conformational form for the six-membered (pyranose) cyclic hemiacetal structure of D-galactose. Show both α- and β-anomers.

Problem 14.9

Draw the open-chain Fischer projections for the following Haworth forms:

(a) CH_2OH, HO, O, OH, HO, OH

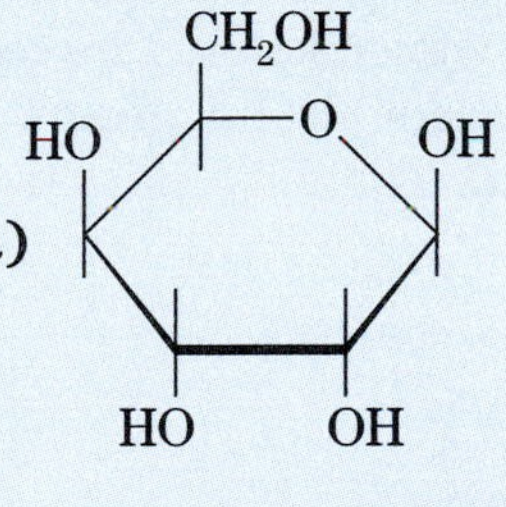

(b) HOH_2C, O, OH, OH, OH

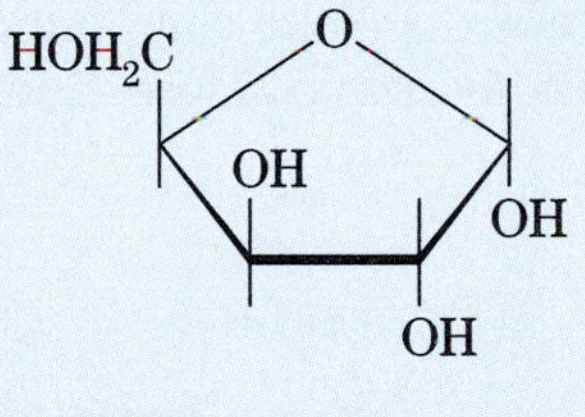

(c) O, OH, OH, HO, HO, CH_2OH

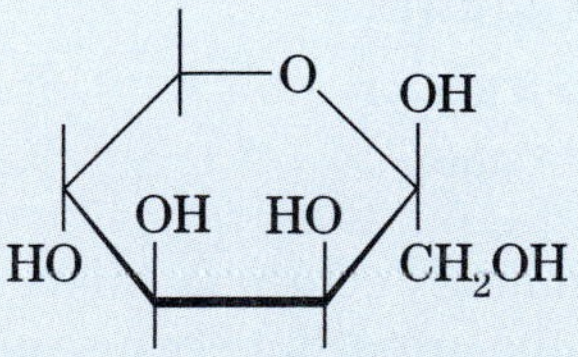

(d) HOH_2C, O, CH_2OH, OH, OH, HO

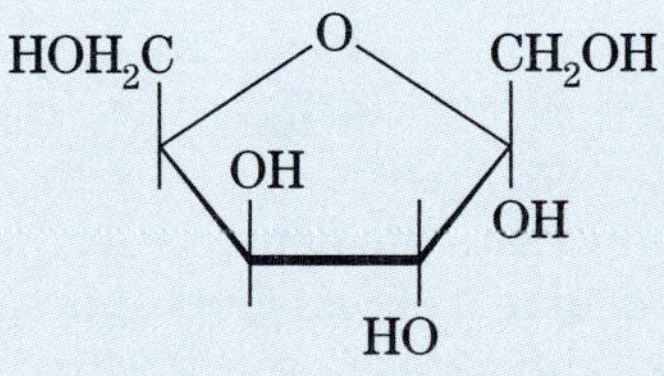

14.4 Some Reactions of Monosaccharides

Since monosaccharides contain carbonyl and alcohol groups, they can undergo the types of reactions that are characteristic of aldehydes, ketones, and alcohols. We will present only a few of these reactions, specifically some of those important to the detection of carbohydrates, their polymerization, and their metabolism.

A. Oxidation of Carbohydrates (Reducing Sugars)

reducing sugar
carbohydrate that has one or more anomeric carbons available for oxidation by a mild oxidizing agent; that is, the carbon contains an alcohol and an ether group on it

Aldehydes are readily oxidized to carboxylic acids by fairly mild oxidizing agents (section 11.4). Therefore, they are good reducing agents. Aldoses have the same chemical property. If a Cu^{2+} complex is used in basic conditions, a precipitate of copper (I) oxide indicates a reducing agent, or in the case of sugars, a **reducing sugar.** This is known as *Fehling's* (tartrate copper complex) or *Benedict's* (citrate copper complex) *test.*

$$\underset{\text{Aldehyde or aldose}}{\mathrm{RCH}(=\mathrm{O})} + 2\mathrm{Cu}^{2+}\ (\text{complex}) + 5\mathrm{OH}^- \longrightarrow \mathrm{RCO}(=\mathrm{O})^- + \underset{\text{Red-brown precipitate}}{\mathrm{Cu_2O}\downarrow} + 3\mathrm{H_2O}$$

Silver ion (Ag^{1+}) may also be used as the oxidizing agent, in which case elemental silver will plate out on the surfaces that the solution contacts. Known as *Tollens test* or the *silver mirror test,* this reaction was used in the past to make silver-backed mirrors.

$$\mathrm{RCH}(=\mathrm{O}) + 2\mathrm{Ag(NH_3)_2}^+ + 3\mathrm{OH}^- \longrightarrow \mathrm{RCO}(=\mathrm{O})^- + \underset{\text{Silver mirror}}{2\mathrm{Ag}\downarrow} + 4\mathrm{NH_3} + 2\mathrm{H_2O}$$

Since both α- and β-anomers of a saccharide are in equilibrium with the open-chain carbonyl, there is no problem in the cyclic forms opening and reacting.

Normal organic ketones do not react with mild oxidizing agents; that is, they give negative Fehling, Benedict, and Tollens tests. However, ketoses are reducing sugars because as 2-oxo compounds with an adjacent alcohol group, they can rapidly tautomerize to aldoses and so be oxidized.

$$\underset{\text{Ketose}}{\mathrm{CH_2OH{-}C{=}O}} \rightleftharpoons \underset{\text{Enediol}}{\mathrm{CHOH{=}COH}} \rightleftharpoons \underset{\text{Aldose}}{\mathrm{HC{=}O{-}CHOH}}$$

For many years the Benedict test was a preliminary screen for diabetes because it is positive if excess glucose is spilled into the urine. But because all common aldoses and ketoses give positive reducing sugar tests, more specific tests have been devised to identify the presence of glucose specifically in body fluids as evidence of diabetes, rather than other sugars excreted due to some other form of abnormal metabolism such as fructosemia. Proteins known as enzymes are usually extremely specific in reacting with compounds. The enzyme glucose oxidase has been mixed with dyes and placed on a paper strip (Test-tape®) so that, when dipped into urine, it will record the presence and relative amount of glucose present. This is the basis of the widely used commercial test kit that both takes a sample of blood by pricking the finger and analyzes it.

The oxidized aldose, or carboxylic acid, is named using the specific monosaccharide stem with the ending **-onic acid.** Glucose becomes gluconic acid and galactose becomes galactonic acid.

If the other end of the molecule, the primary alcohol group, is oxidized (without the carbonyl being oxidized), the resulting product is called a **-uronic acid:** glucuronic acid, galactouronic acid.

Are there any **nonreducing sugars?** The answer is yes; if the anomeric group is no longer free to open, it can't be oxidized. This occurs when the molecule reacts with another mole of alcohol to form an acetal or ketal.

-onic acid
a carbohydrate derivative wherein the aldehyde functional group has been oxidized to a carboxylic acid

-uronic acid
a carbohydrate derivative wherein the last, primary alcohol group has been oxidized to a carboxylic acid

nonreducing sugar
a carbohydrate with all of its anomeric carbons bonded to other groups, unavailable for opening to an aldehyde or ketone carbonyl

GETTING INVOLVED

- ✓ What are the key chemical species in the tests for reducing sugars, that is, in Fehling and Benedict test and in Tollens test?
- ✓ How would the testing solutions appear in the presence and absence of a reducing sugar?
- ✓ Draw the structures of gluconic and glucuronic acids. Which structural features are the same and which are different in the two molecules?

Problem 14.10

Under the basic conditions of reducing sugar tests, fructose tautomerizes to an aldose. There are two possible aldoses. What are they?

Problem 14.11

Draw the structures of mannuronic acid, xylonic acid, and iduronic acid.

See related problems 14.28, 14.29, 14.30.

B. Reduction of Monosaccharides

The aldehyde or ketone group of a monosaccharide can be reduced purposefully or naturally to produce the corresponding sugar alcohol. Reduced glucose is called *sorbitol,* while fructose can be reduced to sorbitol or *mannitol.*

Glucose		Sorbitol		Fructose		Mannitol
O=CH		CH_2OH		CH_2OH		CH_2OH
HCOH		HCOH		C=O		HOCH
HOCH	$\xrightarrow[\text{Cat}]{H_2}$	HOCH	$\xleftarrow[\text{Cat}]{H_2}$	HOCH	$\xrightarrow[\text{Cat}]{H_2}$	HOCH
HCOH		HCOH		HCOH		HCOH
HCOH		HCOH		HCOH		HCOH
CH_2OH		CH_2OH		CH_2OH		CH_2OH

Connections 14.2 www.prenhall.com/bailey

Prevention of Disease and Detoxification

To paraphrase a popular commercial, "Glucose does a body good;" the statement is correct, even if the grammar isn't. Glucose is not only a major body fuel, stored in muscle and liver as its polymeric form, glycogen, but it is also a precursor to oxidation products, ascorbic acid and glucuronic acid, which are essential to metabolism.

Most organisms can produce ascorbic acid through a series of specific enzyme-catalyzed oxidation reactions. Ascorbic acid easily undergoes a reversible oxidation-reduction reaction making it an essential cofactor in metabolism. For example, the formation of collagen and elastin, proteins that comprise our skin, teeth, bones, ligaments, cartilage, and connective tissue, depends upon the oxidation of certain amino acids in those proteins. Ascorbic acid plays an active role in the reactions. In the absence of ascorbic acid, tooth enamel is weakened along with the connective tissue of the gums. Frequent nosebleeds occur as well as bruising; the immune system is generally weakened. This condition is known as scurvy.

Humans, along with other primates, guinea pigs, and the Indian fruit bat, are among the few species that cannot make their own ascorbic acid. Since it must then be provided in the diet, it becomes an essential nutrient or vitamin—vitamin C to be specific. Scurvy was prevalent in the age of sea-faring explorers as it is now in war-torn countries where populations are malnourished. Dietary sources of vitamin C include citrus fruits (lemons, limes, and oranges), strawberries, tomatoes, and green vegetables.

Glucuronic acid, which can be produced by humans, is one of the most important intermediates in the detoxification of the body from regular metabolic wastes as well as foreign toxins including drugs. With its variety of functional groups, it can react, or conjugate, with many kinds of organic molecules to form water-soluble derivatives that can be excreted through the kidneys.

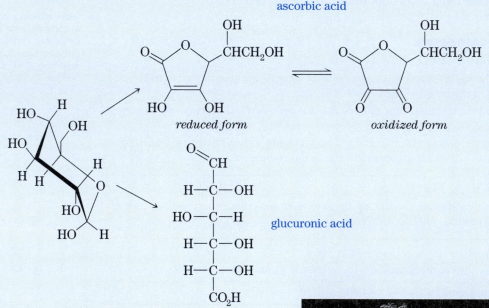

ascorbic acid

reduced form ⇌ *oxidized form*

glucuronic acid

Linus Pauling holding an orange containing vitamin C and an α-helix found in proteins.

Both of these compounds have sweetness and are used as sugar substitutes in candies and chewing gum.

GETTING INVOLVED

Problem 14.12

Xylitol, another sugar substitute, which purportedly can prevent cavities, is produced by the reduction of xylose. Arabitol is the reduction product of arabinose. Draw the structures of xylitol and arabitol. Are they, or the other sugar alcohols, reducing?

C. Esterification

Since they are alcohols, saccharides can condense with acids to form esters (sections 12.6 and 13.5).

$$\underset{\text{Acid}}{R-\overset{\overset{\displaystyle O}{\|}}{C}-\boxed{OH}} + \underset{\text{Alcohol}}{\boxed{H}-O-R'} \longrightarrow \underset{\text{Ester}}{R-\overset{\overset{\displaystyle O}{\|}}{C}-OR'} + H_2O$$

This can occur with organic or inorganic acids, such as phosphoric and sulfuric acids. These inorganic acids are also found in combined forms such as adenosine triphosphate (ATP).

$$\text{Glucose} + \underset{\text{ATP}}{HO-\overset{\overset{\displaystyle O}{\|}}{\underset{\underset{\displaystyle OH}{|}}{P}}-O-\overset{\overset{\displaystyle O}{\|}}{\underset{\underset{\displaystyle OH}{|}}{P}}-O-\overset{\overset{\displaystyle O}{\|}}{\underset{\underset{\displaystyle OH}{|}}{P}}-\text{O-adenosine}} \xrightarrow{\textit{Enzyme}} \text{Glucose-6-phosphate} + \text{ADP}$$

Glucose ATP Glucose-6-phosphate

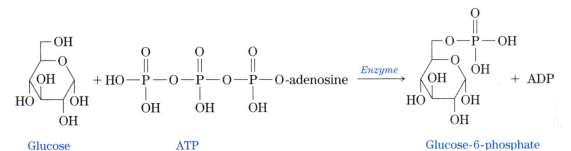

Enzymes are necessary to catalyze the condensation. Phosphate derivatives of saccharides are common as metabolic intermediates in all living organisms.

Sulfate esters of carbohydrates can be found in such biochemically important materials as skin, cartilage, and the lens of the eye.

14.5 Disaccharides and Polysaccharides

So far we have looked only at hemiacetal and hemiketal formation. If another mole of alcohol is available, an acetal or ketal can be made. This is the major means by which monosaccharides polymerize into oligo- and polysaccharides, as well as react with other biochemical molecules.

Keep in mind that a second monosaccharide with its —OH groups is a polyol. If the link is made between the anomeric carbon of one unit and an —OH from a second molecule, a polymerization has begun.

A. Glycosidic Linkages or Bonds

glycoside bond acetal or ketal formed from the reaction of a cyclic monosaccharide molecule with another monosaccharide

The bond made between two monosaccharide units is called a *glycosidic linkage* or **glycoside bond.** Drawn below are a few of the options open to two molecules of glucose reacting together. Notice that the first glucose unit has its hemiacetal

—OH in the α-position (axial). Once it is linked to an alcohol, the glycoside bond will remain in that α-position and the bond will be an α-glycosidic bond. Does it make a difference whether that first unit has the bond in the α- or β- position? Absolutely. We shall discuss more about this as we proceed with this section.

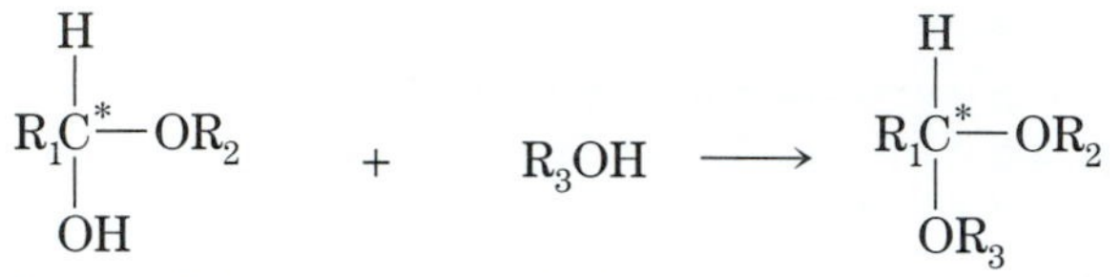

$$R_1C^*H(OH)-OR_2 + R_3OH \longrightarrow R_1C^*H(OR_3)-OR_2$$

hemiacetal *alcohol* *acetal*

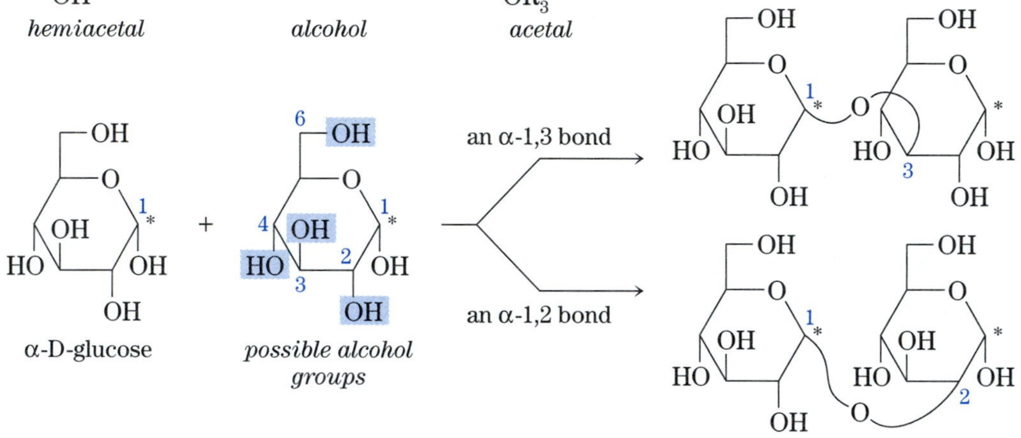

α-D-glucose + *possible alcohol groups* → an α-1,3 bond; an α-1,2 bond

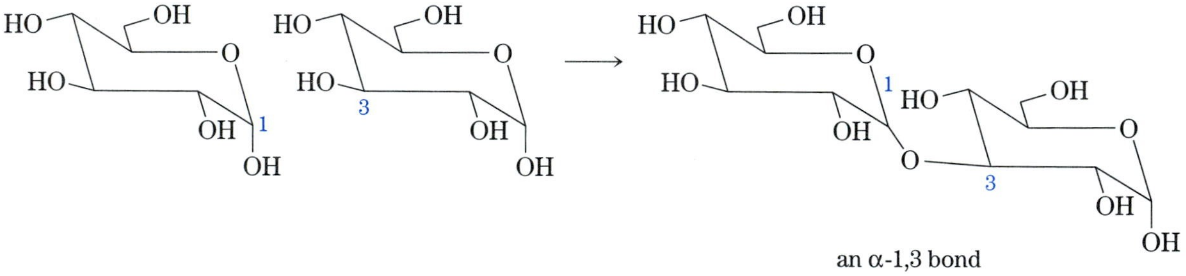

an α-1,3 bond

Notice that the position of the bond at the anomeric carbon in the first sugar unit can be either α- or β-, which is indicated in the name. The orientation (up or down) of the reacting —OH in the second sugar unit is specified by the identity of the monosaccharide (glucose, galactose, etc.) and the hydroxyl position on the ring.

GETTING INVOLVED

✓ Which groups on two monosaccharides react in order to make a glycosidic bond?

✓ Considering two α-D-glucose molecules, list the possible glycosidic bond combinations.

Problem 14.13

Maltose is composed of two glucose units joined in an α-1,4 glycoside bond, while cellobiose has two glucose molecules joined by a β-1,4 bond. Draw the Haworth and conformational structures for the disaccharides maltose and cellobiose. Are these reducing or nonreducing sugars?

Problem 14.14

Identify the type of glycosidic bond found in each of the following disaccharides. Be sure to specify whether the link is α- or β-.

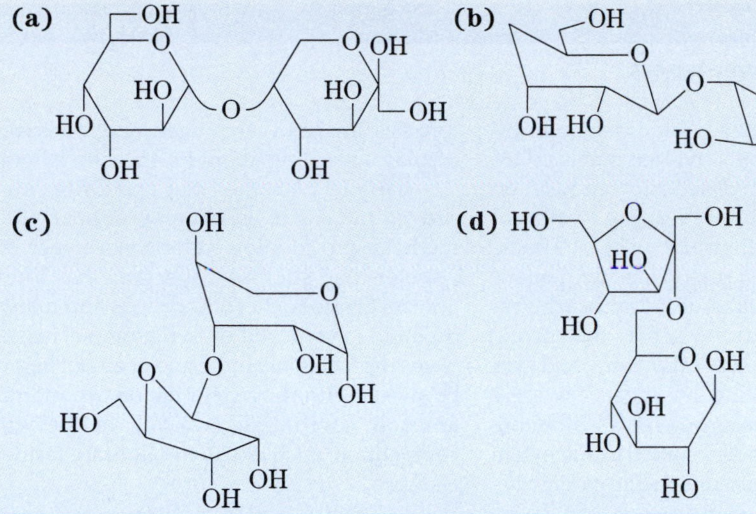

Problem 14.15

Draw the structures of the disaccharides formed from the following monosaccharide units, using the glycosidic linkages specified.

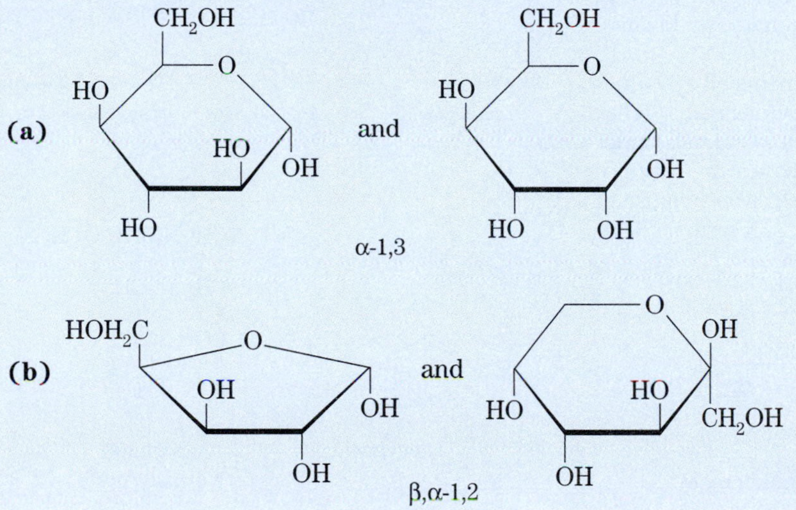

B. Disaccharides

lactose
a disaccharide composed of a galactose and a glucose unit joined by a β-1,4 glycosidic bond

disaccharide
two monosaccharide units linked by a glycosidic bond

1. ***Lactose: Mother's Disaccharide.*** Found exclusively in the milk of mammals, **lactose** (*lac,* Latin for "milk") makes up 4.5% of cow's milk and 6.7% of human milk. This **disaccharide** is composed of galactose and glucose linked by a β-1,4 glycosidic bond.

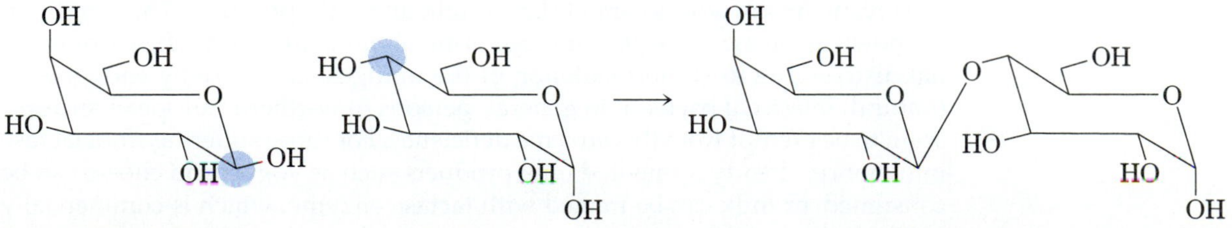

β-D-galactose α-D-glucose β-1,4 bond α-lactose

Connections 14.3

www.prenhall.com/bailey

Low-Calorie Sweeteners

Glucose that is not immediately needed for metabolic energy is either stored as glycogen or converted into lipid for storage in adipose (fat) tissue. Lipid deposits can form on the walls of blood vessels, eventually leading to atherosclerosis and an increased risk of stroke or heart attack. Therefore, for health and cosmetic reasons, many persons have attempted to limit their intake of fat and carbohydrates, especially sucrose. In order to satisfy the "sweet tooth" developed by sugared diets, various natural and synthetic materials have been or are being investigated as sugar substitutes or enhancers. Since these sweeteners either are noncarbohydrate in nature or are not absorbed to any extent in the gastrointestinal tract, they are referred to as *low calorie* or, in some cases, *nonnutritive.*

The sugar alcohols, mannitol and sorbitol, although not as sweet as sucrose, have been used for years as low-calorie substitutes. Their ability to be absorbed in the intestine is minimal, but their capacity for hydrogen-bonding has caused them to be associated with unpleasant laxative action if they are consumed in large quantities.

The use of saccharin, which is approximately 300 times sweeter than sucrose, has come into question because it has been shown to promote cancer in laboratory animals under certain conditions—that is, it can enhance the carcinogenicity of other substances. Under the Delaney clause of the Pure Food, Drug, and Cosmetic Act, saccharin is therefore classified as a carcinogen, and foods containing it must display a warning about its effect on laboratory animals.

Early in 1983, the Food and Drug Administration approved the use of aspartame (L-aspartyl-L-phenylalanyl-methyl ester) as a low-calorie sweetener. About 200 times sweeter than sucrose, aspartame has found its way into gourmet coffees, diet soft drinks, and many other foods. A dipeptide composed of two amino acids, aspartame illustrates the fact that a molecule need not be a carbohydrate to be sweet. With the patent life on aspartame expiring there are new compounds waiting in the wings to offer it competition, such as fructo-oligosaccharides (FOS) and the L-sugars.

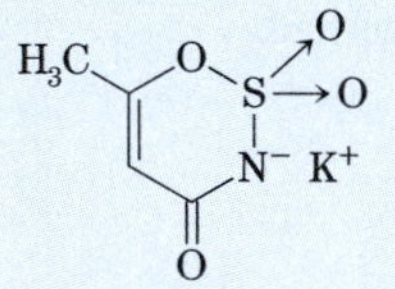

H_3C, O, S, O, O, N^- K^+, O

acesulfame-K

Cl, OH, O, O, HO, ClH_2C, OH, HO, O, $ClCH_2$, OH

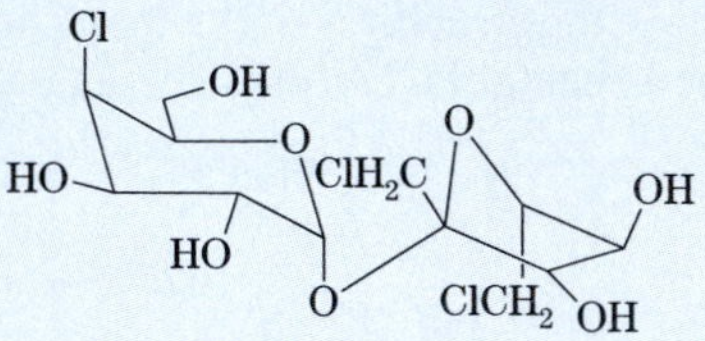

Sucralose (600x the sweetness of sucrose)

$HO_2CCH_2CHCNHCHCNH$ (O, O, NH_2, CH_3), CH_3, CH_3, S, CH_3, CH_3

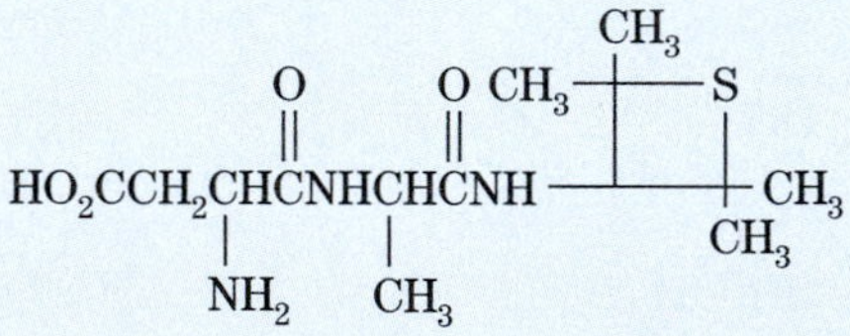

Alitame (2000× the sweetness of sucrose)

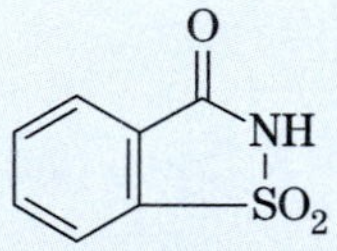

O, NH, SO_2

Saccharin

$H_2NCHCNHCHCOCH_3$ (O, O), CH_2, CH_2, COH, O

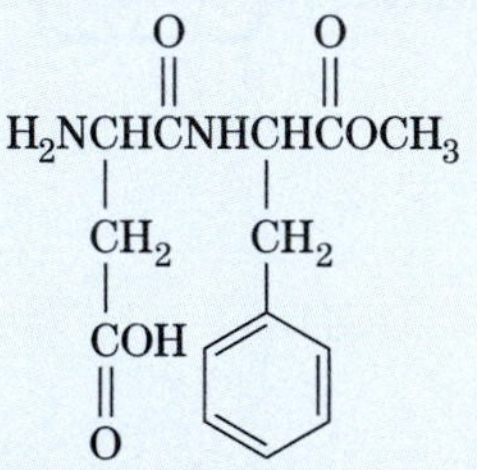

Aspartame (Nutrasweet®)

The second unit of lactose, glucose, has its hemiacetal carbon free, and it can therefore open up to the aldose form. Lactose is thus a reducing sugar and will be oxidized by Fehling, Benedict, and Tollens' reagents.

Lactose requires a special enzyme to break its glycosidic bond. This enzyme is called lactase and it is secreted in the intestines of young mammals. As the infant is weaned, the level of lactase being produced decreases markedly. Very low lactase production is characteristic of 70%–80% of the world's adult population. As a result, many adults cannot digest milk and milk products. This condition is known as *lactose intolerance* and can bring about a great deal of gastrointestinal distress due to the fermentation of the undigested lactose by endogenous (natural) intestinal bacteria. In general, persons of northern European ancestry seem to be exempt from this enzyme deficiency. For those suffering from lactose intolerance, already fermented milk products such as yogurt and cheese can be consumed, or milk can be treated with lactase enzyme, which is commercially available (Lactaid®).

In a product called "sweet acidophilus" a bacterium, *Lactobacillus acidophilus,* is added to regular milk. At refrigerator temperatures, the bacteria are fairly inactive. As the ingested milk is warmed in the gastrointestinal tract, they become active and begin to ferment the carbohydrates of the milk. Since the fermentation does not begin until the milk is ingested, the milk is "sweet," not sour like yogurt and sour cream, which have already been fermented. The benefits of "sweet acidophilus" for those suffering from lactose intolerance are in question, since the lactose could still reach the intestine relatively intact.

2. *Sucrose: The Table Disaccharide.* Everyday table sugar is **sucrose.** It is a disaccharide composed of a glucose unit and a fructose unit linked by a glycoside bond between the two anomeric carbons, an α, β-1,2 bond.

sucrose
a disaccharide composed of a glucose unit and a fructose unit joined by an α, β-1,2 glycosidic bond; a nonreducing sugar

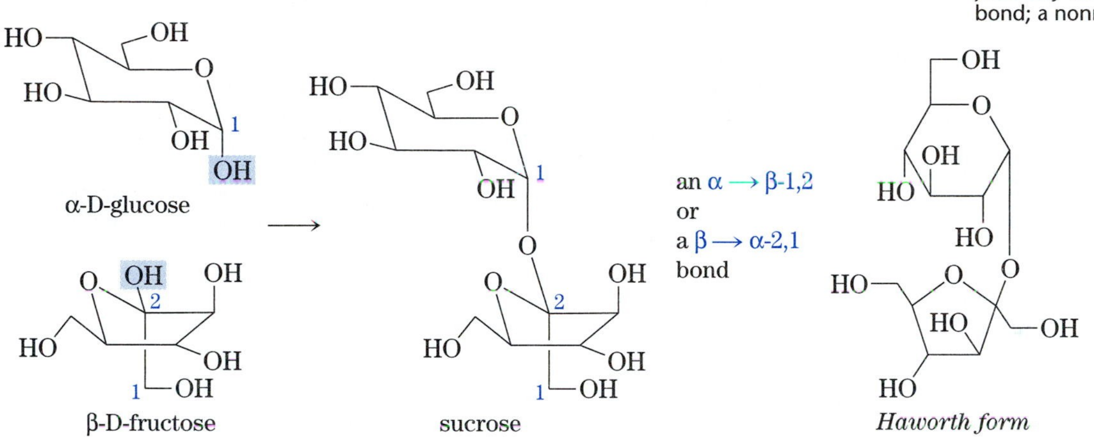

Notice that the anomeric carbons of both units are involved in the glycosidic bond, and therefore neither unit can easily open up to the free aldose or ketose. Because of this, sucrose is a *nonreducing sugar;* that is, it will not give a positive Fehling's test.

Sucrose can be isolated from various sources, including sugarcane (15%–20%), sugar beets (10%–17%), fruits, maple sap, seeds, and flowers. As the disaccharide, sucrose is dextrorotatory. Upon hydrolysis, by either acid or enzyme, the optical rotation changes to levorotatory (+66.5° to −20°), as a result of the release of the fructose, also known as **levulose** ($[\alpha]_D^{20} = -92°$) and glucose, or **dextrose** ($[\alpha]_D^{20} = +52°$) units. Since the sign of the optical activity changes from plus to minus, the hydrolysis process is said to cause *inversion* of the optical rotation, and the mixture of the two monosaccharides is called **invert sugar.** Bees contain an enzyme called *invertase,* which causes this conversion during the production of honey. Our bodies have a similar enzyme called *sucrase.*

levulose
another name for fructose

dextrose
another name for glucose

invert sugar
a mixture of fructose and glucose produced by the breakdown of sucrose

Sucrose is known to be a cause of extensive tooth decay. The material known as plaque that sticks to our teeth is composed of bacterial colonies of *Streptococcus multans* as well as other types of organisms. The bacteria use sucrose both to produce an adhesive with which they stick to teeth and as a food. The end result of their digestion is lactic acid, which causes the corrosion of the mineral deposits (hydroxyapatite) of the teeth and leads to the destruction of the gums. Since most foods are acidic in nature, brushing and flossing teeth frequently are recommended by dental experts. In addition, foods that contain high concentrations of sucrose, especially ones that also stick to the teeth, should be avoided.

Metabolically, fructose and glucose are broken down in much the same fashion and are used for the body's immediate energy requirements. Otherwise, these saccharides are stored as glycogen or converted enzymatically to lipid (fat) and held in adipose tissue. A large dose of sucrose will most likely end up where we need it least—as excess baggage.

Connections 14.4

www.prenhall.com/bailey

Nitrocellulose and Rayon

Long before the human mind had any thought of synthetic polymers, nature had come up with cellulose, in the form of cotton, as an ideal natural fiber. It was indeed fitting that one of the first human experiences with synthetic polymers should entail the inadvertent chemical modification of cellulose.

Serendipity smiled on Christian Schönbein in 1846 when he cleaned up a spill of nitric and sulfuric acids with his wife's cotton apron. After rinsing out the apron with water, he hung it to dry in front of a hot stove. To his utter surprise, the cloth flashed up and disappeared, leaving barely a trace. He had accidentally synthesized cellulose trinitrate, or guncotton.

Cellulose $\xrightarrow[H_2SO_4]{HNO_3}$ Cellulose trinitrate

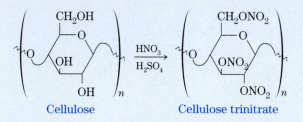

If cellulose is not completely nitrated, another product is formed, called pyroxylin. Pyroxylin and camphor can be combined to form celluloid, a plastic once used for movie film, eyelgass frames, shirt collars, dice, dominoes, and so on. Due to the intrinsic flammability of nitrated cellulose, a good deal of early movie film history has been lost in massive fires. Gradually, of course, this material was replaced by more stable petroleum-based polymers.

$$\xrightarrow{H_2SO_4} \text{cellulose}—\text{OCCH}_3 \;(\text{C=O})$$

The nitration of cellulose converted a very insoluble material, cellulose, into a form that could be dissolved in a solvent and then forced through very small holes to produce a silken thread. However, the material's intrinsic flammability made the resulting fibers short-lived. Less flammable derivatives had to be developed or the solubility of cellulose had to be modified so that it could be formed into threads for cloth.

Acetate rayon is another derivative of cellulose made by acetylating natural cellulose using acetic anhydride. The latter reagent forms ester bonds with the available alcohol groups of the glucose units.

$$\text{Cellulose}—\text{OH} + (\text{CH}_3\overset{O}{\overset{\|}{C}})_2\text{O} + \text{CH}_3\overset{O}{\overset{\|}{C}}\text{OH}$$

Acetic anhydride

The cellulose acetate, in acetone solvent, is extruded through small openings, called spinnerets, to form threads. As the solution leaves the spinneret, hot air flash-evaporates the solvent.

Viscose rayon is actually cellulose that has been derivatized to alter its solubility and then regenerated as it is spun into threads. In the process, carbon disulfide is used to form cellulose xanthate, the sodium salt of which is soluble in basic solution.

The term *viscose* is derived from the fact that the solution of basic cellulose xanthate is very thick, or viscous. An acid bath is used to regenerate the cellulose after it has been forced through the spinnerets. The cellulose thread is used in clothing, carpeting, tire cord, and draperies. If a thin slit rather than spinnerets is used for extrusion, a sheet of cellophane is the result.

$$\text{Cellulose}—\text{O}^-\,\text{Na}^+ + \text{CS}_2 \longrightarrow$$

$$\text{cellulose}—\text{O}—\overset{S}{\overset{\|}{C}}—\text{S}^-\,\text{Na}^+ \xrightarrow{H^+}$$

Cellulose xanthate (soluble)

$$\text{cellulose}—\text{OH} + \text{CS}_2$$

(insoluble)

GETTING INVOLVED

✓ The open chain form of an aldose or ketose is required to give a positive test for a reducing sugar. That is, it must have a hemiacetal or hemiketal structure. An acetal or ketal cannot react. Note that acetals and ketals are *diethers*. Show how the glycosidic link in sucrose contains diethers for both glucose and fructose.

✓ How are the terms fructose, glucose, levulose, dextrose, and invert sugar related?

C. Polysaccharides

There are as many possible polysaccharide structures as there are combinations of monosaccharides and positions of bonding. Those combinations are limitless. We will briefly consider the most abundant homopolymers, those of glucose.

1. ***Starch.*** Plants store their glucose in the form of **starch.** It consists of two related but slightly different polysaccharides, amylose and amylopectin. **Amylose** is polyglucose linked entirely with α-1,4 glycosidic bonds. **Amylopectin** has an amylose-type chain but branches about every 25 glucose units using an α-1,6 glycosidic bond.

Animals possess enzymes that are readily able to cleave the α-bonds in starch and make the glucose available for metabolism.

starch
a natural, complex carbohydrate consisting of the polymers amylose and amylopectin

amylose
a component of starch; linear polymer of glucose units connected by α-1,4 glycosidic bonds

amylopectin
a component of starch; branched polymer of glucose units connected with α-1,4 glycosidic bonds in its linear chains with α-1,6 branching in intervals of about 25 units

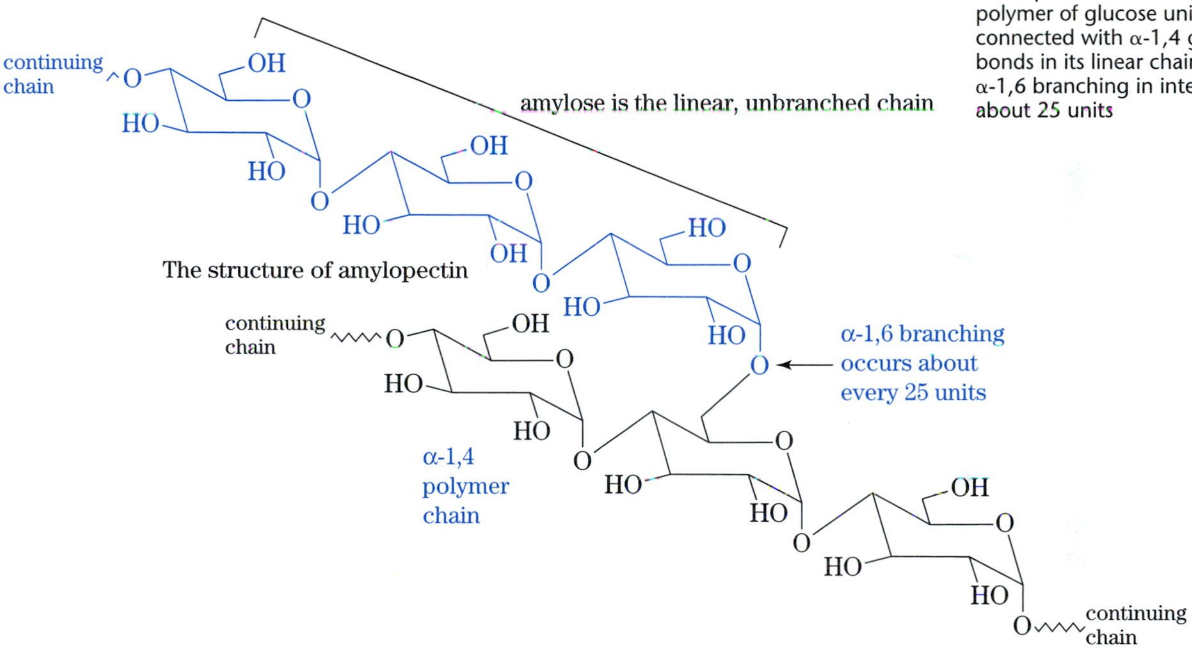

The structure of amylopectin

amylose is the linear, unbranched chain

α-1,6 branching occurs about every 25 units

α-1,4 polymer chain

2. ***Glycogen.*** Animals store glucose using a polymer quite like amylopectin, except that the branching occurs every 8 to 10 glucose units. This adaption produces a more compact structure. A limited amount of **glycogen** is stored in the liver and muscle tissue, where it is a readily available source of energy.

3. ***Cellulose.*** Plants have a rigid exterior that acts as structural support and protection and is composed of a polyglucose-linked β-1,4 called **cellulose.** The

glycogen
branched polymer of glucose units connected with α-1,4 glycosidic bonds in its linear chains with α-1,6 branching in intervals of 8 to 10 units

cellulose
a linear polymer of glucose units linked by β-1,4 glycosidic bonds

The structure of cellulose

β-1,4 bond

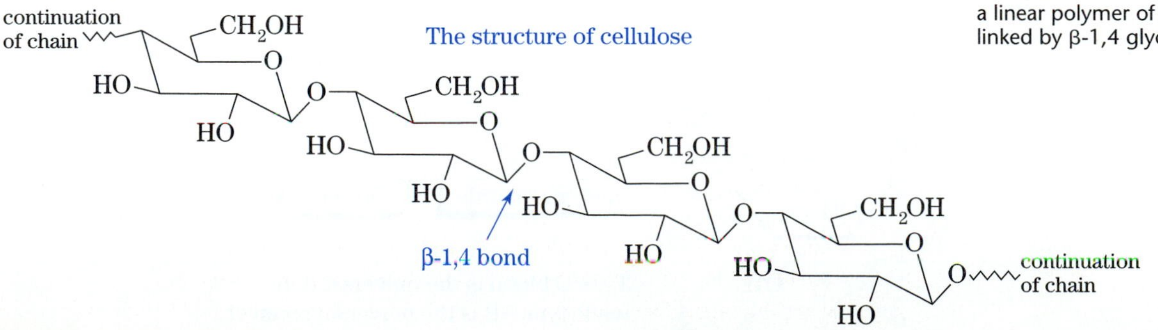

β-bond is not susceptible to animal enzymes, so the cellulose fibers cannot be used as a food source for humans and most animals. However, ruminants such as sheep, goats, and cows have gut bacteria that produce enzymes, cellulases, that digest cellulose, providing glucose for nourishment.

Cellulose is accompanied in its function by other polysaccharides such as hemicellulose (polyxylose) and pectin along with a complex polymer called *lignin.* Cotton, which is more than 90% cellulose, has a tremendous capacity to absorb water because of its large potential for hydrogen bonding.

4. ***Polysaccharide Variations.*** The variations of monomer units as well as of the position and stereochemical orientation of the glycoside bond lead to a tremendous variety of polysaccharides. Add to this the fact that monosaccharide units may be oxidized, derivatized, or otherwise modified, and we have an amazing number of possibilities. A few of these are seen below.

continuation of chain
OH
O
HO
CH_3CNH
O
O continuation of chain

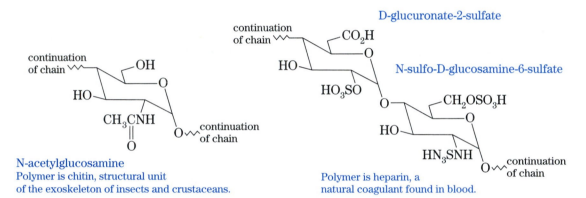

N-acetylglucosamine
Polymer is chitin, structural unit of the exoskeleton of insects and crustaceans.

D-glucuronate-2-sulfate
continuation of chain
CO_2H
O
HO
HO_3SO
N-sulfo-D-glucosamine-6-sulfate
O
CH_2OSO_3H
O
HO
HN_3SNH
O continuation of chain

Polymer is heparin, a natural coagulant found in blood.

Blood group types differ in the presence or absence of a galactose unit or a derivatized galactose unit on the nonreducing end of a polysaccharide chain.

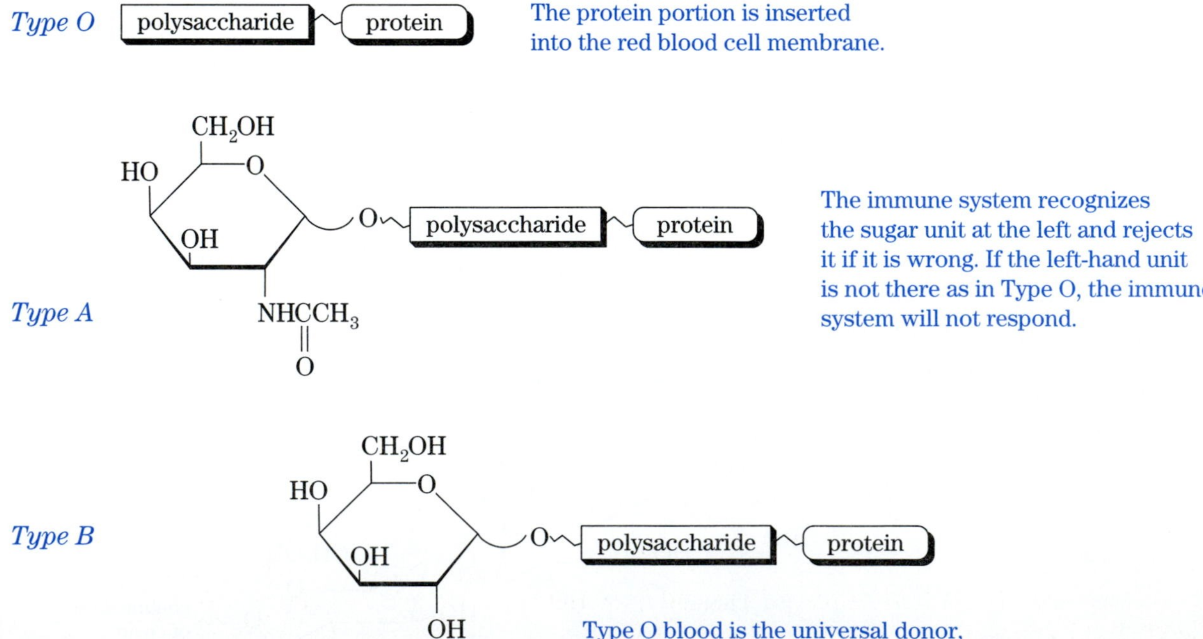
Type O
polysaccharide
protein
The protein portion is inserted into the red blood cell membrane.
CH_2OH
HO
O
OH
O
polysaccharide
protein
Type A
$NHCCH_3$
O
The immune system recognizes the sugar unit at the left and rejects it if it is wrong. If the left-hand unit is not there as in Type O, the immune system will not respond.
CH_2OH
HO
O
Type B
OH
O
polysaccharide
protein
OH
Type O blood is the universal donor, while type AB is the universal receiver.

GETTING INVOLVED

✓ Draw the Haworth structures for at least six polymer units of amylose and glycogen. Be sure to make the correct glycoside bonds.

✓ Can animals digest both starch and cellulose? What structural similarities and differences do these two natural polymers possess?

Problem 14.16

Identify the functional groups in both chitin and heparin. Are these natural polymers acidic or basic?

See related problems 14.26 and 14.27.

Problems

14.17 **Terms:** Distinguish between the members of the following pairs of terms:

(a) hexose, pentose
(b) aldose, ketose
(c) reducing sugar, nonreducing sugar
(d) monosaccharide, polysaccharide
(e) α-D-glucose, β-D-glucose
(f) Haworth formula, Fischer projection
(g) amylose, amylopectin
(h) glycogen, cellulose
(i) Type 1 diabetes, Type 2 diabetes
(j) viscose rayon, acetate rayon
(k) Fehling's and Tollens' tests

14.18 **Structure:** How are the members of the following pairs of saccharides different from each other structurally? Which are reducing, and which are nonreducing? Explain

(a) cellobiose, maltose
(b) lactose, sucrose
(c) α-D-glucose, α-D-galactose
(d) α-D-glucose, α-D-fructose
(e) α-D-xylose, β-D-ribose
(f) maltose, lactose
(g) cellulose, starch

14.19 **Structure:** Draw Haworth formulas for the six-membered ring structures (pyranose forms) of the following:

(a) β-D-fructose **(b)** α-D-idose
(c) β-D-talose **(d)** α-D-lyxose

14.20 **Structure** D-2-deoxyribose is found in DNA, our genetic code. Draw the structure of this monosaccharide.

14.21 **Reactions:** Deoxyribose and ribose (RNA) form esters with phosphoric acid in DNA and RNA. Can both of these monosaccharides react to form the same number of ester combinations? Explain.

14.22 **Structure:** Draw the open-chain forms of the following cyclic saccharides:

(a)

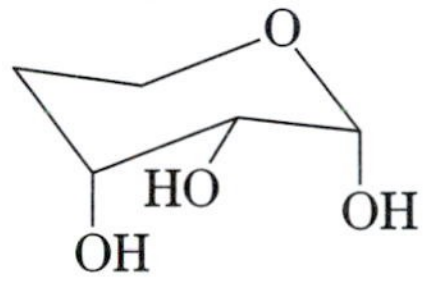

(b)

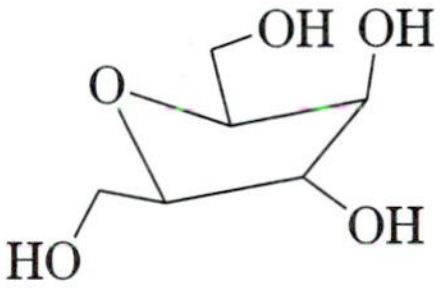

(c)

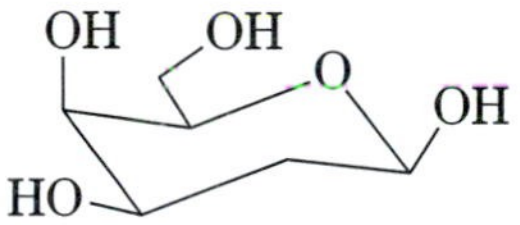

(d)

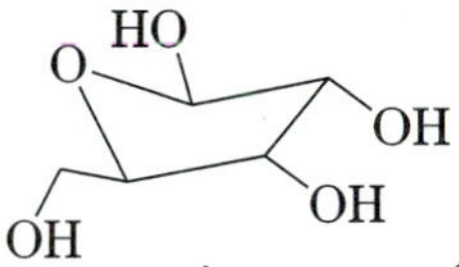

14.23 **Stereoisomers:** Draw the stereoisomers of 3-ketopentose. Which are the enantiomers, diastereomers, and meso compounds?

14.24 **Reactions:** Pure α-D-glucose or pure β-D-glucose in the presence of methanol (CH_3OH) and acid will give a mixture of α- and β-methyl glucosides. Why?

14.25 **Structure:** Specify the type of glycosidic bond that appears in each of the following disaccharides. Also identify the general type of monosaccharide units that appear in each, such as aldopentose.

(a)

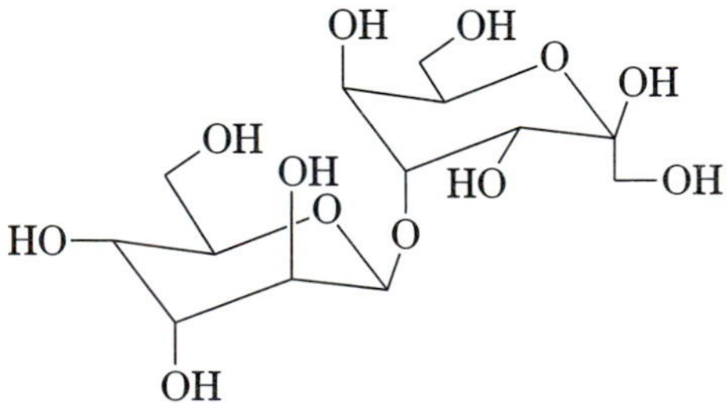

(b)

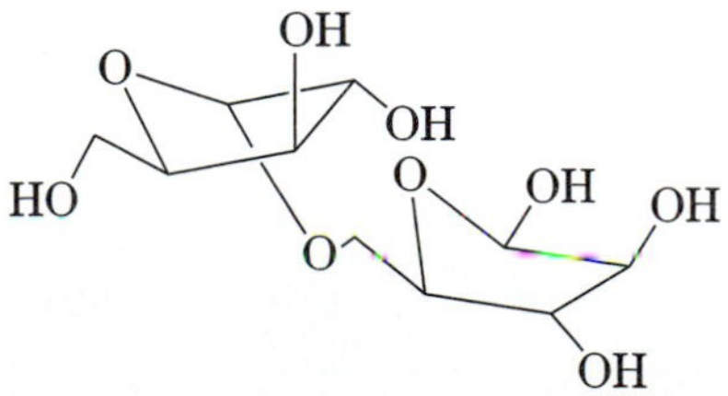

(c)

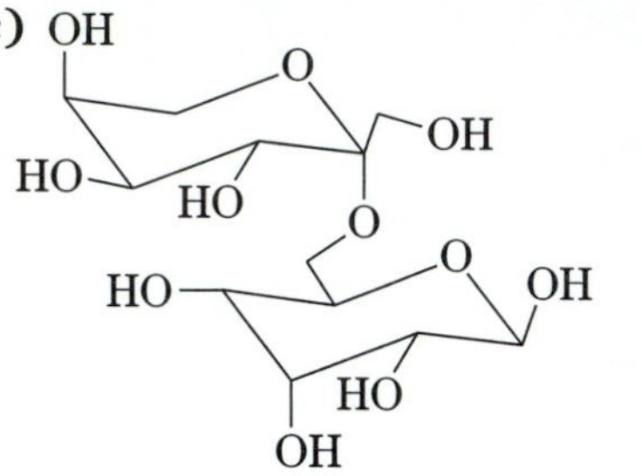

(d)

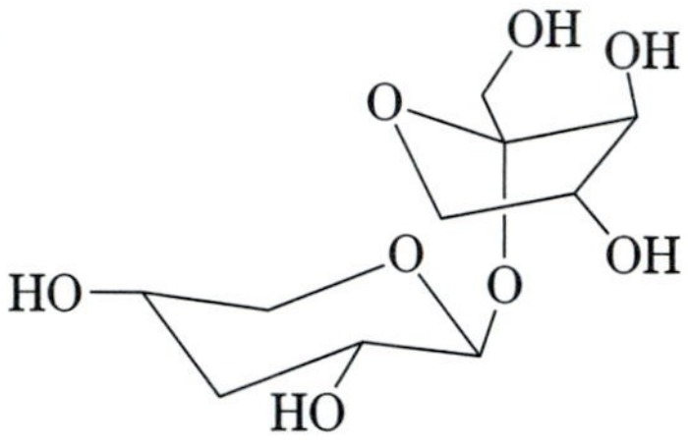

14.26 **Reactions:** Are the disaccharides in the previous question reducing or nonreducing? Explain your answers using the structures.

14.27 **Structure:** Draw the Haworth structures for the following polysaccharides (ring size is indicated in parentheses):

(a) polymannose (pyranose form) linked β-1,3
(b) polyxylose (furanose form) linked β-1,2
(c) polyarabinose (pyranose form) linked α-1,4
(d) polyfructose (furanose form) linked α-2,6 with β-2,4 branching

14.28 **Reactions:** Draw the reactions and products of β-D-galactose in the six-membered ring form with each of the following:

(a) first one mole of methanol and then two moles of methanol
(b) α-D-mannose (six-membered ring also) as a β-1,4-bond
(c) copper (II) in basic solution
(d) α-D-fructose (pyranose) in an α-1,6-bond

14.29 **Reactions:** Why do both glucose and fructose give positive Fehling and Tollens' tests?

14.30 **Reactions:** Is a positive Fehling test for glucose in the urine a direct indication of diabetes? Explain your answer.

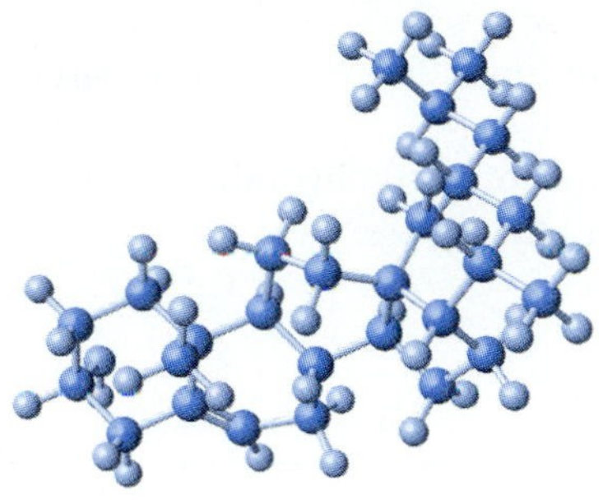

CHAPTER 15

Lipids

15.1 The Nature of Lipids

Although the human body, as well as other organisms, is composed primarily of water, about 70% by weight in animals, the organic biomolecules that constitute the remaining 30% are part of a complex mixture that supports, protects, regulates, directs, and defends the whole entity. Among these is the class known as **lipids.** The functions of lipids range from energy source to membrane formation. Even so, lipids have a very general chemical definition. They are organic molecules soluble to a great extent in nonpolar solvents such as diethyl ether, chloroform, carbon tetrachloride, or benzene. These solvents are used to extract lipids from their more polar neighbors: salts, proteins, carbohydrates, and nucleic acids. A key structural component of all lipids is a large proportion of carbon-carbon and carbon-hydrogen bonds. This makes these compounds hydrophobic (literally "water-fearing") rather than hydrophilic ("water-loving") as are most carbohydrates, proteins, and nucleic acids. Within this loose definition are subcategories such as polar and nonpolar lipids, saponifiable and nonsaponifiable lipids, simple and complex lipids.

lipid
organic biomolecules soluble to a great extent in nonpolar solvents

- ***Nonpolar lipids*** are those with few or no polar bonds. Examples include fats and oils, waxes, and some steroids.
- ***Polar lipids*** have both polar and nonpolar bonds allowing for limited solubility both in polar and nonpolar solvents. Examples are the phospho- and sphingolipids.
- ***Saponifiable lipids*** are those that can undergo hydrolysis in the presence of a base such as NaOH or KOH. Fats, oils, and waxes, as well as phospho- and sphingolipids, are saponifiable.
- ***Nonsaponifiable lipids*** will not be hydrolyzed in the presence of base. Most steroids are nonsaponifiable.
- ***Simple lipids*** have relatively uncomplex structures and either will not be broken down by chemical processes or can be broken down into a limited number of simple compounds. Examples are the steroids and fats and oils.
- ***Complex lipids*** are those with variations in their structures; they can be broken down into several simpler compounds. Sphingolipids are complex.

nonpolar lipid
lipid with few or no polar bonds

polar lipid
lipid with both polar and nonpolar bonds allowing for limited solubility in polar and nonpolar solvents

saponifiable lipid
lipid that can undergo hydrolysis to simpler compounds in the presence of a base such as NaOH or KOH

nonsaponifiable lipid
lipid that cannot be hydrolyzed in the presence of base

simple lipid
lipid with relatively uncomplex structure; it either will not be broken down by ordinary chemical processes or can be broken down into a limited number of simple compounds

complex lipid
lipid that can have variations in its structure and can be broken down into several types of simpler compounds

This chapter will cover most of the major structural variations possible for lipids and their functions.

GETTING INVOLVED

✓ What is a good, generic definition of lipids that includes their chemical properties and structures?

✓ How does a lipid differ in fundamental structure from a carbohydrate or protein?

✓ Differentiate between the members of each pair of lipid types:

- polar and nonpolar lipids
- simple and complex lipids
- saponifiable and nonsaponifiable lipids

15.2 Waxes—Simple Esters of Long-Chain Alcohols and Acids

wax
ester of a long-chain carboxylic acid and a long-chain alcohol

Structurally, **waxes** are defined as esters of long-chain carboxylic acids and long-chain alcohols. They are simple, nonpolar, and saponifiable.

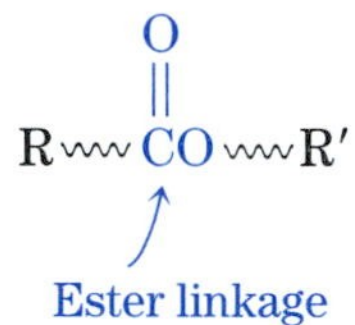

Ski waxing

Natural waxes differ from paraffin wax in that they are high-molecular-weight esters produced directly by living organisms, whereas paraffin wax is a mixture of high-molecular-weight hydrocarbons separated during the fractionation of petroleum. Following are some representative natural waxes and the structures of their principal components.

1. ***Spermaceti.***

$$C_{15}H_{31}\overset{\overset{\displaystyle O}{\|}}{C}OC_{16}H_{33}$$

Spermaceti is a soft wax obtained from the head of the sperm whale; it has a melting range of 42–50°C. It consists largely of cetyl palmitate (above). Because of its softness, it can be used as a base emollient for ointment medications and cosmetics. Also, like paraffin wax, it is used in the production of candles.

2. ***Beeswax.*** Beeswax is taken from the honeycomb and is a mixture of esters of alcohols and acids having up to 36 carbons and some high molecular weight hydrocarbons.

Beeswax has a melting range of 62–65°C and is used in shoe polishes, candles, wax paper, and the manufacture of artificial flowers.

3. ***Carnauba wax.*** Carnauba wax is a very hard wax capable of producing a high polish; it has a melting range of 82–86°C. It is obtained from the leaves of the Brazilian palm tree and is used in automobile and floor waxes and in deodorant sticks. When carnauba wax is hydrolyzed, some hydroxy acids are produced, indicating the presence of large polyesters in the wax; these could contribute to its hardness and durability.

$$CH_3(CH_2)_{24}\overset{\overset{\displaystyle O}{\|}}{C}O(CH_2)_{29}CH_3$$

GETTING INVOLVED

✓ What is the principal organic functional group in waxes?

✓ Why are waxes classified as lipids?

Problem 15.1

Write chemical reactions for the hydrolysis (saponification) of the lipids in beeswax (see sections 13.6(a–b)).

15.3 Fats and Oils—Triesters of Glycerol

Fats and **oils** of either animal or vegetable origin are triesters of the trihydroxy alcohol, glycerol. Consequently they are called **triacylglycerols, glycerides,** or **triglycerides.** The acids making up the triester are known as **fatty acids** because of their length, usually 10–24 carbons in higher organisms.

fat
triester of glycerol wherein the acids are long-chain and highly saturated

oil
triester of glycerol wherein the acids are long-chain and highly unsaturated

triacylglycerol
see **fat** and **oil**

glyceride
see **fat** and **oil**

triglyceride
see **fat** and **oil**

fatty acid
long-chain (10–24) carboxylic acid

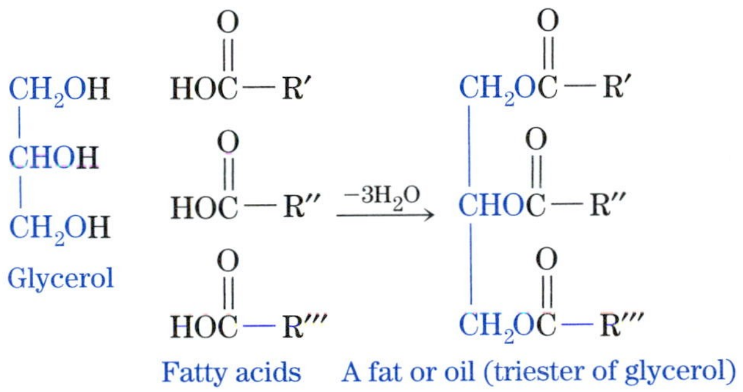

$$\begin{array}{l} CH_2OH \\ | \\ CHOH \\ | \\ CH_2OH \end{array} \quad \begin{array}{l} HO\overset{\overset{O}{\|}}{C}-R' \\ HO\overset{\overset{O}{\|}}{C}-R'' \\ HO\overset{\overset{O}{\|}}{C}-R''' \end{array} \xrightarrow{-3H_2O} \begin{array}{l} CH_2O\overset{\overset{O}{\|}}{C}-R' \\ | \\ CHO\overset{\overset{O}{\|}}{C}-R'' \\ | \\ CH_2O\overset{\overset{O}{\|}}{C}-R''' \end{array}$$

Glycerol   Fatty acids   A fat or oil (triester of glycerol)

The biosynthesis of fatty acids starts with the two-carbon acetate unit, so the final product has an even number of carbons linked in an unbranched chain. Bacteria are known to produce not only fatty acids with an odd number of carbons, but also ones with branched and cyclic chains. The fatty acids in a triglyceride may be the same or may differ.

Fats and oils are structurally alike with one exception; most of the fatty acids in fats are saturated chains, whereas those in oils are unsaturated. A shorthand designation for the various common fatty acids uses a subscript indicating the number of carbons in the chain followed by a colon (:) and the number of double bonds. For example, stearic acid is $C_{18:0}$ and linoleic acid is $C_{18:2}$. See Table 15.1 for other examples.

The position of the double bond can be shown by the symbol delta, Δ, as a superscript with the position of the double bonds. The numbering of the chain begins with the carboxyl group. When more than one double bond is present, the relationship will be allylic; that is, the double bonds will be separated by a methylene group, $-CH_2-$. For example, linoleic acid is $C_{18:2}{}^{\Delta 9,12}$ and arachidonic acid is $C_{20:4}{}^{\Delta 5,8,11,14}$.

Another type of designation that is sometimes used refers to the position of the first double bond starting from the hydrocarbon end or $-CH_3$, called the omega (ω) carbon of the chain. In this case linoleic acid would be an **ω6 fatty acid.** There is currently much study as to the correlation of the amounts of **ω6** and **ω3** triacylglycerols in the diet and decreased risk of heart disease.

omega (ω) 6 fatty acid
an unsaturated fatty acid with its last double bond six carbons in from the methyl end of the chain

omega (ω) 3 fatty acid
an unsaturated fatty acid with its last double bond three carbons in from the methyl end of the chain

Connections 15.1 www.prenhall.com/bailey

Errors in the Metabolism of Fatty Acids—Lorenzo's Oil

In 1992 the movie *Lorenzo's Oil* detailed the real-life struggle of the Odone family, whose young son suffered from an inborn metabolic disorder known as adrenoleukodystrophy (ALD). ALD is an extremely rare condition passed as a recessive X-linked gene through the women in the family. Although carriers may show some mild symptoms of the condition, the actual occurrence is about 1 in 45,000 in the populations of the United States and Europe. At the age of six years, Lorenzo Odone began to show symptoms that could be mistaken for a variety of illnesses: personality disorder and loss of coordination and speech. The film chronicles the efforts of the parents to have their child properly diagnosed and treated for the fatal condition.

The genetic defect is correlated to the accumulation of very-long-chain fatty acids (22–26 carbon range) in the brain and adrenal cortex. One hypothesis is that these fatty acids destroy the myelin sheath of brain nerve fibers, leading to mental and physical deterioration, blindness, seizures, paralysis, and death. The Odones became self-taught experts in the field of lipids and their metabolism, eventually finding that a dietary supplement of a 4:1 mixture of olive oil, which contains oleic acid ($C_{18:1}$) as a major component, and triglycerides of erucic acid ($C_{22:1}{}^{\Delta 13}$) seemed to help slow the progress of ALD in their son. Erucic acid constitutes 40%–50% of the seeds of rapeseed, mustard, and wallflower and up to 80% of nasturtium seeds.

Although controversial in its development and inconsistent in its effects, this mixture, called Lorenzo's Oil, seems to stabilize low blood concentrations of very-long-chain fatty acids and has impeded the development of ALD in some young patients. The normal course of this affliction is about two years from diagnosis to death. Lorenzo Odone has passed adolescence, although he is in an almost vegetative state. Some others who were treated did not have their condition allayed. Recent controlled studies have shown few if any effects on the progress of a milder form of ALD. However, the devotion and studies of Augusto and Michaela Odone have indicated some possible avenues of research. Augusto is the current director of the Myelin Project.

Table 15.1 Common Fatty Acids

Formula	Name
$CH_3(CH_2)_{10}CO_2H$	lauric acid ($C_{12:0}$)
$CH_3(CH_2)_{12}CO_2H$	myristic acid ($C_{14:0}$)
$CH_3(CH_2)_{14}CO_2H$	palmitic acid ($C_{16:0}$)
$CH_3(CH_2)_{16}CO_2H$	stearic acid ($C_{18:0}$)
$CH_3(CH_2)_7CH{=}CH(CH_2)_7CO_2H$	oleic acid ($C_{18:1}{}^{\Delta 9}$)
$CH_3(CH_2)_4CH{=}CHCH_2CH{=}CH(CH_2)_7CO_2H$	linoleic acid ($C_{18:2}{}^{\Delta 9,12}$)
$CH_3CH_2CH{=}CHCH_2CH{=}CHCH_2CH{=}CH(CH_2)_7CO_2H$	linolenic acid ($C_{18:3}{}^{\Delta 9,12,15}$)
$CH_3(CH_2)_4CH{=}CHCH_2CH{=}CHCH_2CH{=}CHCH_2CH{=}CH(CH_2)_3CO_2H$	arachidonic acid ($C_{20:4}{}^{\Delta 5,8,11,14}$)

Table 15.2 lists the composition of familiar triacylglycerols. You can see that oils are usually of plant or marine origin, whereas fats can be commonly found in animal sources. Some triacylglycerols have the same fatty acid at all three positions of esterification. They can be named with that in mind, such as tripalmitin and tristearin. Most naturally occurring fats and oils, however, contain a distribution of fatty acids.

Unsaturation results in a noticeable lowering of the melting point, and hence the formation of an oil, a liquid, at room temperature. This physical difference is due to the structure of the chain. A saturated carbon chain has a staggered, relatively linear nature, which can lead to molecular packing in a semi-solid phase and a higher melting point. The double bonds in oils are geometrically in the *cis* configuration, which "kinks" the chain and makes it difficult to form an organized solid structure. As a result, the melting point is lowered significantly (See Figure 15.1).

Margarines are produced by the partial catalytic hydrogenation of oils. The hydrogenation process used in the United States up to this time causes isomerization of many of the remaining double bonds from *cis* to *trans*. There may be some adverse health effects from these *trans* fatty acids. European methods of catalytic hydrogenation do not seem to cause such isomerization.

Table 15.2 Fats and Oils

Fat or Oil	Iodine Number	Saponification Number	Melting Point, °C	Percent Fatty Acid Composition[a]							
				Saturated Acids				Unsaturated Acids			
				$C_{12:0}$ Lauric Acid	$C_{14:0}$ Myristic Acid	$C_{16:0}$ Palmitic Acid	$C_{18:0}$ Stearic Acid	$C_{18:1}$ Oleic Acid	$C_{18:2}$ Linoleic Acid	$C_{18:3}$ Linolenic Acid	$C_{17:3}$ Eleostearic Acid
Animal Fats											
Beef tallow	31–47	190–200	40–46		3–6	24–32	20–25	37–43	2–3		
Lard	46–66	193–200	36–42		1	25–30	12–16	41–51	3–8		
Butter	36	227	32	1–4	8–13	25–32	8–13	22–29	0.2–1.5	3	
Marine Animals											
Cod liver oil	145–180	180–190			2–6	7–14	0–1	25–31	27–32		
Whale oil	120	195		0.2	9.3	15.6	2.8		35.8		
Vegetable Oils											
Coconut oil	10	255–258	25	44–51	13–18	7–10	1–4	5–8	0–1	1–3	
Corn oil	109–133	187–196	−20		0.1–1.7	8–12	2.5–4.5	19–49	34–62		
Cottonseed oil	105–114	190–198	−1		0–3	17–23	1–3	23–44	34–55		
Olive oil	79–90	187–196	−6			9.4	2.0	83.5	4.0		
Palm oil	54	199	35		1–6	32–47	1–6	40–52	2–11		
Peanut oil	84–102	188–195	−5			8.3	3.1	56	26		
Soybean oil	127–138	185–195	−16		0.3	7–11	2–5	22–34	50–60		
Linseed oil	179	190	−24		0.2	5–9	4–7	9–29	8–29	45–67	
Tung oil	168	193	−3					4–13		8–15	74–91

[a] These percentages do not include short-chain fatty acids or fatty acids present in minute amounts.

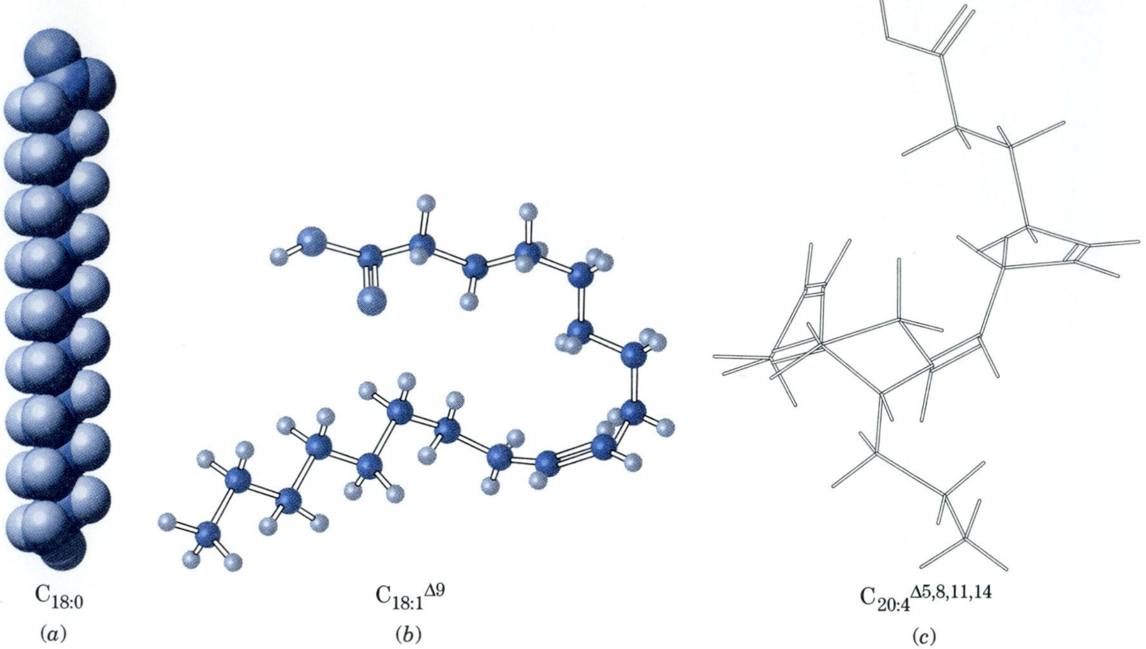

Figure 15.1 Space-filling, ball-and-stick, and wireframe models of (a) stearic acid, (b) oleic acid, and (c) arachidonic acid, respectively. The *cis* orientation of double bonds in naturally occurring unsaturated fatty acids produce a more rigid, curved molecule that interferes with tight packing for semi-solid structures, as occur in membranes.

GETTING INVOLVED

- ✓ What is the structure of a "fatty" acid?
- ✓ Draw the generic structure of a fat/oil.
- ✓ How are fats and oils the same? How are they different?
- ✓ Do naturally occurring fatty acids have a specific chain length?
- ✓ Why are fats and oils classified as lipids?
- ✓ Write and understand the shorthand designation for saturated and unsaturated fatty acids, that is, what are the meanings of the subscripts and superscripts in $C_n{}^{\Delta m}$, $C_{18:1}{}^{\Delta 9}$, for example?

Example 15.1

What is the shorthand designation for the following fatty acid? Is it an ω3 or ω6 fatty acid?

$$CH_3CH_2CH{=}CHCH_2CH{=}CH(CH_2)_6COOH$$

Solution

(a) Count the total number of carbons and the number of double bonds.

$$\overbrace{CH_3CH_2CH{=}CHCH_2CH{=}CH(CH_2)_6COOH}^{C_{14}}$$

It is a $C_{14:2}$ fatty acid.

(b) Number the carbon chain starting with the carboxyl group. Number the double bonds as you would with any unsaturated organic compound.

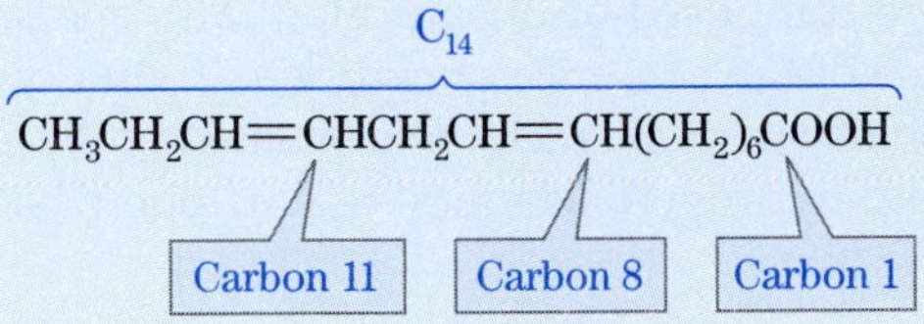

$$\overbrace{CH_3CH_2CH{=}CHCH_2CH{=}CH(CH_2)_6COOH}^{C_{14}}$$

Carbon 11 | Carbon 8 | Carbon 1

This is a $C_{14:2}{}^{\Delta 8,11}$ fatty acid.

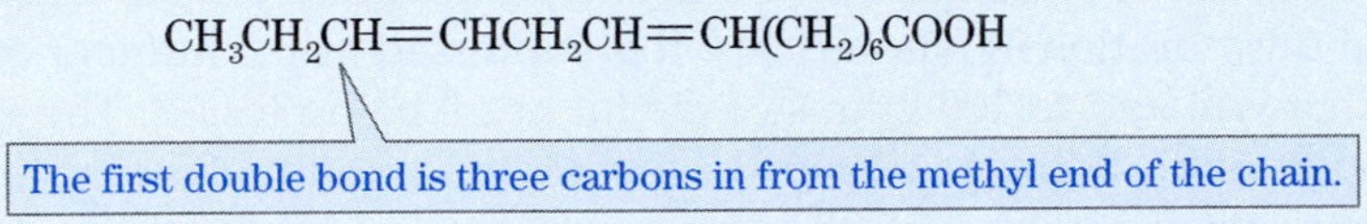

$$CH_3CH_2CH{=}CHCH_2CH{=}CH(CH_2)_6COOH$$

The first double bond is three carbons in from the methyl end of the chain.

It is an ω3 fatty acid.

Example 15.2

Draw the structure of trimyristin, the fat found in nutmeg. Use Table 15.1 for the fatty acid, myristic acid.

Solution

Fats are triacylglycerols and trimyristin must therefore be composed of glycerol and three molecules of myristic acid—$C_{14:0}$. The components are joined by an esterification reaction.

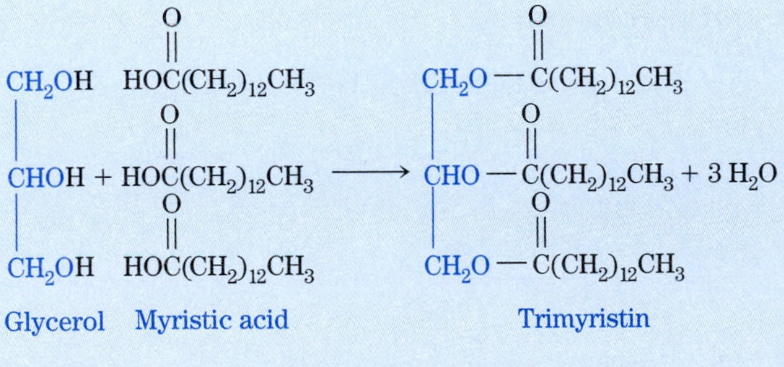

$$\begin{array}{l} CH_2OH \\ | \\ CHOH \\ | \\ CH_2OH \end{array} + \begin{array}{l} HOC(=O)(CH_2)_{12}CH_3 \\ HOC(=O)(CH_2)_{12}CH_3 \\ HOC(=O)(CH_2)_{12}CH_3 \end{array} \longrightarrow \begin{array}{l} CH_2O-C(=O)(CH_2)_{12}CH_3 \\ | \\ CHO-C(=O)(CH_2)_{12}CH_3 \\ | \\ CH_2O-C(=O)(CH_2)_{12}CH_3 \end{array} + 3\,H_2O$$

Glycerol Myristic acid Trimyristin

Problem 15.2

Are fats and oils simple or complex, polar or nonpolar, saponifiable or nonsaponifiable?

Problem 15.3

Draw the structure for the following rare but real long-chain, unsaturated fatty acids:

(a) $C_{28:1}{}^{\Delta 9}$ **(b)** $C_{26:2}{}^{\Delta 5,9}$ **(c)** $C_{24:4}$—an ω6 fatty acid

Problem 15.4

Draw the structure of a triglyceride consisting of $C_{16:0}$, $C_{18:2}{}^{\Delta 9,12}$, $C_{18:0}$.

Problem 15.5

Draw the structure of erucic acid, $C_{22:1}{}^{\Delta 13}$.

See related problems 15.13, 15.14, 15.16, 15.17.

15.4 Reactions of Fats and Oils

A. Addition Reactions

Most fats and oils are composed of unsaturated fatty acids as well as the saturated variety. Because of the presence of carbon-carbon double bonds, fats and oils undergo addition reactions characteristic of alkenes. We consider here the addition of halogens and hydrogen, where X_2 = hydrogen, halogen.

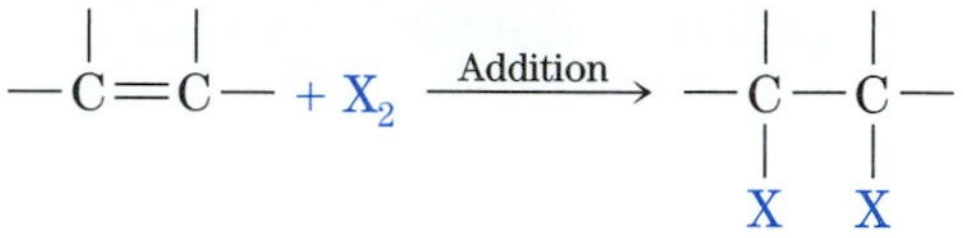

$$-\overset{|}{C}=\overset{|}{C}- + X_2 \xrightarrow{\text{Addition}} -\underset{X}{\overset{|}{\underset{|}{C}}}-\underset{X}{\overset{|}{\underset{|}{C}}}-$$

iodine number a measure of the extent of unsaturation in fats and oils; the number of equivalents of iodine that will add to 100 grams of a fat or oil

1. Addition of I_2 ***(Iodine Number.)*** The iodine number is a measure of the extent of unsaturation in fats and oils. It is expressed as the number of grams of iodine that will add to 100 grams of the fat or oil being tested. The greater the number of double bonds in a lipid, the greater the amount of iodine that adds to 100 grams of it. Thus, high iodine numbers indicate a high degree of unsaturation, and low iodine numbers indicate low unsaturation. In Table 15.2 note that the animal fats have low iodine numbers relative to the more highly unsaturated marine animal and vegetable oils.

hydrogenation of oils the catalytic addition of hydrogen to unsaturated triacylglycerols (oils)

2. Addition of H_2 ***(Hydrogenation.)*** In the presence of a metal catalyst, such as nickel, hydrogen adds to the double bonds of fats and oils, producing more highly saturated glycerides. Consider, for example, the hydrogenation of the following oil, a possible component of soybean oil.

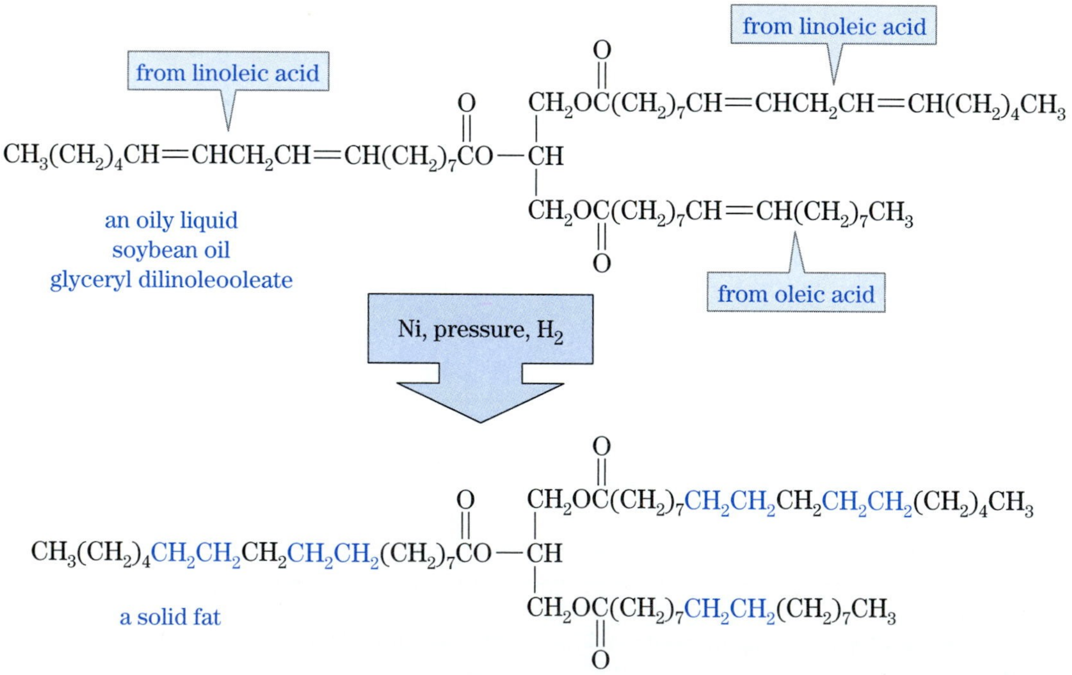

from linoleic acid

$$CH_2O\overset{O}{\overset{\|}{C}}(CH_2)_7CH{=}CHCH_2CH{=}CH(CH_2)_4CH_3$$

from linoleic acid

$$CH_3(CH_2)_4CH{=}CHCH_2CH{=}CH(CH_2)_7\overset{O}{\overset{\|}{C}}O-CH$$

$$CH_2O\underset{O}{\underset{\|}{C}}(CH_2)_7CH{=}CH(CH_2)_7CH_3$$

from oleic acid

an oily liquid
soybean oil
glyceryl dilinoleooleate

Ni, pressure, H_2

$$CH_2O\overset{O}{\overset{\|}{C}}(CH_2)_7CH_2CH_2CH_2CH_2CH_2(CH_2)_4CH_3$$

$$CH_3(CH_2)_4CH_2CH_2CH_2CH_2CH_2(CH_2)_7\overset{O}{\overset{\|}{C}}O-CH$$

$$CH_2O\underset{O}{\underset{\|}{C}}(CH_2)_7CH_2CH_2(CH_2)_7CH_3$$

a solid fat

This general reaction is used in the production of shortenings and margarine. Cooking shortenings and margarine differ from lard and butter in that they are derived from vegetable oils, whereas lard and butter are natural animal fats. In the production of shortening or margarine, liquid vegetable oils are partially hydrogenated in the presence of a catalyst until the desired consistency is achieved. Enough unsaturation is left to create a low-melting, soft product. Complete hydrogenation (as in the example) would produce a hard, brittle fat.

In the manufacture of margarine, these partially hydrogenated vegetable oils (often soybean, corn, and safflower oils) are mixed with water, salt, and non-fat dry milk. Other oils are added to achieve the desired consistency and homogeneity. Vitamins, especially vitamin A, are added along with artificial flavoring and coloring. Diacetyl and methyl acetyl carbinol, which are responsible for the characteristic taste of butter, are common flavorings.

$$\underset{\text{Diacetyl}}{CH_3\overset{\overset{\displaystyle O}{\|}}{C}-\overset{\overset{\displaystyle O}{\|}}{C}CH_3} \qquad \underset{\text{Methyl acetyl carbinol}}{CH_3\overset{\overset{\displaystyle HO}{|}}{C}H\overset{\overset{\displaystyle O}{\|}}{C}CH_3}$$

Finally, preservatives such as potassium sorbate and other salts of carboxylic acids, (Connections 12.1) and antioxidants such as butylated hydroxy toluene and other phenols, (section 9.3.D.2) are added.

B. Oxidation Reactions

1. ***Rancidification.*** Fats and oils, when exposed to air, tend to oxidize or hydrolyze to produce volatile carboxylic acids. These have a sour, unpleasant taste and aroma. The process, called rancidification, makes lard, shortenings, butter, margarine, cooking oils, and milk unpalatable and unusable.

rancidification
oxidation and hydrolysis of fats and oils to volatile organic acids, producing an unpalatable product

Oxidative rancidification involves the oxidation of carbon-carbon double bonds in the alkyl chains of fats and oils to produce carboxylic acids. In hydrolytic rancidification, one or more of the ester units of triacylglycerol are hydrolyzed back to the original acid. Antioxidants (section 9.3.D.2) are added to many edible fat and oil products to retard rancidification.

2. ***Drying Oils.*** When highly unsaturated oils are exposed to air, they undergo an alternative form of oxidation called *drying,* which causes them to harden. This process involves the attack of oxygen at allylic positions (carbons next to double bonds) in the oil to form intermolecular linkages. As oil molecules are drawn into close proximity, the double bonds polymerize, forming a gigantic, interlinking hard mass.

drying oil
an oil that can be hardened by the process of oxidation

This principle governs the drying action of an oil-based paint. Commercial oil-based paints consist of a pigment dispersed in a drying oil, such as linseed oil. When the paint is applied, a volatile thinner such as turpentine evaporates and the oil begins to polymerize, often under the influence of an added catalyst, eventually forming a hard, protective surface.

The drying process is spontaneous and highly exothermic. It can eventually provide enough heat to cause the combustion of cloth and paper. For this reason oily rags should not be stored in closed, unventilated areas.

Linoleum flooring is made from a thick suspension of cork and rosin in linseed oil. The suspension is pressed and allowed to "dry" (the linseed oil oxidizes). A similar process is used to make oilcloth. Tough, durable surface coatings result.

C. Saponification

saponification
the alkaline hydrolysis of esters to produce soaps

soap
sodium and potassium salts of long-chain fatty acids; the hydrocarbon portion is hydrophobic (water-insoluble) but soluble in fats and oils and the ionic part is hydrophilic (water-soluble)

Fats and oils are acid derivatives, triesters of glycerol. When any acid derivative reacts with water, the products are an acid and an alcohol. **Saponification** is the alkaline hydrolysis of esters, resulting in the production of glycerol (the alcohol) and the salts of the constituent fatty acids (since basic conditions are employed, acid salts are formed).

1. ***Production of Soap.*** The term *saponification* means "soap making." The salts of long-chain fatty acids produced by the saponification of fats and oils are **soaps.**

$$\begin{array}{c} \text{R}''\text{∿}\overset{\text{O}}{\overset{\|}{\text{C}}}\text{O}-\underset{\underset{\text{CH}_2\text{O}\underset{\underset{\text{O}}{\|}}{\text{C}}\text{∿R}'''}{|}}{\overset{\overset{\text{CH}_2\text{O}\overset{\overset{\text{O}}{\|}}{\text{C}}\text{∿R}'}{|}}{\text{C}}}-\text{H} \end{array} + 3\,\text{NaOH} \xrightarrow{\text{H}_2\text{O}} \text{HO}-\underset{\underset{\text{CH}_2\text{OH}}{|}}{\overset{\overset{\text{CH}_2\text{OH}}{|}}{\text{C}}}-\text{H} + \begin{array}{c} \text{R}'\text{∿}\overset{\text{O}}{\overset{\|}{\text{C}}}\text{O}^-\text{Na}^+ \\ \text{R}''\text{∿}\overset{\text{O}}{\overset{\|}{\text{C}}}\text{O}^-\text{Na}^+ \\ \text{R}'''\text{∿}\overset{\text{O}}{\overset{\|}{\text{C}}}\text{O}^-\text{Na}^+ \end{array}$$

fat or oil — glycerol — sodium salts of fatty acids

Soap has its origins in antiquity. It was prepared for over two thousand years by mixing fire ashes, which are quite alkaline, with tallow and water. Today, soap is made by two processes.

In the boiling, or kettle, process, up to 50 tons of rendered fat are melted in steel tanks three stories high and then injected with steam and sodium hydroxide solution. Following saponification, brine is added to salt out the soap; this forms an upper curdy layer. The soap is then separated, purified, and cut into bars or chips. Glycerol, for use in the plastics and explosives industries, is recovered from the lower layer, the aqueous salt solution.

The modern continuous soap-making process involves high-temperature water hydrolysis of fats and oils to fatty acids and glycerol. The fatty acids are vacuum-distilled, mixed in specific ratios, and neutralized with alkali to form the soap.

Tallow and coconut oil are frequently the initial glycerides used in the soap industry. Tallow is rendered by heating to produce a liquid. The unmelted protein material is filtered away; the melt is termed *lard.* Tallow or lard produces a good-cleansing but slow-lathering soap. Soaps from coconut oil form better lathers, so some coconut oil is often included in the lipid material to be saponified. Coloring, perfumes, disinfectants, and deodorants can also be added to body soaps. Heavy-duty hand soaps may contain scouring powders, sand, or volcanic pumice, for an abrasive effect. Glycerol confers transparency to bar soaps, while air beaten into the soap will allow it to float. Shaving cream is made by using the potassium salts of fatty acids colloidally dispersed into a foam.

saponification number
the number of milligrams of potassium hydroxide required to saponify 1 gram of a fat or an oil

2. ***Saponification Number.*** The **saponification number** is defined as the number of milligrams of potassium hydroxide required to saponify 1 gram of a fat or an oil. On a molecular basis, one mole of fat or oil requires three moles of KOH for complete saponification because there are three ester linkages in a fat or an oil molecule.

Because a gram of a high-molecular-weight fat has fewer molecules than a gram of a low-molecular-weight one, the weight of KOH needed for saponification will be lower for the high-molecular-weight fat. Thus, high-molecular-weight fats and oils have lower saponification numbers than fats and oils of lower molecular weight. Table 15.2 lists saponification numbers for some common fats and oils.

GETTING INVOLVED

- ✓ What types of chemical reactions are going on during the process of hydrogenation, rancidification, and saponification?
- ✓ Give a practical application of each of the preceding three processes.
- ✓ What is the chemical structure of a soap?

Problem 15.6

Arrange the triglycerides in each of the following groups in order of increasing iodine number: **(a)** trimyristin, triolein, glyceryl oleopalmitostearate; **(b)** stearic, oleic, linoleic, and linolenic acids.

Problem 15.7

Illustrate that you understand the processes of hydrogenation, rancidification, and saponification by reacting a diglyceride that has oleic and linolenic acids at the 1,3-positions of glycerol with the appropriate reagents.

Problem 15.8

Are margarines, the products of the hydrogenation of oils, naturally occurring? Are they "organic" (unprocessed)?

15.5 Soaps and Detergents

A. Structure of Soaps

Dirt adheres to our bodies and clothes by a thin film of fat, oil, or grease. For this dirt to be removed, the oily materials must first be dissolved. The most abundant liquid on earth, and the only one economically feasible for day-to-day washing, is water. But water is a polar liquid—and fats and oils, because of their long hydrocarbon chains, are nonpolar, that is, water-insoluble.

Soaps are structurally capable of solving this dilemma. Recall that soaps are salts of long-chain fatty acids. The long alkyl group has 12–18 carbons, is completely nonpolar, and consequently is soluble in fats and oils but insoluble in water. The other end of the molecule, a carboxylic acid salt, is very polar, in fact, ionic, and is water-soluble. A soap then has two diverse solubility properties—it has a **hydrophilic** end (water-loving), soluble in water, and a **hydrophobic** end (water-fearing), soluble in fats and oils.

hydrophilic water-loving

hydrophobic water-fearing

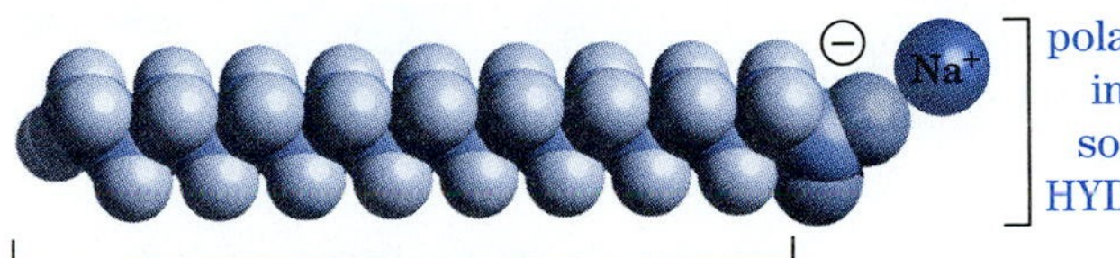

Soap, by simultaneously dissolving in oils and water, removes oil from dirty clothes and emulsifies the droplets in water.

GETTING INVOLVED

✓ What is meant by the terms "hydrophilic" and "hydrophobic"?

Problem 15.9

For each of the following compounds, find the portions that are hydrophilic and hydrophobic.

(a) H_3C CH_2OH H_3C

(b) HO OH O HO OH HO

(c) COO^- $H_3{}^+N$ C CH_2 H

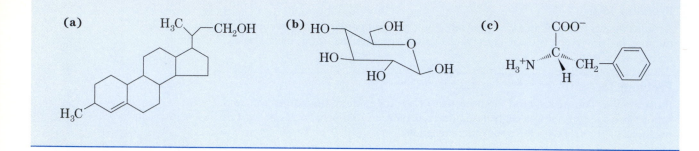

Tyndall light scattering using a soap solution

B. Mechanism of Soap Action

Let us take a closer look at what is happening in a soap solution, on a molecular basis (Figure 15.2). As a soap dissolves in water, the molecules orient themselves on the water's surface with the ionic end submerged and the nonpolar hydrocarbon chain bobbing above the surface like a buoy on the ocean. In this manner, the soap molecule satisfies its opposing solubility characteristics—the water-soluble, hydrophilic end is in the water and the nonpolar, hydrophobic end is not in contact with the water, but is surrounded by nonpolar N_2 and O_2 in air. This molecular orientation lowers the surface tension of water. The liquid surface is no longer made up of strongly associated, hydrogen-bonded water molecules, but of nonpolar, nonassociated hydrocarbon chains, somewhat like gasoline. This gives the water a better wetting capacity, allowing it to spread out

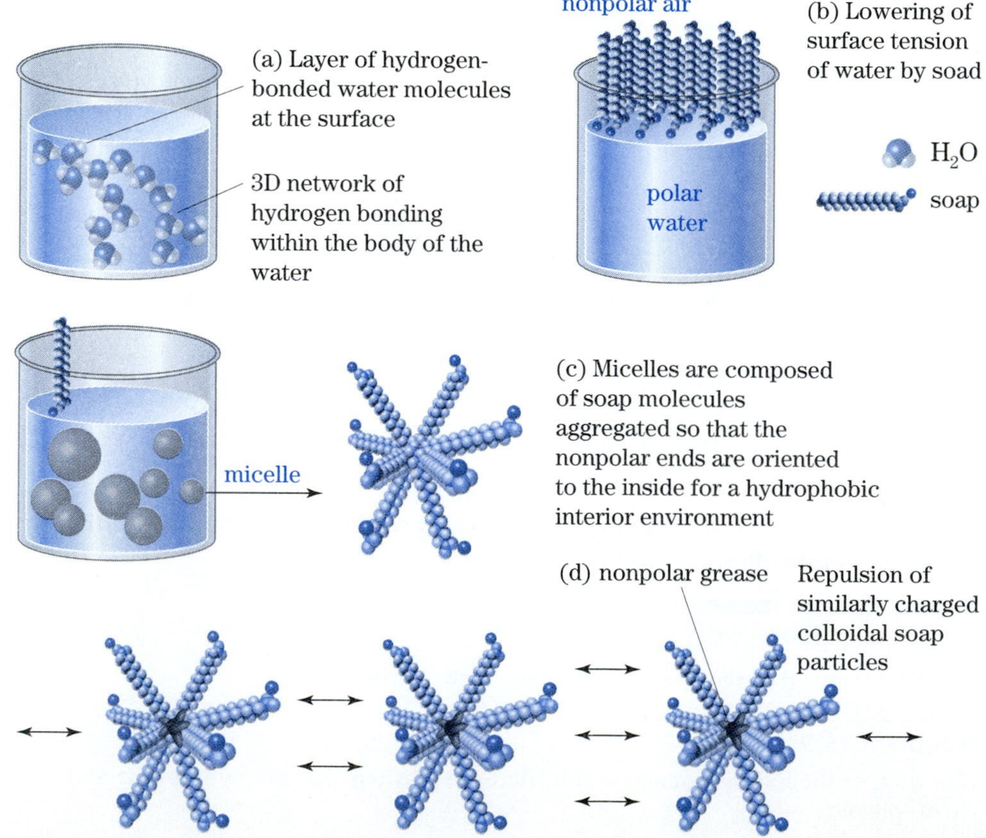

Figure 15.2
Mechanism of soap action.

and penetrate fabrics rather than bead up on the surface. Soaps or detergents are often added to herbicide and pesticide sprays to aid in the emulsification of the active ingredient in the water carrier and to promote better spreading of the solution over the leaves of the treated plants.

What happens to the soap molecules for which there is no room on the water's surface? They will have to orient themselves in such a way beneath the surface that the hydrophobic portions of the molecules have minimal contact with water. The soap molecules achieve this by grouping in three-dimensional clusters, with the nonpolar hydrocarbon chains filling the interior of the cluster and the water-soluble ionic ends composing the outer surface. The molecular conglomerations are called **micelles.** The solubility characteristics of the soap molecules are satisfied in that all the hydrocarbon chains are grouped together away from water (a hydrophobic core) and the ionic portions are in contact with water (Figure 15.2).

micelle
aggregation of polar/nonpolar molecules, like soap, in water such that the nonpolar portions of the molecules are arranged together inside, away from water and the polar portions protrude into the water

If some soiled clothing is submerged in the water, the nonpolar oil films are loosened, and they dissolve in the nonpolar hydrocarbon centers of the micelles. The micelles remain colloidally dispersed in the water, with no tendency to coagulate since there is an ionic repulsion between their charged outer surfaces. The oily films are thus washed away as finely dispersed oil droplets.

C. Detergents

Soaps, the sodium and potassium salts of long-chain fatty acids, have one serious disadvantage; they are insoluble in hard water. Hard water is water containing dissolved salts of calcium, magnesium, and iron picked up as water trickles over and filters through soil, rocks, and sand. Soaps react with these ions to form insoluble scums (the familiar bathtub ring).

$$2R\sim\overset{\displaystyle O}{\overset{\|}{C}}O^-Na^+ + Ca^{2+} \longrightarrow (R\sim\overset{\displaystyle O}{\overset{\|}{C}}O^-)_2Ca^{2+} + 2Na^+$$

Water-soluble Water-insoluble

Detergents, first introduced in 1933, are considerably more effective than soaps in hard water.

detergent
molecules that are not soaps but that have long non-polar, water-insoluble hydrocarbon chains that dissolve fats and oils, and a polar or ionic portion that is water-soluble

Detergents have the same two structural characteristics that soaps do:

1. They possess a long, nonpolar, hydrophobic, hydrocarbon chain that is soluble in fats, oils, and greases.
2. They possess a polar, hydrophilic end that is soluble in water.

Furthermore, in the way they work, detergents are analogous to soaps (as described in section 15.5.B and Figure 15.2).

Synthetic detergents, syndets, fall into three main categories, determined by the structure of the water-soluble portion of the molecule. **Anionic detergents** have an ionic water-soluble end in which the portion attached to the hydrocarbon chain is negative. Alkyl sulfates and alkyl benzene sulfates (ABS) are the two most common anionic detergents.

anionic detergent
cleaning agent with a negatively (−) charged polar end, such as a sulfate or sulfonate

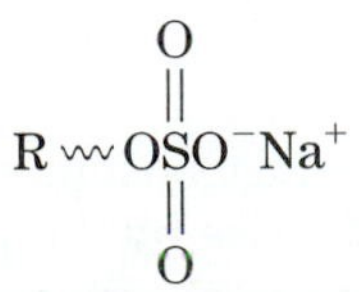

$$R\sim O\overset{\displaystyle O}{\overset{\|}{\underset{\underset{\displaystyle O}{\|}}{S}}}O^-Na^+ \qquad R\sim C_6H_4-\overset{\displaystyle O}{\overset{\|}{\underset{\underset{\displaystyle O}{\|}}{S}}}O^-Na^+$$

A sodium alkyl sulfate 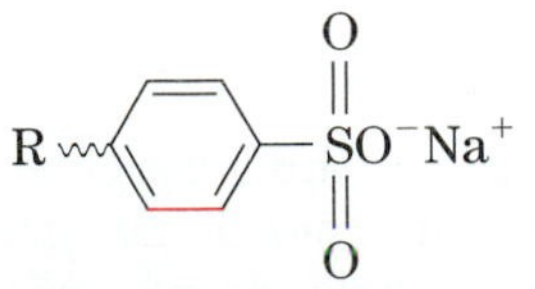 A sodium alkyl benzene sulfonate

cationic detergent cleaning agent with a positively (+) charged polar end, such as an ammonium group

The water-soluble end of a **cationic detergent** is a positive quaternary ammonium salt.

$$R\sim\overset{+}{N}(CH_3)_2-CH_3\ Cl^-$$

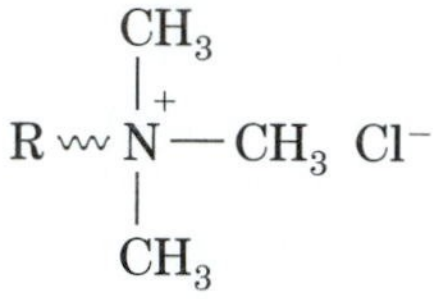

A quaternary ammonium salt

These detergents have significant germicidal properties, and similar compounds such as the ones shown are used in shampoos, mouthwashes, germicidal soaps, and disinfectant skin sprays.

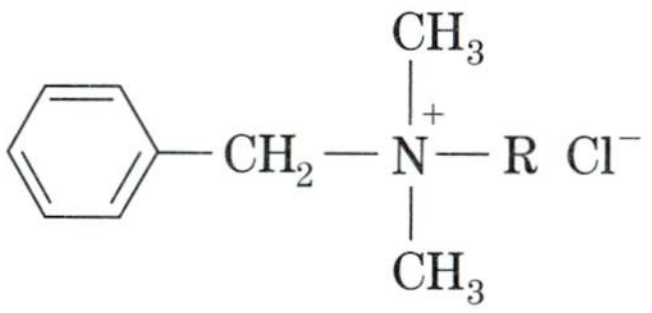

$$C_6H_5-CH_2-\overset{+}{N}(CH_3)_2-R\ Cl^- \qquad CH_3(CH_2)_{15}-\overset{+}{N}C_5H_5\ Cl^-$$

Benzalkonium chlorides (R = 8–14 carbons)

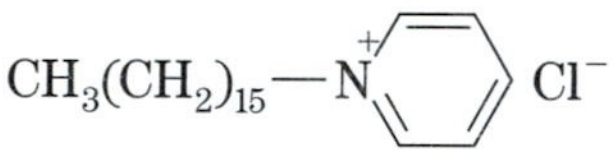

Cetylpyridinium chloride

nonionic detergent cleaning agent with an uncharged, polar end, such as an alcohol

In **nonionic detergents,** the water-soluble end is polar and can hydrogen-bond with water, but it is not ionic.

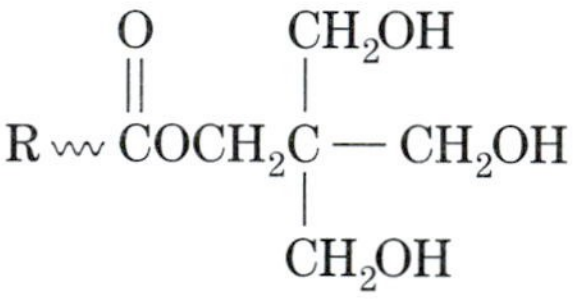

$$R\sim\overset{O}{\overset{\|}{C}}OCH_2C(CH_2OH)_2-CH_2OH \qquad R\sim\overset{O}{\overset{\|}{C}}(OCH_2CH_2)_nOH$$

biodegradable materials that can be metabolized by soil and water bacteria

Most detergents today are **biodegradable.** This means that they can be metabolized by microorganisms in a sewage disposal plant before release into the environment. For a detergent to be biodegradable, the long alkyl chain must be unbranched. Detergents used in the 1950s and early 1960s had branched chains, were not readily biodegradable, and foamed when the water discharged from sewage plants was agitated.

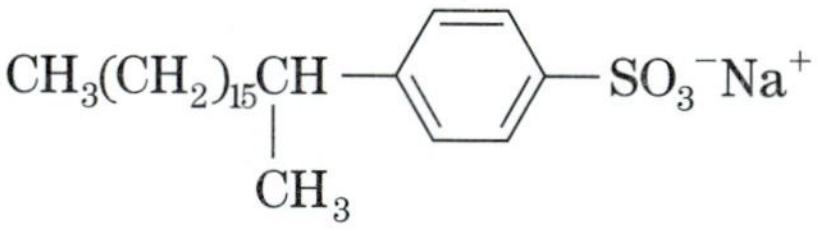

$$CH_3(CH_2)_{15}\underset{CH_3}{\underset{|}{C}H}-C_6H_4-SO_3^-Na^+ \qquad CH_3\underset{CH_3}{\underset{|}{C}H}CH_2\underset{CH_3}{\underset{|}{C}H}CH_2\underset{CH_3}{\underset{|}{C}H}CH_2\underset{CH_3}{\underset{|}{C}H}-C_6H_4-SO_3^-Na^+$$

A biodegradable detergent

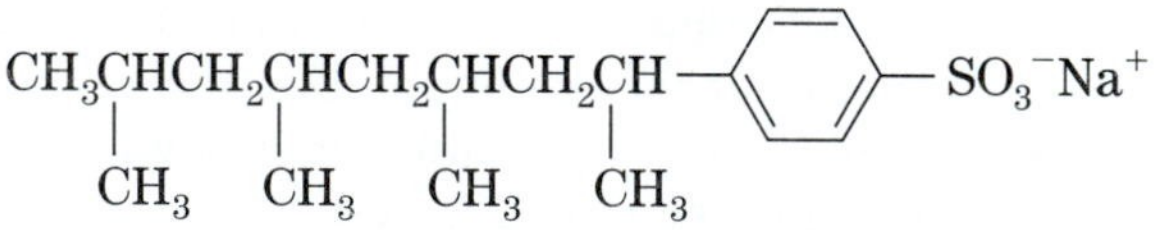

A nonbiodegradable detergent

Chemically pure soaps are regulated by the federal Consumer Product Safety Commission and are considered generally safe for use by the public. Today's hand "soaps" are usually detergents with brighteners, abrasives, and perfumes added. Many are advertised as being germicidal or antibacterial. The detergent composition and advertising claims put them under regulation by the Food and Drug Administration.

GETTING INVOLVED

- ✓ Under what circumstances might it be better to use a detergent than a soap?
- ✓ In terms of chemical structure, how is a detergent similar to a soap and how is it different?
- ✓ Compare the cleaning actions of a soap and a detergent on a molecular level.
- ✓ Be able to identify the functional groups on detergents which classify them as cationic, anionic, and nonionic.
- ✓ Is a soap cationic, anionic, or nonionic?
- ✓ What type of structure might impede the biodegradability of a soap or detergent? **See related problems 15.15, 15.18, 15.19, 15.21.**

15.6 Biolipids—Structures and Functions

As we have seen so far in this chapter, the term *lipid* applies to more than one chemical structure. The common feature of all lipids is a relative insolubility in water of all or a large portion of the molecule. In a living organism, this property allows lipids to serve in many interesting capacities. They are found in cell membranes, as insulating tissue that protects organs from the external environment, as chemical fuel storage depots, and as regulators of metabolism.

We can consider the biolipids in seven general classes depending on both their structural and functional similarities: triacylglycerides (fats), phosphoglycerides, sphingolipids, steroids, prostaglandins (and derivatives), fat-soluble vitamins, and pigments.

A. Triacylglycerols

We need not review the structure of these lipids except to say that they are nonpolar, complex, and saponifiable. Let us continue with a discussion of their functions. Gram for gram, fats and oils deliver about 2.5 as much metabolic energy as do carbohydrates or proteins. Thus fat storage in adipose tissue is a concentrated energy reserve. In addition, we can store more fat than glucose (in the form of the polymer glycogen) because of the insolubility of fats in water. Since fats are hydrophobic, they will not attract water during storage. The hydrophilic nature of glycogen results in hydrogen bonding with water, which increases its weight. The insulating nature of fats keeps our vital organs warm, and their bulk acts as protection against trauma.

B. Phospholipids

phospholipid
complex, saponifiable, polar lipid containing one or more phosphate groups

phosphoglyceride
complex, saponifiable, polar lipid; triester of glycerol in which two acids are saturated and unsaturated long-chain fatty acids and the third acid is phosphoric acid that is further esterified

Phospholipids contain one or more phosphate groups; they are complex, saponifiable, polar lipids. The most common phospholipids are a variation on the triacylglycerol structure with two long-chain fatty acids esterified to the first two positions of glycerol and a phosphate group in an ester linkage at the third position. Thus they are called *glycerophospholipids* (or **phosphoglycerides**). Since phosphate is a triprotic acid, it has three reactive acid groups and can form a second ester with an additional alcohol. This leaves an acid group that can ion-

amphipathic molecule with a polar portion and a nonpolar portion

ize. The versatility of this structure produces a complex lipid with both hydrophilic and hydrophobic properties; this is called an **amphipathic** molecule.

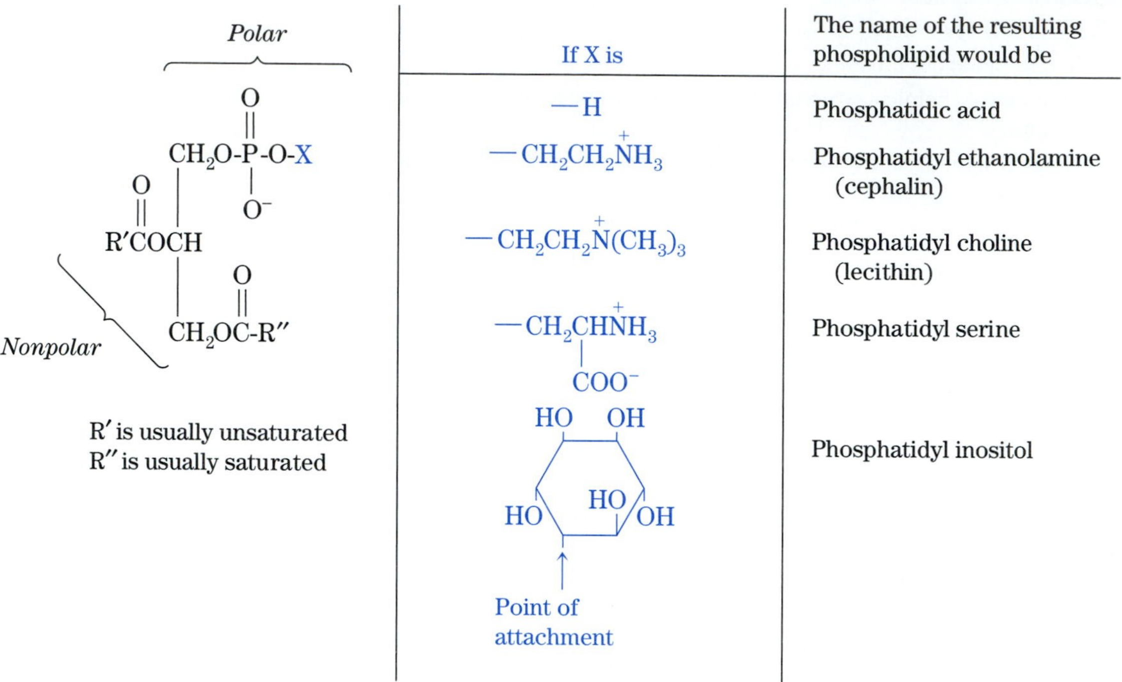

Polar
Nonpolar
CH_2O-P-O-X
O^-
R′COCH
CH_2OC-R″
R′ is usually unsaturated
R″ is usually saturated
If X is
The name of the resulting phospholipid would be
—H
Phosphatidic acid
$-CH_2CH_2\overset{+}{N}H_3$
Phosphatidyl ethanolamine (cephalin)
$-CH_2CH_2\overset{+}{N}(CH_3)_3$
Phosphatidyl choline (lecithin)
$-CH_2CH\overset{+}{N}H_3$
COO^-
Phosphatidyl serine
HO OH
HO
HO
OH
Point of attachment
Phosphatidyl inositol

The amphipathic nature of both the phospholipids and sphingolipids (below) contributes to their function as membrane components in cells.

GETTING INVOLVED

- ✓ Which groups on phospholipids are responsible for their properties of being classified as complex, saponifiable, and polar lipids?
- ✓ Give an example of at least one other type of molecule in this chapter that is amphipathic.
- ✓ Are fats and oils or waxes amphipathic? Explain your answer in terms of their structures.

Problem 15.10

Draw out the structure of a lecithin containing linolenic acid as R' and palmitic acid as R".

Problem 15.11

What compounds would result from the complete hydrolysis of one mole of phosphatidylserine?

C. Sphingolipids

sphingolipid complex, saponifiable, polar lipid; composed of sphingosine linked through an amide bond to a very-long-chain fatty acid and through an ester or acetal linkage to acids or carbohydrates

Sphingolipids are more complex than the phospholipids in that they are derivatives of the amino-alcohol sphingosine. The amine group forms an amide bond with an unsaturated fatty acid, resulting in a ceramide, and the alcohol either can be esterified with an acid, such as phosphoric acid, or can form an acetal or ketal with a carbohydrate molecule or polymer chain. If the phosphate is also esterified with ethanolamine or choline, the molecule is called a *sphingomyelin.* When one or more carbohydrate groups are attached to the ceramide hydroxyl, a glycolipid results. The most common glycolipids are cerebrosides, which contain a single monosaccharide and are important components of the myelin sheath of nerve cells, and gangliosides, which are believed to play a role in nerve signal transmission.

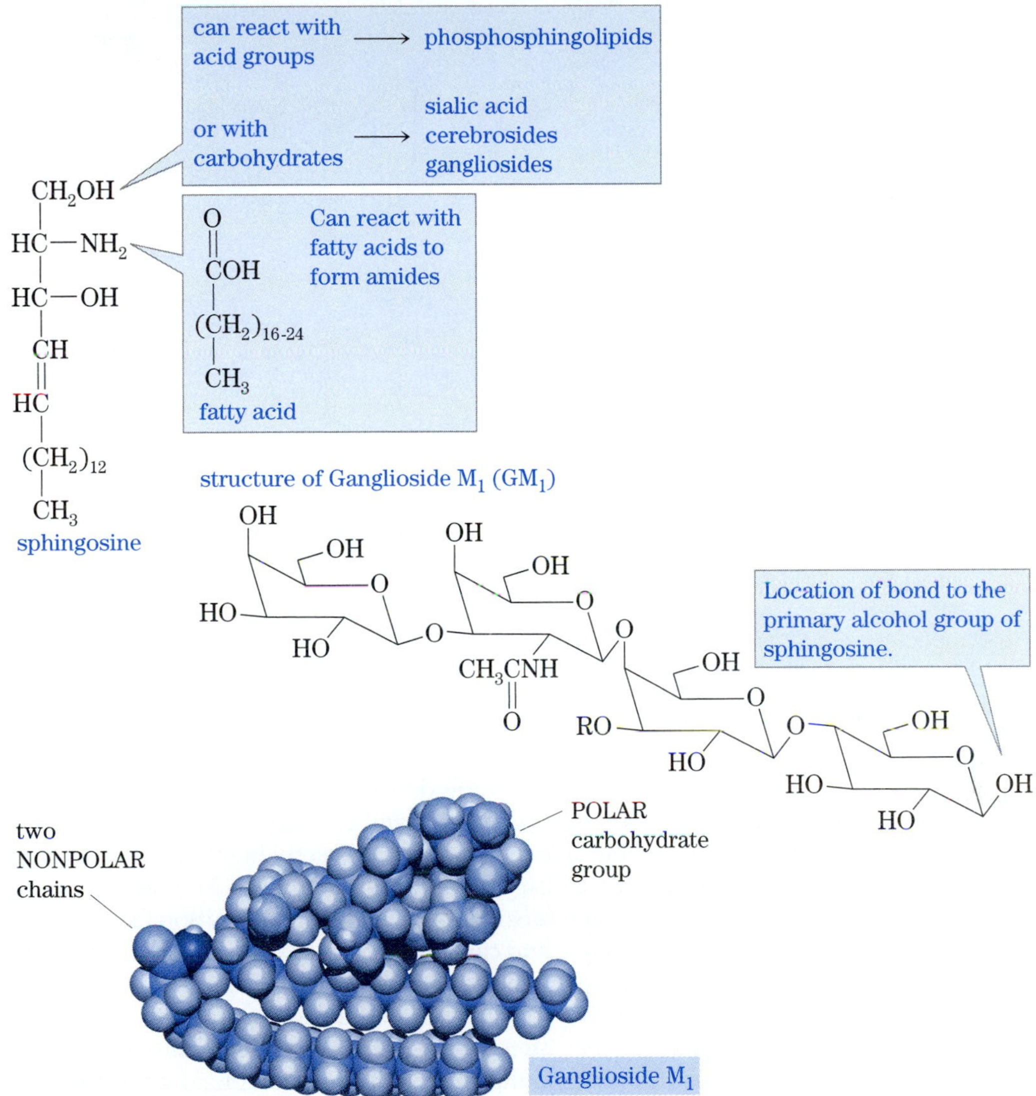

The amphipathic nature of these two types of biolipids causes them to interact with each other and water in a manner that goes one step further than that of the soaps. Rather than forming a monolayer of micelle, the molecules spontaneously arrange into a bilayer that, when extended, can fold back on itself to form a sphere, ellipsoid, or other encapsulated structure. In other words, they can form a cell **membrane,** the semipermeable barrier that allows some substances, but not others, to pass from one side to the other. This helps in the compartmentalization and efficiency of metabolic functions. The cell membrane has many other components, including cholesterol and proteins. Synthetic vesicles, called **liposomes,** can be formed from phospholipids and sphingolipids. Liposomes are used to study membranes and membrane transport and are being studied as

membrane
naturally occurring, semipermeable lipid bilayer composed of phospholipids, sphingolipids, cholesterol, and proteins

liposome
synthetic vesicle with a semipermeable barrier composed of phospholipids

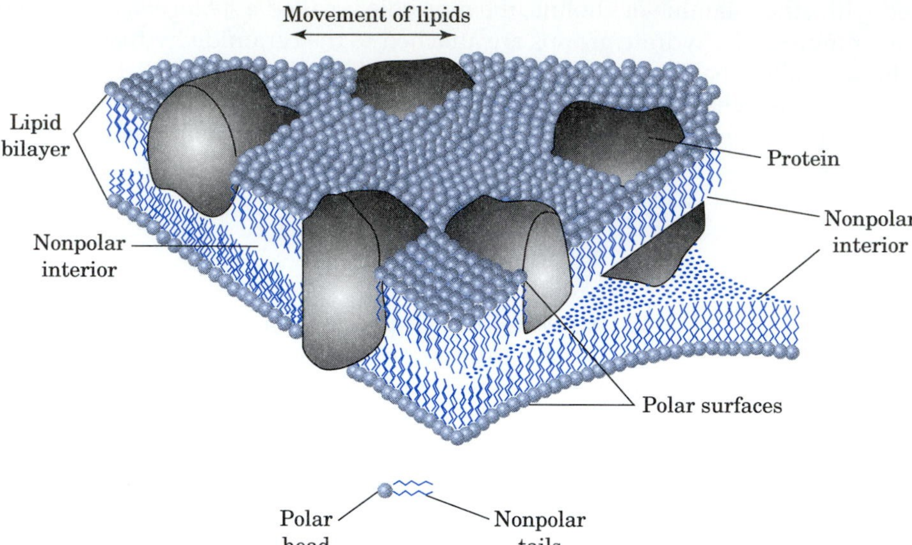

Figure 15.3 Cell membrane diagram showing structural arrangement of lipid bilayer.

a means to deliver drugs inside the body. Figure 15.3 illustrates the model proposed for membrane structure known as the "fluid mosaic" model.

This model views the membrane as a mosaic of lipid and protein, in which the fluidity of the lipid permits both lipid and protein molecules to move laterally. The composition of a membrane varies both in the ratio of protein to lipid and in the percentages of various lipids. Protein content varies from about 20% in the myelin surrounding nerve cells to greater than 70% in the inner membrane of mitochondria. Cholesterol, a steroid, is another important component of the membrane. Having a more rigid structure than the other lipid portions of the membrane, it helps to maintain membrane structure. The concentration of cholesterol in a membrane usually varies directly in relation to the concentration of unsaturated fatty acids present in phospholipids. The carbohydrate portions of gangliosides and cerebrosides can act as cell recognition factors on the outer surface of the membrane and can be attached to proteins, which serve various functions, including that of cell recognition.

Disorders such as ALD, described in Connections 15.1, are frequently fatal, especially when they affect the membranes of nerve tissue. Tay-Sachs disease is a recessive trait that results in the accumulation of ganglioside G_{M2} in the brains of its victims due to the absence of an enzyme, hexoseamidase, to break down G_{M2}. Death occurs within four years after birth following a course of severe brain damage, paralysis, and blindness. Heterozygotes, people who carry only one gene for the disorder, number about 1 in 30 in the northern European population. They can be victims of a late-onset version of Tay-Sachs due to a decrease in hexoseamidase activity.

D. Steroids

steroid
lipid with a four-fused-ring structure, three rings having six members and one ring with five members

Steroids can be saponifiable or nonsaponifiable. All of the **steroids** have a system of four fused rings—three six-membered rings and one five-membered ring. Substituents on this large ring system contribute to functions that range from hormonal regulation to digestion to poison.

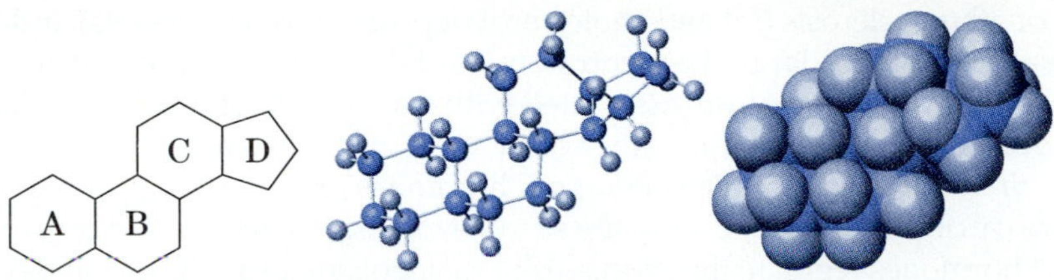

The steroid fused-ring nucleus is shown in three views: two-dimensional; three-dimensional ball-and-stick model; three-dimensional space-filling form. This fused ring system is common to all lipids in this class. Although the system itself is hydrophobic, the presence of hydrophilic side chains can modify the solubility properties of the molecule. The four rings are identified as A,B,C, and D as in the figure.

1. ***Cholesterol.*** The precursor to all steroid endocrine hormones is cholesterol, a simple, nonsaponifiable, nonpolar lipid. The liver is the primary source of its biosynthesis.

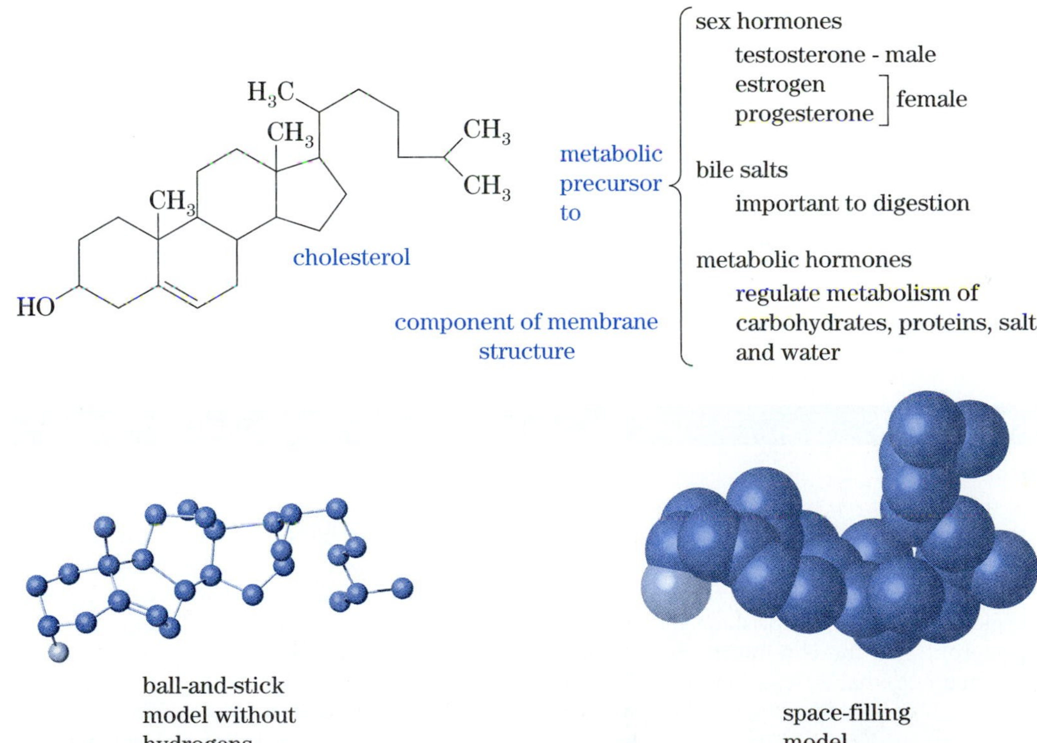

Being hydrophobic, cholesterol must be carried in water-soluble protein complexes through the bloodstream. These complexes contain varying amounts of triglyceride as well as cholesterol and are called *lipoproteins.* The more fat or triglyceride present in the lipoprotein complex, the less dense it will be. The complexes are divided into high-, low-, and very-low-density lipoproteins (as well as other fractions) or HDL, LDL, and VLDL, respectively. Complicated feedback mechanisms, many of which are in the liver, control the plasma concentrations of these complexes. High circulating concentrations of LDL increase the probability of atherosclerosis (fat and cholesterol deposits in blood vessels) and heart disease. Conversely, larger concentrations of HDL, which can result from engaging in regular exercise, are associated with a lower risk of cardiovascular disease. See Connections 5.4 for further discussion of cholesterol.

2. ***Steroid Hormones.*** The endocrine hormones are physiological regulators that are carried by the blood from the site of synthesis to the affected organs. The steroid hormones regulate the processes of metabolism, growth, sexual development, and reproduction. All are derived from cholesterol. The two main organs of secretion are the adrenal cortex and the gonads, that is, the ovaries and the testes.

The adrenal cortex, the outer portion of the adrenal glands which are located above the kidneys, is a primary organ of steroid hormone production. Adrenocortical hormones fall into two general categories: those regulating the metabolism of carbohydrates, proteins, lipids, and so on (glucocorticoids), and those influencing salt and water metabolism (mineralocorticoids). These compounds not only overlap in their functions but also have additional roles in the regulation of the cardiovascular and nervous systems, among others.

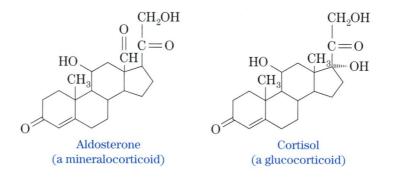

Aldosterone
(a mineralocorticoid)

Cortisol
(a glucocorticoid)

Connections 15.2

www.prenhall.com/bailey

RU-486

Very few chemicals in recent history have generated the political and social argument that RU-486 has in the area of reproductive rights and medical advancement. This compound is a steroid with the generic name of mifepristone. It blocks the action of progesterone and thereby interferes with the gestation of a fertilized egg; that is, it can terminate conception.

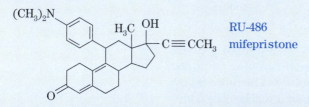

RU-486
mifepristone

Technically, RU-486 is not a contraceptive, because it does not have a chemical mechanism of action similar to that of norethindrone and mestranol, which mimic natural estrogens and progestins. Rather it is an antiprogestin or a contragestive in that it blocks the action of natural progesterone, which is responsible for maintaining the fetus during pregnancy. Without the proper hormonal environment, the fetus cannot survive. Given along with a prostaglandin within 49 days of the last menstrual period, RU-486 has caused a spontaneous abortion in about 95% of women tested. This testing and the current use of the drug have taken place primarily in Europe. Side effects are varied, and the treatment can take place in the privacy of a physician's examination room. There has been strong antiabortion group sentiment against the introduction of RU-486 into the United States for any purpose. In addition to its action as an abortifacient, mifepristone has also shown some promise for the treatment of breast cancer, glaucoma, and other conditions. It has been made available for limited use and research in the United States and will probably generate controversy for many years to come

The glucocorticoids also have antiinflammatory activity, which makes substances like cortisol and the related compounds cortisone and prednisone valuable in the treatment of conditions such as severe allergic reactions and rheumatoid arthritis.

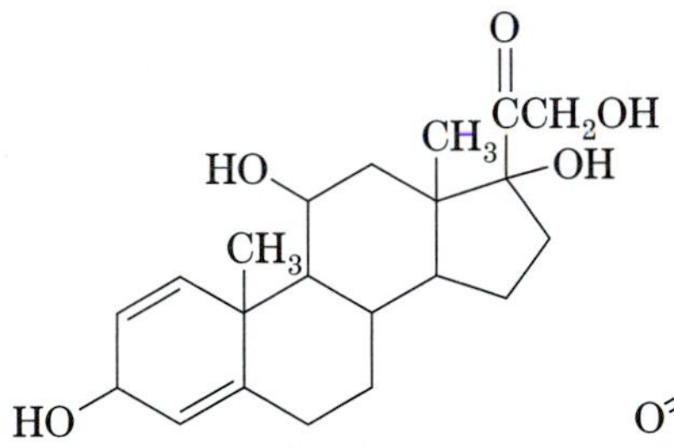

prednisolone
drug used for inflammatory diseases and organ disorders

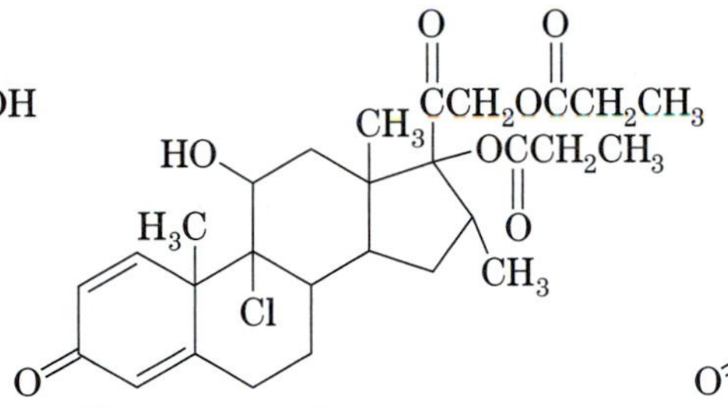

beclomethasone dipropionate
(Beconase®, Beclovent®)

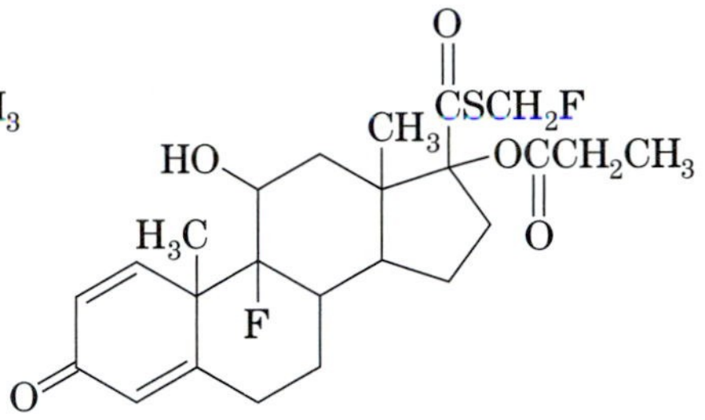

fluticasone propionate
(Flonase®)

aerosol inhalants used to reduce bronchial inflammation found in asthma

The sex hormones, androgens and estrogens, are responsible for the development of the secondary sex characteristics such as the distribution of body fat, protein, and hair, voice timbre, and development of the genital organs. There has been a deep concern over the use of testosterone and its derivatives by athletes to enhance muscle development. Although the use of androgenic (male) steroids does indeed increase muscle protein, their use can also promote the deposition of atherosclerotic plaque, can stimulate skin oil production, and can alter the psychological disposition of the user. Sex hormones and contraceptives were discussed in Connections 3.1.

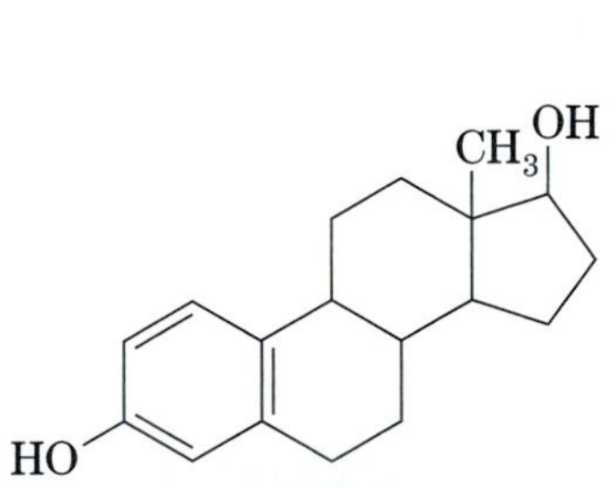

estradiol
an estrogen

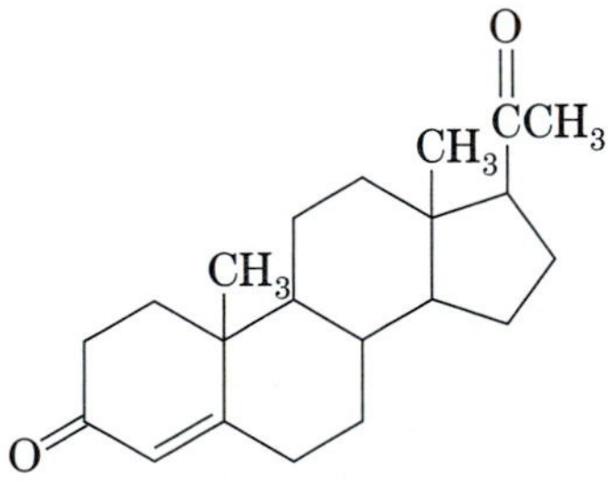

progesterone
a progestin

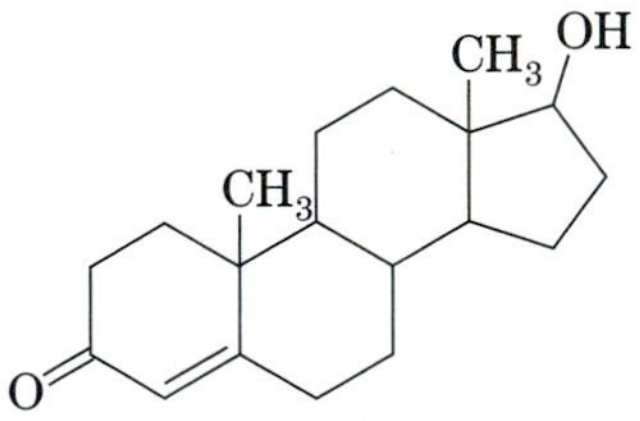

testosterone
an androgen

3. ***Bile Acids.*** The attachment of a polar group to the D-ring of the steroid nucleus leads to an amphipathic molecule called a **bile acid.** The salts of bile acids act in the digestive process as emulsifying agents. Since fats and oils are not water-soluble, the bile acids help to form micelles with them **(emulsification)** in the small intestine so that the triglycerides may be broken down by enzymes before absorption into the bloodstream.

bile acid
amphipathic lipid, a steroid derivative produced in the liver and delivered to the intestines via the bile duct where it acts to emulsify ingested lipids

emulsification
the process of solubilizing polar and nonpolar compounds

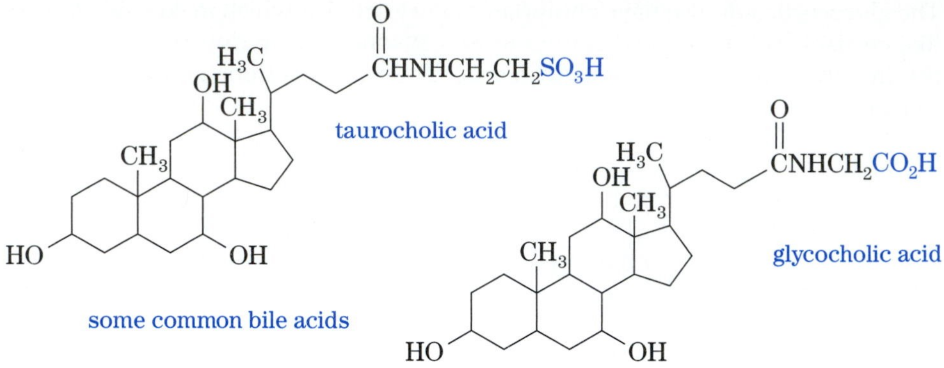

some common bile acids

4. ***Toxins.*** Many steroids are toxic. The nonpolar nature of the molecules allows them to be easily absorbed. Once in the body, circulating in the bloodstream, they have access to any number of sites. Certain Colombian tree frogs produce a variety of toxic steroids that the indigenous population uses as arrow poisons. The foxglove plant produces the complex steroidal mixture known as digitalis. This material affects the strength of heart contractions, helping to relieve the condition known as congestive heart failure but acting as a deadly poison in larger than therapeutic amounts.

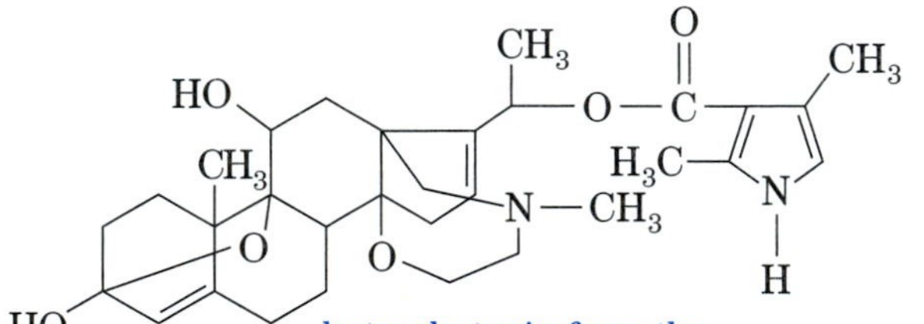

batrachotoxin from the skin of a Colombian tree frog; arrow poison

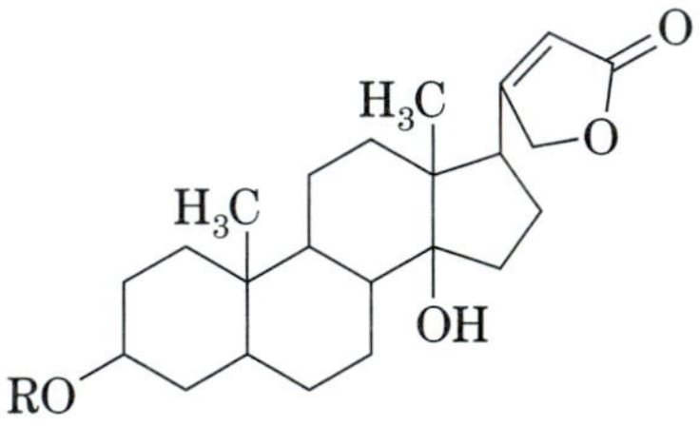

digitoxin
R can be a variety of carbohydrate attachments; cardiac stimulant

GETTING INVOLVED

- ✓ Draw a picture of a membrane lipid bilayer. Justify the arrangement of the phospholipids and sphingolipids in terms of their amphipathic parts: polar and nonpolar.
- ✓ How would cholesterol arrange itself in the bilayer?
- ✓ There are proteins embedded in the membrane and penetrating through the membrane. What would you automatically assume about the polarity of the outsides of these proteins?

See related problems 15.21, 15.23, 15.24, 15.25, 15.26.

E. Eicosanoids—Tissue Hormones

eicosanoid
compound formed from long-chain unsaturated fatty acids

An **eicosanoid** is a compound formed from the $C_{20:4}$ fatty acid, arachidonic acid, or a fatty acid related to it. These unsaturated fatty acids result from the breakdown of the phospholipids in cell membranes during infection or as a reaction to toxic insult. Snake venoms contain enzymes called *phospholipases* that specialize in such breakdown as a means to gain access to blood and tissue cells.

The eicosanoids consist of the prostaglandins, prostacyclins, thromboxanes, and leukotrienes. The first three types of compounds are related to each other both in structure and origin. In fact, the prostacyclins and thromboxanes are biosynthesized from prostaglandins. **Prostaglandins** can cause smooth muscle contraction or relaxation, vasodilation, stimulation of blood clotting, and a variety of other effects.

prostaglandin
lipid tissue hormone synthesized from long-chain fatty acids

An enzyme called *cyclooxygenase* starts the process of arachidonate to prostanoid conversion. It is interesting to note that aspirin, acetyl salicylic acid, inhibits this enzyme, thereby decreasing the formation of prostaglandins. Aspirin also slows blood clotting time, another ramification of its anticyclooxygenase activity.

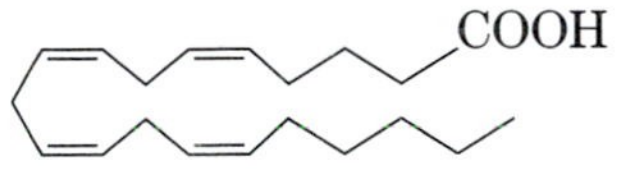

COOH

Arachidonic acid

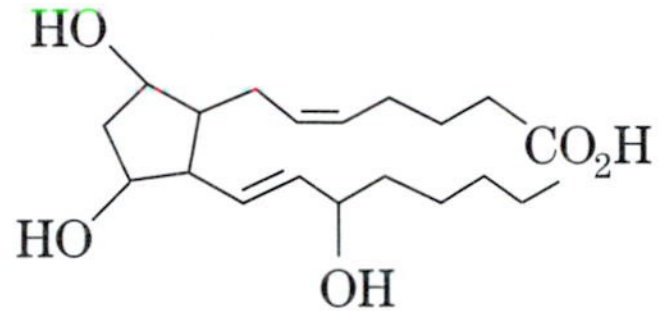

HO

CO_2H

HO

OH

Prostaglandin $F_{2\alpha}$ ($PGF_{2\alpha}$)
Enzaprost® used to induce abortion

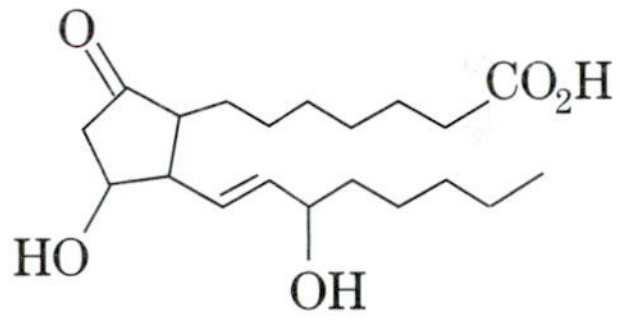

O

CO_2H

HO

OH

Prostaglandin E_1 (PGE_1)
a vasodilator

Prostacyclins are the PGI class of prostanoids. Thromboxanes are another product of prostaglandin metabolism. These two types of compounds are responsible for the prevention and stimulation of platelet cell aggregation, respectively, during the blood clotting process.

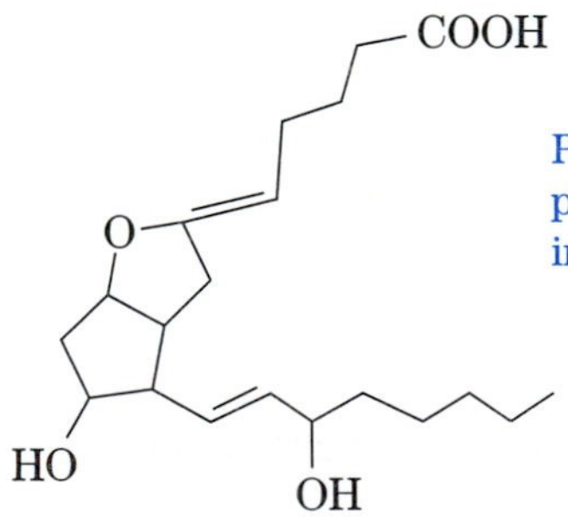

COOH

O

HO

OH

PGI_2 prostacyclin
platelet aggregation inhibitor

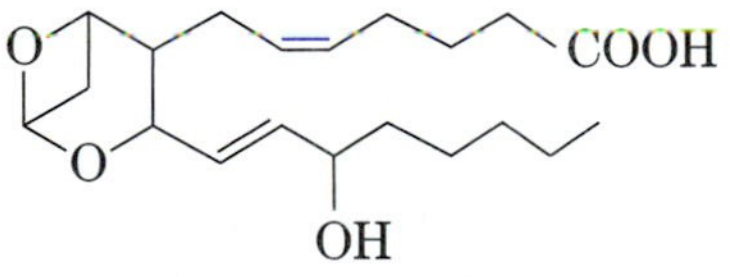

O

COOH

O

OH

Thromboxane A_2 (TXA_2)
inducer of platelet aggregation; also causes contraction of the arteries

Leukotrienes follow a different pathway of biosynthesis catalyzed by the enzyme lipooxygenase. These compounds have potent bronchoconstricting effects and are primarily responsible for the difficulties in breathing experienced by asthmatics and those having a severe anaphylactic experience (shock) due to an insect sting or as a reaction to a drug to which they are allergic.

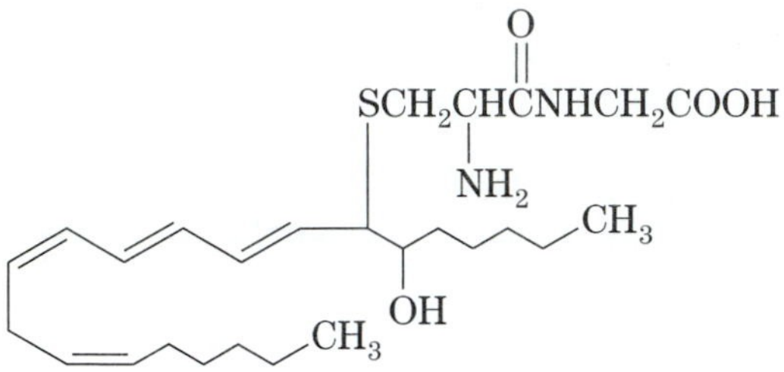

leukotriene D_4 (LTD_4)
along with LTE_4 causes bronchial constriction during an asthma attack

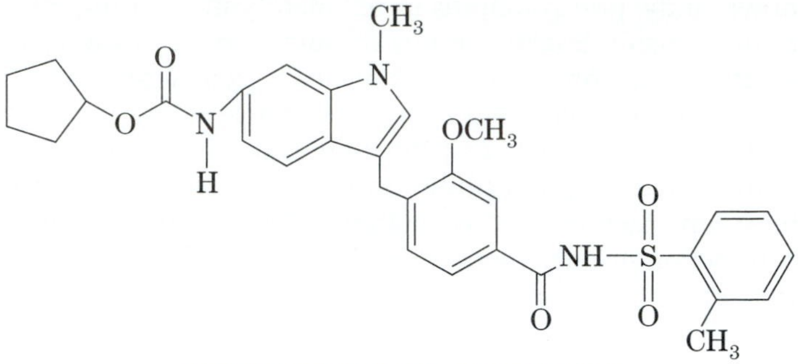

zifirlukast (a leukotriene antagonist)
Accolate®
drug used to prevent and alleviate the brochoconstriction of asthma

F. Vitamins

fat-soluble vitamin nonpolar, nonwater-soluble, essential dietary component; vitamins A, D, E, and K

water-soluble vitamin polar, water-soluble, essential dietary component such as the B complex vitamins and vitamin C

As the term *vitamin* suggests, vitamins are substances essential to life (*vita* is Latin for "life"). They cannot be produced by the normal metabolism of the body. Like hormones, vitamins can take many chemical forms, including those that are water-soluble (the B-complex vitamins and vitamin C) and others that are fat-soluble (A, D, E, and K). Although **water-soluble vitamins** must be supplied frequently, the fat-soluble ones are stored within the body until needed. As a result, it is possible to ingest an overdose of these vitamins.

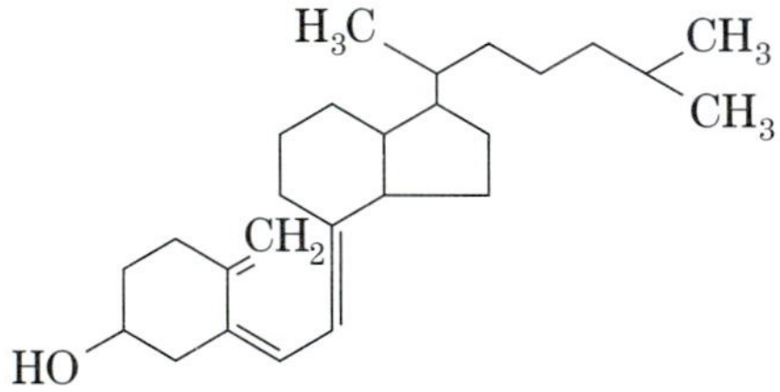

Vitamin D_3

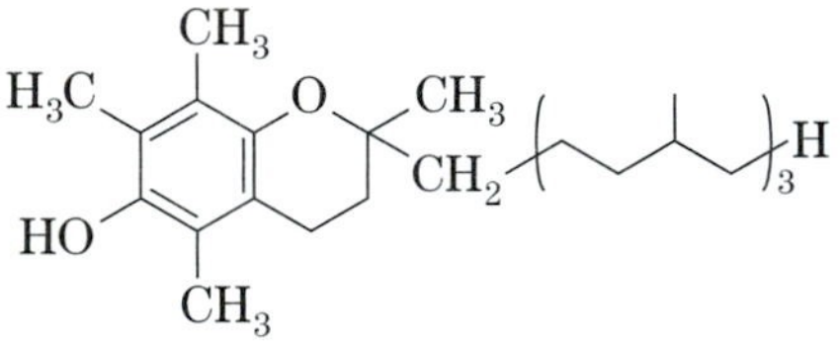

Vitamin E

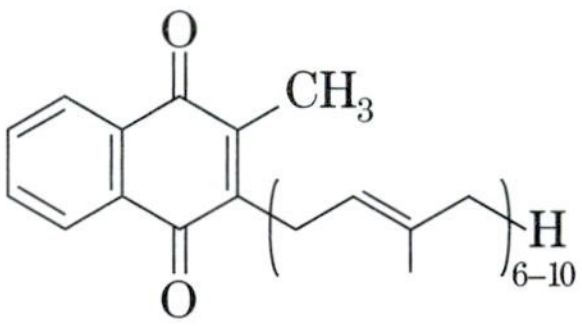

Vitamin K

Vitamin A and its role in the visual cycle and development have been discussed previously (Connections 3.3).

There are several forms of vitamin D, of which D_3 or cholecalciferol is one. It is formed from a precursor in the skin by the action of sunlight. Milk is supplemented with D_3 and D_2, activated ergosterol, obtained from yeast. Vitamin D facilitates the absorption of calcium and phosphorus from the small intestine and their incorporation into bone. A deficiency of vitamin D leads to a condition known as rickets and is evidenced by bone malformations such as bowlegs and extreme tooth decay. Overdoses result in hypercalcification and kidney problems.

Vitamin E, tocopherol, is rarely deficient in diets since it is found in most foods in sufficient quantities. Not much is known about its role in the human body except that it helps to maintain cell membranes by acting as an antioxidant.

The K vitamins are produced by the bacteria inhabiting the intestinal tract. They aid in the complex mechanism of blood clotting; the rare deficiency results in a tendency to hemorrhage. Aspirin and related compounds are antagonistic to the K vitamins.

G. Pigments

Plants and certain algae and bacteria can utilize solar energy for the biosynthesis of their important parts. This process is known as *photosynthesis*, and it requires membrane-bound compounds that can gather light efficiently. The most important of these pigments is chlorophyll. In addition, other compounds, such as the carotenoids and phycobilins, augment the amount of light energy absorbed by an organism. All of these materials are highly conjugated organic molecules and are themselves colored.

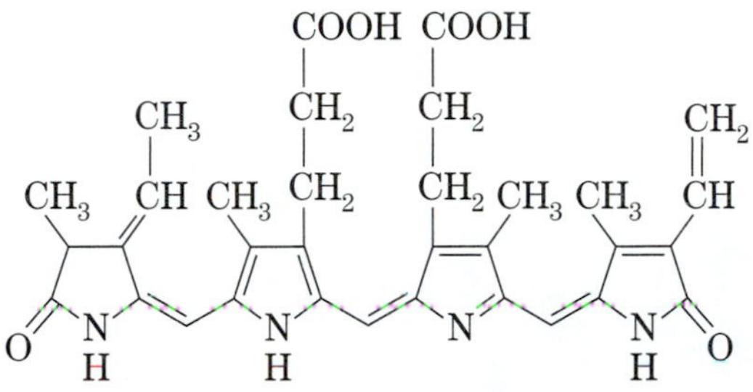

phycoerythrin
(red pigment)

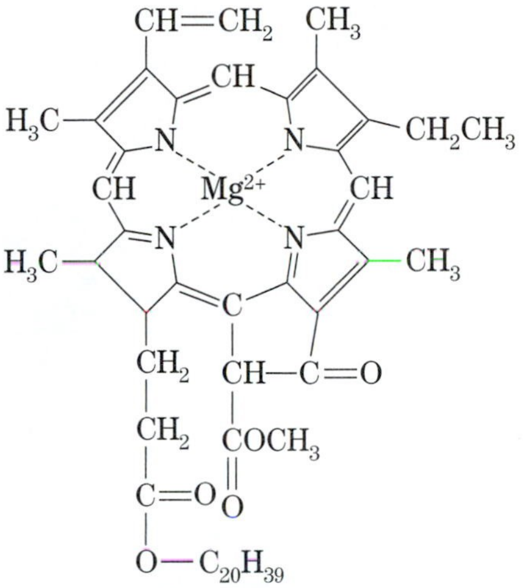

chlorophyll a

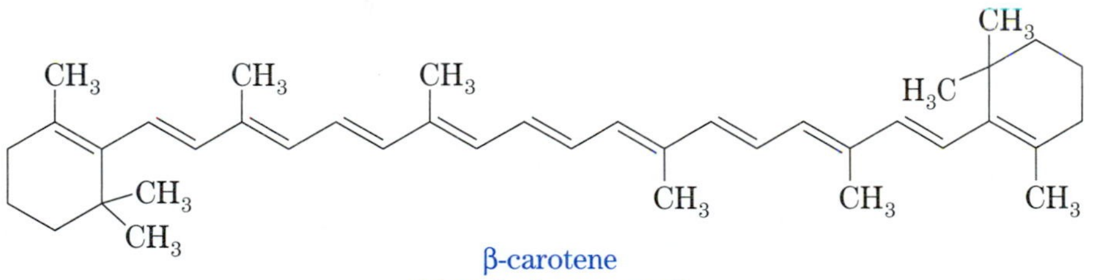

β-carotene
(vitamin A precursor)

H. Other Functions

Various other lipids fulfill many biological functions. For example, squalamine, the first documented steroid antimicrobial to be found in animals, is endogenous (within the organism) to sharks.

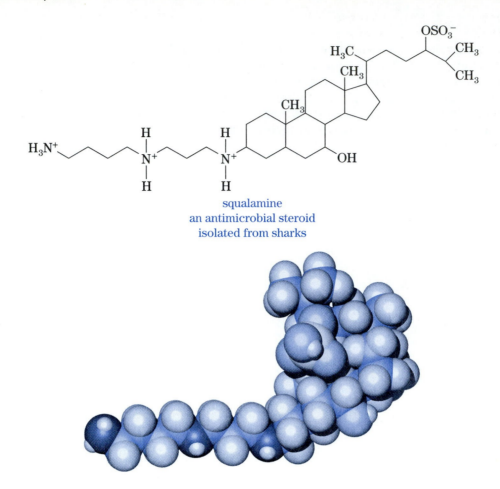

squalamine
an antimicrobial steroid
isolated from sharks

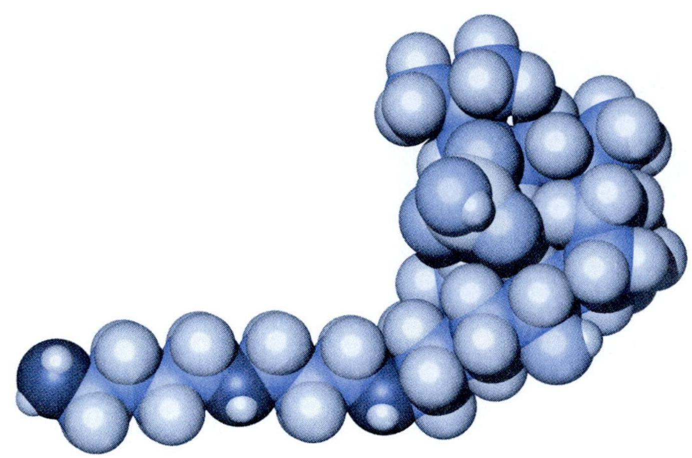

GETTING INVOLVED

Problem 15.12

Is squalamine a polar or nonpolar lipid? Explain your answer using various portions of the molecular structure.

Problems

15.13 **Structures of Fats and Oils:** Draw structures for the following fats and oils:

(a) a glyceride with three lauric acid units, trilaurin

(b) a glyceride with a myristic acid, a palmitic acid, and a stearic acid unit

(c) a glyceride with two myristic acid units and one oleic acid

(d) a glyceride likely to be found in corn oil

(e) a glyceride likely to be found in soybean oil

15.14 **Reactions of Fats and Oils:** Write chemical equations using the following glyceride to describe the reactions indicated:

$$CH_2OC(=O)(CH_2)_7CH{=}CH(CH_2)_7CH_3$$
$$CHOC(=O)(CH_2)_7CH{=}CHCH_2CH{=}CHCH_2CH{=}CHCH_2CH_3$$
$$CH_2OC(=O)(CH_2)_{14}CH_3$$

(a) saponification with NaOH
(b) hydrogenation
(c) I_2/CCl_4

15.15 **Reactions of Soaps:** Write chemical equations showing the reaction of a soap such as sodium stearate with the following:
(a) hard water containing Mg^{2+}
(b) hard water containing Fe^{3+}
(c) an acid solution (HCl)

15.16 **Structure of Fatty Acids:** Draw out the structures of the following fatty acids:
(a) vaccenic $acid_{18:1}{}^{\Delta 11}$
(b) docosahexaenoic acid ($C_{22:6}{}^{\Delta 4,7,10,13,16,19}$)
Are these ω3 or ω6 fatty acids?

15.17 **Structure of Fatty Acids:** At which positions are the following ω6 fatty acids unsaturated?
(a) $C_{14:4}$ (b) $C_{30:5}$ (c) $C_{26:3}$

15.18 **Structures of Soaps and Detergents:** Which of the following would or would not be an effective soap or detergent in water? For each case, explain why.
(a) $CH_3(CH_2)_{14}CO_2{}^-Na^+$
(b) $(CH_3(CH_2)_{16}CO_2{}^-)_2Ca^{2+}$
(c) $CH_3CH_2CO_2{}^-Na^+$
(d) $CH_3(CH_2)_{14}CH_2N(CH_3)_3{}^+Cl^-$
(e) $CH_3(CH_2)_{16}CH_3$
(f) $CH_3(CH_2)_{14}CO_2H$
(g) $CH_3(CH_2)_{14}CH_2OSO_3{}^-Na^+$

15.19 **Properties of Soaps and Detergents:** For whichever compounds you identified as a soap or detergent in the previous question, indicate the hydrophobic and hydrophilic ends of the molecules.

15.20 **Consumer Chemistry:** In a grocery store or drugstore, examine the labels on the following products:
(a) Margarine. Make a list of the vegetable oils used to produce various brands of margarine.
(b) Shortenings. Make a list of the vegetable oils used to produce various brands.
(c) Oils. Make a list of the various types of oils (from different plant sources) available for sale in your local supermarket.
(d) Detergents. Determine if possible the type of detergent, and additives; if the selection is phosphate-based, record the percentage of phosphorus.
(e) Disinfectants. Find some products containing benzalkonium chlorides or cetylpyridinium chloride as antiseptics.
(f) Biolipids. Check various products as to biolipid content and consider the purpose of the compounds noted in that product. Your pharmacist should be able to help you with steroids.

15.21 **Properties of Fats and Oils:** What is the relationship between the melting point of a triacylglycerol and its iodine number?

15.22 **Structure:** How are detergents and phospholipids and sphingolipids alike in structure and function? How do they differ?

15.23 **Structure of Biolipids:** Using the compounds listed below, find the specified organic functional groups and indicate whether those functional groups are polar or nonpolar. If polar, will they donate or receive a hydrogen bond?
(a) testosterone—alcohol, aldehyde, ketone, unsaturation
(b) estradiol—aromatic ring, phenol, amine
(c) aldosterone—ketone, aldehyde
(d) glycocholic acid—amide, carboxylic acid, alcohol

15.24 **Functions of Biolipids:** How would the bile acids act as emulsifying agents for fats and oils in the intestines?

15.25 **Structure of Biolipids:** How many stereocenters are there in squalamine, the shark antimicrobial? How many stereoisomers are possible?

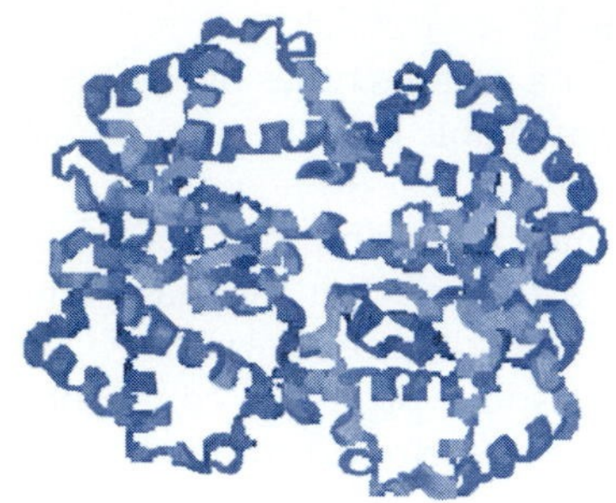

CHAPTER 16

Proteins

By far the most versatile biomolecules in living organisms are the amino acids which make up proteins. Proteins act as catalysts, structural support, protection, transport agents, chemical messengers, and cell recognition factors, to name only a few functions. As with other biopolymers, the organic structure and bonding in proteins give rise to their extraordinary features.

protein
polymer of amino acids

Proteins are polymers composed of monomer units known as *amino acids.* These amino acids are linked by amide bonds in macromolecules with molecular weights ranging from a few thousand to several million atomic mass units. The properties of proteins can be appreciated by considering the characteristics of their constituent amino acids.

16.1 Structure of Amino Acids

A. Fundamental Structure—An Amine and An Acid

amino acids
the monomer units of proteins

As the term *amino acid* suggests, every **amino acid** has an amine group and a carboxylic acid group. Both of these functional groups are attached to the same carbon atom, which usually also has a hydrogen atom and another variable group.

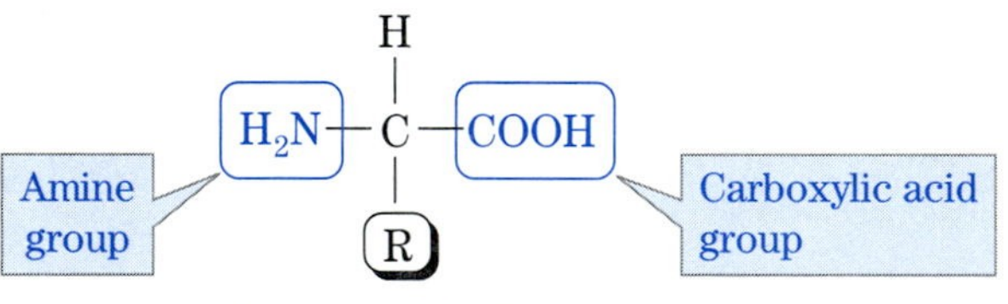

α- (alpha) amino acid
molecule with an amine group on the carbon adjacent to a carboxyl group

These monomers are sometimes referred to as **α- (alpha) amino acids** because the amine is on the carbon next to, or alpha to, the carboxylic acid group or vice versa.

B. Ionization of Amino Acids

Recall that amine and carboxylic acid groups have conjugate acid–base forms in water that are dependent upon the pH of the solution in which they find themselves.

The ionization constants, K_as, for these groups are about 10^{-2} for the carboxyl and about 10^{-9} for the amine group. Therefore, the pK_as are 2 and 9, respectively. This means that at a pH of 2, 50% of the carboxyl groups are in the conjugate acid form and 50% are in the conjugate base form. When the pH is less than 2,

$$-\overset{\displaystyle O}{\overset{\|}{C}}-\ddot{\underset{..}{O}}H \rightleftharpoons H^+ + -\overset{\displaystyle O}{\overset{\|}{C}}-\ddot{\underset{..}{O}}:^-$$

Conjugate acid form — Conjugate base form

$$-\overset{\displaystyle H}{\overset{|}{\underset{\displaystyle H}{\underset{|}{N^+}}}}-H \rightleftharpoons H^+ + -\overset{\displaystyle H}{\overset{|}{\underset{\displaystyle H}{\underset{|}{N:}}}}$$

Conjugate acid form — Conjugate base form

$$-CO_2H \rightleftharpoons H^+ + -CO_2^- \qquad -NH_3^+ \rightleftharpoons H^+ + -NH_2$$

Zero charge — −1 charge — +1 charge — Zero charge

most of the carboxyls are in the uncharged conjugate acid form; above a pH of 2 most are in the −1 charged conjugate base form.

Overall then, an amino acid has several charged forms that are pH-dependent.

$$H_3\overset{+}{N}-\overset{\displaystyle H}{\overset{|}{\underset{\displaystyle R}{\underset{|}{C}}}}-\overset{\displaystyle O}{\overset{\|}{C}}-OH \rightleftharpoons H_3\overset{+}{N}-\overset{\displaystyle H}{\overset{|}{\underset{\displaystyle R}{\underset{|}{C}}}}-\overset{\displaystyle O}{\overset{\|}{C}}-O^- \rightleftharpoons H_2N-\overset{\displaystyle H}{\overset{|}{\underset{\displaystyle R}{\underset{|}{C}}}}-\overset{\displaystyle O}{\overset{\|}{C}}-O^-$$

pH below 2	pH between 2 and 9	pH above 9
amine and carboxyl in conjugate acid forms	amine group as conjugate acid carboxyl group as conjugate base	amine and carboxyl in conjugate base forms
NET CHARGE = +1	NET CHARGE = 0	NET CHARGE = −1

A titration curve for an amino acid shows at least two points of inflection accounting for the titration of the two ionizable groups in the molecule. This is seen in Figure 16.1.

Because amino acids are charged at certain pHs, they move if an electric field, with + and − electrical poles, is applied to the solution. The cationic (+1) form moves to the − pole or cathode, and the anionic (−) form migrates to the + pole or anode. The form with no net charge does not move at all.

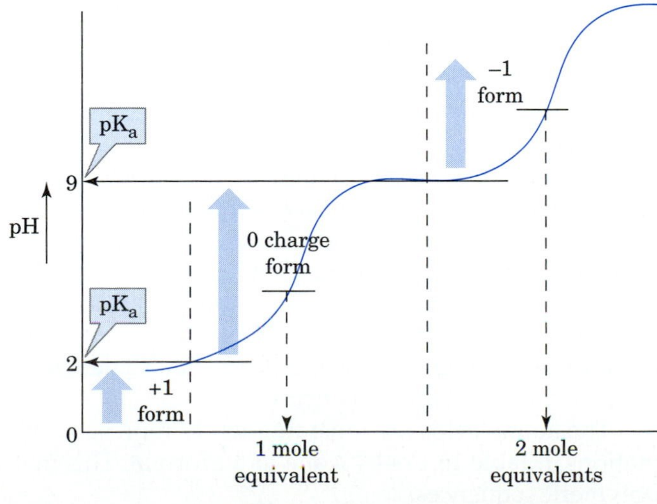

Figure 16.1 Titration curve for an amino acid.

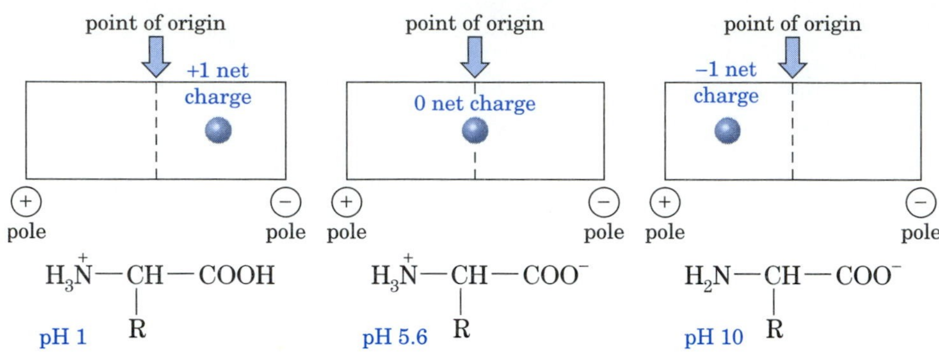

electrophoresis
method of separating charged species in an electric field

zwitterion
the ionized form of an amino acid or peptide that has a net zero charge

isoelectric point (pI) or isoionic pH$_I$
The pH at which an amino acid or protein will not move in an electric field

The process of subjecting amino acids and proteins, or any charged species, to an electric field is known as **electrophoresis.**

The electrically neutral form is called the isoelectric form or **zwitterion** ("zwitter" is German for "both") and the pH at which the isoelectric form exists is called the **isoelectric point** or **isoionic pH$_I$,** the **pI.** A rough idea of the pI can be calculated by averaging the pK_a going from the +1 to the 0 form and the pK_a going from the 0 to the −1 form.

For our generic amino acid $$\text{pI} = \frac{pK_{a(+1 \rightarrow 0)} + pK_{a(0 \rightarrow -1)}}{2} \qquad \text{pI} = \frac{2 + 9}{2} = 5.5$$

This means that at pH 5.5 almost all of our generic amino acid molecules would be in the 0 net charge or zwitterion form, having an equal number of + and − charges.

If the R group contains a functional group that has conjugate acid–base properties, its ionization must be considered along with those of the amine and carboxyl groups. The pI is calculated in the same way as for the generic amino acid; the pK_a values used for the calculation must be those of the +1 --------→ 0 transition and the 0 --------→ −1 transition.

C. The Common Amino Acids

There are 20 amino acids that are commonly found in proteins. Their placement in the protein polymer chain is dependent upon the genetic code, that is, upon the DNA that is present in our genes. The DNA carries the code for the construction of proteins. Table 16.1 illustrates the structures of these 20 amino acids arranged according to the nature of the R group. They are shown in the forms present at very low pH. Table 16.2 lists the pK_a values for the amino, carboxyl, and R groups.

The amino acids designated with a superscript ([a]) are "essential" amino acids; that is, they cannot be made by the normal metabolic processes of the body and must therefore be provided in the diet. Not all food materials supply all of the essential amino acids. For example, corn and grains are deficient in lysine and tryptophan. A poor diet, low in protein and calories, can lead to severe nutritional disorders such as *kwashiorkor* and *marasmus.* These disorders frequently occur in developing or warring nations. Such a deficiency in developed countries can be evidence of anorexia nervosa.

The amino acids are most frequently represented by the three-letter abbreviations in Table 16.1 or by a one-letter format. This makes it easier to write long polymeric sequences.

The two acidic amino acids, aspartic and glutamic, may also have amide forms

Table 16.1 The Common Amino Acids in Their Conjugate Acid Forms

Basic Side Chains	Acidic Side Chains	Alkyl Side Chains	Aromatic Side Chains
$H_3\overset{+}{N}(CH_2)_4CH(\overset{+}{N}H_3)COH$ (C=O) Lysine (Lys)	$HOC(=O)(CH_2)_2CH(\overset{+}{N}H_3)C(=O)OH$ Glutamic acid (Glu)	$HCH(\overset{+}{N}H_3)C(=O)OH$ Glycine (Gly)	$C_6H_5-CH_2CH(\overset{+}{N}H_3)C(=O)OH$ Phenylalanine (Phe)[a]
$H_2\overset{+}{N}=C(NH_2)-NH-(CH_2)_3CH(\overset{+}{N}H_3)C(=O)OH$ Arginine (Arg)[a]	$HOC(=O)CH_2CH(\overset{+}{N}H_3)C(=O)OH$ Aspartic acid (Asp)	$CH_3CH(\overset{+}{N}H_3)C(=O)OH$ Alanine (Ala)	$HO-C_6H_4-CH_2CH(\overset{+}{N}H_3)C(=O)OH$ Tyrosine (Tyr)
$HC=CCH_2CH(\overset{+}{N}H_3)C(=O)OH$ (ring: $H\overset{+}{N}=CH-NH$) Histidine (His)[a]	**Polar Side Chains**	$CH_3CH(CH_3)-CH(\overset{+}{N}H_3)C(=O)OH$ Valine (Val)[a]	indole (N–H)$-CH_2CH(\overset{+}{N}H_3)C(=O)OH$ Tryptophan (Trp)[a]
Sulfur-Containing Side Chains	$HOCH_2CH(\overset{+}{N}H_3)C(=O)OH$ Serine (Ser)	$CH_3CH(CH_3)CH_2CH(\overset{+}{N}H_3)C(=O)OH$ Leucine (Leu)[a]	**Cyclic Side Chains**
$HSCH_2CH(\overset{+}{N}H_3)C(=O)OH$ Cysteine (CysH)[b]	$HOCH(CH_3)-CH(\overset{+}{N}H_3)C(=O)OH$ Threonine (Thr)[a]	$CH_3CH_2CH(CH_3)-CH(\overset{+}{N}H_3)C(=O)OH$ Isoleucine (Ile)[a]	$H_2C-CHC(=O)OH$, H_2C, CH_2, $\overset{+}{N}H_2$ (ring) Proline (Pro)
$H_3CS(CH_2)_2CH(\overset{+}{N}H_3)C(=O)OH$ Methionine (Met)[a]	$H_2NC(=O)CH_2CH(\overset{+}{N}H_3)C(=O)OH$ Asparagine (Asn)		
	$H_2NC(=O)(CH_2)_2CH(\overset{+}{N}H_3)C(=O)OH$ Glutamine (Gln)		

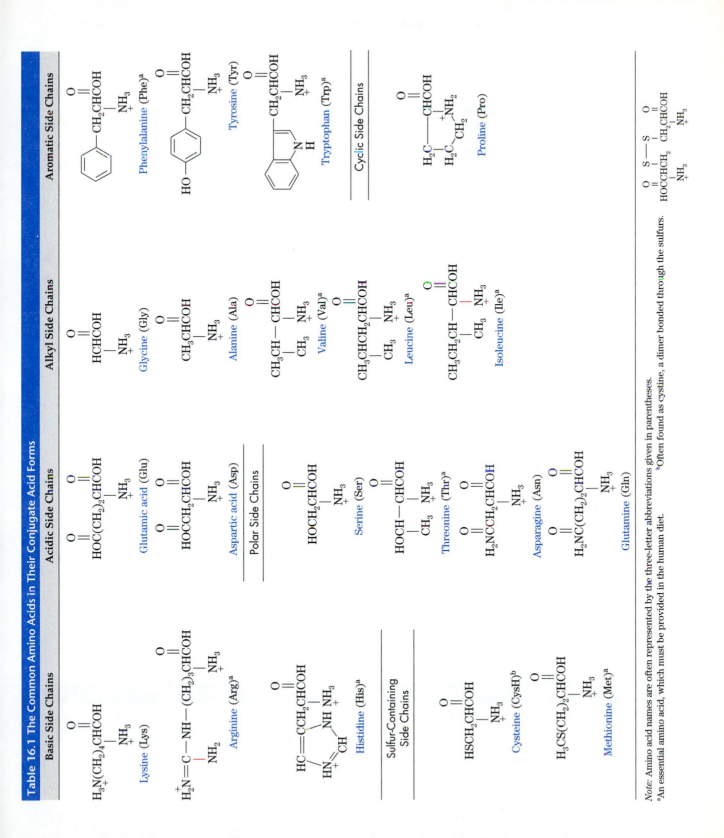

Note: Amino acid names are often represented by the three-letter abbreviations given in parentheses.

[a]An essential amino acid, which must be provided in the human diet.

[b]Often found as cystine, a dimer bonded through the sulfurs. $HOC(=O)CH(\overset{+}{N}H_3)CH_2-S-S-CH_2CH(\overset{+}{N}H_3)C(=O)OH$

on the R side chain—$CONH_2$ rather than COOH. The amino acids are then called asparagine and glutamine, respectively. Amides do not accept or donate a proton under physiological conditions, and therefore the pK_a for the R group no longer exists. However, they are still polar and have the capacity to hydrogen-bond.

GETTING INVOLVED

- ✓ What does the pK_a of an ionizable group tell us?
- ✓ Associate the structures of the ionized forms of a generic amino acid with the titration curve in Figure 16.1.
- ✓ If the carboxyl group was missing from an amino acid, what would be the appearance of the titration curve? Do the same for a structure missing the amino group but retaining the carboxyl group.
- ✓ Where will a zwitterion move in an electric field?
- ✓ Why is it important to know the isoionic pH? What is the other term for this factor?
- ✓ How do you calculate an approximate value for the pI of an amino acid?
- ✓ Where would you find the pI on the titration curve shown in Figure 16.1?
- ✓ Can amino acids have more than three ionized (conjugate acid-base) forms? Briefly explain your answer.

Example 16.1

Draw out the conjugate acid–base forms for aspartic acid. Find its pI and predict the movement of the ionized forms in an electric field at various pH values.

Solution

First draw aspartic acid with all three of its ionizable groups in their conjugate acid forms. Include the pK_a for each ionizable group as found in Table 16.2.

$pK_a = 1.9$

$$\overset{\oplus}{H_3N}-\underset{\underset{CH_2COOH}{|}}{CHCOOH}$$

$pK_a = 9.6$ $pK_a = 3.65$

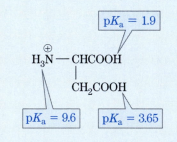

Table 16.2 The pK_a Values for the Common Amino Acids

Amino Acids	pK_as —COOH	—NH_3^+	—R	Amino Acids	pK_as —COOH	—NH_3^+	—R	Amino Acids	pK_as —COOH	—NH_3^+	—R
Ala	2.4	9.9		Gly	2.3	9.6		Pro	2.0	10.6	
Arg	2.2	9.1	11.8	His	1.8	9.0	6.0	Ser	2.2	9.2	
Asn	2.0	8.8		Ile	2.3	9.8		Thr	2.2	9.1	
Asp	1.9	9.6	3.65	Leu	2.4	9.6		Trp	2.4	9.4	
CySH	1.7	10.8	8.3	Lys	2.2	8.9	10.3	Tyr	2.2	9.1	10.1
Gln	2.2	9.1		Met	2.3	9.2		Val	2.3	9.7	
Glu	2.2	9.7	4.3	Phe	2.6	9.2					

Find the net charge on this form and then remove protons in order of increasing pK_a. Calculate the net charge on each new form.

$$\overset{\oplus}{H_3N}-\underset{\substack{|\\ CH_2COOH}}{CHCOOH} \rightleftharpoons \overset{\oplus}{H_3N}-\underset{\substack{|\\ CH_2COOH}}{CHCOO^{\ominus}} \rightleftharpoons \overset{\oplus}{H_3N}-\underset{\substack{|\\ CH_2COO^{\ominus}}}{CHCOO^{\ominus}} \rightleftharpoons H_2N-\underset{\substack{|\\ CH_2COO^{\ominus}}}{CHCOO^{\ominus}}$$

$pK_a = 1.9$ (α-COOH); $pK_a = 9.6$ ($\overset{\oplus}{H_3N}$); $pK_a = 3.65$ (side-chain COOH)

pH below 1.9	pH above 1.9 and below 3.65	pH above 3.65 and below 9.6	pH above 9.6
NET CHARGE +1	NET CHARGE 0	NET CHARGE −1	NET CHARGE −2

Find the zwitterion and use the pK_as on either side of it to calculate the pI.

$$\text{pI of Asp} = \frac{1.9 + 3.65}{2} = 2.8$$

The movement of aspartic acid in an electric field depends upon the pH. At a pH lower than 2.8 most of the molecules are in the cationic (+1) form and migrate to the − electrode. At pH values above 2.8 the aspartic acid molecules take on a negative charge, either −1 or −2, and migrate to the + pole.

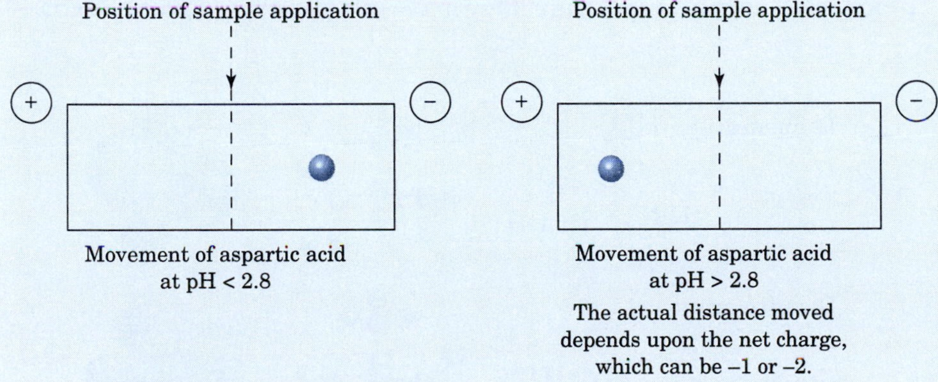

Movement of aspartic acid at pH < 2.8

Movement of aspartic acid at pH > 2.8
The actual distance moved depends upon the net charge, which can be −1 or −2.

Problem 16.1

Look at the R- groups in Table 16.2. Which could go from $+1 \rightarrow 0$ and which could go from $0 \rightarrow -1$?

Problem 16.2

Draw out all of the possible ionized forms for the amino acids lysine, glutamic acid, alanine, and tyrosine. What is the net charge on each form?

Problem 16.3

Construct titration curves for aspartic acid, serine, and arginine. Indicate the pH range in which the various charged forms exist.

Problem 16.4

Toward which pole, + or −, would each of the following amino acids travel at pH 8.7 in an electric field: glutamic acid, arginine, threonine, tyrosine, and histidine?

Problem 16.5

What is the pI for histidine? isoleucine? cysteine?

Problem 16.6

Draw out the charged forms of glutamine and calculate its pI. How does the pI compare to that for glutamic acid?

Problem 16.7

What is the most likely charged form that would exist for histidine at pH 6.8? for tyrosine at pH 13.4?

See related problems 16.25, 16.26, and 16.28.

D. Chirality of Amino Acids

L-amino acid
amino acid with the amine group on its primary chiral carbon oriented in the same way in three dimensions as the —OH in L-glyceraldehyde

If you inspect the structures of all of the amino acids except glycine, you can see that the attachment of a carboxyl, amine, R group, and hydrogen to a central carbon makes that carbon chiral, so the amino acid is optically active. With only one chiral carbon center there are 2^1 or 2 stereoisomers possible, related as nonsuperimposable mirror images or enantiomers. These are referred to as D- and L-amino acids. The genetic code uses only **L-amino acids** in constructing proteins, although D-amino acids may occur as modifications after the genetic code has been transcribed into protein, or they are formed by nongenetically directed processes. D-amino acids occur mainly in lower organisms such as bacteria.

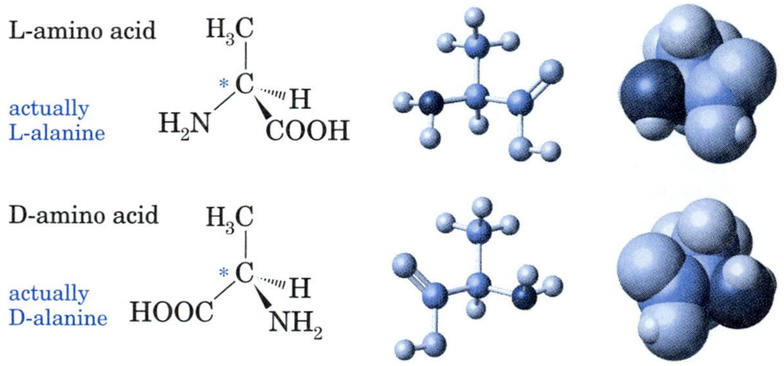

L-amino acid: H_3C, *C, H, H_2N, COOH
actually L-alanine

D-amino acid: H_3C, *C, H, HOOC, NH_2
actually D-alanine

* indicates the chiral carbon center

GETTING INVOLVED

✓ There are two amino acids that have more than one chiral carbon. Identify them and draw out the optical isomers.

Problem 16.8

Use the structures of D- and L-alanine drawn above and determine which is R- and which is S-.

Problem 16.9

Why is glycine optically inactive?

See related problem 16.23.

16.2 The Peptide Bond: Formation of Polypeptides and Proteins

The protein polymer is made by linking together amino acids via an amide, or **peptide bond.** This occurs in a living organism through the transcription and translation of the genetic code. A summary of the reaction follows:

peptide bond
the amide bond formed between the carboxyl and amine groups of two amino acids

$$\mathrm{H_3\overset{+}{N}CH(R_1)\overset{O}{\overset{\|}{C}}O^-} + \mathrm{H_3\overset{+}{N}CH(R_2)\overset{O}{\overset{\|}{C}}O^-} \longrightarrow \mathrm{H_3\overset{+}{N}CH(R_1)\overset{O}{\overset{\|}{C}}-NHCH(R_2)\overset{O}{\overset{\|}{C}}O^-}$$

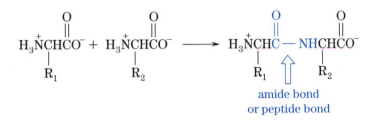

amide bond or peptide bond

The formation of the peptide bond changes the ionization characteristics of the constituent amino acids. The carboxyl group of the first amino acid and the amine function of the second can no longer participate in conjugate acid–base behavior once they are joined by the peptide bond. That leaves the R side chains, as well as the terminal amino and carboxyl groups, as main sources of ionizable groups.

The amino acid chain, called a *polypeptide,* is usually drawn with the free amine group on the left and the free carboxyl group at the right. They are called the N- or amino-terminus and the C- or carboxy-terminus, respectively. As we start to add amino acids to the chain, the complete chemical structure becomes more cumbersome, and we resort to the abbreviations for the amino acids.

Polypeptides and proteins also have isoelectric points or pIs. As with an individual amino acid, if the pH is lower than the pI, the polypeptide has a net + charge; at a pH above the pI, the charge is −. If all of the molecules of a protein have the same net charge, they tend to repel each other. This keeps the protein dispersed in water. However, if the pH is adjusted to the pI, the net charge is zero and the protein molecules can aggregate, precipitating from solution. This is known as **isoelectric precipitation.**

isoelectric precipitation
process of precipitating proteins at their isoelectric points, the pH of minimum solubility

GETTING INVOLVED

✓ Draw normal peptide bonds to link the amino acids serine, phenylalanine, and glutamic acid. There are several sequences of combination; draw only one sequence using the structures. Use the three-letter abbreviations to show the other combinations.

✓ Identify the N-terminus and C-terminus in each combination.

✓ What are the ionizable groups in your tripeptides?

Example 16.2

The structure of bradykinin appears below. This substance is a "tissue hormone" capable of dilating and increasing the permeability of blood vessels. It also causes intense pain. Identify the amino acids and determine the possible charged forms at low and high pH.

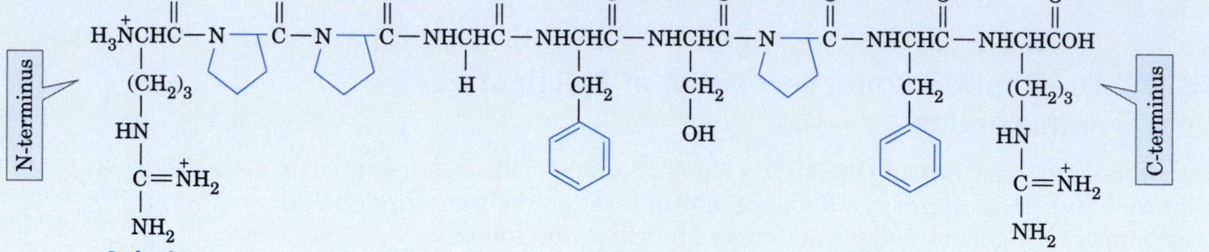

N-terminus

C-terminus

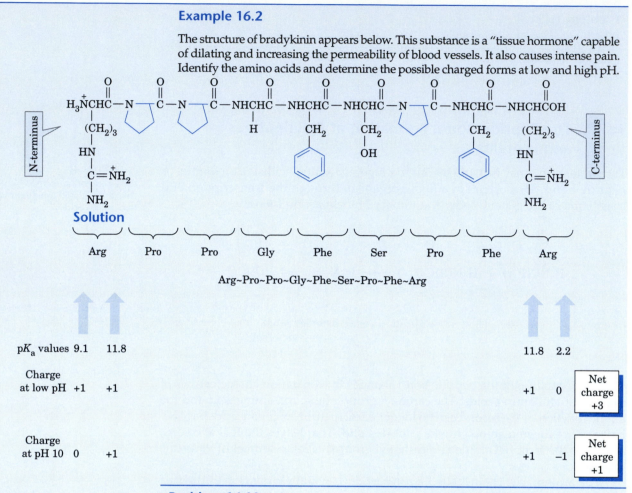

Solution

Arg Pro Pro Gly Phe Ser Pro Phe Arg

Arg~Pro~Pro~Gly~Phe~Ser~Pro~Phe~Arg

	N-terminal Arg		C-terminal Arg		
pK_a values	9.1	11.8	11.8	2.2	
Charge at low pH	+1	+1	+1	0	Net charge +3
Charge at pH 10	0	+1	+1	−1	Net charge +1

Problem 16.10

Find the net charge of the following polypeptide at pH 8.2 (approximate pH in the large intestine):

Ala ~ Lys ~ Asp ~ Tyr ~ Asp ~ His ~ CySH ~ Leu ~ Phe ~ Gln

Problem 16.11

For the polypeptide drawn below, identify the amino acids, Find the pI, and calculate the approximate net charge on the tetrapeptide at physiological pH, that is, pH 7.4.

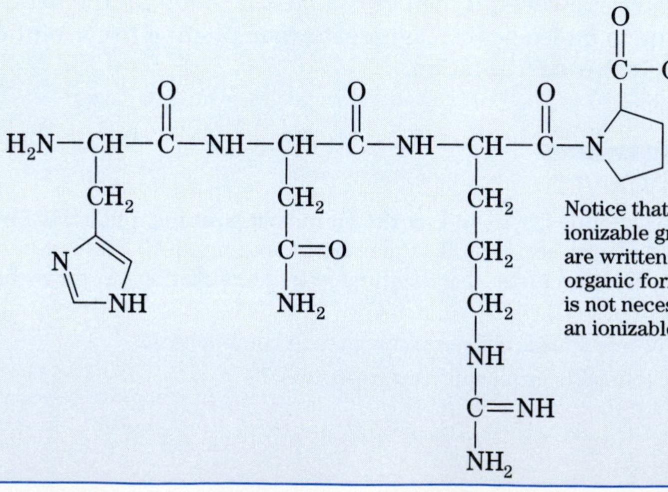

Notice that the ionizable groups are written in an organic form, which is not necessarily an ionizable form.

16.3 The Hierarchy of Protein Structure

Because of their size and chemical nature, proteins exhibit three-dimensional structural organization. There are four formal levels of protein structure, each stabilized by specific intra-molecular interactions (primary, secondary, tertiary, and quaternary).

A. Primary Protein Structure—The Sequence of Amino Acids

The linear arrangement of amino acids in a protein from the free amino end to the carboxyl end is known as its **primary structure.** It is this sequence that is determined by the genetic code and that determines the overall shape and function of the macromolecule.

primary (1°) protein structure the linear sequence of amino acids from N- to C-terminus

B. Secondary Protein Structure—Helices and Pleated Sheets

Secondary structure is the organization of regions or segments of the polypeptide chain that results from hydrogen-bonding between peptide bonds. The hydrogen-bonding produces structures that can be helical or sheetlike.

secondary (2°) protein structure arrangement of a segment of a polypeptide chain into an organized structure, such as an α-helix or β-pleated sheet, stabilized by hydrogen-bonding between peptide bonds

1. ***Alpha and Beta Structures.*** Peptide bond geometry is trigonal planar due to a partial double bond formed by electron delocalization between the carbonyl carbon and the amide nitrogen. This means that there is restricted rotation about the amide bond and geometric isomers can exist. The predominant isomer is *trans;* that is, the oxygen of the carbonyl group and the hydrogen of the amide are across from each other. The α carbons of the attached amino acids have single bonds and are tetrahedral with free rotation about their bonds.

This gives rise to a polymer that looks like a series of flat plates attached by a swivel joint.

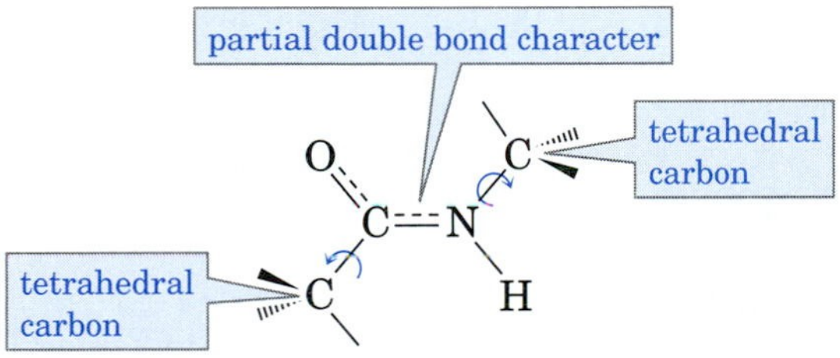

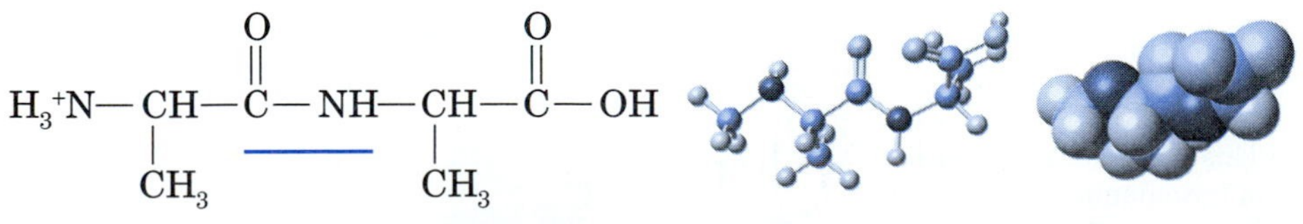

$$H_3^+N{-}CH(CH_3){-}\underline{C(=O){-}NH}{-}CH(CH_3){-}C(=O){-}OH$$

dialanine (Ala~Ala) models
peptide bond is underlined

The hydrogen attached to the amide nitrogen is electropositive (δ+), whereas the oxygen of the carbonyl group is electronegative (δ−). As a result, the amide hydrogen is said to be a hydrogen-bond donor and the carbonyl oxygen is a hydrogen-bond acceptor. The polypeptide chain rotates around the tetrahedral carbons in order to align amide hydrogens with carbonyl oxygens (hydrogen-bond donor–acceptor pairs).

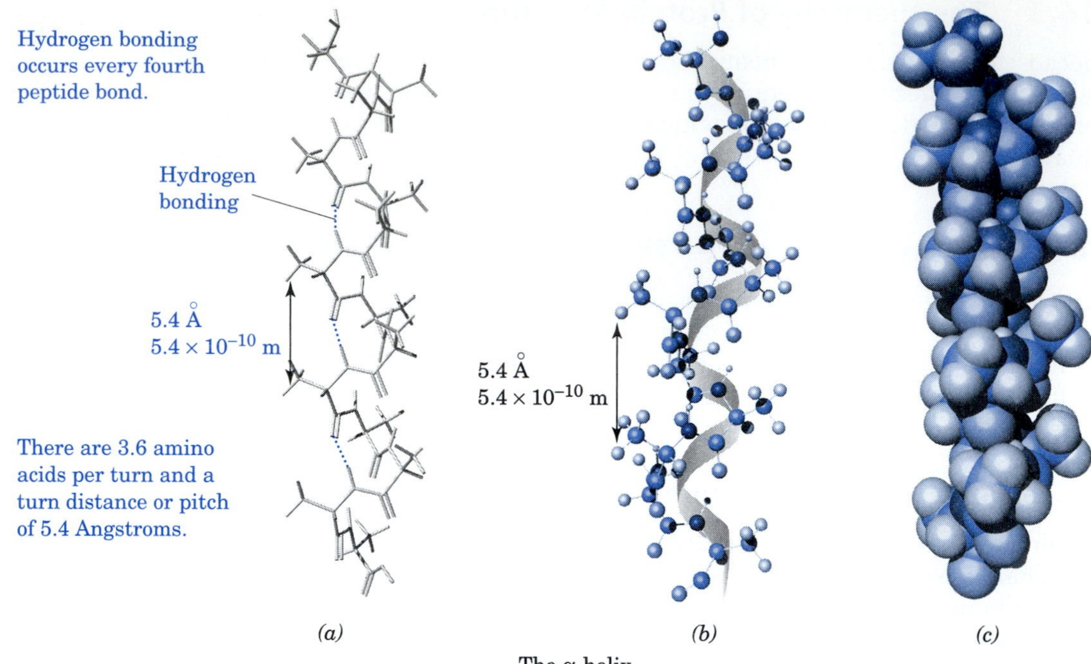
Hydrogen bonding occurs every fourth peptide bond.
Hydrogen bonding
5.4 Å
5.4 × 10^{-10} m
There are 3.6 amino acids per turn and a turn distance or pitch of 5.4 Angstroms.
5.4 Å
5.4 × 10^{-10} m
(a)
(b)
(c)
The α-helix

Figure 16.2 Various representations of the α-helix formed by $(Ala)_{12}$: (a) is a wireframe model with some of the dimensions and hydrogen bonding shown; (b) is a ball-and-stick model overlaid on a ribbon diagram used frequently in biochemistry to represent protein structure; (c) is the same polypeptide in a space-filling model.

α (alpha)-helix
spiral protein secondary structure stabilized by hydrogen-bonding between the peptide bonds of every four amino acids

A partial rotation of about 45° allows the peptide bonds to arrange so that every fourth peptide bond occurs under another (see Figure 16.2). This sets up a spiral or helix, specifically a right-handed helix, known as the **α (alpha)-helix.** (Rotate your right hand in a clockwise direction.) Hydrogen-bonding can occur between the peptide bonds located above and below each other in a direction almost parallel to the long axis of the helix. The R groups protrude from the helix

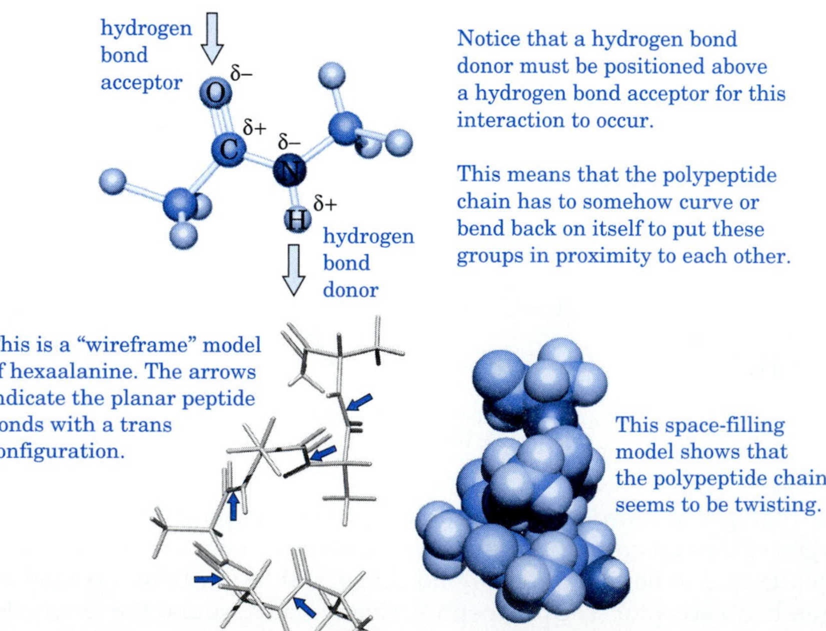
hydrogen bond acceptor
δ−
O
δ+
C
δ−
N
δ+
H
hydrogen bond donor
Notice that a hydrogen bond donor must be positioned above a hydrogen bond acceptor for this interaction to occur.
This means that the polypeptide chain has to somehow curve or bend back on itself to put these groups in proximity to each other.
This is a "wireframe" model of hexaalanine. The arrows indicate the planar peptide bonds with a trans configuration.
This space-filling model shows that the polypeptide chain seems to be twisting.

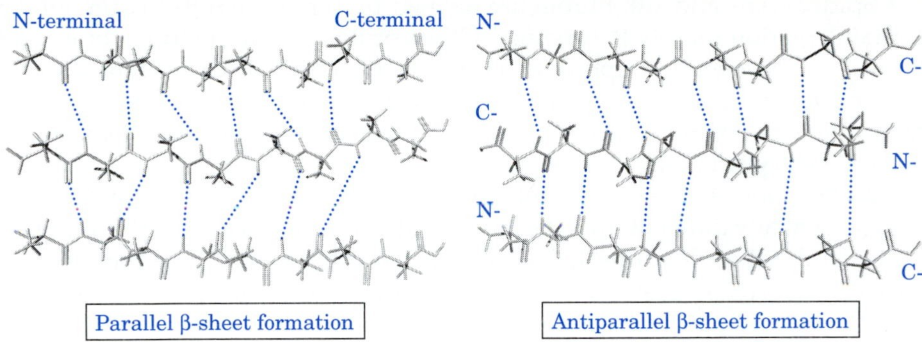

Figure 16.3 Parallel and antiparallel β-sheet formations. Notice the alignment of the N- and C- termini of the chain fragments shown as well as the orientation of the hydrogen bonding. Each chain fragment takes on a pleated ribbonlike appearance with one ribbon hydrogen bonding side-to-side with another.

in a manner analogous to the spokes protruding at almost right angles to the cylindrical hub of a bicycle wheel. (Hair is composed of the protein α-keratin, which is mainly α-helical in nature.)

Full rotation of the bonds to the α-carbons to 180° extends the chain and produces a pleated appearance with the hydrogen bond donors and acceptors located at the sides of the chain and the R groups directed up and down, perpendicular to the chain. If the polypeptide chain itself bends and comes back alongside itself, hydrogen-bonding can occur in a side-to-side arrangement. This is known as a **β (beta)-pleated sheet.**

The polypeptide chain fragments may be oriented such that they are all progressing from N- to C-, called a **parallel sheet,** or they may alternate N- to C- aligned with C- to N-, an **antiparallel sheet.** Figure 16.3 illustrates the parallel and antiparallel β-pleated sheets.

Ribbon cartoons are frequently used to symbolize secondary protein structure. The α-helix is easily recognized as a spiral, whereas β-structure is shown with an arrowhead to indicate the N- to C- orientation of the chain. Figure 16.4 contains examples of such structures.

β (beta)-pleated sheet
layered protein secondary structure stabilized by side-to-side hydrogen-bonding between peptide bonds located in different chains or parts of a chain

parallel β-sheet
β-sheet with its polypeptide strands aligned N- to C-

antiparallel β-sheet
β-sheet with polypeptide strands aligned alternately N- to C- and C- to N-

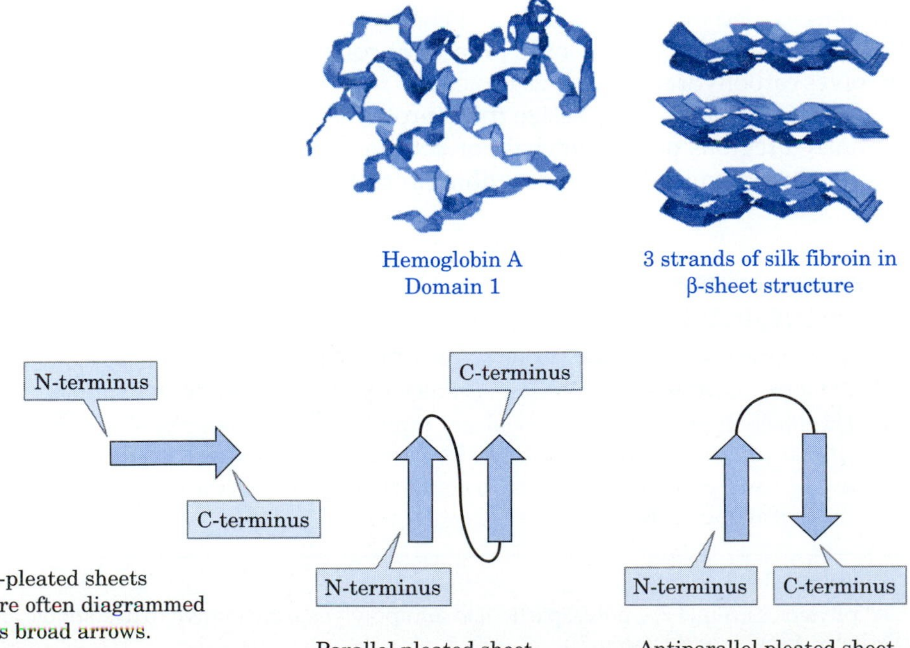

Figure 16.4
Ribbon diagrams of protein supersecondary structures or domains.

Spider webs and silk fibroin are formed by the protein β-keratin, which contains predominantly β structure. Other proteins have mixtures of α- and β-structures depending upon the nature of the amino acids present and the rotation about the α carbons. A limited number of rotational angles occur in proteins due to the presence of the R groups, which can interfere with the stability of a secondary structure.

As you look at the structures of the common amino acids, one stands out as being essentially different from the others in its backbone of amine, chiral carbon center, and carboxyl groups. It is proline, a cyclic amino acid. The constraint of its five-membered ring structure restricts the degree of rotation possible about the α carbon. It will not twist into an α-helix, nor will it extend to form a β-sheet. Rather it "kinks" or bends the polypeptide chain to disrupt potential α- and β-secondary structures.

The only nonoptically active amino acid, glycine, also interrupts α- and β-structures because it has no R group to form any bulk around the polypeptide chain. The R groups can actually help to stabilize or destabilize secondary structure. Glycine, therefore, frequently appears in positions of bends in the chain.

domain
combinations of secondary structures associated into functional units

X-ray structural analysis of proteins has revealed that combinations of secondary structures occur in specific functional groupings known as **domains.** In fact, studies of evolution on a molecular level indicate that new proteins may have evolved by joining, deleting, or modifying the DNA sequences for domain supersecondary structures.

2. ***The Collagen Triple-Helix.*** Collagen is the most abundant protein type in the human body. In its variations collagen contributes to the skin, bones, teeth, ligaments, cartilage, and tendons that cover, support, and hold us together. Collagen is a left-handed, triple, intertwined helix composed primarily of proline and glycine. In this case the structure depends upon the "kink" or bend that proline imposes on the chain. Located at approximately every third position in a chain 1000 amino acids long, one strand of the triple helix looks like an extended paper clip. The lack of a glycine side chain allows three of these strands to come into close proximity, forming a helix composed of three chains.

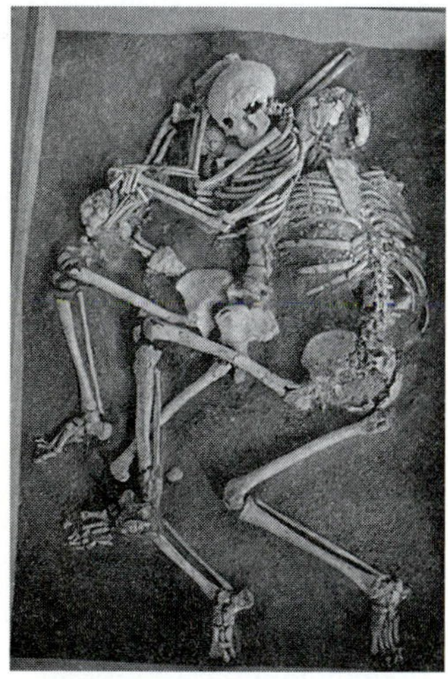

Bones are made of the protein collagen, which holds mineral deposits.

The helix is stabilized by hydrogen-bonding between the peptide bonds of glycines located on different, adjacent chains. The result is a left-handed triple helix (see Figure 16.5). The entire process of collagen assembly is complex and involves carbohydrate as well as protein.

It is important to emphasize that "protein secondary structure" refers to organized regions of protein chain of definite shape, stabilized by hydrogen-bonding between peptide bonds within (intra-) the polypeptide chain.

GETTING INVOLVED

- ✓ In your own words, define secondary structure in proteins.
- ✓ What molecular interaction is responsible for protein secondary structure?
- ✓ The α-helix, β-pleated sheet, and collagen triple helix are all examples of the principal types of protein secondary structure. How are these alike and how do they differ?
- ✓ Are secondary structure interactions limited to those within the same region of the chain of amino acids?

Problem 16.12

At physiological pH 7.4, polyaspartic acid and polylysine are known to destabilize an α-helix. Why does this occur?

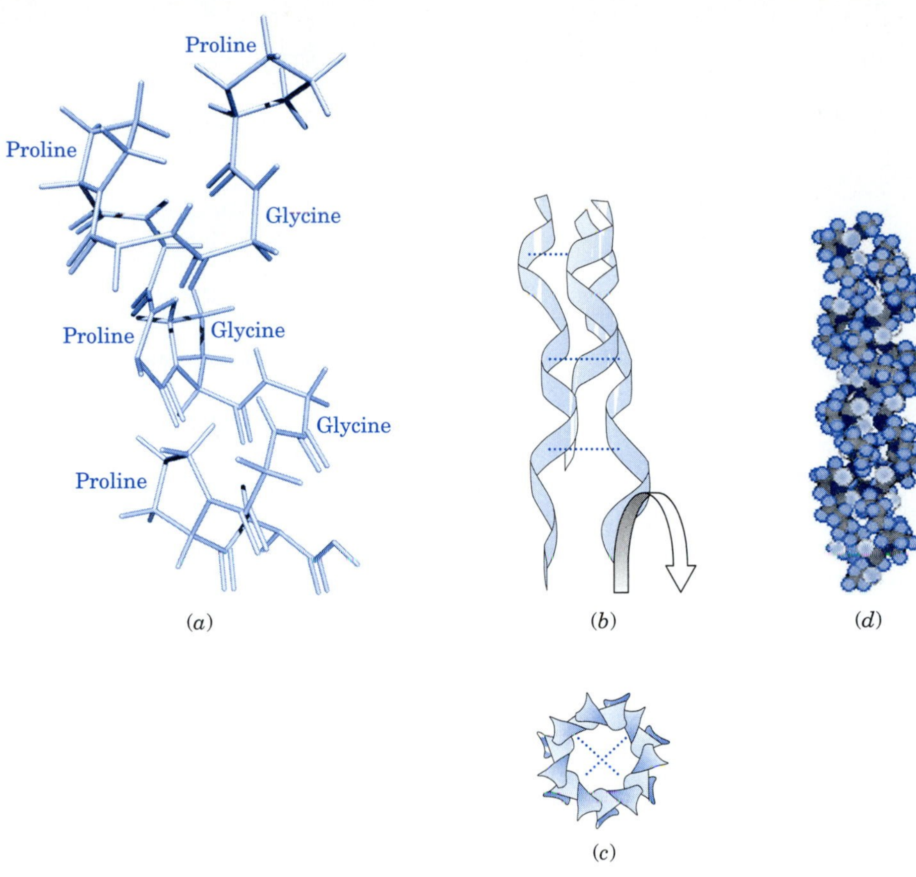

Figure 16.5
Models of collagen.
(a) A single chain with the sequence (Pro~Gly~X)$_3$; the Pro and Gly are highlighted. (b) A ribbon diagram of the collagen triple helix. (c) An end-on view of the triple helix with the dashed lines indicating the directionality of the hydrogen bonding *between* strands. (d) The collagen triple helix in a space-filling model.

Problem 16.13

Are there any special amino acids that encourage or discourage the formation of specific kinds of secondary structure? Explain your answer.

Problem 16.14

Think about the structures of the following amino acids and determine the secondary structures in which they would be "comfortable": Leu, Ala, Ser, Pro, Gly, Tyr, Lys.

C. Protein Tertiary Structure

Proteins can bend and fold into overall structures that may be long and fibrous like hair and bone or more compact, that is, globular, like egg white (albumin). The R side chains participate in both covalent and noncovalent interactions in order to stabilize the protein in its final three-dimensional structure or **tertiary (3°) conformation.**

tertiary (3°) protein structure the folded, completely formed three-dimensional structure of a polypeptide chain that is stabilized by covalent and noncovalent forces

The common covalent side-chain bond that can hold together remote regions of the protein is the **disulfide bond** formed between two cysteine residues. The -SH groups on two cysteines are oxidized to form a covalent disulfide bond or bridge.

disulfide bridge covalent S—S bond formed between the side chains of cysteine residues that may be distant from each other in a polypeptide chain

There are three main noncovalent interactions: hydrogen-bonding, salt bridges (ionic interactions), and hydrophobic interactions. R groups that have a hydrogen atom bonded to an oxygen or nitrogen, such as histidine and serine, can hydrogen-bond with an electronegative group such as the oxygen of a carbonyl or the nitrogen of an amine. This is the same type of force we saw in

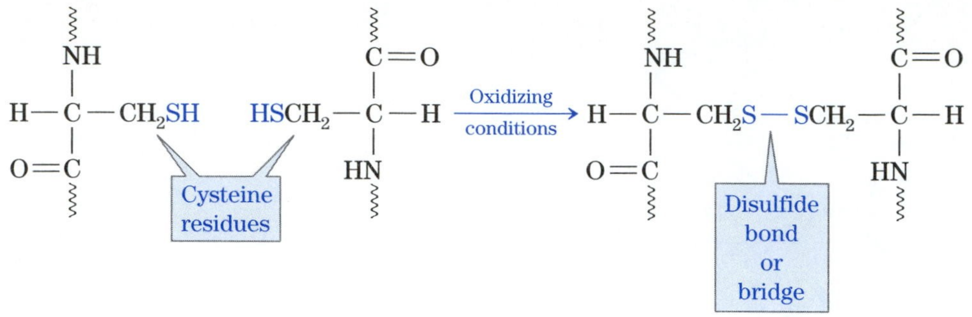

$$\text{H—}\underset{\substack{|\\ \text{O=C}}}{\overset{\substack{\text{NH}\\ |}}{\text{C}}}\text{—CH}_2\text{SH} \quad \text{HSCH}_2\text{—}\underset{\substack{|\\ \text{HN}}}{\overset{\substack{\text{C=O}\\ |}}{\text{C}}}\text{—H} \xrightarrow[\text{conditions}]{\text{Oxidizing}} \text{H—}\underset{\substack{|\\ \text{O=C}}}{\overset{\substack{\text{NH}\\ |}}{\text{C}}}\text{—CH}_2\text{S—SCH}_2\text{—}\underset{\substack{|\\ \text{HN}}}{\overset{\substack{\text{C=O}\\ |}}{\text{C}}}\text{—H}$$

Cysteine residues

Disulfide bond or bridge

secondary structure, but now it is occurring between R side chains rather than between peptide bonds.

salt bridge
ionic interaction (+ to −) between the side chains of acidic and basic amino acids that stabilizes the tertiary and quaternary structures of proteins

Another noncovalent interaction is the formation of **salt bridges** between oppositely charged R groups.

Since most proteins are found in contact with the water that constitutes about 70% of our body weight, the surfaces of these macromolecules should exhibit amino acid side chains that form hydrogen bonds with water or associate via ion–water (ion-dipole) interactions.

hydrophobic interaction
weak attractive, nonpolar interactions between the hydrocarbon side chains of amino acids that stabilize tertiary and quaternary structures of proteins

The hydrocarbon side chains (valine, leucine, phenylalanine) do not interact with water or ions but rather aggregate in a **hydrophobic** environment, often forming a "waxy" core at the inside of a water-soluble protein. While there are very weak interactions between the atoms in these groups, the prevailing force is the avoidance of polarity.

Figure 16.6 illustrates examples of the major tertiary interactions in proteins.

GETTING INVOLVED

- ✓ What kinds of covalent and noncovalent interactions give rise to tertiary structure?
- ✓ Identify any amino acids that would probably not take part in tertiary interactions. Why is this the case?
- ✓ Why is this level of protein structure dependent upon the first two levels of structure?
- ✓ Look at all of the R groups of the common amino acids and identify the types of interactions that could occur. Be sure to consider all possibilities.
- ✓ Explain how repulsive forces could play a role in tertiary structure.

Example 16.3

What type of tertiary interactions could occur between the side chains of the following pairs of amino acids under physiological conditions, that is pH 7.4?

(a) Arg and Asp **(b)** Phe and Leu **(c)** Ser and Gln

Solution

Draw out the side chains of each amino acid. Since we are discussing tertiary interactions, the α amino and carboxyl groups are involved in peptide bonds.

(a) Arg is arginine; its side chain should have a + charge at pH 7.4. Asp is aspartic acid; its side chain should have a − charge at pH 7.4.

ion-ion interaction salt bridge

Arg Asp

$$\text{CH—CH}_2\text{CH}_2\text{CH}_2\text{NH—}\overset{\substack{\text{NH}_2^{1+}\\ \|}}{\text{C}}\text{—NH}_2 \; \cdots\cdots \; {}^{1-}\text{O—}\underset{\substack{\|\\ \text{O}}}{\text{C}}\text{—CH}_2\text{CH}$$

Arg and Asp can form a salt bridge

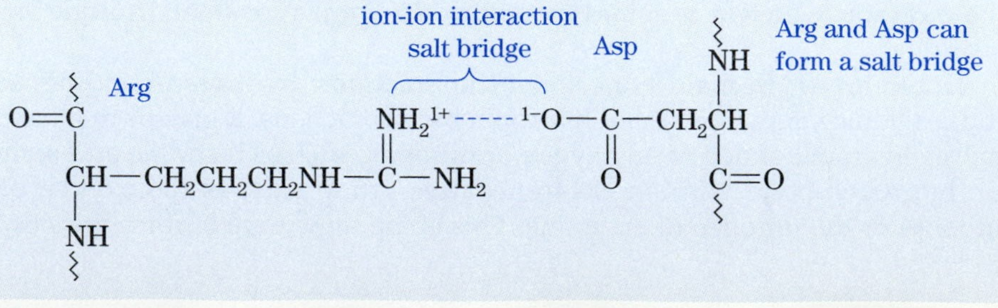

(b) Phe, phenylalanine, has an uncharged, hydrophobic side chain, as does Leu, leucine. Therefore a hydrophobic interaction takes place.

Phe

hydrophobic interaction

Leu

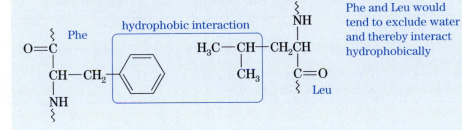

Phe and Leu would tend to exclude water and thereby interact hydrophobically

(c) Ser, serine, and Gln, glutamine, have uncharged, polar side chains. They can form a hydrogen bond.

Ser

hydrogen bond

Gln

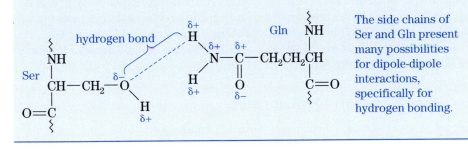

The side chains of Ser and Gln present many possibilities for dipole-dipole interactions, specifically for hydrogen bonding.

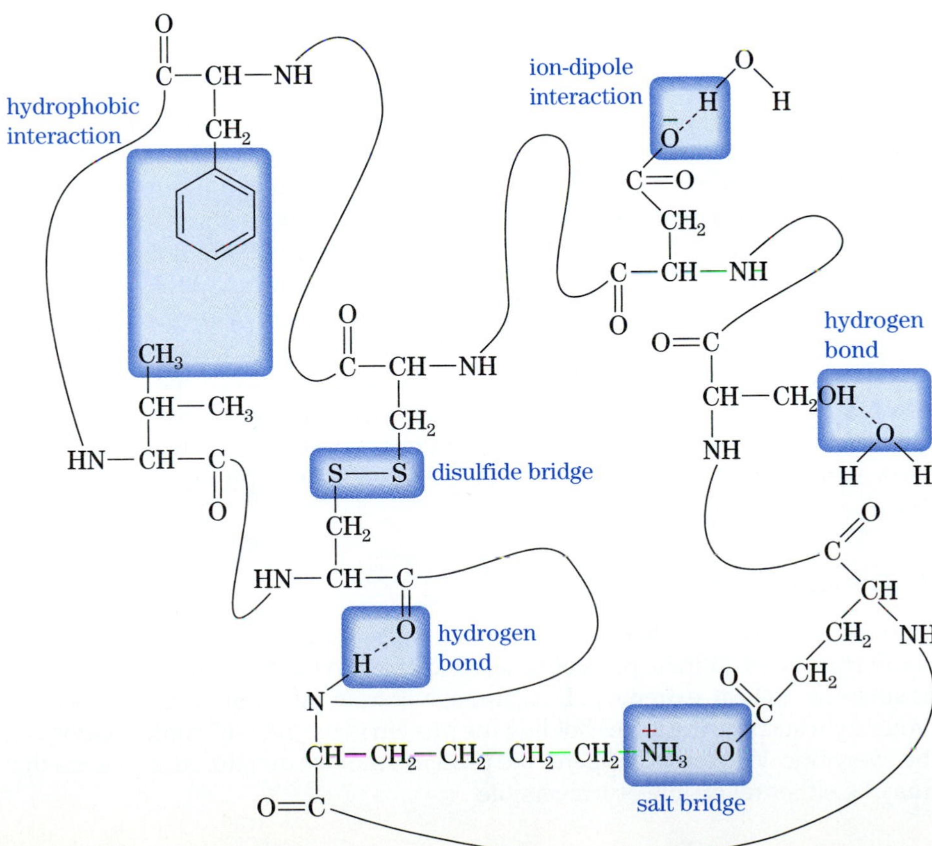

Figure 16.6 Examples of interactions in the tertiary structure of proteins.

Problem 16.15

The serine and glutamine shown in Example 16.3 have more than one possibility for the hydrogen-bonding between their side chains. What other possibilities exist?

Problem 16.16

What type of tertiary interactions could exist between the side chains of the following pairs of species?

(a) Thr and H_2O **(b)** Asn and Trp **(c)** Asp and Glu **(d)** His and Val

See related problems 16.28, 16.29, 16.30.

D. Quaternary Protein Structure—Association of Subunits

subunit of a protein single polypeptide of a protein with tertiary structure that may or may not be functional

quaternary (4°) protein structure noncovalent association of protein subunits to form a functional protein

A significant number of proteins contain more than one polypeptide chain, called a **subunit.** The subunits are held together by the same noncovalent forces of hydrogen-bonding, salt bridges, and hydrophobic interactions that give rise to tertiary conformation. This is called a protein's **quaternary (4°) structure.** The important fact to note is that most multisubunited proteins require all of their subunits in order to be fully functional.

E. Complex Proteins—Proteins Plus

complex protein protein that requires one or more nonprotein portions, such as metal ions or organic groups, in order to function

simple protein protein composed only of polymerized amino acids

Egg white or albumin is a relatively **simple protein** containing nothing but the polypeptide chain folded into its functional 2° and 3° forms. However, proteins sometimes require other types of molecules and ions in order to work. For example, the intestinal enzyme carboxypeptidase requires Zn^{2+} ion. Myoglobin is a muscle protein with one subunit that stores oxygen for use in times of oxygen starvation. It contains Fe^{2+} and a conjugated heterocyclic amine molecule called *heme* which actually binds the O_2 (see Figure 16.7).

Hemoglobin is related to myoglobin in that it, too, is an iron–heme protein, but it is composed of four subunits. The structure of hemoglobin is interesting by virtue of the cooperation that occurs between subunits in order to bind and release oxygen at the appropriate time and place in the body. The red blood cell, or erythrocyte, of a normal adult human contains a large concentration of hemoglobin. The tetramer is made up of two types of protein subunits called α and β; adult hemoglobin has an $\alpha_2\beta_2$ structure. Each subunit has a hole or crevasse in which can be found a heme group complexed with an Fe^{2+}. Molecular O_2 can complex with Fe^{2+} but not with Fe^{3+}. The protein crevasse provides a hydrophobic environment that excludes water. A water environment would facilitate the oxidation of Fe^{2+} to Fe^{3+}. As more O_2 is bound to the hemoglobin tetramer, it becomes easier to oxygenate (add O_2). This means that at the higher oxygen pressure of the lungs, hemoglobin is easily oxygenated, but at the low oxygen tension levels of the veins and capillaries, near respiring cells, it releases the O_2 readily. See Figure 16.9.

F. Denaturation

denaturation process of disrupting the secondary, tertiary, and/or quaternary structures of a protein, usually resulting in irreversible loss of function

Formation of the complete and functional three-dimensional structure of a protein depends on optimal, physiological conditions. What happens when a protein is subjected to heat, extreme pH, organic solvents, or mechanical disturbance? As you might suspect, the forces holding the protein in its "native" conformation can be overcome. When this happens, the protein becomes denatured, a process that may be either reversible or irreversible.

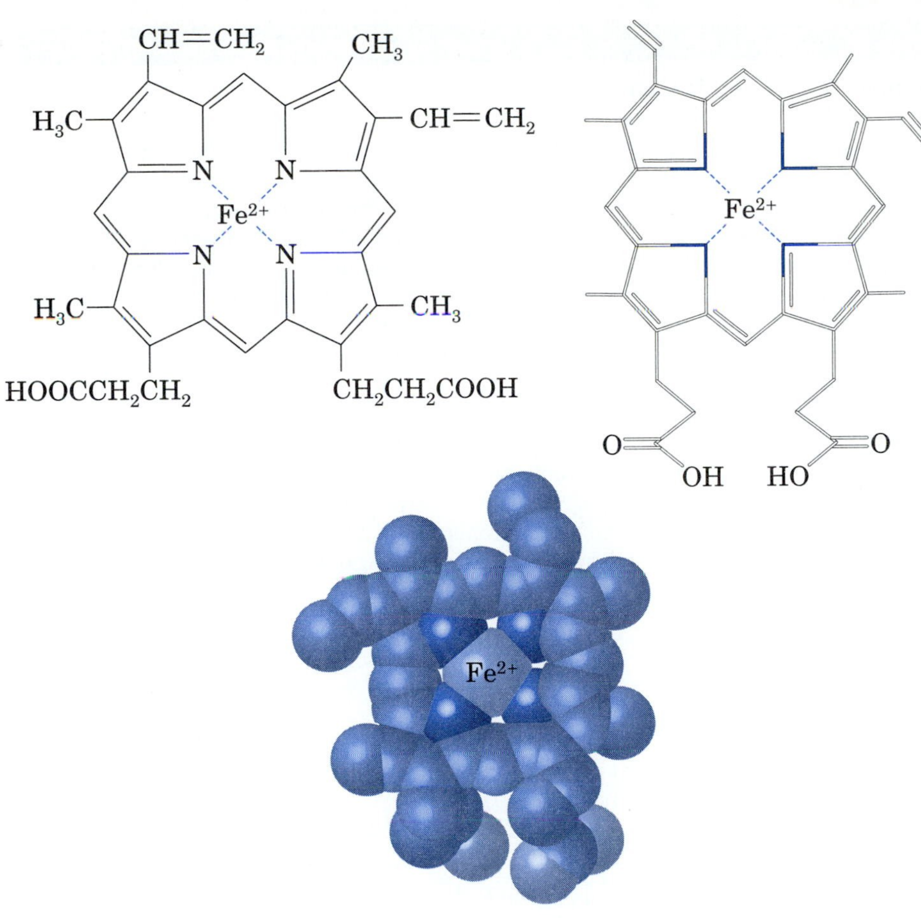

Figure 16.7 Three representations of the heme molecule. Notice that the iron is in the (II) oxidation state and that the heme is a planar structure.

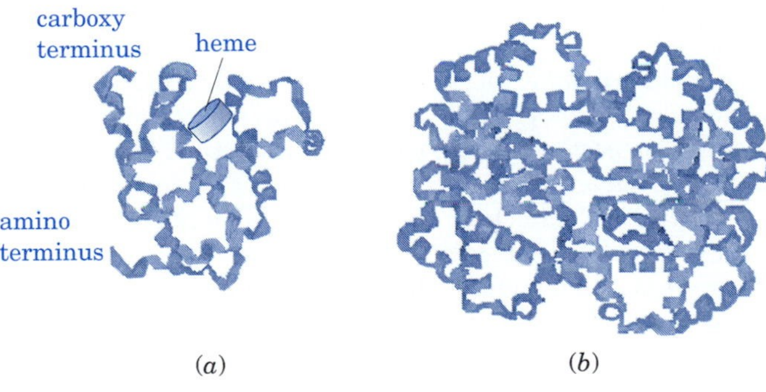

Figure 16.8 Ribbon diagrams of heme proteins. (a) Myoglobin structure illustrates the placement of the heme group in a slot or crevasse made by the secondary and tertiary protein structures. (b) Hemoglobin has four subunits of two different types, α and β, each of which contains a heme group. The subunits act in concert to capture and deliver molecular oxygen.

Consider boiling an egg, for example. As the temperature increases, the molecules of albumin (egg white) begin to vibrate more and more intensely until the tertiary forces as well as many of the secondary ones are negated by the vibrational energy of the unwinding molecule. Once the albumin is opened up, the hydrophobic amino acid core is exposed and aggregates with other exposed cores, forming a solid matrix of associated albumin molecules. We can see this in the conversion of the translucent, gelatinous raw egg white to the opaque hard-boiled egg white.

Connections 16.1 www.prenhall.com/bailey

Sickle Cell Anemia—A Biochemical Disease

More than 400 natural variations in the primary amino acid sequence of hemoglobin are known. Most of these are inconsequential—the genetic code has substituted an amino acid quite similar in structure and properties to the one that should be present; for example, a leucine for an isoleucine.

A devastating condition exists in which hemoglobin, after delivering its oxygen supply to tissues and starting its return trip to the lungs for reoxygenation, polymerizes into thick strands, and literally clogs up the smaller veins and capillaries. The red blood cells (erythrocytes) containing the hemoglobin change from their normal disclike shape to a collapsed, sickled shape. The plugging of blood vessels and the destruction of fragile blood cells lead to gangrene, heart disease, kidney disease, and brain damage. This condition is known as sickle cell anemia, and it occurs in about 0.3% of the African-American population.

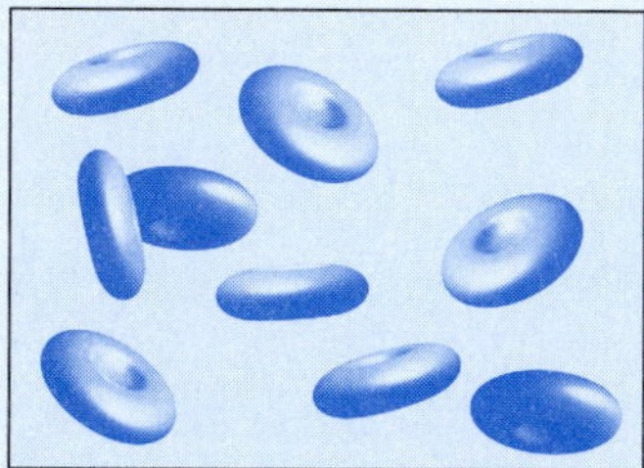

Normal red blood cells

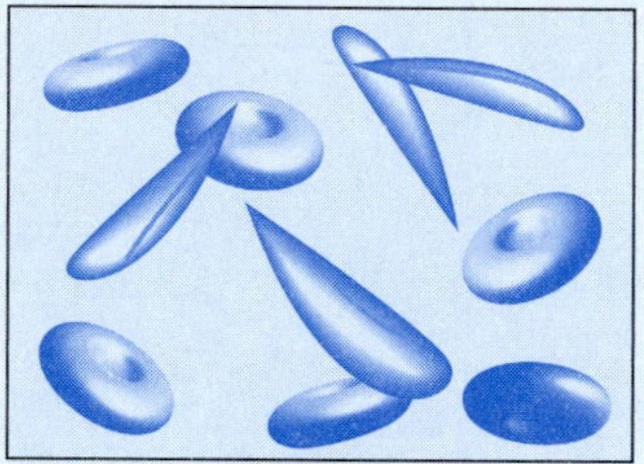

Normal and sickled red blood cells

The cause of this life-threatening condition turned out to be not as complicated as might have been anticipated. It was found by Linus Pauling that sickle cell hemoglobin, or HbS, had a different electrical charge (electrophoretic mobility) at physiological pH compared to normal hemoglobin (HbA). Then Vernon Ingram, using chemical reactivity and chromatography, discovered that there was but one change in the HbS molecule to distinguish it from normal adult hemoglobin, or HbA. The β-subunits of normal HbA have a glutamic acid at the sixth position from the amino end. However, HbS contains a valine at that position. Consider the side chains of glutamic acid and valine, shown in Table 16.1. The substitution of a valine (a hydrophobic amino acid) for a glutamic acid (a hydrophilic amino acid) is what is known as a *nonconservative change.* The HbA Glu is found on the outside of the β-subunit; and while Glu is "content" to be in a water environment, Val is not. Consequently, HbS molecules come together, or aggregate, because Val is attempting to find a compatible environment, away from water.

A person with sickle cell anemia must carry both genes (be homozygous) for HbS. About 10% of African Americans have only one gene for HbS—this is known as the *sickle cell trait.* The gene is also present in populations found in Africa, the Mediterranean, and Middle East countries. This makes the study and treatment of sickle cell disease an international health issue.

Persons with the trait show no overt symptoms of the anemia. The interesting fact about this genetic trait is that two parents, each possessing one gene for HbS, have a 2 in 4 chance of having children with the trait, a 1 in 4 chance of having children with anemia, and a 1 in 4 chance of having children with normal HbA. Is there any advantage in possessing the trait? Indeed, the parasite that causes a certain type of malaria cannot exist for long in the HbA/HbS blood of a trait carrier. The cultural heritage of those exhibiting the trait lies in the tropical, malaria-prone areas of Africa and Asia. It is a survival trait.

GETTING INVOLVED

- ✓ Which forces hold quaternary structures together?
- ✓ What is the difference between a simple and a complex protein?
- ✓ What are the functions of myoglobin and hemoglobin?
- ✓ What structural features do myoglobin and hemoglobin have in common? How do they differ?
- ✓ What is a "conservative" change in primary structure? Give examples of amino acids that could be involved in conservative changes.

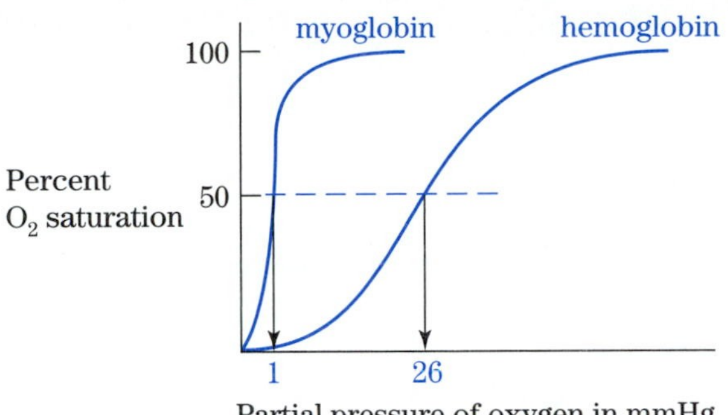

Figure 16.9 Comparative oxygen binding curves for myoglobin and hemoglobin.

Connections 16.2

www.prenhall.com/bailey

Mad Cow Disease

The title of the PBS special read like a science fiction movie ad: "The Brain Eaters." Even Oprah Winfrey got into the fray by stating she would not eat beef anymore for fear of getting something like "mad cow disease."

What was this scourge? Where did it start? How dangerous was it to humans? How could it be stopped?

Mad cow disease occurred in Great Britain after a change in laws permitting cattlefeed to contain meat and bonemeal of other animals such as sheep. The condition seen in the cattle is one of a group called the transmissible spongiform encephalopathies (TSE). Mad cow disease is a bovine spongiform encephalopathy (BSE). The symptoms in cattle are muscle and brain degeneration with premature death. Other types of animals such as minks and mule deer are known to exhibit similar degeneration. Brain material can literally become porous, and animals become incapable of even standing up. Obviously the practice of serving up sheep remains for consumption by cattle has ceased. Luckily, the United States as well as most of the world have not experienced mad cow disease.

BSE was linked to other human TSEs such as "kuru," a fatal dementia known to occur among aboriginal tribes practicing ritual cannibalism, which included eating the brain of a victim infected with the agent. The disappearance of cannibalism has coincided with the elimination of kuru. Crutzfeldt-Jakob disease (CJD), another TSE, was inadvertently spread to the uninfected by the practice of obtaining growth hormone from the pituitaries harvested from human cadavers. CJD and other human TSEs were found to have some hereditary properties.

The TSE in sheep and goats is called "scrapie" because the infected animals will rub themselves incessantly against objects, including barbed wire fences. It takes years or decades to develop symptoms of a TSE, depending upon the organism infected.

At first, the agent of transmission (vector) was thought to be a slow virus that could be passed on by the brain tissue of infected individuals. However, no genetic material could be found in the infectious particles. In the late 1970s and early 1980s Dr. Stanley Prusiner, then at the University of California at San Francisco, discovered that the infectious agent was protein in nature and coined the term *prion* for "proteinaceous infectious particles." The prion hypothesis of Prusiner states that brain tissue naturally produces PrP, a native *Prion Protein*. However, subtle amino acid substitutions in a few PrP molecules eventually cause the alteration of the secondary structure of normal PrP molecules. For example, it was found that the replacement of a leucine an the α-helical native PrP with a proline reoriented the secondary structure to β-sheet. The aberrant PrP can bind to the native, natural prion and slowly cause the conversion of the native into an aberrant form. This could explain the genetic inheritance factors seen in some forms of TSE as well as the long incubation period for the symptoms. These results have been, and still are, under heavy dispute, even though Dr. Prusiner was awarded the Nobel Prize in Physiology/Medicine in 1997 for his discovery of a new kind of infectious agent.

- ✓ What is a "nonconservative" change in primary protein structure? Give examples of nonconservative changes.
- ✓ How does a nonconservative change in hemoglobin that results in hemoglobin S affect the function(s) of this protein?
- ✓ What is a denatured protein?
- ✓ List at least three factors that could lead to protein denaturation and be able to explain how these factors affect protein structure.
- ✓ Are all denaturation processes irreversible?

Problem 16.17

Considering that the molecules in air, O_2 and N_2, are nonpolar, how might you explain the formation of meringue by whipping egg whites?

Problem 16.18

What tertiary interactions in milk proteins would be upset by lowering the pH to about 3, as occurs during souring through the production of lactic acid by *lactobacilli?*

16.4 Functions of Proteins

Catalysis, protection, and regulation are a few of many protein functions.

A. Enzymes—Biological Catalysts

All chemical reactions must proceed through energy barriers, whether slight or huge, in order to form products from starting materials. This energy of activation, E_a, is due to many factors, including the need for the reactants to collide and orient themselves in space correctly and efficiently as well as follow the steps of the mechanism appropriate for the particular reaction. Anything that can enhance one or more of these factors will lower the energy of activation and make it easier for the reaction to occur. We refer to this as **catalysis.** Recall that the process of addition to alkenes, for example, may be acid-catalyzed or may require the presence of a metal such as nickel or platinum. In the case of acid catalysis, the H^+ ion actually participates in polarizing bonds and then is regenerated during the course of the reaction. For metal catalysis, the nickel or platinum provides a surface upon which the reactants may orient themselves to increase the probability of collision as well as to provide an atomic arrangement in space for efficient and productive contact.

catalysis
the process in which a chemical reaction rate is increased due to a lowering of the energy of activation

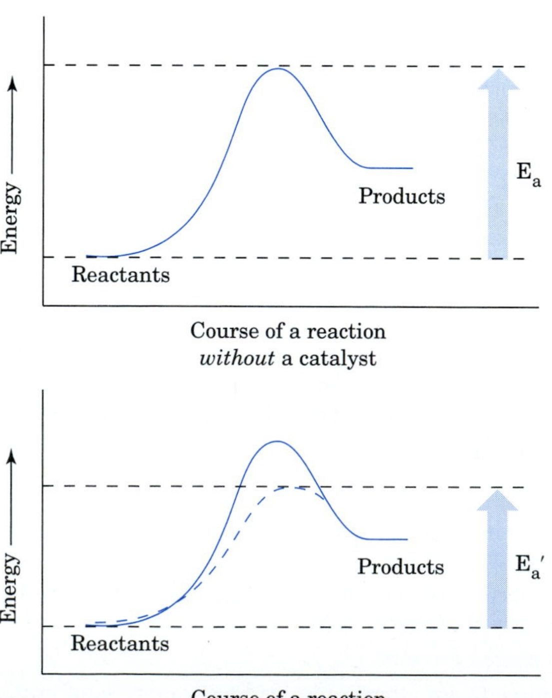

Course of a reaction *without* a catalyst

Course of a reaction *with* a catalyst

enzyme biological catalyst, usually protein in nature

Enzymes are proteins that catalyze biological reactions. Enzymes are classified by the type of reaction that they catalyze: oxidation–reduction, hydrolysis, group transfer, bond breaking, isomerization, or bond making; and according to the reactants with which they interact (see Table 16.3). Technically, an enzyme's name should end in the suffix *-ase*. As an example, the enzyme that catalyzes the following reaction is called maleate *cis-trans*-isomerase.

$$\underset{\text{Maleate}}{(^{-}OOC)HC{=}CH(COO^{-})_{cis}} \rightleftharpoons \underset{\text{Fumarate}}{(^{-}OOC)HC{=}CH(COO^{-})_{trans}}$$

However many enzymes were named before any convention directed such nomenclature, and they retain their common names, such as the stomach enzyme pepsin and the intestinal enzymes trypsin and chymotrypsin. Table 16.4 lists some common enzymes with typical uses.

An enzyme-catalyzed reaction has many advantages over an uncatalyzed reaction or one with a nonenzymatic catalyst. First, enzymes function at a rate thousands, if not millions, of times faster than uncatalyzed or normally catalyzed reactions. Second, enzymes can be very specific not only for the reactants, or substrates, in the reaction, but also for particular stereoisomers of those substrates. Third, enzymes function to produce specific products without the spurious by-products that can occur in organic reactions. These characteristics have led industry to the ever-increasing use of enzymes for the commercial production of natural and synthetic chemicals as well as for the preparation of foods and the cleanup of toxic waste.

Table 16.3 Enzyme Classification by International Enzyme Commission: A Summary

Class 1	*Oxido-reductases* carry out and influence oxidation-reduction reactions with alcohols, carbonyls, carbon-carbon double bonds, amines, etc.
Class 2	*Transferases* facilitate the transfer of certain functional groups, such as carbonyl, acyl, sugar, alkyl, and phosphate groups.
Class 3	*Hydrolases* catalyze the hydrolysis of esters, ethers, peptide bonds, glycosidic bonds, halides, acid anhydrides, and more.
Class 4	*Lyases* allow addition reactions with carbon-carbon double bonds, carbonyls, etc., or form such bonds themselves.
Class 5	*Isomerases* promote isomerization, optical and geometric, and also catalyze various intramolecular reactions, resulting in skeletal isomerization.
Class 6	*Ligases* (synthetases) aid in bond formation between carbon and sulfur, oxygen, nitrogen, or another carbon, and require ATP for energy.

Table 16.4 Some Common Enzymes

Enzyme	Typical Use	Enzyme	Typical Use
Rennin	milk coagulation for making cheese	Collagenase	removes tail from tadpoles when they become frogs
Bromelain	tenderizing meat; chill-proofing beer	Pepsin	begins protein digestion in stomach
Creatine kinase	provides metabolic energy in active muscle tissue	Streptokinase	dissolves blood clots
DNase	breaks up mucus in lungs of cystic fibrosis victims	Reverse transcriptase	responsible for the incorporation of viral genetic material into the host genome

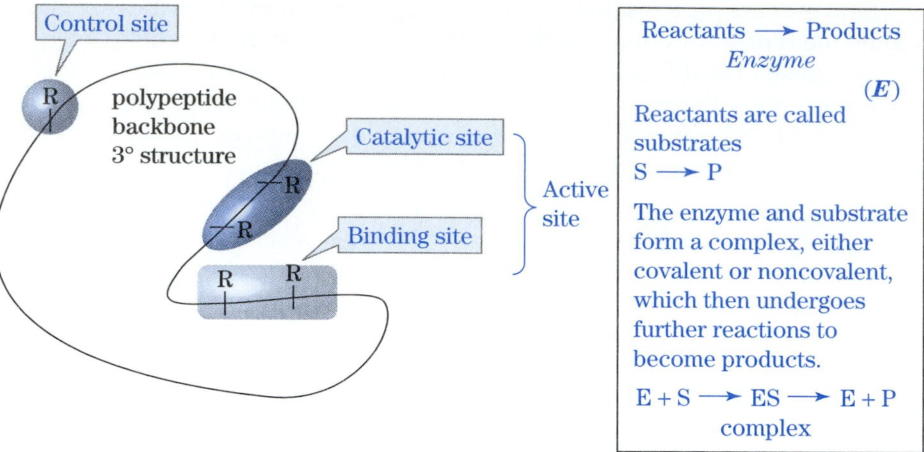

Figure 16.10
An enzyme and the parts contribute to the active and control sites.

Enzymes direct their remarkable feat of catalysis by presenting an interactive, three-dimensional environment to the reactants. Every enzyme molecule has an **active site,** where catalysis takes place. Within the active site is a **binding site,** which attracts and holds the **substrates,** and a **catalytic site,** which participates in the mechanism of the reaction (see Figure 16.10).

active site
functional portion of an enzyme

binding site
portion of an enzyme active site that attracts the substrate

substrates
molecules and/or ions on which an enzyme works

catalytic site
area within the active site of an enzyme that causes catalysis

B. Enzyme Control

The human body contains thousands of different enzymes working on different reactions with different substrates. How are all these reactions coordinated so that a single, coherent organism results? What keeps the body from digesting itself? The answers to these questions are of course very complicated, but we can discuss briefly how some enzymes can be turned on and off.

A common means by which enzymes are prevented from exerting their catalytic effects where they are not needed is their secretion in larger, inactive forms known as **zymogens.** An important example involves enzymes trypsin, chymotrypsin, and carboxypeptidase, which are responsible for protein digestion in the intestines. These proteins are produced in the pancreas as larger proteins; trypsinogen, chymotrypsinogen, and procarboxypeptidase. After biosynthesis, they are secreted through the bile duct into the small intestine, where trypsinogen is changed to trypsin by the action of another enzyme called enteropeptidase. The active trypsin can also convert trypsinogen to trypsin, and chymotrypsinogen and procarboxypeptidase to their active states. Should activation of the zymogens occur before they leave the pancreas, which can happen in certain disease states, then the pancreas will gradually be digested, a condition known as pancreatitis.

zymogen
inactive precursor of an enzyme

Other enzymes can exist in two forms, which differ only in the covalent modification of an amino acid in the protein. For example, the enzyme glycogen phosphorylase is responsible for the first step in the conversion of the storage carbohydrate glycogen to glucose. Glycogen phosphorylase itself needs to be phosphorylated, that is, to be derivatized with two phosphate groups, in order to be enzymatically active. Can you guess what catalyzes the phosphorylation of phosphorylase? That's right—another enzyme, phosphorylase kinase. The active form of phosphorylase can be inactivated (dephosphorylated) by a third enzyme, a phosphatase.

Other materials, natural and synthetic, can slow down or completely stop the action of enzymes. These species are called **inhibitors.** The pancreas, in its role of zymogen secretion, also produces another protein, pancreatic trypsin inhibitor,

enzyme inhibitor
molecule or ion that reversibly or irreversibly slows down or stops the activity of an enzyme

which helps to keep trypsin in check. Heavy metals such as mercury, lead, and arsenic will inhibit enzymes to such an extent that the organism can die. This is the fundamental premise behind the development of many pesticides and poisons.

Because enzymes are organic molecules, they can be manipulated for commercial use. For example, they may be compounded with detergents in order to remove grease or blood stains from clothing or they may be attached to a solid support to convert glucose to fructose in the production of high-fructose corn syrup. The applications of enzyme chemistry are virtually limitless.

GETTING INVOLVED

✓ What are some of the problems that arise from attempting to use enzymes on a large scale for industrial processes, considering that enzymes are proteins?

C. Antibodies—Immune System Protection

The immune system is a complex network of cells, proteins, and chemicals that act in concert to thwart the invasion of anything that is not part of the organism, sometimes referred to as "nonself," or in immunological terms as the antigen. **Antibodies** are part of this protective arsenal.

antibody
glycoprotein produced by the B-cells of the immune system as protection

Connections 16.3 www.prenhall.com/bailey

Testing for Drugs, Pregnancy, and AIDS

Antibodies have proved invaluable in the clinical determination of disease, drug intoxication, and pregnancy. Because they are so specific, antibodies can be generated in an animal to specific substances such as a virus, hormone, or drug. As proteins, the antibodies can be linked chemically to enzymes. The enzymes, in turn, may catalyze a reaction involving a color change, which can be detected by using a single or multiple wavelength spectrophotometer.

This type of assay is called an enzyme-linked immunosorbent assay, or ELISA. The actual process is a little more complex, but the diagram below contains the fundamental concept. This technique can be modified by using radioactive antibody or antigen complexes, which may increase the sensitivity and thereby enhance the limits of detection.

A variety of drugs, from morphine to amphetamine, can be assayed quantitatively in this manner. Pregnancy is determined in over-the-counter kits that detect the presence of the hormone chorionic gonadotropin, which is excreted by a woman during the first few weeks after conception. The AIDS virus has a protein capsule or coat that can be detected via an ELISA assay.

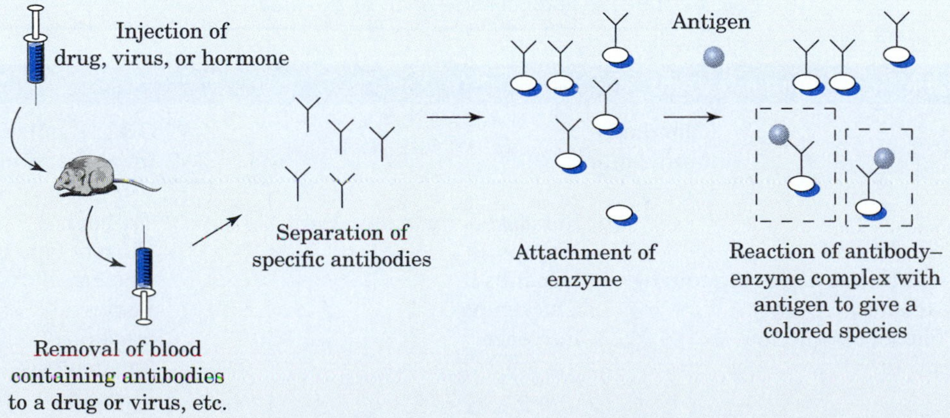

Injection of drug, virus, or hormone

Removal of blood containing antibodies to a drug or virus, etc.

Separation of specific antibodies

Antigen

Attachment of enzyme

Reaction of antibody–enzyme complex with antigen to give a colored species

glycoprotein protein with carbohydrate attached

antigen the material to which the immune system responds

Antibodies are **glycoproteins,** that is, proteins to which carbohydrates are covalently attached. An antibody is produced by the B-cells of the immune system in response to a foreign substance or **antigen.** Formation of an antibody–antigen complex can result in precipitation or in identification to the other immune system components that can help to destroy the invader.

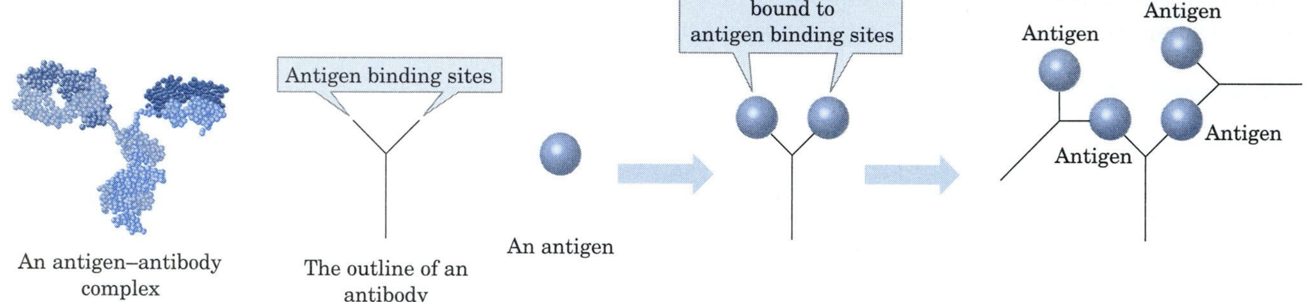

The process of immunization against toxins and disease is based upon the fact that repeated challenges by the same type of antigen result in increasingly intense antibody responses by the cells of the immune system.

Poliomyelitis, a paralytic viral disease, has been almost eradicated from the United States by the immunization of babies with small amounts of the virus, that have been treated to be less dangerous (attenuated). Booster immunizations keep the amount of defensive antibodies high and ready to respond. We can also be immunized against the toxins produced by various bacteria. Antibodies to the venoms of some poisonous animals and insects can be directly administered following the bite or sting. Table 16.5 lists some of the common immunizations available in the United States.

GETTING INVOLVED

- ✓ What is the function of an enzyme? An antibody?
- ✓ Write a generic chemical equation for the action of an enzyme on a chemical reaction.
- ✓ Are enzymes and antibodies simple or complex proteins?
- ✓ How do antibodies interact with antigens?
- ✓ Describe the biochemical basis for immunization.
- ✓ Outline the general steps in an ELISA assay.

Table 16.5 Immunizations

Standard Immunizations		Other Available Immunizations
DPT	Rubella (German measles)	Typhoid
diphtheria	Mumps	Current type of influenza
pertussis (whooping cough)	Hepatitis B	Cholera
tetanus (lockjaw)	Chicken pox	Rabies
Polio (Sabin vaccine)	Influenza	Smallpox
		Tuberculosis
		Hepatitis A
		Pneumonia

D. Polypeptide and Protein Hormones—Metabolic Regulation

The ability of living things to grow, reproduce, and respond to stress is regulated by secretions of biochemicals known as **hormones.** The structures of hormones may be simple, such as those for epinephrine (adrenalin) and cortisone (a steroid), or they may be quite large and complex, such as growth hormone (see Connections 16.4). Most known hormones are steroids (lipids), aminoacidlike molecules, polypeptides, and proteins. Table 16.6 lists some key hormones, their biochemical classes, and one or more primary actions.

hormone
compound secreted by an organ or gland that controls metabolism

16.5 Determination of Protein Structure

The primary structure of a protein can be determined chemically and through molecular biology. The secondary, tertiary, and quaternary structures depend upon instrumental techniques such as X-ray crystallography, nuclear magnetic resonance, and computer modeling for high resolution. The upper levels of protein structure can also be probed chemically, and this information can be used in conjunction with the methods mentioned to get a complete picture of the molecule. For now we shall concentrate on the chemical methods of analysis.

A. Amino Acid Composition

The peptide bond, although stable under physiological conditions, can be broken through the process of acid or base hydrolysis. Subjected to boiling in 6M hydrochloric acid for 18–24 hours, most proteins will break down into their constituent amino acids. The sample of amino acids can then be separated through liquid chromatography. As the separated amino acids leave, or elute from, the column, they can be mixed with a color reagent and assayed with the use of a spectrophotomer (Figure 16.12).

Connections 16.4 www.prenhall.com/bailey

Growth Hormone

Somatotropin or growth hormone is a protein of molecular weight 22,000, having 191 amino acids. It is secreted by the pituitary gland and has its effects on most organs and tissues in the body, notably the muscles and bones. A deficiency in growth hormone results in short stature or the extreme of dwarfism. Hypersecretion will cause elongation of the bones and a coarsening of the skin and facial features, a condition known as gigantism or acromegaly.

Until the early 1980s the only source of functional growth hormone for humans was humans—human cadavers. It took approximately 40 human pituitaries to supply the hormone needed for one child for one year. In contrast to hormones like insulin, which can be harvested from pigs and sheep for human use, growth hormone is very species-specific. In addition there was a risk of developing a fatal brain disorder called Creutzfeldt-Jakob disease from contamination of brain tissue. The advent of recombinant DNA biotechnology opened the door for expressing human proteins in bacterial hosts. Human somatotropin has been available since 1985. It is injected several times per week, causing a child to catch up rapidly and maintain normal growth throughout adolescence.

Three areas of social concern have arisen over the availability of growth hormone. One is its illegal and undetectable use by athletes and a second is its administration to normal children in order to increase stature or athletic potential. The third issue concerns the use of bovine (beef) growth hormone (bGH) to increase milk production in cows. Consumer-group objections have discouraged the introduction of recombinant bGH into agriculture. There is no doubt that advances in the understanding and production of proteins present a challenge not only to science but also to the fabric of society.

Table 16.6 Hormones

Hormone	Source	Type of Biochemical Action
Polypeptide 1–50 amino acids		
• epinephrine and norepinephrine (modified tyrosine)	adrenal medulla	regulate stimulation of heart function, contraction of blood vessels and smooth muscle, control of metabolism
• thyroxine (an iodinated tyrosine dimer)	thyroid	stimulates general cell growth
• releasing and inhibiting factors	hypothalamus	affect secretions of the pituitary
• oxytocin	pituitary	stimulates mammary gland and uterine muscle
• vasopressin	pituitary	regulates blood pressure and water retention
• melanocyte-stimulating hormones	pituitary	control pigmentation
• corticotropin	pituitary	stimulates adrenal steroid synthesis
• calcitonin	thyroid	calcium and phosphorus metabolism
• glucagon	pancreas	increases blood glucose levels
• gastrin	GI tract	stimulates production of acid in stomach and pancreas
• vasoactive intestinal peptide	GI tract	inhibits acid and pepsin secretion
• motilin	GI tract	controls GI muscle
• somatostatin	GI tract	inhibits gastrin and glucagon secretion
• angiotensin	liver	regulates water retention and excretion
Proteins > 50 amino acids		
• insulin	pancreas	lowers blood glucose levels
• growth hormone	pituitary	stimulates general growth and metabolism
• prolactin	pituitary	controls milk secretion
• lutenizing and follicle-stimulating hormones	pituitary	stimulate male and female hormone and cell development
Steroids		
• testosterone	testes and adrenals	regulates male secondary sex characteristics and metabolism
• estradiol	ovaries	regulates female secondary sex characteristics and metabolism
• progesterone	ovaries and placenta	affects egg implantation and pregnancy
• glucocorticoids	adrenal cortex	control protein and carbohydrate metabolism, inflammation
• mineralocorticoids	adrenal cortex	regulate water and salt balance

Overall, the only data obtainable by these means are the types and amounts of the individual amino acids that make up the protein in question.

B. Sequence of Amino Acids—Determination of Primary Structure

There are several organic reagents, such as dansyl chloride, that can react with intact proteins to derivatize the N-terminal amino acid (see Figure 16.11). The "tagged" amino acid can then be separated and identified. However, the procedure destroys the rest of the polypeptide chain and only the N-terminus has been determined. It would be advantageous to have a method in which the rest of the chain remains intact during the course of the experimental procedure.

Edman degradation nondestructive, sequential method of determining polypeptide primary structure

1. ***Edman Degradation.*** Pehr Edman was responsible for developing the sequential method that bears his name. The reagent is phenylisothiocyanate, or

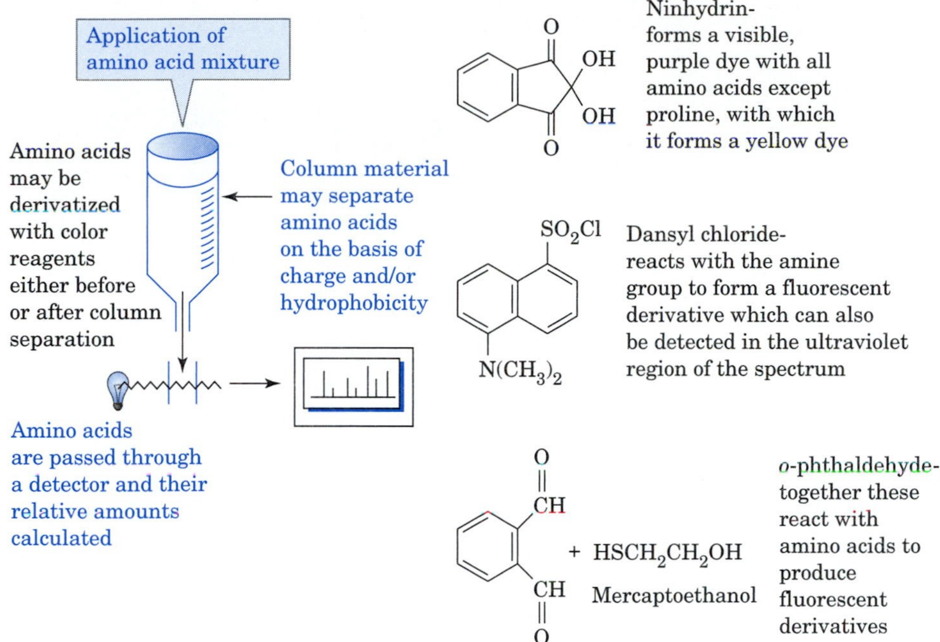
Reagents used for detection
Application of amino acid mixture
Amino acids may be derivatized with color reagents either before or after column separation
Column material may separate amino acids on the basis of charge and/or hydrophobicity
Amino acids are passed through a detector and their relative amounts calculated
OH
OH
Ninhydrin- forms a visible, purple dye with all amino acids except proline, with which it forms a yellow dye
SO2Cl
N(CH3)2
Dansyl chloride- reacts with the amine group to form a fluorescent derivative which can also be detected in the ultraviolet region of the spectrum
CH
CH
+ HSCH2CH2OH
Mercaptoethanol
o-phthaldehyde- together these react with amino acids to produce fluorescent derivatives

Figure 16.11 Schematic of amino acid analysis.

PITC. PITC derivatizes the N-terminal amino acid and leaves the rest of the chain sequence intact. After separating the PTH (phenylthiohydantoin) amino acid, the remaining chain can once again be treated by the Edman reagent.

$$C_6H_5{-}N{=}C{=}S + H_3\overset{+}{N}CH(R_1)C(=O)NHCH(R_2)C(=O)\sim \;\longrightarrow\; C_6H_5{-}N{-}C({=}S){-}NH{-}CH(R_1){-}C({=}O) \text{ (ring)} + H_3\overset{+}{N}CH(R_2)C(=O)\sim$$

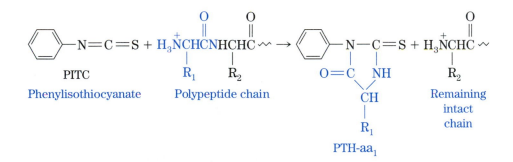

PITC Phenylisothiocyanate; Polypeptide chain; PTH-aa_1; Remaining intact chain

The PTH amino acids can be separated on a chromatographic column and identified by their ultraviolet absorption spectra.

GETTING INVOLVED

Example 16.4

Draw the products of one cycle of the Edman degradation with the polypeptide, Asp ~ Tyr ~ Gly ~ Met.

Solution

The Edman reagent, phenylisothiocyanate, reacts with the N-terminal amino group of Asp and leaves the Tyr ~ Gly ~ Met tripeptide intact.

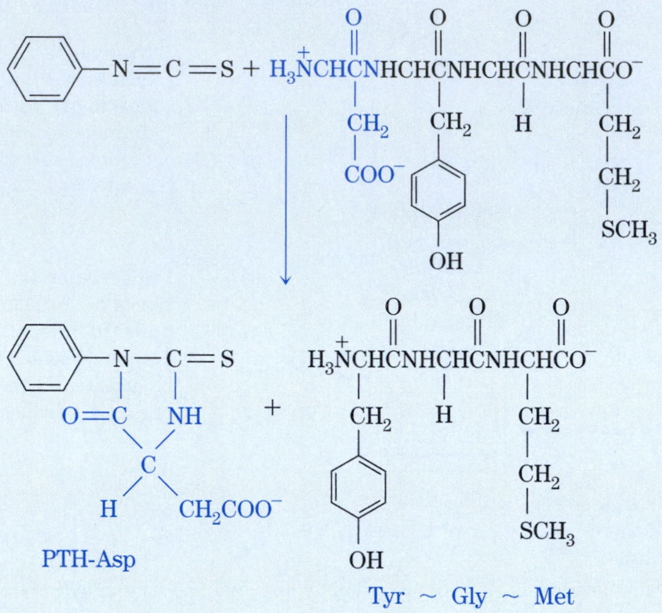

$$C_6H_5{-}N{=}C{=}S + H_3\overset{+}{N}CH(CH_2COO^-)\overset{O}{\overset{\|}{C}}NHCH(CH_2C_6H_4OH)\overset{O}{\overset{\|}{C}}NHCH(H)\overset{O}{\overset{\|}{C}}NHCH(CH_2CH_2SCH_3)\overset{O}{\overset{\|}{C}}O^-$$

$$\downarrow$$

$$\text{PTH-Asp: } C_6H_5{-}N{-}C{=}S,\ O{=}C,\ NH,\ C(H)(CH_2COO^-) \quad + \quad H_3\overset{+}{N}CH(CH_2C_6H_4OH)\overset{O}{\overset{\|}{C}}NHCH(H)\overset{O}{\overset{\|}{C}}NHCH(CH_2CH_2SCH_3)\overset{O}{\overset{\|}{C}}O^-$$

PTH-Asp

Tyr ~ Gly ~ Met

Problem 16.19

Draw the products of two more cycles of the Edman degradation on the tripeptide remaining in Example 16.4.

Problem 16.20

Assume that an N-terminal sequential method has an 85% yield at each of five steps in degrading a polypeptide chain. What is the theoretical yield of the desired amino acid six positions from the N-terminus?

2. ***Fragmenting the Chain.*** The Edman degradation is limited in terms of the length of chain it can successfully sequence, as well as the types of amino acids that can be readily derivatized. Therefore, it is necessary to fragment a long protein chain into pieces manageable for the sequencing routine. There are chemical reagents to do this, such as cyanogen bromide, which breaks the chain at methionine. The easiest and most specific cleavages can be effected by enzymes.

Trypsin, an intestinal protease (peptide bond hydrolase), has a specificity for breaking peptide bonds in a chain at the carboxy end of basic amino acids, that is, lysines and arginines. Chymotrypsin, also found in the small intestine, will hydrolyze peptide bonds contributed by the aromatic, hydrophobic amino acids—phenylalanine, tyrosine, and tryptophan. By performing digestions of the protein to be analyzed with each of these enzymes, perhaps doing an additional chemical cleavage, then separating and finding the amino acid content of the resulting peptides, an overlapping picture of the primary structure can be ascertained.

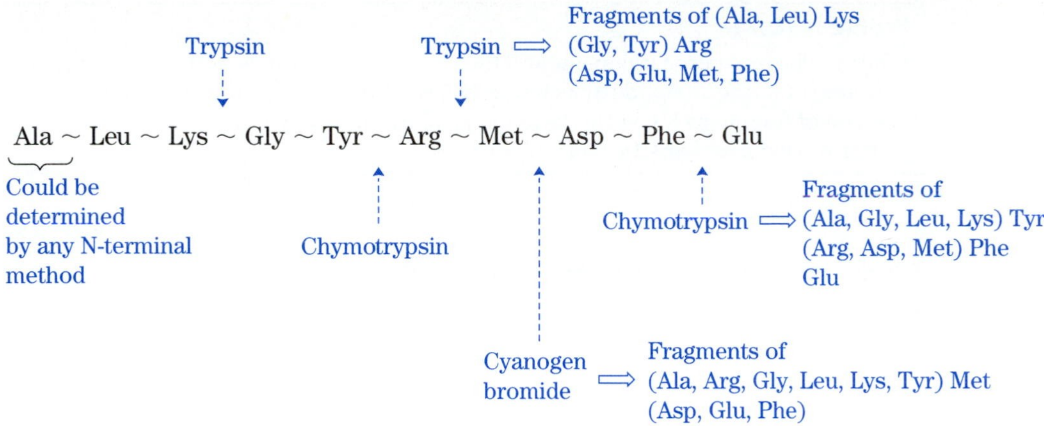

Carboxypeptidase and aminopeptidase are enzymes that will cleave amino acids sequentially from the C- and N-termini, respectively. Timed assays of the released products are taken. This may seem a convenience at first, but incomplete hydrolysis at one or more steps in the breakdown will contaminate subsequent releases of amino acids farther on in the chain.

GETTING INVOLVED

Example 16.5

A hexapeptide was analyzed by using the organic and enzymatic methods described. The results are shown below. Find the primary sequence of the hexapeptide.

(a) Acid hydrolysis and amino acid analysis gave the following content (subscripts refer to the relative quantities of the amino acids; the listing is alphabetical):

$$\text{Gly}_2, \text{His}_1, \text{Ile}_1, \text{Phe}_1$$

(b) One cycle of the Edman degradation produced PTH-Gly.

(c) Digestion of the hexapeptide with trypsin produced two fragments with the following amino acid compositions:

Fragment 1: $\text{Gly}_1, \text{Ile}_1, \text{Phe}_1$ *Fragment 2:* $\text{Gly}_1, \text{His}_1, \text{Lys}_1$

(d) Timed carboxypeptidase digestion of the hexapeptide gave a release of

$$\text{Ile} > \text{Phe} > \text{Gly}$$

Solution

- Part (a) tells us the amino acid content only.
- In (b) we find out the identity of ht N-terminal amino acid—Gly.
- For part (c) remember that trypsin cleaves the chain at the carboxyl end of Lys and Arg. The fact that two fragments were found indicates that the Lys is somewhere inside of the peptide and not at the C-terminal. Notice that Lys cannot be the N-terminal because we have already identified the N-terminal as Gly *and* we did not find free Lys.
- Therefore, we have the sequence of the first three amino acids—Gly ~ His ~ Lys.
- The carboxypeptidase digestion releases amino acids in sequence from the C-terminal. Therefore, Ile must be the C-terminus preceded by Phe; Phe is proceeded by Gly. The complete sequence is Gly ~ His ~ Lys ~ Gly ~ Phe ~ Ile.

Problem 16.21

What are the number of fragments and their amino acid composition if chymotrypsin digestion of the hexapeptide in Example 16.5 is used? What are the number and composition of fragments if trypsin digestion is followed by chymotrypsin?

See related problems 16.31, 16.32, 16.33.

16.6 Organic Synthesis of Polypeptides

The importance of polypeptides and proteins has been an impetus to attempt to synthesize them for practical pharmacological purposes and for study.

A. General Considerations

Making something as simple as the tripeptide Val ~ Tyr ~ Asp does not involve just mixing the three amino acids together. Even if they could react, this would give a mixture of polypeptides: the Val ~ Tyr ~ Asp desired as well as Tyr ~ Asp ~ Val, Asp ~ Val ~ Tyr, Asp ~ Tyr ~ Val, Asp ~Asp ~ Tyr, $(Val)_3$, etc.

Then we must consider that the carboxyl groups are not reactive enough to form peptide bonds readily. Using an activating group such as an acid chloride greatly enhances the carbonyl reactivity. In addition, the amine group of the amino acids that you do not wish to react must be derivatized reversibly. Finally you must add the amino acids sequentially, isolating the first dipeptide product before putting in the third amino acid. A generic scheme for synthesizing the tripeptide might be as follows:

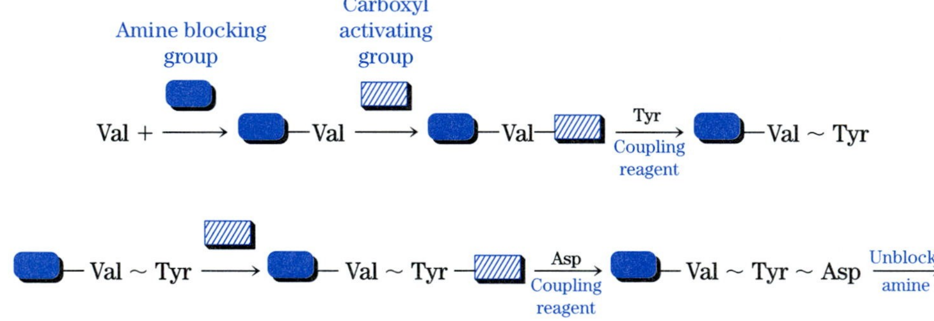

This procedure involves an extensive set of blocking, activating, coupling, and deblocking steps with purification of the desired intermediates along the route. Reactive R groups must be protected and then deblocked in a similar manner. The more amino acids in the polypeptide, the more steps, the lower the yield.

B. Solid-State Synthesis

Merrifield synthesis method of synthesizing polypeptides using a solid phase

Dr. R. Bruce **Merrifield** proposed a novel method of synthesis in 1965. For this discovery, he was awarded the Nobel Prize in 1984. The procedure uses a solid polystyrene resin in which about 10% of the aromatic rings have been derivatized with chloromethyl groups. The carboxyl group of an amino acid reacts with this group via an S_{N2} mechanism to become covalently attached to the resin.

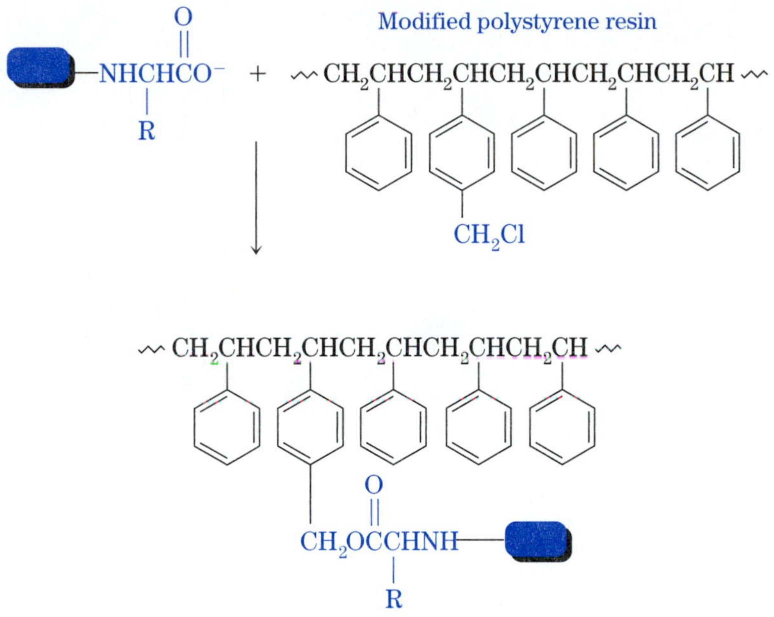

Modified polystyrene resin

$-NHCHRCO_2^- + \sim CH_2CHCH_2CHCH_2CHCH_2CHCH_2CH \sim$ (phenyl rings; one ring bearing CH_2Cl)

↓

$\sim CH_2CHCH_2CHCH_2CHCH_2CHCH_2CH \sim$ (phenyl rings; one ring bearing $CH_2OC(=O)CHRNH-$)

The other desired amino acids are coupled to it; all reagents and solvents can be washed over the growing polypeptide chain as it is held on a solid support. This method works so efficiently that it has been automated. Two common reagents are the *t*-butyloxycarbonyl amino protecting group (Boc) and the coupling agent dicyclohexylcarbodiimide (DCC). The entire process is illustrated in Figure 16.12.

There is a limit to the number and types of amino acids that can be put together in this way. For long chains, that is, for proteins, a biosynthetic process using DNA as a guide is more specific, accurate, and efficient.

GETTING INVOLVED

Problem 16.22

Write out the possible combinations of any four amino acids, assuming that each occurs only once in a tetrapeptide.

Protein chemistry is a complex and fascinating field related to both biology and organic chemistry. It requires research endeavors that involve the talents and coordination of almost every aspect of the physical and medical sciences. We hope the background given here will encourage those interested to study further in the area of biochemistry.

N-blocked C-terminal amino acid is attached to the resin.

Amine group is unblocked.

The next protected amino acid is coupled to the bound amino acid.

The cycle is repeated until the desired polypeptide has been made. Then it is detached from the solid support.

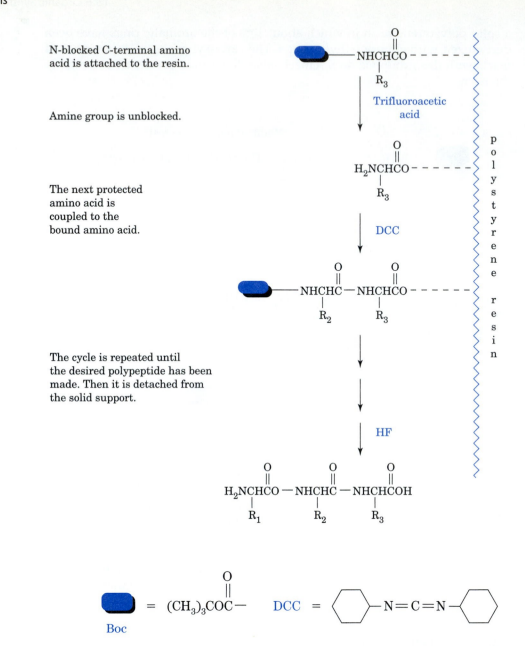

Figure 16.12 The Merrifield solid-phase synthesis of polypeptides.

Problems

16.23 **Structure:** Identify the amino acids having the following characteristics:

(a) optical inactivity
(b) a phenolic group
(c) involvement in covalent bridging
(d) two optical isomers
(e) responsibility for bending a peptide chain and "breaking" a helical structure
(f) hydrogen-bonding through an R side-chain group
(g) more than two possible optical isomers

16.24 **Structure:** α-Amanitine is a polypeptide analogue that is the deadly component of a type of poisonous mushroom, *Amanita phalloides.* From its structure, try to identify the component of amino acids and any novel linkage (besides the α-aminocarboxyl peptide bond).

```
              OH
              |
     H3C     CH—CH2OH
        \   /
         CH   O                  O
         |    ||                 ||
   HN—CH—C—NH—CH—C—NH—CH2—C=O
   |                  |                   |
  C=O              H2C                  NH          CH3
   |                  \   (indole)       |         /
 H  CH                               H—C—CH
  \/  |               O=S    N      OH     |         \
  /\  |                  |   |            C=O        C2H5
HO  N                  CH2  H             |
   |                     \                |
 O=C—CH—NH——C—CH—NH—C—CH2——NH
       |        ||          ||
     H2C—CONH2  O           O
```

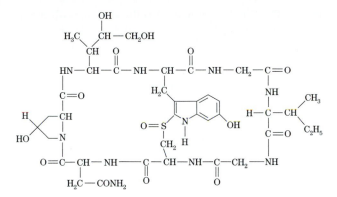

16.25 **Structure:** What are the products of the acid-catalyzed hydrolysis of the following peptide? Name the resulting amino acids, using the three-letter abbreviations.

```
  +     O      O      O      O
        ||     ||     ||     ||
 H3NCHCNHCHCNHCHCNHCHCOH
     |      |       |      |
    CH2   HC—CH3CHOH  CH2
     |      |       |      |
    CH2   CH2     CH3    CH2
     |      |              |
    COH   CH3            CH2
     ||                    |
     O                    CH2
                           |
                          NH3
                          +
```

16.26 **Structure:** Find the pI of the following polypeptides:

(a) met-enkephalin—an opiate neurotransmitter:
Tyr ~ Gly ~ Gly ~ Phe ~ Met

(b) somatostatin (growth hormone inhibiting factor)

```
        ~ Phe ~ Asn ~ Lys ~ Cys ~ Gly ~ Ala
  Phe                        |
  ~                          S
 Trp                         |
  ~                          S
 Lys                         |
  ~                          |
   Thr ~ Phe ~ Thr ~ Ser ~ Cys
```

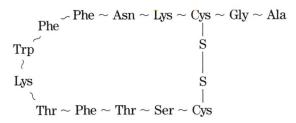

16.27 **Structure:** Histones are proteins associated with nucleic acids. As phosphoric acid derivatives, nucleic acids have a negative charge under physiological pH conditions. What should be the net charge on the histones? Which amino acids should be found to a large extent in the primary structures of histones?

16.28 **Structure:** About 400 variations have been identified in the primary structure of hemoglobin. Some of these variations are conservative, that is, they will not make a difference to the physical properties or functions of the molecule. Others are nonconservative and can be fatal. Shown below is a diagram of a pH 8.0 electrophoresis of normal hemoglobin (HbA) and five variants, including sickle cell hemoglobin. Using the changes listed in the following table and the relative migrations of the variants at pH 8.0, match the variant to its position of the electrophoretogram. HbS is given as an example (remember that there are two α and β chains).

Hb variant	Changes in Primary Sequence				
	Chain	Position from N-Terminus	Amino Acid in HBA	Amino Acid in Variant	
S	β	6	Glu	Val	change of +2
C	β	6	Glu	Lys	
Chesapeake	α	92	Arg	Leu	
Hasharon	α	47	Asp	His	
Koln	β	98	Val	Met	

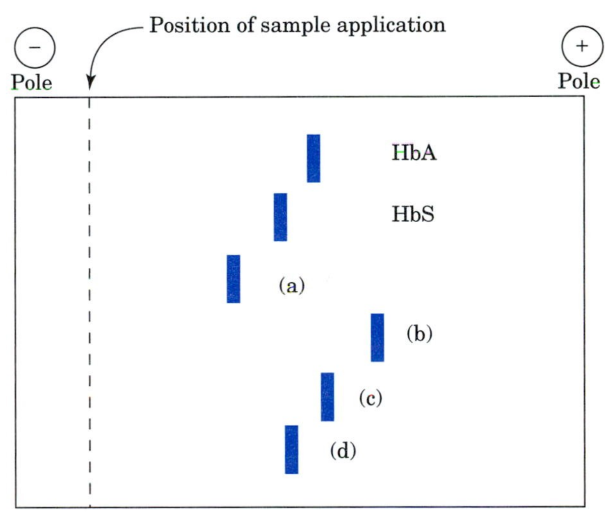

16.29 **Protein Structure:** Where do the following terms fit in a protein's hierarchy of structure (as 1°, 2°, 3°, or 4°)?

(a) the α and β subunits of hemoglobin
(b) Phe-Val side-chain interactions
(c) intrachain hydrogen-bonding
(d) linear sequence of amino acids
(e) salt bridges
(f) disulfide bridges

16.30 **Protein Structure:** What type of tertiary interaction might the side chains of each of the following pairs of amino acids be capable of?

(a) Ser and His **(b)** Phe and Leu
(c) Arg and Glu **(d)** Thr and Val

16.31 **Peptide Sequence:** Draw out the structures for end products of three cycles of Edman degradation for the polypeptide Leu ~ Met ~ His ~ Ser.

16.32 **Peptide Sequence:** A polypeptide, on acid hydrolysis, contained the amino acids Arg (1), Ala (1), Ile (1), Leu (2), Lys (1), Phe (2), Tyr (1). Treating the intact peptide with dansyl chloride and subsequent hydrolysis gave dansyl-Leu. Reaction with carboxypeptidase gave varying amounts of free amino acids, Phe > Leu > Ala. Digestion of the intact polypeptide with trypsin gave the following fragments: Tyr ~ Ile ~ Phe ~ Lys, Leu ~ Arg, and Ala ~ Leu ~ Phe. Chymotryptic treatment of the intact polypeptide produced Ile ~ Phe, Lys ~ Ala ~ Leu ~ Phe, and Leu ~ Arg ~ Tyr. What is the sequence of the nonapeptide? (*Note:* The lines between amino acids represent a peptide bond.)

16.33 **Peptide Sequence:** A polypeptide with the indicated amino acid composition (listed alphabetically) was analyzed as shown below. What is the primary sequence of the peptide?

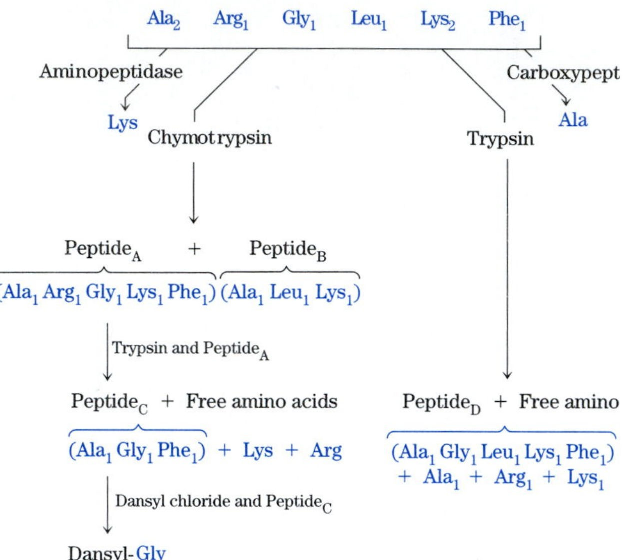

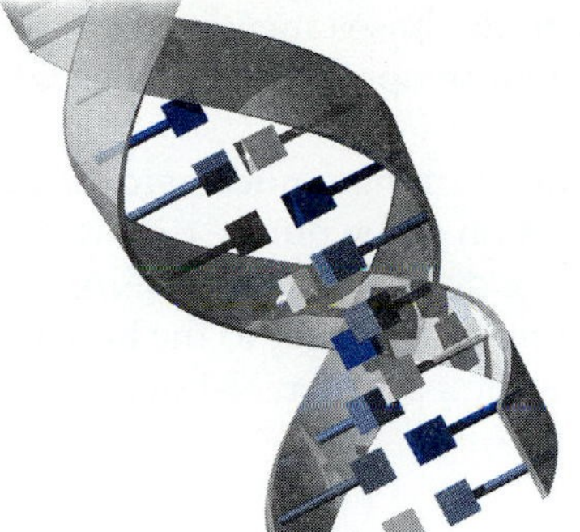

CHAPTER 17
Nucleic Acids

The virtual explosion of biotechnology in the last two decades is ample evidence of the central importance of nucleic acids to chemistry as well as biology. These biopolymers are in the public eye because of their biological and medical promise and have generated serious discussion in the areas of economics, politics, sociology, ethics, and theology.

Nucleic acids are the constituents of our genes. Although their fundamental structures are relatively simple, the process of nucleic acid or gene replication and the translation of the genetic message into tens of thousands of proteins for which it codes is a complex process. This chapter will only touch the surface of a complicated and growing field. Individual monomer units, or nucleotides, as well as dinucleotides also serve as energy carriers and oxidation/reduction agents in metabolism and as chemical messengers relaying life-regulating information within and between cells.

17.1 The Chemical Structure of Nucleic Acids

As the term **nucleic acid** suggests, these biopolymers are acidic in nature and are found in the nucleus of the cell as well as in chloroplasts and mitochondria. The fundamental unit of the polymer is the **nucleotide,** which consists of a heterocyclic base, a sugar, and inorganic phosphate.

Five common heterocyclic bases are found in **DNA** (deoxyribonucleic acid) and **RNA** (ribonucleic acid): two are related to the bicyclic base purine and three to the monocyclic base pyrimidine. Three of the five bases are common to both DNA and RNA, while the two remaining pyrimidines help to distinguish DNA from RNA.

nucleic acid
biopolymer whose monomer unit, a nucleotide, consists of a heterocyclic base, a sugar, and a phosphate group

nucleotide
the monomer unit of a nucleic acid consisting of a purine or pyrimidine base covalently bonded to a ribose or deoxyribose unit, which in turn is bonded to a phosphate group

DNA
deoxyribonucleic acid

RNA
ribonucleic acid

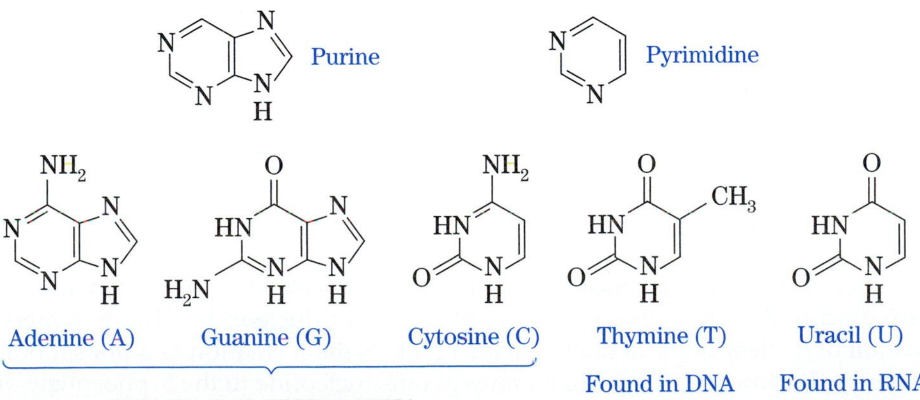

There are variations found in the structures of RNA bases, and DNA can undergo a natural process of methylation. However, the bases mentioned above are the ones found in greatest abundance and are those upon which the genetic code was established.

nucleoside
heterocyclic base bonded to a ribose or deoxyribose unit

These bases are bonded to the monosaccharide, forming a **nucleoside.** Adenine and guanine are attached through the N-9 position of the purine ring system to the hemiacetal group, C-1 or more properly C-1′, of deoxyribose (for DNA) and ribose (for RNA). Note that the glycosidic linkage from the sugar to the base has the β-configuration. The pyrimidines are linked through position N-1 in the ring.

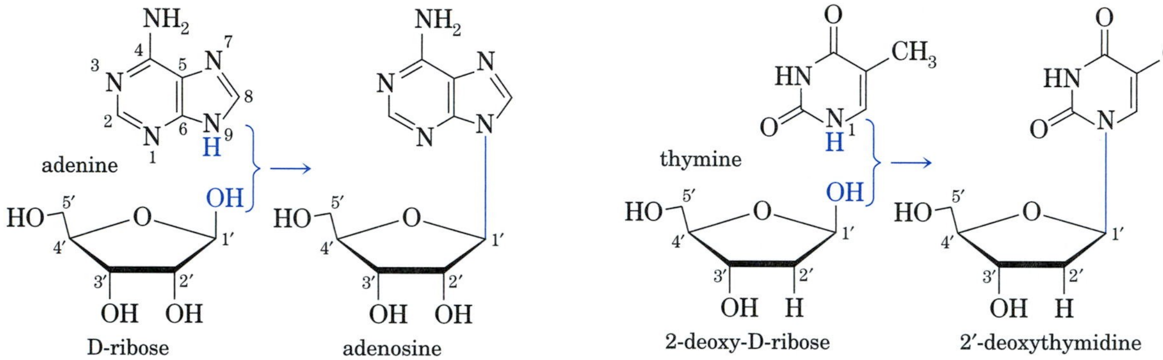

adenine → adenosine (D-ribose)

Formation of a Ribonucleoside

thymine → 2′-deoxythymidine (2-deoxy-D-ribose)

Formation of a 2′-Deoxyribonucleoside

The other nucleosides are named 2′-deoxyadenosine, 2′-deoxyguanosine, and 2′-deoxycytidine for the DNA combinations and adenosine, guanosine, and uridine for the RNA components.

The nucleoside is then esterified through the sugar to a phosphate group to make a nucleotide. Phosphoric acid is a triprotic acid and can react as an acid with each ionizable hydrogen. It can also form one or more ester bonds with available alcohol groups. A nucleotide has a phosphate esterified to position 5′ of the ribose or deoxyribose.

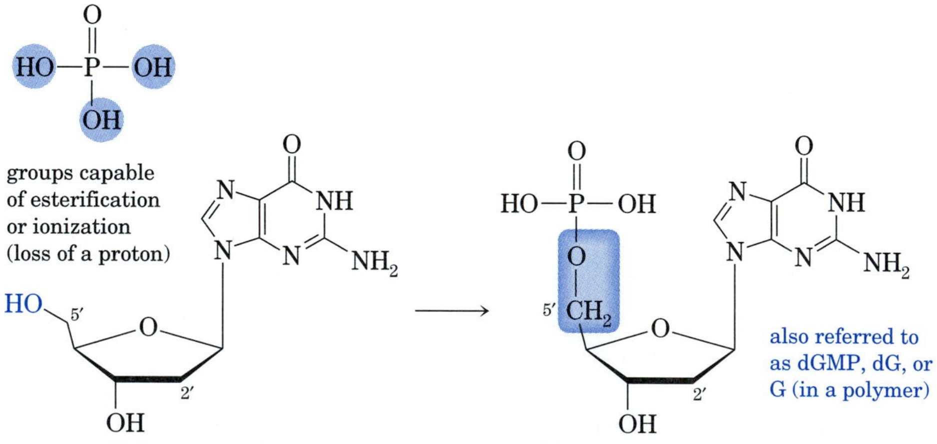

groups capable of esterification or ionization (loss of a proton)

2′-deoxyguanosine → 2′-deoxyguanosine-5′-monophosphate

also referred to as dGMP, dG, or G (in a polymer)

Because of the complicated structure of a nucleotide, shorthand notations are used to designate the bases, nucleosides, and nucleotides. The polymerization of nucleotides into nucleic acids involves the formation of a phosphodiester bridge from the 3′ hydroxy group of one nucleotide to the 5′ phosphate of another.

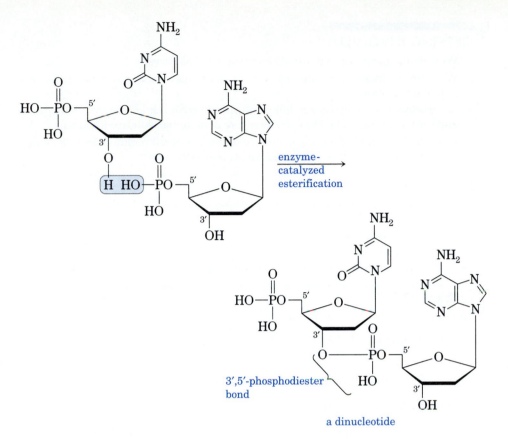

enzyme-catalyzed esterification

3′,5′-phosphodiester bond

a dinucleotide

As this enzyme-catalyzed polymerization proceeds, a regular array develops consisting of a phosphate-sugar "backbone" from which the heterocyclic bases protrude. The backbone can be shown or simply assumed to be the way in which the bases are connected.

Representations of a polynucleotide chain

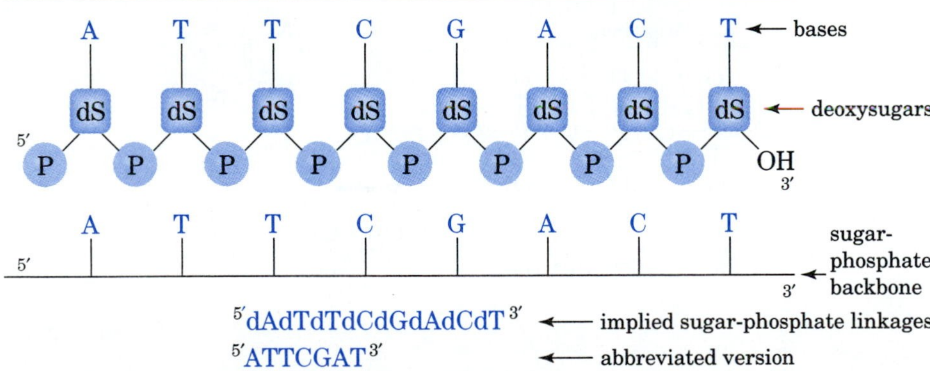

The result is an **oligonucleotide** (just a few units), a **polynucleotide,** or a nucleic acid. You should become familiar with each way to represent polynucleotides. Notice that in the bottom representation above that deoxyribose (dS) is not indicated. The presence of thymine (T) is enough to identify the sequence as DNA.

In all of the phosphates linking the nucleotides, one -OH group remains *underivatized.* The high K_a for this group allows it to deprotonate at physiological pH, producing an anion (−). This means that the phosphate-sugar backbone is highly negatively charged and hydrophilic.

oligonucleotide
polymer containing a few nucleotide units

polynucleotide
polymer containing more than a few nucleotide units

GETTING INVOLVED

- ✓ Why are the purines and pyrimidines categorized as bases?
- ✓ What are two major differences between the structures of a ribonucleotide and a deoxyribonucleotide?
- ✓ Be sure that you know how to link a purine or a pyrimidine base to a ribose unit and then to a phosphate. Draw the structure of mononucleotide with uracil.
- ✓ Draw the structure of a dinucleotide containing a uracil mononucleotide and a cytidine mononucleotide with a 5′, 2′-phosphodiester link.

Problem 17.1

Compare the functional groups in carbohydrates, proteins, lipids, and nucleic acids. How are they alike and how do they differ?

Problem 17.2

Draw three other representations for the structure of the polydeoxynucleotide **GTCC.**

Problem 17.3

Write the complete structure for the polyribonucleotide **UCAG.**

See related problems 17.10, 17.13, 17.14.

17.2 Other Structures Involving Nucleotides

A. Energy Intermediates

ATP, ADP, AMP
adenosine tri-, di-, and monophosphate; energy carriers in metabolism

Nucleotide di- and triphosphates contain high-energy phosphate anhydride bonds that are made during metabolic catabolism (nutrient breakdown) and used during the process of biosynthesis. **Adenosine triphosphate (ATP)** is the best known and most ubiquitous of these molecules, although guanosine and cytidine triphosphates are also important in metabolic processing.

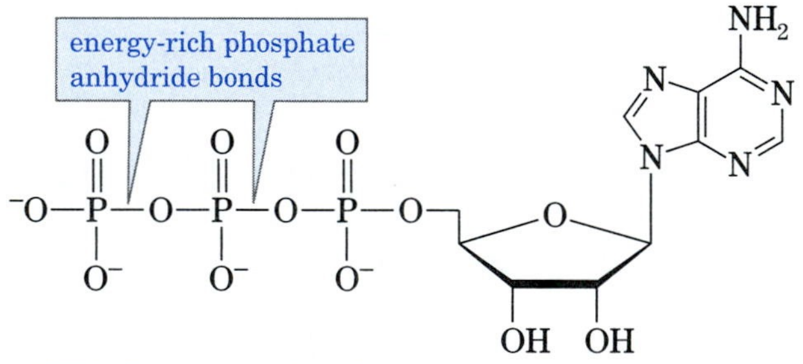

energy-rich phosphate anhydride bonds

ATP (adenosine triphosphate)

ADP

AMP

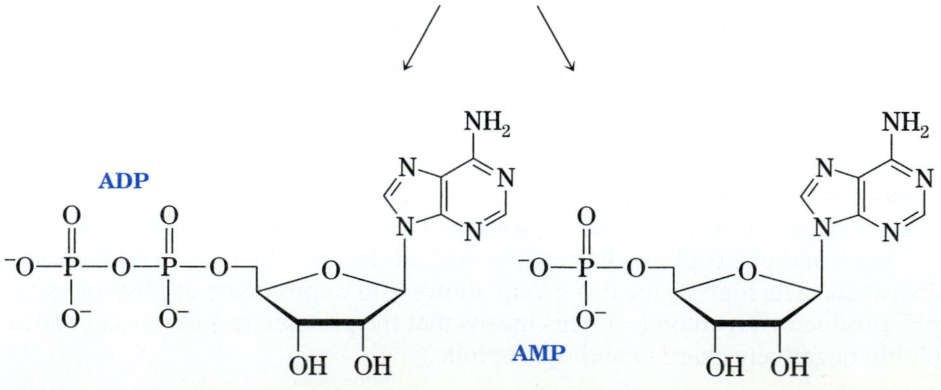

GETTING INVOLVED

✓ Draw the structures for UDP and CMP.

Problem 17.4

A product of the hydrolysis of ATP to AMP is inorganic pyrophosphate (PP_i) or $HP_2O_7^{3-}$, wherein the anhydride bond between the two phosphates has not been broken. Draw the structure of PP_i in its ionized state.

B. Chemical Messengers

The communication of hormone- and nerve-mediated signals can also involve the formation of intracellular messengers known as cyclic nucleotides. 3′,5′-cyclic AMP or **cAMP,** and cGMP are such biomessengers.

cAMP
cyclic adenosine monophosphate; chemical messenger

3′,5′-cyclic adenosine monophosphate

cAMP

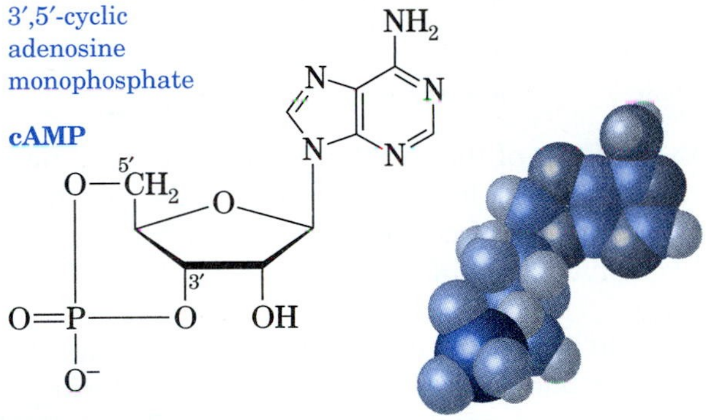

GETTING INVOLVED

Problem 17.5

Draw the structure of 3′,5′-cyclic guanosine monophosphate, cGMP. Should this molecule be acidic, basic, or neutral? Explain your answer.

See related problem 17.15.

C. Redox Factors—Nucleotide Vitamins

Several variations of nucleotides participate in enzyme-catalyzed reactions as cofactors. They contain water-soluble vitamins; that is, they are organic compounds that are essential to life (vitamin), water-soluble, not synthesized within the body, and obtained through the diet. One of these is nicotinamide, not related to nicotine, which is found joined to AMP as nicotinamide adenine dinucleotide or NAD. NAD has two redox forms: **NAD^+** (shown below) and **NADH** in which the oxidized NAD^+ has undergone a hydride reduction at a position para to the nicotinamide ring nitrogen.

NAD^+/NADH
nicotinamide-adenine dinucleotide (oxidized/reduced forms); oxidation-reduction cofactor in metabolism

NAD^+
nicotinamide adenine dinucleotide

nicotinamide

adenine

nucleotide

nucleotide

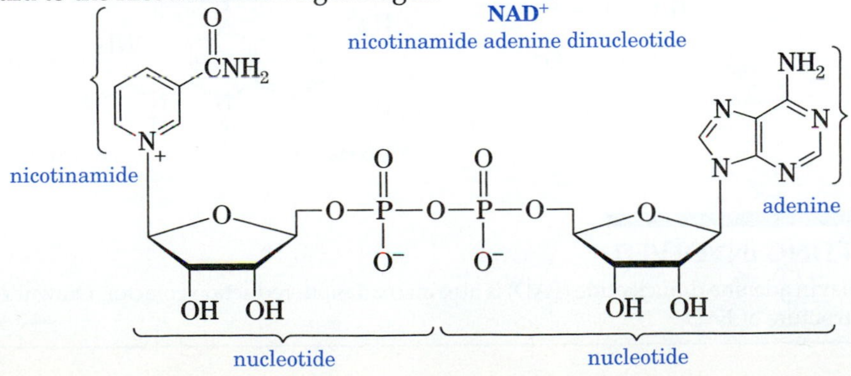

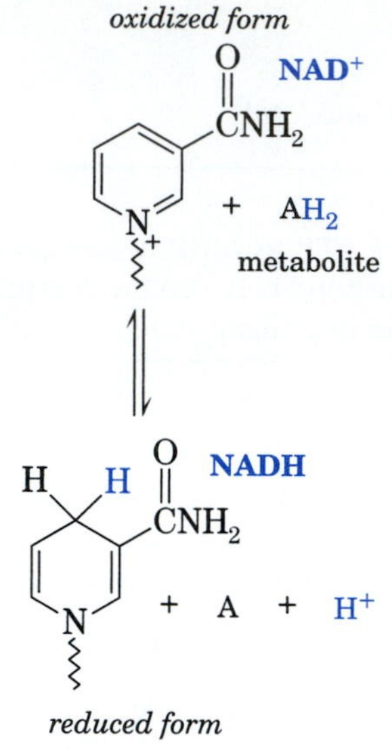

oxidized form

NAD$^+$

CNH_2 + AH_2

metabolite

NADH

CNH_2 + A + H^+

reduced form

Nicotinamide, in the form of the pyridine carboxylic acid niacin, is found in yeast, meats, and wheat germ. Its absence from the diet results in pellagra. Pellagra's symptoms include diarrhea, indigestion, and dermatitis. It can be fatal if left untreated. Excessive intake of niacin causes flushing of the skin and may lead to liver damage.

FAD/$FADH_2$
flavin adenine dinucleotide (oxidized/reduced forms); oxidizing reducing agents

Riboflavin, vitamin B_2, consists of a heterocyclic base called flavin and the reduced form of ribose, ribitol. In the body riboflavin can be joined with a phosphate group to form flavin mononucleotide (FMN). FMN can also be linked with AMP to produce flavine adenine dinucleotide **(FAD).** Both FMN and FAD can undergo reversible oxidation-reduction with the elements of molecular hydrogen adding in a 1,4 manner to the part of the system shown here. The products are abbreviated as $FMNH_2$ and **$FADH_2$.**

A deficiency of B_2 produces dermatitis of the face, an inflammed tongue, and eye disorders.

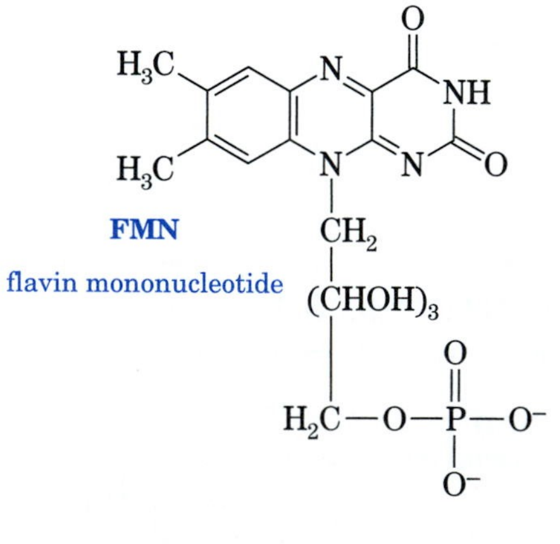

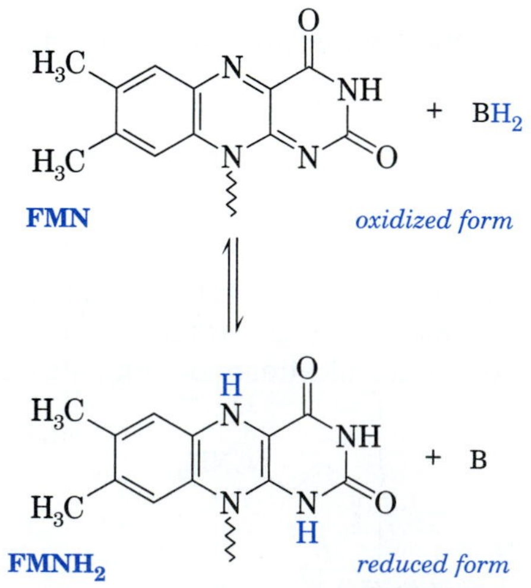

H_3C H_3C NH CH_2 $(CHOH)_3$ $H_2C-O-P-O^-$

FMN

flavin mononucleotide

FMN + BH_2

oxidized form

$FMNH_2$ + B

reduced form

GETTING INVOLVED

✓ Flavin adenine dinucleotide (FAD) is also an oxidation–reduction cofactor. Draw the structure of FAD.

17.3 The Hierarchy of Nucleic Acid Structure

In Chapter 16 we saw that proteins have several levels of superstructure that depend on the ability of various functional groups to participate in covalent and noncovalent interactions. The most important noncovalent intermolecular interaction is hydrogen-bonding. The same types of interactions are exhibited by nucleic acids. The bases establish hydrogen-bonding patterns that result in the well-known double-stranded helix of DNA. Hydrogen-bonding between bases is also used to direct the replication of genetic material and the transcription and translation of the DNA coded message for the production of proteins through RNA.

A. DNA Structure: The Double Helix

It had been known prior to 1953 that the mole ratio of adenine to thymine and guanine to cytosine was usually 1, no matter the source of the DNA. The numbers of the individual bases varied, but that ratio essentially remained the same. The reason for this depends on a recurring phenomenon in chemistry, hydrogen-bonding. Looking at the structures for the bases, we can see that the opportunity for hydrogen-bonding exists since there are electronegative oxygens on the carbonyl groups, electronegative ring nitrogens, and electropositive hydrogens on the amine or imine groups.

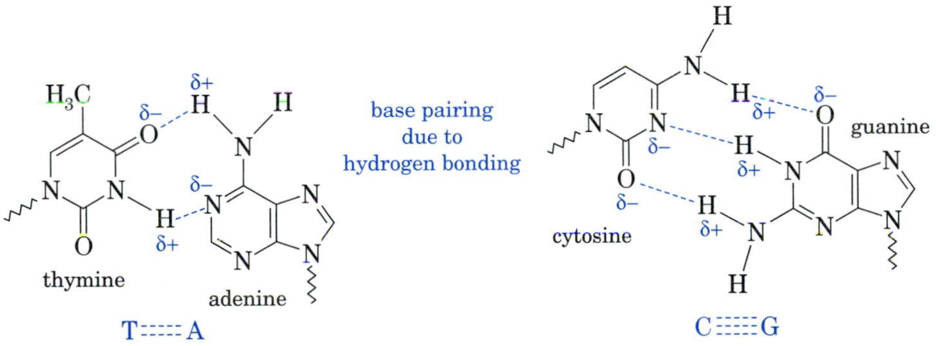

thymine — adenine

T:::::A

base pairing due to hydrogen bonding

cytosine — guanine

C:::::G

The hydrogen-bonding between adenine and thymine, guanine and cytosine, is called complementary **base-pairing.** Maximum base-pairing occurs when A and T are joined by two hydrogen bonds and G and C by three.

base-pairs
complementary bases that can hydrogen-bond to each other; A === T, G ≡≡≡ C, A === U

The actual physical orientation of the entire DNA polymer was not known until 1953, when James Watson, Francis Crick, and Maurice Wilkins interpreted X-ray data (produced by Rosalind Franklin) to indicate a **double-stranded helix.** The Watson–Crick hypothesis, for which the three won the Nobel Prize in 1962, shows two complementary hydrogen-bonded strands aligned in an antiparallel manner.

double helix
Watson–Crick model of DNA in which the heterocyclic bases are oriented toward the interior axis and the sugar-phosphate backbone on the outside of the helix

The two polynucleotide chains are twisted around each other with the bases oriented toward the center axis of the helix and the sugar-phosphate backbone on the outside of the helix, exposed to the aqueous environment of the cell. There are three commonly known forms of the helix called A, B, and Z. These differ in the number of water molecules interacting with the helix as well as in the orientation of the bases to the center of the helix, rise of each turn, and overall handedness of the helix (Figure 17.1). The B helix is the one assumed to exist in water solution and was proposed by Watson and Crick. It is right-handed and rises 34 angstroms per turn ($1\ \text{Å} = 10^{-8}$ cm) or 3.4 nanometers, containing ten nucleotide bases per complete turn. Two grooves appear in the overall structure, one wider than the other. The wider is known as the *major groove,* whereas the smaller is the

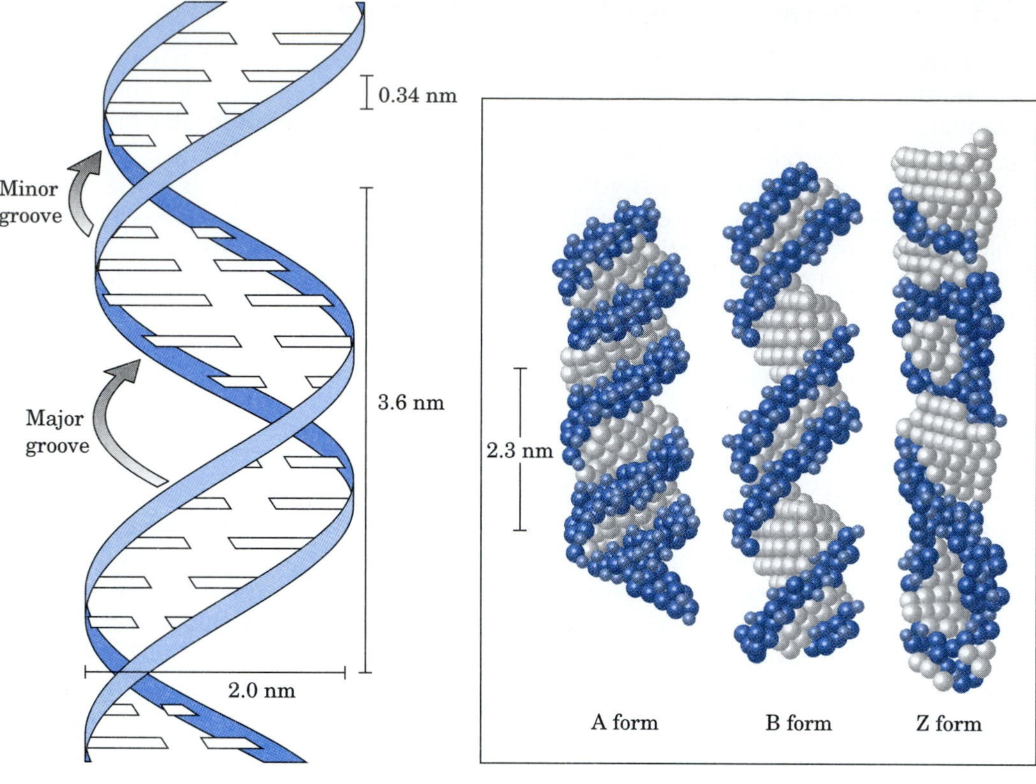

(*a*) General architecture

(*b*) A, B, and Z forms

Figure 17.1 The double helix of DNA. (Adapted from Lehninger, Nelson, and Cox, *Principles of Biochemistry,* 2nd ed. Used with permission.)

minor groove. Other biomolecules interact with DNA in these grooves, helping in its function. For example, basic proteins known as histones stabilize DNA by forming charged (+/−) complexes with it. Single-stranded DNA (ssDNA) does exist in certain organisms but is not common.

genome
the entire genetic makeup of an organism

supercoil
form of compacted DNA

histone
basic protein associated with nucleic acids in eukaryotic organisms

DNA is the material of the **genome,** or hereditary material, of all living organisms from bacteria to human beings. Since the entire human genome, all estimated 100,000 genes, must fit into each cell, there must be much more efficient packing of a double helix. DNA can form into circular rings in lower organisms and is linear in higher organisms. The helices, whether in rings or linear form, can also intertwine, forming **supercoils** (Figure 17.2). In fact, the strain that is induced in superwound coils can aid in transferring genetic information. Supercoiling as well as wrapping around **histone** proteins allow for the compaction necessary to fit the total length of DNA into a single cell.

GETTING INVOLVED

- ✓ RNA contains uracil, which forms base pairs with adenine. Draw a uracil–adenine base pair, indicating the possible sites of hydrogen-bonding.
- ✓ What types of macro- (large) structures are formed by DNA?
- ✓ What is a genome?
- ✓ Briefly mention a few ways in which DNA differs from RNA in its fundamental chemical structure and in its superstructure.
- ✓ Describe the process of base-pairing in DNA
- ✓ How does base-pairing support the theory of Watson and Crick?
- ✓ Is the double helix the largest assembly possible for DNA?

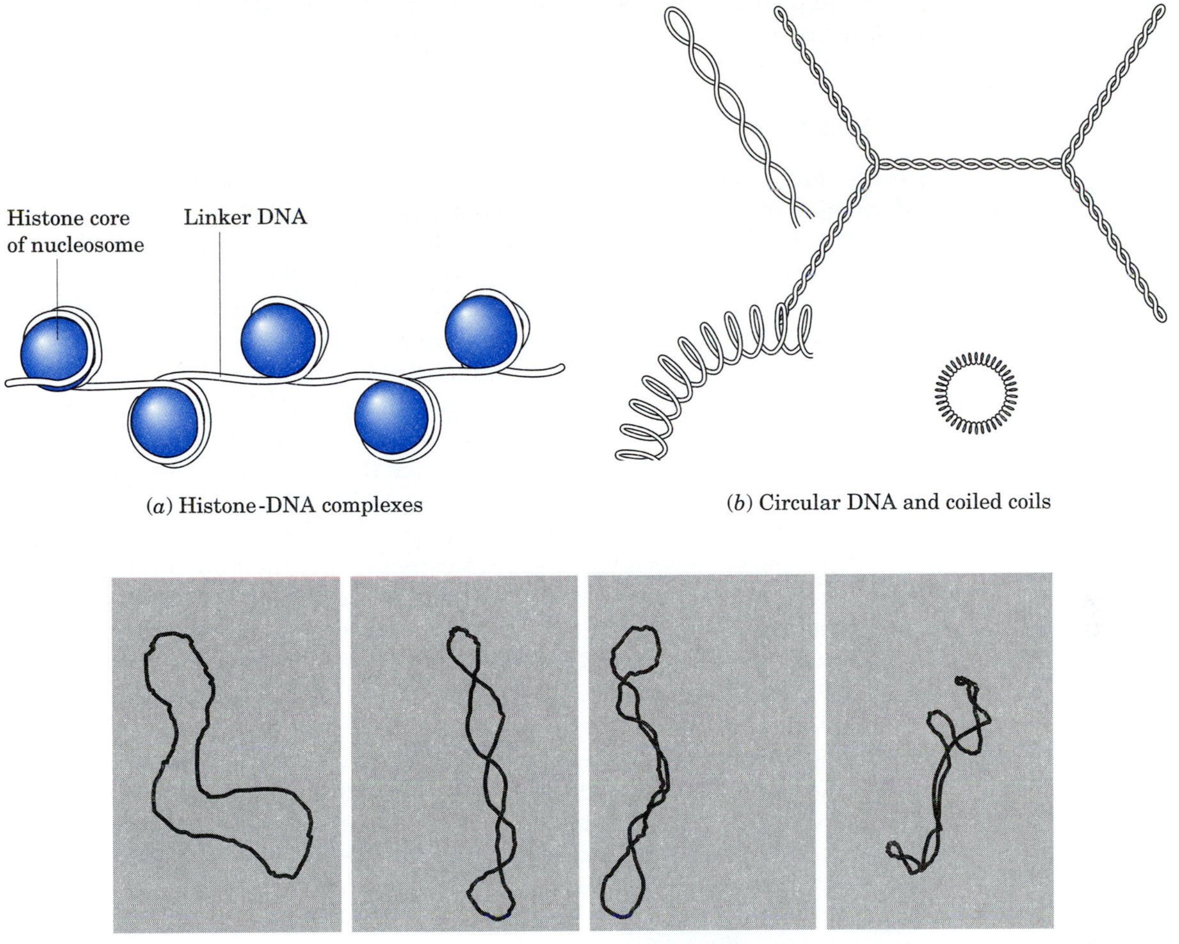

Figure 17.2 Compaction of nucleic acids through coils and histone interactions. (Adapted from Lehninger, Nelson, and Cox, *Principles of Biochemistry*, 2nd ed. Used with permission.)

Problem 17.6

If the histones are basic proteins, which amino acids should occur in large proportion in them? With which negative groups of the polynucleotide will the histones interact?

B. RNA Structure

Although double-stranded RNA (dsRNA) does exist, it is not common. Most RNA is single-stranded (ssRNA), forming a greater variety of superstructures than DNA, which suits its different roles. There are three general types of RNA: messenger RNA **(mRNA)**, ribosomal RNA **(rRNA)**, and transfer RNA **(tRNA)**. We will discuss the overall structures of RNA as we explain their functions.

mRNA
messenger RNA; contains codons for the construction of protein

rRNA
ribosomal RNA; RNA associated with proteins to form the ribosome

tRNA
transfer RNA; brings amino acids to the ribosome for protein synthesis; contains anticodons

17.4 The Genetic Code

DNA, found primarily in the nucleus but to a small extent in mitochondria and chloroplasts, is the ultimate carrier of the genetic code encrypted in the sequence of its bases. RNA acts as a transcribing agent, copying the nuclear DNA message and carrying it to the cytoplasm, where amino acids are assembled into the correct sequence. The proteins that result catalyze the vital reactions and serve in all the other functions mentioned in Chapter 16. The following section gives a brief overview of this process.

A. DNA Replication

replication
process of duplicating DNA

In order for a single cell to grow into a complete organism that passes on genetic information to ensuing generations, it is necessary that DNA be able to reproduce itself. This process, known as **replication,** first involves the uncoiling of the double helix, which occurs in sections. Once uncoiled, a DNA strand is base-paired to the corresponding dNTPs (deoxyribonucleotide triphosphates). Since nucleic acid biosynthesis occurs continuously from the 5′ to the 3′ end, only one strand of the DNA helix is made in one continuous polymer (this will complement the 3′ --------→ 5′ original DNA strand). The second strand of the opened DNA helix is replicated from shorter 3′ --------→ 5′ fragments produced from the base-pairing to the second original DNA strand. Lengths of RNA known as **primers** serve as starting points for DNA formation. After the DNA polymer has been added to the primers, the primers are eventually removed.

primer
lengths of RNA that serve as starting points for DNA formation

semiconservative replication
DNA that is composed of one parent strand (template) and one daughter strand (formed from base-pairing) with a new set of bases

DNA replication is said to be **semiconservative** in that the second generation double helix is composed of one "parent" or original DNA strand and one "daughter" strand (see Figure 17.3). In conservative replication the parent strands would have recombined, and the daughter strands would have formed an entirely new DNA double helix.

ligase
enzyme that connects pieces of polynucleotides

Replication proceeds through a complex interplay of DNA, RNA, deoxyribonucleotide triphosphates, enzymes, and other biochemical factors. Enzymes include DNA polymerases, helicases, primases, and **ligases,** which not only sew together the genetic fragment but also ensure an extraordinary fidelity in replication. Sequences are proofread and can be corrected as the DNA is incorporated into a new copy of the genome. Binding proteins help to keep the helix open.

GETTING INVOLVED
Problem 17.7

Attempt to solve a DNA sequence given the following information from the electrophoresis results of a Sanger-type experiment (See Connections 17.1).

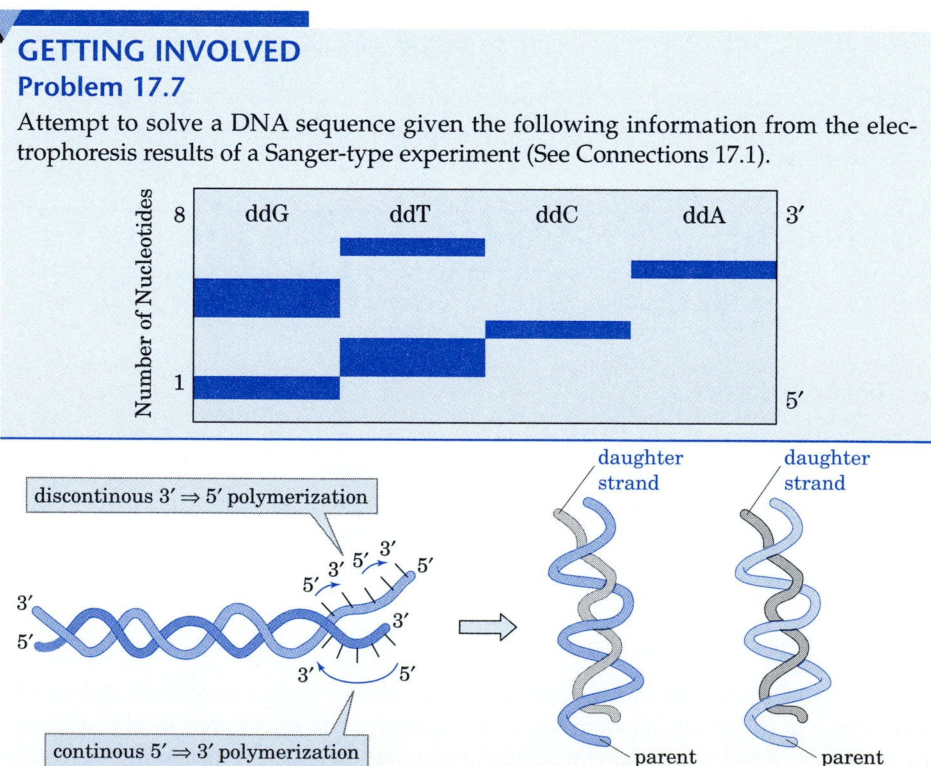

Figure 17.3 Semiconservative mode of replication of DNA.

Connections 17.1 www.prenhall.com/bailey

The Human Genome Project

The scientific community embarked on one of its greatest projects during the last decades of the twentieth century. The quest was to decipher the DNA sequence for the entire human genome—all 3 billion base pairs located in the genes of the 46 chromosomes. It will be left to the scientists of the twenty-first century to determine the functions of the protein products coded in the DNA. As might be expected, the DNA sequences of simpler organisms have already been under investigation and their results will lead the way to the ultimate goal.

The sequencing process is a combination of chemical and enzymatic methods. Two of the earliest procedures were that of Sanger and that of Gilbert and Maxam.

In the Sanger process a template is synthesized that is complementary to a single strand of DNA (see diagram in Problem 17.7). DNA polymerizing enzymes facilitate making the primer copy. This template is then used in further DNA replication; however, at this point small amounts of specific 2′,3′-dideoxyribonucleotides are added (ddATP, ddGTP, ddCTP, ddTTP) along with the normal 2′-deoxynucleotide triphosphates (dNTPs). Without the 3′ hydroxy group the ddNTPs will stop the lengthening chain as they are incorporated. The result is a mix of polynucleotides of varying lengths. These polymers can be separated using electrophoresis on the basis of their molecular weights. The shorter fragments represent the 5′ end of the primer while the longer segments are complements to the 3′ end. In the following diagram, try to coordinate the electrophoretic pattern with the fragments and then the original sequence.

The Gilbert–Maxam method uses specific radioactive labels to "tag" areas of the DNA and then applies restriction endonucleases, enzymes that cleave the DNA chain at specific nucleotides, to break the polymer into smaller fragments. The radioactive markers help in the elucidation of the sequence.

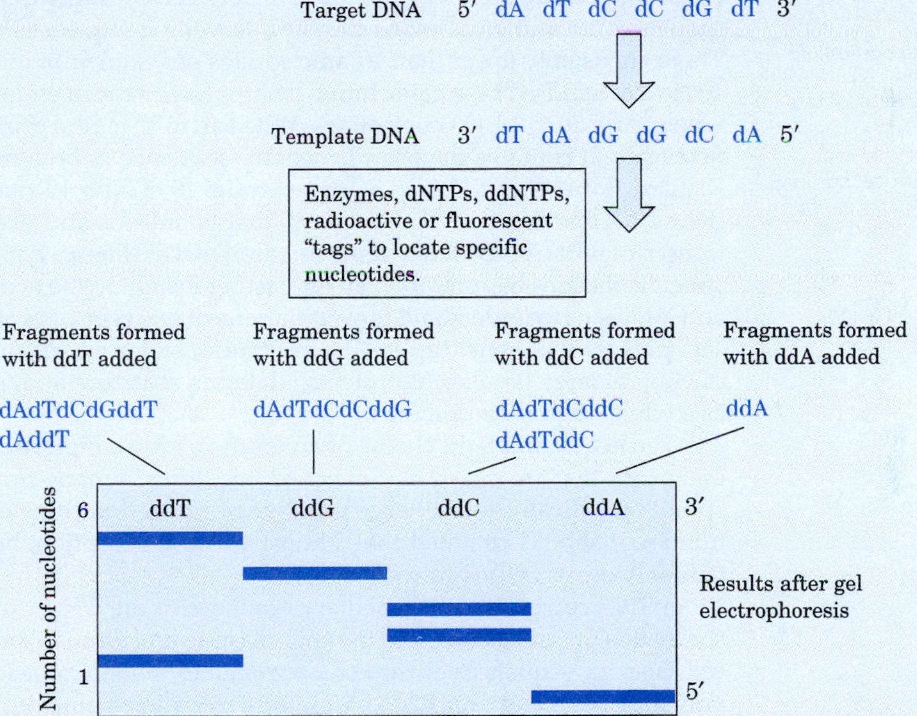

The ability to locate specific loci in chromosomes has produced many diagnostic tests for genetic conditions such as sickle-cell anemia, cystic fibrosis, Huntington's disease, and various types of cancer. Gene therapy techniques are in their infancy and hold great promise for the twenty-first century.

B. Transcription and Translation

transcription
process of making mRNA by complementary base-pairing of ribonucleotides with a piece of DNA chain

translation
process that involves mRNA binding to ribosomes and base-pairing with specific tRNAs holding amino acids; the end product is a protein

sense (+) DNA strand
strand of DNA double helix that is transcribed to mRNA

antisense (−) DNA strand
strand of DNA double helix that is not transcribed to mRNA

exon
expressed sequence; portion of DNA (and mRNA) that is transcribed and translated into protein

intron
an intervening sequence; portion of DNA (and mRNA) that is not transcribed and translated into protein

The processes whereby the genetic code is interpreted to form protein are called **transcription** and **translation.** First the DNA sequence is transcribed into messenger RNA (mRNA) in the nucleus. This process involves the base-pairing of ribonucleotide triphosphates (NTPs) with an unwound portion of the template or (+) or **"sense" strand** of the DNA helix and then enzyme-catalyzed polymerization. Only one DNA strand is transcribed at a time. The untranscribed strand is called the coding, (−), or **antisense strand,** because its sequence will be the same as that for the mRNA produced, with the substitution of a U for a T. The DNA sequence on the template strand is read 3′ to 5′ while the mRNA is synthesized 5′ to 3′. However, when correlations are made between the mRNA and its parent DNA, it is the antisense (−) strand that is usually referred to (see Figure 17.4).

The DNA code for a protein is usually found in several locations, either along one chromosome or on separate chromosomes. Not all of the DNA sequence codes for protein. Some segments are in between coding sequences. The coding sequences are known as **exons** and the intervening sequences are called **introns.** There are usually fewer than 10^3 nucleotides per exon with most in the range of 100 to 200 hundred base pairs. Intron lengths have a much wider variation of anywhere from 50 to 20,000 nucleotides. When an mRNA (the primary transcript) is first made it contains the complementary sequence of both exons and introns. Richard Roberts and Phillip Sharp won the 1993 Nobel Prize in medicine for their 1977 discovery of "split genes," that is, introns and exons. The primary transcript mRNA is edited to remove intron pieces (Figure 17.5). The cutting and splicing of exons has provided an interesting insight to the evolution of proteins and allows us to understand how variations of one type of protein can be found. The processes of gene duplication, mutation, and gene fusion lead to the production of large families of proteins related in structure and/or function either as a whole or in their domains.

The heavy and light chains of antibodies, for example, are made from several exons that are mixed and matched, resulting in many proteins that can respond specifically to the large number of nonself entities encountered by a human. It should be noted that, almost without exception, bacterial cells contain only exons and no introns.

mRNA carries a complementary sequence to the DNA exons onto the ribosomes that are associated with the endoplasmic reticulum in the cytoplasm of the cell. Specific sequences of three bases relate to the amino acid. Since there are only four DNA and four RNA bases, the correlation cannot be one-to-one. Even a two-to-one relationship would produce a code for only 16 of the 20 common

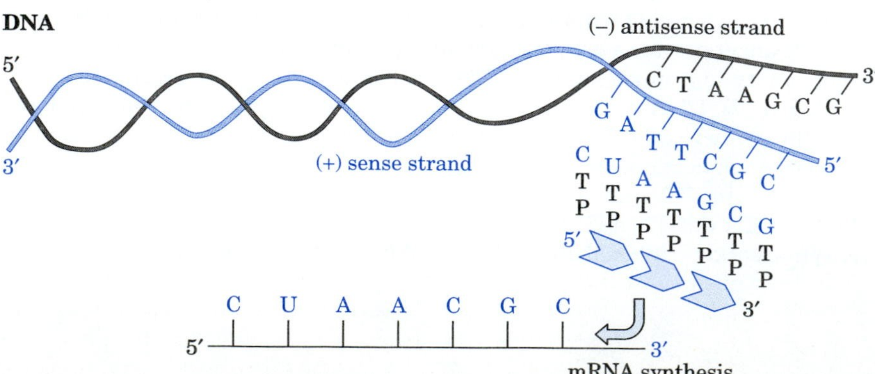

Figure 17.4 The transcription of DNA to RNA.

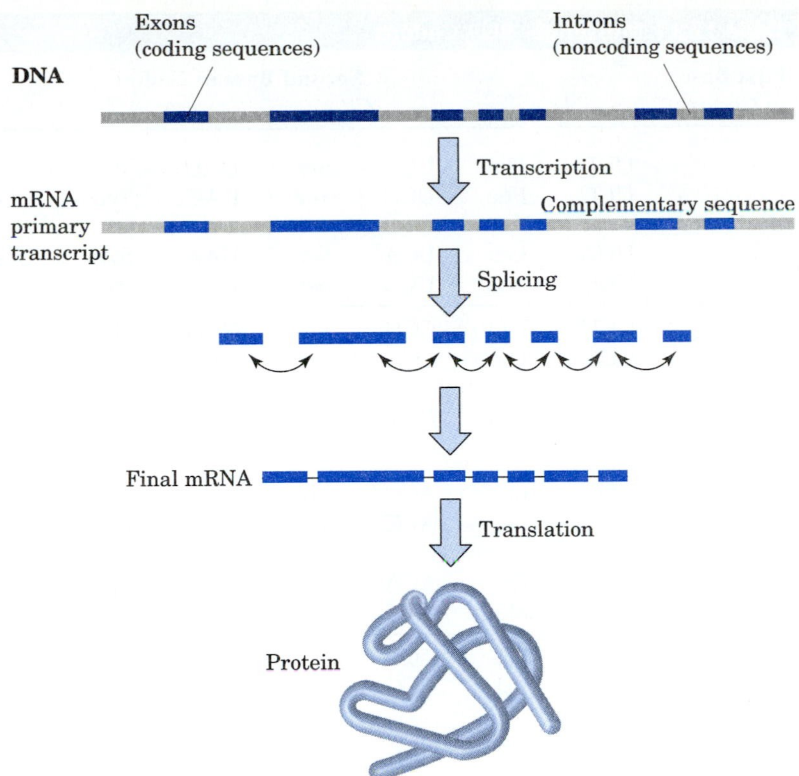

Figure 17.5 The removal of intron sequences from mRNA.

amino acids. However, at three-to-one, 64 different combinations are possible. This means that some amino acids may have more than one "code word." In actuality there are 61 amino acid **codons** (in mRNA) and 3 codons for initiation and termination of the protein chain (sometimes called *nonsense codons*). See Table 17.1 for a list of mRNA codons and their corresponding amino acids or start/stop directions. The availability of more than one codon for most amino acids is called the **degeneracy** of the code. Looking at Table 17.1, one can see that the first base in codons for the same amino acid is usually the same or almost so. The third can be highly variable and is referred to as the "wobble" base.

codon
three-base polynucleotide sequence of mRNA corresponding to an amino acid or protein synthesis directive (start or stop)

degeneracy
characteristic of the DNA code; there is more than one three-base code for most amino acids

The codons found in bacteria are the same as those seen in higher organisms. It seems then that the genetic code is nearly universal for all organisms, whether prokaryotic or eukaryotic.

Ribosomes are large complexes consisting of ribosomal RNA (rRNA), protein, and cofactors. Each ribosome has two major subunits, 50s and 30s in prokaryotes (lower organisms) and 60s and 40s in eukaryotes (higher organisms). Ribosomes provide the environment for translation of the nucleic acid code to a protein amino acid sequence. Ribosomal RNA (rRNA) provides a scaffolding upon which enzymes can interact with the key factors in the manufacture of proteins. mRNA locates itself in a cleft formed by two major portions of the ribosome. From 10 to 100 ribosomes can be associated along one strand of mRNA giving rise to a polysome. More than one protein molecule can thereby be synthesized simultaneously.

ribosome
RNA-protein complexes that serve as the environment for protein synthesis

The third major type of RNA, transfer RNA (tRNA), joins the ribosome-mRNA super complex, carrying with it amino acids for the protein biosynthetic process. tRNA is single-stranded and has a three-dimensional structure that appears elongated and L-shaped, stabilized by hydrogen bonding between base pairs within the molecule. At the 3′ arm of the molecule an amino acid is attached.

Table 17.1 Messenger RNA Codons

First Base in Codon	Second Base in Codon: U		C		A		G	
U	UUU	Phe	UCU	Ser	UAU	Tyr	UGU	Cys
	UUC	Phe	UCC	Ser	UAC	Tyr	UGC	Cys
	UUA	Leu	UCA	Ser	**UAA**	**Stop**	**UGA**	**Stop**
	UUG	Leu	UCG	Ser	**UAC**	**Stop**	UGG	Trp
C	CUU	Leu	CCU	Pro	CAU	His	CGU	Arg
	CUC	Leu	CCC	Pro	CAC	His	CGC	Arg
	CUA	Leu	CCA	Pro	CAA	Gln	CGA	Arg
	CUG	Leu	CCG	Pro	CAG	Gln	CGG	Arg
A	AUU	Ile	ACU	Thr	AAU	Asn	AGU	Ser
	AUC	Ile	ACC	Thr	AAC	Asn	AGC	Ser
	AUA	Ile	ACA	Thr	AAA	Lys	AGA	Arg
	AUG	Met	ACG	Thr	AAG	Lys	AGG	Arg
G	GUU	Val	GCU	Ala	GAU	Asp	GGU	Gly
	GUC	Val	GCC	Ala	GAC	Asp	GGC	Gly
	GUA	Val	GCA	Ala	GAA	Glu	GGA	Gly
	GUG	Val	GCG	Ala	GAG	Glu	GGG	Gly

anticodon
three-base polynucleotide sequence of tRNA that base-pairs with a specific codon

There are more than 30 different tRNAs for the 20 common amino acids. Some tRNAs are more amino acid-specific than others. Located at a polynucleotide loop on the other end of the tRNA is a three-base **anticodon** sequence complementary to the three-base codon on the mRNA (Figure 17.6).

The amino acid-bearing tRNA becomes noncovalently attached to a ribosomal site called the "P" site. For bacteria the "start" codon is for an N-formyl methionine (methionine with a formyl group at the amino N), whereas for eukaryotic organisms the sequence begins with the codon for a methionine. A second tRNA then binds to an adjacent "A" site on the ribosome-mRNA complex.

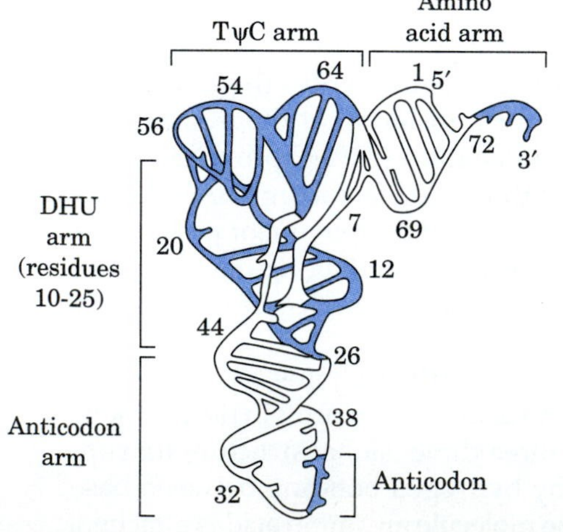

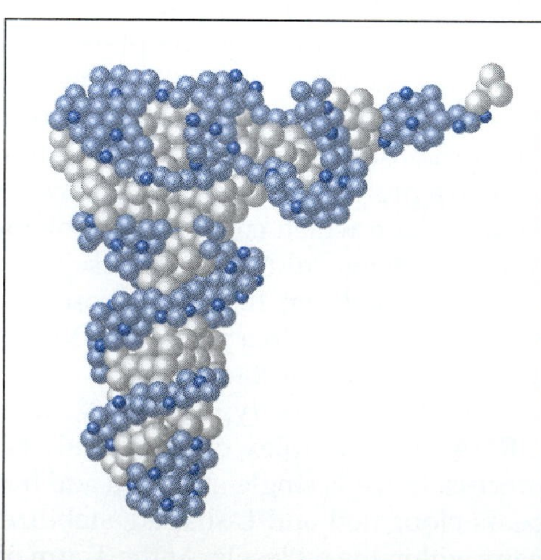

Figure 17.6 Models of transfer RNA (tRNA). (Adapted from Lehninger, Nelson, and Cox, *Principles of Biochemistry*, 2nd ed. Used with permission.)

The first amino acid, which will be the N-terminus of the protein chain, is linked to the second. This results in the liberation of the first tRNA from the P site and the movement of the A site tRNA to the P site, now with a dipeptide attached to the P site. A third amino acid-carrying tRNA binds to the vacant A site and the procedure is repeated. In this manner the protein grows until a "stop" codon is encountered. Then the protein is released from the complex and is transported to cellular areas for modification and/or incorporation into the cell matrix (see Figure 17.7).

Most proteins are biosynthesized with an N-terminal hydrophobic sequence of 15 to 30 amino acids, called the **signal sequence,** which facilitates the targeting of a protein as well as its passage through the membranes of organelles, where storage or posttranslational modification takes place. As the protein is threaded through the membrane, the signal sequence is enzymatically removed.

signal sequence N-terminal protein sequence

Alterations in the nascent protein are called **posttranslational modifications** and may include removal of the N-formyl methionine or methionine that started the chain, addition of carbohydrate (glycosylation), methylation, esterification, phosphorylation, isoprenylation, or cleavage of the single chain into multiple chains to produce a fully functional protein.

posttranslational modifications chemical changes made on a completed protein such as the addition of lipid or carbohydrate or the cleavage of the polypeptide chain

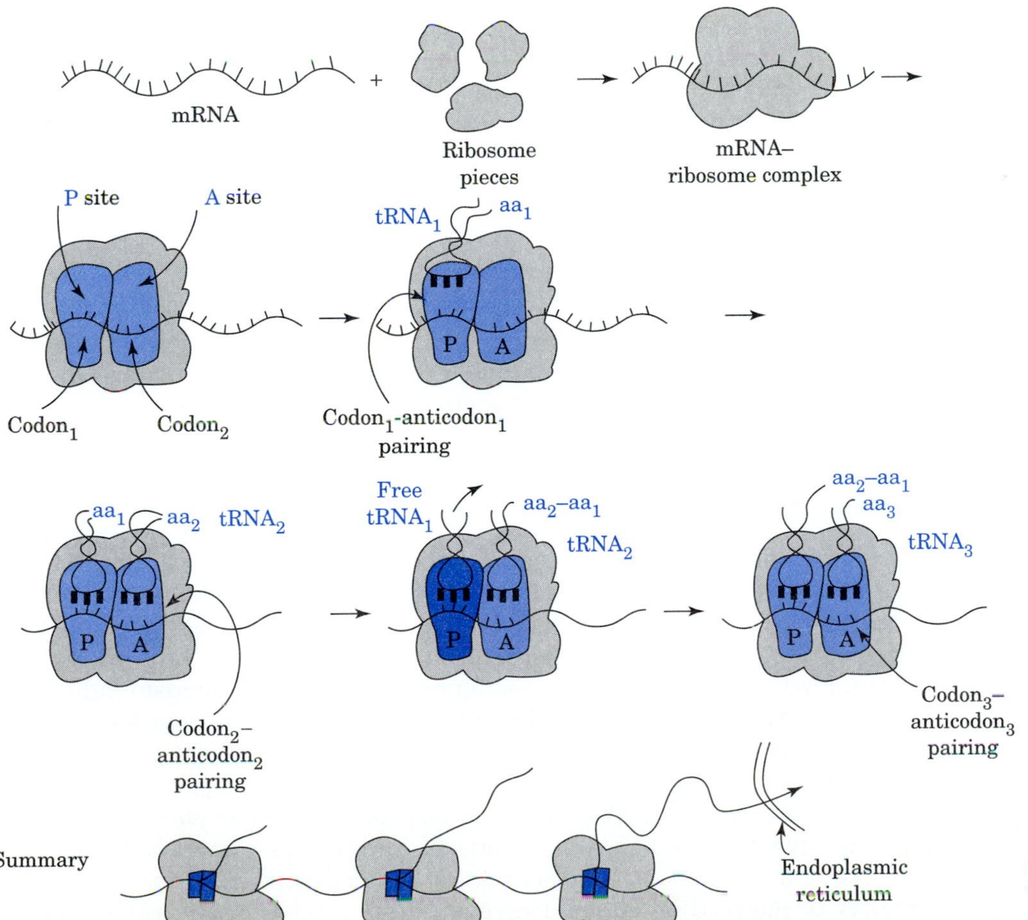

Figure 17.7
Translation of mRNA into protein.

GETTING INVOLVED

✓ What is meant by a codon? anticodon?
✓ What makes the genetic code "degenerate"?
✓ Which amino acid codons are not degenerate?

Problem 17.8

What would be the tRNA anticodon sequences for the following mRNA codon: 5′-GGU ACU CCC UGA-3′? Write the tetrapeptide that is being coded. What is the original DNA sequence for both the sense and antisense strands?

Problem 17.9

Mistakes in making mRNA occur by inserting or deleting bases.

(a) What would happen to the polypeptide sequence coded for in Problem 17.8 if an A were inserted in between the A and C of the second codon or the A were deleted from the second codon?
(b) What would happen if the codon GUG was inserted between the third and fourth codons?

See related problems 17.10, 17.16, 17.17.

Dolly, a sheep cloned from an existing animal

17.5 Characteristics of Transcription and Translation

There are several key points that should be remembered about the genetic code and its direction of protein synthesis.

1. The genetic code is nearly universal. The three-base mRNA codons and their anticodons can be found in prokaryotic as well as eukaryotic organisms.
2. The code is degenerate. Most amino acids have more than one codon.
3. One RNA base sequence is usually read in the same way to produce the same protein in a repeatable manner, that is, there are no overlapping codes. There are a few exceptions to this feature, but it generally holds true.
4. There is a great deal of reliability in the process, but mutations can occur that may or may not lead to viable proteins.
5. The biosynthesis of proteins is energy-consuming.

17.6 Mutation of DNA

mutation
a change in the nucleotide sequence of a DNA molecule which may or may not lead to an alteration in protein structure and/or function

The structure of nucleic acids is sensitive to chemical and physical factors that are present naturally or may be introduced into cells through the environment. During the process of metabolism, free radicals are generated that may affect the reactivity of the nucleotide bases. Dimers of adjacent bases such as thymine occur. Hydrogen-bonding patterns can be altered by tautomerism induced by exposure to radiation or chemical agents. In addition, incorrect purine or pyrimidine bases may be inserted or bases may be deleted during the replication or transcription process, giving rise to stop codons, shifts in the reading frame, or substitutions that change the nature of the protein to be biosynthesized.

An organism has natural mechanisms to deal with many such mutations, for example, excision and replacement of dimers and double-checking the integrity of the reading frame. However, these mechanisms cannot cover all changes and can be overwhelmed when faced with a "flood" of mutational

Connections 17.2

www.prenhall.com/bailey

Acquired Immune Deficiency Syndrome (AIDS)

We are all aware of the challenge that AIDS presents to the entire world community. Caused by the human immunodeficiency retrovirus (HIV), this condition results in the collapse of the immune system. The mode of transmission is usually sexual or involves a mingling of blood between a host and a subject such as occurred in blood transfusions during the 1970s or can still occur in intravenous drug addicts through shared needles. Newborns can be infected with HIV from their mothers. HIV has two major forms, called I and II, in which the former was originally prevalent in the homosexual community in the United States and the latter was spread heterosexually in Third World countries. Today both forms are ravaging many continents.

HIV invades a host organism and, as a retrovirus, incorporates its genome into the host DNA. After a period of time the virus destroys the T_4 cells of the immune system, leaving the body open to opportunistic infections such as pneumonia caused by *Pneumocystis carinii* and/or tuberculosis. Many of those afflicted eventually succumb to a form of cancer known as Kaposi's sarcoma.

The lives of HIV victims have been extended by drug treatment, but an AIDS vaccine is still in the future. As more is learned about HIV, its structures and modes of reproduction and action, research has produced veritable "cocktails" of drug mixtures whose targets attempt to stop the invasion of the virus as well as its incorporation into the host genome. There are three general classes of medications: reverse transcriptase inhibitors (RTIs), non-nucleoside reverse transcriptase inhibitors (NNRTIs), and protease inhibitors. The first two types slow the incorporation of viral RNA into the host DNA. The third type targets a protease enzyme used in the reverse transcription process. Further research into the protein coat of HIV is leading to new methods of approaching drug treatment. The structures of some RTIs and NNRTIs are shown below.

Reverse transcriptase inhibitors

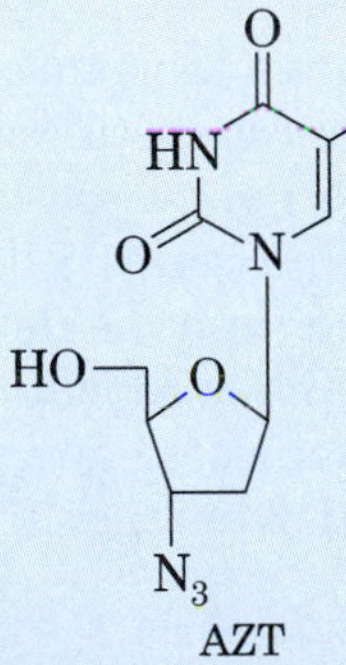

AZT
3′ α-azido-2′3′-dideoxythymidine
(Retrovir®)

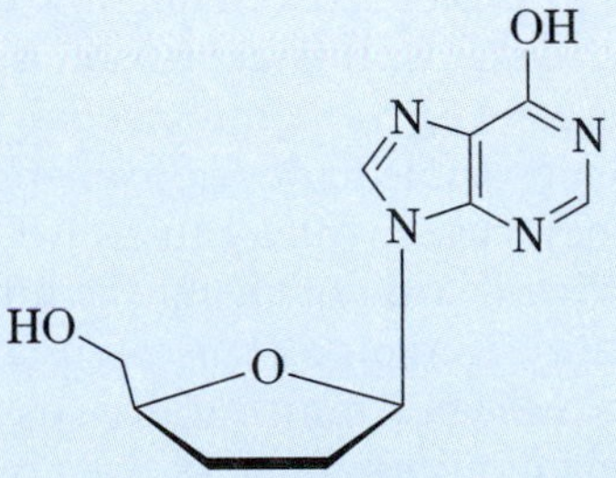

ddI
dideoxyinosine
(Videx®)

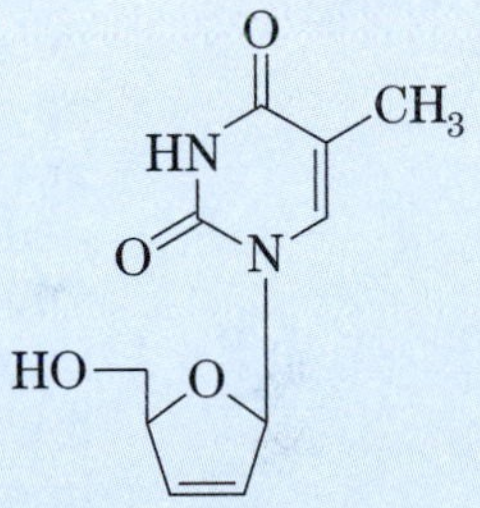

3TC
(Epivir®)

Non nucleotide reverse transcriptase inhibitors

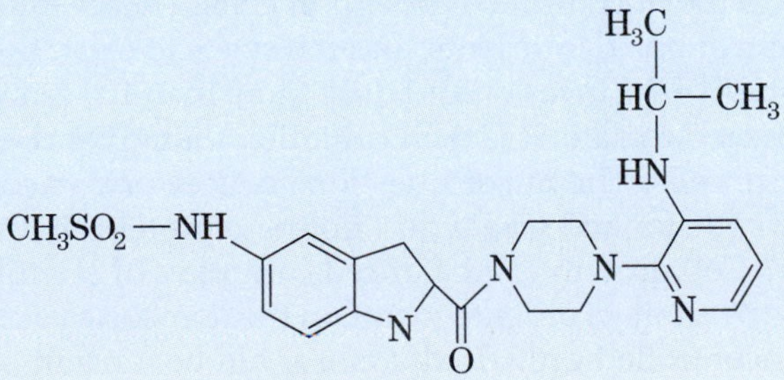

Delaviridine mesylate

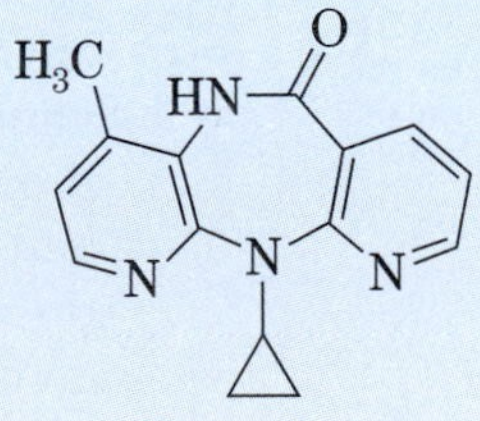

Nevirapine

events. The end result may be positive or negative: positive in that this is a natural way for the evolution and adaptation of an organism; negative in that it can result in the inability of the organism to survive. Site-directed mutagenesis has helped us to understand the roles of various amino acids in the structure of a protein; it is possible to design and produce DNA that will change a single specific amino acid or substitute or delete entire sections of a protein molecule. Protein domains and subunits have been shuffled and recombined into chimeras that retain the properties of the component parts. Animals have had specific genes "knocked out" in order to ascertain the importance of a specific protein in the overall viability of that organism. The possibilities are virtually limitless.

17.7 Viruses

virus
a nonbacterial infectious agent, that consists of DNA or RNA, a few proteins, and a protein coat

ssRNA
single-stranded RNA

dsDNA
double-stranded DNA

ss DNA
single-stranded DNA

reverse transcriptase
enzyme that can incorporate a virus RNA code into host DNA

retrovirus
RNA virus that can encode its genome into the DNA of a host organism, using the enzyme reverse transcriptase

Viruses are unique in that they are mainly nucleic acid with a few enzymes and a protein capsule or coat. The genetic material can be **ssRNA** as well as **ds** and **ss DNA.** The rabies virus, for example, has an ssRNA genome that codes for five proteins: a reverse transcriptase, a nucleoprotein, a phosphoprotein, a matrix protein that lines the inside of the membrane lipid bilayer, and a glycoprotein that constitutes the outer coat. Up to this point we have discussed the passing of genetic information from DNA to RNA to protein. How can a virus be replicated without DNA? The answer lies with a key viral enzyme known as **reverse transcriptase.** It takes the RNA message and puts its complement into the host cell's DNA. The protein synthesis machinery of the host cell is then used to propagate the viral RNA and its proteins. Once the virus particle or virion is assembled and enough virions are present, the host cell is lysed and the virus particles invade other cells. Such viruses are called **retroviruses.** Being RNA-based, they are subject to, and can survive, more mutational events. Therefore, the virus can change its protein coat frequently. This provides a constant challenge to the host defense systems. It is the reason that the flu virus is still with us, as is the rhinovirus causing the common cold. It is also part of the reason that to date it has not been possible to develop an effective vaccine against HIV, human immunodeficiency virus, which causes AIDS (acquired immune deficiency syndrome). Other viruses do not change as much and so have come under control by immunization. Smallpox has essentially been eradicated and poliomyelitis is following. Measles and mumps are controlled wherever there is an effective immunization program.

17.8 Oncogenes

oncogene
gene associated with cancer

"Onco-" refers to cancer. **Oncogenes** are genes connected with cancer, that is, they are related to the uncontrolled growth of cells. These "untamed" cells rob the adjacent tissue of blood, nutrients, and the space to exist. Eventually the organism cannot survive the invasion and dies. More than 100 cancer-related genes have been discovered to date. It is their mutation that gives rise to the proliferation of immature cells. The major questions concerning oncogenes are what proteins do they code for and what effect do these proteins have on cell growth and maturation? Cell growth is a balanced interplay of stimulation and suppression of growth factors in order to maintain the homeostasis or balance within an organism. Uncontrolled growth, therefore, can be a result of direct growth stimulation or the inhibition of suppression. One of the first oncogenes to be discovered and studied extensively is the p53 tumor suppression gene located on human chromosome 17. It has been associated with most types of human cancers from breast cancer to brain tumors. This is because p53 seems to be responsible for controlling the overall mutability of cells in the human genome. Most of the mutations in this gene involve missense codons, that is, changes in one DNA base, giving rise to single amino acid replacements. The alteration of

Connections 17.3 www.prenhall.com/bailey

DNA Fingerprinting

The intron sequences of DNA, far from being merely "junk DNA," are providing a means for the direct identification of individuals. These intron sequences contain regions of short nucleotide segments repeated many times. The length and number of repeats are the keys to identity. Since all cells of the body contain an individual's DNA, minute samples of blood, saliva, semen, hair follicles, or skin can be used for analysis. Of course, small tissue samples mean very small quantities of DNA. The Nobel Prize-winning discovery of an enzyme isolated from thermal hot springs, which will amplify submicroquantities of DNA, has revolutionized molecular biology and its applications, especially forensic science.

The polymerase chain reaction, or PCR, involves isolating nanogram amounts of DNA from tissue (old or new), cleaving it if necessary into manageable fragments, and subjecting it to multiple treatments of a polymerase enzyme in the presence of known DNA primers, with alternate heating and cooling cycles so that the original DNA is uncoiled, replicated, and renatured. The eventual products are separated by electrophoresis and compared. Individuals have unique sets of repeating sequences called polymorphisms that will identify them as reliably as a fingerprint can. Identical twins are the only organisms that have the same genetic sequences. Of course, a drawback to this ultra-sensitive method is the reality of contamination by investigators or others on the scene. Consequently, heightened security and improved sampling techniques are evolving in this new scenario.

DNA fingerprinting has helped to convict rapists and killers as well as exonerate death-row inmates. It has revealed the indiscretions of presidents and located the descendents of ancient populations. It can also be used to diagnose disease states such as Huntington's chorea, cystic fibrosis, and hereditary Alzheimer's, and may eventually help persons with genetic disorders to overcome unfavorable outcomes.

even one amino acid in a protein can severely affect its conformation and function. This is the case with the p53 gene.

17.9 Recombinant DNA and Biotechnology

With a library of gene sequences as well as the technical expertise to analyze, modify, and produce synthetic DNA and RNA, it is now possible and practical to introduce genes not only into simple bacteria but also into animals and eventually into humans.

recombinant DNA
DNA that has been spliced into a foreign host

At its simplest level, natural, semisynthetic, or totally synthetic genes can be introduced into the genome of lower organisms such as the *E. coli* bacteria that inhabit the human gut. The circular plasmid DNA found in many types of bacteria is an easy vehicle to use. It can be removed from its cell, modified, and then returned in order to generate the proteins encoded. In order to introduce a foreign gene, **endonucleases,** which cleave DNA at specific base sequences, are used to open up a plasmid DNA molecule. A synthetic gene can be made with ends that can be annealed or attached by DNA ligases to the opening in the host DNA.

endonuclease
enzyme that cleaves polynucleotides within the chain

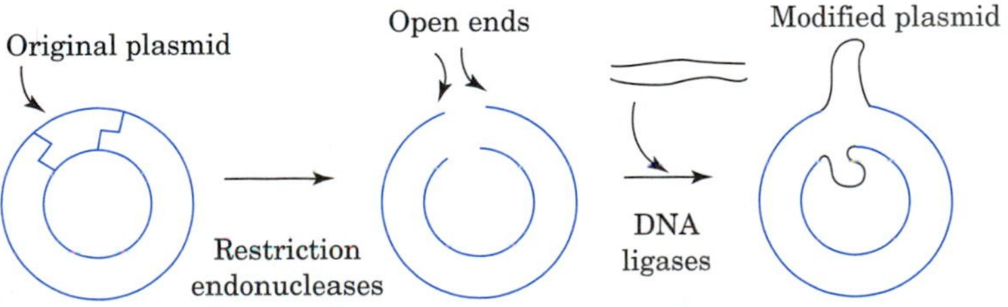

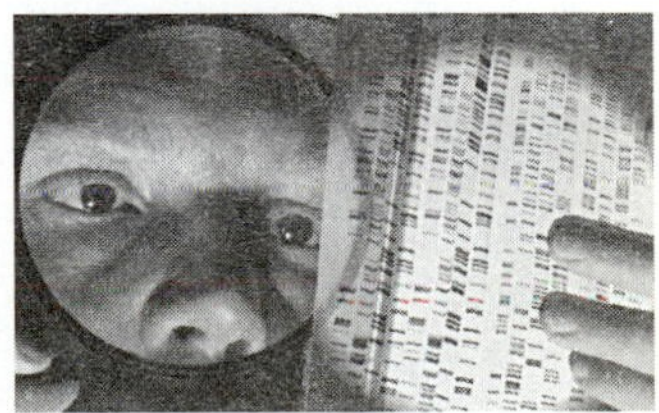

Interpreting DNA

It is in this way that a number of proteins have been mass produced by using the genes from higher animals, including humans. Some of them include human and bovine growth hormones, human insulin, tissue plasminogen activator (dissolves blood clots), and components of the human immune system. The last have then been spliced into mice to produce transgenic animals.

Final observation: The scope of this area is full of promise and, for some, dread. The challenge to society will be to support the knowledge while constraining its potential for abuse. Maybe the challenge is not one to science but rather one to our total humanity.

Problems

17.10 **Structure:** Give three important structural differences between DNA and RNA.

17.11 **Structure:** Uracil hydrogen-bonds to adenine in RNA in place of the thymine found in DNA. Draw the structure of the adenine–uracil hydrogen-bonding pairs.

17.12 **Genetic Code:** Given the following "sense," (−), or template DNA sequence, write the sequences for the corresponding "antisense" DNA, mRNA, and tRNA. Be sure to indicate the 3′ and 5′ ends of the polynucleotides.

G T A A C G T C G C

17.13 **Structure:** If one mole of the polynucleotide in problem 17.12 were completely hydrolyzed, what would be the products and how many moles of each would be produced?

17.14 **Structure:** There are ten nucleotide bases per 360° turn of the DNA molecule. This corresponds to a linear distance of about 34 Å (1 Å = 1 angstrom = 10^{-8} centimeters). How long, in meters, would a DNA molecule be if it contained one million nucleotide base pairs?

17.15 **Energy-Related Nucleotides:** What are the hydrolysis products of one mole of each of ATP, FAD, NADH, and FMN?

17.16 **Genetic Code:** Polypeptides, which have physiological activity such as the hormone glucagon, are derived from much larger protein precursors. What is the minimum number of nucleotide base-pairs that would be needed for the exon coding for glucagon, which has 37 amino acids?

17.17 **Genetic Code:** Two naturally occurring variations discovered in the amino acid sequence of adult hemoglobin involve substituting a lysine for a glutamic acid in the β chain (hemoglobin E) and substituting a tyrosine for a histidine also in the β chain (hemoglobin M_{Boston}). Considering the codons for these amino acids, can you offer some explanation for these natural substitutions?

CHAPTER 18

Spectroscopy

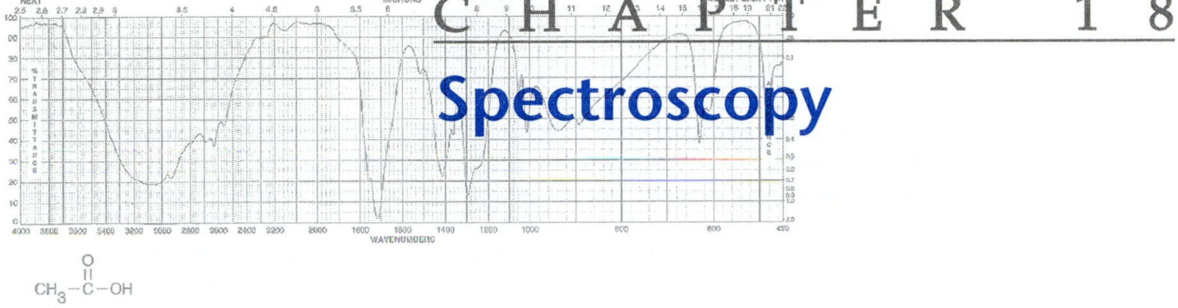

Determining the structures of compounds is central to the science of organic chemistry. In the early part of the twentieth century and before, this could be a tedious and time-consuming task. An array of test-tube assays could give information about functional groups. The unknown compound might be chemically degraded into smaller compounds of known structure or converted into a recognizable derivative for structural information. Independent synthesis of the unknown from known compounds was and often still is used to confirm structures. But in the 1940s instrumental techniques became available for chemical analysis; they have become increasingly sophisticated and important in the past few decades. These electronic instruments greatly shortened the time and increased the capacity for obtaining structural information, and required very small amounts of sample. We will look at some of these techniques in this chapter.

spectroscopy
instrumental method in which the interaction of chemical compounds with electromagnetic radiation is measured

18.1 Spectroscopy

All chemical substances interact with **electromagnetic radiation** in some way. Measuring this interaction can provide valuable information about the substance. When a molecule absorbs energy, a transformation or perturbation occurs that may be either temporary or permanent. Low-energy radiation may merely cause a molecular rotation or a bond vibration. Higher-energy radiation may affect the promotion of electrons to higher energy levels; and radiation of even greater energy can result in bond cleavage and permanent disruption of the molecule.

electromagnetic radiation
various wavelengths of energy

Energy can be visualized as traveling in waves (Figure 18.1). The distance between waves is the **wavelength,** and the number of waves that pass by a point in a given time or the number of waves per unit of distance is the **frequency,** expressed in cycles per second, or hertz (Hz). The relationship between energy ϵ and wavelength or frequency is given by the following equation, where h is a proportionality constant called Planck's constant.

wavelength
the distance between two maxima in an energy wave

frequency
number of waves per unit distance or per unit time [cycles per second, or Hertz (Hz)]

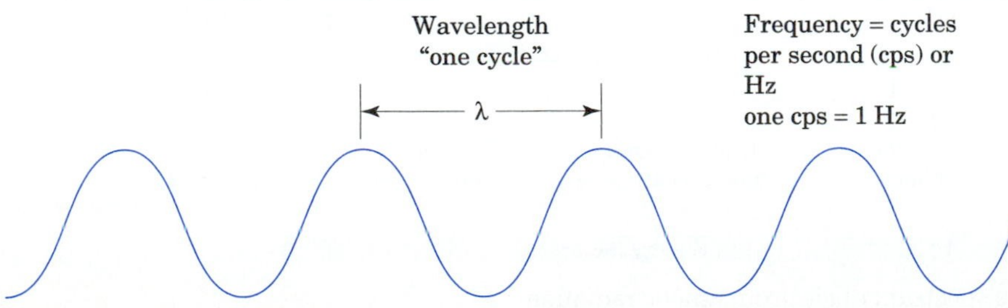

Figure 18.1
Electromagnetic radiation travels in waves characterized by a wavelength λ and frequency υ.

$$\epsilon = h\nu = \frac{hc}{\lambda}$$

h = Planck's constant
c = speed of light
ν = frequency
λ = wavelength

Radiation of a particular wavelength or frequency has a definite, constant amount of energy associated with it. High-energy radiation is characterized by short wavelengths and high frequency, and low-energy radiation by long wavelengths and low frequency. Figure 18.2 shows the spectrum of electromagnetic energy varying in wavelengths from a fraction of an angstrom to thousands of kilometers.

As we have indicated, the interaction of a molecule with electromagnetic radiation causes molecular transformations. Whether the transformation involves molecular rotation, bond vibration, or electronic transition, the molecule absorbs only the wavelength of radiation with exactly the energy necessary for the transition. The processes are quantized and will occur only with specific frequencies or wavelengths. It is not possible either to accumulate radiation of lower energies to attain the total needed for molecular transition or to extract it from higher-energy radiation. The situation is analogous to a vending machine that takes only a single dollar bill. You can obtain your item only if you have a dollar. Trying to insert four quarters is useless. Likewise, a five dollar bill will not be accepted.

Since the absorption of radiation is selective for the particular transition and this transition depends on molecular structure, spectroscopy is invaluable both qualitatively and quantitatively. By measuring the absorption spectra of known compounds, we can correlate the wavelengths of energy absorbed with characteristic structural features. This information is then used to identify structural units in unknowns. The instrument used to measure the absorption of energy by a compound is called a **spectrometer.**

spectrometer
an instrument that measures the absorption of energy by a chemical compound

GETTING INVOLVED

✓ What is the difference between wavelength and frequency in describing electromagnetic radiation?

✓ Is the energy of a wave directly or inversely related to wavelength? To frequency? Does high frequency correspond to short or long wavelengths and high or low energy? Rationalize your concepts with familiar waves presented in Figure 18.2.

✓ What does it mean that a molecular transformation or absorption of energy by a molecule is quantized? What is a spectrometer?

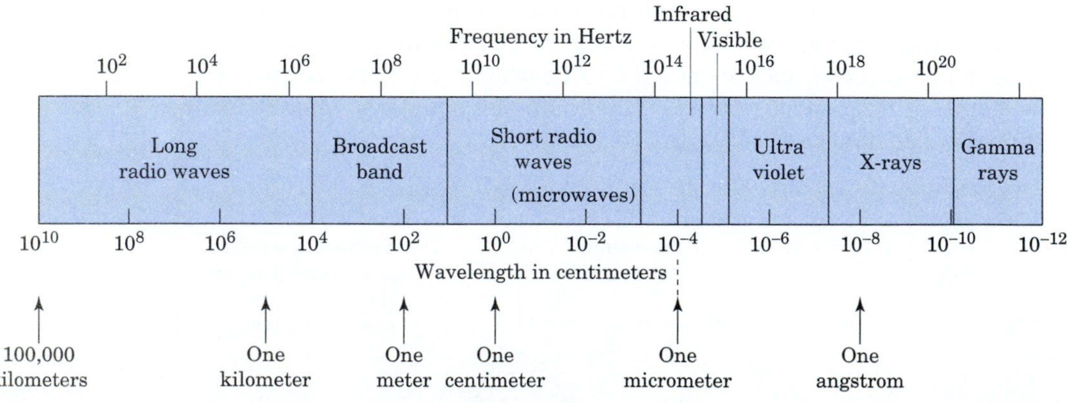

Figure 18.2 The spectrum of electromagnetic radiation.

18.2 Infrared Spectroscopy

An **infrared spectrometer** subjects a sample compound to **infrared radiation** in the 2–15 micrometer (μm) wavelength range. This region is more frequently described in terms of **wavenumber** (frequency), 5000 cm^{-1} to 670 cm^{-1}, which is essentially the number of cycles or waves in a distance of 1 centimeter calculated as $1/\lambda$, with λ in centimeters. Although this radiation is unable to inflict permanent alteration on a molecule, it does supply sufficient energy for bonds in the molecule to vibrate by stretching, scissoring, bending, rocking, twisting, or wagging (Figure 18.3). The atoms of a molecule can be conceived of as linked by springs that are set in motion by the application of energy. As the molecule is subjected to radiation with frequencies in the 5000 cm^{-1} to 670 cm^{-1} range, it absorbs only those possessing exactly the energy required to cause a particular vibration. Energy absorptions are recorded as bands on chart paper.

infrared spectroscopy
spectroscopy using infrared radiation; used to determine bond types and functional groups in organic compounds

infrared radiation
for infrared spectroscopy it is radiation with wavelengths of 2–15 micrometers or frequencies of 5000 cm^{-1} to 670 cm^{-1}

wavenumber
the number of cycles or waves in a distance of one centimeter

Since different bonds and functional groups absorb at different frequencies, an infrared spectrum is usually applicable in qualitative analysis, that is, in determining what types of groups are in a molecule. For example, a carbon-carbon triple bond is stronger than a double bond and requires a higher frequency (greater energy) radiation to stretch. The same considerations apply to carbon-oxygen and carbon-nitrogen bonds.

Increasing bond strength ←

C≡C	C=C	C—C
2100–2260 cm^{-1}	1600–1670 cm^{-1}	800–1200 cm^{-1}

C=O	C—O
1660–1780 cm^{-1}	1000–1300 cm^{-1}

C≡N	C=N	C—N
2210–2260 cm^{-1}	1630–1690 cm^{-1}	1250–1360 cm^{-1}

Increasing energy of absorption (higher frequency) ← to cause bond stretching

Usually one can identify an absorbing bond (group) by the position of the absorption peak. Figure 18.4 illustrates the general area in which various bonds absorb in the infrared.

An infrared spectrum is usually studied in two sections. The area from about 1400 cm^{-1} to 3500 cm^{-1} is the functional group area. The bands in this region are particularly useful in determining the types of groups—alkene, alkyne, aldehyde, ketone, alcohol, acid—present in the molecule. The remainder of the spec-

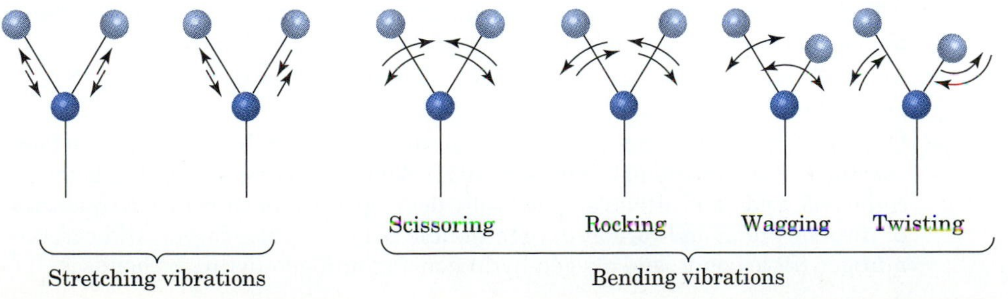

Figure 18.3 Molecular vibrations caused by infrared radiation.

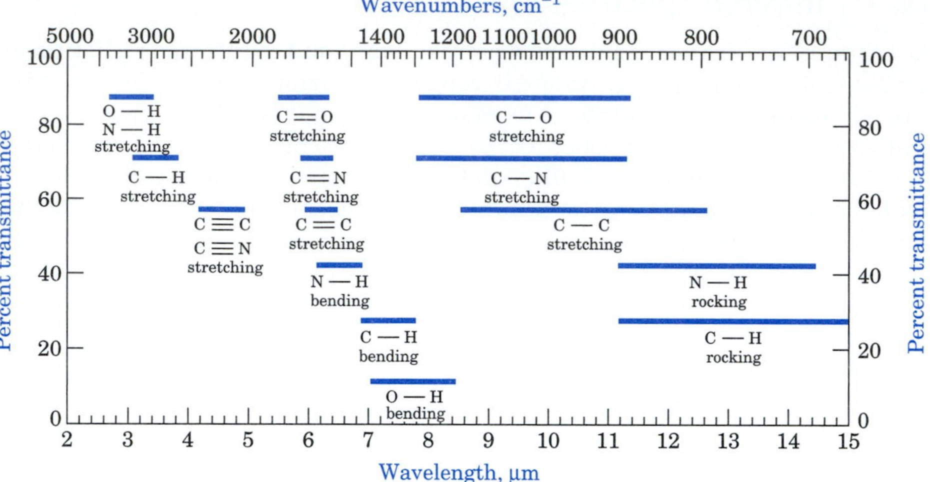

Figure 18.4 Areas of absorption of infrared radiation by various bonds. The lower scale is the wavelength in micrometers (μm). The upper scale is frequency expressed in wavenumbers (the number of waves in 1 cm). The vertical scale describes percentage of transmittance of the sample beam.

trum is called the *fingerprint region.* A peak-by-peak match of an unknown spectrum with the spectrum of the suspected compound in this region can be used, much like a fingerprint, to confirm the unknown's identity. Figure 18.5 contains some sample spectra, and Table 18.1 summarizes some infrared assignments useful in functional group analysis.

You may wish to consider ways to arrange some of the assignments in Table 18.1 in your mind. For example:

1. Alkanes, alkenes, and aromatics show C—H stretches around 2800–3100 cm^{-1}, but for alkynes, the ≡C—H stretch is around 3300 cm^{-1}. O—H and N—H stretches are in the 3000–3500 cm^{-1} range.
2. Double bonds stretch between 1600 cm^{-1} and 1800 cm^{-1} with carbon-carbon double bonds in the lower frequencies and carbon-oxygen in the higher frequencies.
3. Triple bonds, C≡C or C≡N, absorb around 2100–2300 cm^{-1}.

GETTING INVOLVED

- ✓ What frequency range is used in infrared spectroscopy? What does wavenumber mean? How does wavenumber relate to energy of infrared radiation? Does a high wavenumber describe high or low energy radiation?
- ✓ What kinds of perturbations does infrared radiation cause on organic molecules? Why does a triple bond require radiation with a higher wavenumber for stretching than a double bond and a double bond higher than a single bond?
- ✓ In an infrared spectrum, what is meant by the functional group region and the fingerprint region?
- ✓ Using Figure 18.4 and Table 18.1 become familiar with the characteristic infrared absorptions of the functional groups such as alkenes, alkynes, aldehydes, ketones, carboxylic acids, and alcohols. Specifically develop a feel for stretching frequencies of carbon-carbon and carbon-oxygen double bonds, carbon-carbon and carbon-nitrogen triple bonds, and oxygen-hydrogen and nitrogen-hydrogen bonds.

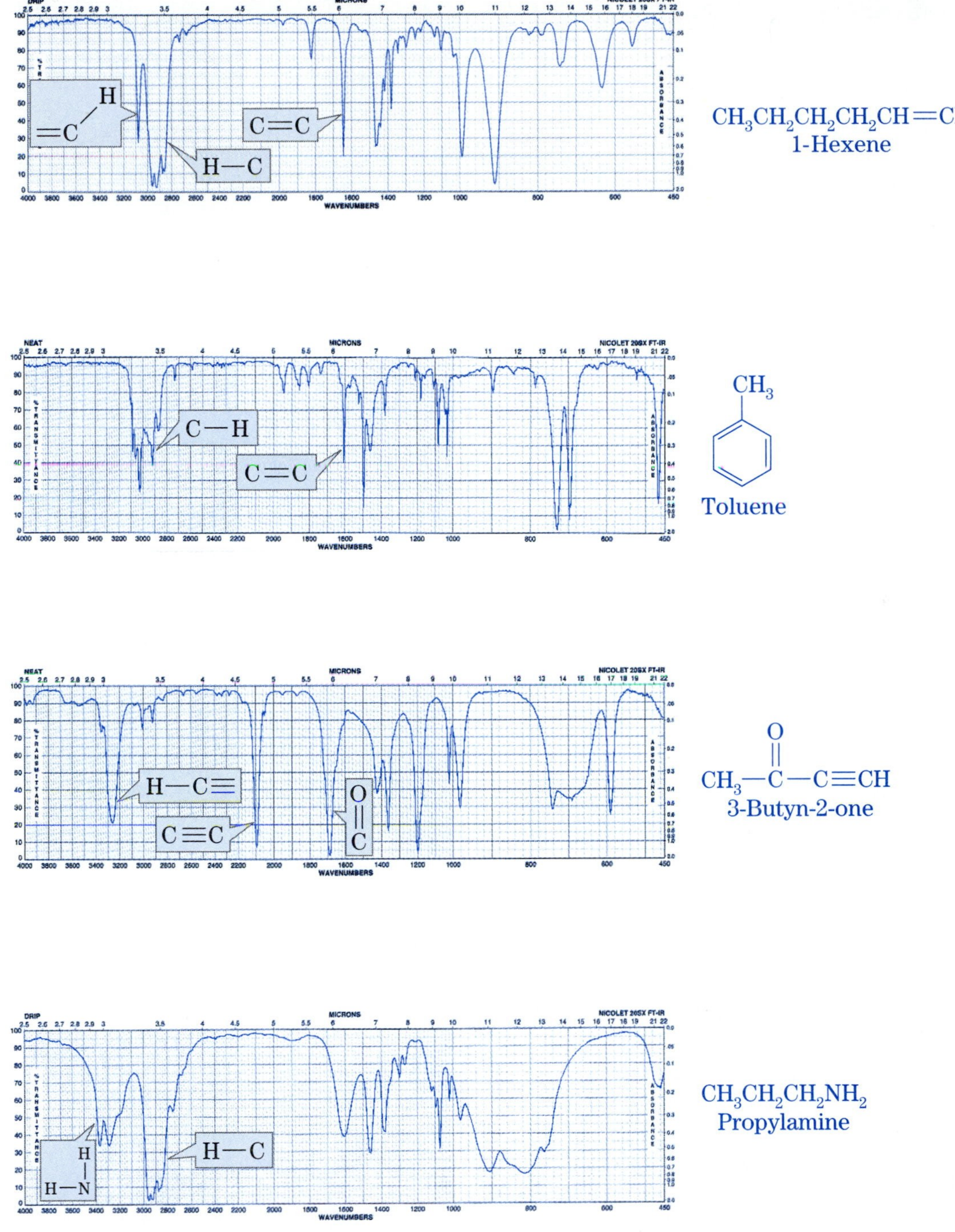

Figure 18.5 Infrared absorption spectra. Compare bands in the 4000–1380 cm^{-1} region with assignments for each functional group in Table 18.1.

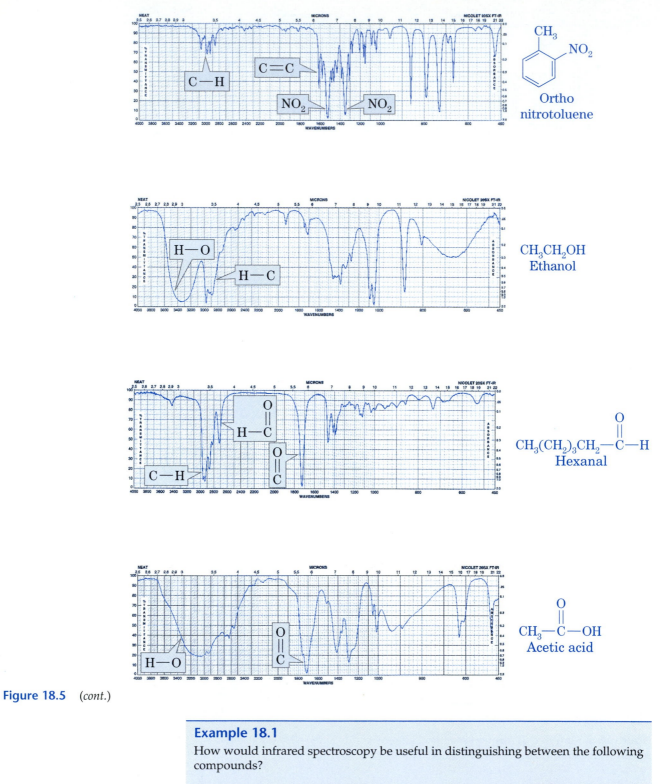

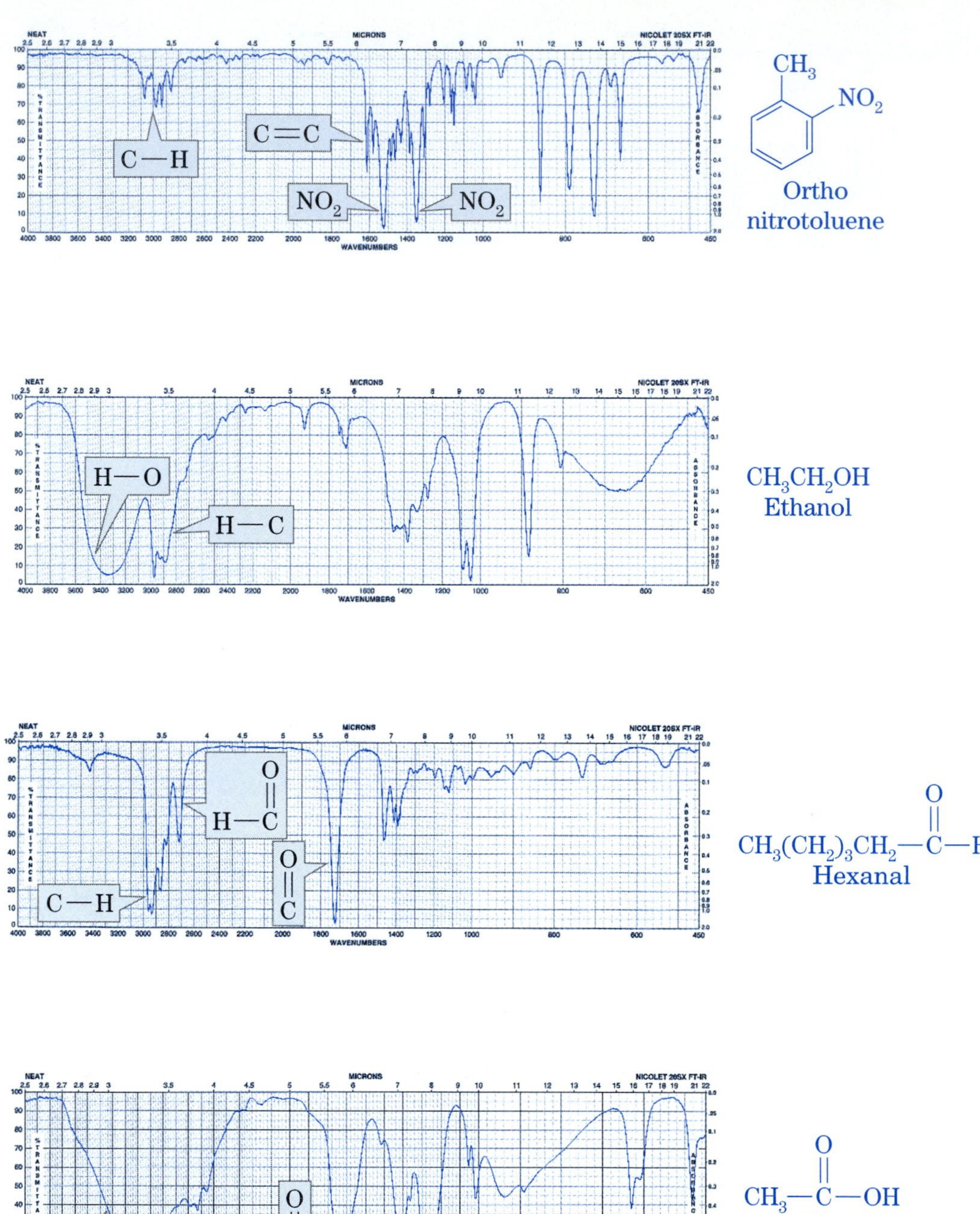

Figure 18.5 *(cont.)*

Example 18.1

How would infrared spectroscopy be useful in distinguishing between the following compounds?

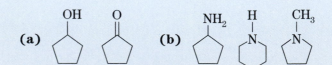

(a) OH O **(b)** NH_2 H N CH_3 N

TABLE 18.1 ◆ Infrared Absorption Assignments

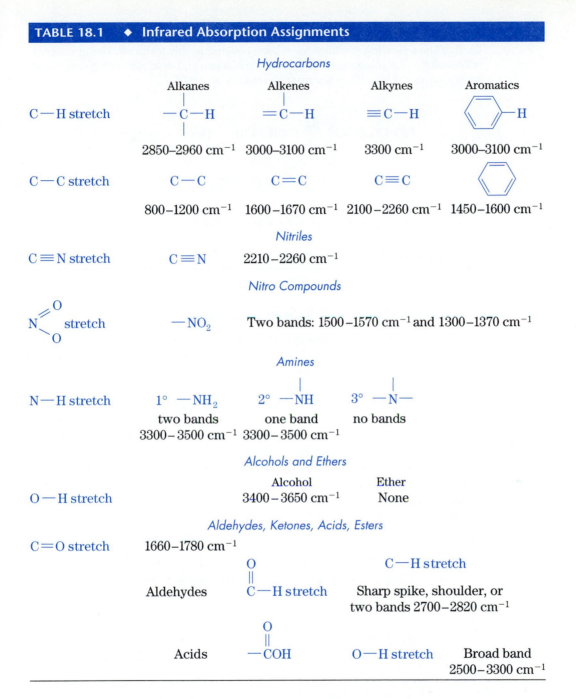

	Hydrocarbons			
	Alkanes	Alkenes	Alkynes	Aromatics
C—H stretch	—C—H (with bonds above and below C)	=C—H	≡C—H	benzene ring—H
	2850–2960 cm^{-1}	3000–3100 cm^{-1}	3300 cm^{-1}	3000–3100 cm^{-1}
C—C stretch	C—C	C=C	C≡C	benzene ring
	800–1200 cm^{-1}	1600–1670 cm^{-1}	2100–2260 cm^{-1}	1450–1600 cm^{-1}
	Nitriles			
C≡N stretch	C≡N	2210–2260 cm^{-1}		
	Nitro Compounds			
N(=O)O stretch	$—NO_2$	Two bands: 1500–1570 cm^{-1} and 1300–1370 cm^{-1}		
	Amines			
N—H stretch	1° $—NH_2$	2° —NH—	3° —N— (three bonds)	
	two bands	one band	no bands	
	3300–3500 cm^{-1}	3300–3500 cm^{-1}		
	Alcohols and Ethers			
		Alcohol	Ether	
O—H stretch		3400–3650 cm^{-1}	None	
	Aldehydes, Ketones, Acids, Esters			
C=O stretch	1660–1780 cm^{-1}			
	Aldehydes	C(=O)—H stretch	C—H stretch: Sharp spike, shoulder, or two bands 2700–2820 cm^{-1}	
	Acids	—C(=O)OH	O—H stretch	Broad band 2500–3300 cm^{-1}

Solution

(a) The difference between the compounds is that the first is an alcohol and the second a ketone. The alcohol will show an O—H stretch around 3400–3650 cm^{-1} but will show no C=O stretch. The ketone will have no O—H stretch, but will show a C=O stretch around 1660–1780 cm^{-1}.

(b) The first compound is a primary amine and will show two N—H stretching bands in the 3200–3500 cm^{-1} region. The second compound is a secondary amine and will show only one N—H stretching vibration in this area. The third compound is a tertiary amine; it has no N—H bonds and will show no such stretching bands.

Problem 18.1

Each of the following reactions have been covered in this textbook. Using infrared wavenumber assignments, describe how the reactant and product in each case could be distinguished to confirm the reaction actually occurred.

(a) $CH_3C{\equiv}N \xrightarrow[H^+]{H_2O} CH_3\overset{O}{\overset{\|}{C}}OH$ **(b)** 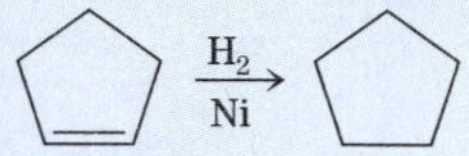 cyclopentene $\xrightarrow[Ni]{H_2}$ cyclopentane

(c) $CH_3CH_2\overset{O}{\overset{\|}{C}}H \xrightarrow[Ni]{H_2} CH_3CH_2CH_2OH$

(d) $CH_3CH_2NH_2 \xrightarrow{2\ CH_3I} CH_3CH_2N(CH_3)_2$

See related problem 18.14.

18.3 Ultraviolet-Visible Spectroscopy

ultraviolet spectroscopy spectroscopy using ultraviolet radiation with wavelengths in the 200–400 nm range

visible spectroscopy spectroscopy using visible light with wavelengths in the 400–750 nm range

In **ultraviolet-visible spectroscopy,** the 200–750-nanometer* region of the electromagnetic spectrum is used. This includes both the visible, 400–750 nm, and near ultraviolet, 200–400 nm. Radiation of these wavelengths is sufficiently energetic to cause the promotion of loosely held electrons, such as nonbonding electrons or electrons involved in a π bond, to higher energy levels. For absorption in this particular region of the ultraviolet, however, there must be conjugation of double bonds. An alternating system of double and single bonds lowers the energy of transition of an electron moving to a higher energy level. If the conjugation is extensive, the molecule may absorb in the visible region and show color (Connections 10.4).

In general, ultraviolet-visible spectroscopy is not used for functional-group analysis as extensively as infrared analysis. Rather, it shows the presence of conjugated unsaturated systems such as the ones illustrated.

$CH_2{=}CHCH_2CH{=}CH_2$

1,4-pentadiene, 178 nm

$CH_2{=}CH{-}CH{=}CHCH_3$

1,3-pentadiene, 223 nm

$CH_2{=}CH\overset{O}{\overset{\|}{C}}H$

Propenal, 210 nm

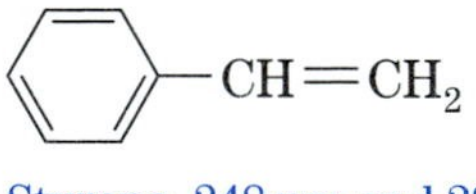

$C_6H_5{-}CH{=}CH_2$

Styrene, 248 nm and 282 nm

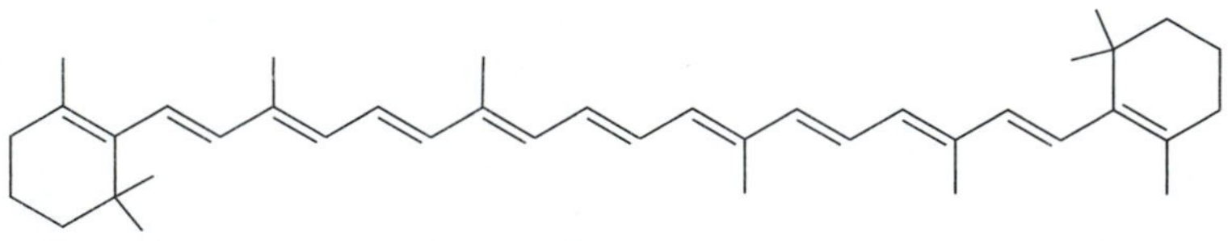

β-carotene (orange color in carrots), 454 nm

Compounds that absorb in this area have characteristic wavelengths of absorption. Thus their presence and concentration in a solution can be detected and measured. This is useful in identifying product ratios and reaction rates and also

*A nanometer is 10^{-9} meter.

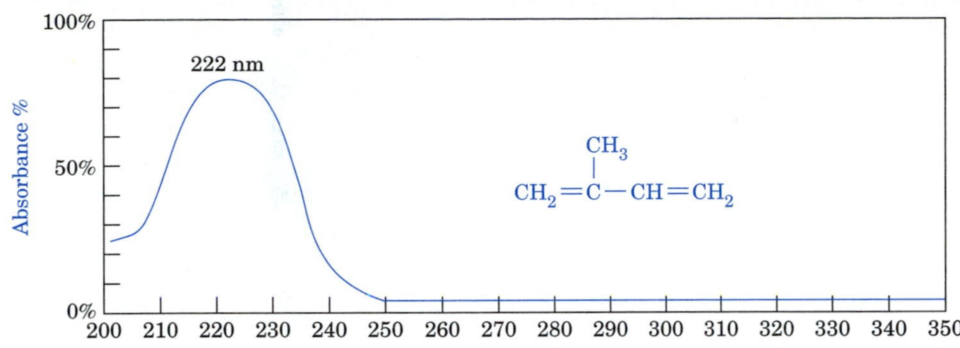

Figure 18.6 Ultraviolet spectrum of isoprene.

in determining other quantitative data. Figure 18.6 shows the ultraviolet spectrum of isoprene.

GETTING INVOLVED

✓ What is the difference in ultraviolet and visible spectroscopy in terms of wavelength of radiation used? What type of structural features are characterized by ultraviolet-visible spectroscopy?

Problem 18.2

Why can the following compounds be distinguished by ultraviolet-visible spectroscopy? (For **c** and **d** refer to Connections 5.3 on terpenes for structures.) **(a)** 1,3-cyclohexadiene and 1,4-cyclohexadiene; **(b)** propanone and propenal; **(c)** menthol (peppermint) and carvone (spearmint); **(d)** squalene (shark liver oil) and vitamin A.

18.4 Nuclear Magnetic Resonance: ^{1}H NMR

The nuclei of some atoms spin. In doing so, they generate a magnetic moment along their axis of spin, acting as tiny bar magnets. The nucleus of the hydrogen atom, mass number of 1 (one proton, no neutrons), exhibits this property and is the one most often analyzed by **nuclear magnetic resonance (NMR) spectroscopy.** If a hydrogen atom is placed in an external magnetic field, its nucleus can align with the field (the more stable arrangement) or against the field (a more energetic, less stable state) (Figure 18.7). Although the energy difference between the states is not great, there is a slightly greater proportion of nuclei in the more stable state in which they are aligned with the external field.

nuclear magnetic resonance spectroscopy
spectroscopy in which compound is placed in a magnetic field and exposed to radio-frequency radiation. It provides information about the carbon and hydrogen structure of an organic compound

To make a nucleus flip from alignment to nonalignment, energy in the radio-frequency range must be applied. For example, a hydrogen nucleus in an external field of 14,092 gauss requires a frequency of 60 million hertz (cycles per second) for the transition. When this frequency is applied, it is absorbed, and the absorption is recorded on chart paper. In practice, either the magnetic field can be held constant and the radio frequency varied, or the radio frequency can be held constant and the magnetic field varied. Student model NMR's are often 60 MHz but modern research instruments are 100–300 MHz or higher.

GETTING INVOLVED

✓ In NMR, hydrogen nuclei are subjected to a magnetic field and then radio frequency energy. What happens before and after the presentation of the field and the radio frequency radiation?

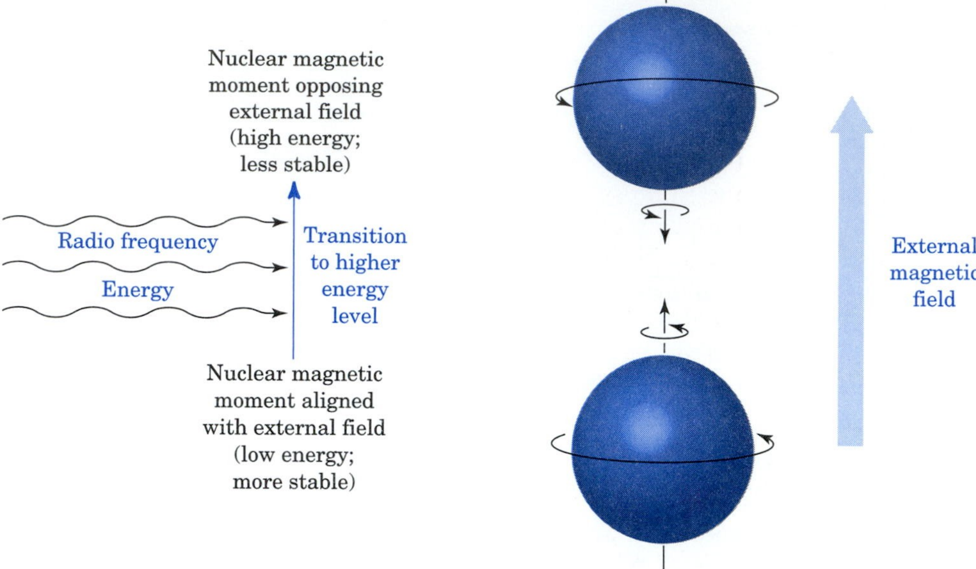

Figure 18.7 The spinning nucleus of the hydrogen atom acts like a tiny magnet that can go into alignment or nonalignment with an externally applied magnetic field. Applying radio-frequency energy can flip protons in the more stable aligned state to nonalignment.

chemical shift the position on NMR chart paper where a carbon or hydrogen nucleus absorbs relative to an internal standard, TMS; measured in δ units

A. Chemical Shift

If NMR's main feat were to detect the presence of hydrogen in a molecule, it would not be worth discussing here. Nuclear magnetic resonance spectroscopy can, however, distinguish between hydrogens in different chemical environments within a molecule. Hydrogens on a benzene ring, on a carbon bearing a chlorine, or on a carbon adjacent to a carbonyl group absorb radio-frequency energies at different applied magnetic fields, which appear at different locations on the recording paper. Furthermore, the position of absorption is relatively constant for hydrogens in a particular chemical or structural environment. Hence, the number of signals recorded on the NMR chart paper indicates the number of different types of hydrogens in a molecule. The position of the peak can give information about the molecular structure in the vicinity of the hydrogens.

To understand fully the value of NMR, then, we must gain a concept of equivalent and nonequivalent hydrogens. Equivalent hydrogens are positioned in structurally and chemically equivalent areas in the molecule. For example, consider the following molecules and convince yourself of the different types of hydrogens shown. The first compound has two methyl groups connected to the same oxygen. The hydrogens on these carbons are chemically equivalent. However, in the second example, bromoethane, the $-CH_2-$ group is bonded to a carbon and a bromine and the CH_3- is bonded to just a carbon. The hydrogens on these two carbons are in significantly different chemical environments and are nonequivalent. In the third and fourth examples, note that the methyl groups are equivalent, but in the fifth example the two methyl groups are in two different chemical environments and are nonequivalent.

(1) $\underset{a}{CH_3}-O-\underset{a}{CH_3}$

1 NMR signal

(2) $\underset{a}{CH_3}\underset{b}{CH_2}Br$

2 NMR signals

(3) $(\overset{a}{CH_3})_2\overset{b}{C}HO\overset{b}{C}H(\overset{a}{CH_3})_2$

2 NMR signals

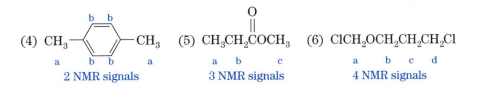

(4) $CH_3-C_6H_4-CH_3$ (a, b b / b b, a) — 2 NMR signals

(5) $CH_3CH_2\overset{O}{\overset{||}{C}}OCH_3$ (a b c) — 3 NMR signals

(6) $ClCH_2OCH_2CH_2CH_2Cl$ (a b c d) — 4 NMR signals

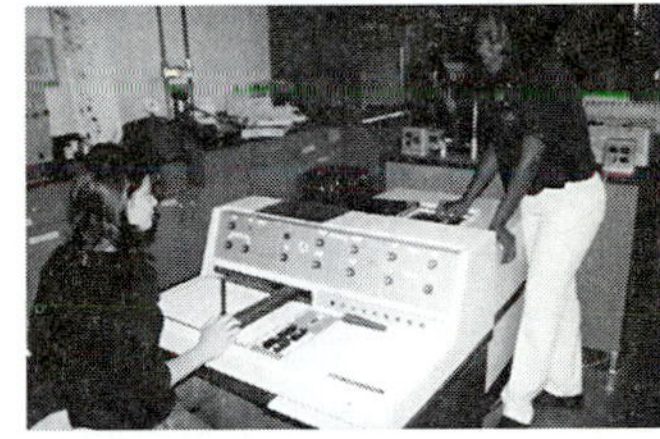

NMR instrumentation

The chart paper for proton nmr is rectangular with a linear scale of so-called δ units across the bottom. Most signals in ^{1}H NMR appear from zero to eight or ten δ units, although peaks at higher values can be recorded easily. To every sample to be analyzed by NMR, a small amount of tetramethylsilane (TMS) $(CH_3)_4Si$, is added as a reference; the TMS signal, caused by the 12 equivalent hydrogens, is defined as δ = 0. The signals of the hydrogens in the molecule being analyzed are compared to TMS; their chemical shift is defined as the number of δ units that the signal is shifted from that of TMS.

The chemical shift of a hydrogen depends on how strongly it experiences the external magnetic field. Electron density in the vicinity of a hydrogen nucleus can shield it from the field. Electron-withdrawing groups can decrease this electron density and shielding. Thus different hydrogens experience the external field to varying degrees and require different amounts of energy to flip from alignment to nonalignment. We see these differences in the NMR as differences in chemical shift. For example, chlorine is strongly electronegative, and the hydrogens on chloromethane are shifted significantly from those on methane. As we add a second and third chlorine, the shift to higher δ values continues by a fairly uniform amount as the hydrogens are progressively deshielded from the external field.

	CH_4	CH_3Cl	CH_2Cl_2	$CHCl_3$
δ-value	0.5	3.1	5.3	7.2
(approximate)				

Knowing the effect of chlorine or any other group on chemical shift is very useful in the interpretation of NMR spectra. Table 18.2 summarizes the characteristic chemical shifts of hydrogens in different types of environments.

Now let us consider a specific example, benzyl alcohol [NMR in Figure 18.8(a)].

$C_6H_5-CH_2OH$ Benzyl alcohol

The NMR spectrum has three distinct peaks at δ = 2.4, δ = 4.6, and δ = 7.3 (δ = 0 is the TMS reference). Examination of the molecule confirms three types of hydrogen present: one hydrogen bonded to an oxygen, two hydrogens bonded equivalently to a carbon, and five essentially equivalent hydrogens attached to the benzene ring. Using Table 18.2, we can now assign each hydrogen type to a signal. Aromatic hydrogens occur between δ = 7 and 8, while hydroxy hydrogens have variable δ values. So the δ = 7.3 must belong to the five benzene hydrogens. The third column in Table 18.2 indicates that the methylene hydrogens, if attached to an alkyl group, $R-CH_2-$, would generate a signal at 0.9–1.6. However, the fourth column specifies the chemical shift due to adjacent groups. Since a benzene ring (Z = aromatic) and an oxygen (Z = 0) are in those positions, their influences will shift the methylene hydrogen's signal 1.4 and 2.8 units, respectively. Added to the normal signal (δ = 1.0), the sum is δ = 5.2, very close to the 4.6 δ value. By elimination, the δ = 2.4 signal must be due to the hydroxy hydrogen.

Table 18.2 Chemical Shifts

Z—C(H)—		Actual δ Value	Chemical Shift Relative to Alkyl H
Z Groups[a]			
Z = alkyl	R—C(H)—	0.9–1.6	0
Z = C=O	—C(=O)—C(H)—	2–2.5	1
Z = aromatic	C_6H_5—C(H)—	2.3–2.9	1.3–1.5
Z = alkene	—C=C—C(H)—	1.8–2.8	0.9–1.2
Z = O	—O—C(H)—	3.3–5	2.4–3.4
Z = Cl	Cl—C(H)—	3.2–4	2.3–2.6
Z = Br	Br—C(H)—	2.7–3.8	1.8–2.2

Z—C(H)—		Actual δ Value	Chemical Shift Relative to Alkyl H
Z = amines	—N—C(H)—	2.2–3	1.3–1.4
Z = NO_2	O_2N—C(H)—	4.4–4.6	3–3.5
Other Groups			
Vinyl hydrogens	—C=C—(H)	4.5–6	—
Aldehydes, acids	—C(=O)—(H), —C(=O)O—(H)	9–12	—
Aromatic hydrogens	C_6H_5—(H)	7–8	—
Alcohol, phenols, amines	—C—O—(H), C_6H_5—O—(H), —N—(H)	variable	—
Tetramethylsilane, TMS reference	$(CH_3)_4S$:	0	—

[a]The table indicates the shift of the circled hydrogen under the influence of the Z group.

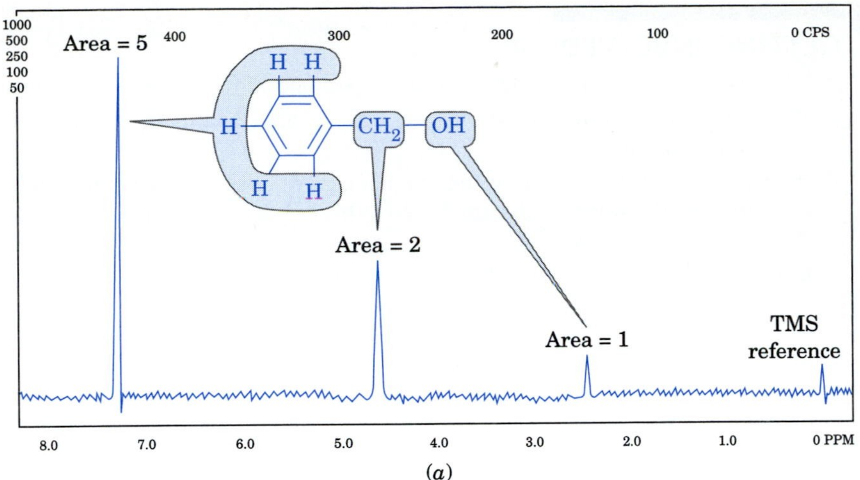

(a)

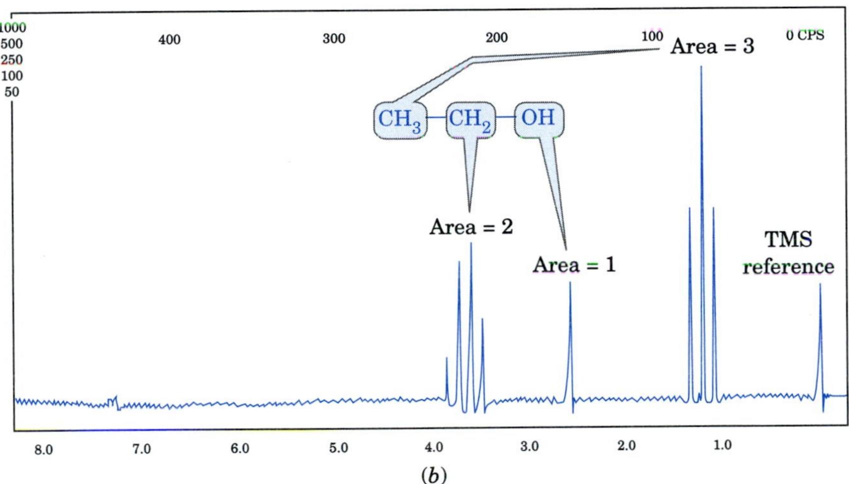

(b)

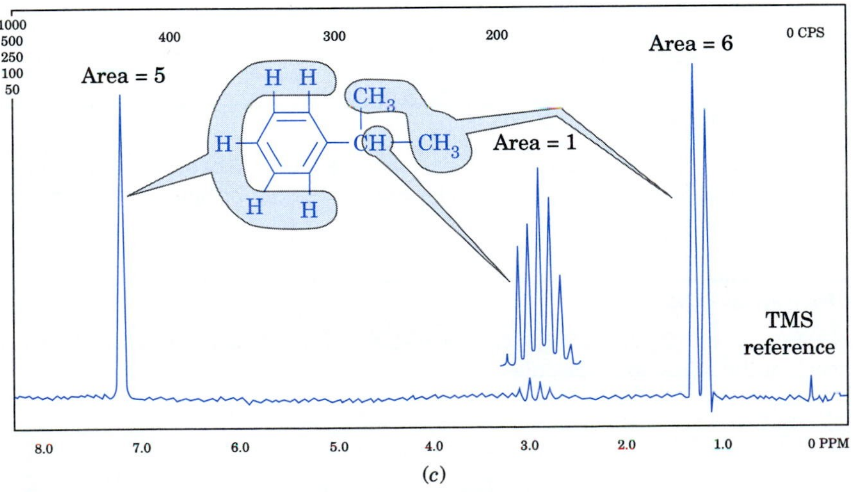

(c)

Figure 18.8 Some NMR spectra. (a) Benzyl alcohol. (b) Ethyl alcohol. (c) Isopropylbenzene NMR spectra, courtesy of Varian Associates.

GETTING INVOLVED

✓ What is a chemical shift? What does it tell you about hydrogens in a molecule?

✓ What are δ-units? What is the purpose of TMS and what is its chemical shift assignment in δ-units?

✓ If a hydrogen is shielded from the magnetic field, does it require more or less energy to flip? What if it is deshielded? Do high δ-values result from shielded or deshielded hydrogens?

✓ Can you use Table 18.2 to predict the value of chemical shifts?

Example 18.2

Predict the chemical shifts of the three types of hydrogens in the following molecule:

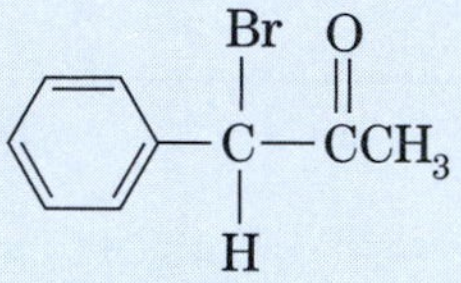

$C_6H_5-CH(Br)-C(=O)CH_3$ (hydrogens labeled a: aromatic, b: CH, c: CH_3)

Solution

(a) aromatic δ = 7–8

(b)
1.0 normal δ value
1.0 adjacent C=O
2.0 adjacent Br
1.4 adjacent $-C_6H_5$
δ = 5.4 total

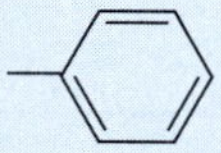

(c) Z = C(=O) δ = 2–2.5

Problem 18.3

How many NMR signals would be produced by each of the following molecules? **(a)** ethanal; **(b)** propanal **(c)** propanone; **(d)** butanone; **(e)** 2-pentanone; **(f)** 3-pentanone.

Problem 18.4

Using Table 18.2 calculate the chemical shifts of the different types of hydrogens in each of the following molecules.

(a) $CH_3\overset{O}{\overset{\|}{C}}CH_2\overset{O}{\overset{\|}{C}}CH_3$ **(b)** $BrCH_2CH_2\overset{O}{\overset{\|}{C}}CHCl_2$ **(c)** 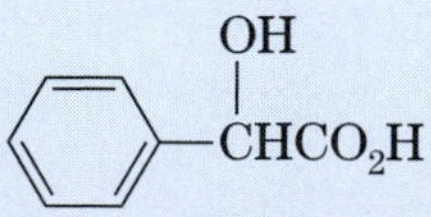 $C_6H_5-CH(OH)CO_2H$

B. Integration

integration
in ^1H NMR a technique that provides the relative numbers of hydrogens in a compound; it is the area under a peak

The relative areas under the various peaks of an NMR spectrum are in proportion to the number of hydrogens contributing to each signal. These areas can be electronically integrated by an NMR spectrometer. Comparison of the areas provides the ratio among the various kinds of hydrogens in the molecule. Consider the NMR spectrum of benzyl alcohol [Figure 18.8(a)], for example. The hydrogens in the molecule are in a 1:2:5 ratio, like the corresponding peak areas in the spectrum.

GETTING INVOLVED

What information does integration in NMR provide?

Problem 18.5

A compound with the formula C_8H_{10} gives three NMR signals with the following chemical shifts and integration values: $\delta=7$ (149); $\delta=2.3$ (58); and $\delta=1.1$ (91). How many hydrogens are represented by each signal?

C. Peak Splitting

peak splitting
in ^1H NMR a phenomenon in which hydrogens on an adjacent carbon split the signal of hydrogens on the other carbon

Hydrogens on adjacent carbons, each with a different chemical shift, can influence the signal of one another. This influence appears as peak splitting. We can generalize the phenomenon by saying that the number of peaks into which a particular hydrogen's signal is split equals one more than the total number of hydrogens on directly adjacent carbons. Assuming that each of the following types of hydrogens is nonequivalent, we should obtain the indicated splitting patterns.

```
                               H                      H
   |   |                  |    |                 |    |
 —C—C—                —C—C—               —C—C—H
   |   |                  |    |                 |    |
   H   H                  H    H                 H    H
Doublet  Doublet       Triplet  Doublet       Quartet  Doublet

   H   H                H         H            H         H
   |   |                |    |    |            |    |    |
 —C—C—H              —C—C—C—          H—C—C—C—H
   |   |                |    |    |            |    |    |
   H   H                H    H    H            H    H    H
Quartet  Triplet          Quintet                 Septet
                          Doublet                 Doublet
```

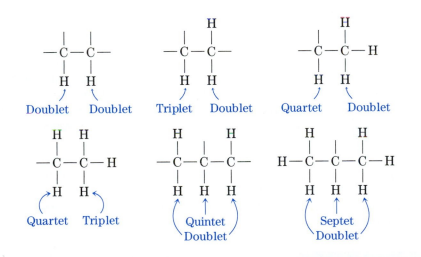

In Figure 18.8(b) (ethyl alcohol), note that the ethyl group is indicated by a quartet and a triplet and that the isopropyl group in Figure 18.8(c) (isopropyl benzene) shows as a septet and a doublet.

Let's use the ethyl alcohol example [Figure 18.8(b)] to explain how splitting occurs. In the presence of an external magnetic field, some hydrogens align with the field and some against. Since the energy difference between the two alignments is not great, the proportion in each state is similar; there is a slightly higher proportion in the aligned state, however. When the hydrogens on a carbon align

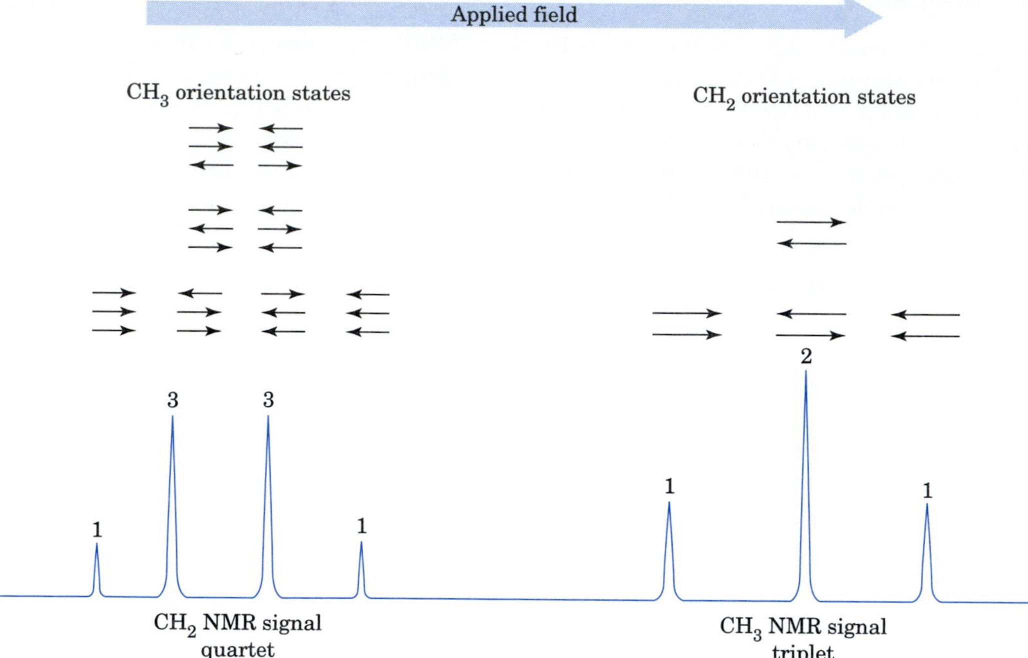

Figure 18.9 Explanation of spin–spin splitting patterns in ethyl alcohol, CH_3CH_2OH.

with the external field, they increase the effective magnetic field felt by the hydrogens on an adjacent carbon. Alternatively, in nonalignment they oppose the external field and decrease the effective field experienced by the hydrogens on an adjacent carbon. These differences cause small but observable differences in chemical shift, which we call splitting.

In the ethyl alcohol example, the three methyl hydrogens have four possible ways to align with the external field: all can be aligned; two can be aligned, one nonaligned; one aligned, two nonaligned; or all three nonaligned. The hydrogens on the adjacent carbon will have different chemical shifts depending on which state the methyl group is in. In the middle two cases, there are three ways to create the alignment/nonalignment possibilities and thus these states are three times as likely as either of the extremes. As a result of the four possibilities, the adjacent hydrogens are split into four peaks, which appear in relative heights of 1:3:3:1. Similar reasoning applies to the effect of the CH_2 group on the CH_3. The two CH_2 hydrogens can be aligned, nonaligned, or one aligned and one nonaligned in two different ways. As a result, the hydrogens on the adjacent carbon are split into three peaks in a 1:2:1 ratio. This is illustrated in Figure 18.9.

GETTING INVOLVED

✓ How are splitting patterns of hydrogens practically recognized? What effect do adjacent hydrogens have on one another?

✓ Can you give a simple explanation of why splitting occurs?

Problem 18.6

Predict the splitting patterns of the hydrogens in the following molecules (ignore aromatic hydrogens and those on oxygens): **(a)** ethanol; **(b)** 1-phenylethanol; **(c)** 2-phenylethanol; **(d)** 1-phenylpropanol; **(e)** 2-propanol.

D. Summary of NMR

The following aspects of NMR provide information.

Chemical Shift. The number of signals corresponds to the number of different types of hydrogens in the molecule. The position of each signal gives information about the structural environment of the hydrogens.

Integration. The relative areas under the signals give the ratio of the numbers of each hydrogen type in the molecule. If the molecular formula is known, the actual number of each type of hydrogen can be determined.

Splitting. The number of peaks into which a signal is split is one more than the total number of hydrogens on directly adjacent carbons.

GETTING INVOLVED

Example 18.3

To conclude our discussion of NMR, let us go through a procedure for identifying the compound with the formula $C_{10}H_{12}O_2$ using the following NMR spectrum.

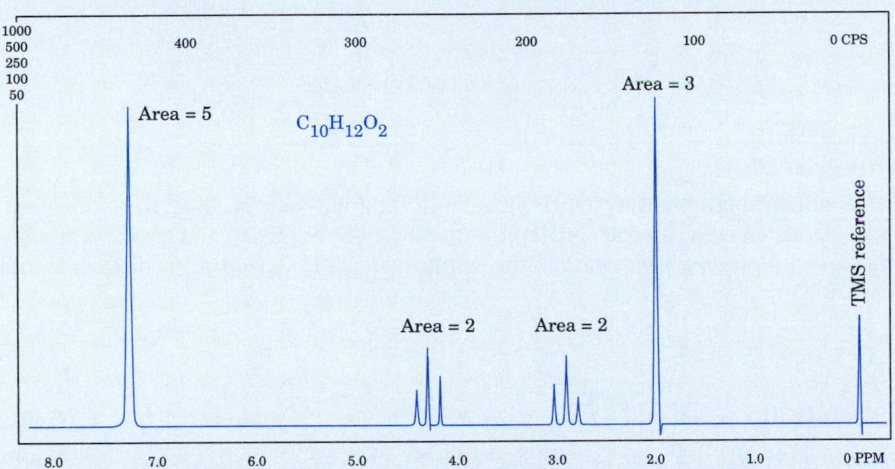

Solution

The four signals indicate four different types of hydrogens. At $\delta = 7.3$, five hydrogens are in the aromatic region—probably a monosubstituted benzene ring. The simplest way of expressing two equivalent hydrogens, indicated by the signal at $\delta = 2.9$ and also at $\delta = 4.3$, is with methylene (CH_2) groups. Finally, the signal at $\delta = 2$ suggests three hydrogens, most simple expressed as a methyl group. Remaining in the formula are a a carbon and two oxygens; these are most simply expressed as $-\underset{\underset{O}{\|}}{C}O-$. Although obviously there are other arrangements of all the groups mentioned, these are the simplest expressions and should be considered first. The pieces of the puzzle are

$\delta = 7.3$ C_6H_5- (benzene ring) $\quad \delta = 4.3$ $-CH_2-$ $\quad \delta = 2.9$ $-CH_2-$

$\delta = 2.0$ $-CH_3$ $\quad -\overset{\overset{O}{\|}}{C}O-$

The spectrum shows that the two methylene groups split each other and thus must be adjacent to each other ($-CH_2CH_2-$). In this arrangement, each methylene splits the other into a triplet (one more peak than the number of hydrogens). Now the puzzle has fewer pieces.

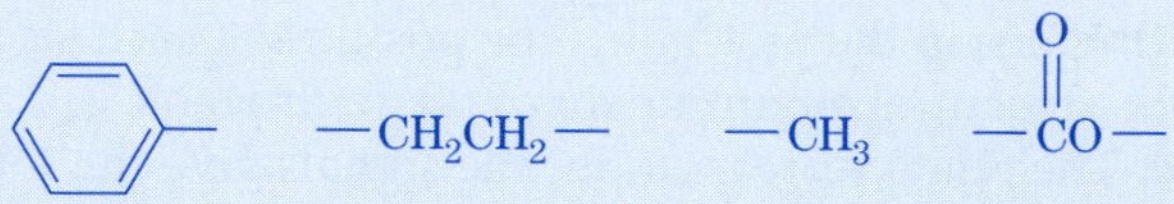

$$C_6H_5{-} \quad {-}CH_2CH_2{-} \quad {-}CH_3 \quad {-}\overset{O}{\overset{\|}{C}}O{-}$$

The methyl group is not split; it must be bonded to the oxygen or the carbonyl carbon. If it were bonded to one of the CH_2's it would be a triplet instead of a singlet. Since its chemical shift is 2.0, it must be bonded to the carbonyl; if it were bonded to the oxygen, the shift would be 3.5 to 5.0. The puzzle now has three pieces.

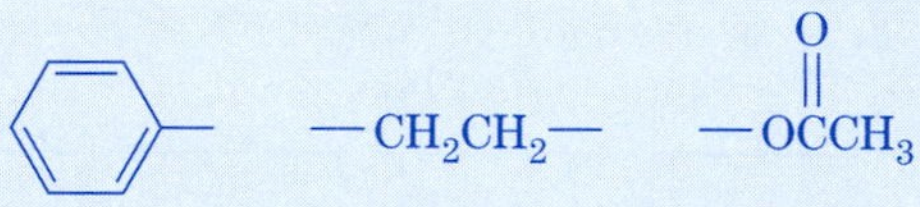

$$C_6H_5{-} \quad {-}CH_2CH_2{-} \quad {-}O\overset{O}{\overset{\|}{C}}CH_3$$

There is only one way these pieces fit together and the chemical shifts of the two CH_2 groups correspond to this structure.

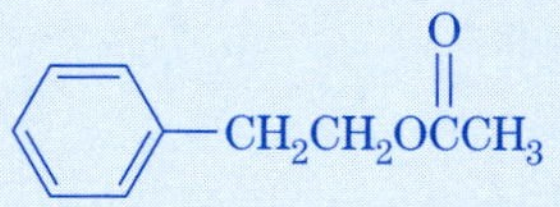

$$C_6H_5{-}CH_2CH_2O\overset{O}{\overset{\|}{C}}CH_3$$

Problem 18.7

^{1}H Nuclear Magnetic Resonance: In each of the following problems, an nmr spectrum is described (chemical shift, splitting, ratio of hydrogens), and two or three isomeric compounds are given. Pick the compound whose spectrum is described and explain your choice.

(a) $\delta = 2.2$ singlet: $CH_3CH_2\overset{O}{\overset{\|}{C}}H$ or $CH_3\overset{O}{\overset{\|}{C}}CH_3$

(b) $\delta = 3.9$ singlet (3), $\delta = 7.8$ (5):

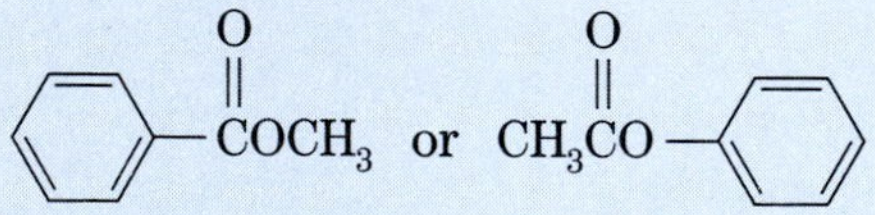

$C_6H_5{-}\overset{O}{\overset{\|}{C}}OCH_3$ or $CH_3\overset{O}{\overset{\|}{C}}O{-}C_6H_5$

(c) $\delta = 1.1$ doublet (6), $\delta = 3.1$ singlet (3), $\delta = 3.5$ heptet (1):

$CH_3CH_2OCH_2CH_3$, or $CH_3O\underset{}{\overset{CH_3}{\overset{|}{C}}}HCH_3$

(d) $\delta = 1.3$ triplet (3), $\delta = 2.7$ quartet (2), $\delta = 7.2$ singlet (2):

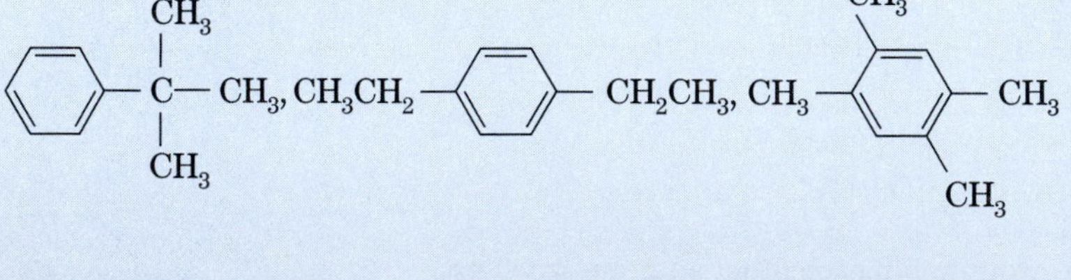

$C_6H_5{-}\underset{CH_3}{\underset{|}{\overset{CH_3}{\overset{|}{C}}}}{-}CH_3$, $CH_3CH_2{-}C_6H_4{-}CH_2CH_3$, $CH_3{-}C_6H_2(CH_3)_2{-}CH_3$ (1,2,4,5-tetramethylbenzene)

(e) $\delta = 1.2$ triplet (3), $\delta = 2.6$ quartet (2), $\delta = 3.7$ singlet (3), $\delta = 7.0$ singlet (4):

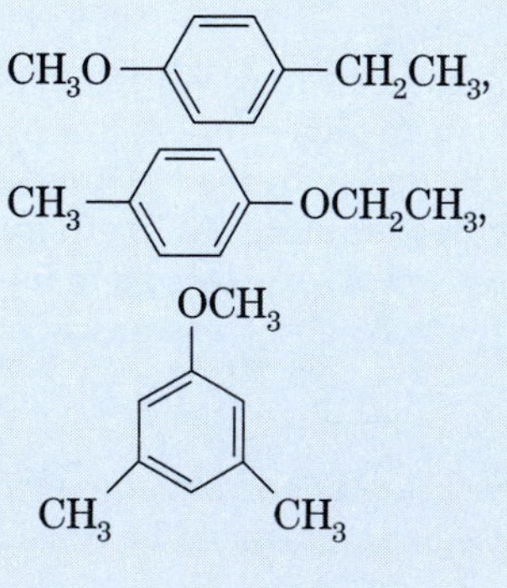

$CH_3O{-}C_6H_4{-}CH_2CH_3$, $CH_3{-}C_6H_4{-}OCH_2CH_3$, 3,5-$(CH_3)_2C_6H_3{-}OCH_3$

See related problems 18.15–18.16.

18.5 Carbon-13 NMR

Like hydrogen (^{1}H), carbon-13 (^{13}C), an isotope of carbon, gives NMR spectra. Since organic chemistry is based on carbon, one can imagine that ^{13}C NMR could be an exciting analytical tool. However, ^{13}C has an isotopic abundance of only 1.1% in nature; only about one in 100 carbon atoms is this NMR active isotope. ^{12}C, normal carbon, is not active in NMR. In a sample of a simple organic compound, most molecules would not even have a carbon-13 as one of the carbons. However, even very small samples have uncountable numbers of molecules and among these are many molecules with a carbon-13, thus providing an analyzable quantity for each position of carbon. Very sophisticated instrumentation is required to record the ^{13}C NMR because of the low concentrations of ^{13}C isotope. This instrumentation became available around 1970 and is in common use today.

^{13}C NMR is useful in the following ways:

1. ***The number of peaks in the spectrum is the number of nonequivalent carbons in the molecule.*** Each different carbon gives a signal. Consider, for example, the xylenes (dimethylbenzenes), which can clearly be distinguished by ^{13}C NMR because of their different substitution patterns on the benzene ring resulting in several nonequivalent carbons (within each structure below, equivalent carbons have equivalent identifying numbers).

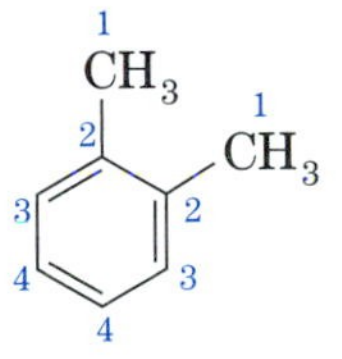

ortho	meta	para
four peaks	five peaks	three peaks

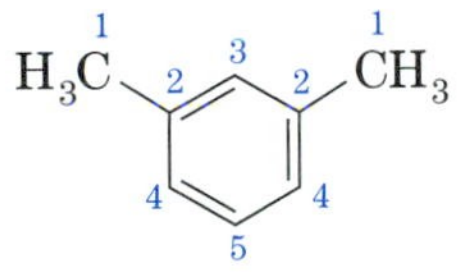

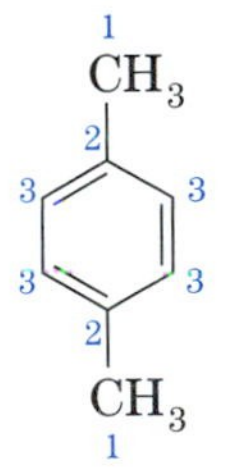

The ortho isomer has four different carbons and the ^{13}C NMR shows four peaks. Using the same reasoning, the meta isomer shows five peaks and the para isomer, only three in the ^{13}C NMR's.

2. ***The chemical shift provides information about the structural environment of each carbon.*** ^{13}C NMR uses tetramethylsilane as a reference and a scale of δ units, as does ^{1}H NMR. The chemical shifts in ^{13}C NMR, however, range over more than 200 δ units rather than the 10–15 units common in ^{1}H NMR. Some representative chemical shifts for carbons in various chemical environments follow.

C—C	C=C,  benzene ring	C≡C	C—N	C—O	C=O	C—Cl, C—Br
10–60	100–150	70–90	30–60	40–80	160–210	30–80

3. ***The number of peaks into which a signal is split is one more than the number of hydrogens bonded to that carbon.*** Because it is unlikely that a simple molecule will have even one carbon-13, much less two side by side, carbon-carbon splitting does not occur. However, splitting of a ^{13}C by attached hydrogens does occur for the reasons described in 18.4.C. A ^{13}C NMR spectrum can be run in a manner that will either show splitting of the carbons by attached hydrogens or not show the splitting. This is illustrated in Figure 18.10(a)–(b), which shows ^{13}C NMR of 2-bromobutane. Figure 18.10(a) shows the spectrum without splitting. A single peak appears for each of the four nonequivalent carbons. Splitting

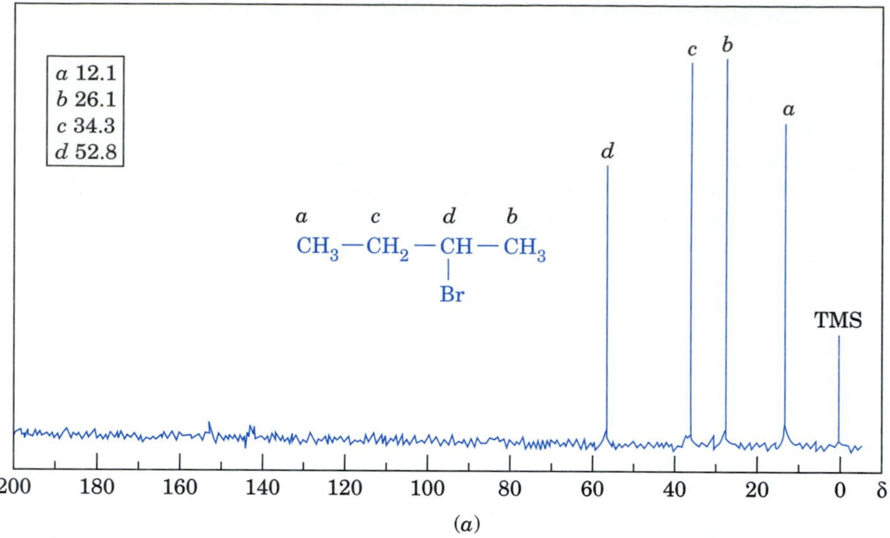

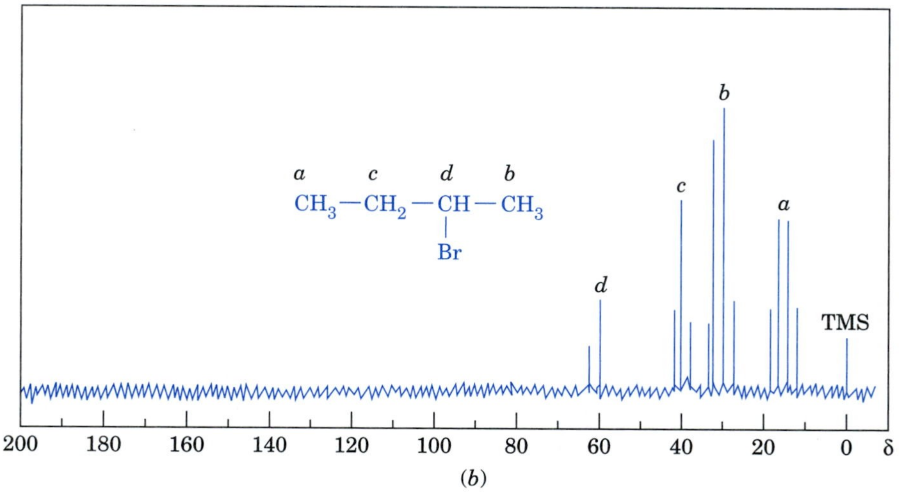

Figure 18.10 ^{13}C nmr spectra of 2-bromobutane. (a) Spectrum without splitting; (b) spectrum showing splitting of ^{13}C signals as a result of coupling with attached hydrogens. The signal at $\delta = 0$ in (a) is the TMS reference.

is shown in Figure 18.10(b). Note that the signal for each carbon is split into one more peak than the number of attached hydrogens.

GETTING INVOLVED

- ✓ Describe the three types of information that can be derived from carbon-13 NMR.
- ✓ What do the number of peaks mean? What is the difference in the chemical shift of a carbon involved in a single bond, triple bond, double bond, or benzene ring? What determines the number of peaks into which a carbon is split?

Problem 18.8

Which of the three trimethylbenzenes gives only three carbon-13 NMR signals at 138, 127, and 21? Which signal(s) correspond to carbons in the benzene ring? How many signals do each of the other two trimethylbenzene isomers show?

Problem 18.9

There are three tetramethylbenzenes. One gives carbon-13 NMR signals at 135, 134, 127, 21, and 16; another at 134, 131, and 19; and the third gives signals at 136, 134, 132, 128,

21, 20, and 15. Identify each of the isomers by the given spectrum. In each spectrum indicate which signals correspond to the methyl groups.

Problem 18.10

How many carbon-13 NMR signals would be predicted for the spectrum of hexamethylbenzene?

Problem 18.11

There are three isomers of C_3H_8O. One gives a carbon-13 spectrum that consists of a quartet and two triplets. Another shows two quartets and a triplet, and the third shows one quartet and a doublet. Draw the structure consistent with each spectrum.

See related problems 18.17–18.20.

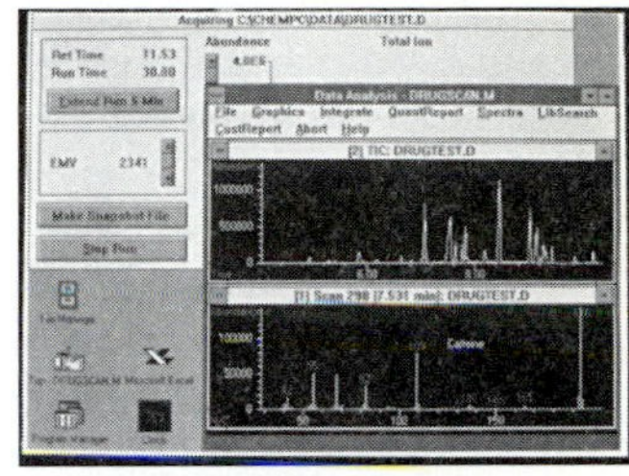

Mass spectrum of an illicit drug

18.6 Mass Spectrometry

By **mass spectral analysis,** it is possible to determine the molecular weight and molecular formula of a compound. The structure of the compound is determined by breaking the molecule into smaller, identifiable fragments and then mentally piecing them back together, like a puzzle.

mass spectrometry
an instrumental analysis in which a molecule is fragmented with radiation and the individual fragment ions are identified for use in determining the structure of the compound analyzed

Mass spectral analysis is initiated by bombarding a vaporized sample with an electron beam. This can cause an electron to be dislodged from the molecule, producing a positive molecular ion. If the electron beam is sufficiently energetic, it may cause the molecule to rupture into a variety of positive fragments.

$$\underset{\text{Molecule}}{\text{ABC}} \xrightarrow[\text{beam}]{\text{Electron}} \begin{array}{ll} \text{ABC}^+ & \\ \text{AB}^+ & \text{C}^+ \\ \text{A}^+ & \text{BC}^+ \end{array}$$

Molecular ion: ABC^+

Fragmentation ions: AB^+, C^+, A^+, BC^+

Connections 18.1 www.prenhall.com/bailey

MRI: Magnetic Resonance Imaging

Magnetic resonance imaging (MRI) is an established diagnostic tool in medicine; it provides information about anatomy and the functioning of cells and organs. MRI is actually a form of nuclear magnetic resonance, ^{1}H NMR (proton NMR) specifically. The human body is largely water (around 70%), which is found in all tissues. Since water has hydrogens, it is NMR active, and, as a result, there is the potential to probe the entire body.

In laboratory NMR used for chemical analysis, a sample is placed in a small, narrow tube and lowered into a powerful magnet. The spinning nuclei of hydrogen atoms act as tiny magnets that align either with or opposite the external magnetic field. The sample is exposed to radio waves, and some hydrogens in alignment are transformed to the less stable nonalignment state. The radio frequency at which this transformation occurs depends on the chemical and structural environment of the hydrogens in the molecules being analyzed. The same principle applies to MRI. The patient is placed in a tunnel surrounded by a large and powerful magnet. The body is magnetically scanned in selected cross-sectional planes. When exposed to radio waves, the body's hydrogen atoms flip from alignment to nonalignment with the external magnetic field, and this transformation is recorded. Hydrogens in different environments, for example,

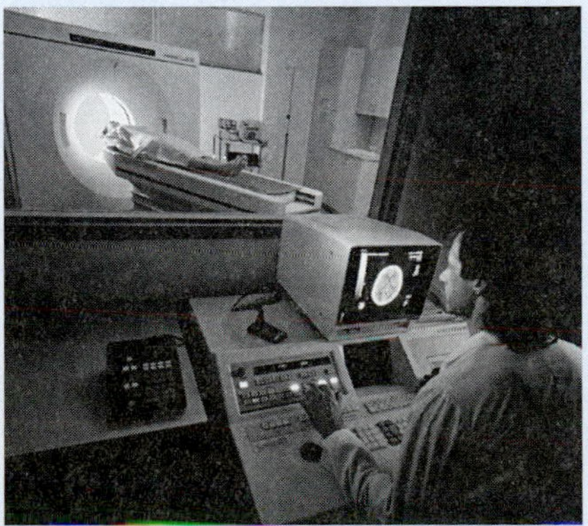

MRI machine

Connections 18.1 *(cont.)*

healthy or diseased tissue or areas of high versus low water content, behave differently in MRI. These differences are recorded and computer analyzed; the result is a high-quality cross-sectional image of body structures and organs.

MRI became available in the early 1980s in some medical facilities. Its use has expanded and diagnostic applications are still being developed. It allows images to be constructed in any plane and is particularly valuable in studying the brain and spinal cord, revealing and diagnosing tumors, examining the heart and important blood vessels, determining blood flow, examining joints, and detecting abnormalities in internal anatomy. MRI is a noninvasive procedure and unlike X-ray radiography, CAT scanning, and radionuclide imaging, MRI does not employ potentially harmful radiation. There are no known risks or side effects and thus it can be used repeatedly.

The ions are then subjected to magnetic and electric fields. Since most of them have a single positive charge, they are separated according to mass (actually, mass to charge ratio, m/e), and the separation is recorded on chart paper. Each ion shows as a peak, the intensity of which describes the relative abundance of that particular ion. Usually the spectrum is then recorded in tabular form, correlating the mass and relative abundance of each ion. For example, the mass spectrum of carbon dioxide is as shown in Table 18.3.

base peak
the most intense peak in a mass spectrum

molecular ion
the peak corresponding to the molecule minus one electron in a mass spectrum

fragment ions
peaks caused by rupture of the molecule into fragments with m/e less than the molecular ion

The most intense peak in the mass spectrum is called the **base peak** and is assigned a value of 100%. The peak formed by the loss of one electron from the molecule is called the **molecular ion** M. In CO_2, the base peak and molecular ion peak are the same. Any peaks of less mass than the molecular ion are called **fragment ions.**

GETTING INVOLVED

What happens to a molecule when it is subjected to mass spectral analysis? What are a base peak, molecular ion peak, and fragment ions? What does it mean that the structure can be determined by piecing together the fragment ions like a puzzle?

A. Molecular Formula Determination

The atomic weights of common elements are averages of the weights of naturally occurring isotopes. For example, the atomic weight of chlorine is 35.5, since there are two abundant isotopes of chlorine in nature: Cl^{35}, 75%; and Cl^{37}, 25%. The mass spectrometer detects each isotope separately, and for chlorine there would be a peak at $m/e = 35$ and a peak at $m/e = 37$, one-third (25/75) as high. By considering such isotopic abundances, one can often determine the elemental composition of a compound.

1. ***Carbon.*** Most natural carbon is ^{12}C but about 1.1% is ^{13}C. For every carbon in the molecular ion M, the next higher ion M + 1 is 1.1% of the M ion. Note in Table 18.3 that the M + 1 ion of CO_2 is 1.11% of the M ion.

Table 18.3 Mass Spectrum of Carbon Dioxide

Mass (m/e)	Relative Abundance, %
28	20
29	0.2
44	100
45	1.11

$$\text{Number of carbons in M ion} = \frac{\text{rel. abund. M} + 1}{0.011 \times \text{rel. abund. M}}$$

C_5H_{12}				$C_{10}H_{24}$			
$C_5^{12}H_{12}$	M	72	100%	$C_{10}^{12}H_{24}$	M	144	100%
$C_4^{12}C^{13}H_{12}$	M + 1	73	5.5%	$C_9^{12}C^{13}H_{24}$	M +1	145	11%

2. ***Chlorine.*** In compounds containing chlorine, the M + 2 ion (two mass units heavier than molecular ion) is about 33% of the molecular ion for each chlorine.

CH_3Cl				CH_2Cl_2			
CH_3Cl^{35}	M	50	100%	$CH_2Cl_2^{35}$	M	84	100%
CH_3Cl^{37}	M + 2	52	33%	$CH_2Cl^{35}Cl^{37}$	M + 2	86	66%

3. ***Bromine.*** Naturally occurring bromine is almost equally abundant in Br^{79} and Br^{81}. So for every bromine in a molecule, the M + 2 ion is approximately 100% of the M ion.

CH_3Br				CH_2Br_2			
CH_3Br^{79}	M	94	100%	$CH_2Br_2^{79}$	M	172	100%
CH_3Br^{81}	M + 2	96	99%	$CH_2Br^{79}Br^{81}$	M + 2	174	198%

4. ***Sulfur.*** For compounds containing sulfur, the M + 2 ion is 4.5% of the M ion for each sulfur, owing to the small isotopic abundance of S^{34} compared with S^{32}.

5. ***Nitrogen.*** If the molecular ion has an odd mass number, there is an odd number of nitrogens in the compound.

6. ***Hydrogen, Oxygen.*** Hydrogen, oxygen, and other common elements must be deduced by elimination after the other elemental components have been determined.

GETTING INVOLVED

- ✓ Describe how one determines the number of carbons in a molecule from mass spectrometry.
- ✓ How are chlorine, bromine, and sulfur detected and the number determined? How is nitrogen possibly identified? How are oxygen and hydrogen deduced?

Problem 18.12

For each of the following compounds, the M, M + 1, and M + 2 ions are given. Calculate the molecular formula.

(a) 96=100%, 97=7.7%, 98=0.1%;
(b) 92=100%, 93=3.3%, 94=33%;
(c) 156=100%, 157=6.6%, 158=98%;
(d) 234=49%, 235=3.3%, 236=98%;
(e) 91=100%, 92=2.2%, 93=4.5%.

See related problem 18.21.

B. Fragmentation Patterns

Using the M, M + 1, and M + 2 ion, we can obtain the molecular mass and either a partial or a complete molecular formula. How do we obtain a structural formula? This is accomplished by analyzing the fragment ions. In mass spec-

Table 18.4 Main Cleavage Patterns in Mass Spectrometry

A. Resonance-Stabilized Ions

$$R \wr C{-}X \longrightarrow {}^{+}C{-}X$$

Prominent cleavage

$X = O$ Alcohols and ethers 31, 45, 59, 73, 87, . . .

Alcohols	$^{+}CH_2OH$	$^{+}CH(CH_3)OH$	$^{+}CH(CH_2CH_3)OH$	$CH_3\overset{+}{C}(CH_3)OH$
	31	45	59	59

Ethers	$^{+}CH_2OCH_3$	$^{+}CH_2OCH_2CH_3$	$^{+}CH_2OCH_2CH_2CH_3$
	45	59	73

$X = N$ Amines 30, 44, 58, 72, 86, . . .

$^{+}CH_2NH_2$	$^{+}CH_2NHCH_3$	$^{+}CH(CH_3)NH_2$	$^{+}CH_2NHCH_2CH_3$
30	44	44	58

$X = C{=}C$ Alkenes 41, 55, 69, 83, 97, 111, . . .

$^{+}CH_2CH{=}CH_2$	$^{+}CH_2CH{=}CHCH_3$	$^{+}CH_2CH{=}CHCH_2CH_3$
41	55	69

$X = -C_6H_5$ (phenyl) Aromatic compounds 91, 105, 119, 133, . . .

$^{+}CH_2C_6H_5$	$^{+}CHCH_3(C_6H_5)$	$^{+}CHCH_2CH_3(C_6H_5)$	$CH_3\overset{+}{C}CH_3(C_6H_5)$	$CH_3\overset{+}{C}CH_2CH_3(C_6H_5)$
91	105	119	119	133

B. Carbonyl Compounds

$$R \wr \overset{O}{\overset{\|}{C}} \wr R$$

Prominent cleavages

Aldehydes and ketones $\quad R_1\overset{O}{\overset{\|}{C}}R_2 \longrightarrow R_1^{+} \quad R_1\overset{O}{\overset{\|}{C}}{}^{+} \quad R_2^{+} \quad R_2\overset{O}{\overset{\|}{C}}{}^{+}$

Carboxylic acids and esters $\quad R_1\overset{O}{\overset{\|}{C}}OR_2 \longrightarrow R_1^{+} \quad R_1\overset{O}{\overset{\|}{C}}{}^{+} \quad {}^{+}\overset{O}{\overset{\|}{C}}OR_2$

R^{+} 15, 29, 43, 57, 71, 85, 99, . . .

$R\overset{O}{\overset{\|}{C}}{}^{+}$ 43, 57, 71, 85, 99, 113, . . . $\quad {}^{+}\overset{O}{\overset{\|}{C}}OR$ 45, 59, 73, 87, 101, 115, . . .

Note: This table does not by any means summarize all important fragmentation patterns; it is a good start, however.

trometry, we take a large molecule whose structure is unknown and break it down with a beam of electrons into smaller, more easily identifiable fragments. The fragments are then pieced back together to obtain the structure of the unknown molecule.

In a mass spectrometer, a molecule can undergo almost any possible cleavage to form all imaginable fragment ions. Fortunately, not all fragment ions form with equal ease. In general, we can say that the probability of fragmentation depends on: (1) bond strengths—almost all important fragmentations in organic molecules are at single bonds rather than at stronger double and triple bonds; and (2) carbocation stability—the fragments are positive, and the more stable the fragment, the greater the ease of formation. Table 18.4 summarizes some of the main cleavage patterns for the types of compounds we have studied in this text. The table also summarizes the numerical sequences of mass numbers associated with the particular fragment types. For example, consider the masses of alkyl groups. The simplest, methyl, CH_3^+, has a mass of 15. Ethyl, $CH_3CH_2^+$, is 29, 14 more, and each of the subsequent alkyl fragments has a mass 14 units more than the one before.

Let us take a specific example, methyl ethyl ketone. From Table 18.4, we see that ketones fragment predominantly on either side of the carbonyl group, giving four principal ions; in this case at $m/e = 15, 29, 43$, and 57. See Example 18.4.

$$CH_3 \wr C(=O) \wr CH_2CH_3 \longrightarrow \underset{15}{CH_3^+} \quad \underset{29}{CH_3CH_2^+} \quad \underset{43}{CH_3C(=O)^+} \quad \underset{57}{CH_3CH_2C(=O)^+}$$

Methyl ethyl ketone

It is evident that both alkyl (R—) and acyl ($R\overset{O}{\overset{\|}{C}}$—) groups have the same numerical sequence. However, if we were to piece this together, we should assign the lowest mass number to the smaller alkyl group and the largest to the larger acyl group.

Another approach to structure determination is to identify the group that must have fallen off the molecule to produce an abundant fragment. To do this, we determine the mass difference between the molecular ion and the important fragment ions. For example, from Table 18.4 we see that alcohols preferentially cleave at the carbon bearing the hydroxy group. See Example 18.5.

$$R_1-C(R_2)(OH)-R_3 \longrightarrow {}^+C(R_2)(OH)-R_3 \quad R_1-\overset{+}{C}(OH)-R_3 \quad R_1-C^+(R_2)(OH)$$

GETTING INVOLVED

✓ What are fragment ions and how do they arise? What two principles largely determine where a fragmentation will occur?

✓ Make sure you understand the fragmentation patterns in Table 18.4 and begin learning the mass sequences that result.

Example 18.4

Now let us identify an unbranched ketone with the formula $C_7H_{14}O$ and principal ions at 29, 43, 57, and 71.

Solution

The smallest fragment, 29, corresponds to an ethyl group $CH_3CH_2^+$ and the largest, 71, to the acyl group $CH_3CH_2CH_2C(=O)^+$. The compound is ethyl propyl ketone.

$$CH_3CH_2C(=O)CH_2CH_2CH_3 \longrightarrow$$

Ethyl propyl ketone

$$\underset{29}{CH_3CH_2^+} \quad \underset{43}{CH_3CH_2CH_2^+} \quad \underset{57}{CH_3CH_2C(=O)^+} \quad \underset{71}{CH_3CH_2CH_2C(=O)^+}$$

Example 18.5

Suppose we have an unknown alcohol that gives a mass spectrum with a molecular ion of 116 and principal ions at 101, 87, and 73.

Solution

First, determine the difference between the molecular ion and fragment ion, M-101, M-87, and M-73. The peak M-101 is 15. One of the R groups then must have a mass of 15 and be a methyl. M-87 is 29, and must represent an ethyl group. Finally, M-73 is 43, either a propyl or an isopropyl group. A possible structure then for the unknown is one in which $R_1 = -CH_3$, $R_2 = -CH_2CH_3$, and $R_3 = -CH_2CH_2CH_3$.

$$CH_3-C(CH_2CH_3)(OH)-CH_2CH_2CH_3$$

Problem 18.13

In each of the following problems, a partial description of an unknown compound and the important ions in its mass spectrum are presented. Write a structure consistent with the information.

(a) a straight-chained ketone with the formula $C_7H_{14}O$ (M=114) with major fragments at 29, 57, and 85; **(b)** a monosubstituted benzene with the formula $C_{10}H_{16}$ (M=134) with major fragments at 77, 105, and 119; **(c)** a straight-chained secondary alcohol with the formula $C_5H_{12}O$ (M=88) and major fragments at 45 and 73; **(d)** a straight-chained ester with the formula $C_4H_8O_2$ (M=88) and major fragments at 15, 43, and 73; **(e)** a straight-chained alkene with the formula C_6H_{12} (M=84) and a major fragment at 69.

See related problems 18.21–18.23.

Problems

18.14 **Infrared Spectroscopy:** How could one distinguish between the members of the following sets of compounds by infrared spectroscopy? Give the wavenumber of an easily identifiable absorption band that would appear in one molecule but not the other. Identify the bond responsible for the absorption.

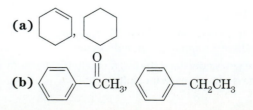

(a) cyclohexene, cyclohexane

(b) $C_6H_5C(=O)CH_3$, $C_6H_5CH_2CH_3$

(c) $CH_3CH_2CH_3$, $CH_3C\equiv CH$

(d) $CH_3\overset{O}{\overset{\|}{C}}OCH_3$, $CH_3CH_2\overset{O}{\overset{\|}{C}}OH$

(e) $CH_3CH_2CH_2NH_2$, $CH_3NHCH_2CH_3$, $(CH_3)_3N$

(f) $CH_3CH_2C\equiv N$, $(CH_3)_3N$

(g) $C_6H_5\overset{O}{\overset{\|}{C}}H$, $C_6H_5\overset{O}{\overset{\|}{C}}CH_3$

(h) CH_3CH_2OH, CH_3OCH_3

(i) CH_3NO_2, $CH_3CH_2CH_3$

(j) $C_6H_5\overset{O}{\overset{\|}{C}}OH$, $C_6H_5\overset{O}{\overset{\|}{C}}CH_3$

(k) $CH_3CH_2NHCH_3$, $CH_3CH_2OCH_3$

18.15 **^1H Nuclear Magnetic Resonance:** Draw ^1H NMR spectra of each of the following compounds. For chart paper, draw a 4-inch line and number from 0 to 8 right to left with 0.5 inch between numbers. Show the chemical shift and splitting of each signal. Also indicate the relative areas of the signals.

(a) CH_4 (b) CH_3OCH_3 (c) $CH_3\overset{O}{\overset{\|}{C}}CH_3$

(d) C_6H_6 (benzene) (e) CH_3Br (f) $CHBr_3$

(g) CH_3OH (h) $C_6H_5-CH_3$

(i) $CH_3-C_6H_4-CH_3$ (para)

(j) $C_6H_5-CH_2O\overset{O}{\overset{\|}{C}}CH_3$ (k) Cl_2CHCH_2Cl

(l) CH_3CHBr_2 (m) $(CH_3)_2CHOCH(CH_3)_2$

(n) $BrCH_2CH_2CH_2Br$

(o) $C_6H_5-CH_2\overset{O}{\overset{\|}{C}}CH_2Cl$

(p) $C_6H_5-CH_2N(CH_2CH_3)CH_2CH_3$

(q) $CH_2{=}C(OCH_2CH_3)_2$

(r) $C_6H_5-CHBrCH_3$

18.16 **^1H Nuclear Magnetic Resonance:** In each of the following problems, a molecular formula is given and the NMR spectrum described (chemical shift, splitting, ratio of hydrogens). Draw a structural formula consistent with the molecular formula and the NMR spectrum.

(a) $C_3H_6O_2$: $\delta = 2.0$, singlet (1); $\delta = 3.7$, singlet (1)
(b) $C_6H_{12}O_2$: $\delta = 1.4$, singlet (3); $\delta = 2.1$, singlet (1)
(c) C_2H_6O: $\delta = 1.2$, triplet (3); $\delta = 3.6$, singlet (1); $\delta = 4.4$, quartet (2)
(d) C_4H_8O: $\delta = 1.1$, triplet (3); $\delta = 2.1$, singlet (3); $\delta = 2.4$, quartet (2)
(e) C_3H_7Br: $\delta = 1.7$, doublet (6); $\delta = 3.4$, heptet (1)
(f) $C_2H_4O_2$: $\delta = 2.0$, singlet (3); $\delta = 11.4$, singlet (1)
(g) $C_2H_3Cl_3$: $\delta = 3.9$, doublet (2); $\delta = 5.8$, triplet (1)
(h) C_7H_8: $\delta = 2.3$, singlet (3); $\delta = 7.2$, singlet (5)
(i) $C_{13}H_{11}Cl$: $\delta = 6.1$, singlet (1); $\delta = 7.3$, singlet (10)
(j) $C_{15}H_{14}O$: $\delta = 2.1$, singlet (3); $\delta = 5.0$, singlet (1); $\delta = 7.0$, singlet (10)
(k) $C_3H_5ClO_2$: $\delta = 1.8$, doublet (3); $\delta = 4.5$, quartet (1); $\delta = 11.2$, singlet (1)
(l) $C_4H_6Cl_2O_2$: $\delta = 1.4$, triplet (3); $\delta = 4.3$, quartet (2); $\delta = 6.9$, singlet (1)
(m) $C_7H_{12}O_4$: $\delta = 1.3$, triplet (3); $\delta = 3.4$, singlet (1); $\delta = 4.2$, quartet (2)
(n) C_8H_{10}: $\delta = 1.3$, triplet (3); $\delta = 2.7$, quartet (2); $\delta = 7.2$, singlet (5)

18.17 **Carbon-13 NMR without Splitting**

(a) There are two isomers with the formula C_3H_7Br. One gives ^{13}C NMR signals at δ values of 36, 26, and 13, and the other at 45 and 28. Draw the structure that corresponds to each spectrum.

(b) There are four isomers of C_4H_9Cl. Draw the one that gives two ^{13}C NMR peaks at 67 and 34 and another that gives three signals at 53, 31, and 20.

(c) There are three dibromobenzenes. One gives ^{13}C NMR signals at 133 and 121; another at 134, 128, and 125; and the third at 134, 131, 130, and 123. Identify the ortho, meta, and para isomers.

(d) There are seven isomers with the formula $C_4H_{10}O$, four alcohols and three ethers. Identify the structures that give four ^{13}C NMR signals, those that give three, and those that give only two.

(e) There are three isomers with the formula C_5H_{12}.

One shows two signals in the ^{13}C NMR, another three, and the third one shows four. Draw the structure of each.

(f) There are three ketones with the formula $C_5H_{10}O$. Which one shows only three signals in the ^{13}C NMR (212, 35, 8)?

(g) Which of the 18 isomers of C_8H_{18} gives the fewest signals in the ^{13}C NMR spectrum? How many signals are there for this compound?

18.18 **^{13}C NMR with Splitting**

(a) There are two isomers of C_2H_6O. One gives a single ^{13}C NMR signal split into a quartet and the other gives a quartet and a triplet. Draw a structure for each.

(b) There are three isomers of C_4H_8O that have carbon-oxygen double bonds. One gives a ^{13}C NMR spectrum that consists of a quartet and two doublets; another gives two quartets, a triplet, and a singlet; and the third gives one quartet, one doublet, and two triplets. Draw a structure consistent with each spectrum.

(c) One isomer of C_4H_{10} shows a quartet and a triplet in the ^{13}C NMR and the other a quartet and a doublet. Draw structures for each.

(d) There are three isomers of C_5H_{12}. One isomer shows a quartet and two triplets in the ^{13}C NMR. Another shows two quartets, a doublet, and a triplet. The third one shows a quartet and a singlet. Draw a structure consistent with each spectrum.

18.19 **^{13}C nmr:** The following compound gives ^{13}C NMR signals at 161, 81, and 28. Identify the carbon responsible for each signal and indicate the expected splitting.

```
O  CH3
‖  |
HCOCCH3
   |
   CH3
```

18.20 **^{13}C NMR:** The following compound gives the ^{13}C NMR spectrum described. Write the chemical shift described for each signal next to the carbon(s) responsible for the signal. Two carbons have already been assigned as an illustration.

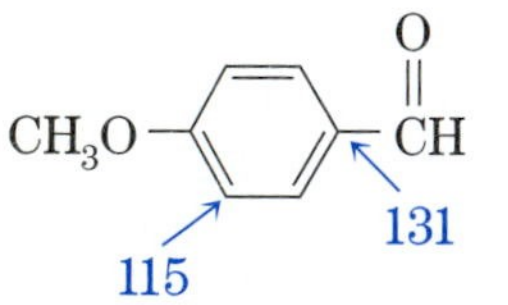

CH_3O–(benzene ring)–$\overset{O}{\overset{\|}{C}}H$ (ring carbons labeled 115 and 131)

190 doublet	165 singlet
131 singlet	115 doublet
133 doublet	55 quartet

18.21 **Mass Spectrometry:** In each of the following problems the M, M + 1, and M + 2 ions of the mass spectra of the unknown compounds are given. Calculate a molecular formula.

(a) 114 = 100%, 115 = 8.8%, 116 = 0.1%
(b) 64 = 100%, 65 = 2.2%, 66 = 33%
(c) 136 = 40%, 137 = 1.3%, 138 = 39%
(d) 48 = 100%, 49 = 1.1%, 50 = 4.5%
(e) 96 = 80%, 97 = 1.8%, 98 = 54%
(f) 234 = 50%, 235 = 3.3%, 236 = 99%
(g) 59 = 75%, 60 = 2.5%, 61 = 0.05%
(h) 140 = 30%, 141 = 2.3%, 142 = 10%
(i) 92 = 70%, 93 = 1.5%, 94 = 6.3%

18.22 **Mass Spectrometry:** In each of the following problems, a partial description of an unknown compound is presented along with the important fragmentations of its mass spectrum. Write a structure consistent with the data for the unknown.

(a) A straight-chained ketone with the formula $C_8H_{16}O$ (M = 128) and major *m/e* ions at 29, 57, 71, and 99.

(b) A straight-chained alkene with the formula C_7H_{14} (M = 98) and major *m/e* ions at 69 and 83.

(c) A monosubstituted benzene with the formula $C_{12}H_{18}$ (M = 162) and major *m/e* ions at 133 and 147.

(d) A straight-chained secondary alcohol with the formula $C_7H_{16}O$ (M = 116) and major *m/e* ions at 59 and 87.

(e) A straight-chained secondary amine with the formula $C_5H_{13}N$ (M = 87) and major *m/e* ions at 58 and 72.

(f) A straight-chained ester with the formula $C_{10}H_{20}O_2$ (M = 172) and major *m/e* ions at 57, 85, and 115.

18.23 **Mass Spectrometry:** Using the concepts in Table 18.4, predict the major fragmentations (one to five ions) in the mass spectra of the following compounds:

(a) $CH_3(CH_2)_5\overset{O}{\overset{\|}{C}}(CH_2)_3CH_3$

(b) $CH_3CH_2\overset{O}{\overset{\|}{C}}OH$ **(c)** $CH_3(CH_2)_5\overset{O}{\overset{\|}{C}}H$

(d) $CH_3\overset{O}{\overset{\|}{C}}OCH_2CH_3$ **(e)** 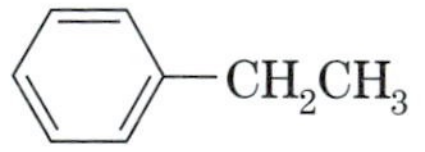 C_6H_5–CH_2CH_3 (benzene ring)

(f) C_6H_5–$C(CH_3)(CH_2CH_2CH_3)CH_2CH_3$ (benzene ring attached to C bearing CH_3, CH_2CH_3, and $CH_2CH_2CH_3$)

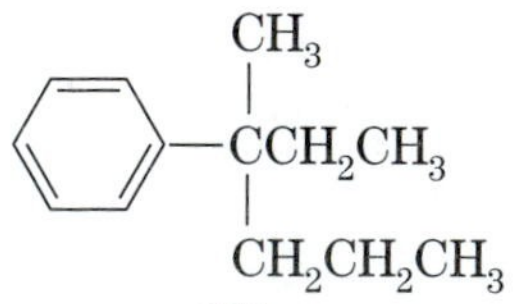

(g) $CH_3CH_2\underset{}{\overset{CH_3}{\overset{|}{C}}}HCH{=}CHCH_2CH_2CH_2CH_3$

(h) $CH_3CH_2\overset{CH_3}{\overset{|}{C}}{=}\overset{CH_3}{\overset{|}{C}}CH_3$

(i) $CH_3CH_2CH_2\overset{CH_3}{\overset{|}{N}}CH_2CH_3$

APPENDIX

Summary of IUPAC Nomenclature

In this appendix the nomenclature of organic compounds presented in the text is compiled and summarized. More detailed presentations of the various aspects of organic nomenclature and examples can be found in the following sections.

I Nomenclature of Nonaromatic Compounds

Nomenclature Rule 1: Naming the Carbon Chain

The base of the name of an organic compound is derived from the Greek name for the number of carbon atoms present in the longest continuous chain of carbons (Table A.1). A cyclic chain is designated by the prefix *cyclo-*. Side-chain alkyl and alkoxy groups are named as shown in Table A.2.

Nomenclature Rule 2: Describing Carbon-Carbon Bonds

If all carbon-carbon bonds in the longest continuous carbon chain are single bonds, this is indicated by the suffix *-ane*. Double bonds are described by the suffix *-ene,* and triple bonds by the suffix *-yne.* See Table A.3a.

Nomenclature Rule 3: Naming Functional Groups

The functional groups in Table A.3b are designated by a suffix when only one such group is present. If more than one such group is present, the group highest in the table is allocated the suffix and the rest are indicated by prefixes. To insert the suffix, drop the *-e* of the parent hydrocarbon and add the suffix.

Nomenclature Rule 4: Naming Substituents

The groups in Table A.3c are named only by prefixes.

Table A.1 Continuous-Chain Hydrocarbons

First Ten Hydrocarbons

CH_4	Methane	$CH_3(CH_2)_4CH_3$	Hexane
CH_3CH_3	Ethane	$CH_3(CH_2)_5CH_3$	Heptane
$CH_3CH_2CH_3$	Propane	$CH_3(CH_2)_6CH_3$	Octane
$CH_3(CH_2)_2CH_3$	Butane	$CH_3(CH_2)_7CH_3$	Nonane
$CH_3(CH_2)_3CH_3$	Pentane	$CH_3(CH_2)_8CH_3$	Decane

Higher Hydrocarbons

$C_{11}H_{24}$	Undercane	$C_{20}H_{42}$	Eicosane	$C_{40}H_{82}$	Tetracontane
$C_{12}H_{26}$	Dodecane	$C_{21}H_{44}$	Heneicosane	$C_{49}H_{100}$	Nontetracontane
$C_{13}H_{28}$	Tridecane	$C_{22}H_{46}$	Docosane	$C_{50}H_{102}$	Pentacontane
$C_{14}H_{30}$	Tetradecane	$C_{23}H_{48}$	Tricosane	$C_{60}H_{122}$	Hexacontane
$C_{15}H_{32}$	Pentadecane	$C_{26}H_{54}$	Hexacosane	$C_{70}H_{142}$	Heptacontane
$C_{16}H_{34}$	Hexadecane	$C_{30}H_{62}$	Triacotane	$C_{80}H_{162}$	Octacontane
$C_{17}H_{36}$	Heptadecane	$C_{31}H_{64}$	Hentriacontane	$C_{90}H_{182}$	Nonacontane
$C_{18}H_{38}$	Octadecane	$C_{32}H_{66}$	Dotriacontane	$C_{100}H_{202}$	Hectane
$C_{19}H_{40}$	Nonadecane	$C_{33}H_{68}$	Tritriacontane	$C_{132}H_{266}$	Dotriacontahectane

Cyclopropane	Cyclobutane	Cyclopentane	Cyclohexane	Cyclooctane

Table A.2 Alkyl and Alkoxy Groups

$CH_3—$	Methyl	$CH_3O—$	Methoxy
$CH_3CH_2—$	Ethyl	$CH_3CH_2O—$	Ethoxy
$CH_3CH_2CH_2—$	Propyl	$CH_3CH_2CH_2O—$	Propoxy
$CH_3(CH_2)_2CH_2—$	Butyl	$CH_3(CH_2)_2CH_2O—$	Butoxy
$CH_3(CH_2)_3CH_2—$	Pentyl	$CH_3(CH_2)_3CH_2O—$	Pentoxy

Branched Alkyl Groups

CH_3CHCH_3 (bond from central C, down)	$CH_3CHCH_2CH_3$ (bond from second C, down)	$CH_3CHCH_2—$ with CH_3 on central C	CH_3CCH_3 with CH_3 above and bond below
Isopropyl	Secondary butyl (*sec*-)	Isobutyl	Tertiary butyl (*tert*-, or *t*-)

Nomenclature Rule 5: Numbering the Carbon Chain

In numbering a carbon chain, the lowest numbers are given preferentially to (1) the functional group in Table A.3b named by a suffix, followed by (2) carbon-carbon multiple bonds (double bonds take precedence over triple bonds when numbering would otherwise give like results), and (3) groups named by prefixes.

TABLE A.3 ◆ Group Nomenclature: Prefixes and Suffixes

Class	Functional Group	Prefix	Suffix
a. Groups indicated by suffix only:			
Alkanes	C—C	. . .	*-ane*
Alkenes	C=C	. . .	*-ene*
Alkynes	C≡C	. . .	*-yne*
b. Groups indicated by prefix or suffix:			
*Carboxylic acids	—COOH, —C(=O)OH, —CO_2H	*carboxy-*	*-oic acid*
Aldehydes	—CHO, —C(=O)—H	*oxo-*	*-al*
Ketones	—C(=O)—	*oxo-*	*-one*
Alcohols	—OH	*hydroxy-*	*-ol*
Amines	—NH_2	*amino-*	*-amine*
c. Groups indicated by prefix only:			
Halogenated compounds	—F	*fluoro-*	. . .
	—Cl	*chloro-*	. . .
	—Br	*bromo-*	. . .
	—I	*iodo-*	. . .
Nitrated compounds	—NO_2	*nitro-*	. . .
Alkylated compounds	—R	*alkyl-*	. . .
Ethers	—OR	*alkoxy-*	. . .

*Nomenclature of carboxylic acid derivatives:

Derivative	*Suffix*	*Example*	
Carboxylic acid RC(=O)OH	*-oic acid*	CH_3C(=O)*OH*	Ethan*oic acid*
Acid chloride RC(=O)Cl	Change *-ic acid* to *-yl chloride*	CH_3C(=O)*Cl*	Ethan*oyl chloride*
Acid anhydride RC(=O)OC(=O)R	Change *-acid* to *anhydride*	CH_3C(=O)OC(=O)CH_3	Ethanoic *anhydride*
Esters RC(=O)OR′	Change *-ic acid* to *-ate;* precede name by name of R′ or M.	CH_3C(=O)OCH_3	*Methyl* ethano*ate*
Salts RC(=O)OM		CH_3C(=O)ONa	*Sodium* ethano*ate*
Amides RC(=O)NH_2	Change *-oic acid* to *-amide*	CH_3C(=O)NH_2	Ethan*amide*

Procedure for Naming Organic Compounds

1. Determine and name the longest continuous chain of carbons (Rule 1, Table A.1).
2. If all carbon-carbon bonds are single bonds, retain the *-ane* suffix of the parent hydrocarbon. For carbon-carbon double bonds, use the suffix *-ene,* and for carbon-carbon triple bonds, *-yne* (Rule 2, Table A.3a).
3. Name the group highest in Table A.3b with a suffix by dropping the *-e* from the end of the parent hydrocarbon and replacing it with the appropriate suffix (Rule 3, Table A.3b).
4. Number the carbon chain, giving preference to the functional group in Table A.3b named by a suffix, then carbon-carbon double or triple bonds, and finally, groups named by prefixes (Rule 5). Complete the suffix by identifying the location of groups described.
5. Name (and locate with a number) all other groups with prefixes (Rule 4, Table A.3b and c).

Example A.1

Give the IUPAC name for

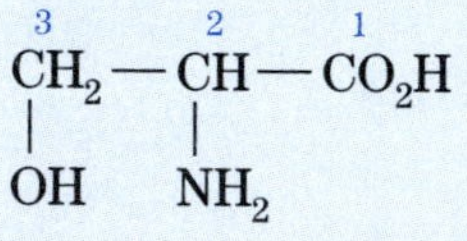

$$\underset{\text{OH}}{\overset{3}{\text{CH}_2}}-\underset{\text{NH}_2}{\overset{2}{\text{CH}}}-\overset{1}{\text{CO}_2\text{H}}$$

Solution

1. There are three carbons in the longest chain: prop.
2. There are no carbon-carbon double or triple bonds: propan.
3. The acid group is highest in Table A.3 and is named with a suffix: propanoic acid.
4. The chain is numbered right to left, with preference given to the acid group.
5. The amino and hydroxy groups are named with prefixes and located. The complete name is

2-amino-3-hydroxypropanoic acid

Example A.2

Give the IUPAC name for

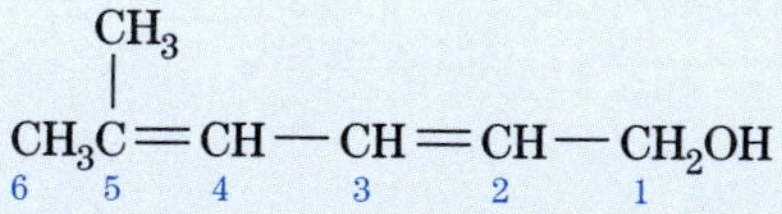

$$\underset{6}{\text{CH}_3}\overset{\text{CH}_3}{\underset{5}{\text{C}}}=\underset{4}{\text{CH}}-\underset{3}{\text{CH}}=\underset{2}{\text{CH}}-\underset{1}{\text{CH}_2\text{OH}}$$

Solution

1. There are six carbons in the longest chain: hex.
2. There are two carbon-carbon double bonds: hexadiene.
3. The alcohol group is named with a suffix: hexadienol.
4. The chain is numbered right to left, with preference given to the alcohol group: 2,4-hexadien-1-ol.
5. The methyl group is named with a prefix and located. The complete name is

5-methyl-2,4-hexadien-1-ol

Example A.3

Give the IUPAC name for

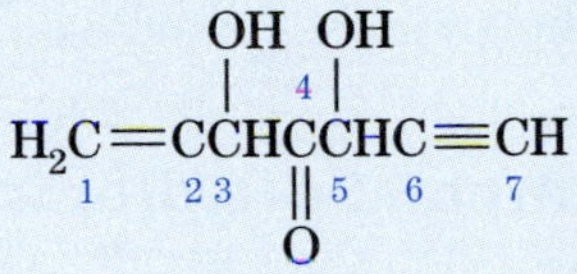

$$\begin{array}{c} \text{OH}\ \ \text{OH} \\ |\ \ {}^{4}\ \ | \\ \underset{1}{\text{H}_2\text{C}}{=}\underset{2}{\text{C}}\underset{3}{\text{C}}\text{H}\overset{}{\underset{\|\atop \text{O}}{\text{C}}}\underset{5}{\text{C}}\text{H}\underset{6}{\text{C}}{\equiv}\underset{7}{\text{CH}} \end{array}$$

Solution

1. There are seven carbons in the longest chain: hept.
2. There is one double and one triple bond: hept-en-yne.
3. The ketone group is highest in Table A.3 and takes a suffix: hepten-yn-one.
4. Numbering in either direction gives the ketone (the highest priority group) the same number (4) and places a multiple bond on carbon-1. In this case, then, the double bond takes precedence and numbering is left to right. All groups are located in the suffix: 1-hepten-6-yn-4-one.
5. The two alcohol groups are named with prefixes. The complete name is

3,5-dihydroxy-1-hepten-6-yn-4-one

II Nomenclature of Substituted Amines and Amides

Nomenclature Rule 6: Locating Substituents on Amines and Amides

If one or more hydrogens on the nitrogen are replaced by substituents, the positions of these substituents are indicated by a capital N.

Example A.4

Give the IUPAC name for

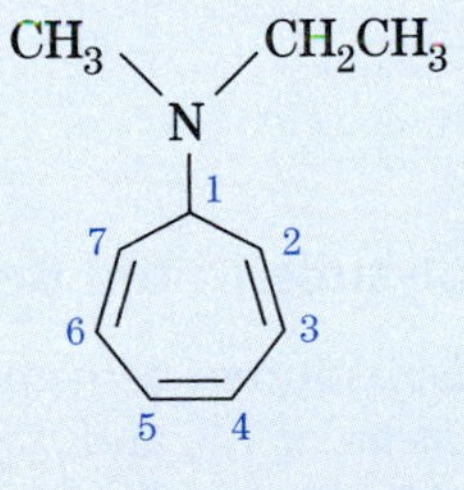

CH_3 and CH_2CH_3 on N; N attached to C-1 of a cycloheptatriene ring (ring positions 1–7, double bonds at 2–3, 4–5, 6–7)

Solution

1. There is a seven-carbon ring: cyclohept.
2. There are three double bonds: cycloheptatriene.
3. The amine group is named with a suffix: cycloheptatrienamine.
4. Numbering begins with the amine group: 2,4,6-cycloheptatrien-1-amine.
5. The alkyl groups are named with prefixes, and N's are used to locate them since they are on a nitrogen. The complete name is

N-ethyl-N-methyl-2,4,6-cycloheptatrien-1-amine

III Nomenclature of Carboxylic Acid Derivatives

Nomenclature Rule 7: Suffix Endings of Acid Derivatives

Carboxylic acid derivatives are named by modifying the suffix ending on the name of the parent acid, as shown in the footnote of Table A.3.

IV Nomenclature of Aromatic Hydrocarbons

Nomenclature Rule 8: Parent Aromatic Ring Systems

Following are common aromatic ring systems and their names:

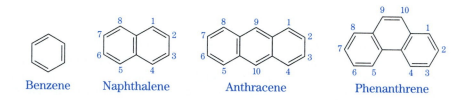

Benzene   Naphthalene   Anthracene   Phenanthrene

Nomenclature Rule 9: Monosubstituted Aromatics

Monosubstituted aromatic compounds are named as derivatives of the parent ring system (such as chlorobenzene). Some monosubstituted benzenes are frequently referred to by common names.

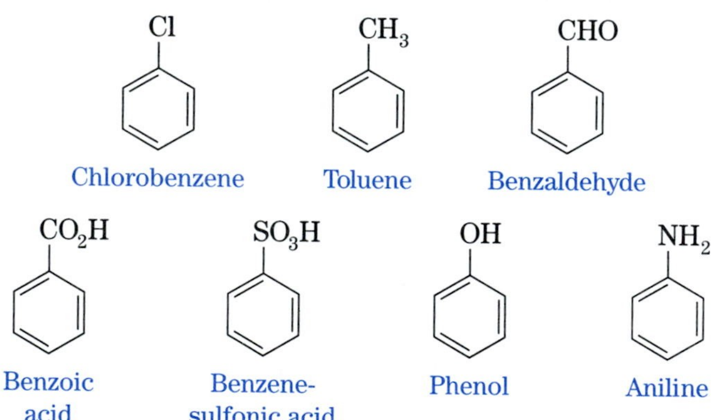

Nomenclature Rule 10: Polysubstituted Benzenes

The positions of groups on disubstituted benzenes can be designated by numbers or *o*-, *m*-, *p*-; *ortho*- or 1,2; *meta*- or 1,3; and *para*- or 1,4. On more highly substituted benzenes, numbers only are used to indicate the relative positions of groups.

o-dichlorobenzene   *m*-bromochlorobenzene   *p*-nitrobenzaldehyde

Nomenclature Rule 11: Aromatic Rings as Prefixes

Aromatic rings can be named using prefixes when this will simplify the overall name.

Phenyl C_6H_5- Benzyl $C_6H_5-CH_2-$

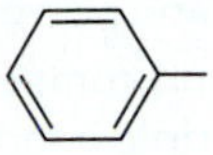

$CH_3CH(C_6H_5)CH_2CH_2OH$

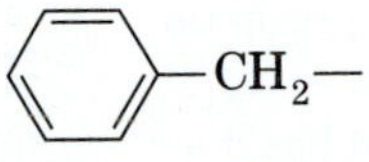

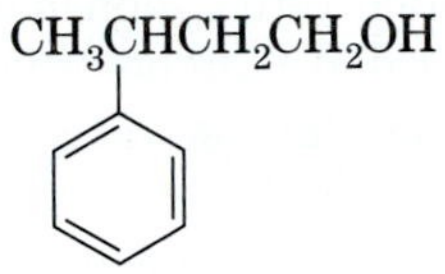

3-phenylbutanol

$CH_3C(C_6H_4NO_2)=CHCHO$

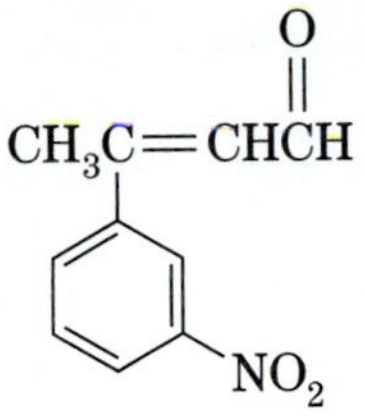

3-(meta-nitrophenyl)-2-butenal

Glossary

(α)-alpha amino acid molecule with an amine group on the carbon adjacent to a carboxyl group.

α-anomer the cyclic monosaccharide form that has the OH group on the new chiral center below the ring; on the right in a Fischer projection.

(α)-alpha helix spiral protein secondary structure stabilized by hydrogen-bonding.

(α)-alpha hydrogen hydrogen on a carbon connected to a carbonyl group.

acetal carbon bonded to two OR groups; the diether product of the reaction between an aldehyde and two moles of alcohol.

acid anhydride functional group in which RCO_2 of one acid molecule is bonded to the carbon-oxygen double bond of another.

acid chloride functional group in which a chlorine is bonded to a carbon-oxygen double bond.

acidity constant K_a, product of the concentrations of the ionized form of an acid divided by the concentration of the un-ionized form.

activating group group that increases the reactivity of an aromatic compound to electrophilic substitution.

active site functional portion of an enzyme.

acyl group RC=O group found in carboxylic acids and derivatives.

addition polymer polymer that results from polymerization of alkenes.

addition reaction reaction in which atoms or groups add to adjacent atoms of a multiple bond.

alcohol ROH, alkane in which a hydrogen is replaced with OH.

aldehyde functional group in which at least one H is bonded to a carbonyl.

aldol condensation base-catalyzed reaction between two aldehyde or ketone molecules to form a product with both alcohol and carbonyl groups.

aldose a polyhydroxy aldehyde.

alkaloids plant-produced nitrogenous bases that have physiological effects on humans.

alkane compound composed of only carbon and hydrogen and single bonds.

alkene compound composed of carbon and hydrogen and at least one double bond.

alkylation introduction of an alkyl group into a molecule.

alkyl group hydrocarbon chain with one open point of attachment.

alkyl halide alkane possessing at least one F, Cl, Br, or I.

alkyne compound composed of carbon and hydrogen and at least one triple bond.

allyl $CH_2{=}CHCH_2$ is the allyl group.

allylic carbocation carbocation in which positive carbon is directly attached to a carbon-carbon double bond.

allylic free-radical free-radical carbon directly attached to a carbon-carbon double bond.

amide functional group in which NH_2, NHR, or NR_2 is attached to a carbon-oxygen double bond.

amine derivative of ammonia in which one or more hydrogens are replaced by organic groups.

amphipathic molecule with a polar portion and a nonpolar portion.

amylopectin a component of starch; branched polymer of glucose units connected with α-1,4 glycosidic bonds in its linear chains with α-1,6 branching in intervals of about 25 units.

amylose a component of starch; linear polymer of glucose units connected by α-1,4 glycosidic bonds.

anion negatively charged ion.

anomer one of two optical isomers formed at the new chiral center produced when an aldehyde or ketone reacts with one mole of an alcohol.

aromatic compounds compounds that resemble benzene in structure and chemical behavior.

arylamine derivative of ammonia in which at least one hydrogen is replaced by aromatic ring.

atom smallest particle of an element.

atomic number number of protons (or electrons) in an atom.

atomic weight weighted average of an element's naturally occurring isotopes.

Aufbau principle the described order of filling atomic orbitals from lowest to highest energy with electrons.

axial bonds bonds on a cyclohexane chair perpendicular to the ring with three up and three down on alternating carbons.

β-anomer the cyclic monosaccharide form that has the OH group on the new chiral center above the ring.

base peak the most intense peak in a mass spectrum.

basicity constant K_b, product of the concentrations of the protonated form of an amine and the remaining anion divided by the concentration of the unprotonated amine.

bile acid amphipathic lipid, a steroid derivative produced in the intestine to emulsify ingested lipids.

bimolecular term that describes a reaction rate that depends on the concentration of two species.

biodegradable materials that can be metabolized by soil and water bacteria.

boat conformation an unstable conformation of cyclohexane with 109° bond angles but in which most bonds are eclipsed.

boiling point temperature at which a liquid becomes a gas.

bond angle angle between two adjacent bonds.

bonding pair outer-shell electron pair involved in a covalent bond.

bond length distance between atoms in a covalent bond (usually in angstroms, 10^{-10} meters).

bond strength energy required to break a covalent bond (usually in kcal/mole).

Bronsted-Lowry acid acid that is a hydrogen ion donor in a chemical reaction.

carbanion a species with a carbon that has only three bonds, eight outer-shell electrons including one nonbonding pair, and a negative charge.

carbocation a species with a carbon that has only three bonds, six outer-shell electrons, and a positive charge.

carbocation stability order of stability is $3° > 2° > 1°$.

carbohydrate a polyhydroxy—aldehyde or ketone; the polymers and derivatives of such compounds.

carbonyl the carbon-oxygen double bond, $C{=}O$

carboxylic acid functional group in which OH is attached to a carbon-oxygen double bond.

catalyst a reagent that influences the course and rate of a reaction without being consumed.

cation positively charged ion.

cationic polymerization addition polymerization of alkenes initiated by an electrophile.

cellulose a linear polymer of glucose units linked by β-1,4 glycosidic bonds.

chain reaction a reaction that sustains itself through repeating chain-propagating steps.

chair conformation the most stable conformation of cyclohexane in which all bonds are staggered and bond angles are 109°.

chemical shift the position on NMR chart paper where a carbon or hydrogen nucleus absorbs relative to an internal standard, TMS; measured in δ units.

chiral carbon atom carbon with four different bonded groups.

chiral compound a compound that is not superimposable on its mirror image; these compounds rotate plane polarized light.

cis isomer geometric isomer in which groups are on the same side of a ring or double bond.

Claisen condensation a method for making β keto esters from esters with α hydrogens.

complex lipid lipid that can have variations in its structure and can be broken down into several types of simpler compounds.

condensed formula structural formula in which not all the bonds or atoms are individually shown.

configuration the orientation of groups around a chiral carbon or around a carbon-carbon double bond.

conformational isomers isomers that differ as a result of the degree of rotation around a carbon-carbon single bond.

conjugate base species formed by loss of a proton from an acid.

conjugation alternating double and single bonds in a molecule.

constitutional isomers isomers that vary in the bonding attachments of atoms.

covalent bond bond formed by the sharing of electrons (in pairs) between two atoms; the bond is formed by the overlap of atomic orbitals.

crossed aldol condensation aldol condensation between two different aldehydes or ketones.

cyanohydrin carbon with both an OH and CN bonded.

cycloalkane cyclic compound containing only carbon and hydrogen.

deactivating group group that decreases the reactivity of an aromatic compound to electrophilic substitution.

dehydration reaction in which the elements of water (H and OH) are eliminated from a molecule.

dehydrohalogenation a reaction in which hydrogen and halogen are eliminated from a molecule.

density weight per unit volume of a substance.

detergent amphipathic molecules that are not soap.

dextrorotatory rotation of plane-polarized light to the right (d or +).

dextrose another name for glucose.

diastereomers stereoisomers that are not mirror images.

diazonium salt compound in which a molecule of nitrogen is bonded to an aromatic ring.

dimer two structural units.

disaccharide two monosaccharide units linked by a glycosidic bond.

double bond bond with two shared pairs of electrons.

D-sugar the most common carbohydrates; D- refers to the right-hand orientation of the chiral OH group farthest from the carbonyl group.

drying oil an oil that can be hardened by the process of oxidation.

E, Z terms used to describe the configurations of alkene geometric isomers.

E1 elimination unimolecular; the two-step elimination mechanism.

E2 elimination bimolecular; the one-step elimination mechanism.

eclipsed conformation around a carbon-carbon single bond in which attached atoms are as close together as possible.

electromagnetic radiation various wavelengths of energy.

electron negatively charged subatomic particle with negligible mass.

electron configuration description of orbital occupancy by electrons of an atom or ion by energy level and number of electrons.

electron dot formula molecular representation using dots to show each atom's outer-shell electrons, both bonding and nonbonding pairs.

electronegative element with electron-attracting capabilities.

electronegativity ability of an atom to attract its outer-shell electrons and electrons in general.

electrophile an electron-deficient species that accepts electrons from nucleophiles in a chemical reaction. Electrophiles are Lewis acids.

electrophilic addition addition reaction initiated by an electron-deficient species (electrophile).

electropositive element with electron-donating capabilities.

elimination reaction a reaction in which atoms or groups are removed from adjacent atoms to form a double or triple bond.

emulsification the process of solubilizing polar and nonpolar compounds

enantiomers stereoisomers that are mirror images.

endothermic reaction a reaction in which energy is released.

energy of activation the energy difference between reactants and a transition state.

enol compound with OH bonded to a carbon-carbon double bond.

enolate resonance-stabilized carbanion resulting from abstraction of an α hydrogen.

epimer one of two diastereomers that differ in the orientation of groups at only one carbon.

epoxide three-membered ring cyclic ether.

equatorial bonds bonds on cyclohexane chair parallel to the ring.

ester functional group in which OR, alkoxy, is attached to a carbon-oxygen double bond.

ether ROR, oxygen with two organic groups.

exothermic reaction a reaction in which energy is absorbed.

fat triester of glycerol wherein the acids are long-chain and highly saturated.

fat-soluble vitamin nonpolar, nonwater-soluble, essential dietary component; vitamins A, D, E, and K.

fatty acid long-chain (10–24 carbons) carboxylic acid.

Fischer projection method for expressing the structure of stereoisomers.

formal charge difference between the number of outer-shell electrons "owned" by a neutral free atom and the same atom in a compound.

free radical a neutral species with a carbon that has only three bonds and seven outer-shell electrons, one of which is unpaired.

free-radical polymerization addition polymerization of alkenes initiated by a free radical.

frequency number of waves per unit distance or per unit time (cycles per second).

functional group a structural unit (grouping of atoms) in a molecule that characterizes a class of organic compounds and causes the molecule to display the characteristic chemical and physical properties of the class of compounds.

functional isomers isomers with structural differences that place them in different classes of organic compounds.

furanose five-membered ring form of a monosaccharide.

gas state of matter with variable volume and shape. Molecules are independent, in random motion, and without intermolecular attractions.

gem diol carbon with two bonded OH groups.

geometric isomers *cis* and *trans* isomers; a type of stereoisomerism in which atoms or groups display orientation differences around a double bond or ring.

glyceride see fat and oil.

glycogen branched polymer of glucose units connected with α-1,4 glycosidic bonds in its linear chains with α-1,6 branching in intervals of 8 to 25 units.

glycosidic bond diether formed from the reaction of a cyclic monosaccharide molecule with another monosaccharide.

Grignard reagent the reagent RMgX developed by Nobel laureate Victor Grignard.

halogenation reaction in which halogen is introduced into a molecule.

Haworth structure two-dimensional five- or six-membered ring representation of the cyclic form of a monosaccharide; OH groups that appear on the right in a Fischer projection are drawn down (below the plane of the ring) in a Haworth structure and those on the left are drawn up.

heat of reaction the difference in energy between reactants and products in a chemical reaction.

hemiacetal carbon bonded to both an OH or an OR group; the alcohol-ether product of the reaction between an aldehyde and one mole of an alcohol.

heterocycle cyclic compound where at least one ring atom is not carbon.

heterolytic cleavage bond cleavage in which the bonding electrons are unevenly divided between the two parting atoms.

homologous series a series in which each compound differs from the one preceding by a constant factor; each of the members of the homologous series methane, ethane, propane, butane, pentane, and so on differs from the preceding by a CH_2 group.

homolytic cleavage bond cleavage in which the bonding electrons are evenly divided between the two parting atoms.

hybridization combination of atomic orbitals to form new orbitals of different shapes and orientations.

hydration reaction in which the elements of water (H and OH) are introduced into a molecule

hydrocarbon compound composed of only carbon and hydrogen.

hydrogenation reaction in which the elements of hydrogen (H_2) are introduced into a molecule.

hydrogenation of oils the catalytic addition of hydrogen to unsaturated triacylglycerols (oils).

hydrogen-bonding intermolecular attractions caused by hydrogen bonded to an electronegative element (O, N, F) being attracted to a lone pair of electrons of another electronegative element.

hydrolysis cleavage of a bond by water.

hydrophilic water-loving.

hydrophobic water-fearing.

hyperglycemia the condition of having a higher than normal amount of sugar, usually referring to glucose, in the blood.

imine compound with a carbon-nitrogen double bond.

infrared radiation for infrared spectroscopy it is radiation with wavelengths of 2–15 micrometers or frequencies of 5000 cm^{-1} to 670 cm^{-1}.

infrared spectroscopy spectroscopy using infrared radiation; used to determine bond types and functional groups in organic compounds.

integration in NMR a technique that provides the relative numbers of hydrogens or carbons in a compound; it is the area under a peak.

invert sugar a mixture of fructose and glucose produced by the breakdown of sucrose.

iodine number a measure of the extent of unsaturation in fats and oils; the number of grams of iodine that will add to 100 grams of a fat or oil.

ionic bond bond between two atoms caused by electrostatic attraction of plus and minus charged ions.

ionic charge sign and magnitude of the charge on an ion.

isomers compounds with the same molecular formula but different structural formulas.

isotope atoms of an element that differ in number of neutrons.

ketone functional group in which two organic substituents are bonded to a carbonyl.

ketose a polyhydroxy ketone.

lactose a disaccharide composed of a galactose and a glucose unit joined by a β-1,4 glycosidic bond.

levorotatory rotation of plane-polarized light to the left (l or −).

levulose another name for fructose.

Lewis acid a substance that can accept a pair of electrons for sharing from a Lewis base in a chemical reaction. Electrophiles are Lewis acids.

Lewis base a substance with an outer-shell nonbonding electron pair that it can share in a chemical reaction with a Lewis acid. Nucleophiles are Lewis bases.

Lewis structure another term for electron dot formula.

line-bond formula molecular representation in which bonding electron pairs are represented by lines.

lipid organic biomolecules soluble to a great extent in nonpolar solvents.

liposome synthetic vesicle with a semipermeable barrier composed of phospholipids.

liquid state of matter with constant volume but variable shape; molecules in random motion but with intermolecular attractions.

malonic ester synthesis a method for preparing mono and disubstituted acetic acids.

Markovnikov's rule rule for predicting orientation of addition of unsymmetrical reagents to unsymmetrical alkenes.

mass number number of protons plus neutrons in an atom.

mass spectrometry an instrumental analysis in which a molecule is fragmented with radiation and the individual fragment ions are identified for use in determining the structure of the compound analyzed.

melting point temperature at which a solid becomes a liquid.

membrane naturally occurring, semipermeable lipid bilayer composed of phospholipids, sphingolipids, cholesterol, and proteins.

meso compounds stereoisomers that are superimposable on their mirror images.

micelle aggregation of amphipathic molecules, like soap, in water such that the nonpolar portions of the molecules are arranged together inside away from water and the polar portions protrude into the water.

molecular formula formula that gives the number of each kind of atom in a compound.

molecular ion the peak corresponding to the molecule minus one electron in a mass spectrum.

molecular weight sum of the atomic weights of the atoms in a compound.

molecule smallest particle of a compound; a bonded group of atoms.

monomer compound(s) from which a polymer is made.

monosaccharide a single carbohydrate unit.

multiple bond a double bond or triple bond.

neutron neutral subatomic particle with mass of 1.

Newman projection a way of representing conformational isomers using an end-on projection of a carbon-carbon bond.

nitrile compound with a carbon-nitrogen triple bond.

nonbonding pair a lone outer-shell electron pair not involved in a bond.

nonpolar lipid lipid with few or no polar bonds.

nonreducing sugar a carbohydrate with all of its anomeric carbons bonded to other groups, unavailable for opening to an aldehyde or ketone carbonyl.

nonsaponifiable lipid lipid that cannot be hydrolyzed in the presence of base.

nuclear magnetic resonance (NMR) spectroscopy spectroscopy in which compound is placed in a magnetic field and exposed to radio-frequency radiation. It provides information about the carbon and hydrogen structure of an organic compound.

nucleophile species with electron availability that donates electrons to electrophiles in a chemical reaction. Nucleophiles are Lewis bases.

nucleophilic acyl substitution nucleophilic substitution in which an atom of group attached to an acyl group, RC=O, is replaced.

nucleophilic substitution substitution reaction in which a nucleophile replaces a leaving group such as a halide.

nucleus center of atom; contains protons and neutrons.

oil triester of glycerol wherein the acids are long-chain and highly unsaturated.

oligosaccharide a polymer of two to ten saccharide units.

omega (ω) 3 fatty acid an unsaturated fatty acid with its last double bond three carbons in from the end of the chain.

omega (ω) 6 fatty acid an unsaturated fatty acid with its last double bond six carbons in from the end of the chain.

-onic acid a carbohydrate derivative wherein the aldehyde functional group has been oxidized to a carboxylic acid.

orbital a defined region in space with the capacity to be occupied by up to two electrons.

organic chemistry the chemistry of the compounds of carbon.

organometallic compound where a metal atom is covalently bonded to a carbon.

oxidation removal of hydrogen from a carbon-oxygen single bond or insertion of oxygen in a molecule.

oxonium ion ion formed by the bonding of a hydrogen ion to the oxygen of an alcohol or ether.

peak splitting in ^1H NMR a phenomenon in which hydrogens on an adjacent carbon split the signal of hydrogens on the other carbon.

phenol ArOH, aromatic ring with bonded OH.

phospholipid complex, saponifiable, polar lipid; triester of glycerol in which two acids are saturated and unsaturated long-chain fatty acids and the third acid is phosphoric acid that is further esterified.

pi bond covalent bond formed by the overlap of parallel p orbitals at both lobes.

pK_a negative logarithm of the acidity constant, K_a.

pK_b the negative logarithm of the basicity constant, K_b.

plane-polarized light light oscillating in only one plane.

polar bond a covalent bond between two atoms of different electronegativities causing one atom to have a greater attraction for the bonding pair(s) and thus charge separation.

polarimeter instrument used to measure the rotation of plane-polarized light.

polar lipid lipid with both polar and nonpolar bonds allowing for limited solubility in polar and nonpolar solvents.

polyamide polymer (large molecule) in which the repeating structural units are connected by amide linkages.

polyatomic ion ion composed of several atoms.

polyester polymer (large molecule) in which the repeating structural units are connected by ester linkages.

polyhydric alcohol alcohol with more than one hydroxy group.

polymer a giant molecule composed of a repeating structural unit.

polysaccharide a polymer with more than ten saccharide units.

p orbital a double-lobed atomic orbital.

positional isomers isomers that differ in the location of a noncarbon group or a double or triple bond.

posttranslational modifications chemical changes made on a completed protein such as the addition of lipid or carbohydrate or the cleavage of the polypeptide chain.

potential energy diagram a graphical depiction of energy changes during a chemical reaction.

primary atom atom with one directly attached carbon (alkyl group).

prostaglandin lipid tissue hormone synthesized from long-chain fatty acids.

proton positively charged subatomic particle with mass = 1.

pyranose six-membered ring form of a monosaccharide.

R, S terms used to describe the configurations of chiral carbons.

racemic mixture 50/50 mixture of enantiomers.

rancidification oxidation and hydrolysis of fats and oils to volatile organic acids, producing an unpalatable product.

reaction equation an equation that shows what happens in a chemical reaction by showing reactants and products.

reaction intermediate an unstable, short-lived species formed during a chemical reaction; examples are carbocations, free radicals, and carbanions.

reducing sugar carbohydrate that has one or more anomeric carbons available for oxidation by a mild oxidizing agent; that is, the carbon contains an alcohol and an ether group on it.

reduction introduction of hydrogen into a molecule, often resulting in the loss of oxygen or conversion of double bonds to single bonds.

regioselective an addition reaction that produces one of two possible positional isomers predominantly.

regiospecific an addition reaction that produces one of two possible positional isomers exclusively.

resolution through diastereomers a method for separating enantiomers.

resonance energy a measure of the degree to which a compound is stabilized by resonance.

resonance forms symbolic, nonexistent structures, differing only in positions of electrons, that are used to describe an actual molecule or ion.

resonance hybrid "average" of the resonance forms used to describe a molecule or ion that cannot be described by a single structure.

R-group R is a generic symbol for an alkyl group.

salt ionic compound composed of cation from a base and anion from neutralized acid.

saponifiable lipid lipid that can undergo hydrolysis in the presence of a base such as NaOH or KOH to simpler compounds.

saponification the alkaline hydrolysis of esters to produce soaps.

saponification number the number of milligrams of potassium hydroxide required to saponify 1 gram of a fat or an oil.

saturated a saturated molecule has all single bonds; each atom has the maximum number of attached atoms possible.

sawhorse diagram a way of representing conformational isomers with stick drawings.

Saytzeff rule in applicable elimination reactions, the most substituted alkene (with alkyl groups) will predominate.

secondary atom atom with two directly attached carbons (alkyl groups).

sigma bond molecular orbital (covalent bond) formed by the head-to-head overlap of atomic orbitals.

signal sequence N-terminal protein sequence.

simple lipid lipid with relatively uncomplex structure; it either will not be broken down by ordinary chemical processes or can be broken down into a limited number of simple compounds.

single bond bond with one shared pair of electrons.

skeletal isomers isomers that differ in the arrangement of the carbon chain.

SN_1 substitution nucleophilic unimolecular; the two-step nucleophilic substitution mechanism.

SN_2 substitution nucleophilic bimolecular; the one-step nucleophilic substitution mechanism.

soap sodium and potassium salts of long-chain fatty acids.

solid state of matter with constant volume and shape; strong attractive forces between immobile molecules in crystal lattice.

solubility the amount of material that will dissolve in a solvent and produce a stable solution.

s orbital a spherical atomic orbital.

sp-hybridization combination of one s and one p orbital to form two sp hybrid orbitals that are linearly oriented.

sp^2-hybridization combination of one s and two p orbitals to form three sp^2 hybrid orbitals that are trigonally oriented.

sp^3-hybridization combination of one s and three p orbitals to form four sp^3 hybrid orbitals that are tetrahedrally oriented.

specific rotation calculated degree of rotation of an optically active compound.

spectrophotometer an instrument that measures the absorption of energy by a chemical compound.

spectroscopy instrumental method in which the interaction of chemical compounds with electromagnetic radiation is measured.

sphingolipid complex, saponifiable, polar lipid; composed of sphingosine linked through an amide bond to a very-long-chain fatty acid and through an ester or acetal linkage to acids or carbohydrates.

stable octet an outer-shell electron configuration of eight electrons (s^2p^6).

staggered conformation around a carbon-carbon single bond in which attached atoms are as far apart as possible.

starch a natural, complex carbohydrate consisting of the polymers amylose and amylopectin.

stereocenter usually a carbon with four different bonded groups.

stereoisomers isomers with the same bonding attachments of atoms but different spatial orientations.

steroid lipid with a four-fused ring structure, three rings having six members and one ring with five members.

strong acid acid that is 100% ionized in water solution.

structural formula formula that provides the bonding arrangement of atoms in a molecule.

structural isomers isomers that vary in the bonding attachments of atoms.

substitution reaction a reaction in which an atom or group on a molecule is replaced by another atom or group.

sucrose a disaccharide composed of a glucose unit and a fructose unit joined by an α,β-1,2 glycosidic bond; a nonreducing sugar.

sulfide RSR, sulfur with two bonded alkyl groups.

tautomerism an equilibrium between two structural isomers.

tautomers two easily interconvertible structural isomers.

tertiary atom atom with three directly attached carbons (alkyl groups).

thiol RSH, alkane in which a hydrogen has been replaced by SH.

transesterification conversion of one ester into another by replacing the OR group.

trans isomer geometric isomer in which groups are on opposite sides of ring or double bond.

transition state a dynamic process of change in which bonds are being broken and formed in a chemical reaction.

triacylglycerol see fat and oil.

triglyceride see fat and oil.

triple bond bond with three shared pairs of electrons.

ultraviolet spectroscopy spectroscopy using ultraviolet radiation with wavelengths in the 200–400 nm range.

unimolecular term that describes a reaction rate that depends on the concentration of one species.

units of unsaturation a unit of unsaturation is expressed as a ring or double bond. A triple bond is two units of unsaturation.

unsaturated An unsaturated molecule has at least one double bond or triple bond.

-uronic acid a carbohydrate derivative wherein the last, primary alcohol group has been oxidized to a carboxylic acid.

valence the number of covalent bonds an atom usually forms.

valence electrons an atom's outer-shell electrons.

vinyl $CH_2{=}CH$- is the vinyl group.

visible spectroscopy spectroscopy using visible light with wavelengths in the 400–750 nm range.

vulcanization process in which rubber is treated with sulfur to improve its properties.

water-soluble vitamin polar, water-soluble, essential dietary component such as the B complex vitamins and vitamin C.

wavelength the distance between two maxima in an energy wave.

wax ester of a long-chain carboxylic acid and a long-chain alcohol.

weak acid acid that is only partially ionized in water solution.

Index

B

C

D

N

Q

R

Organic Functional Groups

FUNCTIONAL GROUP NAME	FUNCTIONAL GROUP	EXAMPLE	USE OR OCCURRENCE OF EXAMPLE
HYDROCARBONS			
Alkane	$-\overset{\mid}{\underset{\mid}{C}}-\overset{\mid}{\underset{\mid}{C}}-$	$CH_3CH_2CH_3$	propane—rural or camping gas
Alkene	$>C{=}C<$	$CH_2{=}CH_2$	ethene—precursor of polyethylene
Alkyne	$-C{\equiv}C-$	$HC{\equiv}CH$	acetylene—used in oxy-acetylene torches
Aromatic	benzene ring	benzene ring—CH_3	benzene, toluene—high octane gasoline components
CARBOXYLIC ACIDS AND DERIVATIVES			
Carboxylic acid	$-\overset{O}{\overset{\parallel}{C}}OH$	$CH_3\overset{O}{\overset{\parallel}{C}}OH$	acetic acid—vinegar acid
Acid chloride	$-\overset{O}{\overset{\parallel}{C}}Cl$	$CH_3\overset{O}{\overset{\parallel}{C}}Cl$	acetyl chloride—organic synthesis
Acid anhydride	$-\overset{O}{\overset{\parallel}{C}}O\overset{O}{\overset{\parallel}{C}}-$	$CH_3\overset{O}{\overset{\parallel}{C}}O\overset{O}{\overset{\parallel}{C}}CH_3$	acetic anhydride—organic synthesis
Ester	$-\overset{O}{\overset{\parallel}{C}}O\overset{\mid}{\underset{\mid}{C}}-$	$H\overset{O}{\overset{\parallel}{C}}OCH_2CH_3$	ethyl formate—artificial rum-flavoring agent
Amide	$-\overset{O}{\overset{\parallel}{C}}-\overset{\mid}{N}-$	$H_2N\overset{O}{\overset{\parallel}{C}}NH_2$	urea—found in urine
ALDEHYDES AND KETONES			
Aldehyde	$-\overset{O}{\overset{\parallel}{C}}H$	$H\overset{O}{\overset{\parallel}{C}}H$	formaldehyde—biological preservative
Ketone	$-\overset{\mid}{\underset{\mid}{C}}-\overset{O}{\overset{\parallel}{C}}-\overset{\mid}{\underset{\mid}{C}}-$	$CH_3\overset{O}{\overset{\parallel}{C}}CH_3$	acetone—fingernail polish remover